INTERNATIONAL UNION FOR CONSERVATION OF NATURE AND
NATURAL RESOURCES

THE IUCN
INVERTEBRATE
RED DATA BOOK

Compiled jointly by

SUSAN M. WELLS, ROBERT M. PYLE and N. MARK COLLINS

of the IUCN CONSERVATION MONITORING CENTRE

with the help and advice of the Species
Survival Commission of IUCN and other
experts throughout the world

Illustrated by

SARAH ANNE HUGHES

Published by IUCN, Gland, Switzerland, 1983
(IUCN Conservation Monitoring Centre, 219(c) Huntingdon Road,
Cambridge CB3 0DL, U.K.)
Prepared with the financial assistance of
THE WORLD WILDLIFE FUND
and
THE UNITED NATIONS ENVIRONMENT PROGRAMME
A contribution to the Global Environment Monitoring System

IUCN

IUCN (International Union for Conservation of Nature and Natural Resources) is a network of governments, nongovernmental organizations (NGOs), scientists and other conservation experts, joined together to promote the protection and sustainable use of living resources.

Founded in 1948, IUCN has more than 450 member governments and NGOs in over 100 countries. Its six Commissions consist of more than 700 experts on threatened species, protected areas, ecology, environmental planning, environmental policy, law and administration, and environmental education.

IUCN

- monitors the status of ecosystems and species throughout the world;
- plans conservation action, both at the strategic level through the World Conservation Strategy and the programme level through its programme of conservation for sustainable development;
- promotes such action by governments, intergovernmental bodies and nongovernmental organizations;
- provides assistance and advice necessary for the achievement of such action.

ISBN No. 2-88032-602-X

Printed by Unwin Brothers Limited,
The Gresham Press, Old Woking, Surrey, U.K.

PREAMBLE

It has always been one of the principal functions of IUCN to collect data on threatened species and habitats in order to promote appropriate conservation action. For many years it has initiated and solicited such information on a global scale and, in the case of species, this has been carried out mainly through the Specialist Groups of the Species Survival Commission and individual consultants.

Some years ago it was recognized that this collection of data should be co-ordinated and centralized and so, with the financial support of the World Wildlife Fund (WWF), the United Nations Environment Programme (UNEP) and others, this led to the establishment of the Conservation Monitoring Centre (CMC) in 1979.

One of the aims of CMC is to process and store all the information received in such a way that it allows for presentation in a variety of forms. This flexibility has been achieved through the use of electronic word- and data-processing equipment and CMC is now rapidly acquiring the ability to produce many different types of output ranging from specialist reports on a geographical or taxonomic basis to the new series of Red Data Books of which this volume is a part. Further invertebrate volumes in preparation will cover papilionid butterflies and molluscs. CMC is also planning volumes on cave ecosystems and coral reefs.

The present work, representing a slight departure from the traditional concept of a Red Data Book, is the result of these activities and has involved the sustained efforts of three compilers. Robert Pyle was instrumental in initiating the project and began compilation in January 1980, concentrating on insects and spiders. In March 1980 Susan Wells was appointed to compile the sections covering molluscs and other non-arthropods and in May 1982 Mark Collins took over the sections on insects and spiders from Robert Pyle. In addition, the project has received support from the Centre's administrative staff and guidance from the SSC's Specialist Group members and consultants and the many other expert associates who help to ensure that our data are both up to date and of the highest quality. I hope that we may continue to rely on such help, and thank them all for their support.

I would like also to express my thanks and congratulations to all those who have contributed towards this work; in particular the joint compilers Susan Wells, Robert Pyle and Mark Collins, Sally Hughes who drew the illustrations, Kathy McVittie who provided technical assistance, Suzanne Vernon who was responsible for handling the text storage and processing equipment and Tony Mence and Chris Huxley, the Centre's managers.

> G.Ll. Lucas
> Chairman
> Species Survival Commission

January 1983

CONTENTS

INTRODUCTION

CONSERVE INVERTEBRATES?

tebrates are conservatively estimated to comprise about three quarters of all
 species. Approximately 1.4 million invertebrates have been described,
ared with 0.25 million flowering plants and only 21 000 vertebrates. About
l million known invertebrate species are insects, although constant new
veries indicate that this may be only a fraction of the world total. The
tebrate faunas of the tropical forest regions and deep seas are very poorly
n and a large proportion of species remain undescribed. Hundreds of new
tebrates are described each year and, although estimates vary widely, there
obably between 2 and 10 million unknown invertebrates, mainly insects and
nids, living in poorly studied biomes. Unfortunately, those areas for which
nowledge is poorest are often the richest in terms of biological diversity, and
under the greatest pressures from growing human populations and
opment. The conversion of rain forest alone is believed to be resulting in a
reduction in the variety of invertebrate life, many species becoming extinct
e they are known to science.

orld Conservation Strategy outlines the rationale for species conservation in
ontext of sustainable development (7). Living resource conservation is seen
e three specific objectives:

maintain essential ecological processes and life support systems (such as
il regeneration and protection, the recycling of nutrients and the cleansing
 waters) on which human survival and development depend;

preserve genetic diversity (the range of genetic material found in the
orld's organisms), on which depend the breeding programmes necessary for
e protection and improvement of cultivated plants and domesticated
iimals, as well as much scientific advance, technical innovation, and the
curity of the many industries that use living resources;

ensure the sustainable utilization of species and ecosystems (notably fish
d other wildlife, forests and grazing lands) that support millions of rural
mmunities as well as major industries.

portance of invertebrates in ecological processes and as a living resource of
 to man should not be under-estimated. Invertebrates are major
ients of food chains, are of primary importance in the cycling of nutrients,
iy a significant role in the maintenance of soil structure and fertility.
 play a vital role in the pollination of many plants including those of
iic importance. Many invertebrates, especially insects, are predators,
oids and parasites of pest species, and exert a natural control which has
een overlooked in the past (see Insecta introduction).

eople think of invertebrates only in terms of those species which are
 of disease, human parasites, and pests of crops and buildings. The many
 which they are beneficial to man are rarely considered. The importance
tebrate fisheries is discussed in the introductions for marine molluscs and
eans, and those for other phyla describe the wide range of other
orates, ranging from worms to insects, used as food in different parts of
ld. Crustaceans and molluscs provide much of the protein in the diet of
 people, and shrimps probably represent the most valuable export trade in
mals.

i

Many invertebrates contain active compounds with medically useful properties, which may serve as models in the development of new drugs. Marine invertebrates have yielded substances with antiviral, antimicrobial, tumour-inhibitory, anticoagulant, cardioactive or neurophysiological properties (14). Marine communities have recently been found to exhibit complex relationships between species and sophisticated organization at the community level, which accounts for the fact that organisms found in environments such as coral reefs contain many biochemical compounds that feature novel forms and functions (2,11).

Invertebrates, particularly arthropods, are used extensively in biological and integrated control programmes and, although considerable care must be taken in the application of such methods, their use is often preferable to unselective methods involving chemicals. The composition of aquatic and terrestrial invertebrate communities can be invaluable as indicators of pollution and other environmental changes (4).

Products derived from invertebrates are as diverse as silk, dyes, lime, and building materials. Invertebrates are also used in basic research in genetics, biochemistry, embryology, evolutionary science and many other disciplines. In many cultures invertebrates have long been valued for aesthetic reasons. Some have been endowed with magical attributes; others, such as corals and shells, are used for personal decoration; and many, through their intriguing and beautiful forms, have inspired artists and craftsmen (see introductory sections on Cnidaria, marine molluscs, Insecta).

INVERTEBRATE CHARACTERISTICS AND CONSERVATION

The ecology and population dynamics of invertebrates differ in several ways from those of vertebrates, and an understanding of invertebrate life histories is essential for their successful conservation. The majority of invertebrates are short-lived and produce far more offspring than will survive the natural forces of predation, parasitism and starvation. With their considerable powers of recolonization, over-exploited invertebrates are more likely to recover than are vertebrates. It will be seen in the following reviews that the invertebrates threatened by exploitation or predation tend to be those either with a reproductive strategy similar to that of vertebrates, producing comparatively few young with a low natural mortality (e.g. land snails), or with a high larval mortality so that recruitment to the adult population is very low (e.g. giant clams).

Invertebrate endemism is not uniform throughout the world and unfortunately the greatest diversity of invertebrate species is found in those areas which are both the least studied and are under greatest threat from man. Although there are many exceptions, the temperate latitudes in general support a less diverse and less specialized fauna, with fewer restricted endemic species. This is largely because such areas have only been recolonized since the last Ice Age ended 10 000 years ago, a short evolutionary period compared with, for example, the 60 million years of relatively stable and favourable conditions in central and southern Africa and the forest refugia of the Amazon basin. At least 50 per cent of the world insect fauna is believed to live within the 6.3 per cent of the earth's land surface that supports tropical moist forest (9, see Insecta introduction).

Some freshwater and terrestrial invertebrates have poor dispersal abilities and are able to survive in extremely small ranges, such as single springs, caves and isolated patches of vegetation. This seems to be the case in tropical rain forest where, despite the great density of species, population densities are often low and ranges may be very small (see Insecta introduction). However, many other invertebrates have extraordinarily wide and often disjunct ranges as a result of their ability to disperse through flight, by the agency of man and other animals, in air currents, and in sea currents as planktonic larvae. Many non-arthropod invertebrates are hermaphroditic or bisexual, individuals having both male and female reproductive organs. In hermaphroditic species in which self-fertilization occurs only a single individual may be needed to extend a species range.

More than half the invertebrate phyla are entirely marine and many of the rest are predominantly so. The remarkable diversity of marine life is evident in the fact that most taxonomic classes are represented in the sea and most basic forms have evolved there, even though more species now inhabit the land, because of the adaptive radiation of certain groups such as the insects (5). Most marine species are more widely distributed than are terrestrial species and are less likely to suffer man-induced extinction. Some 70 per cent of marine invertebrates are dispersed as freely-drifting planktonic larvae but nevertheless, local populations of widely distributed species can become genetically differentiated. Genetic variation in marine organisms is discussed in detail in (13). The extent of geographical differentiation depends on several factors including length of the planktonic phase, the fertility and size of the parent population, and natural environmental influences.

The greatest diversity of species in many marine groups is exhibited in the Indo-Pacific region, in particular in and around South East Asia. This is probably attributable to higher rates of evolution in this region, as well as to stability of the maritime climate and trophic resources, a heterogenous environment and the potential for local isolation of populations. Although there are few cases of localized endemism of marine species, a high genetic variation is found in the animals of coral reefs and certain deep-sea areas, suggesting that the genetic resources of populations in these species-rich communities are particularly extensive. A high degree of endemism has been documented in the off-shore communities of isolated oceanic islands such as Lord Howe Island, the Easter Islands, the Hawaiian Islands and some southern Atlantic islands. Unique forms also evolve in seas with restricted access, such as the Red Sea, where species have diverged from their Indian Ocean relatives. These areas tend to have far fewer species, but may have over 25 per cent endemism within biotopes such as coral reefs.

THREATS TO INVERTEBRATES

The causes of the decline and extinction of all wild species can be broadly divided into natural events and human activities. Natural genetic shift in invertebrate populations, the evolution of new species, and the extinction of unsuccessful species through natural selection, immigration and emigration, are not considered in Red Data Books. Man-induced changes are currently the most important factors affecting most organisms as they operate on a time scale drastically foreshortened from that of natural changes over geological time. Through technological innovation and population pressure, man is creating or destroying landscapes and biotopes within a geological instant, permitting no opportunity for natural selection of adaptable genotypes.

As illustrated in the Red Data Books for mammals and for amphibians and reptiles, the major threat to wild species throughout the world is the destruction and alteration of the habitats on which they depend (3,16). Through increasing population pressure, human activities have steadily modified wilderness areas into man-made landscapes of settlement, agriculture and industry that preclude coexistence with many wild creatures. The invertebrates are no exception. Three other factors recur regularly in the following pages: pollution (which could be seen as a form of habitat alteration, but is here treated separately), exploitation, and the introduction of exotic species. Frequently these threats act together in causing the decline of a species and it is often difficult to determine the factors of greatest impact. A brief discussion of these four general threats is given below, and they are discussed in more detail, along with threats of a more specific nature, in individual accounts.

1. Habitat alteration

1a. Terrestrial species.
Terrestrial invertebrates are affected by many types of habitat alteration, of which the most important are deforestation, agricultural conversion and intensification, alteration of grasslands (through inappropriate grazing, scrub regeneration and other factors), industrialization and urbanization. Many terrestrial invertebrates have ranges so small that they could be eradicated by single events such as the building of a factory or the granting of a timber concession. Deforestation of tropical equatorial regions is widely considered to represent the greatest single threat to insects and probably to many other arthropods and pulmonate molluscs. It has been estimated that tropical moist forests are being converted at the rate of between 73×10^5 and 20×10^6 ha per year, the figure depending on the criteria used (1,10). This represents between 0.8 and 2.1 per cent of the total available area of this biome (15). The loss of insect species is inevitable in the wake of conversion on this scale but the impact is largely undocumented (see Insecta introduction). Terrestrial molluscs are the second group most seriously affected by loss of forest. Land snails have undergone greatest speciation on islands, particularly those of the Pacific Ocean. The widespread reduction of forest cover on Pacific islands, resulting from increasing population pressure and growth in tourism, has led to catastrophic rates of extinction, reaching as much as 50-66 per cent of terrestrial molluscs in the Hawaiian Islands (see terrestrial molluscs introduction). Modification of habitat through conversion to plantation forestry and agriculture, the intensification of agricultural and pastoral methods by use of fertilizers and pesticides, and urbanization and industrialization, generally reduce the diversity of invertebrate communities. Specific instances are discussed in the following reviews (and see Insecta introduction).

1b. Freshwater species.
Like terrestrial invertebrates, many freshwater species have extremely restricted ranges, occurring in single water bodies such as springs, ponds, lakes, rivers, fens and mires. Pollution directly affects the quality of the water (see below), but total alteration and loss of wetlands and water bodies is increasingly occurring through drainage, channelization, impoundment, and the capping or tapping of springs. The effects of such activities are discussed in detail in the introductions for freshwater molluscs, crustaceans, chelicerates, insects and threatened communities. There are several well-documented cases of the loss of invertebrates as a result of alteration of freshwater biotopes. One of the most striking examples involves the drastic changes that have taken place as a result of the impoundment and channelization of the rivers of the south-eastern U.S.A. In North America, of the formerly diverse molluscan fauna of over 1000 named species, some 40-50 per cent is now either extinct or endangered (see freshwater mollusc introduction). The flooding of Lake Pedder in Tasmania is another example of the devastating effect that such activities can have on unique invertebrate communities. Despite these case histories, alteration of freshwater ecosystems continues to accelerate, as illustrated by the current controversy over the proposed damming of the Lower Gordon River, also in Tasmania. Loss of fenland and mires through drainage and changes in agricultural practice have affected animals as diverse as leeches, molluscs, spiders and caddis-flies.

1c. Marine species.
The great majority of marine species spend much, or critical periods, of their lives in shallow or intertidal waters, mainly in tropical seas, and relatively few are pelagic throughout their life histories. A very small section of the oceans therefore harbours a large proportion of species and is most productive of organic matter. It is these coastal areas which support some of the most dense human

populations and are most vulnerable to human activities. It has been predicted, for example, that when the population of the United States reaches 300 million, half of the population will be living in the coastal belts on 5 per cent of the total land area. There is evidence that some of the most diverse communities, such as those of tropical reefs and the continental slope, are particularly susceptible to environmental changes (5,11). Furthermore, many of the commercially important invertebrate fisheries are dependent on coastal habitats. Young prawns and shrimps, for example, are dependent on protected estuarine waters adjacent to marshes, mudflats and mangroves and in some regions the loss of these critical habitats has resulted in a reduction in the yields of offshore fisheries (12). In most maritime countries, alteration of coastal habitat is taking place rapidly as a result of urbanization, industrialization, tourism and recreational development, dredging, damaging fishing methods, and a host of other activities which are made apparent in the subsequent sections (see introductions to Cnidaria, marine molluscs, Crustacea and Echinodermata sections).

2. Pollution

2a. Terrestrial.
Some terrestrial invertebrates may be susceptible to atmospheric pollution and acid rain (see Insecta introduction) but in general this is only of major importance to aquatic species. The use of fertilizers in intensive agriculture tends to reduce the diversity of plants and their dependent invertebrates.

2b. Freshwater.
Pollution of streams, rivers and lakes by acidic mine drainage, domestic, thermal and industrial effluent, siltation (from soil-run off following erosion and deforestation) and acid rain have all been cited in the following reviews as having had adverse effects on a range of freshwater molluscs, crustaceans and insects. Acid rain is an issue of particular concern in northern Europe and North America and could spread to other parts of the world (see freshwater molluscs introduction). Several invertebrates, such as certain species of protozoans, annelids, crustaceans and molluscs, may apparently benefit from organic pollution. For example, tubificid worms and some leeches appear in huge numbers in sites of domestic pollution, but this is generally at the expense of less tolerant species (4). The long-term effects of pollutants on many freshwater invertebrates is still poorly understood and requires further research.

2c. Marine.
The effects of pollutants in the marine environment are equally poorly understood. Fresh oil can cause heavy mortalities of invertebrates and in certain cases, long-term effects on populations are indicated. In other cases marine communities appear to be able to regenerate. Fragile communities such as mangroves and coral reefs are most vulnerable to oil spills and to the clean-up operations which often involve toxic dispersants (6). The effects of other forms of marine pollution, such as industrial and domestic effluents and siltation from soil run-off and mining activities, are also poorly understood, but it is becoming apparent that they often have adverse impacts on filter-feeding and benthic organisms.

3. Exotic introductions

Intentional or accidental introductions of exotic animals and plants will invariably upset the balance of invertebrate communities. Introduced vertebrates have, in numerous cases, caused extinction or serious declines in invertebrate species through predation (see accounts of Lord Howe Island Stick-insect and giant wetas) or alteration of native vegetation by over-grazing or trampling. Other introduced species pose threats through competition for food resources or habitat (see freshwater crayfish), and through hybridization. The ill-considered unintentional introduction of exotic species in biological control programmes can have a disastrous impact on non-target invertebrates. The introduction of the carnivorous land snail <u>Euglandina rosea</u> to numerous islands in the Pacific in an attempt to control the Giant African Snail (<u>Achatina fulica</u>) has caused the extinction or drastic decline in population size of many endemic land snails, and is discussed in detail in the section on terrestrial molluscs.

4. Exploitation

Unlike vertebrates, most invertebrates can withstand a considerable level of harvest because of their high reproductive capacities, and in general exploitation is unlikely to lead to many extinctions. However, over-collection can have serious effects if the population is already critically depleted, if the population is small and there is a high value per individual (such as the rare butterflies and molluscs which become collectors' items), or if the species has a reproductive strategy more like that of a vertebrate, with low juvenile recruitment to the population and low reproductive rates.

In many cases, exploitation of invertebrates is to be encouraged, provided that a rational management strategy is adopted to ensure sustainable utilization. Many marine species provide, or could potentially provide, a substantial source of revenue. Over-exploitation in the sea is a matter of concern, not so much because it threatens individual species but because it alters the structure of the community, particularly predator-prey relationships. In addition, the taking of unsustainable yields will eventually lead to permanent depletion of stocks, and the loss of industries and income. Human population growth has led to excessive subsistence fishing in many areas, and in others, the introduction of modern and sometimes damaging techniques has greatly increased the impact of fishing (5,13). About 25 valuable marine fisheries are seriously depleted worldwide, including those for a number of invertebrate species which are discussed in the introductory sections on marine molluscs and crustaceans.

INVERTEBRATE CONSERVATION - PAST AND PRESENT

1. Documentation of threatened invertebrates

Since the 1970s national interest in invertebrate conservation has greatly increased, primarily for insects (see Insecta introduction) but also for other groups. Australia, New Zealand, the U.S.A. and several European countries have been documenting their threatened invertebrates. The U.S. Fish and Wildlife Service and Office of Endangered Species have instigated numerous surveys and produced extensive documentation on national threatened and rare invertebrates. Included in the 762 species (and populations) presently listed under the U.S. Endangered Species Act are 25 clams (freshwater mussels), 13 insects, nine snails and two crustaceans. Recovery plans are in progress for some species and are being drawn up for others. In addition, many states have drawn up their own legislation and initiated their own conservation projects.

In Europe, national Red Data Books and official threatened species lists for invertebrates have been produced for parts or all of Austria, Belgium, the Federal Republic of Germany, Luxembourg, Poland, Spain, and Switzerland, and others are in preparation. The European Invertebrate Survey of the International Commission for Invertebrate Survey is mapping the past and present distributions of many species and the programme is being expanded to include North America.

Coral reef communities and other marine invertebrates are being given increasing attention, particularly in the Caribbean, Mediterranean, South East Asia and the Pacific (see Cnidaria and marine molluscs introductions).

2. Legislation and international conventions

Invertebrates are included in the wildlife legislation of many countries, notably in North America (see above) and Europe. In the U.K. Wildlife and Countryside Act, 19 of the 39 animals (excluding birds) listed for protection are invertebrates, including 14 insects, three snails and two spiders. Specific invertebrates have been considered in some developing countries such as the Philippines and Papua New Guinea, which have enacted legislation respectively for corals and butterflies. However, legislation to protect individual species of invertebrates is not always useful unless it provides for stringent controls over habitat alteration. Legislation controlling the exploitation of invertebrates is often ineffective because this is rarely a crucial threat. No amount of protective legislation will succeed if a species' habitat is destroyed. In some cases a strict ban on collection may be a positive disadvantage if it prohibits collection for scientific work which could provide vital insight into the reasons for the decline of the species.

Invertebrates are poorly represented on the Appendices of the Convention on International Trade in Endangered Species of Wild Fauna and Flora (CITES) and many of those that are listed are only incidentally, or not at all, involved in trade. Consequently, the extent to which CITES can benefit invertebrates is a controversial issue. A positive aspect of CITES listing is that international attention is drawn to the species concerned, thus assisting conservation efforts at the national level. However, this is not the main purpose of CITES. For species to qualify for inclusion in one of the Appendices they should fulfil a strict set of criteria. The difficulty of obtaining the necessary information for invertebrates, in terms of numbers and rates of change of population size, means that few meet these criteria. Nevertheless, it is generally considered that certain species merit inclusion on Appendix II in order to provide means of monitoring international trade in them.

Other international conventions have paid little specific attention to invertebrates. For example, the Berne Convention on the Conservation of European Wildlife and Natural Habitats, which came into force on 1 June 1982, lists only vertebrates and plants on its Appendices. Some conventions, such as the 1978 Barcelona Convention for the protection of the Mediterranean Sea against pollution and the 1971 Ramsar Convention for wetlands of international importance, may indirectly have beneficial effects. Since many invertebrates are threatened over wide geographical areas, or are migratory, international conventions could play an important role in their conservation.

3. Protected areas

There are few examples of reserves created specifically for invertebrates although many species are afforded protection in national parks and protected areas established for more general reasons. However, in some countries, such as the U.K. and U.S.A. (see Insecta introduction), some sites have been selected for protection or as a high priority for conservation on the basis of their entomological importance, and there are isolated instances of reserves created for other invertebrates.

Marine reserves play an important role in the conservation of marine invertebrates, ensuring the continued existence of breeding populations from which individuals can disperse to recolonize over-exploited or disturbed areas. In many parts of the world where the sea and its resources are valuable to people, marine areas and populations have been protected for centuries (8). Although such measures often arose from spiritual beliefs, many seem to have been introduced for functional purposes, such as the reduction of conflict over limited resources within and between communities of people. Many countries throughout the world are giving priority to the establishment of marine reserves. Management problems and techniques, such as the need to resolve conflict between different uses of the marine environment, were discussed in detail at the Third World National Parks Congress in Indonesia, October 1982.

INVERTEBRATE CONSERVATION - THE FUTURE

Each review in this volume includes a section on 'Conservation Measures Proposed' which recommends specific actions required to improve management and conservation. These proposals are primarily concerned with the protection of the animal and its habitat, the elimination of threats to its survival, and the collection of biological data on which to base more effective remedial measures. An outline of conservation requirements for invertebrates in general is given below:

1. Further research

This is a prime requirement for many invertebrates. The lack of data on the taxonomy, biology and ecology of invertebrates, their distributions, population dynamics and the effects of human activities on their long-term survival has been stressed throughout this introductory section. The surveys being carried out in Europe and North America should be extended with all possible speed to the tropics and other poorly studied areas.

2. Legislation

Legislation for individual taxa is not necessarily a high priority for invertebrates. However, where it involves habitat protection, it may be essential for the future survival of the taxon. Improvement in existing legislation and management is required for a number of species that are commercially exploited, so that size limits and closed seasons relate more closely to the life history of the species concerned. Closer control of international trade in species for luxury or ornamental purposes (such as marine invertebrates for souvenirs) is required where this use conflicts with the benefit that the species could provide in its country of origin.

3. Creation of reserves and protected areas

Reserves are usually based on the requirements of vertebrates or plants but it cannot be assumed that within these the needs of invertebrates will adequately be met. Without detailed knowledge, key invertebrate habitats may be overlooked. Because of their small size and relatively modest needs, invertebrates are able to occupy more and smaller ecological niches than are vertebrates. It is essential that surveys of potential reserve areas include an assessment of the invertebrate fauna, to ensure that human impact on a small scale is not overlooked. The very characteristics that make many invertebrates so vulnerable, namely their small size and restricted distributions, may also facilitate their protection. In some cases, it may be possible to ensure their survival by the extension of existing reserves or by the creation of small reserves that cause little disruption to local people.

4. Education

An increase in public awareness of the need for invertebrate conservation is a high priority. Unless of high economic value, invertebrate species invariably have the lowest priority for action by governments, national and international conservation agencies and funding bodies. In countries where subsistence life styles still predominate, there is often a greater awareness of the importance of these smaller forms of life, but education must be aimed at the rational management of these resources.

5. Encouragement of captive breeding programmes and aquaculture

Aquaculture has been in progress for many years for a variety of marine and freshwater invertebrates of commercial importance. Encouragement must be given to the establishment of similar programmes for other species that may not be of large scale commercial importance but that are a major resource to local people. Despite several attempts at captive breeding of terrestrial invertebrates for commercial purposes or conservation, there are few instances where it has been entirely successful. Rearing of wild-bred invertebrates to increase production is potentially more valuable and has been successful for a number of species including butterflies and many marine invertebrates.

6. Re-introductions to areas where the taxon is extinct or very rare

A few attempts have been made to re-introduce terrestrial invertebrates to areas where they have become extinct, but these have rarely been successful and there is little prospect of the widespread use of such schemes. However, research on re-introductions of invertebrates to over-exploited or under-utilized habitats for commercial aquaculture and hatchery projects appears to be promising.

7. Control of feral or introduced animals

Greater efforts are required to control or eliminate introduced species, particularly those which drastically alter the habitat through their grazing and trampling activities (e.g. sheep, cattle, rabbits), cause population reductions through predation (e.g. the snail Euglandina rosea, fish, rats) and which affect native species through competition (e.g. crayfish). Extensive trials and experimentation must be undertaken before the initiation of any biological control programme that involves the introduction of species that could have an adverse impact on non-target native invertebrates.

* * *

As the newest and most innovative of the IUCN Red Data Books, the Invertebrate RDB is breaking new ground. It aims to increase public awareness of the need for invertebrate conservation and gives suggestions of ways in which this may be carried out. The examples chosen emphasize that invertebrates have enormous ecological and economic importance. Subsequent invertebrate Red Data Books will cover relatively well-documented groups such as papilionid butterflies and molluscs, and vulnerable communities such as caves and coral reefs in which invertebrates constitute a large proportion of the fauna. Concurrently, data on all threatened invertebrates will be collated for the IUCN threatened species database. This information will be available to national and regional endangered species programmes, conservation funding agencies and those involved in international conservation initiatives. It is hoped that this collation of information on the status and conservation requirements of invertebrates will stimulate further interest, awareness and action to ensure the future survival of these organisms and their environments.

REFERENCES

1. FAO/UNEP (1981). Tropical Forest Resources Assessment Project. (3 vols.) FAO, Rome.

2. Grant, P.T. and Mackie, A.M. (1977). Drugs from the sea - fact or fantasy? Nature 267: 786-788.

3. Groombridge, B. and Wright, L. (1982). The IUCN Amphibia-Reptilia Red Data Book. Part 1. IUCN, Gland, Switzerland.

4. Hart, C.W. and Fuller, S.L.H. (Eds). (1974). Pollution Ecology of Freshwater Invertebrates. Academic Press, New York.

5. Holt, S. and Segnestam, M. (1982). The seas must live: why coastal and marine protected areas are needed. Paper prepared for the Marine and Coastal Areas Workshop, Third World National Parks Congress, Bali, Indonesia. October 1982. 11 Pp.

6. IUCN Commission on Ecology, Oil Pollution Working Group, (1982). The impact of oil pollution on living resources. Draft report.

7. IUCN/UNEP/WWF (1980). World Conservation Strategy: Living Resource Conservation for Sustainable Development. IUCN, Gland, Switzerland.

8. Johannes, R.E. (1978). Traditional marine conservation in Oceania. Ann. Rev. Ecol. Syst. 9: 349-364.

9. Myers, N. (1979). The Sinking Ark. Pergamon Press, Oxford, 307 pp.

10. Myers, N. (1980). Conservation of Tropical Moist Forests. Report to the National Academy of Sciences. National Research Council, Washington D.C.

11. Myers, N. (1981). The role of protected areas in conservation of marine genetic resources. Draft paper prepared for the Third World National Parks Congress, Bali, Indonesia, October 1982.

12. Odum, W.E. (1982). The relationship between protected coastal areas and marine fisheries genetic resources. Paper prepared for Marine and Coastal Areas Workshop, Third World National Parks Congress, Bali, Indonesia, October 1982. 12 pp.

13. Polunin, N. (1982). The role of protected areas in conserving marine genetic resources. Paper prepared for the Marine and Coastal Areas Workshop, Third World National Parks Congress, Bali, Indonesia, October, 1982. 19 pp.

14. Ruggieri, G.D. (1976). Drugs from the sea. Science 194: 491-497.

15. Sommer, A. (1976). An attempt at an assessment of the world's tropical moist forests. Unasylva 28: 5-24.

16. Thornback, J. and Jenkins, M. (1982). The IUCN Mammal Red Data Book. Part 1. IUCN, Gland, Switzerland.

METHODS

Collection of information for the Invertebrate Red Data Book began at the Conservation Monitoring Centre in early 1980. It soon became apparent that documentation of threats to invertebrates presented a unique set of issues and problems. The selection of species for inclusion in the vertebrate RDBs is generally based on data that illustrate a decline in the range or population size of the species concerned. For most invertebrates such information is not available, and even for the better-known species the ecological basis on which constructive conservation recommendations could be based is lacking. There is an extreme paucity of data on the invertebrates of South America, Africa and Asia. In the species data sheets the emphasis on temperate species is an artefact which reflects this lack of data from the tropics and not the actual world distribution of threats to invertebrates.

The invertebrates include several groups with life-styles that have not previously been considered in RDBs. It is still extremely difficult to estimate population sizes and map distributions of marine species, although the development of SCUBA diving equipment, underwater cameras and deep-sea submersibles has dramatically increased our knowledge of marine life. Furthermore, the effect of man's activities on marine species is still not fully understood; for example, it is not known how long it takes a damaged coral reef to recover nor how long the effects of a major oil spill will last. As discussed in the Introduction, human activities are unlikely to cause the extinction of marine invertebrates, but many populations are experiencing local depletions. Furthermore, entire communities such as coral reefs, salt marshes and mangroves are suffering extensive damage which will affect their invertebrate life. Consequently, the marine species in this volume have been chosen to emphasize particularly vulnerable communities and species of economic value to man. There is a bias towards this latter group, not so much because these species are in greater jeopardy (in fact the reverse is often true) but because of the need for management of these species for the benefit of local people.

Invertebrate parasites and pests raise moral questions. Enormous amounts of effort and money go into attempts to eradicate such species, but it is a matter for debate whether they should be driven to extinction. Genetic diversity, even in these species, could be an asset in the future. Numerous parasites of endangered species are themselves endangered and one example, the Pygmy Hog Sucking Louse, has been given in this volume. There are undoubtedly many others but parasites cannot be considered a high priority when the needs of obviously beneficial invertebrate species are so great.

These characteristics of invertebrates have resulted in the adoption of a more flexible style of presentation than that traditionally used in the RDBs. This volume contains a set of case histories illustrating human pressures on invertebrate populations and habitats, and the range of species threatened worldwide. No attempt has been made to include all threatened invertebrates, and the selected examples are not necessarily those in greatest danger. In this respect the Invertebrate RDB resembles the Plant RDB, but differs from most of the vertebrate volumes, which aim to include all threatened taxa within their respective classes.

The book is divided into sections on phyla, with the Arthropoda being further divided between the Chelicerata, Insecta and 'Myriapoda'. In some phyla it has been impossible to identify particular taxa in any of the RDB categories. In such instances a general account of the group is given, emphasizing those aspects of its biology that are of importance in terms of conservation, stressing its scientific interest and value to man, and outlining the types of threat to the species. For groups in which threatened taxa have been identified, an introductory section is followed by reviews for example taxa. The introductory section is similar to that for groups with no species accounts, provides a context for the subsequent accounts, and emphasizes those areas not well covered by the succeeding taxa examples.

The species accounts follow the standard IUCN RDB format with a few modifications. A brief description is included since many species will not be familiar to all readers. For the same reason, the Invertebrate RDB is the first volume to be illustrated, and each major group has at least one illustration. These are not scientific figures, but have been designed to provide an accurate picture of the animal in its natural surroundings and have been checked by suitable authorities. Although for many species there is no information on population size, this paragraph has been retained and annotated where possible. The paragraph on captive breeding has been inserted only when applicable. The paragraph on scientific interest and potential value is a development of a similar paragraph in the Plant RDB. Most invertebrates do not have the popular appeal of birds and mammals and attempts at their conservation are sometimes met with derision. In this volume every effort is made to emphasize the reasons why the conservation of invertebrates is a matter of genuine concern.

As with the volumes on vertebrates, only taxa considered to be rare or threatened throughout their ranges are included. For most of the taxa, particularly terrestrial species, the traditional IUCN status categories have been used. However, in some cases the strict criteria required by these categories cannot be applied although there is evidence that the taxon is of special concern and warrants inclusion. The problems encountered in applying status categories to threatened invertebrates and the need for new categories is discussed in the section on IUCN Status Categories (p.xix).

The reviews have been compiled in the same way as those in previous Red Data Books. Data are obtained from published, unpublished and occasionally verbal reports from a variety of sources. The IUCN Species Survival Commission has Specialist Groups for Lepidoptera (butterflies and moths), Odonata (dragonflies and damselflies), Formicidae (ants), Mollusca, freshwater Crustacea, cave ecosystems and coral reefs. They have played important roles in providing information, selecting species for inclusion and refereeing reviews. A variety of people not formally associated with IUCN, including field and museum scientists, wildlife park personnel and amateur naturalists, have also kindly provided information.

A preliminary draft account is compiled from the information gathered. In some cases workers contacted have provided outline drafts which are then amplified or refined as necessary using other data on file. The preliminary draft is then circulated to a number of authorities for comment, correction, and the insertion of additional material. When conflicting opinions arise, both viewpoints are generally represented, although the compiler may make the final choice of category. There may have been one or several review phases, depending on the geographical range of the animal concerned and the number of authorities involved. No reviews are final, and all are subject to the inclusion of new information. It is hoped that this volume will prompt comments on, and corrections of, the information presented.

The length of reviews is very variable. Some are unusually long because several taxa have been included. This system has not been used in previous volumes, and has been introduced to emphasize and describe the large number of species involved. For example, there are 22 extant species of the snail genus Achatinella on Oahu, Hawaii, all of which are endangered by the same factors and can be described in a single review. As with the new volumes of the Mammal and

Amphibia-Reptilia books, the reviews include as much information as possible, contributing further to the differences in length.

Due to limitations in the equipment used to produce camera-ready copy for the printer, certain accents and other pronunciation marks are missing. This applies mainly to words in Spanish, Portuguese, Scandinavian, Russian and East European languages. We apologize for errors or misunderstandings that might arise as a result of this.

ACKNOWLEDGEMENTS

We are very grateful to the many institutions and people who have given generous assistance in the compilation of this volume and without whom the task could not have been attempted. Wherever appropriate, acknowledgements to individuals have been made at the end of each account. In addition we would particularly like to thank the following for providing information and reviewing the accounts:

Chairmen, vice-chairmen and members of the IUCN/SSC Lepidoptera, Odonata, Ant, Cave Invertebrate, Coral Reef and Freshwater Crustacea Specialist Groups for providing information and comments.

The staff of the British Museum (Natural History), in particular the staff of the Entomology Department, of the Entomology, Zoology and General libraries and of the following sections of the Zoology Department: Annelida, Arachnida, Bryozoa, Coelenterata, Crustacea, Echinodermata, Mollusca, Polychaeta and Porifera, and Protozoa.

The staff of the U.K. Nature Conservancy Council and of the Institute of Terrestrial Ecology at Furzebrook and Monks Wood.

The staff of the Balfour Library in the Zoology Department, University of Cambridge.

The staff of the Royal Entomological Society of London.

The members of the European Invertebrate Survey of the International Commission for Invertebrate Survey.

The staff of the regional offices of the U.S. Fish and Wildlife Service and of the Office of Endangered Species, Washington D.C.

We would also like to thank the following for their help in the production of the volume:

Kathy McVittie, who has subedited the entire volume, and prepared the index and taxonomic and geographical lists.

Mary Green, Michael Morton and Lissie Wright, who have provided invaluable technical assistance.

Suzanne Vernon and Carol Hovenden, who were responsible for the typing and handling of the text on the word processor.

Tony Mence, Adrian Friday and Richard Barnes, who reviewed the final manuscript.

Our colleagues at the Conservation Monitoring Centre, who gave enthusiastic encouragement and helpful advice on many matters.

Melanie Collins, who gave support and encouragement.

INFORMATION REQUIRED FOR DATABASE

Report to be mailed to IUCN Conservation Monitoring Centre, 219(c) Huntingdon Rd, Cambridge, CB3 ODL, U.K.

1.   Country

2.   Date

3.   Reporter

      Name:

      Address:

4.   Classification

      Phylum: Class: Family:

      Genus and species: Common Name:

5.   Distribution

      Present: Former:

      If possible, please include a map. Is present range preferred or enforced habitat?

6.   Population

      Estimated numbers in the wild. Indicate date of estimate and describe method of estimation. Are numbers increasing, decreasing or stable?

7.   Habitat and Ecology

      Biome type, altitudinal range or depth, brief notes about feeding habits and diet, reproduction (breeding season, larval stages, number of young, age of sexual maturation), longevity, associated species etc.

8.   Scientific interest and potential value

      Use in scientific or medical research; commercial value in trade or local economy.

9.   Threats to Survival

      E.g. habitat destruction, over-exploitation, pollution, introduced species, natural disasters.

10. Conservation Measures Taken

Legal measures (international conventions, national laws); is law enforced? Protected areas - does it occur in national parks, reserves etc? If so, please list. Management programmes or research programmes in progress.

11. Conservation Measures Proposed

Same as for 10, but measures that are needed for the conservation of the taxon.

12. Captive Breeding

Numbers in captivity. Does it breed readily in captivity? Where and when?

13. References

Including published papers, unpublished manuscripts, personal communications or references to correspondence (cited as In litt.).

The traditional IUCN status categories have been used for most of the taxa considered in this volume. However, as noted in the Introduction, the information necessary to fulfil the criteria for these categories is often lacking for invertebrates. There is generally little difficulty in the case of species with very restricted distributions. Accurate population estimates may even be possible, allowing a more objective application of the category definitions. However, several invertebrates with apparently high population numbers and wide distributions have been included because there is general agreement that some form of conservation action is required (e.g. the Medicinal Leech, some of the crayfish, the Apollo Butterfly and wood ants). The categories 'Indeterminate' and 'Insufficiently Known' have often been used, but the status of some of these species may be better described as 'Of Special Concern', a category which is now increasingly used in national lists of threatened species.

It has become apparent that new categories are required for certain circumstances. 'Commercially Threatened' has been introduced to allow consideration of the many marine invertebrates which are of economic importance to man and for which there is evidence that widespread over-exploitation is taking place (e.g. Queen Conch and precious corals). 'Threatened Phenomenon' has been introduced for taxa which are not in danger of extinction but which have certain threatened populations of major scientific interest (e.g. Monarch Butterfly over-wintering sites in Mexico and U.S.A.). 'Threatened Community' accounts have been introduced to illustrate whole invertebrate communities which are at risk. This concept is discussed in more detail on p.559.

It is emphasized that the application of IUCN status categories to threatened organisms of any kind poses many problems and inevitably involves the subjective judgement of the Red Data Book compilers. The Species Survival Commission is planning to undertake a review of these categories in the near future.

EXTINCT (Ex)

Species not definitely located in the wild during the past 50 years (criterion as used in The Convention on International Trade in Endangered Species of Wild Fauna and Flora - CITES).

ENDANGERED (E)

Taxa in danger of extinction and whose survival is unlikely if the causal factors continue operating.

Included are taxa whose numbers have been reduced to a critical level or whose habitats have been so drastically reduced that they are deemed to be in immediate danger of extinction. Also included are taxa that are possibly already extinct but have definitely been seen in the wild in the past 50 years.

VULNERABLE (V)

Taxa believed likely to move into the 'Endangered' category in the near future if the causal factors continue operating.

Included are taxa of which most or all the populations are decreasing because of over-exploitation, extensive destruction of habitat or other environmental disturbance; taxa with populations that have been seriously depleted and whose ultimate security has not yet been assured; and taxa with populations which are still abundant but are under threat from severe adverse factors throughout their range.

RARE (R)

Taxa with small world populations that are not at present 'Endangered' or 'Vulnerable', but are at risk.

These taxa are usually localized within restricted geographical areas or habitats or are thinly scattered over a more extensive range.

INDETERMINATE (I)

Taxa known to be 'Endangered', 'Vulnerable', or 'Rare' but where there is not enough information to say which of the three categories is appropriate.

OUT OF DANGER (O)

Taxa formerly included in one of the above categories, but which are now considered relatively secure because effective conservation measures have been taken or the previous threat to their survival has been removed. This volume does not include taxa in this category.

INSUFFICIENTLY KNOWN (K)

Taxa that are suspected but not definitely known to belong to any of the above categories, because of lack of information.

N.B. In practice, 'Endangered' and 'Vulnerable' categories may include, temporarily, taxa whose populations are begining to recover as a result of remedial action, but whose recovery is insufficient to justify their transfer to another category.

Threatened is a general term to denote species that are 'Endangered', Vulnerable, 'Rare', or 'Indeterminate' and should not be confused with the use of the same word by the U.S. Office of Endangered Species.

In addition the following three categories have been used in this volume:

COMMERCIALLY THREATENED (CT)

Taxa not currently threatened with extinction but most or all of whose populations are threatened as a sustainable commercial resource, or will become so unless their exploitation is regulated.

This category applies only to taxa whose populations are assumed to be relatively large.

N.B. This category is not used if any of the traditional categories apply. In practice it has only been used for marine species of commercial importance that are being overfished in several parts of their ranges.

THREATENED COMMUNITY (TC)

A group of ecologically linked taxa occurring within a defined area, which are all under the same threat and require similar conservation measures.

N.B. These may include species for which the traditional categories would apply if single species accounts were written. The main purpose of such a category is to deal with the large number of invertebrates that may be found in one place and for which a series of almost identical accounts would be necessary.

THREATENED PHENOMENON (TP)

Aggregates or populations of organisms that together constitute major biological phenomena, endangered as phenomena but not as taxa.

TAXONOMIC LIST OF THREATENED INVERTEBRATES
DESCRIBED IN THIS VOLUME

The individual species accounts in this volume follow the systematic order given below, except in the case of the Mollusca, which are grouped into marine, freshwater and terrestrial sections, within which the accounts are again in taxonomic order.

Within families, taxa are arranged in alphabetical order. The letter preceding each taxon refers to its status category (see page xix).

Invertebrate taxonomy is undergoing constant revision and there are considerable differences of opinion even at the phylum level. As a basis we have used 'Synopsis and Classification of Living Organisms' (McGraw-Hill, New York) edited by S.B. Parker (1982). Where there is serious disagreement with this (e.g. the annelid order Hirudinea and the insect order Phasmatodea) we have adopted suggestions from specialist taxonomists. For this volume we have kept to the traditional view of the Arthropoda as a single phylum. Where relevant, matters of taxonomic dispute are referred to in the review of the taxon in question.

Each taxon that has been reviewed has a common name. Wherever possible we have used an existing name but in some cases it has been necessary to create a new one. We have sought specialist advice and in most cases the new name is an English translation of the scientific name, or describes the geographical distribution, morphology or biology of the species. Common names for individual species have been capitalized.

PROTOZOA

CILIOPHORA

    POLYHYMENOPHOREA

        HETEROTRICHIDA

        Stentoridae

K   Stentor introversus   Tartar's Stentor

EUMETAZOA

CNIDARIA

    ANTHOZOA

        GORGONACEA

        Plexauridae

K   Eunicella verrucosa   Broad Sea Fan

        Coralliidae

CT	Corallium elatius	Boke Coral
CT	C. japonicum	Aka-sango Coral
CT	C. konojoi	Shiro-sango Coral
CT	C. rubrum	Mediterranean Coral
CT	C. secundum	Pink Coral
CT	Corallium sp. nov.	Midway Deep Sea Coral

ACTINIARIA

   Edwardsiidae

V   <u>Nematostella</u> <u>vectensis</u>   Starlet Sea Anemone

   ANTIPATHARIA

CT Antipathidae   Black coral

PLATYHELMINTHES

  TURBELLARIA

    TRICLADIDA

      Kenkiidae

E   <u>Sphalloplana</u> <u>holsingeri</u>   Holsinger's Groundwater Planarian
E   <u>S. subtilis</u>   Biggers' Groundwater Planarian

NEMERTEA

  ENOPLA

    HOPLONEMERTEA

    Prosorhochmidae

V   <u>Antiponemertes</u> <u>allisonae</u>   Terrestrial nemertine worm
R   <u>Argonemertes</u> <u>autraliensis</u>   "
R   <u>A. hillii</u>   "
R   <u>A. stocki</u>   "
R   <u>Geonemertes</u> <u>rodericana</u>   "
R   <u>Katechonemertes</u> <u>nightingaleensis</u> "
R   <u>Leptonemertes</u> <u>chalicophora</u>   "
R   <u>Pantinonemertes</u> <u>agricola</u>   "

MOLLUSCA

  GASTROPODA

    MESOGASTROPODA

    Hydrobiidae

V   <u>Coahuilix</u> <u>hubbsi</u>   Cuatro Ciénegas snail
V   <u>Cochliopina</u> <u>milleri</u>   "
V   <u>Durangonella</u> <u>coahuilae</u>   "
V   <u>Mexipyrgus</u> <u>carranzae</u>   "
V   <u>M. churineanus</u>   "
V   <u>M. escobedae</u>   "
V   <u>M. lugo</u>   "
V   <u>M. mojarralis</u>   "
V   <u>M. multilineatus</u>   "
V   <u>Mexithauma</u> <u>quadripaludium</u>   "
V   <u>Nymphophilus</u> <u>minckleyi</u>   "
V   <u>Paludiscala</u> <u>caramba</u>   "

Strombidae

CT Strombus gigas .... Queen Conch

Cymatiidae

R Charonia tritonis .... Triton's Trumpet

GYMNOSOMATA

Corambidae

K Doridella batava .... Zuiderzee Doridella Sea Slug

BASOMMATOPHORA

Planorbidae

E Ancylastrum cumingianus .... Tasmanian Freshwater 'Limpet'

STYLOMMATOPHORA

Achatinellidae

E	Achatinella abbreviata	Little agate shells
E	A. apexfulva	"
E	A. bellula	"
E	A. buddii	"
E	A. bulimoides	"
E	A. byronii	"
E	A. caesia	"
E	A. casta	"
E	A. cestus	"
E	A. concavospira	"
E	A. curta	"
E	A. decipiens	"
E	A. decora	"
E	A. dimorpha	"
E	A. elegans	"
E	A. fulgens	"
E	A. fuscobasis	"
E	A. juddii	"
E	A. juncea	"
E	A. lehuiensis	"
E	A. leucorraphe	"
E	A. lila	"
E	A. livida	"
E	A. lorata	"
E	A. mustelina	"
E	A. papyracea	"
E	A. phaeozona	"
E	A. pulcherrima	"
E	A. pupukanioe	"
E	A. rosea	"
E	A. sowerbyana	"
E	A. spaldingi	"
E	A. stewartii	"

E	Achatinella swiftii	Little agate shells
E	A. taeniolata	"
E	A. thaanumi	"
E	A. turgida	"
E	A. valida	"
E	A. viridens	"
E	A. vittata	"
E	A. vulpina	"

Partulidae

E	Partula aurantia	Moorean viviparous tree snail
E	P. dendroica	"
E	P. exigua	"
E	P. mirabilis	"
E	P. mooreana	"
E	P. olympica	"
E	P. suturalis	"
E	P. taeniata	"
E	P. tohiveana	"
E	Samoana diaphana	"
E	S. solitaria	"

Pupillidae

V	Leiostyla abbreviata	Madeiran land snail
V	L. cassida	"
V	L. corneocostata	"
V	L. gibba	"
V	L. lamellosa	"

Streptaxidae

V	Gulella planti	Plant's Gulella Snail

Caryodidae

E	Anoglypta launcestonensis	Granulated Tasmanian Snail

Endodontidae

V	Discus defloratus	Madeiran land snail
V	D. guerinianus	"

Camaenidae

R	Papustyla pulcherrima	Manus Green Tree Snail

Helicidae

V	Caseolus calculus	Madeiran land snail
V	C. commixta	"
V	C. sphaerula	"
V	Discula leacockiana	"
V	D. tabellata	"
V	D. testudinalis	"
V	D. turricula	"
V	Geomitra moniziana	"
V	Helix subplicata	"
R	Helix pomatia	Roman Snail

BIVALVIA

UNIONOIDA

Unionidae

Ex	Epioblasma arcaeformis	Sugar-spoon
E	E. biemarginata	
V	E. brevidens	
V	E. capsaeformis	
E	E. curtisi	Curtis Pearly Mussel
Ex	E. flexuosa	
E	E. florentina	Yellow-blossom Pearly Mussel
E	E. haysiana	Northern Acorn Riffle Shell
E	E. lefevrei	Lefevrei's Riffle Shell
Ex	E. lenior	
E	E. lewisi	Fork Shell
E	E. metastriata	
E	E. othcaloogensis	Southern Acorn Riffle Shell
E	E. penita	Penitent Mussel
Ex	E. personata	
Ex	E. propinqua	
E	E. sampsoni	Sampson's Pearly Mussel
Ex	E. stewardsoni	
E	E. sulcata delicata	White Catspaw Mussel
E	E. s. sulcata	Purple Catspaw Mussel
E	E. torulosa gubernaculum	Green-blossom Pearly Mussel
E	E. t. rangiana	Tan-blossom Pearly Mussel
E	E. t. torulosa	Tubercled-blossom Pearly Mussel
V	E. triquetra	Snuffbox
E	E. turgidula	Turgid-blossom Pearly Mussel
E	E. walkeri	Brown-blossom Pearly Mussel

Margaritiferidae

I	Margaritifera auricularia	Spengler's Freshwater Mussel
V	M. margaritifera	Freshwater Pearl Mussel

VENEROIDA

Tridacnidae

I	Hippopus hippopus	Horse's Hoof Clam
I	H. porcellanus	China Clam
K	Tridacna crocea	Crocus Clam
V	T. derasa	Southern Giant Clam
V	T. gigas	Giant Clam
K	T. maxima	Small Giant Clam
I	T. squamosa	Scaly Clam

ANNELIDA

HIRUDINEA

ARHYNCHOBDELLAE

Hirudinidae

I	Hirudo medicinalis	Medicinal Leech

OLIGOCHAETA

HAPLOTAXIDA

Megascolecidae

E	Megascolides americanus	Washington Giant Earthworm
V	M. australis	Giant Gippsland Earthworm
E	M. macelfreshi	Oregon Giant Earthworm

Acanthodrilidae

V	Chilota spp.	South African acanthodriline earthworms
V	Diplotrema spp.	"
V	Microscolex spp.	"
V	Udeina spp.	"

Microchaetidae

V	Microchaetus spp.	South African giant earthworms
V	Tritogenia spp.	"

ARTHROPODA

MEROSTOMATA

XIPHOSURA

Limulidae

K	Carcinoscorpius rotundicauda	Horseshoe crab
K	Limulus polyphemus	"
K	Tachypleus gigas	"
K	T. tridentatus	"

ARACHNIDA

ARANEAE

Theraphosidae

K	Brachypelma smithi	Red-knee Tarantula Spider

Linyphiidae

R	Troglohyphantes gracilis	Kocevje subterranean spider
R	T. similis	"
R	T. spinipes	"

Lycosidae

E	Adelocosa anops	No-eyed Big-eyed Wolf Spider
R	Pardosa diuturna	Glacier Bay Wolf Spider

OPILIONES

Phalangodidae

V   Banksula melones   Melones Cave Harvestman

CRUSTACEA

ANASPIDACEA

Anaspididae

V   Allanaspides helonomus   Tasmanian anaspid crustacean
V   A. hickmani   "
V   Anaspides spinulae   "
V   A. tasmaniae   "
V   Paranaspides lacustris   "

ISOPODA

Cirolanidae

V   Antrolana lira   Madison Cave Isopod

Sphaeromatidae

E   Thermosphaeroma thermophilum   Socorro Isopod

Stenasellidae

R   Mexistenasellus parzefalli   Parzefall's Stenasellid
R   M. wilkensi   Wilken's Stenasellid

DECAPODA

Cambaridae

V   Orconectes shoupi   Shoup's Crayfish

Astacidae

V   Astacus astacus   Noble Crayfish
R   Austropotamobius pallipes   White-clawed Crayfish
V   Pacifastacus fortis   Shasta Crayfish

Parastacidae

V   Astacopsis gouldi   Giant Freshwater Crayfish

Coenobitidae

R   Birgus latro   Coconut Crab

INSECTA

EPHEMEROPTERA

Siphlonuridae

R   Tasmanophlebia lacus-coeruli   Large Blue Lake Mayfly

ODONATA

Hemiphlebiidae

E   Hemiphlebia mirabilis   Hemiphlebia Damselfly

Coenagrionidae

E   Coenagrion freyi   Freya's Damselfly
E   Ischnura gemina   San Francisco Forktail Damselfly

Epiophlebiidae

V   Epiophlebia laidlawi   Relict Himalayan Dragonfly

Cordulegastridae

V   Cordulegaster sayi   Florida Spiketail Dragonfly

Corduliidae

R   Macromia splendens   Shining Macromia Dragonfly
E   Somatochlora hineana   Ohio Emerald Dragonfly

GRYLLOBLATTARIA

Grylloblattidae

V   Grylloblatta chirurgica   Mount St Helens Grylloblattid

ORTHOPTERA

Stenopelmatidae

V   Deinacrida carinata   Herekopare Island Weta
V   D. fallai   Poor Knight's Weta
V   D. heteracantha   Wetapunga
V   D. rugosa   Stephens Island Weta

PHASMATODEA

Phasmatidae

E   Dryococelus australis   Lord Howe Island Stick-insect

DERMAPTERA

Labiduridae

V   Labidura herculeana   St Helena Earwig

PLECOPTERA

Eustheniidae

E   Eusthenia nothofagi   Otway Stonefly

Gripopterygidae

R Leptoperla cacuminis   Mount Kosciusko Wingless Stonefly
R Riekoperla darlingtoni   Mount Donna Buang Wingless Stonefly

ANOPLURA

Haematopinidae

E Haematopinus oliveri   Pygmy Hog Sucking Louse

HOMOPTERA

Cicadidae

V Magicicada cassini   Periodical cicada
V M. septendecim   "
V M. septendecula   "

COLEOPTERA

Carabidae

V Elaphrus viridis   Delta Green Ground Beetle

Cicindelidae
E Cicindela columbica   Columbia Tiger Beetle

Silphidae

E Nicrophorus americanus   Giant Carrion Beetle

Scarabaeidae

V Dynastes hercules hercules   Hercules Beetle
V D. h. reidi   "

Tenebrionidae

R Polposipus herculeanus   Frigate Island Giant Terebrionid Beetle

Curculionidae

V Gymnopholus lichenifer   Lichen Weevil

DIPTERA

Psychodidae

I Nemapalpus nearcticus   Sugarfoot Moth Fly

Blepharoceridae

E Edwardsina gigantea   Giant Torrent Midge
E E. tasmaniensis   Tasmanian Torrent Midge

Tabanidae

E Brennania belkini   Belkin's Dune Tabanid Fly

Drosophilidae

V   Drosophila spp.   Picture-winged flies

TRICHOPTERA

Hydropsychidae

Ex   Hydropsyche tobiasi   Tobias' Caddis-fly

LEPIDOPTERA

Hesperiidae

R   Dalla octomaculata   Eight-spotted Skipper
V   Hesperia dacotae   Dakota Skipper
V   Panoquina errans   Wandering Skipper

Papilionidae

E   Graphium lysithous harrisianus   Harris' Mimic Swallowtail Butterfly
E   Ornithoptera alexandrae   Queen Alexandra's Birdwing Butterfly
E   Papilio aristodemus ponceanus   Schaus' Swallowtail Butterfly
V   P. homerus   Homerus Swallowtail Butterfly
V   Parides ascanius   Fluminense Swallowtail Butterfly
R   P. hahneli   Hahnel's   Amazonian   Swallowtail
                                        Butterfly
V   Parnassius apollo   Apollo Butterfly

Lycaenidae

V   Eumaeus atala florida   Florida Atala Hairstreak Butterfly
V   Maculinea alcon   Alcon Large Blue Butterfly
V   M. arion   Large Blue Butterfly
V   M. arionides   Greater Large Blue Butterfly
E   M. nausithous   Dusky Large Blue Butterfly
V   M. teleius   Scarce Large Blue Butterfly
K   Strymon avalona   Avalon Hairstreak Butterfly

Danaidae

TP   Danaus plexippus   Monarch   Butterfly: Mexican   and
                                          Calfornian winter roosts
K   Idea tambusisiana   Sulawesi Tree Nymph Butterfly

Satyridae

R   Tribe Pronophilini   Andean brown butterflies

Nymphalidae

V   Boloria acronema   Uncompahgre Fritillary Butterfly
E   Euphydryas editha bayensis   Bay Checkerspot Butterfly
E   Heliconius natterei   Natterer's Longwing Butterfly

Sphingidae

E   Euproserpinus wiesti   Wiest's Spinx Moth

HYMENOPTERA

Formicidae

K	Aneuretus simoni	Sri Lankan Relict Ant
K	Aulacopone relicta	Caucasian Relict Ant
R	Epimyrma ravouxi	Ravoux's Slavemaker Ant
V	Formica aquilonia	Wood ants of Europe
V	F. lugubris	"
V	F. polyctena	"
V	F. pratensis	"
V	F. rufa	"
TP	Formica yessensis	Japanese Wood Ant (supercolony)
K	Leptothorax goesswaldi	Goesswald's Inquiline Ant
K	Nothomyrmecia macrops	Australian Nothomyrmecia Ant

Megachilidae

K	Chalicodoma pluto	Wallace's Giant Bee

ONYCHOPHORA

ONYCHOPHORA

V	Peripatidae	Peripatus
V	Peripatopsidae	Peripatus

ECHINODERMATA

ECHINOIDEA

ECHINOIDA

Echinidae

K	Echinus esculentus	European Edible Sea Urchin
CT	Paracentrotus lividus	Purple Urchin

LIST OF THREATENED COMMUNITIES
DESCRIBED IN THIS VOLUME

TROPICAL FORESTS

Usambara Mountains	Tanzania
Rain Forests of Gunung Mulu	Malaysia

METROPOLITAN AREAS

San Bruno Mountain	U.S.A.
Banks Peninsula	New Zealand

XERIC BIOMES

El Segundo Sand Dunes	U.S.A.
Dead Sea Depression	Israel and Jordon

CAVES

Cueva los Chorros	Puerto Rico
Deadhorse Cave	U.S.A.

WETLANDS

Mires of the Sumava Mountains	Czechoslovakia

MARINE BIOMES

Taka Bone Rate Coral Atoll	Indonesia
Roseland Marine Conservation Area	U.K.

LIST OF THREATENED INVERTEBRATES AND THREATENED COMMUNITIES ARRANGED BY COUNTRY AND ZOOGEOGRAPHICAL REGION

The zoogeographical regions first described by Wallace in 1876 are used, slightly modified, as a basis for this index. Boundaries have been adjusted in some cases in order to avoid subdividing a particular political unit.

The Palearctic region is the northerly part of the Old World. It extends over the whole of Europe, U.S.S.R. and China, including the adjacent offshore islands, and includes the Mediterranean coast of Africa. In the present list the entire Arabian Peninsula is assigned to this region.

The Ethiopian region covers central and southern Africa.

The Oriental region covers Asia south of China from the western border of Afghanistan and Pakistan east to the Pacific Ocean including the Philippines and Indonesia.

The Australasian region includes Papua New Guinea, Australia and New Zealand.

The Nearctic region covers all of North America including Greenland, the Aleutian Islands and, in this volume, the whole of Mexico.

The Neotropical region includes the Caribbean Islands, Central and South America.

Oceanic islands have been placed in a separate section.

Non-marine species (including those from brackish water and lagoons) and threatened communities are arranged by country, marine species by broad oceanic region. Since most marine species in this book have wide distributions and records for some are incomplete, countries and islands within oceanic divisions are not given. Readers are referred to the individual accounts for details. Marine threatened communities are listed under the relevant country.

'Ex' following a taxon denotes that it is regarded as extinct in that country and '?' denotes lack of confirmation of its presence.

NON-MARINE SPECIES

PALEARCTIC REGION

EUROPE

Formica aquilonia	Wood ant
F. lugubris	"
F. polyctena	"
F. pratensis	"
F. rufa	"

ANDORRA

Parnassius apollo	Apollo Butterfly
Maculinea arion	Large Blue Butterfly

AUSTRIA

Helix pomatia	Roman Snail
Margaritifera margaritifera	Freshwater Pearl Mussel
Hirudo medicinalis	Medicinal Leech
Astacus astacus	Noble Crayfish
Coenagrion freyi	Freya's Damselfly
Parnassius apollo	Apollo Butterfly
Maculinea alcon	Alcon Large Blue Butterfly
M. arion	Large Blue Butterfly
M. nausithous	Dusky Large Blue Butterfly
M. teleius	Scarce Large Blue Butterfly
Epimyrma ravouxi	Ravoux's Slavemaker Ant

BELGIUM

Helix pomatia	Roman Snail
Margaritifera margaritifera	Freshwater Pearl Mussel
Maculinea alcon	Alcon Large Blue Butterfly
M. arion	Large Blue Butterfly
M. teleius	Scarce Large Blue Butterfly

BULGARIA

Helix pomatia	Roman Snail
Hirudo medicinalis	Medicinal Leech
Astacus astacus	Noble Crayfish
Parnassius apollo	Apollo Butterfly

CANARY ISLANDS (Spain)

Leptonemertes chalicophora	Terrestrial nemertine

CHINA (including Tibet)

Maculinea alcon	Alcon Large Blue Butterfly
M. arion	Large Blue Butterfly
M. arionides	Greater Large Blue Butterfly
M. teleius	Scarce Large Blue Butterfly

CZECHOSLOVAKIA

Helix pomatia	Roman Snail
Margaritifera auricularia	Spengler's Freshwater Mussel (Ex)
M. margaritifera	Freshwater Pearl Mussel
Hirudo medicinalis	Medicinal Leech
Astacus astacus	Noble Crayfish
Parnassius apollo	Apollo Butterfly
Maculinea alcon	Alcon Large Blue Butterfly
M. arion	Large Blue Butterfly
M. nausithous	Dusky Large Blue Butterfly
M. teleius	Scarce Large Blue Butterfly

Mires of the Sumava Mountains

DENMARK

Helix pomatia	Roman Snail
Margaritifera margaritifera	Freshwater Pearl Mussel
Hirudo medicinalis	Medicinal Leech
Astacus astacus	Noble Crayfish
Parnassius apollo	Apollo Butterfly
Maculinea alcon	Alcon Large Blue Butterfly
M. arion	Large Blue Butterfly

FINLAND

Helix pomatia	Roman Snail
Margaritifera margaritifera	Freshwater Pearl Mussel
Hirudo medicinalis	Medicinal Leech
Astacus astacus	Noble Crayfish
Parnassius apollo	Apollo Butterfly
Maculinea alcon	Alcon Large Blue Butterfly

FRANCE (see also Guadeloupe, Martinique, Society Islands)

Helix pomatia	Roman Snail
Margaritifera auricularia	Spengler's Freshwater Mussel
M. margaritifera	Freshwater Pearl Mussel
Hirudo medicinalis	Medicinal Leech
Astacus astacus	Noble Crayfish
Austropotamobius pallipes	White-footed Crayfish
Macromia splendens	Shining Macromia Dragonfly
Parnassius apollo	Apollo Butterfly
Maculinea alcon	Alcon Large Blue Butterfly
M. arion	Large Blue Butterfly
M. nausithous	Dusky Large Blue Butterfly
M. teleius	Scarce Large Blue Butterfly
Epimyrma ravouxi	Ravoux's Slavemaker Ant

GERMANY, FEDERAL REPUBLIC OF

Helix pomatia	Roman Snail
Margaritifera auricularia	Spengler's Freshwater Mussel (Ex)
M. margaritifera	Freshwater Pearl Mussel
Hirudo medicinalis	Medicinal Leech
Astacus astacus	Noble Crayfish
Austropotamobius pallipes	White-footed Crayfish
Coenagrion freyi	Freya's Damselfly
Hydropsyche tobiasi	Tobias' Caddis-fly (Ex)
Parnassius apollo	Apollo Butterfly
Maculinea alcon	Alcon Large Blue Butterfly
M. arion	Large Blue Butterfly
M. nausithous	Dusky Large Blue Butterfly
M. teleius	Scarce Large Blue Butterfly
Epimyrma ravouxi	Ravoux's Slavemaker Ant

GERMANY, DEMOCRATIC REPUBLIC OF

Margaritifera auricularia	Spengler's Freshwater Mussel (Ex)
M. margaritifera	Freshwater Pearl Mussel
Parnassius apollo	Apollo Butterfly
Maculinea nausithous	Dusky Large Blue Butterfly
M. teleius	Scarce Large Blue Butterfly
Epimyrma ravouxi	Ravoux's Slavemaker Ant

GREECE

Helix pomatia	Roman Snail
Hirudo medicinalis	Medicinal Leech
Parnassius apollo	Apollo Butterfly
Maculinea alcon	Alcon Large Blue Butterfly
M. arion	Large Blue Butterfly

HUNGARY

Helix pomatia	Roman Snail
Hirudo medicinalis	Medicinal Leech
Astacus astacus	Noble Crayfish
Parnassius apollo	Apollo Butterfly
Maculinea alcon	Alcon Large Blue Butterfly
M. nausithous	Dusky Large Blue Butterfly
M. teleius	Scarce Large Blue Butterfly

ICELAND

Margaritifera margaritifera	Freshwater Pearl Mussel (?)

IRELAND, REPUBLIC OF

Margaritifera margaritifera	Freshwater Pearl Mussel
Austropotamobius pallipes	White-footed Crayfish

ISRAEL

Dead Sea Depression

ITALY

Helix pomatia	Roman Snail
Margaritifera auricularia	Spengler's Freshwater Mussel
Hirudo medicinalis	Medicinal Leech
Astacus astacus	Noble Crayfish
Austropotamobius pallipes	White-footed Crayfish
Parnassius apollo	Apollo Butterfly
Maculinea teleius	Scarce Large Blue Butterfly
Epimyrma ravouxi	Ravoux's Slavemaker Ant

JAPAN

Maculinea arion	Large Blue Butterfly
M. arionides	Greater Large Blue Butterfly
M. teleius	Scarce Large Blue Butterfly
Formica yessensis	Japanese Wood Ant Supercolony

JORDAN

Dead Sea Depression

KOREA, DEMOCRATIC PEOPLE'S REPUBLIC OF

Maculinea teleius	Scarce Large Blue Butterfly

KOREA, REPUBLIC OF

Maculinea teleius	Scarce Large Blue Butterfly

LIECHTENSTEIN

Parnassius apollo	Apollo Butterfly

LUXEMBOURG

Helix pomatia	Roman Snail
Margaritifera auricularia	Spengler's Freshwater Mussel (Ex)
M. margaritifera	Freshwater Pearl Mussel
Hirudo medicinalis	Medicinal Leech
Astacus astacus	Noble Crayfish

MADEIRA (Portugal)

Leptonemertes chalicophora	Terrestrial nemertine
Leiostyla abbreviata	Madeiran land snail
L. cassida	"
L. corneocostata	"
L. gibba	"
L. lamellosa	"
Discus defloratus	"
D. guerinianus	"
Caseolus calculus	"
C. commixta	"
C. sphaerula	"
Discula leacockiana	"
D. tabellata	"
D. testudinalis	"
D. turricula	"
Geomitra moniziana	"
Helix subplicata	"

MONGOLIA

Maculinea arion	Large Blue Butterfly
M. teleius	Scarce Large Blue Butterfly

MOROCCO

Margaritifera auricularia maroccana	Spengler's Freshwater Mussel

NETHERLANDS

Doridella batava	Zuiderzee Doridella Sea Slug
Helix pomatia	Roman Snail
Margaritifera margaritifera	Freshwater Pearl Mussel
Hirudo medicinalis	Medicinal Leech
Astacus astacus	Noble Crayfish
Parnassius apollo	Apollo Butterfly
Maculinea alcon	Alcon Large Blue Butterfly
M. arion	Large Blue Butterfly
M. nausithous	Dusky Large Blue Butterfly
M. teleius	Scarce Large Blue Butterfly

NORWAY

Helix pomatia	Roman Snail
Margaritifera margaritifera	Freshwater Pearl Mussel
Hirudo medicinalis	Medicinal Leech
Astacus astacus	Noble Crayfish
Parnassius apollo	Apollo Butterfly

POLAND

Helix pomatia	Roman Snail
Margaritifera margaritifera	Freshwater Pearl Mussel
Astacus astacus	Noble Crayfish
Parnassius apollo	Apollo Butterfly
Maculinea arion	Large Blue Butterfly
M. nausithous	Dusky Large Blue Butterfly
M. teleius	Scarce Large Blue Butterfly

PORTUGAL (see also Azores, Madeira)

Margaritifera auricularia	Spengler's Freshwater Mussel
Hirudo medicinalis	Medicinal Leech
Austropotamobius pallipes	White-footed Crayfish

ROMANIA

Helix pomatia	Roman Snail
Astacus astacus	Noble Crayfish
Parnassius apollo	Apollo Butterfly

SPAIN

Helix pomatia	Roman Snail
Margaritifera auricularia	Spengler's Freshwater Mussel
Austropotamobius pallipes	White-footed Crayfish
Parnassius apollo	Apollo Butterfly
Maculinea nausithous	Dusky Large Blue Butterfly
M. teleius	Scarce Large Blue Butterfly
Epimyrma ravouxi	Ravoux's Slavemaker Ant

SWEDEN

Helix pomatia	Roman Snail
Margaritifera margaritifera	Freshwater Pearl Mussel
Hirudo medicinalis	Medicinal Leech
Astacus astacus	Noble Crayfish
Parnassius apollo	Apollo Butterfly
Maculinea alcon	Alcon Large Blue Butterfly
M. arion	Large Blue Butterfly

SWITZERLAND

Helix pomatia	Roman Snail
Margaritifera auricularia	Spengler's Freshwater Mussel (Ex)
Hirudo medicinalis	Medicinal Leech (?)
Astacus astacus	Noble Crayfish
Austropotamobius pallipes	White-footed Crayfish
Coenagrion freyi	Freya's Damselfly
Parnassius apollo	Apollo Butterfly
Maculinea alcon	Alcon Large Blue Butterfly
M. arion	Large Blue Butterfly
M. nausithous	Dusky Large Blue Butterfly
M. teleius	Scarce Large Blue Butterfly
Epimyrma ravouxi	Ravoux's Slavemaker Ant
Leptothorax goesswaldi	Goesswald's Inquiline Ant

TURKEY

Hirudo medicinalis	Medicinal Leech

UNITED KINGDOM (see also St Helena, St Lucia, Tristan da Cunha)

Nematostella vectensis	Starlet Sea Anemone
Helix pomatia	Roman Snail
Margaritifera auricularia	Spengler's Freshwater Mussel (Ex)
M. margaritifera	Freshwater Pearl Mussel
Hirudo medicinalis	Medicinal Leech
Astacus astacus	Noble Crayfish (?)
Austropotamobius pallipes	White-footed Crayfish
Maculinea arion	Large Blue Butterfly

Roseland Marine Conservation Area

U.S.S.R.

Margaritifera margaritifera	Freshwater Pearl Mussel
Hirudo medicinalis	Medicinal Leech
Astacus astacus	Noble Crayfish
Parnassius apollo	Apollo Butterfly
Maculinea alcon	Alcon Large Blue Butterfly
M. arion	Large Blue Butterfly
M. nausithous	Dusky Large Blue Butterfly
Aulacopone relicta	Caucasian Relict Ant

YUGOSLAVIA

Helix pomatia	Roman Snail
Hirudo medicinalis	Medicinal Leech
Troglohyphantes gracilis	Kocevje subterranean spider
T. similis	"
T. spinipes	"
Astacus astacus	Noble Crayfish
Austropotamobius pallipes	White-footed Crayfish
Parnassius apollo	Apollo Butterfly
Maculinea nausithous	Dusky Large Blue Butterfly
M. teleius	Scarce Large Blue Butterfly
Epimyrma ravouxi	Ravoux's Slavemaker Ant

ETHIOPIAN REGION

Phylum Onychophora	Peripatus

LESOTHO

Microscolex spp.	South African acanthodriline earthworms
Udeina spp.	"

SOUTH AFRICA

Gulella planti	Plant's Gulella Snail
Chilota spp.	South African acanthodriline earthworms
Diplotrema spp.	"
Microscolex spp.	"
Udeina spp.	"
Microchaetus spp.	South African giant earthworms
Tritogenia spp.	"

TANZANIA

Usambara Mountains

ORIENTAL REGION

Phylum Onychophora	Peripatus

INDIA

Epiophlebia laidlawi	Relict Himalayan Dragonfly
Haematopinus oliveri	Pygmy Hog Sucking Louse

INDONESIA

Idea tambusisiana	Sulawesi Tree Nymph Butterfly
Chalicodoma pluto	Wallace's Giant Bee

Taka Bone Rate Coral Atoll

MALAYSIA

Rain Forests of Gunung Mulu, Sarawak

NEPAL

Epiophlebia laidlawi	Relict Himalayan Dragonfly

SRI LANKA

Aneuretus simoni	Sri Lankan Relict Ant

AUSTRALASIAN REGION

AUSTRALIA

Argonemertes australiensis	Terrestrial nemertine
A. hillii	"
A. stocki	"
Ancylastrum cumingianus	Tasmanian Freshwater 'Limpet'
Anoglypta launcestonensis	Granulated Tasmanian Snail
Megascolides australis	Giant Gippsland Earthworm
Allanaspides helonomus	Tasmanian anaspid crustacean
A. hickmani	"
Anaspides spinulae	"
A. tasmaniae	"
Paranaspides lacustris	"
Astacopsis gouldi	Giant Freshwater Crayfish
Tasmanophlebia lacus-coerulei	Large Blue Lake Mayfly
Hemiphlebia mirabilis	Hemiphlebia Damselfly
Eusthenia nothofagi	Otway Stonefly
Leptoperla cacuminis	Mount Kosciusko Wingless Stonefly
Riekoperla darlingtoni	Mount Donna Buang Wingless Stonefly

Edwardsina gigantea	Giant Torrent Midge
E. tasmaniensis	Tasmanian Torrent Midge
Nothomyrmecia macrops	Australian Nothomyrmecia Ant

LORD HOWE ISLANDS (Australia)

Dryococelus australis	Lord Howe Island Stick-insect

NEW ZEALAND

Antiponemertes allisonae	Terrestrial nemertine
Deinacrida carinata	Herekopare Island Weta
D. fallai	Poor Knight's Weta
D. heteracantha	Wetapunga
D. rugosa	Stephens Island Weta

Banks Peninsula

PAPUA NEW GUINEA

Papustyla pulcherrima	Manus Green Tree Snail
Gymnopholus lichenifer	Lichen Weevil
Ornithoptera alexandrae	Queen Alexandra's Birdwing Butterfly

NEARCTIC REGION

CANADA

Nematostella vectensis	Starlet Sea Anemone
Epioblasma torulosa rangiana	Tan-blossom Pearly Mussel
E. t. torulosa	Tubercled-blossom Pearly Mussel
Margaritifera margaritifera	Freshwater Pearl Mussel
Magicicada septendecim	Periodical cicada
Nicrophorus americanus	Giant Carrion Beetle
Hesperia dacotae	Dakota Skipper (Ex)

MEXICO

Coahuilix hubbsi	Cuatro Ciénegas snail
Cochliopina milleri	"
Durangonella coahuilae	"
Mexipyrgus carranzae	"
M. churinceanus	"
M. escobedae	"
M. lugoi	"
M. mojarralis	"
M. multilineatus	"
Mexithauma quadripaludium	"
Nymphophilus minckleyi	"
Paludiscala caramba	"
Brachypelma smithi	Red-knee Tarantula Spider
Mexistenasellus parzefalli	Parzefall's Stenasellid

Mexistenasellus wilkensi	Wilken's Stenasellid
Brennania belkini	Belkin's Dune Tabanid Fly
Panoquina errans	Wandering Skipper
Phylum Onychophora	Peripatus
Danaus plexippus	Monarch Butterfly: Mexican winter roosts

U.S.A. (see also Hawaiian Islands)

Stentor introversus	Tartar's Stentor
Nematostella vectensis	Starlet Sea Anemone
Sphalloplana holsingeri	Holsinger's Groundwater Planarian
S. subtilis	Biggers' Groundwater Planarian
Leptonemertes chalicophora	Terrestrial nemertine
Helix pomatia	Roman Snail
Epioblasma spp.	Riffle shells
Margaritifera margaritifera	Freshwater Pearl Mussel
Megascolides americanus	Washington Giant Earthworm
M. macelfreshi	Oregon Giant Earthworm
Pardosa diuturna	Glacier Bay Wolf Spider
Banksula melones	Melones Cave Harvestman
Antrolana lira	Madison Cave Isopod
Thermosphaeroma thermophilum	Socorro Isopod
Orconectes shoupi	Shoup's Crayfish
Pacifastacus fortis	Shasta Crayfish
Ischnura gemina	San Francisco Forktail Damselfly
Cordulegaster sayi	Florida Spiketail Dragonfly
Somatochlora hineana	Ohio Emerald Dragonfly
Grylloblatta chirurgica	Mount St Helens Grylloblattid
Magicicada cassini	Periodical cicada
M. septendecim	"
M. septendecula	"
Elaphrus viridis	Delta Green Ground Beetle
Cicindela columbica	Columbia Tiger Beetle
Nicrophorus americanus	Giant Carrion Beetle
Nemapalpus nearcticus	Sugarfoot Moth Fly
Brennania belkini	Belkin's Dune Tabanid Fly
Hesperia dacotae	Dakota Skipper
Panoquina errans	Wandering Skipper
Papilio aristodemus ponceanus	Schaus' Swallowtail Butterfly
Eumaeus atala florida	Florida Atala Hairstreak Butterfly
Strymon avalona	Avalon Hairstreak Butterfly
Boloria acrocnema	Uncompahgre Fritillary Butterfly
Euphydryas editha bayensis	Bay Checkerspot Butterfly
Euproserpinus wiesti	Wiest's Sphinx Moth
Danaus plexippus	Monarch Butterfly: Californian winter roosts

Deadhorse Cave
El Segundo Sand Dunes
San Bruno Mountain

Tribe Pronophilini   Andean brown butterflies
Phylum Onychophora   Peripatus

BERMUDA (U.K.)

Pantinonemertes agricola   Terrestrial nemertine

BOLIVIA

Tribe Pronophilini   Andean brown butterflies

BRAZIL

Graphium lysithous harrisianus   Harris' Mimic Swallowtail Butterfly
Parides ascanius   Fluminense Swallowtail Butterfly
P. hahneli   Hahnel's   Amazonian   Swallowtail
                                         Butterfly
Heliconius nattereri   Natterer's Longwing Butterfly

COLOMBIA

Tribe Pronophilini   Andean brown butterflies

COSTA RICA

Dalla octomaculata   Eight-spotted Skipper

DOMINICA

Dynastes hercules hercules   Hercules Beetle

ECUADOR

Tribe Pronophilini   Andean brown butterflies

GUADELOUPE (France)

Dynastes hercules hercules   Hercules Beetle

JAMAICA

Papilio homerus   Homerus Swallowtail Butterfly

MARTINIQUE (France)

Dynastes hercules reidi   Hercules Beetle

PANAMA

Dalla octomaculata   Eight-spotted Skipper

PERU

Tribe Pronophilini   Andean brown butterflies

PUERTO RICO

Cueva los Chorros

ST LUCIA (U.K.)

Dynastes hercules reidi   Hercules Beetle

VENEZUELA

Tribe Pronophilini   Andean brown butterflies

OCEANIC ISLANDS

AZORES (Portugal)

Leptonemertes chalicophora   Terrestrial nemertine

ST HELENA (U.K.)

Labidura herculeana   St Helena Earwig

TRISTAN DA CUNHA ISLANDS (U.K.)

Katechonemertes nightingalensis   Terrestrial nemertine

RODRIGUES (Mauritius)

Geonemertes rodericana   Terrestrial nemertine

SEYCHELLES

Polposipus herculeanus   Frigate Island Giant Tenebrionid Beetle

HAWAIIAN ISLANDS (U.S.A.)

Achatinella spp.	Little agate shells
Adelocosa anops	No-eyed Big-eyed Wolf Spider
Drosophila spp.	Picture-winged flies

SOCIETY ISLANDS (France)

Partula aurantia	Moorean viviparous tree snail
P. dendroica	"
P. exigua	"
P. mirabilis	"
P. mooreana	"
P. suturalis	"
P. taeniata	"
P. tohiveana	"
Samoana diaphana	"
S. solitaria	"

MARINE SPECIES

Marine areas are arranged in the following order: Mediterranean Sea, Atlantic Ocean (three areas), Caribbean Sea, Indian Ocean, Red Sea, Pacific Ocean (four areas). The species listed occur in at least one or more of the countries given under each oceanic region, but may also be found in the waters around other political units. The reader is referred to the data sheets for details.

MEDITERRANEAN SEA
Algeria, France, Italy, Malta, Morocco, Spain, Tunisia.

Eunicella verrucosa	Broad Sea Fan
Corallium rubrum	Mediterranean Coral
Paracentrotus lividus	Purple Urchin

NORTH-EAST ATLANTIC OCEAN
Belgium, Canary Islands (Spain), Cape Verde Islands, Denmark, France, Ireland, Netherlands, Norway, Portugal, Spain, Sweden, U.K.

Eunicella verrucosa	Broad Sea Fan
Corallium rubrum	Mediterranean Coral
Echinus esculentus	European Edible Sea Urchin
Paracentrotus lividus	Purple Urchin
Roseland Marine Conservation Area	U.K.

NORTH-WEST ATLANTIC OCEAN
Bermuda (U.K.), U.S.A.

Family Antipathidae	Black coral
Strombus gigas	Queen Conch
Limulus polyphemus	Horseshoe crab

SOUTH-WEST ATLANTIC OCEAN

Family Antipathidae	Black coral

CARIBBEAN SEA
Antigua and Barbuda (U.K.), Bahamas, Barbados, Belize, British Virgin Islands, Colombia, Cuba, Dominica, Dominican Republic, Grenadines, Haiti, Jamaica, 'Lesser Antilles', Mexico, Netherlands Antilles, Panama, Puerto Rico, St Lucia (U.K.), Trinidad and Tobago, Turks and Caicos (Jamaica), Venezuela, U.S.A., U.S. Virgin Islands.

Family Antipathidae	Black coral
Strombus gigas	Queen Conch
Limulus polyphemus	Horseshoe crab

INDIAN OCEAN
Aldabra (Seychelles), Andaman Islands (India), Chagos Archipelago (Seychelles), Christmas Island (Australia), Cocos (Keeling) Islands, Kenya, India, Indonesia, Madagascar, Malaysia, Maldives, Mauritius, Mozambique, Nicobar Islands (India), Seychelles, South Africa, Singapore, Sri Lanka, Tanzania, Thailand, Vietnam.

Family Antipathidae	Black coral
Charonia tritonis	Triton's Trumpet
Hippopus hippopus	Horse's Hoof Clam
H. porcellanus	China Clam
Tridacna crocea	Crocus Clam
T. derasa	Southern Giant Clam
T. gigas	Giant Clam
T. maxima	Small Giant Clam
T. squamosa	Scaly Clam
Carcinoscorpius rotundicauda	Horseshoe crab
Tachypleus gigas	"
T. tridentatus	"
Birgus latro	Coconut Crab

RED SEA
Egypt, Saudi Arabia

Family Antipathidae	Black Coral
Charonia tritonis	Triton's Trumpet
Tridacna maxima	Small Giant Clam
T. squamosa	Scaly Clam

NORTH-EAST PACIFIC OCEAN
Hawaii (U.S.A.)

Corallium secundum	Pink Coral
Corallium sp. nov.	Midway Deep Sea Coral
Family Antipathidae	Black coral
Charonia tritonis	Triton's Trumpet

NORTH-WEST PACIFIC OCEAN
China, Japan (including Ogaswara-Gunto Islands and Ryukyu Islands), Philippines, Taiwan.

Corallium elatius	Boke Coral
C. japonicum	Aka-sango Coral
C. konojoi	Shiro-sango Coral
C. secundum	Pink Coral
Corallium sp. nov.	Midway Deep Sea Coral
Family Antipathidae	Black coral
Charonia tritonis	Triton's Trumpet
Hippopus hippopus	Horse's Hoof Clam
H. porcellanus	China Clam
Tridacna crocea	Crocus Clam
T. derasa	Southern Giant Clam
T. gigas	Giant Clam
T. maxima	Small Giant Clam
T. squamosa	Scaly Clam
Carcinoscorpius rotundicauda	Horseshoe crab
Tachypleus gigas	"
T. tridentatus	"
Birgus latro	Coconut Crab

SOUTH-EAST PACIFIC OCEAN
Cook Islands (New Zealand), Henderson Island (U.K.), Pitcairn Island (U.K.), Tuamotu (France)

Corallium spp.	Precious corals
Family Antipathidae	Black coral
Tridacna maxima	Small Giant Clam
T. squamosa	Scaly Clam
Birgus latro	Coconut Crab

SOUTH-WEST PACIFIC OCEAN
American Samoa, Australia, Fiji, U.S.A. Guam (U.S.A.), Kiribati (Caroline, Gilbert, Line, and Phoenix Islands), Marianas (U.S.A.), Marshall Islands (France), New Caledonia, Papua New Guinea, Solomon Islands, Tuvalu (Ellice Islands), Vanuatu, Wake Island (U.S.A.), Western Samoa.

Corallium spp.	Precious corals
Family Antipathidae	Black coral
Charonia tritonis	Triton's Trumpet
Hippopus hippopus	Horse's Hoof Clam
Tridacna crocea	Crocus Clam
T. derasa	Southern Giant Clam
T. gigas	Giant Clam
T. maxima	Small Giant Clam
T. squamosa	Scaly Clam
Birgus latro	Coconut Crab

PROTOZOA

INTRODUCTION The protozoa are a diverse assemblage of over 65 000 minute, mobile, single-celled organisms which, although not a natural group, have been placed together for convenience. Traditionally they were classified as a single phylum, but it is now considered that seven major groups should each have the status of phylum reflecting the polyphyletic origin of the assemblage, the term 'protozoa' being retained as a convenient name for the whole assemblage (10,11,12). Their taxonomy is complex and probably only half the number of existing species have been described so far. Apart from their unicellular nature, other shared characteristics are their small, generally microscopic, size and their mobility. Moisture is their prime ecological requirement and the majority are found in the sea or freshwater, although large numbers are parasitic or mutualistic. Others, particularly flagellates, small amoebae and ciliates, are ubiquitous in soil (13). Many so-called protozoa have plant characteristics and are able to photosynthesize. Others are clearly animals in that they depend on organic substances for food. Some are even capable of surviving in purely inorganic environments.

The main groups are characterized by their different forms of locomotion and the kind of encystment undergone in response to unfavourable conditions (3). Encystment is an adaptation shown by many species whereby they form a hard protective external covering under adverse conditions. As cysts they are dispersed huge distances by wind, water currents and other organisms and can survive for years until conditions become favourable again. The ease with which protozoa may be dispersed accounts for the extraordinarily wide distributions shown by some species, many occurring throughout much of the world. However, the ecological requirements of most species are poorly known and difficult to study; a species may be abundant in a particular spot one week but absent the next week under apparently identical conditions. Many appear to be able to tolerate wide fluctuations in environmental conditions such as salinity, and some survive equally well in fresh or sea water (3,15).

SARCOMASTIGOPHORA This phylum includes two main groups, traditionally called the Sarcodina and the Mastigophora. The Sarcodina are a large group of protozoans that have pseudopodia or flowing extensions of the body which are used for capturing prey and for locomotion. They include the well known Amoeba and the benthic foraminiferans and deep-water radiolarians which both have skeletons or shells of secreted calcium carbonate or sometimes of foreign mineral particles. A few are parasitic, including those causing amoebic dysentery in humans. The Mastigophora possess flagella, long whip-like processes which beat to propel the organism through the water (2). They include many parasitic forms as well as a number of alga-like species. The dinoflagellates, for example, occur in huge numbers in plankton and certain luminescent species contribute to the bioluminescence of the sea (1,3,7).

CILIOPHORA This is the second largest group whose members have cilia (minute hair-like filaments), used for locomotion and feeding. They include the relatively well known Paramecium, Stentor and the colonial Vorticella.

The five other phyla contain mainly parasitic forms, including species which cause human malarias and many economically important diseases of domestic animals.

SCIENTIFIC INTEREST AND POTENTIAL VALUE Protozoa play an extremely important role in many ecological processes. Protozoa and other microbial species make up a large portion of the biomass of many aquatic systems. Foraminiferan and radiolarian shells sink to the sea bottom and form the primary constituent of many ocean bottom sediments (3,4). Protozoa form the basis of most food chains

1

and the planktonic photosynthetic flagellates are important primary producers of organic matter. Planktonic ciliates and flagellates are an important element in food chains (3,15). Other species are decomposers of organic matter and play a role in soil breakdown and the formation of humus (13).

Most animals serve as hosts for protozoa in one form or another, and in a number of cases such relationships are beneficial or essential to the host as well as to the protozoan. Lower termites and wood roaches are dependent on the mutualistic flagellates in their guts which digest the wood eaten by these insects; the products of the digestive process are shared by the flagellate and its host. Marine organisms such as corals and giant clams are dependent on green (zoochlorellae) or brown (zooxanthellae) flagellates which live within the host tissues, using waste nitrogenous materials and in return supplying the host with the products of their photosynthesis. Ciliates occur in huge numbers in the rumen of ruminants such as cattle and sheep. Although it is possible under controlled conditions to rear perfectly healthy animals entirely free of such protozoa, it is probable that the latter make a major contribution to the organic acids and protein in the diets of these herbivores in the natural state.

Some protozoa cause disease: for example Entamoeba histolytica causes dysentery; the flagellate Trypanosoma causes African sleeping sickness and is transmitted by tsetse flies; and the sporozoan blood parasite, Plasmodium, is responsible for malaria and is transmitted by mosquitoes. Species of the flagellate genera Gymnodinium and Gonyaulax are responsible for outbreaks of so-called red tides when, under certain conditions, their planktonic populations increase to such enormous densities that the water becomes coloured (8). At such times concentrations of certain toxic metabolic substances reach such high levels that other marine life may be killed. The occurrence of red tides is restricted to certain regions, many of which coincide with oceanic upwellings. Since these outbreaks occur mainly in coastal waters, land drainage and pollution are thought to be responsible in many cases (14). Ciguatera fish poisoning is also caused by a protozoan, the dinoflagellate Gambierdiscus toxicus which is consumed by fish feeding on corals and algae. Many hundreds of cases of fish poisoning are reported annually throughout the Caribbean and Indo-Pacific, and recent research has suggested that in some cases outbreaks may be correlated with damage to coral reefs, either naturally through hurricanes or as a result of human activities (6).

It is therefore not surprising that protozoa are often thought of in the context of disease. However, many are essential to human life since they help the body to digest food and synthesize important vitamins. Many feed on bacteria, and ciliates play an important role in the purification and removal of pathogenic organisms from polluted water. Sewage processing is dependent on ciliated Protozoa that, by their predatory activities on bacteria, clarify the effluent (9,15).

Protozoans are used for a variety of scientific purposes including environmental, ecological and medical research. They are small and easily handled, multiply rapidly, can be grown on synthetic media, and through asexual reproduction can provide batches of uniform test organisms. Since many species can be cultured, 'libraries' can be maintained and the test organisms used at any time of year. Ciliates are often used for water quality testing and have proved to be as sensitive as other invertebrates or fish for this purpose. Members of the Terahymena pyriformis species complex are used widely as microassay organisms in the detection of essential amino acids, vitamins and proteins and in testing the toxicity of various substances (3,15).

THREATS TO SURVIVAL Clearly it will be a long time before sufficient information is available to ascertain the degree to which protozoan species may be threatened. Their distribution patterns differ from those of other animals and many are cosmopolitan, so that few species are known with very small ranges, although some appear to be highly specialized and limited to habitats such as hot springs. However, research has shown that the requirements of microbial species and communities may be as complex, or nearly so, as those of taxonomically higher organisms, and it is to be expected that disruption of such communities could affect the foodwebs that depends on them.

Although some species are able to survive adverse conditions by encysting, research suggests that others are readily destroyed. Aquatic communities are likely to be vulnerable to pollution, either thermal or toxic. The effects of pollution on marine protozoa are fully reviewed in (8). Some marine ciliates, for example, are very sensitive to sudden increases in temperature as occur at power plants and other industrial discharge sites (5). In such locations the vast majority of ciliates in the immediate vicinity could be rapidly eliminated. The results of pollution studies are often conflicting. One study has shown that foraminiferan species diversity is depressed close to a sewage outfall although at some distance offshore there is an anomalously high diversity compared to unpolluted sites (6,17).

CONSERVATION Work on microalgae has shown that pollutants, particularly in the sea, can have considerable adverse impacts on populations. Comparatively little work in this field has been carried out on protozoa, although there are indications that protozoan populations will react in the same way to pollution stress as do microalgae. Further research on marine protozoa is essential for a full understanding of the impact of pollution on the sea. More controlled field data are required and more laboratory experimental work on food chain effects, food preferences and uptake rates of protozoans by consumers needs to be carried out (8).

REFERENCES
1. Jones, A.R. (1974). The Ciliates. Hutchinson University Library, London.
2. Cairns, J. (1974). Protozoans (Protozoa). In Hart, C.W. and Fuller, S.L.H. (Eds), Pollution Ecology of Freshwater Invertebrates. Academic Press, London. Pp. 1-28.
3. Westphal, A. (1976). Protozoa. Blackie, Glasgow and London.
4. Murray, J.W. (1979). British Nearshore Foraminiferids. Synopses of the British Fauna N.S. 16. Linn. Soc. London, Academic Press.
5. Martinez, E.A. (1980). Sensitivity of marine ciliates (Protozoa, Ciliophora) to high thermal stress. Estuarine and Coastal Marine Science 10(4): 369-381.
6. McMillan, J.P., Granade, H.R. and Hoffman, P. (1980). Ciguatera fish poisoning in the United States Virgin Islands: preliminary studies. J. Con. Virgin Is. 6: 84-107.
7. Barnes, R.D. (1980). Invertebrate Zoology, 4th Ed. Saunders College, Philadelphia.
8. Curds, C.R. (1982). Pelagic protists and pollution - a review of the decade. Ann. Inst. Océanogr. Paris 58(S): 117-136.
9. Curds, C.R. (1975). Protozoa. In Curds, C.R. and Hawkes, H.A. (Eds), Ecological Aspects of Used-water Treatment. Vol. 1. The Organisms and their Ecology. Academic Press, New York. Pp. 203-268.
10. The Committee on Systematics and Evolution of the Society of Protozoologists (1980). A newly revised classification of the Protozoa. J. Protozool. 27(1): 37-58.
11. Parker, S.B. (1982). Synopsis and Classification of Living Organisms. McGraw-Hill Book Co., New York.
12. Corliss, J.O. (1981). What are the taxonomic and evolutionary relationships of the Protozoa to the Protista?

Biosystems 14: 445-459.

13. Stout, J.D. (1974). Protozoa. In Dickinson, C.H. and Pugh, G.F. (Eds), Biology of Plant Litter Decomposition. Vol. 2. Academic Press, London and New York. Pp. 385-420.

14. Iwasaki, H. (1981). Physiological ecology of red tide flagellates. Ch. 12 in Levandowsky, M. and Hutner, S.H. (Eds), Biochemistry and Physiology of Protozoa. Vol. 2. Pp. 357-393.

15. Corliss, J.O. (1979). The Ciliated Protozoa. Pergamon Press.

16. Schafer, C.T. (1982). Foraminiferal colonization of an offshore dump site in Chaleur Bay, New Brunswick, Canada. J. Foram. Research 12(4): 317-326.

17. Schafer, C.T. (1973). Distribution of foraminifera near pollution sources in Chaleur Bay. Wat. Air Soil Poll. 2: 219-233.

We are very grateful to J.O. Corliss, C.R. Curds, J.J. Lee and F. Page for their assistance with this section.

TARTAR'S STENTOR   INSUFFICIENTLY KNOWN

Stentor introversus Tartar, 1958

Phylum   CILIOPHORA   Order   HETEROTRICHIDA

Class   POLYHYMENOPHOREA   Family   STENTORIDAE

SUMMARY A structurally unique, relatively large protozoan, Tartar's Stentor is
known from a single locality in Washington State, U.S.A. Wildlife management
activities have apparently destroyed this population and perhaps the taxon, as it
has not been located elsewhere.

DESCRIPTION Species in the genus Stentor are trumpet-shaped unicellular
organisms which attach themselves to a hard substrate by an adhesive holdfast
organelle. The body is contractile and covered with longitudinal rows of cilia.
The oral cilia round the wider end of the body form a spiral feeding funnel. S.
introversus is between 100 μm and 300 μm long, extending to about 450 μm when
feeding. The endoplasm is opaque and brownish yellow in colour and the body is
longitudinally striped blue-green. The feeding organelles and adoral zone can be
fully retracted and introverted (2,3).

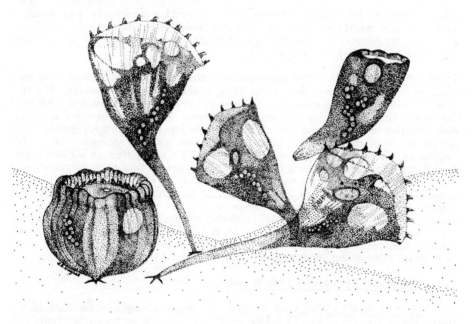

DISTRIBUTION Known only from a single locality, an impounded goose pond at
the headquarters of Willapa National Wildlife Refuge on the eastern shore of
Willapa Bay in south-western Washington State, U.S.A. (2). It was collected there
first in 1952 and on numerous subsequent occasions (2) until about 1970, when it
ceased to appear. Searches in apparently suitable habitats elsewhere around
Willapa Bay have been unsuccessful in locating other colonies (4).

POPULATION No information, and the animal may now be extinct.

5

HABITAT AND ECOLOGY   A freshwater species; the pond where it was known to occur empties over a spillway directly into the bay but there is no evidence of contamination by sea water. However, the original situation may have been that of a minor estuary. Two other freshwater Stentor species occurred in the same pond (2). Species of Stentor normally remain attached by their holdfast, but they can free themselves and swim to a new location. The spiral feeding funnel draws a current of water into the animal carrying particulate matter which provides food (3).

SCIENTIFIC INTEREST AND POTENTIAL VALUE   Thirteen species of Stentor have been described, some of which have been found quite recently, suggesting that others are yet to be discovered. S. introversus is unique in being able to retract the oral end of the body completely. The genus Stentor is potentially of major importance in biomedical research, since a wider range of microsurgery and cell growth experiments can be performed on these organisms than on any other unicellular animal or tissue cell. This is due to a number of factors including their large size, the consistency of their endoplasm which facilitates grafting and the high degree of visible cytoplasmic differentiation within the cell so that fixing and staining during experimentation are not necessary. Much of the research on Stentor has been supported by the American Cancer Society. A fundamental need in the cancer problem is to learn precisely what incites the cell to divide and the great amenability of stentors to manipulation suggests they could be instrumental in providing the answer (3).

Since grafts or cell fusions between different species persist, as well as those within one species, the more species that are available for interspecific grafts the greater is the possibility of exploring species differences. S. introversus has been fused with S. coeruleus and S. niger (4).

THREATS TO SURVIVAL   The destruction of the one known population may be a result of management practices in the Willapa National Wildlife Refuge. The U.S. Fish and Wildlife Service attracts wild geese onto the refuge by feeding them near the headquarters as an attraction for visitors. The large numbers of geese concentrated at the pond have resulted in water pollution through the accumulation of rotted wheat, used as feed, and goose droppings. The cattail vegetation around the margins of the pond has been killed through a combination of polluted water and trampling by geese. It is suggested that these circumstances may have caused the disappearance of the S. introversus population. If other colonies exist around Willapa Bay, they may be threatened by the many activities planned for what the National Estuarine Survey has called the most significant unpolluted West Coast estuary. Non-brackish habitats are rare around the Bay and many have already been disrupted by logging activities. Timber management, highway and residential development, and other uses of the bay margins could threaten any remaining populations of Tartar's Stentor (1).

CONSERVATION MEASURES TAKEN   None.

CONSERVATION MEASURES PROPOSED   A survey of suitable habitats in the region should be undertaken to locate additional populations that might exist. If any are found, they should be protected by careful management of the freshwater bodies in which they occur, or perhaps acquired by a suitable agency or organization for safeguarding. If any additional populations are found on the Willapa Bay National Wildlife Refuge, the U.S. Fish and Wildlife Service should be informed and petitioned to manage it for their well-being.

REFERENCES   1.   Pyle, R.M. (1970). Willapa Bay. Audubon 72 (6): 145.
              2.   Tartar, V. (1958). Stentor introversus n. sp. J. Protozool. 5: 93-95.

3. Tartar, V. (1961). The Biology of Stentor. Pergamon Press, Oxford, 413 pp.
4. Tartar, V. (1982). In litt., 4 February.

We are very grateful to V. Tartar for information supplied for this data sheet, and to F.C. Page and J.O. Corliss for helpful comments.

PORIFERA

Sponges

INTRODUCTION About 5000 living sponge species are known of which about 150 occur in freshwater and the rest in the sea. Sponges are the most primitive multicellular animals and have neither true tissues nor organs, the cells displaying a considerable degree of independence. All species are sessile and exhibit very little detectable movement, which convinced early naturalists that they were plants and it was not until 1765 that their animal nature was finally established (3,4).

They vary greatly in size, from over a metre in height and diameter to less than a centimetre. Although some may exhibit a certain symmetry, the majority are irregular in shape and have massive, erect, encrusting or branching growth patterns. The type of growth pattern displayed is influenced by environmental variables such as nature and inclination of substratum, availability of space, and velocity and type of water current. Where water movements are strong, sponges often grow as round or flattened clumps, but in calmer waters they may assume branching shapes. As a result the same species can have a different appearance under different conditions, a factor which has contributed to the confusion in the taxonomy of this group. Green, yellow, orange, red and purple sponges are often found particularly in shallow tropical waters but in deep waters most species are white, pale-yellow or green. The significance of coloration is not known although protection from solar radiation and a warning function have been suggested (3,4,49).

Sponge architecture is unique in being constructed around a system of water canals termed the aquiferous system. This comprises the incurrent system, which leads from the choanocyte chambers to the ostia (small pores) and the excurrent system which leads from the choanocyte chambers to the oscula (larger openings). The choanocytes are specialized flagellated cells which generate a steady water current throughout the aquiferous system supplying oxygen and food and removing waste products. The volume of water pumped through a sponge is remarkable; for example one species, 10 cm in height and 1 cm in diameter, pumps 22.5 l of water in a day. The interstitial flesh, or choanosome, consists of various types of cells lying mostly without order within a secreted gelatinous matrix. Since the external form of sponges is so variable, the type and arrangement of spicules and skeletal material supporting the body wall is used in identification, and sponges are divided into four classes accordingly. The Calcarea have calcareous spicules; the Hexactinellidae (which include the deep water glass sponges) have six-rayed siliceous spicules; the Demospongiae (the largest group) may have siliceous spicules (but not six-rayed), a skeleton of a fibrous material called spongin or spongin embedded with spicules; and the Sclerospongiae have a massive limey skeleton composed of calcium carbonate, siliceous spicules and organic fibres. Sponges have no gut and no noticeable sense organs or nervous system (3,4,18).

Marine sponges abound in all seas wherever rocks, shells, submerged timbers, plants or coral provide suitable substrates, although many are also found on sand or mud bottoms. Most species prefer relatively shallow water, but some groups such as the glass sponges favour deep water (3,18). In some shallow marine waters sponges may make up 80 per cent of the total biomass of benthic habitat (39) and in the Antarctic they are quantitatively the dominant group in some areas (40). Commercial sponges occur only in tropical or semitropical waters where the temperature does not drop below 50°F (10°), and they are usually only found in abundance on hard substrates, although some will tolerate mud or sand (38). In Tunisia the preferred environment of commercial sponges is characterized by the Posidonia-Caulerpa community, interspersed with sand and rocks outcrops. Since the Tunisian coastline in generally quite exposed, a minimum depth of about 10 m is required to assure sufficient protection from wave action (30). Freshwater sponges grow on almost any substrate including vegetation, and are generally encrusting.

Sponges are highly efficient filter feeders. Species on corals in Jamaica were found to extract very fine particulate matter of which about 20 per cent was bacteria, dinoflagellates and other fine plankton, and 80 per cent was dissolved organic matter of a size below that which can be resolved by light microscopy. In tropical water this latter portion contains seven times as much available carbon as the planktonic portion and the ability of sponges to use this food source probably accounts for their long success as sessile animals, particularly in the tropics (27). Sponges are eaten by a large number of invertebrates and some fish in spite of the presence of toxins and spicules (40,45) and predation may be an important factor in their distribution (40). They support a rich epifauna and endofauna on the surface and in the water canals and choanosome, and have been described as "living hotels". Epibionts and endobionts range from blue-green algae to fish and a higher number of specimens per unit area may be present than in other habitats such as sea grass or rock, making sponges very efficient ecological niches (30).

Most sponges are hermaphroditic, eggs and sperm being produced synchronously or at slightly different times. Sperm are released into the water canals and may fertilize eggs in the same sponge or be drawn into the canal system of another specimen. The fertilized eggs develop into larvae which leave through the oscula and after a short free-swimming existence settle on the bottom and develop into the adult form (3,18). Hippospongia lachne (the wool sponge) needs a minimum temperature of 80°F (27°C) for larval production and it has been suggested that the concentration of mature sponges per acre must be greater than 10 or 11 individuals (about 25 per ha) for maximum larval production to occur (38). Reproduction can also take place by asexual budding or the splitting of the body into smaller parts and in some species, particularly fresh water sponges, asexual reproductive bodies or gemmules are formed which can withstand severe environmental conditions. Sponges have remarkable powers of regeneration, small broken pieces being capable of developing into a new sponge (3,18,19).

SCIENTIFIC INTEREST AND POTENTIAL VALUE   The basic, simple sponge organization has proved a successful and persistent one in evolution. In Palaeozoic times, sponges in reef locations exceeded the combined biomass totals of other benthic animals (17). There is still a great deal to be learnt about these animals, although modern techniques such as electron microscopy, histochemistry and SCUBA diving have proved extremely useful. Sponges provide model systems in tissue culture and cell reaggregation, as they differ from all other groups of invertebrates which occupy similar ecological niches in the almost protozoan independence of their constituent cells. At the same time they ensure that the whole cell mass pumps sufficient water to effect all essential exchanges (4,62).

The endemic sponge faunas in Lake Malawi and Lake Tanganyika, Lake Posso (Sulawesi) and Lake Tiberias (Israel) are of particular taxonomic interest. They have been termed 'thalassoid sponges', since they occur in lakes isolated from the sea in relatively recent times (44,54). Furthermore, with the determination of the base-line environmental parameters for extant North American spongillid species, freshwater sponges will be increasingly used in the palaeolimnological definition of ancient lacustrine habitats (54,63).

Sponges were the first aquatic invertebrate group to be studied in any comprehensive way for new biochemical compounds. An amazing diversity of novel compounds has been found including collagens, nucleotides, nucleosides, amino acids and glycoproteins. Many of these are of potential use in medicine, and compounds with respiratory, cardiovascular, gastrointestinal, anti-inflammatory and antibiotic activities have been identified. The only drug being developed so far is D-arabinosyl cytosine, an important synthetic antiviral agent, which has been produced as a result of the discovery of spongouridine, a compound isolated from a Jamaican sponge, Tethya crypta; three derivatives of this compound have been patented as antiviral and anti-cancer drugs (3,4,31,47).

Sponges from the genera Spongia and Hippospongia have been used for centuries for personal and household purposes. In the first half of the 19th century sponge fishing started on a commercial basis in the Mediterranean and by the end of the last century was well established in Florida, U.S.A., and some of the Caribbean islands. Although only about a dozen species are used commercially, some 400 different 'types' are recognized by the trade. The use of trade names has made it difficult to determine exactly which species are used but probably the following have been or still are the most valuable commercially (5):-

Scientific name	Commercial name	
Spongia graminea	Glove, Grass	Gulf of Mexico, Caribbean
S. barbara	Yellow	"
S. zimocca	Zimocca, Fine dure	Mediterranean
S. officinalis	Turkey cup, Turkey solid, Fine Levant, Elephant ear	"
S. tubulifera		Caribbean
S. petusa		"
Hippospongia lachne	Wool, Sheepswool	Gulf of Mexico, Caribbean
H. gossypina	Velvet	"
H. communis	Honeycomb Horse Sponge	Mediterranean

FAO recorded a world production of 130 tonnes in 1980, of which 74 tonnes came from Tunisia, 45 tonnes from Cuba and 10 tonnes from Turkey (64). Mediterranean sponges are regarded as being of the highest quality and are fished commercially east of the Gulf of Lyon in the north and Algeria in the south, their growth to the west being restricted by the colder waters of the Atlantic. The commercial fishery operates to depths of about 80-90 m (21). Tunisia is currently the major producer, most of the yield being H. communis with smaller quantities of S. officinalis also taken. In 1979, 95 tonnes were collected, mainly by trawlers but also incidentally by coastal fishing boats (6). Greece is the second major Mediterranean producer and traditionally is the centre of the Mediterranean sponge fishing industry. As Greek waters became depleted at the beginning of this century the Greek fleet took to travelling further afield and with their superior diving and fishing techniques exploited the waters of other Mediterranean countries such as Egypt, Libya, Tunisia and Cyprus (5,7). Recently these countries have been making an attempt to expand their own local industries and Libya and Egypt both prohibit foreign boats operating in their water (5) although as yet these countries do not have deep water diving techniques (16). As a result the Greek fishery has declined. Eighty to a hundred Greek fishing boats currently operate off the coasts of Crete, the Peloponesus and some of the Aegean and Ionian waters and the annual yield is about 40 tonnes, most of which are exported (11). Greece still controls much of the international market and now imports large quantities of raw sponges from other Mediterranean countries, which are then prepared for export. Other Mediterranean countries exporting sponges include Turkey, Libya, Yugoslavia, Lebanon and Syria (42,56).

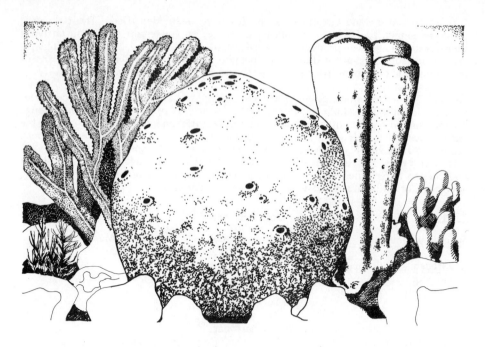

Velvet Sponge (<u>Hippospongia</u> <u>gossypina</u>)

Sponges of the eastern Gulf of Mexico and West Indies are regarded as being of second best quality. By 1954 this industry had practically disappeared as a result of a blight which swept the sponge beds and uncontrolled fishing. However, since then sponge stocks have been increasing (37), although the commercial fishery has still not fully revived. The sponge industry has tended to decline in many areas as a result of the development of artificial sponges, and the difficulty of attracting new sponge divers (14,21). In many ways natural sponges are still superior to the artificial product as they are tougher, last longer, can be more readily cleaned, and the very fine types are still essential for the pottery industry and some medical and scientific purposes. Natural sponges now fetch high prices, and with the current vogue for natural products, it is becoming increasingly worthwhile for younger divers to join the industry (6). In several areas sponges could be a major natural resource if the industry was developed and managed efficiently (39).

Sponges have a variety of other uses. It has been reported that on the Adriatic coast of Yugoslavia, the Leather Sponge <u>Chondrosia</u> <u>reniformis</u> is collected and eaten locally (21). The Finger Sponge <u>Axinella</u> <u>polycapella</u>, which has spicules and a spongin skeleton, is used for applying liquid shoe polish (39). The Egyptians may have used freshwater sponges as a pottery temper (52), and some of the Amazonian Indians still use these, crushed and burnt, for the same purpose. To ensure a regular supply the Indians scatter powdered sponge over the water at the beginning of the flooding period which assists in the propagation of gemmules (46). It has been suggested that sponges could be used to purify water through their filtration activities. A number of studies have demonstrated the removal of microbial pollutants such as the bacterium <u>Esherichia</u> <u>coli</u> and the fungus <u>Candida</u> <u>albicans</u> from waste effluents by the Red Beard Sponge, <u>Microciona</u> <u>prolifera</u>. Since this species concentrates large quantities of bacteria it has been suggested that it could be used in estuaries to combat microbial pollution from faecal contamination (8,10,23,32). This application may not be generally practical in view of the sensitivity of most sponges to highly polluted water (19).

THREATS TO SURVIVAL The complicated taxonomy of sponges and the sparse data on distribution and ecology make it very difficult to identify threatened sponge species. However, sponges are likely to be subject to the same problems as other sessile organisms. Many marine sponges probably have wide ranges, but it is possible that some are restricted to small areas as a result of very specific ecological requirements. The marine sponges of the Carolinian biogeographic province have been well studied and several species appear to be restricted to very small areas although further research may reveal that they have wider ranges. These include Pleraplysilla minchini (reported only three times), Adocia tubifera (from Beaufort Harbor, N.C.), Axinella bookhouti (New River, N.C.), Teichaxinella grayi (New River, N.C.), Phakellia folium (Florida and New River, N.C.) and Dorypleres carolinensis (New River, N.C.) (60,61).

A number of freshwater sponges have very limited distributions and therefore are vulnerable to water pollution or habitat alteration such as damming or channelization. Rare spongillid faunas are found in South American rivers and East African lakes and rivers (20,35). In North America a number of species such as Corvomeyenia carolinensis (South Carolina), Heteromeyenia longistylis (Pennsylvania); Spongilla heterosclerifera (two sites in Lake Oneida, New York), and Spongilla johanseni (New Brunswick, Canada) are known only from areas surrounding their type localities (19). Spongilla heterosclerifa is considered to be endangered since in parts of the lake near more heavily populated areas of the shore, the sponge fauna in general is completely absent and the range of this species may soon be affected (59).

1. Pollution a. Marine Shallow water and cave species are particularly vulnerable to pollution, and some sponges may be the first organisms to disappear from polluted sites (29,33). For example around Marseilles, certain species have disappeared around the sewage outfall, although still occurring elsewhere (41). The genus Spongia has been found to be an important element of the coralligenous community of sessile suspension feeders, with up to 100 species recorded. This community is known to be fragile and very vulnerable to pollution (26). However, other species may withstand pollution and be the first to recolonize damaged areas (43). In general, sponges are probably more sensitive to inorganic chemical pollutants than to organic pollutants, particularly bacterial (57).

In some cases sponges may be sensitive to the effect of fine sediments and other materials that could clog the canal system. In a survey off Edison's nuclear power plant at San Onofre, California, it was found that the high turbidity created during offshore installation of the outfall line completely eliminated the sessile invertebrate community including sponges, although the community re-established itself within three years (19). Densely organized sponges such as Verongia gigantea are less able to cope with sediment loads since their amoebocyte cycling system (the method by which particles are captured) is easily saturated. V. gigantea is restricted to clean water habitats of the outer reef (19).

b. Freshwater In highly polluted waters, species diversity is low although particular species may be abundant. It has been found that almost all members of the genus Heteromeyenia are eliminated in the presence of pollution. Pollution has been shown to cause modification of spicules in some sponges and such species could be useful as pollution indicators (19,53,54). The thalassoid sponge faunas of Lake Baikal and Lake Ochrid are comprised largely of endemics such as Baikalospongia and Ochridaspongia (20,35). It is known that Lake Baikal is threatened by pollution from industrial development (22) and Lake Ochrid could be equally vulnerable. Acidification could be a serious threat to all thalassoid sponges although they are normally protected as they occur in rather deep basins and at great depths. It may also be a threat to some spongillid species, although many of these will tolerate great pH fluctuations, reacting only with deformed or otherwise uncharacteristic spicular assemblages. However, some species inhabiting strictly alkaline localities may fail to reproduce under acidic conditions, their gemmules failing to germinate below certain pH values. For example, studies on Ephydatia mulleri from the Montreal-Quebec region have shown that gemmule hatching in buffered waters below pH 6.0 exhibit slower hatching rates or significantly lower hatching success. If this is a widespread phenomenon it could pose a serious threat for the distribution of some species over North America with the increasing acidification of lakes and streams (50). Siltation may be less of a threat to freshwater sponges than to marine sponges. The majority of freshwater spongillids cope extremely well with large amounts of suspended matter, as found in Australian streams and rivers, where the habitats are strongly affected during seasonally occurring floods. Such sponges have specially adapted canal systems for this purpose (53).

2. Disease From 1938 to 1940 and in 1947-1948 an epidemic destroyed 90-95 per cent of all commercial sponges in most areas of the Caribbean and Gulf of Mexico. The disease was identified as a fungus Spongiophaga communis, which attacks the sponge interior and spreads rapidly towards the periphery until the entire sponge rots away. New areas of sponge bed were affected in sequence down-current from infected areas, but after a time there was little further spread of the disease since it was limited by the mainland to the west, deep ocean to the east, colder waters to the north and possibly the freshwater Amazonian discharge to the south. The disease resulted in a major decline in the wool sponge Hippospongia lachne and the disappearance of the velvet sponge H. gossypina (24,39). Only a few specimens of the velvet sponge have been taken since then off the coast of Florida (5), and off the coast of Colombia (36).

3. Exploitation Overfishing has been reported a number of times in the commercial sponge industry. In 1908 Moore considered all exploited sponge beds to be showing signs of depletion and in some cases conspicuously. The Aegean and Adriatric were noted as being particularly over-exploited (25). Over-exploitation has continued to be reported: in Greece in shallow waters; in Tunisia, where the commercial quality now obtained is said to be poor; and in Turkey (15,28,41). In the U.S.A. the peak of the industry occurred in 1936-37, but by 1951 serious overfishing had occurred as a result of a combination of factors including disease, too many boats and lack of conservation methods (5,38). The decline in demand for sponges has taken pressure off the West Atlantic beds, however, which are reported to be gradually recovering.

Some of the methods used by the commercial fishery are damaging to the sponges themselves, and to their habitat. Bottom trawling and dredges are particularly destructive, and up to 25 per cent of the sponges collected by this method may be damaged. A number of countries now prohibit such methods but in the 1960s dredges were still being used off the coasts of Sicily and have also been used recently in Tunisian and Turkish waters (5,6,14,21).

4. Other factors Hurricanes and heavy rain on the sea are known to kill sponges or cause withdrawal or loss of the outer layer of living material (37). For example Hurricane Allen in 1980 almost destroyed the northern Jamaican macro-sponge fauna of the reefs (48). Low temperatures may also cause death; for example the cold winter of 1977 in the Bahamas killed off the Red Branching Sponge Higginsia strigilata in protected areas of shallow water (39).

Removal of mangroves threatens the whole community of sessile invertebrates, including sponges which grow on the roots. Since very diverse communities of mangrove invertebrates are rather restricted, mangrove destruction could threaten some sponge species (20).

Marine sclerosponges occur locally in underwater caves and tunnels and are collected as curios by SCUBA divers. Since these probably grow very slowly the habitat would probably take a long time to be restored to its original form (20). Axinella polypoides has become rarer around Marseilles where it is collected by divers and dried for decoration (41).

Heavy bottom trawling will effect all benthic fauna including sponges. For example, the Northwest Shelf off the west coast of Australia has been denuded of benthos by prawn trawling, and scallop dredging in many parts of the world, such as Tasman Bay, New Zealand, has had similar effects (48).

Fluoridation of water has a detrimental effect on sponges, causing skeletal abnormalities and the disappearance of these organisms from water pipes. It is now becoming apparent that this may be disadvantageous, since sponges in drinking water pipes may play an important role in the removal of harmful bacteria (19,52). Competition with other species may pose a threat to freshwater sponges. Corvomeyenia carolinensis, known only from its type locality in Carolina, could be threatened by aggressive colonisation of the pond by the sponge Heteromeyenia baileyi (58).

CONSERVATION For many years the commercial sponge fishery has been urged to implement conservation measures. Certain restrictions such as minimum size limits were imposed in the early days of the industry but these were generally recognized as insufficient. In 1908, as a result of a study of the fishery in the Gulf of Mexico it was recommended that (25):
1. The minimum size limit should be increased to 5 inches diameter (13 cm).
2. 'Hard-hat' diving should be banned below 20 fathoms (37 m) to improve safety conditions for the divers.
3. Dredges should be banned above 30 fathoms (55 m).
4. Close seasons should be set.
5. Sponge culture should be encouraged.

In 1964 a reappraisal of the U.S. sponge industry resulted in a similar set of recommendations (38):-
1. A minimum size limit of 6 inches diameter (15 cm) should be enforced since it was considered that gamete production did not start until the sponge had reached a diameter of 5.5 inches (14 cm).
2. Fishing techniques should be improved, with the introduction of lightweight diving suits and other safety measures.
3. Further efforts should be made to develop commercially viable methods of sponge culture including research into embryology and larval production, and surveys for suitable habitats.

In Florida there is currently a minimum size limit of 5 inches (13 cm) for the wool sponge. However, maturation time varies with temperature and although in the northern sponge grounds of the Gulf of Mexico the minimum size for larval production is 6 inches (15 cm), in Honduras maturation may be reached when the sponge is only 2 inches (5 cm) in diameter (39). In the Mediterranean, minimum size limits have been set in some countries, such as Syria and Italy, but apparently these are not properly enforced (15,55). In Tunisia it has been recommended that dredging should be banned at depths less than 50 m, and it is recognized that the sponge industry requires improved management (6).

Some methods of sponge fishing are definitely preferable and should be encouraged. In Tunisia sponges are still collected in shallow water by wading and finding specimens with the toes. Early simple fishing methods had some inbuilt conservation practices; for example fishermen would squeeze the sponge before removing it from the water, thus releasing the gametes in their natural environment (6). The Greeks mainly use diving methods so that a certain degree of selection is possible and it is hoped to convert all divers to using the aqualung (15). The disadvantages of the aqualung are that it is expensive and bottom time is severely limited (21). In the U.S.A. the hooking method may tear the sponge but it leaves the centre piece attached to the substrate, from which the sponge can regenerate. The disadvantage of this method is that only shallow beds can be worked and therefore tend to be fished more often. In the Florida Keys, diving is not allowed and sponges are taken with a rope loop, the small poorly formed specimens being left as breeding stock. Dredges should be banned in those countries which still permit this fishing method (5). Close seasons would be difficult to enforce, but the most effective type would be to close sponge beds in rotation for one or two years. Unfortunately there are very few data available on the rate of regeneration of sponge beds, or of regeneration of sponges from cut bases on which to base management plans.

Sponges are well suited to cultivation since small cut pieces will regenerate and in the early part of this century considerable research was carried out (1,2,12,13). The first attempt was in Florida in 1879, and a sponge farm has been in existence on Andros Island in the Bahamas for many years. Belize is reported to have had a successful sponge culturing programme, and in 1939 the Japanese carried out sponge culture experiments in the Marshall and Caroline Is. which, although not entirely successful, provided useful information (9). Attempts made at sponge culture in Yugoslavia earlier this century were not successful. In the early 1960s, experiments in Greece showed that Spongia officinalis regenerated well if cut pieces were attached to plastic cords (34). Experimentation has shown that sponges cut and tied to an artificial substrate such as cement discs or wire will grow into specimens of commercial size in 5-7 years, and that the total growth in volume of cultured sponges can greatly exceed the original over the same period. Choice of site was found to be important and shallow water where bad weather could easily destroy the beds should be avoided. The planting operation is relatively simple, and could be concentrated in a small area to save time in harvesting. Harvesting need only be carried out if an order has to be filled, and if the sponges are removed carefully, the remaining base could act as a fresh cutting. To a certain extent the quality, size and shape of the sponges can be controlled. The heavy local concentration will lead to the production of large quantities of larvae which will disperse to the surrounding environment if this is suitable. There are a few disadvantages, such as the susceptibility of such high concentrations of sponges to disease, and the fact that there are no returns until the fourth year. Furthermore the beds will not be fully productive until the seventh, although work in the Bahamas suggested that cuttings could reach a commercial size in two years (38). Since the current market for sponges appears to be larger than can be supplied by present production efforts, it is suggested that further investigation into sponge culture is warranted. Tunisia, for example, is interested in culturing sponges but lacks the necessary expertise to set up a programme (6).

If managed effectively, the sponge industry could probably provide a significant income in a number of countries. Species would be unlikely to become threatened as small specimens are not taken, deep populations are not exploited intensively and the species involved have wide distributions. In Tunisia usable species are common far beyond the established fishing grounds (30) and there may be a number of countries which have sponges of commercial value and as yet do not exploit them. For example, the Philippines produce a significant quantity now, most of which are exported to Japan (42).

For species which have no commercial value, further research on taxonomy, distribution, ecology and effects of factors such as pollution and sedimentation is required. Areas with diverse sponge faunas or endemic species should be protected, as is Runaway Bay cave in Jamaica (20). Attention should be paid to the deep shelf (200-300 m) faunas of the southern oceans which are at risk from trawling activities (51). Protection of the endemic freshwater 'thalassoid' sponges of Lakes Malawi, Tanganyika, Baikal, Ochrid, Posso and Tiberias may also be necessary. The only known conservation measures taken for a freshwater sponge is in the U.S.A. where the type-locality of <u>Corvomeyenia carolinensis</u> has been established as a state-patrolled wildlife sanctuary (57).

REFERENCES
1. Allemand-Martin, M.A. (1914). Contribution à l'etude de la culture des éponges. Les essais de spongiculture de Sfax. C.r. Ass. Fr. Avanc. Sci. 42: 375-377.
2. Allen, E.J. (1896). Report on the sponge fishery of Florida and the artificial culture of sponges. J. Mar. Biol. Assoc. (N.S.) 4.
3. Barnes, R.D. (1980). Invertebrate Zoology 4th Ed., Saunders College, Philadelphia.
4. Bergquist, P.R. (1978). Sponges. Hutchinson, London.
5. Bergquist, P.R. and Tizard, CA. (1969). Sponge Industry. In Firth, F.E. (Ed.), Encyclopaedia of Marine Resources, Van Nostrand Reinhold, N.Y. Pp. 665-670.
6. Boulhel, M. (1981). Amenagement des ressources vivantes de la zone littorale tunisienne. In GFCM, Management of living resources in the Mediterranean coastal area. Stud. Rev. Gen. Fish. Counc. Mediterr. 58: 350 pp.
7. van Buren, H.T. (1949). The Sponge Fishing Industry in Libya. Fishery Leaflet 341, USDI Fish and Wildlife Service, Washington D.C.
8. Burkholder, P.R. and Ruetzler, K. (1969). Antimicrobial activity of some marine sponges. Nature 222: 983-984.
9. Cahn, A.H. (1948). Japanese Sponge Culture Experiments in the South Pacific Islands, Fishery Leaflet 309, USDI, Fish and Wildlife Service, Washington D.C.
10. Claus, G., Madri, P. and Kunen, S. (1967). Removal of microbial pollutants from waste effluents by the redbeard sponge. Nature (London) 216: 712-714.
11. Costakopoulos, D. (1980). In litt., 21 April, to Min. of Ag., Athens.
12. Cotte, J. (1907). La Spongiculture. Rev. Scient. (5) tome 8.
13. Crawshay, L.R. (1939). Studies in the market sponges. I. Growth from the planted cuttings. J. Mar. Biol. Ass. U.K. 23(2): 553-574.
14. Dracos, C.S. (1969). Natural Sponge. Chap. 5 in Natural Organic Materials and Related Synthetic products. Vol. 5. Material and Technology, Longman and J.H. de Brussy.
15. Dracos, C.S. (1978). Pers. comm.
16. Dracos, C.S. (1981). Pers. comm.
17. Finks, R.M. (1970). The evolution and ecologic history of sponges during Paleozoic times. Symp. Zool. Soc. Lond. 25: 3-22.
18. George, J.D. and George, J.J. (1979). Marine Life. Harrap, London.
19. Harrison, F.W. (1974). Sponges (Porifera: Spongillidae). Chap. 2 in Hart, C.W. and Fuller, S.L.H. (Eds), Pollution Ecology of Freshwater Invertebrates. Academic Press, N.Y.
20. Hartmann, W.D. (1980). In litt., 16 July.
21. Hunnam, P. (1980). Mediterranean species in possible need of protection. Report to IUCN, Aquatic Biological Consultancy Services Ltd., Chelmsford, U.K.
22. Galaziy, G.I. (1980). Lake Baikal's ecosystem and the problem of its preservation. J. Mar. Tech. Soc. 14(7): 31-38.
23. Kunen, S., Claus, G., Madri, P. and Peyser, L. (1971). The ingestion and digestion of yeast-like fungi by the sponge Microciona prolifera. Hydrobiologia 38: 565-576.

24. de Laubenfels, M.W. and Storr, J.F. (1958). The taxonomy of American commercial sponges. Bull. Mar. Sci. Gulf and Caribbean 8(2): 99-117.
25. Moore, A.H.F. (1908). The Commercial Sponges and the Sponge Fisheries. Washington Bureau of Fisheries 27.
26. Peres, J. (1978). Vulnerabilité des ecosystems méditérranéens à la pollution. Ocean Management 3 (3/4).
27. Reiswig, H.M. (1971). Particle feeding in natural populations of three marine demosponges. Biolog. Bull. 141: 568-591.
28. Ruetzler, K. (1979). In litt., 8 January.
29. Reutzler, K. (1980). In litt., 16 June.
30. Ruetzler, K. (1975/1976). Ecology of Tunisian commercial sponges. Tethys 7(2-3): 249-264.
31. Ruggieri, G.D. (1976). Drugs from the sea. Science 194: 491-497.
32. Sarà, M. (1973). Animali filtratori ed autodepurazione nel mare: il ruolo dei poriferi. Atti del III Simposia Nazionale Sulla Conservazione della Natura. Organizzato dall Istituto di Zoologia dell Universita di Bar. Vol. I: 35-52.
33. Sarà, M. (1980). In litt., 25 August.
34. Serbetis, C.D. (1964). Experimental sponge culture in Greece. Proc. tech. Pap. gen. Fish comm. Mediterr. 7: 487-489.
35. van Soest, R.W.M. (1980). In litt., 30 June.
36. van Soest, R.W.M. (1981). In litt., 28 April.
37. Storr, J.F. (1976). Field observations of sponge reactions as related to their ecology. Aspects of Sponge Biology. Academic Press Inc.
38. Storr, J.F. (1964). Ecology of the Gulf of Mexico commercial sponges and its relation to the fishery. U.S. Fish Wild. Serv. Spec. Sci. Rep. Fish. 466: 1-73.
39. Storr, J.F. (1980). In litt., 5 August.
40. Vacelet, J. (1979). La place des Spongiaires dans les systèmes trophiques marins. Colloques internationaux du C.N.R.S. Biologie des spongiares. 291: 259-
41. Vacelet, J. (1980). In litt., 10 June.
42. Wells, S.M. (1978). The sponge trade. Unpub. report to TRAFFIC (International).
43. Wiedenmayer, F. (1980). In litt. 29 June.
44. Volkmer-Ribeiro, C. and Rosa-Barbosa, R. de (1978). Neotropical freshwater sponges of the family Potamolepidae Brien 1967. Colloques internationaux du C.R.N.S. 291:503-511.
45. Volkmer-Ribeiro, C. and Grosser, K.M. (1981). Gut contents of Leporinus obtusidens "sensu" von lhering (Pisces, Characoidei) used in a survey for freshwater sponges. Rev. Brasil. Biol. 41 (1):175-183.
46. Volkmer-Ribeiro, C. (1976). A new monotypic genus of neotropical freshwater sponges (Porifera-Spongillidae) and evidence of a speciation via hybridism. Hydrobiolgia 50(3): 271-281.
47. Bergquist, P.R. (1978). Sponge chemistry - a review. Colloques internationaux du C.N.R.S. 291: 383-392.
48. Bergquist, P.R. (1981). In litt., 29 July.
49. Litchfield, C. (1976). What is the significance of coloration in sponges? In Harrison, F.W. and Cowden, R.P. (Eds), Aspects of Sponge Biology. Academic Press N.Y. Pp. 28-32.
50. Reiswig, H.M. (1981). In litt., 11 May.
51. Bergquist, P.R. (1968). The marine fauna of New Zealand,

Porifera, Demospongiae. Part I. (Tetractininomorpha and Lithistida). N.Z. Dep. Sci. Ind. Res. Bull. 188: 1-106.

52. Racek, A.A. (1982). In litt., 9 January.

53. Racek, A.A. (1969). The freshwater sponges of Australia. Aust. J. Mar. Freshw. Res. 20: 267-310.

54. Racek, A.A. (1974). The waters of Merom. A study of Lake Huleh. Spicular remains of freshwater sponges. Arch. Hydrob. 74(2): 137-158.

55. Belloc, G. (1949/50). Inventory of the Fishery Resources of Greek Waters. Unpub. Report, FAO, Rome.

56. Gamulin-Brida, H., Pozar-Damac, A. and Simunovic, A. (1974). Benthic biocoenoses of the Adriatic as new sources of food and various raw materials, with special consideration for sponges and corals. Acta Adriatica 16(4): 71-95.

57. Harrison, F.W. (1982). In litt., 29 September.

58. Harrison, F.W. (1977). The taxonomic and ecological status of the environmentally restricted spongillid species of North America. 3. Corvomeyenia carolinensis Harrison 1971. Hydrobiologia 56: 187-190.

59. Harrison, F.W. and Harrison, M.B. (1979). The taxonomic and ecological status of the environmentally restricted spongillid species of North America. 4. Spongilla heterosclerifa Smith 1918. Hydrobiologia 62: 107-111.

60. Wells, H.W., Wells, M.J. and Gray, I.E. (1960). Marine sponges of North Carolina. J. Elisha Mitchell Sci. Soc. 76: 200-245.

61. Harrison, F.W., Calder, D.R., Coull, B.C. and James, F.C. (1976). Status report: lower invertebrates and miscellaneous phyla. In Forsythe, D.M. and Ezell, W.B. (Eds). Proceedings First South Carolina Endangered Species Symposium: 34-38.

62. Harrison, F.W. and Cowden, R.R. (Eds) (1982). Developmental Biology of Freshwater Invertebrates, Alan R. Liss (chapter on freshwater sponges).

63. Harrison, F.W., Gleason, P.J. and Stone, P.A (1979). Palaeolimnology of Lake Okeechobee, Florida: an analysis utilizing spicular components of freshwater sponges (Porifera: Spongillidae). Notulae Nat. Acad. Nat. Sci. Philadelphia 454: 1-6.

64. FAO. (1982). Fishery statistics for sponges. Fisheries Division, FAO, Rome.

We are very grateful to P.R. Bergquist, C.S. Dracos, F.W. Harrison, W.D. Hartman, A.A. Racek, H.M. Reiswig, K. Ruetzler, M. Sarà, R.W.H. van Soest, S. Stone, J.F. Storr, J. Vacelet, C. Volkmer-Ribeiro, and F. Wiedenmayer for assistance in the compilation of this data sheet.

CNIDARIA

INTRODUCTION Cnidarians are predominantly a marine group, and include hydroids, jellyfish, sea anemones, sea fans and corals. Although very variable in appearance, they all have a radially symmetrical body plan. The sac-like body has a central stomach cavity with a single opening, which serves as both a mouth and anus and is usually surrounded by food-capturing tentacles. Stinging capsules (nematocysts) on the tentacles narcotise the prey before it is drawn into the mouth, and in some cases can inflict powerful stings on humans. Most species are carnivorous but some are suspension feeders.

Individuals occur in two different structural grades: a sedentary polyp form and a medusa form which is usually free-swimming. Some groups occur only as polyps and some only as medusae, but others pass through both phases during the life cycle. Polyps have the mouth and tentacles directed upwards and usually have either an external or an internal skeleton. Many form colonies, and reproduce asexually by budding from the parent polyp. Medusae are bell-shaped with the mouth centrally placed on the underside and tentacles hanging down from the rim. They are only occasionally colonial, and reproduce sexually, the fertilized eggs developing into small larvae, often planktonic, which are covered with cilia.

There are approximately 10 000 species of cnidarians, and four classes are currently recognized:

The Hydrozoa characteristically have both a polyp and a medusa phase and most species are marine. They include the complex floating colonies which make up the Portuguese Man-of-War, and the fire corals, found on coral reefs, which produce painful rashes if touched. A few species such as the well-known Hydra are found in freshwater.

The Scyphozoa include all the true jellyfish in which a medusa is the predominant phase. They are entirely marine.

The Cubozoa are also marine medusae, with a characteristic shape having four flattened sides. They include the Australian sea wasps, renowned for their potent stings.

The Anthozoa, the largest class with over 6000 species, is entirely marine and has no medusa phase. There are three subclasses. The Alcyonaria (Octocorallia) include soft corals, sea fans and sea whips (gorgonians), organ-pipe coral, the precious red and pink corals and sea pens. They are mostly colonial and the individual polyps have a skeleton of spicules and eight tentacles. The Ceriantipatharia are generally found in deep waters and include the semi-precious black corals and some of the whip corals. The Zoantharia may be solitary or colonial and generally have six, or multiples of six, tentacles. When present the skeleton is an external calcareous mass lying outside the polyp. This group includes sea anemones, a number of solitary cup corals occurring in temperate waters, and the stony or reef-building corals, found only in tropical or sub-tropical waters where the temperature never drops below 70°F (21°C). Most of the stony corals depend on symbiotic algae (zooxanthellae) in their tissues, which need sufficient light to flourish. Stony corals make up the structural framework of coral reefs. These are built up over a period of many years as old corals die and new colonies form on top. Reefs provide a habitat for other cnidarians such as octocorals, sea anemones, zoanthids and hydrozoans, and for a wide variety of other invertebrate, fish and plant life. Reef organisms are often dependent on each other and form a complex network of relationships (54,55).

SCIENTIFIC INTEREST AND POTENTIAL VALUE Cnidarians play an immensely important role in marine communities and, like sponges and bryozoans, often provide a habitat as well as food for other invertebrates and fish. Most important in this respect are the reef-building corals which provide the basis of one of the most productive marine ecosystems in the world. Coral reefs play a number of important roles. They protect the coastline from erosion by wave action; they support fisheries of major importance to many subsistence economies; and they are often the nursery grounds for species of commercial importance (68,75). The reefs are becoming increasingly important as a tourist attraction. SCUBA diving is now a popular sport, and many countries, such as the Philippines, are gearing their tourist industry to attract and cater for divers and snorkellers (25). Even in temperate waters, sea anemones and soft corals can occur locally in large numbers, where their influence on the ecology of the area must be considerable (5).

Cnidarians have a number of important uses to man. Black and red corals (see reviews) have been used traditionally as medicines, and jellyfish were used in China for treatment of high blood pressure and bronchitis (14). Many species have been found to contain compounds of potential value in modern medicine. Antimicrobial substances occur in various gorgonian corals and antitumour compounds have been found in Hawaiian zoantharians. Aequorin, the bioluminescent protein present in the jellyfish Aequorea aequorea, glows in the presence of calcium or strontium and is sensitive enough to enable detection of minute fluctuations in calcium concentrations in biological fluids. Calcium changes in man often reflect cellular malfunctions and aequorin may prove useful in diagnosing cardiac irregularities and other disease processes (3). Prostaglandins have been isolated from the gorgonian Plexaura homomalla (13). An extract of the sea anemone Rhodactis howesi has an anticoagulant factor and the stony coral Porites has been used as a template for artificial bone (13). Anthozoans are ideal subjects for the study of simple nervous systems and work in this field is a major current interest (10). Recent studies of several anemone species indicate that they are capable of exhibiting complicated behaviour patterns and interacting aggressively, a type of behaviour often difficult to discern in invertebrates(1).

Cnidarians are also used as food. For over a thousand years jellyfish have been commercially exploited for food along the coasts of China, Japan and South Korea. Since 1960 several species, particularly Rhopilema esculentum, have been collected in South East Asia for export to China and Japan (14). In 1980, over 63 000 tonnes of Rhopilema were landed, mainly in Thailand, but also in the Philippines and Indonesia (11).

Stony corals are often used as a building material and for industrial purposes such as the preparation of calcium carbide and cement (6,56,57,58). The main species exploited for these purposes are the slow-growing massive corals such as Porites, Favia, Favites, Leptastrea and Platygyra. The smaller branching species such as Acropora, Seriatopora and Pocillopora are used for the ornamental coral trade. In the 1970s most of this came from the Philippines (in 1976 over 1800 tonnes were exported, 75 per cent of which went to the U.S.A.), but several other countries are involved to a smaller extent (12). The precious and semi-precious corals in the orders Gorgonacea and Antipatharia are used extensively for jewellery (see following reviews). Other precious corals of commercial importance include the gold corals (Gerardia and Primnoa) and the bamboo corals (Lepidisis and Acanella) and there is probably potential for the exploitation of many other deep water gorgonians (30). Sea fans are used as ornaments. Besides Eunicella verrucosa (see review), the Mediterranean species E. cavolinii and E. singularis (29), as well as many tropical species, are collected for this purpose. The hydroids Sertularia and Hydrallmania, known as 'white weed', are fished commercially in Germany and irregularly in the U.K. Having been raked-up from the sea bed, they are processed, dyed and used for decorative purposes, mainly in the U.S.A. (7).

THREATS TO SURVIVAL Like most marine species, very few cnidarians are likely to be in any danger of extinction, since they generally have wide distributions and deep, inaccessible populations which can act as reservoirs. However, several of man's activities are injurious to cnidarians. Certain species have shown a decline in U.K. waters in recent years, including the sea anemones Adamsia carciniopados and Aiptasia mutabilis, the jellyfish Aurelia aurita, some hydroids, gorgonian sea fans and cup corals. In many cases it is difficult to know if such species are truly declining and, if so, what the causal factors are (2). Amphianthus dohrnii, a small anemone which attaches itself to gorgonians and hydroids, was formerly common in the English Channel and Mediterranean, but now seems to be absent from the latter and very scarce in the former (2,5). In British waters it used to occur on the sea fan Eunicella verrucosa which itself has undergone a decline in abundance in some areas recently (see review). The high value of precious and semi-precious corals has led to over-exploitation in many areas (see reviews). The scouring of the sea bed by trawls, oyster dredges and other fishing gear probably affects populations of sea fans, sea pens, alcyonarians and sea anemones, but few data are available on this subject.

A number of cnidarians have restricted distributions and may be vulnerable if they occur in intertidal, marsh or lagoon habitats. The sea anemone Nematostella vectensis is one example (see review). Another is Edwardsia ivelli, known only from its type locality in Sussex, U.K., a brackish lagoon which could be affected by development. Further research may reveal such species to be more common subtidally, but until this is known care must be taken to protect known littoral populations (5). Freshwater cnidarians are likely to be vulnerable to pollution. In the U.S.A., both thermal and chemical pollution have been shown to have dramatic effects on the freshwater hydrozoan Cordylophora lacustris, although there is no evidence that this species is currently declining (18).

The greatest documented threat to cnidarians is the destruction of coral reefs, which is occurring worldwide on an unprecedented scale. The literature on this subject is considerable and no attempt has been made to summarize it here. The IUCN Coral Reef Specialist Group has been reviewing the overall situation and several countries have documented the status of their reefs at the national level (60,61,62).

The greatest destroyers of stony corals may well be natural catastrophes, such as storms, hurricanes and rapid drops in temperature, and it has been shown that given time, reefs have the ability to recover from either natural or human-induced damage (8,26,33,34,46). Several studies have shown coral reefs to have a natural ecological resilence (31,32), and that moderate storms actually contribute to the high species diversity of corals on a reef (34). Regeneration rates for reefs and growth rates for coral appear to be highly variable and probably depend on a wide range of factors. It was originally thought that corals had very slow growth rates and this may be true of some species. Studies of dynamite damaged reefs in the Philippines indicated a recovery time of 38 years to 50 per cent area coral cover (16). Porites compressa was shown to have a growth rate of 2 mm a year in the field (74) although rates of up to 20 mm a year have been recorded in the laboratory (70). A growth rate of 6 mm a year has been recorded for Pocillopora damicornis in the laboratory (70) although rates of 2-4 cm a year have been estimated for P. meandrina (69). Other species grow quite rapidly, and damaged reefs have been recolonized within two to three years (53). In some of the faster-growing species, such as Acropora cervicornis, which break easily, the fragments re-anchor and may have growth rates of up to 3-5 cm or even 12 cm a year (9,21,26).

Over the past decade much publicity has been given to the problem of Acanthaster planci, the Crown-of-thorns Starfish, (see Echinodermata). The extent to which outbreaks of this species have damaged coral reefs has been extensively debated. Certain areas of the Great Barrier Reef are being subjected to renewed infestation but there is still controversy over whether A. planci populations are out of control as a result of yet unidentified influences, conceivably brought about by man's activities, or whether the outbreaks are a natural cyclical phenomenon. Since no satisfactory method has been established for controlling A. planci, further research is necessary to locate vulnerable stages in its life cycle and to determine whether outbreaks are indeed cyclical (28,53). One study suggests that outbreaks are linked with periods of low salinity (i.e. high rainfall) (71).

Although reefs seem capable of surviving considerable natural damage it appears that the human-perturbed reef does not always return to its former configuration (8). Changes made by man have longer term gradual effects which tend to negate the reconstructional capability of damaged reefs and selectively reduce sensitive coral communities (46). Often several of these factors impinge jointly. The IUCN Coral Reef Specialist Group has listed man's more destructive activities (61) and a brief outline of those directly affecting corals is given.

1. Pollution   A review of the effects of pollutants is given in (73). Studies of some types of pollution, including heavy metals from tin smelting and tin dredging (36,37) and a Kaolin clay spill (37), have shown that the effects on reefs are not yet clear and may be negligible. However, silt can have a major impact on stony corals, preventing the settlement of the planula larvae, and hindering the photosynthetic activity of the symbiotic algae within the corals. In many parts of the world, siltation rates are increasing rapidly as a result of soil run-off from the land, caused by erosion in areas of extensive deforestation. Certain sewage treatments and effluent concentrations greatly enhance the growth of algae at the expense of corals. Under experimental conditions it has been found that coral mortality is not directly related to effluent toxicity, but is the result of competition with algae for space and light (42). A study of corals affected by dredging at Miami Beach revealed that gorgonians were most tolerant to siltation- and dredging-induced turbidity. However, although scleractinians were tolerant to a few days' siltation, prolonged exposure caused loss of zooxanthellae, polyp swelling or death (43). Reefs have been damaged by sedimentation in Thailand (35) and in Hawaii, where the disappearance of corals from Kaneohe Bay is well documented (44). By-products of oil-drilling, in the form of mud containing toxic materials, have been shown in the laboratory to affect the stony coral Montastrea annularis (38). However, studies in the field suggest that although branching species suffer, massive corals are not affected (72).

There have been very few detailed quantitative field studies of the effects of oil pollution on coral reefs. Studies in the Red Sea suggest that oil affects the reproductive potential of corals, their reproductive development and the settling behaviour of the planula larvae (49) and that it is a least partially responsible for recolonization failure (48,49). A high mortality of corals was observed in the Gulf of Aqaba during the years in which an oil terminal and a mineral and phosphate loading harbour were developed. This was probably caused by eutrophication and algal growth on the corals resulting from frequent oil spills and deposition of phosphate dust (48). Laboratory experiments indicate a range of possible responses to oil including abnormal mouth-opening and feeding behaviour, mucus secretion, decreased growth rates and increased tissue death rates (65,66). Upper reef corals in shallow water are most likely to suffer severe damage, especially if oil slicks coat the corals during extreme low tides (67). In some cases it appears that crude oil may not itself cause permanent damage, but that clean-up operations after oil spills, involving toxic dispersants or mechanical methods, damage the corals (17).

Coral growth may be impeded by heated effluents, and fast-growing species seem to be less tolerant of thermal enrichment than slow-growing species (45). Four widespread Indo-Pacific species of stony coral (Pavona frondifera, Pectinia lactuca, Leptoseris gardineri and Montipora sp.) are found in Guam exclusively in

Apra Harbour (51) which is under serious threat from development for a power plant and generating and industrial facilities. Pavona frondifera is already being affected by heated water discharges and it is feared that the other species could be similarly vulnerable (4).

2. Disturbance Reefs near tourist resorts and industrial and urban developments are particularly vulnerable. Corals are easily damaged by boats and anchors (39,47), trampling, and fishing with explosives and chemicals such as chlorine (40). In a few cases habitat destruction could have a serious effect on a coral species, since some stony corals appear to have limited ranges. For example, Oculina robusta, O. tenella, Fabellum fragile and Caryophyllia horologium are endemic to Florida (19) and in Guam three species (Euphyllia sp., Plerogyra sinuosa and Tubastrea aurea) are thought to be vulnerable on account of their rarity (20,51).

3. Exploitation The collection of stony corals for building and industrial purposes has caused the loss of large tracts of reef in countries such as the Philippines (25), Sri Lanka (56), Malaysia (57) and Indonesia (58). The ornamental coral trade has also had localized impacts in several countries, including the Philippines (59,60) and Hawaii (23). The main problem seems to be that small immature colonies are preferred and there is intensive localization of collecting in certain areas. Colonies of Seriatopora, for example, have virtually disappeared from some areas in the Philippines (25). On Hawaiian reefs collection of coral colonies by tourists had a local impact in the 1970s. Fungia scutaria, one of the rarer corals found principally in Kaneohe Bay in Oahu, declined as a result of the combined effects of over-collection and pollution. Its limited distribution and low abundance made it particularly vulnerable (23,24).

CONSERVATION For many cnidarians, few data are available to assess the effects of human activities although for corals there is evidence that action must be taken rapidly to preserve a resource of great ecological and economic importance. Although there is little danger of any coral species becoming extinct it is clear that many coral reefs are seriously threatened. Fortunately, steps are already being taken to control some of man's more damaging activities. The Reefwatch project, run from the Tropical Marine Research Unit at York, U.K., is assembling data on the current condition of reefs worldwide using amateur and professional divers. Handbooks have been and are being produced by IUCN, the South Pacific Commission and UNESCO to provide guidelines for monitoring and managing coral reef ecosystems (15,41).

Coral reef reserves have been established in many countries and survey work is being carried out to locate further potential sites. Consideration is being given to the type of reserve implemented, so that local people may still exploit and benefit from the resources of the reef. The Great Barrier Reef Marine Park is one example. The Australian government has established a management regime which provides for multiple use of the reef while ensuring that the important qualities of the ecosystem are protected (50). Numerous examples of parks and reserves for coral reefs could be cited, and methods of establishing and managing these were discussed in detail at the Third World National Parks Congress in 1982. However, the proliferation of such reserves is not a reason for complacency, as methods for maintaining these areas, and conflict with fishermen and other users, still present considerable problems. In temperate countries little interest has been shown in cnidarians, although reserves may be required to protect species with limited distributions or restricted to vulnerable habitats such as estuaries and salt marshes.

Exploitation and trade in corals and other cnidarians should be managed. In several countries e.g. U.S.A. (Florida) (19), Bermuda (64), Bahamas (63) and Guam (20), the taking of corals, sea fans and other cnidarians is prohibited or restricted. Fishery management plans are being drawn up for coral resources in the Gulf of Mexico (30), and for precious coral in the western Pacific (see reviews). Corals have been added to the U.S. Lacey Act (22) which means that corals exported illegally from their country of origin may not be imported into the U.S.A. The Philippines has a total ban on the export of all coral although, as in many countries, this has proved difficult to enforce (23,59). At present exploitation of corals for the ornamental trade and particularly for the building trade should be banned, especially in species which contribute to the structure of the reef, until techniques for managing this resource have been determined (25).

Further research on the effects of pollution on corals and coral reefs is also required. It has been suggested that clean-up operations after oil spills should be carried out using suitable mechanical rather than chemical methods (17). If slicks are sighted near coral reefs they should be treated while still in deep water or diverted away from shallow reef and lagoon areas (67).

Transplantation of stony coral colonies has been shown to be feasible (4,27) and when techniques are perfected may assist in shortening the recovery time of coral reefs damaged by human activities, as well as preserving species endemic to certain localities (4). The creation of artificial reefs using, for example, old motor vehicle tyres is a method that has been used in many areas to promote coral growth and to increase the habitat available to other reef organisms (52). 'Farming' of transplanted colonies has been postulated as a means of managing corals of value to man (25), but considerable research is required before economically viable methods are available. Further research is needed into fecundity, settlement and growth rates to determine recovery rates and sustainable yields.

REFERENCES

1. Brace, R., Pavey, J. and Quicke, D. (1979). Intraspecific aggression in the colour morphs of the anemone Actinia equina: the 'convention' governing dominance ranking. Animal Behaviour 27: 553-561.
2. NERC (1973). Marine Wildlife Conservation. The Natural Environment Research Council, Publications Series 'B', No.5.
3. Bayer, F.M. and Weinheimer, A.J. (1974). Prostaglandins from Plexaura homomalla: ecology, utilization and conservation of a major medical marine resource: a symposium. Studies in Tropical Oceanography 12. Univ. of Miami Press. Coral Gables, Florida.
4. Randall, R.H. and Birkeland, C.E. (1981). Preservation of rare coral species by transplantation and examination of their recruitment and growth. Project proposal.
5. Manuel, R.L. (1981). British Anthozoa. Synopses of the British Fauna (N.S.) 18. Kermack, D.M. and Barnes, R.S.K. (Eds). The Linnaean Society of London and Estuarine and Brackish-Water Sciences Association, Academic Press, London.
6. Pillai, C.S.G. (1973). Coral resources of India with special reference to Palk Bay and Gulf of Mannar. Proc. Symp. Living Resources of the Seas around India. ICAR. Special Pub., Central Marine Fisheries Research Institute, Cochin.
7. Hancock, D.A., Drinnan, R.E. and Harris, W.N. (1956). Notes on the biology of Sertularia argentea. J. Mar. Biol. Ass. U.K. 35: 307-325.
8. Loya, Y. (1976). Recolonization of Red Sea corals affected by natural catastrophes and man-made perturbations. Ecology 57: 278-289.
9. Shinn, E.A. (1976). Coral reef recovery in Florida and the Persian Gulf. Env. Geol. 1: 241-254.
10. Shelton, G.A.B. (1981). Electrical Conduction and

Behaviour in 'Simple' Invertebrates. Oxford University Press.

11. FAO (1981). Yearbook of Fishery Statistics 1980 Catches and Landings, Vol. 50. FAO, Rome.
12. Wells, S.M. (1981). International Trade in Ornamental Corals. IUCN Conservation Monitoring Centre, Cambridge.
13. Ruggieri, G.D. (1976). Drugs from the sea. Science 194: 491-497.
14. Omori, M. (1981). Edible jellyfish (Scyphomedusae: Rhizostomeae) in the Far-East waters. A brief review of the biology and fishery. Bull. Plankton. Soc. Japan 28(1): 1-12.
15. UNESCO (in prep.). Handbook on Coral Reef Management.
16. Alcala, A.C. and Gomez, E.D. (1979). Recolonization and growth of hermatypic corals in dynamite-blasted coral reefs in the Central Visayas, Philippines. Proc. Symp. Ecol. Biogeogr. Southern Hemisphere: Pp. 654-661.
17. Elgershuizen, J.H.B.W. and de Kruijf, H.A.M. (1976). Toxicity of crude oils and a dispersant to the stony coral Madracis mirabilis. Marine Pollut. Bull. 7(2): 22-25.
18. Prochnow, D. (1982). In litt., 23 January.
19. Jaap, W.C. (1980). In litt., 20 May.
20. Hedlund, S.E. (1977). The extent of coral, shell and algal harvesting in Guam waters. Sea Grant Publication UGSG-77-10, University of Guam Marine Laboratory, Technical Report 37.
21. Robinson, C.J. (1980). Cambridge Coral Microstructure Project Jamaica 1980. Dept. of Geography, Cambridge University.
22. Anon. (1981). Lacey Act Amendments of 1981. Congressional Record.
23. Johannes, R.E. (1978). Stony coral harvesting in Oahu. Unpub. report.
24. Grigg, R.W. (1976). Fishery management of precious and stony corals in Hawaii. UNIHI-SEAGRANT TR-77-03. University of Hawaii Sea Grant College Program, Honolulu.
25. Ross, M.A. (in prep). Commercial coral collection in Cebu. Paper to be given at the 15th Pacific Science Congress, Dunedin, New Zealand, February 1983.
26. Tunnicliffe, V. (1981). Breakage and propagation of the stony coral Acropora cervicornis. Proc. Nat. Acad. Sci. U.S.A. 78(4): 2427-2431.
27. Maragos, J.E. (1974). Coral transplantation: a method to create, preserve, and manage coral reefs. UNIHI-SEAGRANT-AR-74-03. University of Hawaii Sea Grant College Program, Honolulu.
28. Kenchington, R.A. (1978). The crown-of-thorns crisis in Australia: a retrospective analysis. Envir. Cons. 5(1): 11-20.
29. Hunnam, P. (1980). Mediterranean species in possible need of protection. Report to IUCN, Aquatic Biological Consultancy Services Ltd., Chelmsford, U.K.
30. Center for Natural Areas (1981). Draft fishery management plan for coral and coral reefs of the Gulf of Mexico and South Atlantic. Gulf of Mexico and South Atlantic Fishery Management Councils.
31. Colgan, M.W. (in press). Succession and recovery on the coral reef after predation by Acanthaster planci (L.). Proc. 4th Int. Coral Reef Symp. Manila, Philippines, May 1981.

32. Pearson, R.G. (1981). Recovery and recolonisation of coral reefs. Mar. Eco. Prog. Ser. 4: 105-122.

33. Antonius, A. (in press). Coral reefs under fire. Proc. 4th Int. Coral Reef Symp. Manila, Philippines, May 1981.

34. Dollar, S.J. (in press). Storm stress and coral community structure in Hawaii. Proc. 4th Int. Coral Reef Symp. Manila, Philippines, May 1981.

35. Chansang, H., Boonyanate, P. and Charuchinda, M. (in press). Effect of sedimentation from coastal mining on coral reefs on northwestern coast of Phuket Island, Thailand. Proc. 4th Int. Coral Reef Symp. Manila, Philippines, May 1981.

36. Brown, B.E. and Holley, M.C. (in press). The influence of tin smelting and tin dredging on the intertidal reef flats of Phuket, Thailand. Proc. 4th Int. Coral Reef Symp., Manila, Philippines, May 1981.

37. Grigg, R.W. and Dollar, S.J. (in press). Coral reef pollution: a major spill causes neglible impact. Proc. 4th Int. Coral Reef Symp. Manila, Philippines, May 1981.

38. Szmant-Froelich, A., Adams, R., Hoehn, T., Johnson, V., Parker, J., Battey, J., Smith, J., Fleischmann, E., Porter, J. and Dallmeyer, D. (in press). The physiological effects of oil-drilling muds on the Caribbean coral Montastrea annularis. Proc. 4th Int. Coral Reef Symp. Manila, Philippines, May 1981.

39. Davis, G.E. (1977). Anchor damage to a coral reef on the coast of Florida. Biol. Conserv. 11: 29-34.

40. Campbell, D.G. (1977). Bahamian chlorine bleach fishing: a survey. Proc. 3rd Int. Coral Reef Symp. Miami, Florida, May 1977. Pp. 593-595.

41. Dahl, A.L. (1981). Coral Reef Monitoring Handbook. South Pacific Commission Publications Bureau, Noumea, New Caledonia.

42. Marszalek, D.S. (in press). Effects of sewage effluents on reef corals. Proc. 4th Int. Coral Reef Symp. Manila, Philippines, May 1981.

43. Marszalek, D.S. (in press). Impact of dredging on a subtropical reef community, southeast Florida coast, U.S.A. Proc. 4th Int. Coral Reef Symp. Manila, Philippines, May 1981.

44. Smith, S.V. (1977). Kaneohe Bay: a preliminary report on the responses of a coral reef/estuary ecosystem to relaxation of sewage stress. Proc. 3rd Int. Coral Reef Symp. Miami, Florida, May 1977.

45. Neudecker, S. (in press). Growth and survival of scleractinian corals exposed to thermal effluents in Guam. Proc. 4th Int. Coral Reef Symp. Manila, Philippines, 1981.

46. Raymond, B. (in press). Bombs, dredges and reefs. Florida and the Caribbean. Proc. 4th Int. Coral Reef Symp. Manila, Philippines, May 1981.

47. Tilmant, J.T. and Schmahl, G.P. (in press). A comparative analysis of coral damage on recreationally used reefs within Biscayne National Park, Florida, U.S.A. Proc. 4th Int. Coral Reef Symp. Manila, Philippines, May 1981.

48. Fishelson, L. (1973). Ecology of coral reefs in the Gulf of Aqaba (Red Sea) influenced by pollution. Oecologia 12(1): 55-68.

49. Rinkevich, B. and Loya, Y. (1977). Harmful effects of chronic oil pollution on a Red Sea scleractinian coral

population. Proc. 3rd. Int. Coral Reef Symp. Miami, Florida, May 1977.

50. Kelleher, G. and Kenchington, R. (1982). Australia's Great Barrier Reef Marine Park: making development compatible with conservation. Paper given at Third World National Parks Congress, Bali, Indonesia, October 1982.

51. Stojkovich, J.O. (1977). Survey and species inventory of representative pristine marine communities on Guam. Sea Grant Publication UGSG-77-12, Univ. of Guam Marine Laboratory, Technical Report 40.

52. Alcala, A.C., Gomez, E.D., Alcala, L., Cowan, M.E. and Yap, H.T. (in press). Growth of certain corals, molluscs, and fish in artificial reefs in the Philippines. Proc. 4th. Int. Coral Reef Symp. Manila, Philippines, May 1981.

53. Cameron, A.M. and Endean, R. (in press). Renewed population outbreaks of a rare and specialised carnivore (the starfish Acanthaster planci) in a complex high diversity system (the Great Barrier Reef). Proc. 4th. Int. Coral Reef Symp. Manila, Philippines, May 1981.

54. Barnes, R.D. (1980). Invertebrate Zoology. Saunders College, Philadelphia.

55. George, J.D. and George, J.J. (1979). Marine Life. Harrap, London.

56. Hoffman, T.W. (1977). Lime from coral - the tragic folly. Wildlife and Nature Protection Soc. of Ceylon Newsletter 40: 4-6.

57. Wood, E.M. (1977). Coral reefs in Sabah: present damage and potential dangers. Malayan Nature Journal 31 (1): 49-57.

58. Anon. (1980). The coast of Bali in danger. Conservation Indonesia 4(5): 8.

59. Wells, S.M. (1982). Marine conservation in the Philippines and Papua New Guinea with special emphasis on the ornamental coral and shell trade. Report to Winston Churchill Memorial Trust, London.

60. Marine Sciences Center (1979). Investigation of the coral resources of the Philippines. Phase II. Final report. University of the Philippines, Manila.

61. Salvat, B. (1982). Coral Reef Newsletter 4. IUCN Commission on Ecology.

62. White, A. (1982). Vulnerable marine resources, coastal reserves and pollution: a southeast Asian perspective. Paper given at Third World National Parks Congress, Bali, Indonesia, October 1982.

63. Anon. (1975). Fisheries Act 1969 (13 of 1969). Continental Fishery Resources (Declaration) Notice 1975, Bahamas.

64. Anon. (1981). Bermuda Division of Fisheries (Legislation). Fisheries Office.

65. Ray, J.P. (1981). The effects of petroleum hydrocarbons on corals. In Proc. PETROMAR 80 Conference. Petroleum and the Marine Environment. Graham and Trotman Ltd., London. Pp. 705-726.

66. Loya, Y. and Rinkevich, B. (1980). Effects of oil pollution on coral reef communities. Mar. Ecol. Prog. Ser. 3: 167-180.

67. IUCN Commission on Ecology, Oil Pollution Working Group (1982). The impact of oil pollution on living resources. Draft.

68. Salvat, B. (in press). Preservation of coral reefs: scientific

whim or economic necessity? Past, present and future. Proc. 4th Int. Coral Reef Symp. Manila, Philippines, May 1981.

69. Maragos, J.E. (1975). Coral colonization in Honokohau Harbour. In A Three-Year Environmental Study of Honokohau Harbour, Hawaii. Oceanic Institute, Waimanalo, Hawaii.

70. Clausen, C.D. and Roth, A.A. (1975). Estimation of coral growth-rates from laboratory ^{45}Ca-incorporation rates. Mar. Biol. 33: 85-91.

71. Birkeland, C. (1982). Terrestrial runoff as a cause of outbreaks of Acanthaster planci (Echinodermata: Asteroidea). Mar. Biol. 69: 175-182.

72. Hudson, H., Shinn, E.A., and Robbin, D. (1982). Effects of offshore oil drilling on Philippine reef corals. Bull. Mar. Sci. 32(4):

73. Brown, B.E. and Howard, L.S. (in prep). Assessing the effects of stress on reef corals - a review.

74. Shinn, E.A. (1966). Coral growth-rate, an environmental indicator. J. Palaeont. 40: 233-240.

75. Odum, W.E. (1982). The relationship between protected coastal areas and marine fisheries genetic resources. Paper prepared for Marine and Coastal Areas Workshop, Third World National Parks Congress, Bali, Indonesia, October 1982.

We are very grateful to F.M. Bayer, B. Brown, P. Cornelius, B. Salvat and R. Williams for assistance with this section.

BROAD SEA FAN   INSUFFICIENTLY KNOWN

Eunicella verrucosa (Pallas, 1766)

Phylum   CNIDARIA   Order   GORGONACEA

Class   ANTHOZOA   Family   PLEXAURIDAE

SUMMARY Eunicella verrucosa is an attractive sea fan found in moderately deep
coastal waters in the East Atlantic, ranging from the Mediterranean to north-west
Ireland. Populations have been depleted in some areas by collection for the curio
trade by professional divers. Recent conservation education efforts among the
diving community have resulted in a decrease in collection pressure.

DESCRIPTION Like all sea fans, Eunicella verrucosa colonies consist of a main
stem, attached to a hard surface, with plant-like branches which spread in one
plane only. The brown horny skeleton is covered with a soft tissue ranging in
colour from white or yellow to deep orange-pink. Specimens from the English
Channel are almost invariably pink; in west coast localities both pink and white
forms occur and Mediterranean examples are always white. Colonies reach a
height of up to 30 cm and a breadth of up to 40 cm (6,9).

DISTRIBUTION East Atlantic including U.K., France, Spain, Portugal and
north-west coast of Africa as far as Mauritania and possibly the Gulf of Guinea
(9). Occurs as far north as the west coast of Ireland, and perhaps west Scotland
(6). A distribution map is being prepared for the west coast of France (8). Found
in the west Mediterranean, off the coasts of Corsica, Marseille (France), and
Genoa (Italy) (9).

POPULATION Unknown, but populations may be abundant in some remote
locations. Rare off the Mediterranean coast of France but more abundant in
Spanish waters (10).

HABITAT AND ECOLOGY It rarely occurs in water shallower than 10 - 20 m but in France is found on the Brittany coast from 2 - 60 m depth (4). In the Mediterranean it is found at depths of 35-200 m (9). It is found in shady areas and occasionally serves as a support for large numbers of sessile organisms such as ascidians, anthozoans and polychaetes (4,9). It is very slow growing (about one cm per year) and has very short-lived larvae (a few hours at the most) which settle close to the parent colony (5,7). Colonies are usually but not invariably orientated across the prevailing water currents (6).

SCIENTIFIC INTEREST AND POTENTIAL VALUE Colonies are popular as souvenirs, sold dried and mounted on wooden blocks. In 1979 a 15-20 year old specimen cost ₤0.80 (2).

THREATS TO SURVIVAL Although there is no threat of extinction, and populations may be abundant in deep water and on inaccessible parts of the coast, this species is threatened locally by over-collecting. The collection of large numbers of Eunicella colonies for the souvenir trade by professional skin divers in the late 1960s and early 1970s in the Isles of Scilly and on the Cornish, Dorset and Welsh coasts in the U.K. is claimed to have caused local depletion (1,3,5,7). Although these claims are based on subjective evidence, a similar situation was found on the Brittany coast of France where colonies were found to be much more abundant in areas where there was no diving pressure (4). The species is potentially vulnerable, since recovery of exploited populations will be slow on account of the slow growth rate and short-lived larvae, which would make recruitment from distant stocks a slow process. A colony worth collecting is likely to be over 20 years old. There is little collecting pressure on the species on the Mediterranean coast of France, where other gorgonians, E. cavolinii, E. singularis and Paramuricea clavata are in greater demand (10).

CONSERVATION MEASURES TAKEN No cases of commercial collection have been reported in the U.K. in the last few years. There is also probably little amateur diver collection in U.K. waters as a result of increasing diver education by organizations such as the Underwater Conservation Society. Most divers are probably now aware of the slow growth of the species and the permanent damage which would be done by intensive collection (3,5).

CONSERVATION MEASURES PROPOSED There is still a real danger of local extinctions if a commercial operator decided to supply the curio trade or if present conservation awareness faded. The potentially vulnerable status of the species should continue to be made clear to divers, who should also be alert to discourage other people collecting colonies.

REFERENCES
1. Mitchell, R. (1980). In litt., 13 August.
2. Hunnam, P. (1980). Marine curio trade. Unpub. draft fact sheet, Underwater Conservation Society.
3. Hiscock, K. (1981). In litt., 10 February.
4. Lafargue, F. (1969). Peuplements sessiles de l'Archipel de Glenan. Vie et Milieu 20(2-B): 415-436.
5. Hiscock, K. (1981). In litt., 25 September.
6. Manuel, R.L. (1981). British Anthozoa. Synopses of the British Fauna, N.S. 18, Academic Press, London and New York. 241 pp.
7. Natural Environment Research Council (1973). Marine Wildlife Conservation. Publication Series 'B', No. 5.
8. Castric, A. (1982). In litt., 21 December.
9. Carpine, C. and Grasshoff, M. (1975). Les gorgonaires de la Méditerranée. Bull. Inst. Océanogr. Monaco 71(1430): 140 pp.
10. Harmelin, J.G. (1983). In litt., 13 January.

We are very grateful to A. Castric, J.G. Harmelin, K. Hiscock, F. Lafargue, R. Manuel, R. Mitchell and S. Weinberg for information supplied for this data sheet.

PRECIOUS CORALS   COMMERCIALLY THREATENED

Corallium elatius Ridley, 1882   Boke, Maguy, Momoiro-sarigo
C. japonicum Kishinouye, 1903   Aka-sango, Oxblood Coral
C. konojoi Kishinouye, 1903   Shiro-sango, White Coral
C. rubrum (Linnaeus, 1758)   Mediterranean or Noble Coral
C. secundum Dana, 1848   Pink Coral, Angelskin Coral
C. sp. nov.   Midway Deep Sea Coral

Phylum   CNIDARIA   Order   GORGONACEA

Class   ANTHOZOA   Family   CORALLIIDAE

SUMMARY The precious corals of the genus Corallium have been highly valued
since antiquity as a carving material and for making jewellery. C. rubrum is found
in the Mediterranean which, until the 19th century, was the centre of the coral
trade. Overfishing of Mediterranean stocks combined with the discovery of
Corallium species off Japan, Taiwan and eastwards to the Hawaiian Archipelago,
resulted in Pacific corals becoming increasingly important on the world market in
the late 19th and 20th centuries. Precious corals are slow-growing and tend to
occur in small, easily exploitable beds which makes them very vulnerable to
over-collection, and there is concern that the Pacific stocks could become as
depleted as those in the Mediterranean. Management of the fishery should be
possible using quotas, limited fishing areas and selective fishing gear. A
management plan is being implemented for the Western Pacific and further
research is needed to provide base-line data for similar plans for other areas.

DESCRIPTION Precious corals grow as plant-like, branching colonies, with the
main stem firmly attached to a hard surface by a holdfast or base. The stem
consists of cemented calcareous spicules which form a hard skeleton that can be
used as a carving material. Colours vary from white to dark red and are specific
to localities and species. C. japonicum is red; C. elatius is pink to dark pink; C.
konojoi is white; C. secundum is pink to white; C. sp. nov. is light pink with darker
spots (12). C. rubrum varies in colour from pink to shades of dark red (8); colonies
in the region of Marseille are predominantly red (28). Colonies of the largest
Pacific species may reach 1 m in height, but C. rubrum grows to only about 25-50
cm (14) and in the region of Marseille most colonies are 8-15 cm in height (28).

Twenty species of Corallium have been described but only six of these are of
commercial importance. The Midway Deep Sea Coral was discovered in 1978, and
has yet to be described but appears to be a new species. The Japanese and
Taiwanese recognize a further variety in the Pacific but this has not been
identified as a valid species (12).

DISTRIBUTION In the Pacific, four major areas have been located as coral
producing regions, all north of 22°N. C. secundum and the Midway Deep Sea Coral
are found in the Hawaiian Archipelago including Emperor Seamounts, and their
range extends as far north as 36°N. C. secundum is found as far south as Hawaii
(19°N); the southern limit of Midway Deep Sea Coral is unknown but it has been
harvested from Hancock Seamount (30°N). C. japonicum and C. konojoi are found
in the Ogaswara-Gunto (Bonin) Is., from Japan south to about 25°N. A third area
covers the Continental Shelf and offshore islands south-west of Japan (35°N) to
Okinawa, where C. japonicum, C. konojoi and C. elatius is found. C. elatius is
found from Okinawa south to the Philippines (12,16,36).

A systematic search for precious corals, especially Corallium spp. has been carried
out in the South Pacific under the auspices of CCOP/SPAC (Committee for
Co-ordination of Joint Prospecting for Mineral Resources in South Pacific

Offshore Areas). The genus is now known to occur in the Solomon Islands, Guam, Vanuatu, Fiji, Tonga, Samoa, the Commonwealth of the Northern Marianas, and the Cook Islands, although little is known of its distribution and abundance (13,15). Two species are known to occur off Guadalupe Island off the west coast of Mexico but are found at depths which are probably too great for exploitation to be practical (33).

C. rubrum used to occur throughout large areas of the Mediterranean, particularly off the coasts of southern France, Corsica, Sardinia, Sicily and North Africa, from Tunis to the Straits of Gibraltar (4). It has been recorded from many localities around the coast of Italy (9), and is still widely distributed in the western Mediterranean particularly off the coast of Provence and Corsica (27,28,29). Although it used to occur in the Aegean Sea, it now appears to be absent, and the reason for its scarcity in the eastern Mediterranean is still unknown (31). It has been reported from Cape Verde and the Canary Islands (17).

POPULATION Unknown.

HABITAT AND ECOLOGY All known species of precious corals occur in greatest abundance on stable, cleanly swept, hard substrates where relatively strong bottom currents, from 0.5-3.0 knots, prevail. The preferred substrate is limestone or volcanic material, on a rise, seamount or gently sloping terrace, free of sediment which smothers young corals and abrade the stalks of older ones causing toppling (13,23). Recent work at Midway Island has shown that there are both 'shallow' and 'deep sea' precious corals found at about 400 m and 1200 m respectively. Pacific species are found at the following depth ranges: C. japonicum 200-300 m; C. elatius 250-400 m; C. konojoi 100-200 m; C. secundum 350-450 m; C. nobile 150-330 m; Midway Deep Sea Coral 1000-1500 m (12). C. rubrum is found attached to rocks in poorly lit areas at depths of 20-200 m, and colonies are more abundant at lower light intensities (9,28). It occurs most often in caves and under overhangs between 30 and 60 m. At greater depths it is found on vertical surfaces and steep inclines, its greater scarcity here probably due to lack of suitable substrate. Colonies are frequently found in dense aggregations, aligned parallel to each other (27,28). Food supply and temperature seem to have little effect on the species distribution (9,13) but studies of C. rubrum have shown that temperature has an important effect on sexual reproduction. The reproductive cycle of Corallium is annual. C. rubrum usually produces gonads in March and spawns in August and September (9), although in warmer, southern Mediterranean waters, spawning occurs earlier in the year and extends over a longer period (10). The Pacific species spawn in June or July (16). Larval life span is unknown.

Precious corals are characterized by great longevity, slow growth, and relatively low rates of mortality and recruitment (16). In C. secundum, colony height increases about 0.9 cm a year to an age of about 30 years. The largest colonies found at Makapua in Hawaii are rarely more than 60 cm in height, but colonies may live for 100 years. In the absence of fishing, mortality approximately equals the recruitment rate and is about 6 per cent a year. Many colonies are killed by boring organisms which cause toppling. C. secundum in Hawaii is reproductively mature at a size of about 12 cm (13 years old). Growth rates in Japan have been estimated at about 0.5 cm/year and observations by fishermen suggest that about 50 years are required before an exhausted bed is productive again (23).

SCIENTIFIC INTEREST AND POTENTIAL VALUE Commercial use of precious coral from the Mediterranean can be traced back to the Neolithic period. In the Iron Age precious coral was used to decorate shields and swords (18). It was used by the Persians, Indians, Chinese and Celts as jewellery and for inlaid decorative work and the Greeks and Romans used it more as a talisman. It was highly valued in Tibet and India at the time of Marco Polo and coral necklaces are still among the most cherished possessions of wealthy Tibetans and are included among the sacred treasures of monasteries (4). Indians believed red coral protected them from evil and cured sterility (5). Precious coral was considered an effective antidote to poison in the Middle Ages and was worn as a talisman against enchantments, witchcraft and Satan. Ground coral was taken internally and grains were put into babies' milk to protect them from fits. Coral jewellery was particularly fashionable in the middle of the 19th century when the Naples industry expanded enormously. Subsequently it became less fashionable, then underwent a revival in popularity in the 1920s and is currently once again in great demand (2).

The Mediterranean fishery was initially controlled by the Italians. For a period from the 17th to early 19th century, the Spanish and French had control, but it then reverted to the Italians. Large quantities were exported to India, but coral also went to many other countries including China (for buttons), Germany, Russia, Tibet and the U.S.A. (5). Torre del Greco is the centre for high quality carving, where many of the carvers learn their craft at the school of coral carvers (8,34). The principal Mediterranean fisheries were off the southern coast of France, Corsica, Sardinia, Sicily and North Africa from Tunis to the Straits of Gibraltar. In the early part of this century a small fishery was set up off the Cape Verde Islands (4). Very small quantities are still collected in the Mediterranean and it is difficult to obtain reliable estimates of harvests. In Tunisia coral fishing takes place in waters between Tabarka and Bizerte, mainly in the winter. In 1979, 5551 kg were reported to have been gathered (3). In Corsica, coral is collected commercially off Bouches-de-Bonifacio (30), and along the west coast (34); in Sardinia it is collected near Alghero (34).

In the 19th century precious corals were found off Japan and the industry became centred on the Western Pacific, dominated by the Japanese and Taiwanese. By the 1960s annual production was about 10 000 kg, mainly from grounds near Japan, Okinawa and Taiwan. In 1967-68 the Hawaiian beds of C. secundum were discovered and landings increased to over 150 000 kg in 1969. In 1978 the Midway Deep Sea Coral bed was found and in 1980 approximately 200 000 kg was harvested from this area by the Taiwanese and Japanese coral fleets. Statistics for annual landings of precious coral in Taiwan and Japan are difficult to obtain because coral fishermen are not licensed and no offical government records are maintained in either country. However, it has been estimated that 226 000 kg of Corallium spp. were landed in 1980 by Taiwanese and Japanese fishermen (12). In 1970 the majority of the catch was landed in Japan but by 1980 approximately two thirds of the total was landed in Taiwan (12,35). This shift is reflected in the two countries' export statistics, exports of precious coral from Taiwan having increased whilst those from Japan have decreased. Japan now depends to a large extent on imports from Taiwan to supply its carving industry (19).

In Japan, much of the coral comes from waters off Tosa Bay, Shikoku (11). The centres of the Taiwanese coral fishermen are the Penghu islands or Pescadores, where about 170 tonnes are collected a year (22), and Suao, Taiwan. In 1981 the Pacific precious coral fishery was being worked by about 200 Taiwanese boats and 100 Japanese boats (12). Precious corals are generally collected by tangle-net dredging using large river boulders. The Taiwanese and Japanese boats with up to 12 lines over the side, drift with currents. The lines are handled by mechanical line haulers and are kept on the bottom for up to an hour at a time (12).

The total industry in Taiwan and Japan is probably worth over U.S.$50 million. A two-tiered market has developed; one is low priced and stable, consisting of the currently abundant Midway Deep Sea Coral. The other is characterized by rapidly escalating high prices for the darker varieties which are becoming rarer from the traditional Western Pacific grounds (12). Most of the raw coral is sold to international buyers through a system of closed auctions in Japan that are operated by coral fishing associations (15). Processing is labour intensive and there are about 1000 small factories in Taiwan, which produce jewellery and complex hand-carved figures. The latter sell in Paris for up to U.S.$50 000 (12). African, Middle Eastern and Asian demand is for large natural pieces whereas demand in Europe and the U.S.A. is mainly for carved pieces (35). World jewellery production today is dominated by Japanese and Italian manufacturers (15). Saudi Arabia is said to be the main customer for coral, with Japan the second largest consumer (22), but it is also reported that Italy imports 60 per cent of Taiwan's coral (35).

Average 1981 prices for Pacific precious corals are (U.S.$/kg): C. japonicum 600-4800, C. elatius 300-3000, C. nobile 100-2500, C. konojoi about 100, C. secundum 75-250 (12). However, the recession is reported to be affecting the coral trade in Taiwan. The price of raw coral in the Pescadores has dropped from an average of U.S.$52.60/lb ($116/kg) in spring 1981 to less than U.S.$18/lb ($37/kg) (for lowest grades) in March 1982, and sales have continued to be very low (1).

A small domestic fishery operated in Hawaii in the 1970s using a submersible but this has ceased operations due to rising costs (12).

THREATS TO SURVIVAL   Slow-growing, long-lived precious coral is very vulnerable to over-exploitation. At the beginning of the Christian era it was already being reported that coral was becoming rare in parts of the Mediterranean as a result of trade between the Mediterranean and India (5). Although there are probably still a few isolated sites that have never been worked, over-collecting of C. rubrum has made the commercial grade scarce in most accessible areas (14) although small colonies are still abundant, particularly in caves (27). Coral stocks off the western coast of Greece are reported to have been exhausted by Italian coral collectors at the beginning of the century (20). In Tunisia the catch remained fairly constant from 1975 to 1979 although catch effort increased; in shallower waters stocks are reported to be almost fished out but surveys have shown that exploitation could be profitably extended to depths greater than 100 m (3). The rich bed of C. rubrum at Coralli di Alghero off Sardinia was being so heavily exploited in the 1970s that it was disappearing at an alarming rate (7). A study of coral populations off the coast of Corsica has shown that the size of colonies in areas undergoing commercial exploitation is significantly smaller than those in unfished areas (30). Coral divers are now travelling further afield to find fresh stocks. The Sardinians go as far as Morocco and Gibraltar and French divers are increasingly turning to Tunisian waters. In Tunisia, foreign divers may keep 60 per cent of their catch but the rest is retained by the host country (34). Studies on C. rubrum have not indicated optimum or maximum sustainable yields for this species.

The history of the Japanese and Taiwanese coral fishery is characterized by a cyclical pattern of discovery of new grounds, heavy exploitation and subsequent depletion of stocks. The fishery has gradually changed from one based on many small boats to one in which most of the coral is collected by fewer, larger boats capable of remaining at sea for periods of up to six months and going much further afield (11,23). Beds that were especially rich in the past but which are now nearly exhausted include the Penghu islands near Taiwan; the continental slope areas near the southern tip of Taiwan; Oxa Bank near Miyako Island; the Danjo Islands near Kyushu, Japan; banks of Tosa, in Shikoku, Japan; Smith Bank, Torishima Island and Sofu Island located almost mid-way between Japan and the Bonin Islands (23). Off Tosa Bay, coral used to be found at depths of near 100 m but it is now only found between 200 and 400 m and the catch has dropped from 8259 kg in 1977 to 5250 kg in 1981 (11). Many Taiwanese coral collectors have reportedly been fined for fishing in Japanese waters (22).

The beds found in 1969 off Hawaii were also rapidly overfished, resulting in declining catches through the early 1970s, until the discovery of the Midway Deep

Sea Coral. Yields of the darker coloured varieties from the traditional grounds have continued to decline. This has resulted in rapidly escalating prices which are primarily responsible for beds remaining viable in spite of declining production. The trend can be expected to continue unless significant new coral beds are discovered. If the Midway grounds are also depleted, the industry's major sources of raw material will be jeopardized and the economic consequences will be incurred by Taiwan and Japan as well as countries such as Italy and the U.S.A. which now depend on their exports (12).

Studies in the Mediterranean have indicated that C. rubrum is very susceptible to silt and pollution and along the coast of Provence it is only found in unpolluted areas (28).

CONSERVATION MEASURES TAKEN Attempts have been made to prohibit coral fishing in several places but invariably these have been unsuccessful. When recovery of the beds has occurred it has usually been accidental, for example, as a result of interruption of fishing by war (15). Between 1879 and 1890 grounds off the north coast of Africa were fished on a nine or ten year rotation but lack of enforcement eventually led to severe depletion of the beds. The selection of this length of time for recovery was based on observations by fishermen and early research (5,15,25). In Spain there is a permit system for divers and the Meda Islands have been proposed as a marine conservation area in order to preserve and possibly reintroduce C. rubrum and gorgonians (14).

Before 1868 coral fishing in Japan was managed inadvertently, since the Shoguns confiscated any coral collected by fishermen, but after this date no management was attempted. In 1963 the government of Okinawa attempted to regulate the newly discovered beds south of Okinawa by limiting entry into the fishery by a permit system. However, too many permits were issued and the beds were rapidly depleted (15,23). In the Philippines, Presidential Decree 1219 of 1977 bans the export of unprocessed precious coral (6).

Extensive research on the Hawaiian precious coral stocks (16,24) has led to the formulation of a Fishery Management Plan for the Precious Coral Fisheries of the Western Pacific Region (FMP). This establishes a set of conservation and management measures designed to achieve a balance between protecting coral resources and allowing a harvest on a sustainable yield basis for the Fishery Conservation Zone (FCZ) of Hawaii, Guam and American Samoa. Four categories of management areas have been designated: 'Established Beds' for which estimates of maximum sustainable yield (MSY) are reasonably precise and where selective gear would be used; 'Conditional Beds' for which the MSYs could be estimated by comparing their size relative to the established beds; 'Refugia' where coral harvesting would be prohibited and which would serve as baseline study areas and possible reproductive reserves; and 'Exploratory Permit Areas' in which coral beds almost certainly exist but have not yet been located. Only one of the Hawaiian beds, Makapuu, has been studied enough to be classified as an Established Bed. Five other beds off the Hawaiian islands are classified as Conditional, and the Wes Pac bed, situated between Nihoa and Neckar islands, has been proposed as a Refugium. As new beds are discovered and more accurate MSYs are determined, the categorization of the beds may be upgraded. Quotas and size limits have also been set for different species (the plan covers gold and bamboo, as well as pink, corals) in different beds. The MSY for C. secundum in the Makapuu bed has been calculated at 1000 kg/year (15). In the U.S.A. corals are now listed under the Lacey Act which means that the import of corals obtained illegally in the country of origin is prohibited.

CONSERVATION MEASURES PROPOSED Unless some form of management is instituted, supplies of precious coral may become even more unpredictable. Precious coral can be managed by limiting harvest levels, areas or gear but a biological basis for further guidelines is required. The research carried out in Hawaii and the Fishery Management Plan for the Western Pacific are major advances in this direction. Further survey work is required in the South Pacific to determine whether any of the <u>Corallium</u> now known to occur there is commercially valuable (13). Management of the precious coral resources of the Emperor Seamounts outside the U.S. FCZ could come about by extension of U.S. jurisdiction over this area or by a UN multilateral treaty based on the principle of 'common heritage'. Future action regarding jurisdiction over what are now international sea bed resources will probably depend on the final text negotiated at the Law of the Sea Conference. In the interim, some degree of 'management' over future supplies of precious coral may come about as a result of stockpiling and controls imposed by cartels in Taiwan and Japan. In the longer term, supplies will depend on the discovery of new grounds and ultimately on the success of an international management regime (12).

In the Mediterranean, possible measures for conservation include restricting participation in the fishery through permits; imposing size limits and catch quotas; establishing and managing reserve areas and restricting trade (14). Current research on growth rates of <u>C. rubrum</u> should provide data on recovery times for exploited populations, which appear to be very variable (32). Marine reserves proposed for Italy include two areas where rich beds of <u>C. rubrum</u> can still be found: the Coralli di Alghero, and the Arcipelago della Maddalena, both in Sardinia (7). La Scandola, part of the regional Park of Corsica, south of Calvi, and the reserve at Cap l'Abeille near Banyuls in France also have rich beds of this species. However, it is absent from the reserve at Port-Cros, off Marseille, France, although this area has not been fished for 15 years (32).

Tangle-net and dredge methods of fishing should be strictly regulated. Historically the primary method used to collect precious coral both in the Mediterranean and in the far western Pacific has been dredging with tangle-nets. With the exception of the Mediterranean the distribution of all major known coral beds is below the depth range of conventional SCUBA equipment. However, dredging is both destructive and inefficient, and recently attempts have been made to develop other methods. A submersible has been used off Hawaii, and remotely controlled vehicles are currently being developed by separate companies in Hawaii and Taiwan. Use of a submersible permits selective harvest, mature colonies can be avoided, other benthic species are not disturbed, and size limits can be enforced. Furthermore nearly all the coral dislodged from the bottom is brought to the surface. Dredges only recover about 40 per cent of what is initially 'knocked down', although they can be dragged repeatedly over the same area. However a submersible is costly and the depths and areas which can be worked are limited. It is unlikely that many Pacific countries could afford such a system (13,15,21). A comparison of the relative impacts of dredging versus selective harvest suggests that if dredging quotas are 0.2 of selective harvest quotas, the impact to the coral bed and associated benthic organisms will be the same (15).

REFERENCES 1. Anon. (1982). Recession hits the coral gatherers. <u>Fishing News International</u> March 1982.
2. Anon. (1977). Coral returns to fashion. <u>Retail Jeweller</u> 13 October : 8.
3. Boulhel, M. (1981). Aménagement des ressources vivantes de la zone littorale tunisienne. In GFCM, Management of living resources in the Mediterranean coastal area. <u>Stud. Rev. Gen. Fish. 1981 Counc. Meditterr.</u> (58): 95-130.
4. Hickson, S.J. (1924). <u>An Introduction to the Study of Recent Corals.</u> University Press, Manchester.
5. McIntosh, W.C. (1910). A brief sketch of the red or precious coral. <u>Zoologist</u> 14: 1-22.
6. Marcos, F.E. (1977). Presidential Decree No. 1219, Manila,

Philippines.

7. Cassola, F. and Tassi, F. (1973). Proposta per un Sistema di Parchi e Riserve Naturali in Sardegna. Bull. Soc. Sarda Sci. Nat. 13: 5-83.

8. Ritchie, C.I.A. (1970). Carving Shells and Cameos. Willmer Bros. Ltd., Birkenhead.

9. Barletta, G., Marchetti, R. and Vighi, M. (1968). Ricerche sul Corallo Rosso: 4 Ulteriori Osservazioni Sulla Distribuzione del Corallo Rosso nel Tirreno. Istituto Lombardo (Rend.Sc.) B 102: 119-114.

10. Vighi, M. (1972). Etude sur la reproduction du Corallium rubrum (L.). Vie Milieu 23(1) A: 21-32.

11. Anon. (1982). Coral prices skyrocket as sources are depleted. Chikyo no Koe (Journal of Environmental and Energy Issues in Japan), April issue, Friends of the Earth, Tokyo.

12. Grigg, R.W. (1982). Precious coral in the Pacific. Economics and development potential. Infofish Marketing Digest 2: 8-11.

13. Grigg, R.W. and Eade, J.V. (1981). Precious corals. Report of the Inshore and Nearshore Resources training workshop. Suva, Fiji, July 1981.

14. Hunnam, P. (1980). Mediterranean species in possible need of protection. Report to IUCN, UNEP/UNESCO UNEP/16.20/INF.6.

15. Western Pacific Regional Fishery Management Council (1980). Fishery Management Plan and Proposed Regulations for the Precious Coral Fisheries of the Western Pacific Region. Honolulu, Hawaii.

16. Grigg, R.W. (1976). Fishery Management of Precious and Stony Corals in Hawaii. Sea Grant Technical Report UNIHI-SEAGRANT-TR-77-03. Univ. of Hawaii Sea Grant Program.

17. Bayer, F.M. (1964). The genus Corallium (Gorgonacea: Sleraxonia) in the western North Atlantic Ocean. Bull. Mar. Sci. Gulf Caribb. 14(3): 465-478.

18. Brailsford, J. (1978). Early Celtic Masterpieces from Britain. British Museum Publications Ltd.

19. Wells, S.M. (1981). International Trade in Corals. IUCN Conservation Monitoring Centre, Cambridge, U.K.

20. Belloc, G. (1950). Inventory of the Fishery Resources of Greek waters. FAO Report, Rome.

21. Grigg, R.W., Bartko, B., and Brancart, C. (1973). A new system for the commercial harvest of precious coral. Sea Grant Advisory Report UNIHI-SEAGRANT-AR-73-01.

22. Nerbonne, J.J. (1981). Coral... a treasure from the sea. The Asia Magazine 26 April : 21-25.

23. Grigg, R.W. (1971). Status of the Precious Coral industry in Japan, Taiwan and Okinawa: 1970. UNIHI-SEAGRANT -AR-71-02. The University of Hawaii Sea Grant Program.

24. Poh, K. (1972). Economics and Market Potential of the Precious Coral Industry in Hawaii. UNIHI-SEAGRANT -AR-71-03.

25. Lacaze Duthier, H. de. (1864). Histoire Naturelle de Corail. Paris.

26. Grigg, R.W. (1982). In litt., 1 September.

27. Laborel, J. and Vacelet, J. (1958). Etude des peuplements d'une grotte sous-marine du golfe de Marseille. Bulletin de l'Institut Océanographique 1120: 20 pp.

28. Laborel, J. and Vacelet, J. (1961). Répartition bionomique du Corallium rubrum Lmck. dans les grottes et falaises

sous-marines. Rapp. Comm. Int. Mer. Médit. 16(2): 465-469.

29. Laborel, J. (1961). Contribution à l'étude directe des peuplements benthiques sciaphiles sur substrat rocheux en Mediterranée. Rec. Trav. Stat. Mar. Endoume 33(20): 117-173.

30. Marin, J. and Reynal de Saint-Michel, L. (1981). Premières donnés biométriques sur le corail rouge, Corallium rubrum Lmck. de Corse. Rapp. Comm. Int. Mer Médit. 27(2): 171-172.

31. Zibrowius, H. (1978). A propos du corail rouge en Méditerranée Orientale. XXVIe Congres - Assemblée plénieré de la C.I.E.S.M., Antalya 1978, Comité de Benthos: 2 pp.

32. Laborel, J. (1982). In litt., 16 September.

33. Bayer, F.M. (1982). In litt., 6 October.

34. Mouton, P. (1982). La fièvre du corail. Océans 107:3-19

35. Westbrook, J.R. (1981). Taiwans 'Gem' Fishermen. Fish. News Int. 20(1): 20-21.

36. Kishinouye, K. (1903). Preliminary note on the Coralliidae of Japan. Zool. Anz. 26: 703.

We are very grateful to F.M. Bayer, F. Boero, R.W. Grigg, J. Laborel, D. Pessani, L. Rossi and J. Vacelet for help with this data sheet.

STARLET SEA ANEMONE   VULNERABLE

Nematostella vectensis Stephenson, 1935

Phylum   CNIDARIA   Order   ACTINIARIA

Class   ANTHOZOA   Family   EDWARDSIIDAE

SUMMARY A brackish-water sea anemone found in salt marshes of the northern hemisphere. It is endangered in England, U.K. where it has been recorded from ten sites at seven different localities. Only four populations now definitely survive, at least one of which (Pagham) appears to be unstable. It is vulnerable in North America although populations seem stable at present.

DESCRIPTION A small (up to 6 cm but usually less than 2 cm long), elongated, burrowing anemone with up to 20 (typically 16) tentacles in two cycles. In contraction it is corrugated and opaque, but when extended becomes smooth and translucent. The general colour appears grey to whitish, modified by the internal mesenterial filaments (greenish, brown, black or pinkish depending on previous food) and the opaque pigmentation; the tentacles are flecked with white (15,18). The markings on the column are occasionally absent altogether (4,18). Like other burrowing anemones, the body column is rounded posteriorly forming a physa. Nematosomes (tiny spherical bodies incorporating stinging cells) are visible with a hand lens through the translucent body wall; peristaltic waves continually pass down the column when the animal is exposed (11,18). Detailed description with photographs in (18); field recognition data in (12).

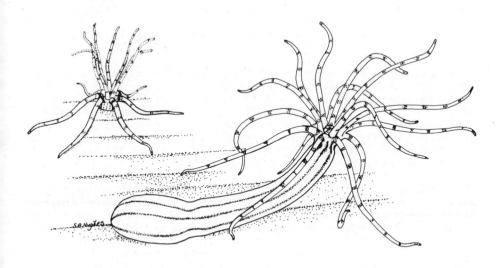

DISTRIBUTION Known only from England, U.K. and North America. English populations are currently known from East Anglia (Shingle Street, Suffolk) and on the south coast (Pagham, Sussex; Gosport, Hampshire; Langton Herring, Dorset)

(14,19). Only known Canadian locality is in Nova Scotia, comprising at least nine adjacent saltmarsh pools at Minas Bay, Kingsport (1,19). In the U.S.A. found on the Pacific coast (several localities are known in California (7,8) one in Oregon and one in Washington (2,19)) and on the Atlantic coast (one locality in Massachusetts (4) (originally described as Nematostella pellucida), one in Rhode Island (13), one in New York (9), one in Delaware (10), one in Chesapeake Bay (3) and one in Florida (22)). As with many marsh species, accurate plotting of geographical distribution is difficult due to the paucity of research on appropriate habitats. It is possible that Nematostella vectensis will eventually be found in the extensive marshes and brackish waters of north-west Europe and that its known distribution in North America will be extended.

POPULATION No estimate of the world population can be made; densities are very variable throughout the year and at different localities. In the New York locality, one to twenty specimens have been found after searches at different times of year; other attempts to find anemones in the same place have yielded no specimens (9). Similar results have often been obtained in California (8) and in English localities, but very high densities sometimes occur: in a Norfolk pool in September, 1974, a maximum density of 12 572 anemones per m^2 was recorded, and an estimate of over 5 million individuals in a single pool has been made (21).

HABITAT AND ECOLOGY Lives in fine, soft muds at the edges of creeks on salt marshes and in brackish pools (19) but never on the shore-line proper. It is usually found buried upright in the mud with only the tentacle crown and part of the column exposed but it is sometimes found lying extended on the mud or attached to sea grasses, such as Ruppia or Zostera, and algal masses of Cladophora and Chaetomorpha (11,15,19). It cannot burrow in sand (21). Markedly euryhaline (i.e. able to tolerate a wide salinity range and has been recorded in England at salinities from 8.96 to 51.54°/oo) and eurythermous (i.e. tolerant of a wide temperature range and recorded in England at temperatures from -1 to 28°C) (21). Individuals frozen in ice for several hours can be revived on thawing (11). The anemone survives well in unaerated aquaria for several weeks (11,21), although the oxidized layer at the mud surface should ideally be at least 1 cm deep at all seasons (19).

Limiting ecological factors are probably heavy wave action, extreme hypoxia and desiccation. Extreme shelter is necessary to allow build-up of fine mud which Nematostella requires to burrow in; burrowing offers a short term (less than one week) means of surviving desiccation (19). The species is unselective in feeding, taking mainly small copepods and midge larvae (19), corixids and gastropods (particularly Hydrobia) (6) but also a great range of other prey. Reproduces sexually in summer and autumn, having non-planktonic larvae (5,18), and asexually by transverse fission, probably throughout the year (6,11,19). No known predators.

The original habitat of this anemone was probably in shallow pools of high marshes and at sides of creeks at marsh edges in estuaries and bays (19). North American records reflect this natural distribution pattern which facilitates the spread of the species from established to developing marsh, probably by tidal and storm action sweeping anemones from pool to pool and possibly to a lesser extent by wildfowl fortuitously transporting vegetation with adhering anemones. The English distribution is atypical and is the reason for the species being endangered.

SCIENTIFIC INTEREST AND POTENTIAL VALUE N. vectensis is one of only three species known in this unusual genus; the other two, of whose ecology nothing is known, occur in the Arctic Ocean (18). It is not only an intrinsically valuable species, but also the only one easily accessible for studying nematosomes, which are found only in this genus and are thus unique in the animal kingdom; their function is still unknown (20). It is one of very few Actiniaria in the world inhabiting saltmarshes, and its discontinuous distribution merits further study.

Being apparently so sensitive to hypoxia it may well act as an indicator species giving early warning of pollution (19).

THREATS TO SURVIVAL In England some populations have been exterminated by pollution or through pools drying up. The present localized distribution, in isolated pools bounded by sea walls or shingle banks, is probably the result of centuries of sea wall construction and marsh reclamation, increasing the vulnerability of the resulting fragmented populations to climatic and further human influences and preventing natural spread (19). In North America, the marsh habitats of the known Nematostella populations are coming under pressure from human use and pollution, particularly on the east coast (9). It should be emphasized that the number of localities in the world is more significant than the number of individuals because, should a marsh be polluted or a relict pool dry up, the whole population, regardless of its size, would be exterminated, as has occurred several times in England (19).

CONSERVATION MEASURES TAKEN A long-term study on the biology of English populations of N. vectensis was initiated in 1971 and has provided much information necessary to take effective conservation measures (16-21). At Cley, Norfolk, Half-Moon Pond was cleared of rubbish in 1974 and further dumping was prohibited to prevent pollution. In 1975, a year of drought, the pond dried up, but a remnant of the population was saved and transferred to a nearby open marsh pool. Other studies have since been undertaken on marsh habitats in North America (9,10) and in Florida the species occurs in St. Mark's Wildlife Refuge near Tallahasee (22).

CONSERVATION MEASURES PROPOSED Further reintroductions in England should be made on open marshes unbounded by sea walls and free from other human influence, so that natural spreading can occur. Sites should be selected near to where Nematostella exists or previously existed and should be adequately protected. The best time for transfer would be late summer to autumn, when both sexual and asexual reproduction are in progress and natural prey is plentiful. Known natural populations should be protected.

CAPTIVE BREEDING The species can be maintained satisfactorily in aquaria and will undergo asexual reproduction under such conditions (6,11,19).

REFERENCES 1. Bailey, K. and Bleakney, J.S. (1966). First Canadian record of the brackish water anthozoan, Nematostella vectensis Stephenson. Can. Fld Nat. 80(4): 251-252.
2. Bleakney, J.S. (1980). In litt. to R. B. Williams.
3. Calder, D. (1972). Phylum Cnidaria. In Wass, M.L. (Compiler). Check list of the biota of Lower Chesapeake Bay. Spec. scient. Rep. Va Inst. mar. Sci. 65: 97-102.
4. Crowell, S. (1946). A new sea anemone from Woods Hole, Massachusetts. J. Wash. Acad. Sci. 36(2): 57-60.
5. Frank, P. and Bleakney, J.S. (1976). Histology and sexual reproduction of the anemone Nematostella vectensis Stephenson 1935. J. nat. Hist. 10: 441-449.
6. Frank, P. and Bleakney, J.S. (1978). Asexual reproduction, diet, and anomalies of the anemone Nematostella vectensis in Nova Scotia. Can. Fld Nat. 92: 259-263.
7. Hand, C. (1957). Another sea anemone from California and the types of certain California anemones. J. Wash. Acad. Sci. 47(12): 411-414.
8. Hand, C. (1982). In litt., 16 March.
9. Hotchkiss, F.H.C. (1976). In litt. to R.B. Williams.
10. Jensen, L.D, (1974). Environmental responses to thermal discharges from the Indian River Station, Indian River, Delaware. Electric Power Research Institute, publication

no. 74-049-00-3, Palo Alto.

11. Lindsay, J.A. (1975). A salt marsh anemone. Mar. Aquarist 6(8): 43-48.
12. Manuel, R.L. (1981). British Anthozoa. In Kermack, D.M. and Barnes, R.S.K. (Eds)., Synopses of the British Fauna, (N.S.) 18. Linn. Soc. London, Academic Press.
13. Nixon, S.W. and Oviatt, C.A. (1973). Ecology of a New England salt marsh. Ecol. Monogr. 43: 463-498.
14. Sheader, M. (1981). In litt. to R.B. Williams.
15. Stephenson, T.A. (1935). The British Sea Anemones Vol. II. London, The Ray Society.
16. Williams, R.B. (1973). Euplotes alatus, a hypotrichous ciliate new to Great Britain. Trans. Norfolk Norwich Nat. Soc. 22: 383-386.
17. Williams, R.B. (1973). The significance of saline lagoons as refuges for rare species. Trans. Norfolk Norwich Nat. Soc. 22: 387-392.
18. Williams, R.B. (1975). A redescription of the brackish-water sea anemone Nematostella vectensis Stephenson, with an appraisal of congeneric species. J. nat. Hist. 9:51-64.
19. Williams, R.B. (1976). Conservation of the sea anemone Nematostella vectensis in Norfolk, England and its world distribution. Trans. Norfolk Norwich Nat. Soc. 23: 257-266.
20. Williams, R.B. (1979). Studies on the nematosomes of Nematostella vectensis Stephenson (Coelenterata: Actiniaria). J. nat. Hist. 13: 69-80.
21. Williams, R.B. (unpublished).
22. Heard, R.W. (1982). Guide to Common Tidal Marsh Invertebrates of the Northeastern Gulf of Mexico. Mississippi - Alabama Sea Grant Consortium. 82 pp.

We are very grateful to R.B. Williams for compiling this data sheet and to C. Hand and R. Manuel for reviewing it.

| Phylum | CNIDARIA | Order | ANTIPATHARIA |
| Class | ANTHOZOA | Family | ANTIPATHIDAE |

SUMMARY Although data on the abundance and distribution of Black Coral throughout the world is scanty there have been many reports of depletions of local populations in shallow waters due to over-collecting for the jewellery and curio trade. The trade should be monitored and management plans set up for Black Coral fisheries. Since identification is difficult, all species in the order Antipatharia have been included in this sheet.

DESCRIPTION Classification and identification of the 150 known species of Black Coral is complicated, some species assigned to the genus Leiopathes have been referred to a distinct family Leiopathidae. There are few data on their natural history because of the depth at which they occur. Black Coral forms branched plant-like colonies varying in height from a few centimetres to several metres, which often aggregate to form 'forests' or 'beds'. The slender branches are strengthened by a brown or black skeleton of horny material; polyps which are situated in the living tissue around this skeleton are short and cylindrical with six tentacles which cannot be retracted. Colonies off Trinidad were found to adopt fan-like growth forms which are probably nutritional adaptations to unidirectional current flow. Branches of many species of Black Coral appear to bear polyps on one side only, often on the leeward side, also probably a nutritional adaption. The order also includes whip or wire corals which exhibit coiling shapes. The lower end of the colony is attached to some firm object by a flattened base or may simply extend into the sediment. They are urticant (stinging) (7,10,21).

A number of studies on the growth of Black Corals are being carried out. Average rates of increase in height for Hawaiian species are 6.4 and 6.1 cm per year; patterns of growth are probably constant and indeterminate, and periodicity of ring formation in Antipathes dichotoma appears to be annual (10). Black Coral from Curaçao has daily growth rings, and a coral piece of 1 cm diameter is probably 15 years old (15).

DISTRIBUTION Mainly in tropical and subtropical waters in all major oceans, although a few species are found in shallow waters in temperate regions (7). The distributions of individual species are little known at present. Some appear to have wide distributions: Antipathes dichotoma occurs throughout the Indo-Pacific in deep waters and in the Red Sea (7); A. japonica is common at about 10 m depth from Taiwan to central Japan; Cirrhipathes spiralis and C. anguina (whip corals) occur from 30-500 m throughout the Indo-West Pacific (13). Others have localized and restricted distributions: A. aperta is found in all fiords of the west coast of southern New Zealand (30). A. panamensis is known from Panama, Colombia, Ecuador and the Galapagos Is.; A. galapagensis is endemic to the Galapagos Is. (33).

POPULATION Unknown, and as for all colonial species is very difficult to estimate.

HABITAT AND ECOLOGY Found in deep water, usually between 30-110 m although some have been recorded from depths of 2000-3000 fathoms (4000-6000 m) (12); in the Caribbean the greatest density is between 40-200 m (8); in Hawaii adult colonies were found to tolerate depths of 1-146 m (9). Colonies are frequently associated with terraces and undercut notches, which suggests that abundance is related to habitat space (10). In New Zealand, A. aperta is found in shallow, cold waters, on the steep rock walls of fiords (30). Light and water turbulence play important roles in the distribution of A. grandis off the coast of Hawaii. Adult colonies can withstand light intensities of up to 60 per cent of the surface incident light, but it appears that larvae cannot settle and survive under a light intensity greater than 25 per cent of the surface light. This would account for the greatest densities being found below 35 m. Colonies were found at shallower depths only in turbid water or in shaded areas (9). Off Curaçao, Black Corals are found as shallow as 10 m and in the shade of rocks occasionally shallower; the greatest density of colonies is found between 15 and 50 m (15). In the Galapagos, A. panamensis and A. galapagensis are found throughout the archipelago where there are steep vertical rocky walls (34). They occur at depths from 3-50 m; the unusually shallow depths at which they occur may be due to the cool waters surrounding the Galapagos. It appears that A. panamensis, like most black corals, settles preferentially in dark areas (caves, underhangs) from which the colony grows outward into the current. A. galapagensis appears to tolerate a wider range of substrates, often more exposed. A. panamensis is rarer compared with A. galapagensis (33). Black Corals may be scarce where sediment covers terraces (10). Only animal material has been observed to be ingested by A. grandis (9). The normal method of feeding is probably trapping of animal plankton by the nematocysts (21). Polyps are either male or female but colonies may be hermaphroditic (7). Reproductive maturity in A. dichotoma is probably reached between 10 and 12.5 years and the reproductive cycle may be annual (10). Asexual reproduction occurs naturally by fragmentation of branch ends. Colonies may have a life span of 70 years (10).

SCIENTIFIC INTEREST AND POTENTIAL VALUE Antipatharians are of interest for their unusual body structure compared with other anthozoans, and they are important in providing a habitat for many other small invertebrates of systematic value (13,16). In the Middle and Far East Black Coral has long been accredited with medicinal and magical properties (26). It was once abundant in the Persian Gulf from where it was exported to India (25). It has recently become a popular material for souvenir jewellery and is sold to tourists in many tropical countries, providing income to local people (3,4,6,10). Synthetic 'Black Coral' is often sold to unsuspecting tourists (28). An unknown but probably fairly large amount is involved in international trade. For example, in June 1982 the U.S.A. imported 24,000 pieces of Black Coral jewellery from Taiwan and 359 pieces from the Philippines (27). In 1981, prices for A. dichotoma on the international market were 20-50 U.S. $/kg (28).

Black Coral is harvested commercially in Hawaii; the estimated standing crop of A. dichotoma in the Auau Channel is 84,000 colonies and an optimum sustainable yield of 5,000 kg can be taken over an annual period (10). It is collected by divers who remove colonies with an axe or sledge hammer and float them to the surface with air bags (28). Small quantities of A. dichotoma are collected commercially on Guam, off Orote Point and in Umatac Bay at 30-70 m (22). A. pennacea and A. dichotoma are used commercially in the Caribbean as well as Cirrhipathes lutkeni, although the latter is less valuable due to its limited girth (8). Black Coral has been exploited commercially in Mexico off the coast of Quintana Roo, for 10-12 years (24) and also in a number of other sites such as Plancar reef, around the coast of Cozumel, Puerto Angel in Oaxaca, and the Bay of Cabo San Lucas in South Baja California (31). At Bajo de Punta Gorda, in the Bay of Cabo San Lucia the Black Coral industry has been run as a co-operative since 1974. Coral is collected at depths of 50-75 m and a diver collects on average 5.5 kg a day and is paid 1500 pesos per kg. The coral is made into jewellery for export to the U.S.A. It takes about 3 months to process 22 kg, so the rate of utilisation is not very high. Collecting is carried out under the system devised for Hawaii (10) and colonies with a height of less than 1.20 m are not taken (31). The increasing number of tourists visiting the Galapagos Is. in recent years has resulted in the development of a small black coral jewellery industry in the islands. Black coral collection is mainly incidental to the lobster fishery. There are indications that the industry could expand to provide exports to markets in the U.S.A. and Ecuador(33).

THREATS TO SURVIVAL Locally depleted in several areas particularly in shallow water, but unlikely to be exploited to extinction due to its inaccessibility and the fact that other populations may yet be discovered (11,17). The Caribbean has been particularly overfished. Over-collection is reported to be severe in Barbados (8,19). Black Coral is exploited commercially in Jamaica at Montego Bay and Ocho Rios (8) and stocks may be threatened. In St. Lucia the Ministry of Fisheries believes Black Coral to be all but fished out. The locals deny this but carvings are reported to be noticeably smaller than they were 4-5 years ago (6). Attempts to regulate trade in Black Coral in the U.S. Virgin Islands have proved unsuccessful and it is thought that local populations could be destroyed within the next ten years (5). In the Netherlands Antilles, Black Coral exploitation is reported to be severe off Curaçao and Aruba and moderate off Bonaire (8,15). Over-exploitation has occurred in Mexican waters, particularly off the island of Cozumel, where divers are now having to go to much greater depths (24). In the other areas in Mexico where commercial collecting occurs, stocks have been severely depleted. In the Bay of Cabo San Lucas, South Baja California, attempts are being made to manage the fishery, but these are frustrated by foreign and 'pirate' divers who take undersized colonies and sell them at lower prices (31). However, in other areas such as the Chinchorro Bank, Black Coral stocks are under-exploited (24). In New Zealand A. aperta, the only commercially valuable species, is under threat throughout its range from collection for the jewellery market (16). Black Coral in waters around the Chagos Archipelago could be vulnerable to collectors (18). In the Galapagos, the shallow depths at which Black Corals occur make them especially vulnerable to over-collection (33). Japanese Black Corals are reported to be unsuitable for carving and are not collected (13).

Black Corals may also be affected by habitat disturbance. In New Zealand large numbers of colonies were killed in Deepwater Basin, at the head of Milford Sound, as a result of the diversion of the Cleddau River during construction of an airstrip (16).

CONSERVATION MEASURES TAKEN The order Antipatharia is listed on Appendix II of the Convention on International Trade in Endangered Species of Wild Fauna and Flora (CITES). Trade in Black Coral between parties is therefore allowed only if a valid export permit has been issued by the country of origin.

Black Coral is protected under national legislation in several countries. The order Antipatharia is listed under the Netherlands legislation for protected non-native species (23). It was declared endangered in the British Virgin Islands under the Endangered Animals and Plants (Protection of Black Coral) Order, 1979; trade is forbidden except under licence and import and export is banned (14). Collection is apparently allowed only under licence in Antigua, Belize, Trinidad and Tobago (14); in the U.S. Virgin Islands, Black Coral is defined as a living manageable resource and can be collected only with a permit. In the Netherlands Antilles all corals and gorgonians are officially protected off Curaçao and Bonaire, and in 1974 a similar law was in preparation for Aruba. Enforcement has been difficult but collecting is now controlled by the Bonairean Handicrafts Foundation and tourists rarely collect Black Coral now (8, 36). In the Bonaire Marine Park a survey of Black Coral colonies, A. pennacea and A. cf. dichotoma, is being carried out (20). At Quintana Roo in Mexico, regulations exist to control the commercial Black Coral fishery but these are poorly enforced. In theory, colonies below a certain size may not be collected (24). In the Bay of Cabo San Lucas attempts are being made to manage the fishery along the same lines as the Hawaiian fishery, as described above (31). In Jamaica it is illegal to collect coral without a licence, but again policing is difficult (8); in Barbados it is protected in a limited area of an underwater park on the west coast by the Marine Areas (Preservation and Enhancement) Act of 1976 (19). In Hawaii permits are issued for limited commercial harvesting and minimum size limits are in force (10); in Florida collecting permits are issued only for educational and scientific purposes (17). Black Coral has been protected in New Zealand since December 1980 (16).

CONSERVATION MEASURES PROPOSED   Further research is required to determine the effects of harvesting on Black Coral populations. Studies are needed on the distribution, ecology and reproductive patterns of species used commercially and the current trade requires monitoring. Management plans could then be drawn up for Black Coral fisheries similar to the one implemented in Hawaii. For Hawaiian species a minimum size limit of 1.2 m in height and/or 2.5 cm in stem diameter for harvesting has been established. It was also recommended that trade in 'display' colonies should be prohibited as this encourages collection of small immature colonies (10). In 1982 the Black Coral harvest in Hawaii was about 50 per cent lower than in previous years on account of a depressed market and past stockpiling. Present levels are significantly below the estimates of a maximum sustainable yield, and currently in this area at least there is no danger of over-collection (29). Similar recommendations are being considered by the Western Pacific Regional Fishery Management Council for their Fishery Management Plan for the precious coral fisheries of the Western Pacific Region (37). It is recommended that collection of A. dichotoma on Guam should be more restricted and that further studies to determine the impact of harvesting should be carried out (22). In New Zealand studies have been started on growth, feeding and distribution of A. aperta to provide information which it is hoped will lead to management practices (16). Recommendations for the management of the Mexican Black Coral fishery are made in (24). In some areas, such as Arrowsmith Bay, Mexican Black Coral stocks could support a substantial fishery. However this would have to be planned so that a balance is achieved between conservation and exploitation and the local inhabitants receive maximum benefit. A study is required to determine the socio-economic characteristics of the industry and the sustainable yield of the coral stocks (32). The biology, distribution and abundance of Black Corals in the Galapagos islands are currently being studied in order to assess the ability of local populations to sustain limited harvest (33). It is hoped that the new marine extension plan for the Galapagos National Park will create a mechanism whereby black corals may be conserved perhaps through regulated collection and the establishment of refuge areas of relatively small size where collection would be prohibited (35).

REFERENCES

1. Anon. (1978). Black Coral Regulations. <u>U.S. Virgin Is.</u>
<u>Department of Conservation and Cultural Affairs Newsletter.</u>
2. Anon. (1980). Amendment to Appendices I and II of CITES.
SAA Paper (80)(18) (6).
3. Barber, J. (1980). Report to IUCN/WWF on a visit to the
Caribbean. Unpubl. manuscript.
4. Brendel, M. (1980). In litt.
5. Canoy, M.J. (1981). In litt., 7 April.
6. Cherfas, J. (1980). In litt., 15 April.
7. George, J.D. and George, J.J. (1979). <u>Marine Life.</u> Harrap
and Co., London.
8. Goldberg, W. (1981). In litt., 27 February. <u>Caribbean Black</u>
<u>Coral Survey</u> (information supplied by I. Kristensen, J.
Woodley, Dr. Sander, R.L. Colman).
9. Grigg, R.W. (1965). Ecological studies of black coral in
Hawaii. <u>Pacif. Sci.</u> 19: 244-260.
10. Grigg, R.W. (1976). <u>Fishery Management of Precious and</u>
<u>Stony Corals in Hawaii.</u> Sea Grant Technical Report
UNIHI-SEAGRANT-TR-77-03. Univ. Hawaii.
11. Grigg, R.W. (1980). In litt., 6 June.
12. den Hartog, J.C. (1981). In litt, 9 January.
13. Ikenouye, O. (1981). In litt., 4 March.
14. Ministry of Natural Resources and Public Health (1979).
Endangered Animals and Plants (Protection of Black Coral)
Order, 1979. <u>Government Information Service News Release.</u>
15. Noome, C. and Kristensen, I. (1975). Necessity of
conservation of slow growing organisms like Black Coral.
<u>Stinappa</u> II: 76-77 C.C.A. Ecology Conference, Bonaire,
Netherlands Antilles National Parks Foundation.
16. Richardson, J. (1981). Black Coral Programme. Unpubl.
account of study programme.
17. Robertson, G.W. (1980). In litt., 18 July.
18. Sheppard, C.R.C. (1981). In litt., 8 April.
19. Sheppard J. (1980). In litt., 30 May.
20. van't Hof, T. (1981). Progress Report Jan-April 1981.
Bonaire Marine Park, IUCN/WWF project 1496.
21. Warner, G.F. (1981). Species descriptions and ecological
observations of black corals (Antipatharia) from Trinidad.
<u>Bulletin of Marine Science</u> 31(1): 147-163.
22. Hedlund, S.E. (1977). The extent of coral, shell and algal
harvesting in Guam waters. Sea Grant Publication
UGSG-77-10.
23. Anon. (1980). Ananhangsel A van het beslait ter Uitvoering
van Artikel 3 van de wet bedreigde uitheemse diersoorten.
24. De la Torre, A.R. (1978). Coral negro: Un recurso o una
especie en peligro. In Hignian, J.B. (Ed.), <u>Proc. Annual Gulf</u>
<u>and Caribbean Fisheries Institute</u> 31: 158-163.
25. Pillai, C.S.G. (1973). Coral resources of India with special
reference to Palk Bay and Gulf of Mannar. <u>Proc. Symp.</u>
<u>Living Resources of the Sea around India.</u> ICAR. Special
Pub. Central Fisheries Research Institute. Cochin.
26. Hickson, S.J. (1924). <u>An Introduction to the Study of Recent</u>
<u>Corals.</u> University Press, Manchester.
27. Mack, D. (1982). In litt., 30 July.
28. Grigg, R. and Eade, J.V. (1981). Precious corals. In Report
on the Inshore and Nearshore Resources Training Workshop,
Suva, Fiji. 13-17 July 1981, CCOP/SOPAC.
29. Grigg, R.W. (1982). In litt., 5 August.
30. Grange, K.R., Singleton, R.J., Richardson, J.R., Hill, P.J.

51

and Main, W. de L. (1981). Shallow rock-wall biological associations of some southern fiords of New Zealand. New Zealand Journal of Zoology 8: 209-227.

31. Castorena, V. and Metaca, M. (1979). El Coral Negro, una riqueza en peligro. Tecnica Pesquera 22-27.

32. Castorena, V. (1979). Coral Negro - una possible estrategia. Tecnica Pesquera : 20-21.

33. Robinson, G. (1982). Antipatharian Corals (Black corals) of the Galapagos Islands: Biology, Distribution and Abundance. Unpubl. Project Proposal, Charles Darwin Research Station, Galapagos.

34. Wellington, G. (1975). The Galapagos marine environments. A resource report to the Department of National Parks and Wildlife. Quito, Ecuador.

35. Robinson, G. (1982). In litt., 20 October.

36. Van't Hof, T. (1982). Bonaire Marine Park: an approach to coral reef management in small islands. Paper given at the IUCN Third World National Parks Congress, Bali, October 1982.

37. Western Pacific Regional Fishery Management Council. (1980). Fishery Management Plan and Proposed Regulations for the Precious Coral Fisheries of the Western Pacific Region. Honolulu, Hawaii.

We are very grateful to F.M. Bayer, M.J. Canoy, J. Cherfas, P. Cornelius, W. Goldberg, R.W. Grigg, J.C. den Hartog, O. Ikenouye, J. Richardson, G.W. Robertson, G. Robinson and G.F. Warner and for information provided for this data sheet.

PLATYHELMINTHES

INTRODUCTION The Platyhelminthes are a group of wormlike soft-bodied animals comprising about 25 000 species. They have no body cavity or special respiratory structures and respiration occurs throughout the body surface. The gut usually has only one opening which serves as both mouth and anus. The majority are parasitic, including the flukes (class Trematoda), the monogeneans (Class Monogenea) and the tapeworms (class Cestoda), but there is a fourth class, (Turbellaria) which consists of about 4000 free-living species. Many platyhelminths are dorso-ventrally flattened which accounts for their common name, flatworm.

The trematodes are a major group of endoparasites in all major vertebrate groups. Their complex life cycles normally involve a molluscan host. Well-known trematodes include the liver flukes, those which cause fascioliasis in domestic animals, and the blood flukes which cause human diseases such as schistosomiasis. The Cestoda are highly specialized endoparasites, the most well-known being the tapeworm Taenia, some species of which infect humans (1).

The Turbellaria are given more attention here, as several species could be rare or threatened. Most are aquatic, living on the bottom in sand or mud as part of the interstitial fauna, or on algae, and a few pelagic marine species are known. The majority are marine, but there are many freshwater species and some terricolous species. The latter are found largely in the tropics, confined to very humid areas and hiding beneath logs and leaf mould during the day. Some land species are comparatively large, one species attaining a length of 60 cm. The best known group, on account of its size and ubiquity in freshwater, is the Tricladida (planarians).

The majority are carnivorous, preying on various small invertebrates or scavenging dead animal matter. In triclads and some related forms, the prey is pierced by the pharynx and food is then pumped into the body. A few feed on algae and diatoms, and some are commensal; for example, three species are found living on the gills of Limulus, a horseshoe crab, feeding off scraps of the latter's food. They are generally more active at night, moving by muscular undulation or ciliary activity. Turbellarians are hermaphroditic, and either a free-swimming larva is produced after copulation or the eggs hatch into juveniles which undergo direct development. Many flatworms also reproduce asexually by transverse fission of the adults and freshwater planarians, like sponges are noted for their ability to regenerate from small pieces. Some species are perennial, whereas others are annual, dying after reproduction has taken place.

Although many species have wide or even cosmopolitan distributions, flatworms are sometimes restricted to fairly specific environmental conditions. Streams in Europe generally show a characteristic succession, apparently linked with temperature, of different species from the headwaters to the coast (2,4).

SCIENTIFIC INTEREST AND POTENTIAL VALUE Platyhelminths are generally thought of in terms of their parasitic members and are usually considered to be of no very great benefit to man. However, the free-living forms are proving to be of great scientific interest. Planarians are the simplest animals to possess a true brain and have been found to have a simple form of memory. They are used extensively in research on the molecular basis of memory although much of the work is controversial. Freshwater planarians are also renowned for their ability to withstand prolonged experimental starvation when their bodies may be reduced to as little as one fifth of the original (4). Planarians are good indicators of palaeogeographical relationships as they tend to disperse very slowly, being fragile and having no resistant or specifically adapted dispersal phases (4). Freshwater species occur mainly in clean unpolluted waters and their study may be of some importance in the analysis of water pollution. For example, the European species Crenobia alpina is widely used as an indicator of unpolluted waters (2).

THREATS TO SURVIVAL Very little information is available on specific threats to platyhelminths. As in many other invertebrates, it is extremely difficult to obtain population estimates. Planarians are usually collected by putting out bait which means that the numbers of specimens collected cannot be used for estimates. The main concern is for the free-living freshwater and terrestrial Turbellaria, many of which may be confined to specific habitats and restricted ranges. For example, some 50 species of freshwater flatworms in the Tricladida have been described from North America, the majority of which appear to be limited to subterreanean habitats or narrow geographical ranges (2), such as Dendrocoelopsis hymanae, Planaria occulta and several species in the genera Sphalloplana, Phagocata and Kenkia (3,15). Some are limited to single caves, such as Kenkia rhynchida, endemic to Malheur Cave in Oregon, U.S.A. (6) (see introduction to Chelicerata) and a species known only from Devil's Icebox Cave and a small spring in Rock Bridge State Park near Colombia (8). Romankenkius pedderensis is known only from Lake Pedder, Tasmania (12) and is possibly now extinct as a result of a hydro-electric power scheme which necessitated flooding of the lake (13). In other cases apparently limited distributions may be due to lack of field work; Dendrocoelopsis spinosipenis for example has so far only been recorded from Yugoslavia and South Sweden but almost certainly will be found elsewhere in Europe (11).

Planarians generally appear to be intolerant of pollution and this is likely to be the main threat to many groundwater cave dwelling species. Triclad flatworms usually require well-oxygenated water, although some tolerate temporary oxygen depletion. Acid waters are also unfavourable. Some of the so-called 'ancient lakes' such as Lake Baikal and Lake Ohrid have large numbers of endemic species (9,14) which could be affected by pollution. In Europe, planarians are often absent from polluted streams although other faunal groups (tubificid oligochaetes, Ephemeroptera larvae, and gammarid amphipods) may still be found. However, if pollution leads to an increase in the abundance of food organisms, the population density of some flatworms may increase (2).

CONSERVATION No conservation measures are known to have been taken specifically for flatworms, although in some cases species are now protected as a result of protection of an entire ecosystem. For example, Sphalloplana mohri, a species now known to be widespread in subterranean waters in Texas, is protected in Ezell's Cave as a result of the acquisition of the cave by the Nature Conservancy in order to protect the Texas Blind Salamander, Eurycea rathbuni (7). The species known from the Devil's Icebox Cave and the small spring in Rock Bridge State Park has been studied by the Greater Ozarks Endangered Species Task Force with the aim of determining its distribution within the cave and to ascertain whether it occurs outside its limited known range (8).

For cave species, the maintenance of conditions in the environment to which the species is adapted is of prime importance. For example Kenkia rhynchida has survived in Malheur Cave despite regular vists by large numbers of people using it as a meeting place for a Masonic order. To date this has not affected the flatworm's habitat, but should the water in the cave lake ever be pumped to the surface, this would present a direct threat (5,10).

REFERENCES   1.   Barnes, R.D. (1980). Invertebrate Zoology, 4th Ed., Saunders College, Philadelphia.
2.   Kenk, R. (1974). Flatworms (Platyhelminthes: Tricladida). In Hart, C.W. and Fuller, S.L.H. (Eds), Pollution Ecology of Freshwater Invertebrates. Academic Press, New York, San

Francisco and London. Pp 67-80.

3. Kenk, R. (1979). Freshwater triclads (Turbellaria) of North America. 9. The genus Sphalloplana. Smithsonian Contributions to Zoology 246: 1-38.

4. Ball, I.R. and Reynoldson, T.B. (1981). British Planarians. Synopsis of the British Fauna (N.S.) No. 19, Linn. Soc., Cambridge University Press.

5. Benedict, E.M. (1981). Untitled. Draft manuscript for North American Biospeleology Newsletter 26.

6. Kawakatsu, M. and Mitchell, R.W. (1981). Redescription of Kenkia rhynchida, a troglobitic planarian from Oregon, and a reconsideration of the family Kenkiidae and its genera. Annotationes Zoologicae Japonenses 54: 125-141.

7. Anon. (1968). Ezell's Cave. Information Sheet.

8. Love, J. (1979). Pink Planararia (sic) Project. Ozark Guardian September 1979: 3.

9. Kenk, R. (1978). The Planarians (Turbellaria: Tricladida, Paludicola) of Lake Ohrid in Macedonia. Smithsonian Contributions to Zoology 280: 56 pp.

10. Benedict, E.M. (1973). Malheur Cave as a biospeleological resource. The Speleograph 9(3): 47-48.

11. Dahm, A.G. (1960). Dendrocoelopsis spinosipenis (Kenk) from Yugoslavia and Sweden, and Digonoporus macroposthia An Der Lan. (Turbellaria, Tricladida, Paludicola). Lunds Universitets Arsskrift N.F. Ard. 2 Bd 56(8): 1-39.

12. Ball, I.R. (1982). In litt., 15 October.

13. Australian Conservation Foundation (1972). Pedder Papers: Anatomy of a Decision. Australian Conservation Foundation, Victoria.

14. Kozhov, M. (1963). Lake Baikal and its Life. Monographiae Biologicae, W. Junk, The Hague.

15. Norden, A. (1982). In litt., 22 December.

We are very grateful to I.R. Ball, J.E. Cooper, A. Dahm, D.I. Gibson, R. Kenk, A. Norden and R.M. Reiger, for assistance with this section.

Sphalloplana holsingeri Kenk, 1977   Holsinger's Groundwater Planarian
S. subtilis Kenk, 1977   Biggers' Groundwater Planarian

Phylum   PLATYHELMINTHES   Order   TRICLADIDA

Class   TURBELLARIA   Family   KENKIIDAE

SUMMARY These two eyeless, unpigmented flatworms are known only from a single spring in Virginia, U.S.A. Both are probably now extinct as a result of development of the locality into a parking area, in spite of the fact that notification of the importance of the spring had been made to the relevant authorities.

DESCRIPTION Both species are eyeless and unpigmented. Sphalloplana subtilis reaches a length of 16 mm, S. holsingeri a length of 15 mm (1). Detailed description in (1).

DISTRIBUTION Both species are known only from a walled spring on a private property in Fairfax County, Virginia, U.S.A. (1).

POPULATION Unknown. Ten specimens of S. subtilis were collected in 1977; S. holsingeri appeared to be more abundant since 80 specimens were collected but baiting techniques were used (1).

HABITAT AND ECOLOGY The species inhabit shallow groundwater that comes to the surface by way of a small spring. The spring basin is enclosed by a brick structure with a removable concrete cover (1).

SCIENTIFIC INTEREST AND POTENTIAL VALUE Both species are highly specialized animals, restricted to subterranean ground water. Their nearest relatives are found in caves further west in the Appalachians and Interior Low Plateaux (2).

THREATS TO SURVIVAL Both planarians are probably now extinct. In 1978 plans were under way for the construction of a federally financed housing project, which included a parking lot in the area where the well is situated. The ground was to be bulldozed to a depth of about 14 ft (4 m) and no provision had been made for preserving the well. In spite of reports notifying the Office of Endangered Species of the existence of this threat, no action was taken and the area was developed as planned (3).

CONSERVATION MEASURES TAKEN None.

CONSERVATION MEASURES PROPOSED Adjacent springs have been searched for these species but without sucess (4). It is possible that they have retreated to the surrounding underground waters, but with the elimination of springs and wells in the area it would now be very difficult to verify the species' existence (3). If it is confirmed that they are extinct, this case provides an example of how easily invertebrate species may be lost.

REFERENCES   1. Kenk, R. (1977). Freshwater triclads (Turbellaria) of North America. 9. The genus Sphalloplana. Smithsonian Contributions to Zoology 246: 1-38.
              2. Holsinger, J.R. (1979). Subterranean freshwater planarians (Order Seriata). In Linzey, D.W. (Ed.), Endangered and Threatened Plants and Animals of Virginia. Virginia Polytechnic Institute and State University, Blacksburg,

Virginia Pp.187-188.

3. Dodd, C.K. (1981). Probable extinction of two species of flatworms in northern Virginia. Assoc. Southeast. Biol. Bull. 28(2): 79.

4. Dodd, C.K. (1982). In litt., 30 September.

We are very grateful to S. Chambers and C.K. Dodd for providing the information for this data sheet.

NEMERTEA

Ribbon Worms

Nemertines are unsegmented slender worms, capable of extreme elongation and contraction. They range in size from a few mm to several cm or occasionally longer. Most are more or less uniformly coloured but several species are strikingly and characteristically patterned with stripes, bands or geometric shapes and many bathypelagic species are brilliantly coloured in red, orange or yellow.

The total number of species is estimated to be about 900, many of which are known from only single or damaged specimens. Most are free-living marine organisms, occurring intertidally or sublittorally beneath boulders or in crevices, burrowing into soft sandy or muddy sediments or in association with algae or colonial sessile invertebrates. Some species occur on land (see data sheet), some in brackish water and there are six, mostly monotypic, freshwater genera. A few are commensal or parasitic and one group is pelagic.

Nemertines are related to the Platyhelminthes. Characteristically they have a long unsegmented muscular proboscis with a piercing stylet lying above a simple gut. There is an elementary closed blood system of capillary vessels, and the surface epithelium of the body is ciliated. They have a variable number of eyes; the terrestrial species Argonemertes australiensis, for example, may have as many as 170. Nearly all species are active predatory carnivores or scavengers, frequently using the proboscis to catch prey. The proboscis can also be used for rapid movement, when it is ejected forcibly and the tip anchored to the substrate; the nemertine then retracts the rest of its body forwards over the proboscis. Ciliary and creeping movement is used under normal circumstances, the nemertine gliding along in its own mucus.

Little is known of the conservation status of any of the aquatic nemertines. However, in the small group of terrestrial species, several appear to have very small ranges, and these are considered in the following data sheet.

TERRESTRIAL NEMERTINE WORMS

Pantinonemertes agricola (Willemoes - Suhm, 1874)	RARE
Geonemertes rodericana (Gulliver, 1879)	RARE
Argonemertes australiensis (Dendy, 1889)	RARE
A. hillii (Hett, 1924)	RARE
A. stocki (Moore, 1975)	RARE
Antiponemertes allisonae (Moore, 1973)	VULNERABLE
Leptonemertes chalicophora (Graff, 1879)	RARE
Katechonemertes nightingaleensis (Brinkmann, 1947)	RARE

Phylum	NEMERTEA	Order	HOPLONEMERTEA
Class	ENOPLA	Family	PROSORHOCHMIDAE

SUMMARY Twelve terrestrial nemertines have been described, all of which are restricted to moist, shady habitats. Although a few species are widespread, many are restricted to oceanic islands or have been found only from limited areas. In these cases they tend to be highly vulnerable to habitat disturbance.

DESCRIPTION The terrestrial nemertines were originally assigned to a single genus, Geonemertes, but have recently been divided into a number of genera (6). They make up a convergent assembly of forms separately evolved from marine ancestors, having evolved parallel but distinctive adaptations, in particular of the circulatory and excretory organs, to cope with a terrestrial life-style (4,7). Twelve species have been described, of which eight are considered to be rare or vulnerable.

Pantinonemertes agricola Unstriped, varies in colour (white, pink, orange, grey, black). Length varies from 15 to 150 mm, the larger specimens being found in water (6).

Geonemertes rodericana Dark green with a single white mid-dorsal stripe and four white spots around the eyes (6); 27-55 mm long (5).

Argonemertes australiensis Variable in colour; white, yellow or brown with median dark brown stripe, with or without additional slight lateral stripes (3,6). The large cream specimens have only been found from rain forest while dark forms inhabit both forests and more open areas. Males up to 60 mm long; females reach 84 mm (10).

A. hillii Purple brown worms with two bright red lateral bands in the anterior two thirds of the body; in the male these are orange red, in the females deep red. The posterior end is dark; the ventral surface is a light mottled brown. Length varies between 23 mm and 43 mm and there are about 80 eyes over the anterior tip of the worm (1,2,3,5).

A. stocki Dark brown dorsally and dorsolaterally with abrupt transition to lateral and ventral cream colour; fragments spontaneously. The only known specimen was 9.5 mm long (3).

Antiponemertes allisonae Dorsal surface mottled brown with clear stripe over proboscis; only two eyes; 10 mm long (6,9).

Leptonemertes chalicophora White or pink with red anterior end (6). Up to 15 mm long and fragments easily (8).

Katechonemertes nightingaleensis Olive grey or yellowish white with two broad brown stripes dorsally (6).

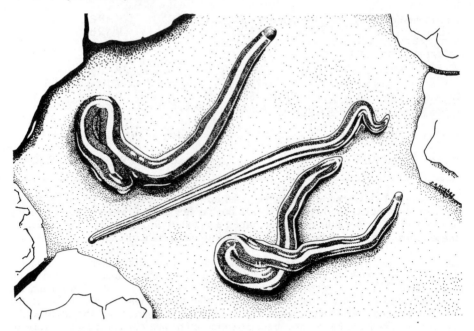

DISTRIBUTION Land nemertines occur naturally only on oceanic islands or in Australia and New Zealand, although some species have been introduced to Europe via plants. The reason for such a distribution remains obscure but could be explained by the colonization of land by marine or littoral ancestors occurring independently in widely separated localities (4,6).

Pantinonemertes agricola Bermuda (6).

Geonemertes rodericana Known only from Rodrigues Island in the Indian Ocean where it was discovered in 1879 (5).

Argonemertes australiensis Found in a number of sites in Australia, from Queensland through the Canberra Hills to Victoria and Tasmania. Recorded from North Gippsland and other sites in Victoria; used to be relatively abundant in Tasmania (3,10); in New South Wales found in the Brindabellas and on Mt Kosciusko; in Queensland found on the Lamington plateau (3).

A. hillii New South Wales, Australia, east of the crest of the Great Dividing Range. First recorded from Barrington Tops (2). Since then, has been found at Point Look Out, New England National Park at 5000 ft (1525 m); at Dorrigo National Park at 2400 ft (732 m); and at the foot of Dorrigo Mountain at 500 ft (153 m) on the upper reaches of the Bellinger river (3). A specimen described from the Richmond river may also be A. hillii (1). Range appears to extend eastwards to the edge of the coastal plain, and the western limit appears to be the crest of the mountain chain along the eastern coast of New South Wales. Has also been found on the Lamington plateau, Queensland near Binna Burra (3).

A. stocki Only one specimen found in 1972 at Point Look Out, New South Wales, 50 miles (80 km) east of Armidale (3).

Antiponemertes allisonae Confined to a single locality near Menzies Bay, Banks Peninsula, South Island, New Zealand (9).

Leptonemertes chalicophora Madeira, Azores, Canary Islands, and as an immigrant in European glass houses. Details given in (8). Has also recently been recorded from a garden in California (13).

Katechonemertes nightingaleensis Has only been recorded from Nightingale Island in the Tristan da Cunha group in 1937 (5).

POPULATION Unknown for all species.

HABITAT AND ECOLOGY Terrestrial nemertines have rather more precise environmental requirements than terrestrial flatworms. They are found from below the low-tide level up to an altitude of 4000-5000 ft (1200-1525 m) and to a considerable distance inland. They live in conditions of high humidity and reduced light intensity, such as among leaf-litter, beneath stones or fallen logs and under the bark of decaying trees. They often remain immobile for considerable lengths of time, enclosed in a thick covering of mucus; this behaviour may be related to the conservation of water and energy (7,10). Eggs are laid in gelatinous capsules deposited in very damp places. In Argonemertes australiensis as many as seven capsules may be laid in two months, the larger capsules containing several hundred eggs. Egg laying occurs throughout the year but is most frequent in April and May (4,10).

Pantinonementes agricola Upper shore-line; also occurs below low-water level in mangrove swamps, under stones or among matted algae. On land occurs in moist earth under stones on hillsides (5).

Geonemertes rodericana: Inhabits damp woods (5).

Argonemertes australiensis Occurs in wet sclerophyll in the Brindabellas and near Nothofagus trees at Lamington (3). In Tasmania it is found from near sea level to 1500 m, in cool and moist situations in shady gullies and on the margins of forest clearings. It is found under decaying logs, moss-covered stones, fallen leaves and the loose bark of dead trees. During warm and dry periods of the year it often retreats into the crevices of decaying logs and under deeply embedded stones. It feeds on young myriapods and small insects, especially Collembola.

A. hillii Occurs in damp woodlands, but is found at greater altitudes and in drier conditions than other land nemertines. At Barrington Tops it was not uncommon under logs if the soil was good, and was found also in scrub (2). At Dorrigo found in tropical and subtropical rain forest (3). At Point Look Out, found under fallen timber in dry sclerophyll forest where there are severe frosts in winter (3).

A. stocki The only known specimen was found in dry sclerophyll forest at 1575 m, under a log (3).

Antiponemertes allisonae Found in open bush (9).

Leptonemertes chalicophora Occurs in damp, stony, shaded habitats where endemic plant species predominate, but also found in areas where there are virtually no native plants. Found from near sea level to an altitude of 1000 m (8).

Katechonemertes nightingaleensis Found in the upper littoral, just above the tide level (5).

SCIENTIFIC INTEREST AND POTENTIAL VALUE The fifteen species of land nemertine are of interest since they occur in an almost exlusively marine phylum and because several of the species occur in isolation on oceanic islands, providing an interesting case of convergent evolution. They are often associated with a rich and varied soil fauna; for example Antiponemertes allisonae is one of a number of endemic invertebrate species found on Banks Peninsula. There is still much to be learnt about their physiology and ecology (4,5).

THREATS TO SURVIVAL Since many species are restricted to moist forest habitats, they are particularly vulnerable to habitat disturbance such as logging and forest clearance.

Pantinonemertes agricola Could be affected by the development of tourism on Bermuda (11).

Argonemertes australiensis Could be vulnerable to deforestation, since the forest is being rapidly logged by the Japanese for chipboard (11). Many localities in Tasmania are rapidly being laid bare and several sites in Victoria have been obliterated by the expansion of Melbourne (3).

Antiponemertes allisonae Its habitat is being destroyed by shrub clearance (11,12).

Leptonemertes chalicophora Could be threatened on Madeira by the growth of tourism, since sites are being destroyed by hotel building (14), but it is probably relatively safe in other localities (12).

Geonemertes rodericana, Argonemertes hillii, A. stocki and Katechonemertes nightingaleensis are all limited to single localities or their known ranges are very restricted.

CONSERVATION MEASURES TAKEN Both Argonemertes hillii and A. stocki occur in the New England National Park and A. hillii has also been found in Dorrigo National Park (3).

CONSERVATION MEASURES PROPOSED In a number of places land nemertines occur in areas of great scientific interest and rich soil fauna; one such example is on Madeira at Ribeira Seco (8); another is Banks Peninsula in New Zealand (see Threatened Community sheets). Consideration should be given to setting aside such areas as reserves. Species known only from single localities should be given special protection. It is possible that the conservation proposals drawn up for Madeira (see data sheet for Madeiran land snails) may help to protect L. chalicophora (14).

REFERENCES   1.   Fletcher, J.J. (1891). Notes and exhibits. Proc. Linn. Soc. N.S.W. 6: 167.
2.   Hett, M.L. (1924). On a new land nemertean from New South Wales Geonemertes hillii sp.n. Proc. Zool. Soc. Lond. 1924: 775-787.
3.   Moore, J. (1975). Land nemertines of Australia. Zoological Journal of the Linnean Society 56 (1): 23-43.
4.   Pantin, C.F.A. (1961). Geonemertes: a study in island life. Proc. Linn. Soc. London 172 (2): 137-151.
5.   Pantin, C.F.A. (1969). The genus Geonemertes. Bulletin of the British Museum (Nat. Hist.) Zoology 18 (9): 263-310.
6.   Moore, J. and Gibson, R. (1981). The Geonemertes problem (Nemertea). J. Zool. Lond. 194: 175-201.
7.   Gibson, R. (1972). Nemerteans. Hutchinson University Library, London.
8.   Moore, J. and Moore, N.W. (1972). Land nemertines of Madeira and the Azores. Boletim do Museu Municipal do Funchal 26(113): 31-44.

9. Moore, J. (1973). Land nemertines of New Zealand. <u>Zool. J. Linn. Soc.</u> 52: 293-313.
10. Hickman, V.V. (1963). The occurrence in Tasmania of the land nemertine <u>Geonemertes australiensis</u> Dendy with some account of its distribution, habits, variation and development. <u>Pap. Proc. R. Soc. Tasm.</u> 97: 63-75.
11. Moore, J. (1980). In litt., 28 March.
12. Johns, P. (1981). In litt., 29 May.
13. Moore, J. (1982). In litt., 13 February.
14. Bramwell, D., Montelongo, V., Navarro, B. and Ortega, J. (1981). Informe sobre la conservacion de los bosques y la flora de la isla de Madeira. Jardin Botanico Canario 'Viera y Clavijo' and IUCN.

Data for this sheet were supplied by J. Moore and we are most grateful for her assistance, and to R. Gibson for his comments on the draft.

MOLLUSCA

After insects, molluscs are probably the most familiar invertebrates. In numbers and diversity, they are second only to the arthropods with probably more than a 100 000 species living to-day and a long fossil history. Unlike some invertebrate groups they are relatively well known taxonomically. This is partly because over the centuries large shell collections have been built up which provide a good basis for taxonomic work. Of all the invertebrates, molluscs probably have been and still are most valued by man. They are a major food source in many parts of the world, their shells provide a variety of products, and the diverse and beautiful forms that these may take have led to molluscs becoming important elements in the art, culture and traditions of many races.

Molluscs have managed to exploit every type of habitat and lifestyle and consequently the body plan is extremely variable throughout the phylum. There are seven classes: - Aplacophora, Monoplacophora, Polyplacophora, Gastropoda, Bivalvia, Scaphopoda and Cephalopoda. The Aplacophora (the worm-like solenogasters), the Monoplacophora (the most primitive living molluscs) and the Scaphopoda (elephant tusk shells) are small, entirely marine groups. The Polyplacophora, or chitons, are more familiar, occurring throughout the world on rocky shores. The Gastropoda is the largest and most diverse class, comprising some 65 000-75 000 species and including marine, freshwater and terrestrial forms, many of which are important to man. The Bivalvia are also well known since they include many of the commercially important edible molluscs such as oysters, mussels, clams and scallops. Most are marine, but a few important groups occur in freshwater lakes and rivers. The Cephalopoda are all marine and include squid, octopus and cuttle fish.

The majority of molluscs are characterized by the presence of a shell which provides protection against predators and, in terrestrial species, against desiccation. The shells are composed mainly of calcium carbonate and many have an inner layer of nacre or mother-of-pearl which is composed of tiny blocks of calcium carbonate arranged in layers. The shells of a few species, such as the Chambered Nautilus Nautilus pompilius and the pearl oyster Pinctada sp, consist almost entirely of mother-of-pearl and are highly valued by man. Pearls are another important product of molluscs and are formed when sand grains or other particles become lodged between the shell and the part of the body known as the mantle. Gastropods usually have spirally coiled shells, although some, such as the nudibranchs, have no shell or only a very reduced one. Bivalves have a hinged shell in two parts, and unlike gastropods which are generally mobile, are sedentary, living buried in mud or attached to rocks. Many molluscs have a radula, a unique molluscan structure covered with rows of teeth, and used for feeding. It is adapted for grazing on vegetation or algae in some species, and in others for a carnivorous way of life. Bivalves have no radula and are filter feeders, straining plankton and other minute organisms from the surrounding water. The cephalopods are the most highly developed molluscs and have the largest brains of all invertebrates. Their shell is usually reduced or absent, or may be adapted as a buoyancy organ. They are carnivorous and can swim at great speeds using a form of jet propulsion.

In view of the diversity to be found within the molluscs this chapter has been divided into three subsections, and marine, freshwater and terrestrial species are treated separately.

INTRODUCTION Marine molluscs are found mainly in the intertidal zone and in shallow waters where they sometimes occur in huge numbers. All seven mollusc classes have marine representatives; these are most abundant among the gastropods, bivalves and cephalopods. Gastropods include a wide range of marine species such as abalone, limpets, top shells, winkles, whelks, conchs, cowries, cones, and volutes as well as some with reduced shells such as the sea hares and sea slugs. The majority are bottom dwellers, moving along by the creeping action of their muscular foot. Although many, such as the whelks, are carnivorous and feed on other molluscs and invertebrates, the majority feed by scraping bacterial, fungal and algal films off vegetation, stones or the sea bottom. Their shells are extremely varied and are often colourful and highly ornate particularly those from the Indo-Pacific. The nudibranchs or sea slugs include some of the most beautiful molluscs, often flamboyantly coloured and matching their seaweed backgrounds. They feed on sessile animals such as corals, sea anemones and sponges (38).

Most marine bivalves are filter feeders, requiring water with a high content of organic matter, and consequently they are particularly numerous in coastal waters. Some, such as mussels and oysters, are surface dwellers,and are attached to a firm substrate, often in dense beds where they depend on the cleansing action of tidal currents to prevent them from becoming buried in their own waste products. Others, such as cockles and some clams, are burrowers and have siphons to the surface (38).

Cephalopods, which include squid, octopus and cuttlefish, are generally large, the most famous being the Giant Squid Architeuthis which may attain a length of 18 m and is the largest living invertebrate. All are active predators and have tentacles for locating and capturing prey. Squid are pelagic, swimming by a form of jet propulsion, and feed on fish and crabs. Octopus are benthic and crawl on the sea bottom; they inject their prey with poison (38).

Most marine molluscs produce planktonic larvae after spawning. In cases where these larvae are long-lived, the species may have a very wide distribution. Many Indo-Pacific species, such as the well known Tiger Cowry Cypraea tigris, are found from the coast of East Africa east to the west and central Pacific. Others, including the Giant Clams (see data sheet), have larvae with shorter planktonic lives and their distributions are correspondingly smaller. The juveniles of some marine gastropods, such as the volutes (Volutidae) from Australia, undergo direct development in which case the species may have an even more restricted distribution. This, combined with the tendency for such species to have a long life-span, may make them vulnerable to the adverse effects of human activities. Growth of marine molluscs is often fast in the first few years of life, but then slows down. For example, Trochus niloticus may take only 1-2 years to reach a size of 8 cm in diameter, but a further 8 years may be required before it reaches 12 cm in diameter (13,15). This may be of consequence for species collected by man, if the large specimens are the most valuable. In contrast bivalves of commercial importance have comparatively rapid growth rates, mussels, for example, reaching marketable size in one to three years (44).

SCIENTIFIC INTEREST AND POTENTIAL VALUE Many marine molluscs play an important role in food chains. Bivalves convert large quantities of suspended organic matter in coastal waters and estuaries into animal protein and provide food for water fowl and fish. Commercially important bottom-living fish often depend on bivalves, and pelagic opisthobranchs and cephalopods are eaten by whales and fish. Sperm whales eat very little else except squid and it has been estimated that whales and seals in the Southern Oceans alone consume about 10 million tonnes of squid a year (41). Several gastropods, such as limpets and top shells, play an important role in controlling intertidal growth of algae, sponges and other encrusting organisms (38).

It is only possible to give the most cursory overview of the importance of molluscs in the world's fisheries. Molluscs provide primary protein for many people, especially in coastal areas if fish are not abundant (2). Even so, optimum utilization of molluscan food resources has not yet been achieved in some countries. In 1980, world nominal catches of molluscs totalled 4 951 718 tonnes, including 972 885 tonnes of oysters, 613 965 tonnes of mussels, 364 173 tonnes of scallops, 1 176 771 tonnes of clams and 1 572 098 tonnes of cephalopods (8). Annual oyster consumption worldwide is projected to increase to more than 2.3 million tonnes by the year 2000 (3) and oysters constitute 25 per cent of world mollusc production (39). At least 12 species of gastropod, 50 bivalves and 13 cephalopods are currently exploited in the Mediterranean alone. About 30 566 tonnes of gastropods and bivalves were produced from this region in 1978 and over 45 000 tonnes of cephalopods were also caught for use as food and bait (4). Some 38 species of bivalves, four of chitons and four of gastropods are collected commercially for food in the Caribbean and western central Atlantic. In this region most of the fishing is for subsistence and is not reported, but in 1978 a total of about 260 000 tonnes, mainly of bivalves, was recorded. The major commercial fisheries are for oysters (Crassostrea virginica) in the U.S.A. and Mexico, scallops (Argopecten gibbus and A. radians) in the U.S.A. and Ark Clams (Arca zebra) in Venezuela (7). The Queen Conch, Strombus gigas, is particularly important in the Caribbean (see data sheet). Similar figures could be cited for the other major fishing areas of the world.

In several developing countries, cephalopods are now the most important fishery export item after shrimp, and in many areas this resource is still under-utilized. Although there is little demand for cephalopods in North America and northern Europe they are highly valued in southern Europe, South East Asia and South America. The world catch fluctuates between 0.8 and 1.2 million tonnes, of which 70 per cent is made up of squid, 17 per cent of octopus and 10 per cent of cuttlefish. The Japanese account for half of the catch, Spain, Thailand and the Republic of Korea taking most of the remainder (9).

Molluscs have many other important uses besides food. They are used in scientific research; for example, Aplysia californica, a sea hare, is currently being used in studies on memory because of its large nerve cells (5). Antiviral and antibacterial substances have been extracted from species such as the abalone, oyster, Queen Conch, and squid. Mercenene, an extract of the whole body of the clam Mercenaria mercenaria contains an antitumour factor, and a wide range of other species have been found to contain pharmacologically active compounds (6).

Some molluscs are pests, such as the muricid, Urosalpinx, which feeds on bivalves and causes considerable damage to oyster beds (40). Byssate bivalves often cause fouling in harbours and power station effluent and intake pipes. Another pest is the well-known shipworm (Teredinidae) which bores into timber and mangrove roots, using the excavated sawdust as food. Although shipworms damage piers, boats and other wooden structures, they play a useful role in the reduction of sea-borne wood (38).

Molluscan shells have been used by man for centuries. They have been valued as purely ornamental objects by many societies and cultures, used as currency, and endowed with magical attributes. Ground shells produce the finest lime for pottery glazes, poultry food additives and, in South Asia, for chewing with betel nuts. Where shells can be dredged in large quantities either dead or alive, they are used for road construction and as foundations for buildings (20).

Mother-of-pearl is used for decorative inlay work, buttons, jewellery and other ornamental articles. The most valuable mother-of-pearl is obtained from four species; Pinctada maxima, P. margaritifera, Trochus niloticus and Turbo marmoratus, although a number of others (Pinctada spp., Haliotis spp. and some freshwater mussels) are also used. The mother-of-pearl trade declined with the development of plastics, but recently it has been undergoing a revival; Japan, Taiwan and South Korea are the centres of the carving industry. Currently there is an annual demand of about 6000 tonnes of unprocessed Trochus shell alone (13). The Philippines and Indonesia export large quantities of Pinctada spp., and Indonesia, the Solomon Islands and other Pacific islands export Trochus niloticus and Turbo marmoratus (14). In the South Pacific the export of mother-of-pearl may be the only source of cash income for people on some of the remoter islands (13) and the button industry is being revived in some areas (45). Japan is the main producer of pearls, techniques for culturing these valuable molluscan products having been developed in this country first. It is thought that there is potential for pearl culture industries in other countries (18).

Shell collecting has been a popular hobby for centuries. The ornamental shell trade is now a major industry, centred in the Philippines which exported about 3500 tonnes of ornamental shells in 1979. The main consumer is the U.S.A., whose imports rose from just over 1000 tonnes in 1969 to about 4400 tonnes in 1978. There are currently an estimated 1000 shell dealers in the U.S.A., and in Florida alone there are probably between 5000 and 10 000 retail outlets such as gift shops, hotels and department stores (33). Other major consumers are Japan and many European countries. The species most in demand are the big colourful gastropods from the Indo-Pacific and Caribbean, such as the Queen Conch Strombus gigas, the Tiger Cowry Cypraea tigris, the Pearly or Chambered Nautilus Nautilus pompilius and helmet shells, cones and volutes (33). The jewellery industry has made use of shells for many centuries. Naples in Italy has long been important for cameos, carved from the shell of the Indo-Pacific gastropod Cassis. Recently shell jewellery and other shellcraft articles have undergone a revival in popularity. Small cowries, dove shells and a variety of other shells are used in large quantities to make necklaces, lampshades, decorated boxes and figurines. The Philippines is the main supplier and is also the centre of the capiz shell (Placuna placenta) industry. Nearly three and half million capiz shell articles were exported in 1979 (14).

THREATS TO SURVIVAL As for many marine invertebrates there are few examples of molluscs that are threatened with extinction. The Giant Clams (see data sheet) are possible exceptions, but even they may still occur in large numbers on some of the remoter Pacific atolls. There are, however, many examples of commercially important species which have been locally over-exploited; some examples are given below. Less information is available on the effects of pollution and habitat disturbance, but in some cases these factors could have serious effects.

1. Exploitation Abalone has been overfished in many countries including Japan, China, South Africa, New Zealand, south Australia and Mexico (43). Haliotis tuberculatus has practically disappeared from the English Channel (16), and the Pacific species were heavily overfished on the coast of California, U.S.A. earlier this century (22). Oysters were once the food of poor people but are now considered a luxury in most countries. Natural oyster grounds have been destroyed through coastal development, domestic and industrial pollution, poor management and over-exploitation. In some cases, where pollution has not harmed the bivalves themselves, they have been rendered unfit for human consumption. By 1964, 12 per cent of the total active shellfish-producing area in the U.S.A. had been closed for health reasons and oyster production on the Atlantic coast had been reduced from 43 million kg in 1920 to less than half that amount. Oyster fisheries have also been lost in the Philippines, Australia, Great Britain and on the Pacific coast of the U.S.A. (21,39,40,43). Scallops were fished to virtual commercial extinction

off Tasmania earlier this century (39). In Chile the giant mussel Choromytilus chorus and the gastropod Concholepas concholepas have both been overfished (23). Even cephalopods, a resource under-utilised in many regions, have been overfished in some areas. Catches of cuttlefish in the eastern central Atlantic have declined drastically and the north-west Pacific catches, which still represent 60 per cent of the total world production, have been declining steadily since 1968 (9).

Commercial mother-of-pearl species have been over-collected in many areas (18). In Papua New Guinea catches of Trochus niloticus decreased progressively during the first half of this century, until 1956 when regulations were introduced (26). In the Andaman and Nicobar Islands, the introduction of new diving techniques by the Japanese in the 1930s caused rapid depletion of T. niloticus; whereas in 1933 over 20 shells could be collected in an hour, by 1935 only two or three were found in the same length of time (27). In New Caledonia, populations have been periodically over-collected (13) and by the 1950s there was evidence of serious overfishing in Palau (15). Intensive exploitation has led to the disappearance of the pearl oyster Pinctada margaritifera from some Polynesian atolls (2).

There is evidence that the ornamental shell trade is causing depletion of local populations of some of the more popular shells, although it is unlikely to cause the biological extinction of any species (14). Shells have been heavily collected on the Kenyan coast for many years by amateur shell collectors, tourists and commercial operators. This has resulted in the extreme scarcity of large colourful species on the more accessible parts of the coastal reef (1,30). Similar situations are to be found in Guam, Florida, the Seychelles and other regions where tourism has recently become a major industry (31,32,33). In the Philippines certain species are becoming rare (29), particularly the giant clams Tridacna spp. and the Giant Triton Charonia tritonis (see data sheets).

Marine species with restricted ranges are particularly susceptible to over-collection. Easter Island and Mauritius have several endemic cowries such as Cypraea cribellum and C. esontropius; with the growth of tourism, these species could be over-collected (11). Cymbiola rossiniana, a volute endemic to the south-west coast of New Caledonia, is highly sought after, fetching high prices on the international market and there are fears that it could be over-collected (28).

The rapid growth of SCUBA diving as a hobby and the development of new fishing techniques could pose a threat to some species since previously inaccessible populations can now be reached. Certain deep-sea molluscs have always been rare and have fetched high prices on the shell market; for example the Glory-of-the-Sea Cone Conus gloriamaris, fetched U.S.$2000 in 1964 but it is now being found in comparatively large numbers, particularly in the Philippines, and sells for U.S.$200-300 (29). Harpa costata is a very valuable, apparently rare species, endemic to Mauritius. Specimens have been collected in considerable numbers in the past (11) but the impact of exploitation on this and similar species is unknown since nothing is known of their ecology and population dynamics.

2. Pollution Very little is known about the effects of pollution on marine molluscs. Oil on the shore and in the intertidal zone can kill molluscs by smothering or poisoning them, and filter feeders and deposit feeders may ingest and accumulate dispersed or sedimented oil. Moderate and chronic exposure of bivalves to low concentrations can have an important commercial impact. It has been recognized that there is a need to control pollution from oil and chemicals discharged from ships in Dongonab Bay in the Sudan to prevent damage to Pinctada margaritifera which is cultured in the area (25). Bivalves filter a large volume of water, and ingest hydrocarbons which accumulate, particularly in the fatty tissues, at a faster rate than they are eliminated. However, following the cessation of pollution it is usual for hydrocarbons in the tissues to be reduced to low levels within 30 days. Like many other marine invertebrates, molluscs may be

more vulnerable to toxic dispersants than to the oil spill itself (36,37).

Pollution could be particularly serious for brackish water and estuarine species. Almagorda newcombiana, a gastropod restricted to Humboldt Bay in California, U.S.A., nearly disappeared in the 1960s as a result of sawdust pollution from newly built sawmills. Fortunately this species now seems to be recovering, since several of the mills have been abandoned and others have been converted to operations which do not produce sawdust (24). The sea slug Doridella batava, (see data sheet) is another brackish water species that has been affected by habitat modification. The Atlantic Geoduck, Panopea bitruncata, the largest western Atlantic bivalve, has been designated as Rare by the state of Florida. Although it may be more abundant in offshore or deep water, it is currently only known from a few populations in estuarine environments between North Carolina and Texas. These are liable to be destroyed by changes such as encroaching sand bars, sedimentary deposits and harbour pollution (35).

CONSERVATION Marine reserves have now been established in many countries. These may protect populations from which individuals can migrate to recolonize depleted areas. However, in very few cases have reserves been created specifically to protect threatened molluscs. An exception is the reserve created around the island of Kouaré, New Caledonia for the rare volute Cymbiola rossiniana (which unfortunately is not properly supervised) (28). Further work is required to improve knowledge of the distribution of such species so that they can be taken into consideration in the establishment of future reserves.

Most countries have some form of legislation for their edible molluscs, such as closed seasons, minimum size limits for collecting, and commercial quotas. In many cases controls are only brought into force after evidence of over-collection has become obvious (see data sheet for Queen Conch). However, where there is a strong economic incentive, appropriate management techniques have been devised and should provide examples for other situations (47). For example in the Channel Islands, fishing for the ormer Haliotis tuberculatus is controlled. In Jersey, skin-diving and SCUBA gear are prohibited and there is a closed season during the breeding period from May to August (16). Strict legislation governs the abalone industry in California, U.S.A. and includes size limits, closed seasons and a prohibition on abalone export in any form (43). Choromytilus chorus is now protected in Chile (23) and is listed on Appendix II of CITES (Convention on International Trade in Endangered Species of Wild Fauna and Flora). This means that trade in this species between countries which are parties to CITES is only allowed if a valid export document has been supplied by Chile. Collecting is only permitted from authorized breeding places.

Many countries now have legislation for their mother-of-pearl fisheries. In Papua New Guinea there is a minimum size limit for the collection of Trochus niloticus (26,29); in New Caledonia there is a minimum size limit and fishing is only permitted under licence (13). In Palau, the collecting season is limited to one month or less a year, and shells smaller than 76 mm may not be taken. In addition, 16 sanctuaries have been designated from which collection is prohibited (15). Similar legislation exists for Guam (31) and Vanuatu (46).

In a few countries ornamental species are protected under national wildlife legislation and their collection is prohibited. In Bermuda the following species are protected: Queen Conch (Strombus gigas), Harbour Conch (Strombus costatus), Bermuda Cone (Conus bermudensis), Netted Olive (Oliva reticularis), Bermuda Scallop (Pecten ziczac), Calico Scallop (Argopecten gibbus), Atlantic Pearl Oyster (Pinctada imbricata) and helmets and bonnets of all species (Cassididae) (10). In Australia the Giant Triton (Charonia tritonis), Giant Helmet (Cassis cornuta), Giant Clam (Tridacna gigas) and the Black Cowry (Cypraea friendi thersities) are protected (12). The collection of Harpa costata is prohibited in Mauritius (33). In

Kenya, in the early 1970s, a licensing system was introduced to control the shell trade (30). In 1979 a complete ban on the export of shells was introduced, and specimens could only be sold through licensed retailers within the country. However, enforcement has been difficult and posters have now been produced by the East African Wildlife Society for display in hotels and tourist resorts, to inform visitors of the damage that shell collecting can cause (17).

A shell trade managed on a sustainable yield basis could provide a valuable source of income in many developing countries. In Papua New Guinea an attempt has been made to set up a shell business, run by the government, on an ecologically sound basis, and would-be collectors are issued with a booklet instructing them how to collect shells (29). Amateur shell collectors are gradually becoming aware of the need for conservation. The Hawaiian Malacological Society has produced a shellers' creed with 'four rules to shell by', which include leaving coral untouched, avoiding disturbance of eggs and breeding groups of molluscs, returning stones and loose coral to their original positions, and leaving imperfect and immature shells to ensure a future breeding stock (19).

Aquaculture or mariculture is being intensively developed in many areas for a variety of molluscs (3,15,43) and eventually may be the answer to some of the problems of over-exploitation. As filter feeders, bivalves are well suited to cultivation because they can be grown at very high densities and are economic plant protein transformers. Some examples are given to illustrate the importance of marine mollusc aquaculture.

Mussels are particularly easy to cultivate, yields varying according to the methods used, the water quality, and environmental factors such as climate. Good grounds can yield up to 150 tonnes per ha. Spain, France and Holland are the main producers using various substrates including rafts, suspended ropes and the sea bed for spat settlement (21,39,43). Bivalves, including mussels, oysters and venus shells, are cultured in the Mediterranean in coastal lagoons and brackish water lakes. About 20 molluscs are cultivated in the Indo-Pacific region. In Indonesia the bivalves Pinctada, Anadara and Mytilus are cultured at a low intensity in marine and brackish water environments by providing improved conditions for spat settlement and growth. The method relies on natural populations for recruitment and on natural food (34). Crassostrea gigas, the Pacific oyster, is cultured extensively in Japan in fully controlled hatcheries and interest in this species is developing in Europe. The spat is used to re-seed depleted native oyster beds (39,40). Small beds of Crassostrea angulata in the Po Delta region, Italy, were reported to be recovering from undisciplined overfishing, after restocking for cultivation (4). The mangrove oyster C. rhizophora is cultured in Venezuela (3). Scallop farming started in Japan in 1963 in response to a sudden decrease of more than 80 per cent in the production of natural stocks along the coast of Hokkaido. By 1972, 23 000 tons were being produced (3), and other countries are now investigating the culture of scallops for farming and re-seeding depleted areas (42).

Consideration is being given increasingly to the cultivation of gastropods such as abalone, limpets, periwinkles and conches. In Japan large populations of abalone are supported in semi-enclosed areas with sea urchins, also a valuable resource. Predators, and algae that are unsuitable for food, are removed and algae favoured by the two species are planted. Large scale aquaculture of cephalopods still poses problems, although some species (Sepia esculentus, Loligo bleekeri and Octopus vulgaris) have been commercially cultivated in Japan (3).

Considerable care needs to be taken in the establishment of aquacultural enterprises. Marine monocultures are subject to the same problems of increased risk of disease as are terrestrial monocultures, they may affect detrimentally the natural fauna in surrounding areas and, in the case of mussels, can cause fouling problems. Mariculture with the aim of reseeding depleted natural stocks, as well as providing cultured stock is now being attempted for a number of species including Strombus gigas, Tridacna spp. (see data sheets) and Trochus niloticus, and laboratories in the Caribbean and Pacific, such as the Micronesian Mariculture Demonstration Center, are very active in this field (15). Clearly many marine molluscs are valuable resources if properly managed and for such species, conservation efforts should be directed towards finding ways of exploiting them on a sustainable yield basis. For the comparatively few species with small populations or restricted ranges, more research is required to determine the conservation measures required.

REFERENCES

1. Evans, S.M., Knowles, G., Pye-Smith, C. and Scott, R. (1977). Conserving shells in Kenya. Oryx 13(5): 480-485.

2. Salvat, B. (1967). Importance de la faune malacologique dans les atolls Polynésiens. Cah. Pacif. 11: 7-49.

3. Kline, O. (1977). Marine Ecology. Vol 3. Cultivation, Part 3. John Wiley and Sons, U.K.

4. Hunnam, P.J. (1980). Mediterranean species in possible need of protection. Report to IUCN, UNEP/UNESCO, UNEP/16.20/INF.6.

5. Anon. (1982). Science and technology brief: memory as a brain teaser. The Economist 284 (7250): 72-73.

6. Ruggieri, G.D. (1976). Drugs from the sea. Science 194: 491-497.

7. Stevenson, D.K. (1981). A review of the marine resources of the Western Central Atlantic Fisheries Commission (WECAFC) region. FAO Fish. Tech. Pap. 211: 132 pp.

8. FAO. (1981). 1980 Yearbook of Fishery Statistics. Catches and Landings. Vol. 50, FAO, Rome.

9. Hotta, M. (1976). Production, trade and consumption in cephalopods and cephalopod products. FAO Fisheries Circular No. 340.

10. Anon. (1981). Bermuda Division of Fisheries. (Legislation). Fisheries Office.

11. Dance, P. (1969). Rare Shells. Faber and Faber, London.

12. Coleman, N. (1976). Shell collecting in Australia. A.H. and A.W. Reed, Australia.

13. Bouchet, P. and Bour, W. (1980). The Trochus fishery in New Caledonia. South Pacific Commission Fisheries Newsletter 20: 9-12.

14. Wells, S.M. (1981). International Trade in Ornamental Shells. IUCN Conservation Monitoring Centre, Cambridge, UK: 22 pp.

15. Heslinga, G.A. and Hillmann, A. (1980). Hatchery culture of the commercial top shell Trochus niloticus in Palau, Caroline Islands. Aquaculture 22: 35-43.

16. Bossy, S.F. (1976). A survey of the Channel Island ormer (Haliotis tuberculata) industry. Unpub. rept.

17. Anon. (1981). Marine conservation poster launched. Swara 4(6): 7.

18. Salvat, B. (1980). The living marine resources of the South Pacific past, present and future. In UNESCO, Population-environment relations in tropical islands: the case of eastern Fiji. MAB Technical Notes 13: 131-148.

19. Anon (undated). A sheller's creed. Hawaiian Malacological Society, Honolulu.

20. Saul, M. (1974). Shells. Hamlyn Publishing Group, London.

21. Iverson, E.S. (1976). Farming the edge of the sea. Fishing News Books Ltd., Farnham, U.K.

22. Robilliard, G. (1975). Comments on a 'Draft report on recent

abalone research in California with recommendations for management.' Festivus 6(4): 19-24.

23. Castilla, J.C. and Becerra, R.M. (1975). The shell fisheries of Chile: an analysis of the statistics 1960-1973. In Proceedings of the International Symposium on Coastal Upwelling, Coquimbo, Chile.

24. Keen, A.M. (1970). American Malacological Union Symposium: rare and endangered molluscs. 8. Western marine molluscs. Malacologia 10(1): 51-53.

25. George, T.T. (1976). Water pollution in relation to aquaculture in Sudan. Supplement 1 au rapport du Symposium sur l'Aquaculture en Afrique Accra Ghana 1975. 4, FAO Comm. for Inland Fisheries of Africa, FAO, Rome, pp 753-763.

26. Barletta, G. (1976). I molluschi e la legge. Conchiglie 12(9-10): 394-398.

27. Rao, H.S. (1937). On the habitat and habits of Trochus niloticus Linn. in the Andaman Seas. Rec. Ind. Mus. Calcutta 39: 47-82.

28. Bouchet, P. (1979). Coquillages de collection et protection des recifs. Report to O.R.S.T.O.M., Centre de Noumea, New Caledonia.

29. Wells, S.M. (1982). Marine conservation in the Philippines and Papua New Guinea with special emphasis on the ornamental coral and shell trade. Report to Winston Churchill Memorial Trust, London.

30. Wells, S.M. (1978). The Kenyan shell trade. Report to TRAFFIC International.

31. Hedlund, S.E. (1977). The extent of coral, shell and algal harvesting in Guam waters. Sea Grant Publication UGSG-77-10, University of Guam Marine Laboratory Technical Report 37.

32. Salm, R.V. (1978). Conservation of marine resources in Seychelles. IUCN Report to the Government of Seychelles, Morges, Switzerland.

33. Abbott, R.T. (1980). The Shell and Coral Trade in Florida. Special Report 3, TRAFFIC (USA), Washington D.C.

34. Soegiarto, A and Polunin, N. (1982). The marine environment of Indonesia. IUCN Report to the Government of the Republic of Indonesia.

35. Franz, R. (Ed) (1982). Invertebrates. Vol. 6. In Pritchard, P.C.H. (Ed.), Rare and Endangered Biota of Florida. University Presses of Florida.

36. R.C.E.P. (1981). Oil Pollution of the Sea. Eighth report of the Royal Commission on Environmental Pollution, October 1981. H.M.S.O. London.

37. IUCN Commission on Ecology (1982). The impact of oil pollution on living resources. Draft report by the Oil Pollution Working Group.

38. Barnes, R.D. (1980). Invertebrate Zoology, 4th Ed. Saunders College, Philadelphia.

39. Gulland, J.A. (Ed.) (1971). The Fish Resources of the Ocean. Fishing New Books Ltd., Farnham, U.K.

40. Davidson, P. (1976). Oyster Fisheries of England and Wales. M.A.F.F. Laboratory Leaflet 31. Lowestoft, U.K.

41. Laws, R.M. (1977). Seals and whales of the Southern Ocean. Phil. Trans. Roy. Soc., London B 279: 81-96.

42. Franklin, A., Pickett, G.D., Conner, P.M. (1980). The scallop and its fishery in England and Wales. M.A.F.F. Laboratory

Leaflet 51. Lowestoft, U.K.

43. Bardach, J.E., Ryther, J.H. and McLarney, W.O. (1972).
    Aquaculture. Wiley-Interscience, New York, London.
44. Dare, P.J. (1980). Mussel Cultivation in England and Wales.
    M.A.F.F. Laboratory Leaflet 50, Lowestoft, U.K.
45. Cox, R. (1981). Buttonholing Pacific progress. Development
    Forum 9(7): 1 pp.
46. Anon. (1957). Joint regulation to provide for the control of
    the removal, sale or export of sea shell. 11 of 1957.
    Condominium Gazette 197: 641.
47. Korringa, P. (1980). Management of marine species.
    Helgoländer wiss. Meeresunters 33: 641-661.

We are very grateful to J. Taylor for comments on this section.

QUEEN or PINK CONCH

COMMERCIALLY THREATENED

Strombus gigas Linnaeus, 1758

Phylum   MOLLUSCA

Order   MESOGASTROPODA

Class   GASTROPODA

Family   STROMBIDAE

SUMMARY Strombus gigas is a major food resource in the Caribbean where it has been exploited by subsistence and commercial fisheries for centuries. It is generally found in sea grass beds, preferring shallow waters. In recent years overfishing has depleted populations in many areas, giving rise to international concern over the future of the Conch fishery. A few countries have introduced legislation in an attempt to manage the fishery but there are considerable problems with enforcement. Intensive research into Conch mariculture is underway in several countries aimed at restocking depleted Conch habitats.

DESCRIPTION Strombus gigas has a heavy, solid shell, reaching a maximum of 20.5 cm in length at maturity (28). The exterior is light pink to white and the interior of the aperture is a rich pink, yellow or peach. There is a row of blunt spines on the whorls below the suture. The periostracum is thick and brown, but becomes brittle and flakes off when the animal dies. The operculum is large and clawlike and is used for locomotion and protection. Live adult shells may weigh over 2.5 kg (56,61) and males are smaller than females (17). Juveniles do not have the large flaring apertural lip of the adults. Very occasionally a moderately valuable pink pearl is produced in the fleshy mantle (52). S. gigas is readily distinguished from the other five species of Strombus (apart from S. goliath) by the large size of the adults and the deep pink of the aperture (20,61). S. samba is a synonym of S. gigas, often used for thick-lipped forms (20).

DISTRIBUTION Bermuda, south-east Florida (has been recorded as far north as Georgetown, South Carolina (54)), West Indies and throughout the Caribbean and Southern Gulf of Mexico, down to Belize, Panama, Colombia and Venezuela. It is found in coastal waters around the following islands: Bahamas, Cuba, Jamaica, Turks and Caicos, Haiti, Dominican Republic, Puerto Rico, Leeward Islands, Windward Islands, Barbados, Trinidad and Tobago, Netherlands Antilles (1,55).

POPULATION Actual numbers are unknown. In general it is now rare around areas of high human population density but may still be common elsewhere. It used to be particularly abundant in the coastal waters of Belize, the Turks and Caicos, Antigua and Barbuda, the Bahamas, the Grenadines and some Venezuelan offshore islands (La Orchila, Los Roques, Las Aves) but is now less common in all these areas (18,23). In 1979 the average Conch density in an area off the Turks and Caicos Islands was estimated at one conch per 9.4 m² (26). In Venezuela a few Conchs are still to be found in shallow waters around the offshore islands including Los Testigos, Los Frailes, La Blanquilla and La Tortuga. La Orchila still has a large population since it is a military zone and only fishing for sport is allowed. Los Roques and Las Aves Barlovento y Sotavento also still have extensive Conch resources (23). It is still fairly common is some parts of the Grenadines. St Lucia has many suitable areas but Conchs are now only found occasionally (23). In the British Virgin Islands, Anegada is now the only island left with Conch stocks (24). Reportedly they have never been abundant in the Dominican Republic (which has a very narrow band of shallow coastal water), Dominica (which has limited stable shallow coastal waters) Trinidad, and Barbados (23).

HABITAT AND ECOLOGY Found in sea grass beds around islands and coral reefs. Generally occurs in shallow water where light penetration is sufficient to permit growth of large quantities of benthic algae and sea grasses such as <u>Thalassia</u> <u>testudinum</u>, <u>Syringodium</u> and <u>Cymodocea</u> <u>manatorum</u>. Conchs have been recorded from depths of 200 feet (60 m) but are rarely found as deep as 100 feet (30 m), perhaps because of the lack of food at such depths, although in shallow waters they are sometimes found on sandflats, gravel, coral rubble and hard coral rock bottoms (20). Juveniles are found on shallow sandflats and grass beds where the sea grass is not too thick, the smaller ones (less than 3 inches (7 cm) long) remaining buried during the day (27). Both adults and juveniles are predominantly nocturnal (20, 26). Adults as well as juveniles may remain partially or completely buried for several weeks in the winter. This was thought originally to be for shell deposition (20) but it is more likely that the animals become dormant over the colder months (26). Like other strombids, <u>S. gigas</u> is herbivorous and grazes on microscopic algae found on sand and detached sea grass leaves (but not on sea grass itself), and considerable quantities of sand may be swallowed (15,25). A variety of species of algae are consumed, and preliminary studies have shown that food preferences change over the course of the year (13), and differ according to geographical locality (25,60).

Growth rate has been investigated a number of times (2,28,27,58). The shell length of yearling Conchs ranges from 7.6 to 10.8 cm, that of two year olds from 12.6 to 17.0 cm and that of three year olds from 18.0 to 20.5 cm. Mean longevity is about six years and reproductive maturity is reached by 3-4 years (2,28). A study in Cuba showed a growth rate of 4-8 cm a year (58).

The Queen Conch has a number of characteristic behaviour patterns. Most well known is its method of moving by rapid leaps, digging its operculum into the sand and pulling the body rapidly forward using its muscular foot. This may be of importance in escaping from predators and also enables the animal to right itself (15,16,25). Predation on juveniles is heavy, primarily by Tulip Murex (<u>Fasciolaria</u> <u>tulipa</u>), Apple Murex (<u>Murex</u> <u>pomum</u>), octopi, hermit crabs, Loggerhead Turtles and fish (20,27) but adults are probably only preyed on by Loggerheads, the Florida Horse-conch (<u>Pleuroploca</u> <u>gigantea</u>), octopi (20) and rays. Conchs are migratory, moving in groups from deeper to shallower water to spawn. They also appear to migrate to deeper offshore water with sparse algae and sand in winter, adult Conchs moving greater distances than juveniles. Occasionally Conchs are be found clumped together, possibly for protection against wave action in storms (26), although females are found in similar clusters during spawning (26,34).

Queen Conchs come into shallow waters to breed during the summer months. The spawning season in Florida is late May to September (41); in the Bahamas and Turks and Caicos it is March to October (20,32) and in Venezuela it is July to November (27). Fertilization is internal and each female lays several spawn masses per season which may contain as many as 500 000 eggs. The egg mass consists of a single continuous tube which is sticky when extruded so that sand grains adhere to it, probably camouflaging the eggs and deterring predators from eating them. The veligers hatch about four to five days later and have a pelagic life of about two to three weeks (17,20,60).

<u>S. gigas</u> sometimes lives commensally with the Conch Fish, <u>Apogon</u> <u>stellatus</u>, which is 3-6 cm long and lives in the mantle cavity of adult Conchs, leaving at night to feed (20,50). Small crabs of the species <u>Porcellana</u> <u>sayana</u> are also found living within the shell (51).

<u>SCIENTIFIC INTEREST AND POTENTIAL VALUE</u> Queen Conch has long been an important part of the diet of Caribbean people and as a protein resource it has been second only to finfish for at least a hundred years (13). It is now the second most valuable Caribbean fishery resource after the Spiny Lobster, <u>Panulirus</u> <u>argus</u> (13,20). Traditionally, wooden sailing sloops were used to reach the fishing grounds and Conchs were collected in small dinghies using a hook on the end of a

pole. This method is still used to a limited degree in the Bahamas and Turks and Caicos (2,6,13), but small boats with outboard motors are more popular for getting quickly to distant Conch grounds in the Turks and Caicos. Elsewhere, as in Venezuela, Colombia and the Dominican Republic, traditional native boats with small inboard engines are still used. Most fishermen are free divers, using a face mask and fins, but as the remaining Conch stocks occur predominantly in deep water, many are turning to SCUBA gear which is used to depths down to 20 m (2,13). A good Conch diver can collect up to 600 specimens in four hours diving in depths of 1-5 m (2). The meat is generally removed from the shell at sea. Conchs to be consumed locally are sold uncleaned at village markets. Processing plants, the majority of which are in Belize and the Turks and Caicos, prepare frozen Conch for export to the U.S.A., usually via Miami (13). In 1980, 424 000 kg were imported through Miami, of which 20 per cent came from Belize and 58 per cent from the Turks and Caicos (42).

Broadly speaking, two types of fishery exist: highly productive fisheries in which most of the marketing is done through centralized facilities usually under government control, and less productive fisheries under individual control (22). Growing populations and tourism have increased demand in those countries that have traditionally consumed Conch (13,28).

U.S.A., Florida The fishery was traditionally located on the south coast. Very little is now landed but the high price of imported Conch meat (US$ 7.15-8.13/kg) leads occasionally to collection restrictions being ignored. Even if the bag limit is adhered to, the total undetermined quantity collected by tourists and sports divers may be very large since over one million tourists visit the Florida Keys yearly (13). Florida is also the centre of the huge shell collecting industry (7).

Bahamas Landings have declined since 1959 when over 150 tonnes were collected by the Bahamas fishery and a further 100-200 tonnes consumed in the outer islands (5). Conchs are now still collected around islands such as Great Abaco, Andros, Berry, Exuma and Eleuthera for delivery to Miami, Freeport and Nassau (13,23).

Cuba Does not export Conch but it is still used for bait and is an important source of food in coastal villages (13,23,58).

Turks and Caicos Beginning in the late 1800s the Turks and Caicos exported large quantities of dried Conch to Haiti where it was a major source of protein. In the 1950s an annual average of 1.7 million Conchs were exported (33). The industry subsequently declined but underwent a revival after 1973 when the frozen Conch export industry to the U.S.A. began; 1 900 000 Conchs were landed in 1976 (13,57). In 1979 it was reported that 3 million Conchs were being collected annually on the Caicos Bank (26) and nearly 3 million Conch were exported in 1981 (32). Divers may collect as many as 400 Conch a day (13). There are four Conch cleaning and freezing plants, three in South Caicos and one in Providenciales (23,32).

Mexico In 1962 there was a fishery at Quintana Roo but Conch was being taken in only small quantities due to lack of marketing facilities. It was thought that canning might be feasible in the future (59). Mexico is reported to fish in waters off Belize (38).

Haiti In spite of severe overfishing, some local markets are still supplied and substantial amounts are even exported to Miami. Haitians account for a large portion of the sales of Conch meat in Miami and New York (13), and such is local demand for the product that even when there was still considerable local production in 1943, 3 900 000 dried Conchs were imported into Haiti from the Turks and Caicos (33).

Dominican Republic In 1977, 132 339 kg were caught, mainly from the offshore banks (Mouchoir, Navidad, Silver and Southern Bahamas). A hookah air system is used so that Conch can be gathered from deeper waters (13).

U.S. Virgin Islands In the 1960s Conch was the third most important fishery after fish and lobster (29). Now that shallow waters have been exhausted, SCUBA gear is used for collection. By 1974 to satisfy demand in restaurants and hotels, 35 000 Conch a year were being imported from Florida and Puerto Rico (24).

Antigua and Barbuda Conch used to be sent from Barbuda to Puerto Rico before stocks were depleted. Local demand is still being satisfied in Antigua and small quantities are exported to Guadeloupe. Collecting is by native SCUBA divers since shallow water beds are now barren (13,23).

Dominica Small quantities are gathered by spearfishermen although Conchs have never existed in significant numbers (13,23).

St Lucia Conch is seldom found in less than 9 m of water, but is taken occasionally by SCUBA divers and sold locally (13).

St Vincent As a result of overfishing, the Union Island fishery now only produces a limited quantity for export to Martinique (13).

Grenada The Grenadines used to be a major supplier of Conch meat to Trinidad and Tobago and to Martinique. The main Conch grounds are at 40-50 feet (12-15 m) in the north at Conference Point, Bedford Point, Carriacou, Petit Martinique and in the shallow waters surrounding the small islands in that area. Trade with Trinidad almost came to an end, although domestic consumption in Grenada remained fairly steady at 7000-9000 kg a year in spite of prices having doubled. The market is under-supplied by several tonnes, but frozen Conch is exported through Union Island to Martinique. Some fishermen keep Conch in natural corrals, bringing them in when the market is favourable. Most use SCUBA gear, collecting Conch from 12-15 m (13,31). In 1975, a catch of 100 tonnes was recorded by FAO for Grenada, and 20 tonnes for Union Island (61). Recently the export trade to Trinidad has been revived (62).

Trinidad and Tobago Trinidad has never been very productive but Bombshell, Chacachare and Scotland Bays were exploited until they were fished out. Demand is still high and Conch is occasionally imported from the Grenadines or Los Testigos (13).

Netherlands Antilles The fishery is concentrated in the waters around Bonaire and Lac Bay. In the past, Bonaire exported to Curaçao, but imports now come from Mexico and Los Roques (39).

Belize Exports of Conch meat rose from 50 000 kg in 1965 (representing 2.2 per cent of the value of fishery exports) to 561 000 kg in 1972, when it was the second most important commercial fishery after lobster (36,37). After 1972, exports dropped to 211 000 kg in 1978. In 1976 Conch accounted for 23.7 per cent of the value of total fishery exports and in spite of the decline in catch, prices (rising from Belize $0.59/kg in 1965 to Belize $3.30/kg in 1976) have maintained the value of export sales. Although local demand is high, the good export prices mean that most of the meat is frozen and sold abroad. Poaching and non-co-operative sales in the marketing area of the five Conch fishery co-operatives mean only 10 per cent of the meat reaches the local people (13,23). Vessels from Honduras are reported to fish off Belize (38).

Panama Conch is important in local diets especially for the Cuna Indians, and most fishing is carried out around the San Blas Islands. Small quantities are exported to Miami (13).

Colombia There is a fishery in the San Bernardo Archipelago (47).

Venezuela Production dropped from 180 000 kg in 1972 to 10 000 kg in 1975 as a result of restrictions on fishing. The reported catch for 1978 was 5363 kg but actual figures are probably nearer 20 000 kg since there is a considerable amount of poaching. Los Roques - Las Aves supplies 90 per cent of the catch, most of which goes to Bonaire, Curaçao and Martinique (2,13,23,27) but a few Conch are still available in shallow waters around Los Testigos, Los Frailes, La Blanquilla, La Tortuga and La Orchila (13).

Queen Conch was once used extensively for bait in fish traps, and the viscera are still used as bait in many areas. In Victorian times Conch shells were used for making cameos, and they have been used as decorative ornaments for many centuries. Pulverized Conch shell was used for making porcelain and for quicklime for the sugar factories, for which purpose it was said to be superior to coral (19,20,33). Currently it is one of the most popular species in the shell trade, most of the specimens in trade being by-products of the meat industry (7). Occasionally the Queen Conch produces a pearl, which fetched high prices in the past but is not particularly valuable now (52). Conch has a high content of assimilable copper which may account for the low incidence of copper-deficiency anaemia in the Bahamas (20). It may also help to increase resistance to polio (20) and could be a valuable source of drugs (30,53).

THREATS TO SURVIVAL In all Caribbean countries heavy fishing pressure for local use and the export market has severely depleted stocks in areas close to island population centres and fishing villages, to the point where catches have decreased and contain smaller specimens. Recently overfishing has become so severe that the price of Conch has increased to almost more than that of fish, with demand still unsatisfied. As a result, an increasing number of consumers in the Caribbean are having to import Conch meat. In many areas Conch populations seem to be incapable of recuperating naturally even if fishing is curtailed completely (22,23,28). Deeper areas are being exploited by SCUBA divers although these seem to be less productive, possibly because growth rates are slower in deeper water. In the past, such areas may have provided a refuge from which, through immigration, more heavily exploited inshore waters were restocked (24). Small island fisheries are particularly vulnerable to over-exploitation through the activities of individuals (62).

In Florida, in the Lower Florida Keys, adult specimens are now rarely found (3,7). Conch populations are said to be no longer declining, but despite protective legislation they do not seem to be recovering particularly fast. The heaviest pressure comes from the ornamental shell trade and tourists (48).

In Bermuda stocks are seriously depleted (49).

In the Bahamas, exports have dwindled and virtually all the extensive Conch flats no longer have any fishable stocks (23). By 1959 Conch had already become fairly scarce in the Bimini area as a result of over-exploitation (17), and in the 1960s Conchs had to be imported (3,5,29). Fishermen have to range further to find stocks, although in many places Conchs are still common and are a staple part of the diet (6,13). In other areas Conch is regarded as an expensive delicacy (6).

In the Turks and Caicos, accessible stocks are being depleted rapidly although good catches are still made as fishermen have faster boats and now go out to the southern edges of Caicos Bank (23). Poachers from other countries are also contributing to the decline. It is thought that the human population will increase

83

dramatically in the near future as a result of immigration from the Bahamas and this could lead to a significant increase in resource exploitation (14).

Severe overfishing has been reported in Haiti (13).

In the British Virgin Islands fishing is said to be very heavy on Horseshoe Reef, Anegada (9,24). The U.S. Virgin Islands are also said to be facing an eventual overfishing problem (10).

Conchs are now hard to find in the shallow waters of western Barbuda and in Antigua (23).

There was excessive fishing of conch on Guadeloupe and Martinique in the last century (19).

In Dominica Conchs accessible to free divers have long since been taken (23).

On St Lucia Conchs are seldom found in less than 30 feet (9 m) of water now and are only infrequently taken by SCUBA divers (23). Overfishing has been reported in St Vincent (13).

In the Grenadines Conchs are becoming scarcer in waters less than 7 m deep and divers may have to go to about 22 km offshore. The Grenadines export fishery has almost folded and it is difficult to get Conch even at Grenville, the principal landing place (23,31).

In Trinidad stocks at Bombshell, Chacachare and Scotland Bays were fished out several years ago. Increased demand in both Trinidad and Tobago has led to severe overfishing of Tobago stocks (13,23).

In Belize the catch has steadily declined; for example exports totalled 593 000 lbs (268 934 kg) in 1977 compared with 1 052 000 lbs (477 098 kg) in 1971. Conch fishing effort has increased slightly every year since 1973 but the catch per diver effort has steadily declined from about 36 kg in 1975 to 11 kg in 1978. Large numbers of juveniles, about 12 cm long, are now being taken (13,23).

In Colombia Conchs have been fished commercially in the Rosarios and Bernardos Islands with the result that the basic food source for the local inhabitants has almost been eliminated (35).

In Venezuela the Conch is considered to be Vulnerable and conservation measures have been implemented (see below, 8). Prior to this the relatively small stocks had virtually been eliminated (27). Considerable poaching still occurs, however, especially in the isolated Las Aves islands (23).

CONSERVATION MEASURES TAKEN With increasing evidence of overfishing, several attempts have been made to introduce controls, although these are often very difficult to enforce. Research is being carried out on Conch biology and fishery management in many countries. IUCN and IO CARIB (the Intergovernmental Oceanographic Commission for the Caribbean) have been involved in conservation strategies to include overall management of Conch resources (4,12) although no action at the international level has been taken. Conch mariculture projects are now underway in the Bahamas, Bonaire, Belize, Grenada, Miami, Puerto Rico, Turks and Caicos, and Venezuela (22) and are discussed below. More specifically, conservation action has been taken in the following countries:

Florida has placed a moratorium on commercial Conch fishing and there is a bag limit of ten specimens per person per day (4,7).

In the Bahamas, the Fisheries and Conservation Act protects this species and ensures maximum local utilization by banning the export of edible Conch and whole shells. Conch products, if edible Conch constitutes 40 per cent or less of the product, crushed shells for terrazzo manufacture, shells with cut lips and shell jewellery may be exported (13).

The Queen Conch is protected in Bermuda (11).

In the Turks and Caicos, as a result of recommendations arising from a study of the Conch fishery (57), a minimum size limit of 7 inches (17.8 cm) has been imposed and licences are required for fishing and exporting Conch. SCUBA gear is prohibited (40). However, further measures are required to ensure the future of the industry (57).

Puerto Rico periodically has a closed season (4).

In the Netherlands Antilles a minimum size limit of 20 cm has been in effect since the 1950s but has never been enforced (39).

The Belize Fisheries Department has had a comprehensive Conch fishery research programme underway since 1974, with support from the Canadian International Development Research Centre. There is a closed season from 1 July-30 September, a minimum legal size limit of 18.8 cm (Conchs this size yield 100 g of marketable meat)(13,23) and the use of SCUBA gear is prohibited. Management and control is effected through the fisheries cooperatives, which are conservation minded but unfortunately have no powers of enforcement. Illegal landings may account for half the catch and as much as 32 000 kg a year of this may be exported to Mexico (38).

In Venezuela the National Fisheries Office imposed a closed season and required Conch fishermen to be licensed. It is thought that the drop in production from an estimated 180 tonnes in 1972 to 10 tonnes in 1975 was due to these measures. Through these controls Venezuela has succeeded in maintaining its natural stocks but pressure is intense and if controls were to be relaxed the populations could be destroyed very rapidly (23).

CONSERVATION MEASURES PROPOSED   Since Conch is the single most important staple food in Caribbean islands (when available at reasonable prices) and is also bringing high prices on the export market, great efforts must be made to save the fishery for the small scale fishermen of the Caribbean. Some grounds (for example in Belize, the Turks and Caicos, the Bahamas and Venezuela), far from human settlement, still contain healthy Conch populations and do not appear to be overfished. But effective management and enforcement programmes must be instituted with great haste if these stocks are to continue providing a sustainable yield (23).

In the Bahamas, there are plans for greater restrictions on exports (44) and it is hoped that the large sea area with low population density and suitably enforced protective legislation will ensure that further overfishing does not occur (6). The Bahamas together with the Turks and Caicos have a greater area for Conch fishery than the rest of the Caribbean combined, with in excess of 100 000 sq. miles (260 000 km^2) of shallow water suitable for this species (6). In the Turks and Caicos it is necessary to relieve pressure on those populations that are already overfished. The most promising alternative sources of income are thought to be fin fisheries and agriculture, but public education and awareness programmes as well as the updating of fisheries legislation are also important (12). In Venezuela, in spite of controls, pressure is intense and Conch populations could be virtually destroyed in a very short period of time if controls were relaxed (23). For Belize it is thought that management of natural populations would be a more satisfactory solution than reseeding Conch grounds with hatchery reared individuals (38). In the Netherlands Antilles it has been suggested that Conch fishing should be prohibited for a two year period and that subsequently commercial fishing should be restricted to a very small number of licensed local operators. Only shells with a fully developed lip should be taken, a public education programme should be initiated and a reseeding project developed (39).

In some countries it may be necessary to terminate Conch fishing completely and it is clear that there should be no intensification until adequate stock assessments can be made and sound management plans developed (23, 24). The use of SCUBA to collect Conch should be generally discouraged, to protect Conch living in deep water which could act as residual breeding populations (45). There is a wide diversity among Caribbean countries in the size of their Conch resources, their fisheries and their regulatory options but it is generally agreed that improvement of the use of the stocks is feasible provided adequate support is supplied to those countries requiring it. Different management strategies will be necessary in different countries. Regulation may be comparatively simple in the case of large fisheries with centralized marketing. For smaller fisheries operating on an almost individual basis, mariculture may be the only way to slow down and eliminate the over-exploitation of stocks and improve local income and employment through the use of Conch resources. For larger fisheries, commercial hatcheries could increase economic yield (22,28).

A meeting on fisheries management and mariculture of the Queen Conch was held in January 1981 in the Bahamas to review the existing information on this species and to consider further research. Comprehensive recommendations were made for all aspects of fishery management and research (45) and are summarized here.

1. A statistical data base is needed as an aid to management. The statistics available are very poor (for example there are considerable discrepancies between FAO statistics and those reported in other literature), and most are probably under-estimates. The most accurate figures are for the Turks and Caicos, the Bahamas and Belize, where there are centralized fisheries, but these figures are far from complete because of unreported sales (42).

2. Further data is required on behaviour, habitat requirements, food preferences and the general biology of juvenile Conchs. The mechanisms that trigger migration must be determined accurately, and this behaviour pattern must be taken into consideration when fisheries are being monitored, to ensure that fluctuations in populations due to natural causes are not confused with those due to over-exploitation (43). Studies of age-specific food preferences and the types of foods that can be assimilated need to be further investigated, bearing in mind the possibility of cultivation of key foods or production of artificial diets (45). The locations of breeding populations for each island should be identified.

3. Efforts must be made to co-ordinate the work of mariculture research groups to avoid duplication. Specific topics which require further attention have been identified (45).

4. More information is needed on the two types of juvenile rearing that are being considered: intensive-controlled rearing and extensive-uncontrolled rearing in natural areas (45).

5. Conch fishery management should be studied in the countries where it is required, to define the methods which would be most appropriate in each case. Fisheries data for Belize and the Turks and Caicos Islands, where attempts have been made to manage the Conch fishery for some years, should be

analysed as the methods used might serve as models for other countries. Education programmes should be established.

A meeting took place at the Gulf and Caribbean Institute in Nassau, November 1982, to discussed progress in these new initiatives and to propose future directions (21).

CAPTIVE BREEDING Research on the mariculture of Queen Conch has been in progress since the 1960s (2,27,28,41) and the results so far indicate that Conchs could be raised in very simple systems. Five separate laboratories are currently rearing Conch in the Caicos Islands, Venezuela, Bonaire and Puerto Rico and a combined project is underway in the Bahamas and Miami.

At Pine Cay, Turks and Caicos Islands, a pilot hatchery has been set up with funds from PRIDE (a foundation for the Protection of Reefs and Islands from Degradation and Exploitation) and the Nixon Griffis Foundation. At the Wallace Groves Aquaculture Foundation in the Bahamas, studies on hatchery-rearing are being undertaken in collaboration with the University of Miami, and in Venezuela work is being carried out at the Foundacion Los Roques Research Station (13,23). In Puerto Rico a joint programme with the University of Puerto Rico and the U.S. National Marine Fisheries Service has been in progress for a year. In 1981, 5000 juvenile S. gigas and S. costatus were raised, and the prediction for 1982 was twice as great. The largest Conch hatchery is being established at Puerto Morelos in Quintana Roo, Mexico and is intended to have an annual production capacity of about 0.5 million Conchs (60). In the Grenadines, Conch studies at Carriacou have been carried out with the support of the Rockefeller Brothers Fund. Several other countries are interested in setting up mariculture and reseeding programmes, including the Bahamas (6), Belize (23) and the Dominican Republic, where potential sites for Conch raising have been identified by two government agencies, Caza y Pesca and INDOTEC, and include Catalinita, Bahia San Lorenzo, Puerto Viejo and Bahia Luperon (13,23).

Research suggests that hatcheries could be established with a relatively low capital investment and that most of the raising activities could be carried out by fishermen and/or technician's aides (23), thus providing employment as well as Conch products. Most energy needs for the hatchery could be satisfied by wind and solar power, which are unlimited natural resources in the Caribbean islands (32). It appears that water quality and suitable nutrition are the most important single factors in a successful system (22). Juveniles have been found to grow well in tanks with rich growths of algae, and Conchs have reached 31.7 mm in length 171 days after hatching (2,27). Growth rates in captivity may be higher than in the wild, since juveniles feed during the day as well as at night in culture systems. Their phytoplankton food requirements seem to be very specific and have not been fully studied (27). The greatest proportionate yields of meat are from large juveniles. An acceptable market size of 190 mm length with a total weight of 850 g and a meat yield of 100 g has been obtained with individuals of 2.5 years (28). The greatest mortality rate appears to be during the initial planktonic veliger stage and it is suggested that mariculture operations should raise Conch to 4 cm in length and then release them onto subtidal algal flats, rather than culture them to market size (2,28). With the overfishing of Conch and the decline of Green Turtle populations, sea grass beds are now an under-utilised resource. Reseeding with Conchs from hatcheries would make use of this highly productive habitat (27,28). However, considerable research is still needed to define the correct techniques for commercially viable mariculture (13).

REFERENCES   1.   Clench, W.J. and Abbott, R.T. (1941). The genus Strombus in the Western Atlantic. Johnsonia 1: 1-15.
              2.   Brownell, W.N., Berg, C.J. and Haines, K.C. (1977). Fisheries and aquaculture of the Conch, Strombus gigas in

the Caribbean. In Stewart, H.B. (Ed.), Cooperative investigations of the Caribbean and adjacent regions, Caracas, Venezuela, 12-16 July 1976. FAO Fish. Rep. 200: 59-69.

3. Abbott, R.T. (1973). The Kingdom of the Seashell. Hamlyn.

4. Brownell, W.N. and Berg, C.J. (1978). Conchs in the Caribbean. Sea Frontiers 24(3): 178-185.

5. Boss, K.J. (1969). Conchs. In Firth, F.E. (Ed.), Encyclopaedia of Marine Resources. Van Nostrand Reinhold, New York. Pp. 135-140.

6. Attrill, R. (1980). In litt., 14 January.

7. Abbott, R.T. (1981). The shell trade in Florida: status, trade and legislation. Special Report 3, TRAFFIC U.S.A. Washington D.C.

8. Sanger, P., Hegerl, E.J. and Davie, J.D.S. (Eds) (1981). 1st Report on the Global Status of Mangrove Ecosystems. IUCN Commission on Ecology, Working Group on Mangrove Ecosystems.

9. Canoy, M. (1981). In litt., 7 April.

10. Ogden, J.C. (1981). In litt., 15 April.

11. Anon. (1981). Bermuda Division of Fisheries. Fisheries Office, Southampton.

12. Anon. (1979). A strategy for the conservation of living marine resources and processes in the Caribbean region. IUCN Report.

13. Brownell, W.N. and Stevely, J.M. (1981). The biology, fisheries and management of the queen conch, Strombus gigas. Marine Fisheries Review 43(7): 1-12.

14. Anon. (1979). Turks and Caicos Islands, Caribbean: Development of institutional infrastructure and capacity to manage marine resources. Project proposal to IUCN.

15. Berg, C.J. (1975). A comparative ethological study of strombid gastropods. Behaviour 51: 274-322.

16. Parker, G.H. (1922). The leaping of the stromb (Strombus gigas Linn.). J. Exp. Zool. 36: 205-209.

17. Robertson, R. (1959). Observations on the spawn and veligers of Conchs (Strombus) in the Bahamas. Proc. Malac. Soc. London. 33(44): 164-171.

18. Flores, C. (1964). Contribution to the knowledge of the genus Strombus in the coastal waters of Venezuela. Memorias de la Sociedad de Ciencias Naturales La Salle, Caracas 24(69): 261-276.

19. Beau, M. (1858). De l'utilité de certains mollusques marins vivants sur les côtes de la Guadeloupe et de la Martinique. J. Conchyliol. 7: 25-40.

20. Randall, J. (1964). Contributions to the biology of the queen conch, Strombus gigas. Bull. Mar. Sci. Gulf. Caribb. 14(2): 246-295.

21. Goodwin, M.H. (1982). Overview of conch fisheries and culture. Paper presented to 35th Gulf and Caribbean Fisheries Institute, November, 1982.

22. Goodwin, M. and Taylor, S. (1981). Queen Conch Fisheries and Mariculture Meeting. In Ref. (46).

23. Brownell, W. (1978). Report on the status of conch fisheries and related research in Belize, Turks and Caicos, Dominican Republic, Antigua, Dominica, St. Lucia, Barbados, Grenada, Trinidad and Tobago, and Venezuela. Inter-regional Project for the Development of Fisheries in the Western Central Atlantic (WECAF).

24. Anon. (undated). Marine Environments of the Virgin Islands. Technical supplement No.1, Office of the Governor, Virgin Islands Planning Office, Coastal Zone Management Program.

25. Robertson, R. (1961). The feeding of Strombus and related herbivorous marine gastropods. Notulae Naturae 343: 1-9.

26. Hesse, K. (1979). Movement and migration of the queen conch, Strombus gigas, in the Turks and Caicos Islands. Bull. Mar. Sci. (Miami) 29(3): 303-311.

27. Brownell, W.N. (1977). Reproduction, laboratory culture and growth of Strombus gigas, S. costatus and S. pugilis in Los Roques, Venezuela. Bull. Mar. Sci. 27(4): 668-680.

28. Berg, C.J. (1976). Growth of the queen conch, Strombus gigas, with a discussion of the practicality of its mariculture. Mar. Biol. 34(3): 191-199.

29. Damman, A.E. (1969). Study of the fisheries potential of the Virgin Islands. Spec. Rep. Caribb. Res. Inst. 1: 1-204.

30. Sigel, M.M., Wellham, L.L., Lichter, W., Dudeck, L.E., Gargus, J.L. and Lucas, A.H. (1969). Anticellular and antitumoractivity of extracts from tropical marine invertebrates. In Youngken, H.W. (Ed.), Proc. Symp. Food and Drugs from the Sea 1969: 281-294. Marine Technological Society, Washington D.C.

31. Adams, J.E. (1970). Conch fishing industry of Union Island, Grenadines, West Indies. Trop. Sci. 12(4): 279-288.

32. Davis, M. (1982). In litt., 30 August.

33. Doran, E. (1958). The Caicos Conch trade. Geogr. Rev. 48(3): 388-401.

34. Percharde, P.L. (1970). Further underwater observations on the molluscan genus Strombus as found in the waters of Trinidad and Tobago. Carib. J. Sci. 10(1-2): 73-77.

35. Anon (1977). Coral reefs - Caribbean coast of Colombia. IUCN/WWF Full Project Outline, 1416/1977.

36. FAO (1973). Report to the Government of British Honduras on fisheries management and potential. Based on the work of R.H. Baird, FAO/TA Marine Biologist. Rep. FAO/UNDP (TA), (3203): 49 pp.

37. Blakesley, H.L. (1977). A contribution to the fisheries and biology of the Queen Conch, Strombus gigas L. in Belize. Abstract of paper presented at American Fisheries Society, 107th Annual Meeting, Vancouver, B.C. 15-17 September 1977.

38. Egan, B. (1981). Belize - management and replenishment of Conch stocks. In Ref. (46).

39. Hensen, R. (1981). Netherlands Antilles - management and replenishment of stocks. In Ref. (46).

40. Hesse, C. (1981). Turks and Caicos - management and replenishment of stocks. In Ref (46).

41. D'Asaro, C.N. (1965). Organogenesis, development and metamorphosis in Queen Conch Strombus gigas with notes on the breeding habits. Bull. Mar. Sci. 15: 359-416.

42. Stevely, J.M. (1981). Current status of Queen Conch fisheries. In Ref. (46).

43. Berg, C.J. (1981). Conch biology. In Ref. (46).

44. Higgs, C. (1981). Bahamas - management and replenishment of stocks. In Ref. (46).

45. Berg, C.J. (1981). Summaries, conclusions and recommendations. In Ref. (46).

46. Berg, C.J. (Ed.) (1981). Proceedings of the queen conch fisheries and mariculture meeting. The Wallace Groves

Aquaculture Foundation, Jan. 8-10 1981, Discovery House, Freeport, Bahamas.

47. Moncaleano Archila, A. (1978). Delineamientos estadisticos para la evalucion del stock y la actividad pesquera sobre el caracol de pala Strombus gigas L. en el Archipelago de San Bernardo, Mar Caribe, Colombia. Divulg. Pesq. Inst. Desarr. Recurs. Nat. Renov. Bogota, 13(5).

48. Stevely, J.M. (1981). Florida, U.S.A. - management and replenishment of stocks. In Ref. (46).

49. Burnett-Herkes, J. (1981). Bermuda - management and replenishment of stocks. In Ref. (46).

50. Breder, C.M. (1948). Observations on coloration in reference to behaviour in tide-pool and other marine shore fishes. Bull. Am. Mus. Nat. Hist. 92: 281-311.

51. Telford, M. and Daxboeck, C. (1978). Porcellana sayana Leach (Crustaceana: Anomura) symbiotic with Strombus gigas (Linnaeus) (Gastropoda: Strombidae) and with three species of hermit crabs (Anomura: Diogenidae) in Barbados. Bull. Mar. Sci. 28: 202-205.

52. Coomans, H.E. (1973). Pearl formation in gastropod shells. Sb. Nar. Mus. Praze 29B (1-2): 55-64.

53. Prescott, B. and Li, C.P. (1966). Antimicrobial agents from sea food. Malacologia 5: 45-46.

54. Shoemaker, A.H. (1971). Strombus range extensions. Nautilus 85: 72.

55. Coomans, H.E. (1958). A survey of the littoral Gastropoda of the Netherlands Antilles and other Caribbean islands. Stud. Fauna Curaçao Other Caribb. Isl. 8(31): 42-111.

56. Little, C. (1965). Notes on the anatomy of the queen conch, Strombus gigas. Bull. Mar. Sci. 15: 338-358.

57. Hesse, C. and Hesse, K. (1977). Conch industry in the Turks and Caicos Islands. Underwater Naturalist 10(3): 4-9.

58. Alcolado, P.M. (1976). Crecimiento, variaciones morfologicas de la Concha y algunos datos biologicos del cobo, Strombus gigas. Serie Oceanologica 34, Habana, Cuba.

59. Carranza, J. (1962). Survey of the marine fisheries and fishery resources of the Yucatan Peninsula, Mexico. Sc.D. Thesis, Univ. Michigan, Ann Arbor. 193 pp.

60. Chanley, P. (1982). In litt., 1 August.

61. Fischer, W. (Ed.) (1978). FAO Species identification sheets for fishery purposes. Western Central Atlantic (Fishery Area 31) Vol. 6. FAO Rome.

62. Goodwin, M. (1983). In litt., 4 January.

We are very grateful to R. Attrill, C.J. Berg, W. Brownell, J. Burnett-Herkes, M. Canoy, P. Chanley, C.N. D'Asaro, M. Davis, M.H. Goodwin, K. Hesse, J.C. Ogden and J.M. Steveley, for providing information for this data sheet.

TRITON'S TRUMPET or GIANT TRITON   RARE

Charonia tritonis (Linnaeus, 1758)

Phylum	MOLLUSCA	Order	MESOGASTROPODA
Class	GASTROPODA	Family	CYMATIIDAE

SUMMARY Charonia tritonis has a large attractive shell which is popular with
shell collectors and has had a long history of use in the Indo-Pacific as a trumpet.
It is found on coral reefs and appears to be rare throughout much of its range. In
some places this may be a result of overcollection.

DESCRIPTION Charonia tritonis has a large, very attractive shell reaching 40 cm
or more in length. It has a high pointed spire, and is creamy white with purple and
brown markings and an orange interior to the aperture (1).

DISTRIBUTION Indo-west Pacific (1) to the Red Sea (19), including the Great
Barrier Reef, Australia (4), the Philippines (6), Indonesia, Seychelles, Mozambique
(22), Guam (14), the Marshall Islands, Hawaii (23) and Fiji.

POPULATION Unknown, but densities appear to be naturally low in many areas.
In a three month study in part of the Sudanese Red Sea only two specimens were
found; over the same period local shell collectors had taken five and they reported
that specimens were only found occasionally (19). Reported to be uncommon in
Guam (14) and on the Great Barrier Reef where in a study of 81 reefs, only 24
specimens were found (20). In a survey of 133 sites on 92 reefs in Indonesia only
two specimens were recorded (22). Population densities may actually be higher as
specimens are difficult to find on account of their nocturnal habits and
camouflaged shell (19).

HABITAT AND ECOLOGY No detailed ecological study has been carried out on the Giant Triton. It is usually found among corals on coral reefs. On the Great Barrier Reef it tends to occur on seaward slopes or on coral pinnacles in lagoons (20). It is usually nocturnal (19) but in Indonesia has been recorded from sand near coral in broad daylight (22). It feeds principally on starfish including Culcita novaeguinea, the blue starfish Linckia laevigata and the Crown-of-thorns, Acanthaster planci, but also occasionally on holothurians (2,15,20,21). When preying on A. planci it takes mainly large juveniles and adults on which it may feed whole (4,5,15). It may take up to six years to reach its maximum size (3). It has been reported that the female lays clumps of sausage-shaped egg capsules under protective rocks. The larvae are long lived and have considerable dispersal abilities (3,23,24).

SCIENTIFIC INTEREST AND POTENTIAL VALUE The shells of this species have been used as a trumpet in the South Sea islands, Indonesia, Hawaii and the Seychelles for centuries (3). More recently it has become one of the more sought-after shells in the ornamental shell trade and is collected in large numbers in the Philippines and other countries (6,17). It has been collected extensively in Queensland waters since World War II and available evidence indicates that thousands of specimens were collected on the Great Barrier Reef during the 1950s and sold to tourists and to the shell trade via the commercial trochus shell dealers on Thursday Island in the Torres Straits (7). Shells were collected intensively at several islands in the Marianas and Carolines in the 1960s (18). In 1970 a thousand specimens were seen on sale in Suva market, Fiji said to have come from remote islands (10) and in 1971 specimens were seen on sale at several localities on Manus Island, Papua New Guinea (7). Heavy collection was reported in this area in the 1970s. In Hawaii in 1971, shells were on sale for U.S.$65 each. The value in 1978 was put at U.S.$7-30 depending on size and quality (16). In 1981 specimens in the Philippines sold for U.S.$5-20 (17).

It has been suggested that C. tritonis may be a key species in the coral reef community. As a predator of the Crown-of-thorns Starfish it was thought to be important in keeping this species in check (7). Its role in this context was hotly disputed at the peak of the Acanthaster outbreaks in the Indo-west Pacific, and no final conclusion has been reached, as several other factors are almost certainly involved. Specimens of C. tritonis do not appear to be sufficiently common to exert any control over A. planci populations, and starfish attacked by the mollusc have been found to survive (7,15). Furthermore, C. tritonis may prefer other species of starfish (21). However, areas with particularly low numbers of C. tritonis have been found to correlate with infestations of the Crown-of-thorns (4,5,7,20).

THREATS TO SURVIVAL Over-collection may have been responsible for reports of depletion of this species in some areas. Furthermore, the large numbers taken strongly suggest that populations may have declined. Populations have been severely decimated by over-collecting in Guam waters (8) and the species is said to be vulnerable to collecting in southern Japanese waters (9). Some shell dealers in the Philippines reported Giant Tritons to have become scarce in 1981 (17). At the end of the 1960s C. tritonis was reported to be common in Palau and Rota in the Pacific, areas which were seldom visited by collectors (15). In the Caribbean, a similar species, the Variegated Triton C. variegata, has also been depleted through over-collection (13). However, despite these reports, this species, although certainly rare, may not be threatened at present. Its wide distribution and prolific long-lived larvae would suggest that it may be able to support current collecting pressure (24).

Like many other species C. tritonis populations could have been reduced as a result of the accumulation of toxic residues from pesticides and other pollutants in the tissues of specimens, particularly since this species is a predator of high trophic status (4,7) but there is no data to substantiate this.

CONSERVATION MEASURES TAKEN In 1970 a ban was imposed in Fiji on the collection and export of C. tritonis (10). It is protected in the Seychelles (11). It has been protected throughout Queensland waters since 1969 (although legislation is poorly enforced) (2), and is now protected throughout the rest of Australia (12). In 1970 Taiwanese fishermen were fined by the Australian government for collecting Giant Tritons on Swain Reefs near the southern end of the Great Barrier Reef (7).

CONSERVATION MEASURES PROPOSED Legislation for the protection of this species is required in Guam (8), and better enforcement of existing legislation in Australia is necessary. Further research is required on the population dynamics of this species. Attempts to rear it in captivity have failed (23).

REFERENCES   1. Oliver, A.P.H. (1975). The Hamlyn Guide to Shells of the World. Hamlyn, London.
2. Endean, R. (1980). In litt., 14 November.
3. Abbott, R.T. (1973). The Kingdom of the Seashell. Hamlyn, London.
4. Endean, R. (1969). Report on investigations made into aspects of the current Acanthaster planci (Crown-of-thorns) infestations of certain reefs of the Great Barrier Reef. Fisheries Branch, Queensland Dept. of Primary Industries.
5. Pearson, R. and Endean, R. (1969). A preliminary study of the coral predator Acanthaster planci (L.) on the Great Barrier Reef. Queens. Fish. Notes. Dep. Harbours. Mar. 3:27.
6. Wells, S.M. (1982). Marine conservation in the Philippines and Papua New Guinea with special emphasis on the ornamental coral and shell trade. Report to Winston Churchill Memorial Trust, London.
7. Endean, R. (1973). Population explosions of Acanthaster planci and associated destruction of hermatypic corals in the Indo-west Pacific region. In Jones, O.E. and Endean, R. (Eds), Biology and Geology of Coral Reefs Vol II. Biology I. Academic Press, New York.
8. Hedlund, S.E. (1977). The extent of coral, shell and algal harvesting in Guam waters. Univ. Guam Marine Lab. Tech. Rep. 37. 34 pp.
9. Ikenouye, O. (1981). In litt., 4 March.
10. Owens, D. (1971). Acanthaster planci starfish in Fiji: a survey of incidence and biological studies. Fiji Agr. J. 33: 15.
11. Salm, R. (1978). Conservation of marine resources in Seychelles. IUCN Report to the Government of Seychelles, Morges, Switzerland.
12. Coleman, N. (1976). Shell collecting in Australia. A.H. and A.W. Reed, Australia.
13. Ogden, J.C. (1981). In litt., 15 April.
14. Stojkovich, J.O. (1977). Survey and species inventory of representative pristine marine communities on Guam. Univ. of Guam Marine Laboratory, Sea Grant Publication UG SG-77-12. Technical Report 40.
15. Chesher, R.H. (1969). Destruction of Pacific corals by the sea star Acanthaster planci. Science 165: 280.
16. Wagner, R.J.L. and Abbott, R.T. (1978). Standard Catalog of Shells. 3rd Ed. American Malacologists, Inc., Delaware.
17. Wells, S.M. (1981). International Trade in Ornamental Shells. IUCN Conservation Monitoring Centre, Cambridge.
18. Chesher, R.H. (1969). Acanthaster planci: impact on Pacific coral reefs. Doc. No. PB 187 1631. Westinghouse

93

Electric Corporation Report to U.S. Dept. Interior.

19. Ormond, R.F.G. and Campbell, A.C. (1970). Observations on *Acanthaster planci* and other coral reef echinoderms in the Sudanese Red Sea. *Symp. Zool. Soc. London.* 28: 433-454.

20. Endean, R. and Stablum, W. (1973). A study of some aspects of the crown-of-thorns starfish (*Acanthaster planci*) infestations of reefs of Australia's Great Barrier Reef. *Atoll Res. Bull.* 167.

21. Vines, P.J. (1970). Field and laboratory observations of the crown-of-thorns starfish, *Acanthaster planci*. *Nature (London)*. 228: 341.

22. Salm, R. (1982). In litt., 30 April.

23. Berg, C.J. (1971). The egg capsule and early veliger of *Charonia tritonis* (Linnaeus). *Veliger* 13: 298-299.

24. Kay, A. (1982). In litt., 20 September.

We are very grateful to R. Endean, A. Kay and R. Salm for information provided for this data sheet.

ZUIDERZEE DORIDELLA SEA SLUG

INSUFFICIENTLY KNOWN

Doridella batava (Kerbert, 1886)

Phylum	MOLLUSCA	Order	GYMNOSOMATA
Class	GASTROPODA	Family	CORAMBIDAE

SUMMARY This small brackish water sea slug is believed to be endemic to an area centred on the former Zuiderzee in the Netherlands. The closure of the Zuiderzee caused its disappearance from most known localities, including its type locality, and it may now be extinct.

DESCRIPTION A small sea slug reaching only about 5 mm in length, Doridella batava is variable in colour, ranging from yellow with black patches to dark brown or black all over. It has a circular disc-shaped mantle and a flat, almost circular foot (4). The shape of the mantle, which is complete at the posterior end of the animal, separates this species from its nearest relatives (1).

DISTRIBUTION Believed to be endemic to the Netherlands. The type locality is at Durgerdam near Amsterdam beyond the IJ, a bay of the former Zuiderzee. It has been recorded in many localities on the coast of the former Zuiderzee from Amsterdam to Den Helder (1), and in 1947 from an area near Oudeschild in Texel (3,4).

Although D. batava is considered to be endemic to Dutch waters, it is possible that it was introduced by shipping during the last century to the Zuiderzee which has existed as a brackish area for only 400 years. D. batava has no close relatives in other European waters but similar species are common on the western side of the Atlantic (7). However it is included here to illustrate the vulnerability of species restricted to estuarine and brackish-water habitats.

POPULATION Unknown.

HABITAT AND ECOLOGY Found in brackish water generally in salinities from 7-25°/°°, but it can probably withstand salinities up to 30°/°° since large numbers have been found in Den Helder harbour. The lowest salinity in which it has been found is 6.7°/°°. It is restricted to shallow water on or near Cordylophora caspia on which it may feed. It has also been found on Membranipora (1,2).

SCIENTIFIC INTEREST AND POTENTIAL VALUE When first discovered in 1886, it was described as Corambe batava and caused great interest since its only known relative, Corambe sargassicola, occurred in the Sargasso Sea. A third Corambe species was described as C. testudinaria from Arcachon in France in 1889 and a new family was erected for these species in view of their considerable differences from other nudibranchs (5). Currently its nearest known relative is Corambella baratariae Harry, a species recently described from the Gulf of Mexico. It has now been assigned to the genus Doridella.

THREATS TO SURVIVAL The closing of the Zuiderzee resulted in the disppearance of all the brackish water fauna in the area, and D. batava is almost certainly now extinct throughout its original range and definitely extinct in its type locality (1). The Zuiderzee was closed in May 1932, following which this species continued to be found in the northern part of the enclosed area until July of that year. Since then numerous surveys have failed to reveal its presence in Holland, and its brackish-water habitat has been destroyed at Texel (6).

CONSERVATION MEASURES TAKEN None.

CONSERVATION MEASURES PROPOSED It could possibly survive still in the harbour at Den Helder or in brackish water on some of the Wadden Sea Islands, such as Texel although both these areas are severely polluted. It is also conceivable that the species may be found in France or England (5). Surveys should be carried out in the Wadden Sea and Den Helder harbour to ascertain if the species is still extant. If it is found, steps should be taken to ensure that it is protected in a suitable unpolluted area of brackish water. Considerable work has been carried out directed towards the management and improvement of conditions in the Wadden Sea which would be beneficial to D. batava if it is found (8).

REFERENCES
1. Butot, L.J.M. (1977). Het Kaaskenswater en het Natuurwetenschappelijk belang van Typelocaliteiten. RIN-rapport. Rijksinstituut voor Natuurbeheer, Leersum. 15 pp.
2. Swennen, C. (1961). Data on distribution, reproduction and ecology of the nudibranchiate molluscs occurring in the Netherlands. Neth. J. Sea. Res. 1(1-2): 191-240.
3. van Benthem Jutting, W.S.S. (1922). Zoet - en Brakwater mollusken. In Redeke, H.C. (Ed.), Flora en Fauna der Zuiderzee: 391-410.
4. van Benthem Jutting, W.S.S. (1936). Mollusca (I) B. Gastropoda; Opisthobranchia; Amphineura et Scaphopoda. Fauna van Nederland afl. VIII, Leiden: 106 pp.
5. Butot, L.J.M. (1982). In litt., 24 February.
6. Swennen, C. (1982). In litt., 1 June.
7. Swennen, C. (1982). In litt., 23 June.
8. Wolff, W.F. and Zijlstra, J.J. (1980). Management of the Wadden Sea. Helgoländer wiss. Meeresunters 33: 596-613.

We are very grateful to L.J.M. Butot and C. Swennen for providing information for, and reviewing, this data sheet.

GIANT CLAMS

Hippopus hippopus (Linnaeus, 1758)	Horse's Hoof, Bear Paw or Strawberry Clam	I
H. porcellanus Rosewater, 1982	China Clam	I
Tridacna crocea (Lamarck, 1819)	Crocus, Saffron-coloured or Boring Clam	K
T. derasa (Röding, 1798)	Southern Giant Clam	V
T. gigas (Linnaeus, 1758)	Giant Clam	V
T. maxima (Röding, 1798)	Small Giant Clam	K
T. squamosa Lamarck, 1819	Scaly or Fluted Clam	I

Phylum	MOLLUSCA		Order	VENEROIDA
Class	BIVALVIA		Family	TRIDACNIDAE

SUMMARY The largest bivalves in the world, the giant clams are restricted to limited areas of the Indo-Pacific. They have long been an important component of subsistence fisheries and more recently there has been large scale commercial exploitation, particularly by the Taiwanese, of the larger species for both shells and meat. Certain aspects of their biology and the ease with which they can be harvested make them highly vulnerable to over-exploitation and populations have declined in many areas. Recent research has suggested that it may be feasible to culture giant clams.

DESCRIPTION The family contains seven species:

Hippopus hippopus reaches 39.7 cm in length. The shell is elongate to ovate, triangular, or sub-rhomboidal in shape and the valves are heavy and thick, usually coloured with strawberry blotches and often obscured by encrusting organisms. The mantle is irregularly mottled along the edges and in the centre with deep yellowish-green (22).

H. porcellanus has a thinner and smoother shell than that of H. hippopus, usually lacking the strawberry coloration, and more semi-circular in outline. The mantle is a sombre olive green colour (45).

Tridacna crocea is the smallest giant clam, reaching only 15 cm in length. The valves are greyish-white, often tinged with pinkish-orange or yellow both inside and out, and are usually quite smooth. They are strongly triangularly-ovate in shape (22).

T. derasa, the second largest tridacnid, is distinguished by its low primary and radial sculpture, variable shape, lack of spiny projections and its white shell. It reaches 51.4 cm or more in length (22,32).

Tridacna gigas is the largest living shelled mollusc and may weigh over 200 kg, of which 55-65 kg is the weight of the living tissue. The shell may grow to 137 cm in length, is sub-oval to fan-shaped and white, and bears a number of radiating ribs (22).

T. maxima, probably the most widespread species, also reaches 35-40 cm in length, and has the brightest mantle colouring of all the giant clams. Shell shape and mantle colour are very variable (16,22).

T. squamosa has an elongate shell with about five strong, low, rounded radial ridges, each carrying fluted scales growing large towards the edge of the valve. The valves are white, tinged with lemon yellow towards the margin, and may reach 35-40 cm in length (22). In the southern Philippines and Indonesia some individuals are found with deep orange and yellow shells (24,46).

Mantle colour in most species varies geographically and ranges through brilliant blue, green, purple and brown depending on incident light. The vivid pigmentation is caused by iridophores and may help to protect the tissues against the effects of the intense light to which they are continually exposed (34). It may also serve to confuse potential predators or to warn them of potential danger (16). When the mantle is fully expanded, the shell margins are entirely obscured by the scalloped, coloured siphonal tissues. The exhalant aperture, at the end of a tubular extension, is in the middle of the upper surface. The mantle tissues are briefly withdrawn when a shadow falls on them and only withdraw permanently during periods when animals are left exposed at the extreme low water of spring tides (33). The basic biological features of tridacnids are reviewed in (48).

DISTRIBUTION Restricted to limited areas of the Indo-Pacific; their ecological requirements (shallow waters of coral reefs) and short larval life (preventing long range dispersal) may confine them to their present ranges (22).

Hippopus hippopus occurs from the Malay Peninsula to eastern Melanesia. There are records for at least the following countries: Singapore, Ryukyu Islands, Philippines, Indonesia, North Borneo, Western Australia and Queensland, Papua New Guinea, Solomon Islands, Vanuatu, (New Hebrides), Fiji, New Caledonia, Caroline Islands, Marshall Islands, Gilbert Islands, Tonga (22).

H. porcellanus appears to have a more restricted range than H. hippopus but overlaps with it. It is known from the Sulu Archipelago in the southern Philippines, one specimen has been recorded off Masbate Island, central Philippines (45), and specimens have been recorded recently from Indonesia (46).

Tridacna crocea ranges from the western coast of the Malay Peninsula east to Micronesia and north to the Ryukyu Islands, Japan, including Thailand, Singapore, Philippines, Indonesia, North Borneo, Papua New Guinea, the Solomon Islands, the Caroline Islands (22). It is reported to be common on the Great Barrier Reef, Australia (19) and in 1969 was the most abundant giant clam in Palau in the Caroline Islands (10).

T. derasa is known from the Philippines, Indonesia (Irian Jaya), Guam, Cocos-Keeling Islands, Australia, Papua New Guinea, New Caledonia, Caroline Islands (22).

T. gigas occurs from the Philippines to Micronesia. Records exist for the Ryukyu Islands, Philippines, Indonesia, western Australia and Queensland, Papua New Guinea, Solomon Islands, Vanuatu, (New Hebrides), Caroline Islands, Marshall Islands, Gilbert Islands (22).

T. maxima and T. squamosa are the most widespread, extending from the Red Sea and the East African coast to the Tuamotu Archipelago and Pitcairn Island, and from southern Japan to the coast of New South Wales, Australia (16,22). Both species have been recorded from Mozambique, Kenya, Madagascar, Saudi Arabia, Egypt, Seychelles, Mauritius, Maldives, Chagos, Thailand, Malaysia, Japan, Philippines, Indonesia, North Borneo, Australia, Papua New Guinea, New Caledonia, Fiji, Marianas, Caroline Islands, Marshall Islands, Gilbert Islands, Tuvalu (Ellice Islands), Samoa, Tonga, Tuamotu. T. maxima has also been recorded from South Africa, Sri Lanka, Andaman Islands, China, Taiwan, Solomon Islands, Vanuatu (New Hebrides), Lord Howe Island, Wake Island, Line Islands and Henderson Island (22).

POPULATION Unknown. The smaller species are still found abundantly in many areas. In certain areas of the Great Barrier Reef T. crocea density regularly exceeds 100 animals per m^2 (30). In closed lagoons at Tuamotu, T. maxima has been found at densities of 63 per m^2, that is up to 90 000 individuals per hectare (31). In Indonesia near Vogelkop as many as 30-50 large specimens of T. derasa and T. gigas have been found on certain reefs (29). In many areas populations are now much reduced (see below), although in shallow reef areas Tridacnidae are often the commonest bivalves, if not in actual numbers, nearly always in biomass (33).

HABITAT AND ECOLOGY Restricted to shallow waters of coral reefs. T. gigas is found on sand and among corals on reefs from about 1 to 20 m in depth. Some or all of the shell may be exposed at low tides (22). T. derasa generally occurs on the outer edges of reefs at about 4-10 m (22); it appears to be restricted to oceanic environments and is not found on reefs adjacent to large land masses (47). H. hippopus is found on sandy substrates on coral reefs down to 6 m in depth and on sea grass beds near the reef (22). T. maxima is essentially a reef-top inhabitant living on the surfaces of the reef or sand, or partly embedded in coral (16). T. crocea burrows by mechanical, and possibly chemical, means into coral boulders on the reef-top and lives with only the valve margins visible; it is found most often on top of small, detached coral boulders on the interior reef flat (30). T. squamosa usually occurs on coral reef surfaces in depths less than 15 m, most often in protected environments such as reef canyons and fissures, sheltered lagoons and marine lakes (19). It is uncommon in atoll environments (47). Juveniles of all six species are attached to the reef by a gelatinous byssus. This is retained in small species but is gradually lost in T. gigas, T. derasa and H. hippopus (33).

Giant clams feed by filtering plankton from the sea with their gills. A supplementary source of food is provided by zooxanthellae of the species Symbiodinum (= Gymnodinum) microadriaticum, which are specialised dinoflagellates that live symbiotically in the mantle tissues and produce glycerol, a sugar-like substance, during photosynthesis. Waste products of the clam are utilized by these algae. Special transparent columns of cells in the mantle surface, called hyaline organs, focus light deeply into the mantle tissue and the zooxanthellae are found in greatest numbers around them. Zooxanthellae also occur in corals and a number of other reef dwelling coelenterates. Their need for light contributes to the fact that giant clams are limited to shallow water. This supplementary source of food for the clams may account for the large size attained by some individuals, since the algae may contribute more than 50 per cent of the clams' metabolic carbon requirements. Algae do not appear in the mantle tissues until about 2-3 weeks after metamorphosis and seem to enter at the veliger stage via the alimentary canal. After this, growth rates increase sharply. In laboratory cultures survival and growth of veligers and juveniles with zooxanthellae is better than those without. No adult clams are found without these algae (14,22,27,33,35,36). Considerable research has been carried out to investigate the extent to which clams might gain additional food by digesting the zooxanthellae but it now seems likely that the greatest benefit to the clam is from their photosynthetic activity (33,37). No positive evidence has yet been forthcoming concerning the degree to which tridacnids are autotrophic. However on Motupore Island, Papua New Guinea, several large specimens of T. gigas have been maintained in tanks for four years and have grown slightly; other species, kept for shorter times, have grown at normal rates and have been induced to spawn. Since the tanks have a 50 per cent daily turnover of water the degree of sustenance available from primary production is very slight and the clams must therefore be capable of a high degree of autotrophy. Respiration studies suggest a progressively greater degree of autotrophy with increasing size, and juveniles may be dependent on filter feeding to meet their nutritional requirements (17). Recently it has been shown that the algae that are often found in the stomach of clams may be filtered from the water, and that under some circumstances they may originate from corals. Reef building corals have been reported to expel their algal symbionts in response to heat stress (43).

Giant clams may live for up to 100 years, but further research is needed to confirm this (6). All species tend to grow rapidly during the early years of life, but slow down with increasing size and age, and growth eventually ceases (3,16,19,25), although calcification of the main body of the shell may continue (17). T. gigas has been found to have a comparatively fast growth rate. Estimates vary from up to 55 cm in six years (4); 8-12 cm a year for 12-25 cm length specimens (c. 2-4 years old) (3); and from a shell length of 10 cm and weight of 180 g to a shell length of 50 cm and weight of 29 kg in six years (16). Slower growth rates have been indicated for T. squamosa (10) and T. maxima (16,21). Field growth studies on specimens of 12-25 cm in length indicated growth rates of 3-6 cm per year for T. derasa, 3-5 cm per year for H. hippopus and 2-4 cm per year for T. squamosa (3).

Giant clams are protandrous hermaphrodites, i.e. the male gonad matures before the female gonad, and sperm are released before the eggs (3,14,23,27). Unlike most hermaphrodite bivalves, giant clams rarely self-fertilize. Reproductive success may be largely a function of the population density of breeding adults, and the probability of fertilization will decrease rapidly as the distance between the spawning adults increases (23). Consequently there may be a minimum threshold density below which reduced populations would be unable to recover. T. maxima in Tonga mature as males at about 55 mm and 50 per cent are fully mature (producing both eggs and sperm) at about 105 mm. All were found to be fully mature at more than 140 mm long. T. squamosa were probably fully mature at a minimum size of 300 mm (15). T gigas may not reach reproductive maturity for more than five years after recruitment to the reef. Factors influencing gonad development and initiation of normal spawning behaviour in situ are not known although seasonality, temperature, moon phase (affecting tides) and water motion have all been implicated. Seasonality is the only strong influence but different species seem to have different breeding seasons, and different environmental cues may trigger gonad maturation and spawning (3,14,36). There also seems to be considerable variation between years within species. H. hippopus has a peak breeding season from January to March on the Great Barrier Reef and in Palau, Caroline Islands; this species and T. crocea are probably summer breeders. T. squamosa and T. maxima may have peak breeding seasons in the winter or late summer. T. maxima has been reported spawning in June and July (25), and from November to March (14) on Guam. On Fiji T. maxima and T. crocea spawn in June and July (27). On Palau H. hippopus spawns in June and T. crocea in July (14). In Tonga T. deresa may spawn in early summer (15,25).

The pelagic larval life of the veliger is much shorter than that reported for most tropical molluscs. It has been calculated (from laboratory experiments) at 12 days for T. crocea, 11-12 days for T. maxima, 10 days for T. squamosa and 9 days for H. hippopus (10,14,25,27) but under natural conditions itcould be even shorter (27). After settlement and metamorphosis, juveniles seek a suitable permanent settling spot that will give them maximum protection (14). Changes in shell morphology during larval growth are described in (49). T. crocea larvae aggregate during settlement (26) and this is likely to be the case with other tridacnids. Recruitment to the reef is poor, despite the large numbers of eggs produced. Rarity of juveniles and abundance of large specimens suggest that aggregations are maintained by having low adult mortality, slow adult but fast juvenile growth, and spasmodic juvenile recruitment (19,25).

SCIENTIFIC INTEREST AND POTENTIAL VALUE   Giant clams are a dominant feature of most Indo-Pacific coral reefs (16), but there is still much to be learnt about these unique, highly specialized animals (18). They are the largest bivalves in the world (T. gigas is the largest shelled mollusc known) and are major tourist attractions in coral reef areas (25). Large specimens are thought to be very old, possibly well over 100 years (7). T. maxima has been used in studies on genetic variation (38).

All species have been used extensively by local people from South East Asia, the Gilbert Islands, Marshall Islands, and across the Pacific to the Tuamotu Islands, for tools, washbasins and food (6,31). Giant clams have long been taken from reefs in Tonga as part of the subsistence fishery. In recent years a commercial fishery has developed there, most clams being taken around the island of Tongatapu. Estimates of minimum landings were 24 tonnes in 1974 increasing to 153 tonnes in 1978 (whole weight, including shells). T. maxima is the most common species, making up 94 per cent of landings. T. derasa makes up about one per cent of landings by number and probably 10-15 per cent by weight. Its shells are commonly used to decorate graves and gardens. T. squamosa makes up about one per cent of landings around Tongatapu but is much more common in the sheltered waters around Vava'u (15). In Manus Province, Papua New Guinea, clams are collected from the reef and placed in clam 'gardens' on the reef flat, where they continue to grow, and can be harvested when required or used in emergency if bad weather prevents fishing (24).

Giant clam meat is popular in Asian countries and is used in Japanese Sushi cooking (13). It is now fished commercially in large quantities particularly by the Taiwanese. Usually only the large white cylindrical adductor muscles are removed (although all the meat is edible), the remainder being discarded and the valves left on the reef; dried adductor muscle is reported to fetch HK$440-770 per kg (US$82-143) (17). The Philippines exported 17 tonnes of frozen clam meat to Hong Kong and Japan in 1979 compared with 3 tonnes in 1978. The commercial clam fishery in this country increased its catch from 243 tonnes in 1976 to 2861 tonnes in 1979 (figures which may or may not include shell weights) (40).

Giant clams are collected in large numbers for their shells. Tridacna spp. are among the most popular ornamental shells in Florida (1), and a substantial fishery for shells existed in the Palau Islands in 1969 (10). In the Philippines there is still a major trade in the shells which are used for bird baths, as washbasins in hotels and restaurants and for other decorative purposes. In 1981 many warehouses in Cebu and Zamboanga were stocked with shells of the larger species of giant clam awaiting export. T. squamosa is also sought by shell collectors on account of its attractive, thin, coloured shells with prominent fluted scales (24). In Indonesia giant clam shells are used to make a smooth white flooring called teraso. Dead shells are dug from reef flats near Seribu and sent to Jakarta for processing (29).

THREATS TO SURVIVAL   Giant clams, particularly the larger species, have proved to be very vulnerable to over-exploitation. Juvenile recruitment of giant clams appears to be erratic and poor, so extensive cropping of adults could cause local extinctions, and populations may not recover without re-introductions (25). Reefs from the northern Marianas to the southern tip of the Great Barrier Reef have been stripped, and on many no breeding stock remains. This is largely due to the ease with which the clams can be harvested and the fact that they are in demand for their meat and shells (7,22).

Surveys of giant clams on Helen Reef in the Palau Islands were carried out in 1972, 1975 and 1976. The large species were being heavily collected by the Palauans and foreign fishermen, mainly for their meat. In 1972 moderate populations of T. gigas and T. derasa were found and it was thought that they could support a small controlled fishery (11). By 1975, over-exploitation of these species as well as H. hippopus was obvious. T. crocea and T. maxima had been least affected (5). In 1976, all six tridacnid species could still be found in the region but populations of the large species had been greatly reduced, and most clams observed appeared to be smaller than the mean size reported in 1972. Fewer dead shells were seen than in earlier surveys, possibly because they were being taken for the tourist market (12). In Ponape, empty shells are the only indication that T. gigas once abounded.   Recent extinctions have also been reported from Truk and Kosrae in the eastern Caroline Islands and from Guam (47).

T. derasa and T. gigas have been heavily exploited by Taiwanese fishing vessels in Australian waters and between 1969 and 1977 about 156 000 giant clams were taken from Swain Reefs. From what is known of their reproductive strategy, it has been estimated that an increase in adult mortality caused by overfishing would be disastrous to existing stocks on the Great Barrier Reef (19).

In Ryukyu in the southern part of Japan, fishermen have been collecting T. crocea and H. hippopus, and the populations of these two species, along with T. gigas and T. squamosa, are threatened. T. crocea and H. hippopus can be found no longer on shallow reefs, and it is thought that giant clams in this area could disappear within ten years (13).

T. derasa and T. gigas are now probably extinct throughout much of western Indonesia; off Jakarta they have been eliminated from Kepulanan Seribu as a result of over-collection for their meat and shells (29). There are fears that the increased demand for shell floor tiling could further jeopardize stocks (44).

In the Philippines, the shell trade has apparently had a considerable impact on populations, particularly of the larger species. These are now only readily available in the southern Philippines, and increasingly collectors are having to go as far as the South China Sea to find shells (24). A study in the Philippines revealed that the total standing stock of four species (H. hippopus, T. crocea, T. squamosa and T. maxima) was much higher (79-260 kg per ha) in a protected sanctuary, than in a neighbouring unprotected area (41-43 kg per ha). T. squamosa was taken frequently but was found to grow faster than the other three species and T. maxima with the slowest growth rate was particularly vulnerable as it is easily visible (2).

Subsistence harvesting of giant clams by native peoples has also had a marked effect in reducing clam populations on accessible reefs (20). Demand for clam meat in Tonga is greater than supply, and loan schemes for fishermen, for the purchase of boats and outboard motors, have increased pressure on the stock. Fishermen are having to travel to reefs at increasingly greater distances to obtain reasonable catches. Furthermore the development of export markets in Samoa may result in greater pressures on the relatively untouched areas in the Ha'apai and Vava'u groups. H. hippopus has been recorded from Tonga but in a 1979 study only dead shells were found, and it has probably been fished to such an extent that it is now quite rare there (15). Adult specimens of T. maxima are virtually absent from some Marshall Island atolls where they were very abundant twenty years ago (39). In Fiji Tridacna species were reported to be rare in 1971 near human habitation, as a result of over-collection for food (42). H. hippopus seems to be very vulnerable to exploitation, perhaps partly due to the fact that it is not attached to the substrate and so is easy to take. A midden on Motupore Island, Papua New Guinea, reveals that H. hippopus was collected in huge numbers in that area about one thousand years ago. It is now very rare there; clearly a fairly dramatic event occurred to the local population and there is the possibility that over-exploitation was the reason for its decline (18).

CONSERVATION MEASURES TAKEN Giant clams occur in a number of marine reserves but detailed documentation is lacking. In Indonesia, T. gigas is reported to occur in the proposed Karimun Jawa Marine Reserve and may occur in the following protected areas: Karimun Togian (Central Sulawesi), Karimun Seribu (West Java), Pulau Moyo, and Pulau Komodo (44). In Australia all species of giant clams are totally protected on the Great Barrier Reef (20), and T. gigas is protected in all Australian states (6). Regular surveillance of north Australian and Great Barrier Reef waters is co-ordinated by the Australian Coastal Surveillance Centre and makes use of Defence Forces and contract civilian surveillance aircraft. The contract for the Great Barrier Reef Region provides for six days general surveillance flying each week, in addition to daily coastal flights and surveillance by military aircraft. The introduction of the 200 mile Australian Fishing Zone and increased surveillance have almost eliminated clam poaching on the Great Barrier Reef by Taiwanese (7,20,28). In Papua New Guinea, Taiwanese boats caught poaching giant clams have been confiscated, their captains fined and the boats subsequently sold back to the owners (24).

Studies on the feasibility of giant clam mariculture are in progress in several countries (3,17,36, and see below). An attempt to re-introduce T. gigas to Guam began this year with a successful shipment of 500 juvenile clams maricultured in Palau (47). The export of wild Tridacna meat and the harvest of wild clams except for local consumption are prohibited in Palau (41).

CONSERVATION MEASURES PROPOSED The role of giant clams in the coral reef ecosystem is still not fully understood and the consequences of uncontrolled exploitation of the larger species cannot be predicted. Further research on ecology, growth rates and reproductive behaviour would lead to informed decisions regarding the future conservation and possible controlled exploitation of this resource. The survival of unharvested clams in deep water could be significant in the successful recovery of exploited populations (19). Some of the clams, now scattered, could be consolidated into breeding units to increase the probability of successful fertilization (3). Research into mariculture of clams should be encouraged and intensive cultivation investigated to determine economic feasibility. Studies at Palau have indicated that reef reseeding should be possible, and that a self-sustaining fishery could be developed, increasing local employment and cash revenue (8). Breeding populations could be introduced to areas where clams are now extinct, and regulations could be imposed to restrict harvesting.

Juveniles reared in the laboratory could be used to reseed selected reefs (3). The possibility that the larger species will be introduced to coral reef areas outside their known range must be recognized and the potential consequences considered very carefully at an early stage (47).

T. gigas and T. derasa are being proposed for listing on Appendix II of CITES (Convention on International Trade in Endangered Species of Wild Fauna and Flora). If listed, commercial trade in these species between party states will only be permitted provided an appropriate export permit has been issued by the country of origin. Listing on Appendix II would not interfere with mariculture efforts or attempts to improve harvests for local people, but would enable international trade in shells to be monitored and controlled.

Meanwhile existing stocks should be protected. New legislation should be adopted in Palau to designate and protect additional Tridacna sanctuary sites (8). In Indonesia a marine reserve has been proposed to include the Auri Archipelago and Anggremios in the western part of Cenderawasih Bay, behind Vogelkop, to protect the rich marine resources found there, which include large numbers of T. gigas and T. derasa (29).

Recommendations have been made to the Tongan Government for the management of clam stocks. If it is decided to manage the fishery on a sustainable yield basis, a size limit (around 115 mm shell length for T. maxima) should be established as well as several measures aimed at controlling and then reducing fishing effort (15).

CAPTIVE BREEDING   Giant clams would appear to be suitable animals for mariculture, and prospects for their commercial cultivation are reviewed in (47). They have low mortality, no major predators and in large specimens a high degree of autotrophy is shown (17). Laboratory studies have shown that they may spawn prolifically, have a short larval life and require little maintenance while growing (3). The introduction of macerated or freeze-dried gonad into the surrounding water seems to be the most reliable method for inducing spawning (9,14,23,27) although other methods have been used (3,14). Larvae of all six species have been reared to juveniles recently, but with low survival rates (3,14,21,27). Growth rates are reviewed in (47). On Palau, laboratory-spawned T. derasa and T. gigas have been reared to male-phase maturity, in mixed culture with the herbivorous gastropod Trochus niloticus. Laboratory-reared T. gigas at 17 months had a mean length of 8.3 cm. However, mortality was 99 per cent from egg to juvenile stage. T. derasa had a lower mortality (but it was still high at 97 per cent from veliger to juvenile stage) and at 5 months had a mean length of 1.7 cm. T. squamosa reared in the laboratory reached a mean length of 6.7 cm after two years. Sunlight may be the single most critical factor influencing growth (3). Larvae have been found to settle on tank bottoms without a special substrate and, after metamorphosis, grow remarkably fast. Although young clams attach by byssal threads they readily reattach if removed carefully and could, at a suitable size, be transferred to a clam 'farm' in the sea where they might have higher growth rates (3).

T. gigas, with its fast growth rate, seems to have particular potential, and although induction of spawning has been a major difficulty with this species (17), recently large scale rearing of this species has been successful on Palau (41). At present, the main factors limiting giant clam cultivation include high larval mortality, uncertainty about spawning, and filamentous overgrowth in rearing tanks. The latter problem may be solved by culturing Trochus niloticus juveniles (which graze on algae) with tridacnid juveniles. Hatchery operations may be rather complex and extended, but studies on larval nutrition requirements and improved rearing techniques should result in increased survival rates. After settlement and attainment of a size of about 1 cm it is likely that juveniles could be reared in floating trays until they are large enough (3-4 cm) to be placed in suitable benthic habitats (14). Active support and interest of local Pacific island governments will be important prerequisites in the development of a successful giant clam mariculture industry (3,17).

REFERENCES

1. Abbott, R.T. (1980). The Shell Trade in Florida. Status, Trade and Legislation. Special Report 3. TRAFFIC (USA), Washington D.C.
2. Alcala, A.C. (in press). Standing stock and growth of four species of molluscs (family Tridacnidae) in Sumilon Island, Central Visayas, Philippines. Proc. 4th Int. Coral Reef Symp. Manila, Philippines, May 1981.
3. Beckvar, N. (1981). Cultivation, spawning and growth of the giant clams Tridacna gigas, T. derasa and T. squamosa in Palau, Caroline Islands. Aquaculture 24: 21-30.
4. Bonham, K. (1965). Growth rate of giant clam Tridacna gigas at Bikini Atoll as revealed by radioautography. Science 149: 300-302.
5. Bryan, P.G. and McConnell, D.B. (1976). Status of giant clam stocks (Tridacnidae) on Helen Reef, Palau, Western Caroline Islands, April 1975. Mar. Fish. Rev. 38: 15-18.
6. Coleman, N. (1976). Shell collecting in Australia. A. H. and A.W. Reed, Australia.
7. Endean, R. (1980). In litt., 14 November.
8. Perron, F.E. (1982). Conservation of giant clams in Palau. Project Proposal to WWF/IUCN. January 1982.
9. Gwyther, J. and Munro, J.L. (1981). Spawning induction and rearing of larvae of tridacnid clams (Bivalvia: Tridacnidae). Aquaculture 24: 197-217.
10. Hardy, J.T. and Hardy, S.A. (1969). Ecology of Tridacna in Palau. Pacific Sci. 23: 467-472.
11. Hester, F.J. and Jones, E.C. (1974). A survey of giant clams, Tridacnidae, on Helen Reef, a western Pacific atoll. Mar. Fish. Rev. 36: 17-22.
12. Hirschberger, W. (1980). Tridacnid clam stocks on Helen Reef, Palau, Western Caroline Islands. Mar. Fish. Rev. 42: 8-15.
13. Ikenouye, O. (1981). In litt., 4 March.
14. Jameson, S.C. (1976). Early life history of the giant clams T. crocea, T. maxima and H. hippopus. Pacific Sci. 30: 219-233.
15. McKoy, J.L. (1979). Giant clams in Tonga under study. South Pacific Commission Fisheries Newsletter 19: 1-3.
16. McMichael, D.F. (1975). Growth rate, population size and mantle coloration in the small giant clam Tridacna maxima (Röding) at One Tree Island, Capricorn Group, Queensland. In Proc. 2nd Int. Coral Reef Symp. 1, Great Barrier Reef Committee, Brisbane, Australia. Pp. 241-254.
17. Munro, J.L. and Gwyther, J. (in press). Growth rates and maricultural potential of tridacnid clams. Proc 4th Int. Coral Reef Symp. Manila, Philippines, May 1981.
18. Munro, J.L. (1981). Pers. comm.
19. Pearson, R.G. (1977). Impact of foreign vessels poaching giant clams. Australian Fisheries 36 (7): 8-11, 23.
20. Pearson, R. (1980). In litt., 23 June.
21. Richard, G. (1978). Quantitative balance and production of Tridacna maxima in the Takapoto lagoon. Proc. 3rd. Int. Coral Reef Symp. Univ. of Miami, Miami. Pp. 599-605.
22. Rosewater, J. (1965). The family Tridacnidae in the Indo-Pacific. Indo-Pacific Mollusca 1 (6): 347-396.
23. Wada, S.K. (1954). Spawning in the tridacnid clams.

Japanese J. Zool. 11: 273-285.

24. Wells, S.M. (1981). Giant clams - a case for CITES listing. TRAFFIC Bulletin 3(6): 60-64.

25. Yamaguchi, M. (1977). Conservation and cultivation of giant clams in the tropical Pacific. Biol. Conserv. 11: 13-20.

26. Hamner, W.M. (1978). Intraspecific competition in Tridacna crocea, a burrowing bivalve. Oecologia (Berl.) 34: 267-281.

27. La Barbera, M. (1975). Larval and post-larval development of the giant clams Tridacna maxima and Tridacna squamosa (Bivalvia: Tridacnidae). Malacologia 15 (1): 69-79.

28. Kenchington, R.A. (1982). In litt., 16 December.

29. Salm, R.V. (1981). Heads we win, tails we lose. Conservation Indonesia. 5(3-4): 12-14.

30. Hamner, W.M., and Jones, M.S. (1976). Distribution, burrowing and growth rates of the clam Tridacna crocea on interior reef flats. Oecologia (Berl.) 24: 207-227.

31. Salvat, B. (1969). Dominance biologique de quelques mollusques dans les atolls fermés (Tuamotu, Polynesie): phénoméne récent-conséquences actuelles. Malacologia 9: 187-189.

32. Braley, R. (in prep.). Observations on occurrence and natural spawning of the giant clams Tridacna gigas and T. derasa along the northern Great Barrier Reef.

33. Goreau, T.F., Goreau, N.I., and Yonge, C.M. (1973). On the utilization of photosynthetic products from zooxanthellae and of a dissolved amino acid in Tridacna maxima f. elongata (Mollusca: Bivalvia). J. Zool. 169: 417-454.

34. Kawaguti, S. (1966). Electron microscopy on the mantle of the giant clam with special references to zooxanthellae and iridophores. Biol. J. Okayama Univ. 12(3-4): 81-92.

35. Trench, R.K., Wethey, D.S. and Porter, J.W. (1981). Observations on the symbiosis with zooxanthellae among the Tridacnidae (Mollusca, Bivalvia). Biol. Bull. 161: 180-198.

36. Fitt, W.K. and Trench, R.K. (1981). Spawning, development and acquisition of zooxanthellae by Tridacna squamosa (Mollusca, Bivalvia). Biol. Bull. 161: 213-235.

37. Fankboner, P.V. (1971). Intracellular digestion of symbiotic zooxanthellae by host amoebocytes in giant clams (Bivalvia: Tridacnidae) with a note on the nutritional role of the hypertrophied siphonal epidermis. Biol. Bull. 141: 222-234.

38. Ayala, F.J., Hedgecock, D., Zumwalt, G.S. and Valentine, J.W. (1973). Genetic variation in Tridacna maxima, an ecological analog of some unsuccessful evolutionary lineages. Evolution 27: 177-191.

39. Johannes, R.E. (1975). Pollution and degradation of coral reef communities. In Ferguson Wood, E.J. and Johannes, R.E. (Eds), Tropical Marine Pollution. Elsevier, Amsterdam.

40. Anon. (1980). Fisheries Statistics of the Philippines 29: 1979. Bureau of Fisheries and Aquatic Resources, Manila.

41. Heslinga, G.A., Perron, F.E., and Orak, O. (in press). Mass culture of giant clams (F. Tridacnidae) in Palau. Aquaculture.

42. Owens, D. (1971). Acanthaster planci starfish in Fiji: survey of incidence and biological studies. Fiji Agr. J. 33: 15-23.

43. Fankboner, P.V. and Reid, R.G.B. (1981). Mass expulsion of zooxanthellae by heat-stressed reef corals: a source of food for giant clams? Experientia 37(1): 251-252.

44. Usher, G. (1982). Indonesian species data sheet for Tridacna gigas. Unpub. report.

45. Rosewater, J. (1982). A new species of Hippopus (Bivalvia: Tridacnidae). The Nautilus 96(1): 3-6.
46. Usher, G. (1982). Pers. comm.
47. Munro, J.L. and Heslinga, G.A. (1982). Prospects for the commercial cultivation of giant clams. Paper presented at the 35th meeting of the Gulf and Caribbean Fisheries Institute, Nassau, Bahamas, Nov. 1982.
48. Yonge, C.M. (1980). Functional morphology and evolution in the Tridacnidae (Mollusca: Bivalvia: Cardiacea). Records Aust. Mus. 33(17): 737-777.
49. Rosewater, J. (1981). Changes in shell morphology of post-larval Tridacna gigas Linne (Bivalvia: Heterodonta). Bull. Am. Mal. Union 1980 (1981): 45-48.

We are very grateful to R. Braley, R. Endean, J. Gwyther, W.M. Hamner, G. Heslinga, O. Ikenouye, R. Kenchington, J.L. Munro, R.G. Pearson, J. Rosewater, R. Salm, and G. Usher for information and helpful comments on this data sheet.

INTRODUCTION The Gastropoda and Bivalvia are the only two mollusc classes with freshwater representatives. Freshwater gastropod snails in the order Mesogastropoda, like their marine counterparts, use gills for respiration and are predominantly tropical, although there are a number of well-known temperate families such as the Pleuroceridae and the Hydrobiidae. Snails in the order Basommatophora (subclass Pulmonata) have lungs like most terrestrial snails and generally come to the surface to breathe, although some have evolved a secondary gill. The order includes several well known families such as the Lymnaeidae and the Planorbidae. Few bivalves have succeeded in colonizing freshwater, and except in North America, the number of species in any area is not large. They include the Sphaeriidae (pea mussels), Corbiculidae, Margaritiferidae, Unionidae, Etheriidae (tropical freshwater oysters from the southern hemisphere) and Mutelidae (freshwater mussels from the southern hemisphere).

North America has the richest freshwater mollusc fauna in the world, of which the prominent members are the Unionidae (naiads) and the gill breathing snails in the family Pleuroceridae. Over 1000 endemic naiads and pleurocerids have been described, many of which are inhabitants of the Tennessee, Cumberland and other river drainages of the south-eastern U.S.A. The reasons for this extraordinary diversity are complex but probably reflect a benign climate, varying topography and abundant calcium. This account is heavily biased towards the North American fauna, largely because of the vast quantity of data available which illustrate the full range of threats to freshwater molluscs and the type of conservation action that can be taken to ensure their future survival. Information on threatened freshwater molluscs in other parts of the world is poor, even in European countries, but there is evidence that there is a similar need for conservation.

Freshwater molluscs tend to prefer hard water which provides calcium for shell growth. The gastropods are herbivorous or detrital feeders and live in aquatic vegetation or on the bottom of lakes and rivers. The bivalves are generally sedentary bottom dwellers or burrowers and like marine members of their class, are remarkably lethargic, showing long periods of apparent complete inactivity enclosed within their shells. It has been calculated that it may take a freshwater mussel a year to travel one mile. The larger bivalves require a fairly firm bottom in which to embed themselves but smaller species such as Sphaerium and Pisidium, which may only reach 1.5 mm in diameter, attach themselves to aquatic vegetation or the surface film of water by slime threads. The small size of the Sphaeriidae and their resistance to desiccation permit dispersal of individuals over long distances and may explain the widespread distribution and abundance of this family, and the fact that few threatened sphaeriid species have been identified (33). Many freshwater mollusc species have distinct phenotypes (i.e. morphological forms) which live under different ecological conditions. This leads to much confusion in taxonomy, particularly in the North American fauna, and may have important consequences when conservation measures are being considered (32).

The Unionacea is probably one of the best known freshwater bivalve groups and includes the naiads of North America and the freshwater mussels of Europe. Species in the Unionacea have a unique form of reproduction, involving a parasitic larval stage. Sperm is shed by the male into the water where it is dispersed by currents. It is drawn into the female and fertilizes the eggs in the gills in chambers called marsupia. Unique larval forms called glochidia, some armed with hooks or spines, are formed and are finally shed in large numbers. Most perish but some attach to, or are eaten by, host fish, and spend a period of time (possibly dependent on temperature) as obligate parasites. During this period of encapsulation the glochidia metamorphose into juvenile mussels, eventually rupturing and dropping from the host. After settling on a suitable substrate they develop into adults. The host fish species are still unknown for most species of mussel, although the Centrarchidae (sunfish and bass) may account for one fifth of the known hosts in the U.S.A. and their distribution must have had a major influence on mussel distribution. The most successful and widespread naiads parasitize several different species of host fish. The host fish permit dispersal of larvae and provide them with nourishment and protection against bacterial infection. Adult fish usually suffer no harm and infection appears to lead to the development of a resistance, although young fry occasionally die from secondary infections and infection may cause problems in fish hatcheries.

Little is known of the glochidium-host relationship or of the ecological requirements of many of the rarer naiads, but the requirements for successful reproduction seem to include sand- and silt-free riffles (shallow areas of fast-flowing water), abundant food and dissolved oxygen. A commercially exploited population can take many years to replace itself as individuals may be four or more years old before producing eggs or sperm. However, they often continue to breed for the rest of their lives (1,2,33). Thick-shelled river species normally live for 20-40 years and some individuals have been reported surviving for as long as a century (see review of Margaritifera margaritifera).

SCIENTIFIC INTEREST AND POTENTIAL VALUE Freshwater molluscs play an important role in the ecology of streams, rivers and lakes (21). In North America, naiad beds containing hundreds of thousands of individuals of twenty or more species, packed tightly together shell to shell, served in the past to stabilize the bottom and cleanse the river by collective pumping of water through their bodies. Naiads supported numerous animals such as snails, fish and mammals (3) and may still be the staple winter food of muskrats and mink, and an important food for racoons and river otters in many areas of the U.S.A. (6).

The vast radiation and speciation of the North American freshwater molluscs has long been of scientific interest, and has provided good material for the study of evolution. Furthermore, some of the striking river confluences shown by the distribution of freshwater naiads clearly verify the events postulated by Pleistocene geologists (6). Substantial collections of fossil, subfossil and recent shells have been built up which provide a basis for many studies. Middens, burial sites and rock shelter sites often yield good subfossil material as naiads were used extensively by prehistoric man for food and the shells for tools and decoration. Later crushed naiad shells were used as a tempering agent in Indian pottery (45). Many freshwater molluscs are good general indicators of river quality and can be used to monitor streams for the presence of metals and pesticides (21). Naiads may eventually play an important role in the monitoring of heavy metal pollution since this produces a disturbance in the pattern of their annual growth rings (31). The potential value of naiads is described in detail in (6).

Freshwater pearl mussels and naiads have been valued for centuries for their pearls and nacre or mother-of-pearl. In the Margeritiferidae, Margaritifera margaritifera is the most economically important species (see review) but the North American Unionidae were also used heavily by the U.S. pearl button industry in the early part of this century. Over-collection and the development of plastics led to a decline in the U.S. industry in 1937-40 but renewed commercial interest in species such as Actionaias ligamentina carinata and Megolonias giganti came about in the 1950s when the Japanese started to use crushed naiad shells to seed oysters for commercial pearls. The eastern U.S.A. has been almost the sole supplier (5) and currently produces an average of 1500-2000 tonnes a year (29). Surprisingly, some of the commercially valuable species are often able to flourish in impounded lakes (30,33). Naiad shells are still used in small quantities for buttons, while unused shell material is used as fertilizer (4).

THREATS TO SURVIVAL Evidence is growing that freshwater molluscs are widely threatened. In the U.S.A. the status of species in many states has been documented, revealing a staggering decline in population numbers and many extinctions. At least a dozen naiad species are presumed to have become extinct in this century and about 20 per cent of those remaining are seriously endangered (1). For example, in Ohio, the Ohio River, Lake Erie and their respective tributaries had a total of 78 different species and subspecies when the pioneers arrived. Roughly 200 years later, almost 25 per cent can no longer be found and nearly half of the remaining 59 species are uncommon or rare (50). In many cases the changes were apparent by the end of the last century, and are clearly related to the expansion of urban and industrial development. For example, as early as 1858, the naiad fauna had started disappearing from the Scioto River in Ohio, presumably related to the disappearance of riffle habitat (24). The Mississippi River, considered one of the least damaged North American rivers, has lost 27 per cent of its original naiad species (44). In the Coosa and Tennessee Rivers, two endemic freshwater gastropod genera, Tulatoma in the family Viviparidae and Apella in the Pleuroceridae, and some 30 other endemic pleurocerid species have become extinct (3). Many mollusc populations and species throughout the U.S.A. were extinct as early as 1944 (19).

The situation in other parts of the world is less well known but work in Europe is suggesting that many species are vulnerable. In Italy, the most threatened mollusc species are thought to be those living in rivers and springs, such as Unio elongatulus, Microcondylaea compressa, Sphaerium corneum and Theodoxus fluviatilis (20). Several species are considered to be vulnerable in the U.K., including Theodoxus fluviatilis, Myxas glutinosa, Segmentina nitida (21,42) and Margaritifera margaritifera (see review). S. nitida is particularly rare and several of its remaining localities in ponds and ditches are threatened by pollution, infilling and drainage (21). A study in South Yorkshire has shown that if one per cent of the freshwater habitats of a 10 km grid square is destroyed, the total freshwater fauna can be impoverished by nearly 40 per cent (22).

The major causes of the decline of freshwater mollusc populations are pollution and alteration of watercourses, although in the past over-exploitation of commercially valuable species has occurred. A detailed account of factors adversely affecting North American freshwater bivalves and gastropods is found in (33,34) and a brief discussion of the main threats worldwide is given below.

1. Habitat alteration The North American freshwater mollusc fauna has been seriously affected by alteration of watercourses. In the early part of this century increased demands for electrical power and improved river transportation led to the construction of a series of locks and dams on many of the major North American rivers . In particular, in 1933 the Tennessee Valley Authority (TVA) was created as a regional resource development agency to construct a system of dams that would provide navigation of the Tennessee River, control floods and produce electric power. By 1967, nine dams had been built on the main river leaving only 22 miles (35 km) of free-flowing water. Dams cause major alterations to the environment and few species, particularly naiads, can withstand the fluctuations in water level created each time a hydro-electric plant starts or stops generating. Riffles, the favoured habitats for juvenile naiads, are drowned and silt covers the beds. Impoundment affects various aspects of reproductive behaviour. The deeper, cooler water behind dams delays maturation, water currents are reduced, decreasing the chance of fertilization as the released sperm have less chance of encountering a female, and host fish disappear. Light penetration may be reduced as a result of siltation, and production of food material is diminished. Even if adults continue to survive in impounded waters, they may cease to reproduce; for example the population of Plethobasus cicatricosus in the impounded Tennessee River is now known to be non-reproducing (2,9,33). Mussel Shoals, on the Tennessee River, Alabama, used to have a remarkable assemblage of 68 naiad

species including 10 species of Epioblasma. This area is now inundated by Wilson Dam and a survey in 1963 revealed that only 24 species remained, about 52 per cent of the naiad fauna having disappeared (43). Impoundment has been cited as one of the major factors in the decline of the Spiny River Snail Io fluvialis which is endemic to the Tennessee River system. Like many naiads it is confined to riffles and the smoother stretches below these. It is now known only from one site in the Nolichucky River and from the Powell and Clinch River (23).

Channelization, which is often a major part of dam projects, usually results in a lifeless ditch, the bottom covered in mud. Dredging tends to eradicate all bottom life and even when it ceases, the bottom may take a decade or more to recover and may return with only 50 per cent of its original diversity. Furthermore, such measures tend to create motionless pools alternating with stretches where silt and sand scud along the bottom, both types of habitat being unsuitable for many molluscs (3,33). Modern methods of agriculture also lead to heavy siltation, creating similarly unsuitable substrates for larval settlement, and clogging the gills of mussels, which prevents respiration. Sand grains may have an abrasive effect on the shell, eroding the periostracum which normally protects the shell from the erosive effects of acids (3,33).

Species endemic to springs or similarly restricted localities are generally very vulnerable to drainage and alteration of groundwater level, as well as to the factors mentioned above. Twelve freshwater hydrobiid snails, apparently endemic to springs in Ash Meadows, a valley located about 110 km north-west of Las Vegas, U.S.A., are threatened by groundwater pumping, and are being considered for listing under the U.S. Endangered Species Act (18). A second example concerns two freshwater hydrobiid gastropods listed as Endangered in Florida. The Loose-coiled Snail, Aphaostracon chalarogyrus is known only from Magnesia Spring near Hawthorne and the Enterprise Spring Snail Cincinnatia monroensis only from a small seepage run on the south edge of Enterprise. The type locality of the latter has already been destroyed and the population density of the snail is extremely low (15). The review of the snails of the Cuatro Cienégas basin, Mexico, illustrates similar problems.

2. Pollution Pollution from a variety of sources affects freshwater molluscs (17,21). Acid rain, a result of emissions of sulphur dioxide and nitrogen oxides from electricity generation, smelting and refining, causes a lowering of pH values in lakes and rivers. This has occurred increasingly in north-west Europe and eastern North America over the past two decades. In Sweden alone about 20 000 lakes are thought to have been affected, largely as a result of pollutants originating in Central Europe and the U.K. (13). Hydrogen ion levels below a pH of 5.5 are deleterious to most benthic organisms (35). In Norway sphaerid mussels from lakes with low pH were found to have very thin and often quite soft shells. Where the pH drops to about 6.0, five sphaerid species disappear (Pisidium nitidum, P. subtruncatum, P. milium, P. conventus, Sphaerium corneum and S. nitidum) (14) and snails are similarly affected (16). Gastropods are very sensitive to zinc, copper, mercury and silver pollution. For example, the Ystwyta River in Wales was still devoid of molluscs 35 years after lead mining had ceased in the region (36). In Europe the decline of many freshwater species is attributable to pollution (17,21) as illustrated by Margaritifera margaritifera in the following review.

Pollution has had a major effect on the North American freshwater mollusc fauna. In 1962, it was estimated that 48 000 miles (76 800 km) of river in the U.S.A. were affected by the draining of acid wastes from areas where mining for minerals has been carried out (34). In Pennsylvania, many streams in the Ohio River drainage, especially in coal mining areas, have lost all their benthic fauna due to acid water pollution (28). Pollution may cause the loss of the remaining populations of Io fluvialis as municipal sewage treatment facilities replace former septic tank systems and acid mine draining effects other areas in the Clinch River watershed (23).

Pesticides may also be detrimental. In this context it is important to mention the relationship of bilharzia eradication programmes to mollusc conservation. In tropical climates, bilharzia (schistosomiasis) is a serious health problem. The vectors of this disease are freshwater snails and most attempts at control of the disease involve the application of molluscides. Such efforts could have serious effects on non-vector snails (37). The increasing numbers of water impoundment schemes in developing countries may result in the spread of bilharzia to new areas, since vector snails thrive in the still waters created by dams (46).

3. Introduced species In many cases where introduced species are apparently affecting native freshwater molluscs, habitat alteration is generally also a factor and may play the more important role. Native freshwater molluscs in North America may be under pressure from competition with the introduced Asian clam, Corbicula manilensis. Unlike naiads, this species has a free-swimming larva and is able to exploit virtually any substrate. Its range and population size have expanded dramatically since its introduction to the west coast in the late 1940s (33). An introduced snail Melanoides sp. may be a similar threat to the endemic snails of Ash Meadows (18).

4. Exploitation Mussels are particularly susceptible to over-collecting because of their low reproductive rate and long life-span, as illustrated by Margaritifera margaritifera (see data sheet) and the North American unionids. The breeding stock of the latter was rapidly reduced at the beginning of this century to the point where reproduction no longer offset mortality. The dredging methods used destroyed the stream bed, gravid females aborted as a result of disturbance, there was incidental death of juveniles below the useful and legal size limit, and adults died as a result of being unable to re-bury themselves once they had been uprooted. The recent introduction of SCUBA diving gear allows whole beds to be collected, and river after river has been depleted. Methods of collection, including dredges, are still generally unselective and invariably lead to large scale destruction of the river bed and the loss of uncommercial, often less abundant species (1,33,52).

CONSERVATION The U.S.A. is one of the few countries to have taken major steps towards the conservation of its freshwater mollusc fauna. It is not possible to fully review the considerable body of literature available but a brief discussion of some of the efforts in progress is given here. Several naiads are listed under the U.S Endangered Species Act and a number of recovery programmes (e.g. for the Higgins' Eye Mussel Lampsilis higginsi (51)) and conservation projects have been implemented. The status of freshwater molluscs in many states has been documented. For several years the American Malacological Union has taken an active role in conservation efforts (7). In 1971 the following recommendations were made for the conservation of the North American naiads (6):

1. Improve pollution control.
2. Greatly reduce or stop the rate of stream and river impoundment.
3. Greatly reduce or stop stream channelization.
4. Greatly reduce or stop commercial harvesting until populations recover their former abundance.
5. Determine the ecological requirements for rare and endangered species.
6. Establish sanctuaries in rivers known to harbour relict populations of such species.

To a greater or lesser extent these recommendations have been taken up, but unless full support and encouragement are given to the programmes underway and initiatives taken to implement others, many species will remain under serious threat. It should be noted that critical habitat has not been designated for any species of mollusc, and that molluscs have lowest priority (after vertebrates, plants and insects) for conservation effort and research funds within the U.S. Department of the Interior (8). The following cases provide an example of the type of measures that are being taken and which, in some instances, may be applicable in other countries.

The U.S. Fish and Wildlife Service's Virginia Cooperative Fishery Research Unit, along with the Biology Department of the Virginia Polytechnic Institute and State University are studying the nine naiad species listed as Endangered in Virgina. Work is concentrating on those in south-western Virginia waters but is expected to have applications to other naiads of the Tennessee River Basin. Surveys have been carried out to determine current distributions, and quantitative measurements were taken where several individuals of a listed species were found. The fish at these sites have been studied and it has been confirmed that non-game fish are important hosts. This has implications for certain traditional fishery management practices, such as the stocking of streams with game species which could change fish community compositions. It has been recommended that fisheries management operations in Endangered mussel habitat be carefully reviewed. The Tennessee Valley Authority has recently completed similar studies in the upper Tennessee River drainage to evaluate the status of Cumberlandian mussels throughout their range (4).

A two year mollusc study is underway as part of the Environment Impact Research Program at the U.S. Army Engineer Waterways Experiment Station for the Upper Mississippi, Tennessee and Alabama River systems. The study is centred on federally listed Endangered molluscs but is designed to be useful to biologists working on other species. The aims of the study include the collection of information on sampling techniques, the collection of biological and ecological data on federally listed species and the analysis of techniques used to relocate or create habitat for molluscs. Information on sampling techniques will be used to encourage studies to estimate population sizes, an aspect which has not been dealt with in many studies of endangered freshwater molluscs. A field guide to federally listed freshwater molluscs is in preparation, and may include other uncommon species or species proposed for listing (38).

Until more is known of the particular needs of individual freshwater molluscs, general conservation of habitat and pollution control will be the best approaches. Few reserves or sanctuaries have been created specifically for freshwater molluscs although this is clearly a prime requirement for species reduced to single populations. Naiad sanctuaries have been established in Tennessee, U.S.A., to protect the State's endangered naiads and it has been recommended that similar areas should be designated in Virginia (4). It is thought that one third of the endangered molluscs in the U.S.A. could be restored to viable populations fairly inexpensively by the creation of small reserves or by extending the boundaries of existing ones to include critical habitats (9). For freshwater species greater care may need to be taken in establishing boundaries than is the case for terrestrial species. For example, for the naiads, care must be taken to ensure that a reserve includes at least part of the range of the host fish.

Research on the North American mollusc fauna reveals that some regions merit special attention. Some of these are indicated in the data sheet for Epioblasma. The Meramec River Basin in Missouri, where most rivers and streams are relatively unpolluted, rapid-flowing, have silt-free hard waters and plentiful host fish, could be considered (25). The Green River along the upper one third of its length still contains favourable mollusc habitat, particularly in the vicinity of Munfordville. It probably has the finest representative Ohioan mollusc fauna in existence, 64 species and forms having been collected from the entire system. Particular attention should be paid to its protection (26). The Big Walnut Creek system, Franklin Co., Ohio, has been proposed for preservation and habitat improvement because of the relatively high species diversity still found there (27).

In some instances it may be feasible to improve habitats for threatened species and even to translocate them to more suitable localities. A gravel bar habitat for molluscs and other benthic organisms is being designed in the Tombigbee River near Colombus, Mississippi (38). Translocation of endangered populations may be the only solution for very rare species threatened by large scale commercial waterway developments. A population of Lampsilis higginsi is being considered for relocation from a proposed bridge construction site in the Mississippi River (11) and transplant sites are being selected for Conradilla caelata (= Lemiox rimosus) and Quadrula intermedia before the completion of the Columbia Dam which would impound much of the Duck River (12). It is hoped that re-introductions can be carried out for Fusconaia cuneolus and F. edgariana in Virginia, to broaden their very limited ranges (4). Other studies have recommended areas for restocking of commercially valuable species. A stretch of the Allegheny River, Pennsylvania has been recommended for restocking with Actiononaias carinata, since the only populations with potential for commercial harvest are those in French Creek and these could easily be eliminated through unregulated collecting (28).

Several North American freshwater molluscs are listed on the Appendices to CITES (Convention on International Trade in Endangered Species of Wild Fauna and Flora). Countries which are parties to CITES may not trade commercially in species listed on Appendix I; species on Appendix II may be traded commercially between parties provided a valid export permit has been issued by the country of origin. However, the listed naiads are probably involved in trade only incidentally, when dredged up with commercially valuable species, and controls are difficult to enforce because of the need for experts to identify many of the species. It has been proposed that a minimum size limit of a length of three inches (7 cm) should be set on all shells taken commercially (52). The suggestion has been made that, if naiads in North America could be restored to their original numbers, some of the associated industries could be revived. As a result of the rapid depletion of naiad resources earlier in the century by the pearl button industry, the U.S. Bureau of Fisheries started to investigate their possible artificial propagation. However, their efforts were abandoned when the industry declined (39). Juvenile mussels have been grown experimentally using trout fry food; this was successful although growth was slow (40). Renewed studies involving the artificial infection of host fish and transplantation of mussels are currently in progress at the Virginia Polytechnic Institute (41).

In Europe, conservation of freshwater molluscs is not as far advanced as in the U.S.A. but interest is increasingly being taken in the issues involved. The work of the European Invertebrate Survey on non-marine molluscs is discussed in the section on terrestrial molluscs, and conservation action for the Freshwater Pearl Mussel Margaritifera margaritifera in Europe is described in the following review. The European aquatic snail Myxas glutinosa, known in Britain from only two sites (53), is listed under the U.K. Wildlife and Countryside Act of 1981 (48). The national Red Data Book for West Germany (47) and the regional RDB for the province of Steiermark in Austria (49) include several freshwater molluscs. In Austria, a reserve has been created for three small snails, (Bythinella pareissi, Theodoxus prevostianus and Fagotia acicularis audebartii) which are endemic to a hot spring near Bad Vöslau and came under threat when a bottling factory was built nearby. The establishment of this tiny 120 m long reserve means that activities that would have an effect on the spring may not be undertaken without the consent of the Molluscan Department of the Natural Museum in Vienna (10). Increasing efforts must be made to improve pollution control. Attempts are being made to counteract the effects of acid rain in Scandinavia by adding neutralizing materials such as calcium carbonate to affected water bodies and in Sweden about 1000 lakes have been treated in this way (13).

In other regions, little action has been taken and further research and documentation of the species involved is a high priority. The reviews of the Cuatro Ciénegas snails and the Tasmanian Freshwater 'Limpet' illustrate that similar problems to those encountered in Europe and the U.S.A. occur in other parts of the world.

REFERENCES   1. Stansbery, D.H. (1971). Rare and endangered mollusks in the Eastern United States. In Jorgensen, S.E. and Sharp, R.W. (Eds), Proc. Symp. on Rare and Endangered Mollusks (Naiads). U.S. Dept. of the Interior, Fish and Wildlife Service.

2. Stansbery, D. (1970). Eastern freshwater mollusks. (1). The Mississippi and St. Lawrence River systems. In Clark, A.H. (Ed.), Papers on the rare and endangered mollusks of North America. Malacologia 10(1): 9-20.

3. Davis, G.M. (1977). Rare and endangered species: a dilemma. Frontiers 41(4): 12-14.

4. Anon. (1982). Virginia's endangered mussels studied by State's Co-op Fishery Research Unit. Endangered Species Technical Bulletin 7(3): 6-7.

5. Coker, R.E. (1919). Freshwater mussels and mussel industries of the United States. Bur. Fish. Bull. 36: 13-89. Bur. Fish Doc. 865 (quoted in 2).

6. Stansbery, D.H and Stein, C.B. (1971). Why naiades (pearly freshwater mussels) should be preserved. Stream Channelization (Part 4) Hearings before a subcommittee of the Committee on Government Operations, House of Representatives, 92nd Congress, 1st Session, 14 June: 2177-2179.

7. Clarke, A.H. (1981). Introduction. In Proceedings of the Second American Malacological Union Symposium on the Endangered Mollusks of North America. Bull. Am. Mal. Union, Inc. 1981: 41-42.

8. Chambers, S.M. (1981). Protection of mollusks under the Endangered Species Act of 1973. Bull. Am. Mal Union, Inc. 1981: 55-59.

9. Imlay, M.H. (1977). Competing for survival. Water Spectrum 9(2): 7-14.

10. Paget, O.E. (1981). In litt., 27 March.

11. Nelson, D. (1982). Relocation of Lampsilis higginsi in the Upper Mississippi River. In U.S. Army Engineer Waterways Experiment Station, CE, Report of Freshwater Mollusk Workshop, Vicksburg, Mississippi.

12. Jenkinson, J.J. (1982). Cumberlandian mollusks conservation program. In U.S. Army Engineer Waterways Experiment Station, CE, Report of Freshwater Mollusks Workshop, Vicksburg, Mississippi.

13. Rodhe, W. (1981). Reviving acidified lakes. Ambio 10(4): 195-196.

14. Okland, K.A. and Kuiper, J.G.J. (1980). Smamuslinger (Sphaeriidae) i ferskvanni Norge - utbredelse, okologi og relasjon til forsuring. Internal Report IR 61/80. SNSF Project 'Acid precipitation - effects on forest and fish', Aas-NLH, Norway.

15. Franz, R. (Ed.) (1982). Invertebrates. Vol. 6 In Pritchard, P.C.H. (Ed.), Rare and Endangered Biota of Florida. University Presses of Florida.

16. Okland, J. (1980). Factors regulating the distribution of

freshwater snails (Gastropoda) in Norway. Haliotis 10(2): 108.

17. Ant, H. (1976). Arealveränderungen und gegenwärtiger Stand der Gefährdung mitteleuropäischer Land-und Süsswasser Mollusken. Schriftenreihe für Vegetationskunde 10: 309-339.

18. U.S.D.I. Fish and Wildlife Service (1982). Emergency protection approval for two Ash Meadows fishes. Endangered Species Technical Bulletin 7(6): 1,3-4.

19. Goodrich, C. (1944). Pleuroceridae of the Coosa River basin. Nautilus 58: 40-48.

20. Giusti, F. (1981). In litt., 14 March.

21. Kerney, M. and Stubbs, A. (1980). The Conservation of Snails, Slugs and Freshwater Mussels. Nature Conservancy Council Interpretive Branch, Shrewsbury, U.K.

22. Lloyd-Evans, L. (1975). The biogeography of snails in Yorkshire. The Naturalist (1975): 1-12.

23. Stansbery, D.H. and Stein, C.B (1976). Changes in the distribution of Io fluvialis (Say, 1825) in the Upper Tennessee River System (Mollusca: Gastropoda: Pleuroceridae). Bull. Am. Mal. Union, Inc. 1976: 28-33.

24. Stansbery, D.H. (1961). A century of change in the naiad population of the Scioto River system in central Ohio. Ann. Rep. Am. Mal. Union: 20-22.

25. Buchanan, A.C. (1980). Mussels (Naiades) of the Meramec River Basin, Missouri. Missouri Dept. of Conservation, Aquatic Series 17, Jefferson City, Missouri.

26. Stansbery, D.H. (1965). The naiad fauna of the Green River at Munfordville, Kentucky. Ann. Rep. of Am. Mal. Union: 13-14.

27. Stansbery, D.H. (1976). The occurrence of endangered species of naiad molluscs in Lower Alum and Big Walnut Creeks. Report to Ohio Dept. of Transportation, OSUMZ Report 17.

28. Dennis, S. (1969). Pennsylvania mussel studies. Final Report, Eastern Michigan University, Ypislanti, Michigan.

29. FAO (1982). Statistics for freshwater pearl mussels. Fisheries Division, FAO, Rome.

30. Williams, J.D. (1981). Distribution and habitat observations of selected Mobile Basin unionid mollusks. In U.S. Army Engineer Waterways Experiment Station, CE, Report of Freshwater Mollusks Workshop, Vicksburg, Mississippi.

31. Imlay, M.J. (1982). Use of shells of freshwater mussels in monitoring heavy metals and environmental stresses: a review. Malacologial Review 15: 1-14.

32. Clarke, A.C. (1982). The recognition of phenotypes in Unionidae. In U.S. Army Engineer Waterways Experiment Station, CE, Report of Freshwater Mollusks Workshop, Vicksburg, Mississippi.

33. Fuller, S.L.H. (1974). Clams and mussels (Mollusca: Bivalvia). Ch. 8 in Hart, C.W. and Fuller, S.L.H. (Eds), Pollution Ecology of Freshwater Invertebrates. Academic Press, New York.

34. Cairns, J., Crossman, J.S., Dickson, K.L. and Herricks, E. (1971). The recovery of damaged streams. Ass. Southeast Biol. Bull. 18: 49-106.

35. Schwartz, J. and Meredith, D.G. (1962). Mollusks of the Cheat River watershed of West Virginia and Pennsylvania, with comments on present distributions. Ohio J. Sci. 62:

203-207.

36. Wurtz, C.B. (1962). Zinc effects on freshwater mollusks. Nautilus 76: 53-61.
37. Harman, W.N. (1974). Snails (Mollusca: Gastropoda). Ch. 9 in Hart, C.W. and Fuller, S.L.H. (Eds), Pollution Ecology of Freshwater Invertebrates. Academic Press, New York.
38. Miller, A.C. (1982). The mollusk study. In U.S. Army Engineer Waterways Experiment Station CE. Report of Freshwater Mollusks Workshop, Vicksburg, Mississippi.
39. Lefevre, G. and Curtis, W.C. (1912). Studies on the reproduction and artificial propagation of freshwater mussels. Bur. Fish. Bull. 30: 103-201, Bur. Fish. Doc. 756 (quoted in 1).
40. Imlay, M.H. and Paige, M.L. (1972). Laboratory growth of freshwater sponges, unionid mussels, and sphaerid clams. The Progressive Fish Culturist 34(4): 210-216.
41. Smith, D.G. (1981). In litt., 25 August.
42. Kerney, M. (1982). The mapping of non-marine Mollusca. Malacologia 22(1-2): 403-407.
43. Stansbery, D.H. (1964). The Mussel (Muscle) Shoals of the Tennessee River revisited. Ann. Rep. Am. Mal. Union: 25-28.
44. Havlik, M.E. and Stansbery, D.H. (1977). The naiad mollusks of the Mississippi River in the vicinity of Prairie du Chien, Wisconsin. Bull. Am. Mal. Union Inc.: 9-12.
45. Stansbery, D.H. (1966). Utilization of naiads by prehistoric man in the Ohio valley. Ann. Rep. Am. Mal. Union: 41-43.
46. Brown, J. (1979). Schistosomiasis in Saudi Arabia - a review. J. Saudi Arabia Nat. Hist. Soc. 24: 7-15.
47. Blab, J., Nowak, E. and Trautmann, W. (1977). Rote Liste der gefährdeten Tiere und Pflanzen in der Bundesrepublik Deutschland. Kilda Verlag, Greven.
48. Anon. (1981). Wildlife and Countryside Act 1981. Chapter 69. H.M. Stationery Office, London.
49. Gepp, J. (Ed.) (1981). Rote Listen Gefährdeter Tiere der Steiermark. Österriechischen Naturschutzbundes, Landesgruppe Steiermark.
50. Anon. (1974). Endangered wild animals in Ohio. Ohio Dept. of Natural Resources, Div. of Wildlife.
51. Stern, E.M. (1981) Higgin's Eye Mussel recovery plan: problems and approaches. In U.S. Army Engineer Waterways Experiment Station CE, Report pf Freshwater Mollusks Workshop, Vicksburg, Mississippi.
52. Bogan, A.E. and Parmalee, P.W. (in press). The mollusks. Vol.2. In Tennessee's Rare Wildlife, Tennessee Heritage Programme, Nashville, Tennessee.
53. Kerney, M.P. (1976). Atlas of the Non-marine Mollusca of the British Isles. European Invertebrate Survey/Conchological Society of Great Britain/Biological Records Centre, Institute of Terrestrial Ecology, Huntingdon, U.K.

We are very grateful to S. Chambers, M. Kerney and S. Morris for reviewing this account and to A.H. Clarke, D.G. Smith, D.H. Stansbery and others who reviewed earlier drafts.

Cochliopina milleri Taylor, 1966
Coahuilix hubbsi Taylor, 1966
Durangonella coahuilae Taylor, 1966
Mexipyrgus carranzae Taylor, 1966
M. churinceanus Taylor, 1966
M. escobedae Taylor, 1966
M. lugoi Taylor, 1966
M. mojarralis Taylor, 1966
M. multilineatus Taylor, 1966
Mexithauma quadripaludium Taylor, 1966
Nymphophilus minckleyi Taylor, 1966
Paludiscala caramba Taylor, 1966

Phylum   MOLLUSCA   Order   MESOGASTROPODA

Class   GASTROPODA   Family   HYDROBIIDAE

SUMMARY This group of recently described aquatic snails is part of a unique fauna, many species of which are still to be described, which is endemic to the intermontane desert basin of Cuatro Ciénegas, Coahuila State, northern Mexico. The snails, some of which are subterranean, are found in a variety of habitats including rivers, marshes and lagunas or salt lakes. The area is threatened by drainage for irrigation, increased recreational use and development of industry. The species considered on this data sheet are listed on CITES Appendix II but the only effective conservation measure will be habitat protection.

DESCRIPTION In 1966 18 molluscs were described from the Cuatro Ciénegas basin, Mexico, twelve of which are endemic hydrobiids and are listed above (2). A more recent study, made on living animals as well as shells, has resulted in slight alterations to the taxonomy (12). It appears that most of the endemic genera are closely related to the non-endemics found in the valley. The six nominal species of Mexipyrgus (see above) have been put into a single variable species, M. churinceanus, and three new taxa have been described: Nymphophilus acarinatus, Mexiostiobia manantiali and Coahuilix landyei (12).

The snails are generally small and have brown sculptured shells. Six of the genera (Coahuilix, Mexipyrgus, Nymphophilus, Mexithauma, Paludiscala and Mexiostiobia) are endemic to the basin. In most cases they differ morphologically from other members of their groups. Colour banding is rare in the Hydrobiidae but is found in the genus Mexipyrgus where its function is unknown. Mexipyrgus is also unusual in exhibiting marked sexual dimorphism and remarkable divergence in form. Interestingly the morphological similarities and differences between the species are not correlated consistently with their geographical location. The other genera are unusual in their elaborate sculpture, their shape, and in some cases their large size (2).

Cochliopina milleri shell 3 mm in diameter, broadly conical in shape; sculpture consisting of about 12-15 spiral cords covered by dark brown periostracal colour bands. The most closely similar species are found in Guatemala (2).

Coahuilix hubbsi has a minute shell (1.0 x 0.5-0.6 mm in adults) with two and a quarter whorls. Aperture has flared lip; sculpture consists of raised riblets irregularly spaced. Only dead shells were found in 1966 (2), but live specimens found more recently show that the animal is blind and unpigmented (12).

<u>Durangonella coahuilae</u> shell 3-3.5 mm long with about five and a half whorls (2).

<u>Mexipyrgus carranzae</u> a comparatively large species, shell length reaching 7.0-7.2 mm with six and a half to six and three quarter whorls. Shell is sculptured, light to dark brown in colour with 6-8 wide brown bands (2).

<u>M</u>. <u>churinceanus</u> shell 6.5-7.0 mm long with seven whorls. Sculptured, light brown in colour with 5-14 (8-10 commonly) sharply defined wide brown bands of variable spacing. Similar in size and colour to <u>M</u>. <u>lugoi</u> but narrower with fewer, wider bands (2).

<u>M</u>. <u>escobedae</u> shell length 4.5 mm, with five and a half whorls; 20-30 indistinct dark brown bands on dark brown background; sculpture is dominated by axial ribs; similar to <u>M</u>. <u>carranzae</u> in its heavy pigmentation and ribbing (2).

<u>M</u>. <u>lugoi</u> shell 7.3 mm long with six whorls and 25-35 narrow dark brown bands on a light brown background (2).

<u>M</u>. <u>mojarralis</u> shell 4.5-5.0 mm long with five and a half to six whorls; 2-4 brown bands on pale yellowish-brown background. Although geographically isolated from <u>M</u>. <u>churinceanus</u>, it is most similar to this species although occurring in closer proximity to <u>M</u>. <u>multilineatus</u> (2).

<u>M</u>. <u>multilineatus</u> shell 5.0-5.1 mm long with six whorls and 0-7 (usually 2-4) narrow brown bands at the beginning of the body whorl, and often with up to 20 narrow bands present on the outer lip (2).

<u>Nymphophilus minckleyi</u> unusually large compared with other members of this family, shell reaching 10 mm in length, and unusual in being conical in shape. Shell and body are light ochre in colour, and may be covered with algae (2).

<u>Mexithauma quadripaludium</u> globose shell, 7.5 mm long with four and a half whorls; sculpture consists of 10-12 spiral cords between which the periostracum is brown. Shell often covered with diatoms (2).

<u>Paludiscala caramba</u> turriform shell, 2.1-2.5 mm long with six and a half to seven and a half whorls, resembling the marine Epitoniidae in form. Only dead shells which were white and sculptured, were found in 1966 (2). Live animals collected more recently were found to be blind and unpigmented (12).

New taxa: <u>Nymphophilus acarinatus</u> has a smaller shell than <u>N</u>. <u>minckleyi</u> and no peripheral keel. <u>Mexistobia manantiali</u> has a small globose shell. <u>Coahuilix landyei</u> is much larger than <u>C</u>. <u>hubbsi</u> (12).

<u>DISTRIBUTION</u> The valley of Cuatro Ciénegas is an area of 40 x 30 km enclosed by ranges of the Sierra Madre Oriental in central Coahuila, north-eastern Mexico, and is part of the Chihuahuan Desert (2).

<u>Cochliopina milleri</u> known only from its type locality in the Rio Mesquites (2).

<u>Coahuilix hubbsi</u> known only from its type locality in the northernmost pool of Posos de la Becerra, 14 km south-west of Cuatro Ciénegas (2).

<u>Durangonella coahuilae</u> known for certain only from Laguna Grande close to the mouth of the Rio Churince. Empty shells of <u>Durangonella</u> sp. were found at Rio Mesquites and a spring tributary to El Mojarral but these may represent a different species (2).

<u>Mexipyrgus carranzae</u> found in Laguna Tio Candido only, 14 km south of Cuatro

Ciénegas (2).

M. churinceanus recorded from Laguna Churince, the type locality, and Posos de la Becerra, 14 km south-west of Cuatro Ciénegas (2).

M. escobedae known only from Laguna Escobeda, 12 km south of Cuatro Ciénegas (2).

M. lugoi known only from Rio Mesquites, 9 km south-west of Cuatro Ciénegas (2).

M. mojarralis known only from West Laguna in El Mojarral, 1.7 km east-northeast of the northern tip of the Sierra de San Marcos (2).

M. multilineatus known only from East Laguna in El Mojarral, 1.9 km east-northeast of the northern tip of the Sierra de San Marcos (2).

Nymphophilus minckleyi recorded from Rio Mesquites, Laguna Churince, an unnamed laguna west of Rio Churince, Posos de la Bercerra, West Laguna and East Laguna in El Mojarral, Laguna Escobeda, and Laguna Tio Candido (2).

Mexithauma quadripaludium recorded from Laguna Tio Candido, Laguna Churince, Posos de la Becerra, West and East Lagunas in El Mojarral, Laguna Escobeda (2).

Paludiscala caramba found in a spring tributary in the area of marshes and lagunas called El Mojarral (2).

New taxa: Nymphophilus acarinatus is found only in the south-east of the basin. Mexiostiobia manantiali is widespread and usually sympatric with Durangonella coahuilae. Coahuilix landyei is frequently sympatric with C. hubbsi (12).

POPULATION Unknown for all species.

HABITAT AND ECOLOGY The floor of the basin is at 740 m, the surrounding bare limestone mountains reaching 3000 m in altitude. The region is climatically arid with an average precipitation of less than 20 mm a month. Water originates from thermal (25°-35°C) springs, most of which are now drained by canals but which in the past terminated in closed shallow basins in which permanent saline lakes occurred. There are at least five separate surface drainages, of which only the one involving the Churince Laguna, Rio Churince and Laguna Grande is relatively undisturbed by man. It consists of a subterranean water source with a stream draining from it through marshy terrain to a closed lake where extreme concentration via evaporation occurs. In the other drainage systems the closed lakes are no longer connected with the water sources. These drainage systems result in several clearly defined types of aquatic habitat. Posos are pits formed by the foundering of subterranean channels and are fed by travertine-lined natural pipes from the underground water source. Characteristically these develop into larger lagunas, or lake springs, which may have a slow water current passing through them and range in size from 50-100 m in diameter. The bottom is usually covered with about 3-6 cm of flocculent ooze. The Rio Mesquites is the largest river and has travertine banks, with a marl rubble or sand bottom in areas of current, and mud and plant debris in the bays. The streams and channels produce extensive marshy areas with clear shallow water over soft mud. Throughout the area the water tends to be crystal clear, ranging in temperature from 16.3-31.0°C, and is highly mineralized, depositing salts in abundance. The yellow water lily Nymphaea is abundant in quiet areas of the rivers and in deep water in the lagunas (6,7).

No field study of the living snails has been made, and collecting efforts have been concentrated mainly in the large springs. Based on conditions at collecting localities, the ecological requirements of the species concerned are given below but it is possible that further field studies will reveal these to be broader or different (2).

Cochliopina milleri occurs in soft mud in backwaters and sheltered places along the edge of the stream, never in the main current. Often associated with Mexipyrgus lugoi (2).

Coahuilix hubbsi and C. landyei have both been found live in the small groundwater outlets of the basin and are obviously adapted to this environment and possibly to the subterranean waters of the basin as well (12).

Durangonella coahuilae commonly found in the upper 5-10 mm of the soft bottom mud of the laguna. None were found within about 4-5 m of the mouth of the Rio Churince, possibly because of the lower salinity (2).

Species in the genus Mexipyrgus live in the upper 1-2 cm of the soft, flocculent ooze or mud of the lagunas in all depths of water. This somewhat discontinuous habitat may have contributed to isolation between populations and the resultant tendency towards speciation. Individuals probably feed on micro-organisms in the ooze and move little since they are always found buried (2).

Mexipyrgus carranzae found in soft bottom mud of Laguna Tio Candido. Although irrigation schemes have lowered the water level slightly this does not seem to have had any drastic effect on the snail fauna (2).

M. churinceanus found in typical Mexipyrgus habitat (2).

M. escobedae found abundantly in the soft flocculent ooze on the bottom of the laguna, which is a conical pool about 20 m in diameter and 10 m deep (2).

M. lugoi unlike Cochliopina milleri whose distribution it shares, M. lugoi is found in the main current of the stream as well as in the soft mud of the backwaters (2).

M. mojarralis abundant in the upper 1-2 cm of the soft flocculent ooze on the bottom of West Laguna. This is up to 7 m deep and fed by springs which emerge through vents up to 1 m in diameter at the north-western end (2).

M. multilineatus found in similar habit to M. mojarralis although East Laguna is shallower (1-2 m) than West Laguna (2).

Nymphophilus minckleyi found characteristically on Nymphaea (water lily) leaves, generally on the undersides at a depth of about 0.5 m. In the Laguna west of Rio Churince where there is no Nymphaea it was found only on the lower surfaces of limey algae masses and tufa. It may feed on lily leaves. Compared to other hydrobiids this species crawls at a moderate speed and moves smoothly. Egg capsules are smooth, hemispherical and contain a single embryonic snail (2).

Mexithauma quadripaludium usually found on a hard substrate such as stones or tufa, but sometimes on a firm shelly mud bottom. Probably feeds on algae or eats organic matter or micro-organisms in the superficial layers of mixed detritus in shallow waters (2).

Paludiscala caramba found in the small groundwater outlets of the basin; ecology probably similar to that of Coahuilix (12).

SCIENTIFIC INTEREST AND POTENTIAL VALUE The Cuatro Ciénegas molluscs make up one of the most spectacular endemic faunas of freshwater snails known in the Western Hemisphere. Nowhere else in North America are there six genera restricted to such a small area and among freshwater snails such a close grouping of related species evolved from a common ancestor is unknown. Unlike many freshwater snails, but like marine gastropods, most of the species are strikingly distinct in their shape or coloration. They provide a remarkable opportunity to study speciation, since in some cases a number of species are restricted to a single spring whereas others found with them are widespread in the basin. Mexipyrgus is proving to be a particularly interesting genus, many isolated populations having been found which show varied degrees of morphological differentiation correlated to degree of isolation, spring size, temperature and bottom type (12). Other significant findings are still possible since the subterranean valleys have yet to be examined, and a number of species additional to those mentioned in this data sheet have yet to be described. For example, shells from a species of Assiminea were collected from Posos de la Becerra, a rare inland occurrence of a usually coastal marine family (2).

The snails appear to provide an important source of food for the cichlid fish. The genus Cichlasoma includes a number of cryptic endemic species, one of which is adapted to feeding entirely on snails. Probably some 75 per cent of snails reaching maturity are eaten by fish (2). The Cuatro Ciénegas basin has within a few hundred kilometres a range of aquatic habitats unexcelled in any other known desert bolson (6), and a large number of other endemic species are known from the area. Three forms of turtle are endemic (5), one of which, the Aquatic Box Turtle, Terrapene coahuila, is listed in the IUCN Amphibia-Reptilia Red Data Book as Vulnerable (3). About half of the twenty fish species known there are restricted to the valley, four endemic subterranean isopods have been described (1,4) and there are a number of endemic scorpions (7). Relationships differ markedly from group to group but some species in each taxon are highly differentiated from their living relatives and must date from an early age, particularly the molluscs and crustaceans (7).

THREATS TO SURVIVAL   For many centuries, possibly since the 16th century, canals have been dug for irrigation purposes. Canalization lowers the water level to that of the standing head of the springs, which can effect the marshes seriously but does not seem to have such a serious impact on rivers and the bigger deeper lakes (7,10). In most drainages, the water sources are no longer connected with the rest of the system (6), and a few snail populations have already been lost.

Before 1964, Posos de la Becerra was one of the largest and most complex of the aquatic habitats in the Cuatro Ciénegas basin. It suffered some modification in 1961 through the construction of a bathing pool but in 1964 was drastically modified by the construction of a canal which caused the water level to fall 46 cm in two days. By April 1965 the extensive marshes associated with the spring had been drained and the overall surface of water and marsh had been reduced from about 10 km^2 to less than 0.2 km^2. In addition, swimmers had greatly disturbed the silty bottom of the Laguna and increased turbidity. Snails which had previously been common (Mexipyrgus, Mexithauma and Nymphophilus) had become rare and very few were found alive. An additional effect was the concentration of the snail-eating fish Cichlasoma into a small area, thus increasing competition for food. In December 1965 no snails could be found there (1). However, a more recent estimate suggests that only 0.25-0.5 km^2 of original habitat was lost (5). Laguna Escobeda, the only known locality for Mexipyrgus escobedae, has also been drastically altered. Its natural level was evidently 2 m higher before the present outlet was dug. M. escobedae is now the only snail found alive there, the aquatic vegetation is almost nil, fish are scarce and the biological diversity and productivity of the Laguna has been drastically reduced by the elimination of large areas of shallow vegetated water (2).

Current pressures operating on the aquatic habitats of the basin are not entirely clear. Some reports say that since the late 1960s the basin has undergone very little additional destruction (8,10). However, it has also been reported that pressure on the springs and seeps in the valley continues from agricultural usage, and that within the last five years the area around El Mojarral has become increasingly dry (9,13). Canals are being dug or deepened in some areas to provide adequate water flow in the Rio Salado de los Nadadores, which is the primary supply for a major steel mill complex at Mondova which uses large quantities of water (10). In 1979 a pipeline was being installed from Monterrey, the large industrial city to the south-east of Cuatro Ciénegas, which could interfere with the basin's drainage system (9). Many undescribed springs with Mexipyrgus churinceanus populations have had their outflows channelized to increase flow into the Rio Mesquites. A slower consequence of the canal system will be the mixing of streams and populations that were once naturally separate, a change which has already begun to affect the endemic fish and turtles (2,11).

Human use of the area for recreation is said to be increasing enormously and in 1979 the Rio Mesquite, which had previously been accessible only over a very bad dirt road, became the site of a commercial swimming resort (9). Laguna Churince was slightly dredged in 1981 and has been used as a swimming facility (13). The construction of the Cuatro Ciénegas to San Pedro de las Colonias highway now permits access to the basin from both the east and the south, bringing many more visitors to the area, and placing additional pressure on the basin (10). Another threat is the potential for commercial exploitation of the deposits associated with the springs and extensive dunes of the basin (7).

CONSERVATION MEASURES TAKEN The snails are listed on Appendix II on CITES (Convention on International Trade in Endangered Species of Wild Fauna and Flora), which means that trade in these species between countries which are party to the Convention requires a valid export permit from Mexico. However, there is no evidence of any trade in the snails or their shells.

CONSERVATION MEASURES PROPOSED An overall strategy is required for the conservation of the unique aquatic habitats of the Cuatro Ciénegas basin. Restrictions should be placed on the indiscriminate construction of canals that could result in the drainage of any of the areas, and steps should be taken to investigate fully the possible effects of any proposed irrigation projects in the surrounding areas (5).

REFERENCES
1. Cole, G.A. and Minckley, W.L. (1966). Speocirolana thermydronis, a new species of cirolanid isopod crustacean from central Coahuila, Mexico. Tulane Stud. Zool. 13: 17-22.
2. Taylor, D.W. (1966). A remarkable snail fauna from Coahuila, Mexico. The Veliger 9(2): 152-228.
3. Groombridge, B. and Wright, L. (1982). IUCN Amphibia-Reptilia Red Data Book Part 1., IUCN, Gland, Switzerland.
4. Cole, G.A. and Minckley, W.L. (1970). Speocirolana, a new genus of cirolanid isopod from northern Mexico, with descriptions of two new species. Southwest. Natur. 15: 71-81.
5. Brown, W.S. (1974). Ecology of the aquatic box turtle Terrapene coahuila (Chelonia, Emydidae) in northern Mexico. Bull. Florida State. Mus. Biol. Sci. 19(1): 1-67.
6. Minckley, W.L. and Cole, G.A. (1968). Preliminary limnologic information from the Cuatro Ciénegas area, Coahuila, Mexico. Southwest. Natur. 13: 421-431.
7. Minckley, W.L. (1969). Environments of the bolson of Cuatro Ciénegas, Coahuila, Mexico, with special reference to the aquatic biota. Sci. Ser. 2. Texas Western Press, Univ. of Texas, El Paso. 65 pp.

8. Brown, W.S. (1981). In litt. to B. Groombridge, 21 March.
9. Garstka, W.R. (1981). In litt. to B. Groombridge, 9 February.
10. McCoy, C.J. (1981). In litt. to B. Groombridge, 26 January.
11. Smith, H.M. and Smith, R.B. (1979). <u>Synopsis of the Herpetofauna of Mexico</u>. 6: Guide to Mexican Turtles. John Johnson, North Bennington, Vermont.
12. Hershler, R. (1982). In litt., 21 July.
13. Hershler, R. (1982). In litt., 1 July.

We are very grateful to R. Hershler and M. de la Garza for assistance with this data sheet.

TASMANIAN FRESHWATER 'LIMPET'   ENDANGERED

Ancylastrum cumingianus (Bourguignat, 1854)

Phylum   MOLLUSCA   Order   BASOMMATOPHORA

Class   GASTROPODA   Family   PLANORBIDAE

SUMMARY A large freshwater mollusc in a genus endemic to Tasmania, Australia. Little is known of its biology and ecology and predation by introduced trout could have a major impact on the remaining small population.

DESCRIPTION The pale yellow shell, 8-12 mm in height, is patelliform (limpet-shaped) and asymmetrical with a coiled spire over the right posterior quarter. The spire has fine spiral striae and the main part of the shell has 7-10 long longitudinal angular ribs (2,5).

DISTRIBUTION Confined to a few lakes and streams of the mountain region of central Tasmania, Australia (5,7). Specimens have been taken from the south end of Great Lake, below Weio at Meina (2) and from some of the adjacent smaller lakes (8).

POPULATION It is practically extinct, although isolated populations may still survive (4).

HABITAT AND ECOLOGY Found in cold, oligotrophic waters, living on rock surfaces or under stones in still or slowly flowing water. General biology not known (7).

SCIENTIFIC INTEREST AND POTENTIAL VALUE Ancylastrum is a genus endemic to Tasmania (3,6). Although it is not taxonomically a limpet, it is the only patelliform (limpet-like) member of the major freshwater gastropod family Planorbidae (4).

THREATS TO SURVIVAL The species has been reduced to its current status probably as a result of the introduction of trout as a sport fish. A well preserved specimen was recovered from a trout stomach in Great Lake in 1978 (5), and limpets were found to be of some importance as trout food in an early study (1).

CONSERVATION MEASURES TAKEN None.

CONSERVATION MEASURES PROPOSED Further information is required on the distribution and ecology of this species. Populations in lakes without introduced trout should be protected.

REFERENCES   1. Evans, J.W. (1942). The food of trout in Tasmania. Salmon and Freshwater Fisheries Commission, Hobart.
             2. Hubendick, B. (1964). Studies on Ancylidae. The subgroups. Medd. Goteborgs Musei Zool. Avd., 137: Goteborgs K. Vet.-Vitt.-Samk. Handl., Ser. B, 9(6): 1-72.
             3. McMichael, D.F. (1967). Australian freshwater molluscs and their probable evolutionary relationships. A summary of present knowledge. In Weatherley, A.H. (Ed.), Australian Inland Waters and their Fauna. Australian National University Press, Canberra.
             4. Smith, B. (1980). In litt., 8 September.
             5. Smith, B.J. and Kershaw R.C., (1979). Field guide to the

non-marine molluscs of south eastern Australia. ANU Press, Canberra.

6. Lake, B.S. (1974). Conservation. In Williams, W.D. (Ed.), Biogeography and Ecology in Tasmania. W. Junk, The Hague.
7. Smith, B.J. and Kershaw, R.C. (1981). Tasmania Land and Freshwater Molluscs. Fauna of Tasmania, Univ. of Tasmania.
8. Smith, B.J. (1982). In litt., 28 October.

We are very grateful to B.J. Smith for information provided for this data sheet.

Epioblasma Rafinesque, 1831 (= <u>Dysnomia</u>)

E. arcaeformis (Lea, 1831)	Sugar-spoon	Ex
E. biemarginata (Lea, 1857)		E
E. brevidens (Lea, 1831)		V
E. capsaeformis (Lea, 1834)		V
E. curtisi (Utterback, 1915)	Curtis Pearly Mussel	E
E. flexuosa (Rafinesque, 1820)		Ex
E. florentina (Lea, 1857)	Yellow-blossom Pearly Mussel	E
E. haysiana (Lea, 1834)	Northern Acorn Riffle Shell	E
E. lefevrei (Utterback, 1915)	Lefevrei's Riffle Shell	E
E. lenior (Lea, 1842)		Ex
E. lewisi (Walker, 1910)	Fork shell	E
E. metastriata (Conrad, 1840)		E
E. othcaloogensis (Lea, 1857)	Southern Acorn Riffle Shell	E
E. penita (Conrad, 1834)	Penitent Mussel, Southern Comb Naiad	E
E. personata (Say, 1929)		Ex
E. propinqua (Lea, 1857)		Ex
E. sampsoni (Lea, 1861)	Sampson's Pearly Mussel	E
E. stewardsoni (Lea, 1852)		Ex
E. sulcata delicata (Simpson, 1900)	White Catspaw Mussel	E
E. s. sulcata (Lea, 1829)	Purple Catspaw Mussel	E
E. torulosa gubernaculum (Reeve, 1865)	Green-blossom Pearly Mussel	E
E. t. rangiana (Lea, 1839)	Tan-blossom Pearly Mussel	E
E. t. torulosa (Rafinesque, 1820)	Tubercled-blossom Pearly Mussel	E
E. triquetra (Rafinesque, 1820)	Snuffbox	V
E. turgidula (Lea, 1958)	Turgid-blossom Pearly Mussel	E
E. walkeri (Wilson and Clark, 1914)	Brown-blossom Pearly Mussel, Tan Riffle Shell	E

Phylum	MOLLUSCA	Order	UNIONOIDA
Class	BIVALVIA	Family	UNIONIDAE

SUMMARY The Riffle Shells are freshwater naiads or unionids, and are members of the unique molluscan fauna endemic to North America, most of which is under threat from pollution, large scale projects involving waterways, and commercial exploitation. Epioblasma is characteristic of riffles or shoals, i.e. shallow areas of streams with a sandy gravel substrate and rapid currents. Most species are restricted to the Tennessee and Cumberland River systems and are threatened by loss of habitat as a result of impoundment, channelization and pollution. Several species are probably extinct as they have not been collected within the last 50 years. E. triquetra is the most abundant and widespread species. The three least abundant species, now considered extinct, are E. arcaeformis, E. stewardsoni and E. biemarginata. Several species are listed under the U.S. Endangered Species Act, on CITES, and on state lists of threatened species. There are no recovery plans specifically for Epioblasma but some of the conservation measures being implemented for naiads in general should benefit the remaining members of this group.

DESCRIPTION The shells are small to medium in size (up to 8 cm in length), solid, and usually ovate to triangular in shape. The hinge teeth are well developed and vary in thickness. The periostracum is greenish-yellow to brown, usually with green rays which may or may not be interrupted. The shells are strongly sexually dimorphic and in most species shells of females have a greatly expanded projection covering the marsupial area of the gills (3,32).

There is much disagreement in the current literature as to which generic name should be used for the group. Dysnomia (Agassiz, 1852) was the name most frequently used from 1922 to the present time but recently the older but formerly rejected name Epioblasma (Rafinesque, 1831) has often been used. In 1978 it was proposed that the name Plagiola (Rafinesque, 1891) should be used (32). Until the problem is considered by the International Commission on Zoological Nomenclature, it is generally agreed however, that the name Epioblasma should still be used (41). Within the genus there is further taxonomic confusion between species and subspecies. In many cases this is due to their morphological similarity and the fact that many exhibit ecophenotypic variation, that is, shell form and size vary according to environmental conditions. The same species may therefore have differently shaped shells depending on whether individuals belong, for example, to a river population or to a lake population (32). Many of the species are extremely difficult to tell apart, but there are a few exceptions, which are mentioned below:

E. triquetra, the most widely distributed and most primitive species exhibits little morphological variation. It does not resemble any other member of the genus and may be distinguished by its long triangular outline, sharply truncated posterior end and rows of green mottling (32,35). The shells are highly sexually dimorphic (21). Adults are commonly 2-5 cm long.

E. brevidens male shells can usually be distinguished from shells of other species by the tendency for the green rays, which are often present on the entire surface, to be broken into dots (32).

E. arcaeformis may be distinguished by the extreme inflation of both male and female shells (32). It is medium sized, reaching 7 cm in length (28).

E. lenior is distinguished by its thin shell and delicate green rays which are restricted to the posterior end (32).

E. torulosa exhibits considerable ecophenotypic variation. Individuals in small streams tend to lack tubercles and to be more compressed, whereas those in large streams show strong tubercle development. Since considerable variation also occurs between lake and stream specimens there has been a proliferation of names for this species (32). Three subspecies are generally recognized. E. torulosa torulosa seems to integrate with E. t. rangiana in the upper Ohio tributaries and with E. t. gubernaculum in the upper Tennessee tributaries (22,37).

E. sampsoni is similar to E. torulosa and has been regarded as a variant of E. torulosa rangiana (18), although more recently it has been recognized as a full species (32).

There is little morphological difference between E. florentina and E. curtisi, and the latter is often considered a subspecies of the former (22,32). E. walkeri has also been considered a subspecies of E. florentina since the two forms intergrade and are not geographically isolated. The former is regarded by Johnson as an ecophenotypic variation of E. florentina, with no taxonomic status (32).

E. haysiana is easily distinguished from other members of the genus by its polished, tawny to chestnut periostracum, small size, and unusually thick and heavy shell with purple nacre (32).

E. flexuosa is strikingly dimorphic. The female's marsupial swelling may be dark green, although specimens from the Ohio River do not usually have this coloration (32). It is probably a sibling species of E. stewardsoni as the two species show a number of similarities. E. lewisi is also very similar and has been considered a phenotypic variant of E. flexuosa.

The nacre colour of E. sulcata exhibits ecophenotypic variation, changing from purple to white through its range (32).

The other species have no particular distinguishing features. Full descriptions and details of taxonomy are given in (3,19,22,32).

DISTRIBUTION All but one species (E. penita from the Mobile-Alabama-Coosa river system in Alabama) occur in the Tennessee River system, U.S.A., and fourteen are also found in the Cumberland River system, U.S.A. (32). Past and present distributions of individual species are given below but it is recognised that these may not be as complete or as accurate as might be hoped for, due to taxonomic confusion in the literature.

E. triquetra Former distribution:- Upper White River system, Missouri; Missouri River drainage, Kansas and Missouri; Mississippi River system, Wisconsin and Iowa; Illinois River drainage, Illinois; Tennessee and Cumberland River systems; Green River drainage, Kentucky; Ohio River system from Indiana to Pennsylvania; St Lawrence River system: Lake Michigan and Lake Erie (32).

In Missouri it used to occur in the Meramec, Bourbeuse and St. Francis Rivers (26,35). By 1980 was thought to exist only at seven scattered sites in the Bourbeuse although a 1964 survey reported it from Dry Fork of the Meramec, Big River, and Huzzah, Courtois and Terre Bleue creeks (35).

It may be extinct in Kansas although there are old records from Marais des Cygnes River, Franklin Co. and the Wakarusa River dating from the early part of this century (27).

It is the only member of the genus found in the Mississippi River system (Upper Mississippi) (32) but has never been widely recorded there. Appropriate habitats are not plentiful and have now almost gone, and it may be extinct in this area (39).

Was collected from the Illinois River, Illinois until 1911 but it was probably never common here (36).

In Alabama restricted to the Tennessee River system (22) including Bear Creek (10,15); it was found in Mussel Shoals in 1925 but not in 1963 (55). Was found in the Duck River in Tennessee at least until 1968 (11,14). In 1972, recorded from the Clinch River above Norris Reservoir in Virginia and Tennessee (49).

Recorded from Green River, at Munfordville, Kentucky in the early 1960s (44).

In Pennsylvania found recently (1969) only in French Creek. It was once widely distributed although uncommon, occurring in the Allegheny River (Armstrong Co.), French Creek and Crooked Creek; also found rarely in Monongohela and the Beaver drainage (30). Still occurs in the Scioto River system, central Ohio (51) and the Muskingum River in eastern Ohio (64).

In Michigan it has been collected alive from only four sites in the last 20 years, but it is probably more widespread (21,43). It occurs in the upper mainstream and lower North Branch of the Clinton River, Michigan but has disappeared from the lower mainstream (24,43). Has been recorded from Lake Erie, Lake St Clair, Lake Huron and from scattered localities in rivers throughout southern Michigan (21,43,62), such as Little Portage River (in south-eastern Michigan) (21).

E. brevidens Former distribution:- Tennessee River system, Virginia, Tennessee and Alabama; Cumberland River system, Kentucky and Tennessee (32).

It may still be extant in the following localities: the headwater tributaries of the Tennessee, including the following: the Powell River from Virginia (Lee Co.) south to Tennessee (Claiborne, Hancock and Campbell Cos.); the Clinch River above Norris Reservoir from Virginia (Lee and Scott Cos.) to Tennessee (Hancock and Anderson Cos.) (28,49); the North Fork Holston River, Virginia (Lee Co.) downstream to Tennessee (Knoxville, Knox Co.); the Nolichucky River (63). It has been found in the main channel of the Tennessee River at Knoxville (Tennessee) and down river as far as Mussel Shoals (Alabama) (15), but no longer occurred in the latter in 1963 (55). The Elk River (Tennessee) contained a population which probably extended into Alabama (15). The species may still occur in the Duck River, Tennessee (Marshall and Maury Cos.) (14). In the Cumberland drainage it used to occur in the Stones River (Rutherford and Davidson Cos.), the Caney Fork River (Putnam Co.) and the main stream of the Cumberland River at Nashville (Davidson Co.) (28).

E. penita Former distribution:- Mobile-Alabama-Coosa River system, Georgia, Alabama and Mississippi (3,22,32).

It may have occurred in the lower Tombigbee and lower Black Warrior Rivers before their impoundment (33). The only recent specimens are those from the Tombigbee where it is found in a stretch of about 25 miles (40 km) between the Aberdeen impoundment, Mississippi, and Gainesville, Alabama (22,33,58).

E. arcaeformis Former distribution:- Tennessee River system, Tennessee and Alabama; Cumberland River system, Kentucky and Tenessee (32).

Has never been found in great numbers but was reported to be locally abundant in the Holston River drainage of the Tennessee River system. Used to occur in the lower section of the Clinch River which is now impounded (49). Recorded from Mussel Shoals, Tennessee River, Alabama in 1925 but not in 1963 (55). Has not been seen for 50 years and is presumed extinct, since its entire range is under a series of impoundments (22,32).

E. lenior Former distribution:- Tennessee River system, Virginia, Tennessee and Alabama; Cumberland River system, restricted to Stones River, Tennessee (32).

It was recorded from the upper Clinch River (Virginia and Tennessee) early this century (63) but had disappeared by 1972. The last known population is now covered by the Priest Reservoir on the Stones River in Tennessee, and the species is considered extinct, having not been seen since 1918 (22,32).

E. torulosa Former distribution:- Tennessee River system, Tennessee and Alabama; Ohio River system, from Illinois to Pennsylvania, including Wabash Green, Licking and Kentucky River drainages; St Lawrence River system: Lakes Michigan, Huron and Erie (32).

The E. torulosa complex was once common across Ohio (a midden at the McGraw Site revealed E. torulosa to be the second most abundant shell present) but collecting in the 1960s revealed that it was living only in three very restricted localities within the State (53). In Michigan it was originally present in scattered localities but has not been collected alive for over 40 years, although it could persist in the relatively undisturbed drainages of the south-east (43). <u>E</u>. torulosa was recorded from Green River at Munfordville, Kentucky in the early 1960s (44) but had disappeared from Mussel Shoals, Tennessee River, Alabama by 1963 (55).

Of the three subspecies recognized, <u>E</u>. <u>t</u>. gubernaculum and <u>E</u>. <u>t</u>. torulosa are considered to be almost extinct.

<u>E</u>. <u>t</u>. gubernaculum was described from southern headwaters in eastern Tennessee and western Virginia including the Clinch and Powell drainages and the Holston River, North Fork; some specimens have been found in the southern Appalachians. The most recent records are from the Clinch River above Norris Reservoir in 1972 (49) and the subspecies may now be extinct (37,59).

<u>E</u>. <u>t</u>. torulosa used to be abundant in the Tennessee River in northern Alabama, up past Mussel Shoals into the lower Clinch and Holston Rivers, the lower Wabash, lower New, lower Scioto and lower Nolichucky Rivers (18,22,37,55). In the Ohio River drainage it was found in tributaries in Indiana, Ohio, Pennsylvania, west Virginia and Kentucky (3,37); in the Great Lakes system it occurred in Lake Erie, possibly Lake St Clair, Michigan, Ontario (Canada), Ohio and Indiana (3,37). It was absent from the Scioto R. in central Ohio by 1961 (51). It was said to have been collected in commercial operations on the lower Ohio River (Kentucky-Illinois) in 1967 (16) and from the Nolichucky River near its mouth in western Tennessee in 1970 (18), but it was not found above the impoundment on the Nolichucky in 1974 (37) and it may now be extinct (22,37).

<u>E</u>. <u>t</u>. rangiana is restricted to the Ohio River and Lake Erie-St Clair drainage systems (18,22,37); it used to occur in the section of the Clinton River (a tributary of Lake St Clair, north of Detroit, Michigan) at Pontiac but was not found there in 1978 (24). Its original range in the Great Lakes area included the basins of Lake Erie and Lake St Clair, where it was always very rare and is now very scarce (18,62). In 1959 it was recorded at Alum Creek and Big Walnut Creek, Franklin Co., Ohio (50). Since 1964 fresh shells or living specimens have been seen at Big Darby Creek, Ohio; St Joseph River, Ohio and Indiana; (possibly Mohican-Walhonding Rivers, Ohio); Allegheny River (especially French Creek); Elk River, western Virginia; Green River, Kentucky; Sydenham River, Ontario, Canada; Bass Is., Lake Erie (37). In 1961 it still occurred in the Scioto River system, central Ohio (51). The Big Darby, Sydenham and French Creek populations are probably the only remaining breeding populations (37), the most healthy of which is probably that of the Sydenham River (25).

<u>E</u>. sampsoni Former distribution:- Tennessee River system, Tennessee; Ohio River system, lower Wabash River drainage, Indiana; Ohio River drainage to Cincinnati, Hamilton Co., Ohio (32).

It was formerly found in fair numbers in the lower Wabash River but is considered reproductively extinct now (18,31,32).

<u>E</u>. propinqua Former distribution:- Tennessee River system, Tennessee and Alabama; Cumberland River system, Tennessee; Ohio River system from lower Wabash River drainage, Indiana to Ohio River, Cincinnati, Hamilton Co., Ohio (32).

Considered extinct as it has not been seen for 50 years although it was once common on the shoals of the Tennessee River in northern Alabama (22), and used to occur in the lower section of the Clinch River which is now impounded (49).

<u>E</u>. biemarginata Former distribution:- Tennessee River system, Tennessee and Alabama; Cumberland River system, Big South Fork, Kentucky (32).

133

Had disappeared from Mussel Shoals in the Tennessee River, Alabama by 1963 (55). Now considered extinct since the last known population, in the Elk River, was smothered by quarry washings (18,22).

E. capsaeformis Former distribution:- Tennessee River system, Virginia, Tennessee and Alabama; Cumberland River system, Kentucky and Tennessee (3,32).

Had disappeared from Bear Creek (10) and Mussel Shoals (Tennessee River, Alabama) by 1963 (55). Recorded from Duck River in 1968 (11), and the Clinton and Powell River drainages and Holston River North Fork in 1979 (23). Recorded from the Clinch River above Norris Reservoir in 1972 (49).

E. florentina Former distribution:- Upper White system, Missouri; Tennessee River system, Virginia, Tennessee and Alabama; Cumberland River system, Kentucky and Tennessee (32).

It is now restricted in the Tennessee River system to the South and Middle Forks of the Holston in Virginia (18,23) although it might still occur in the Clinch River in Virginia and Tennessee. It had disappeared from Mussel Shoals, Tennessee River, Alabama by 1963 (55).

E. curtisi was known only from Black River (19) and Castor River in Missouri where it was very rare (26). It was collected from the former in small numbers in 1964 (22) and from the Castor River in the 1970s (57) although it has been listed as extinct (22).

E. walkeri Former distribution:- common in the Cumberland and Tennessee River systems (1).

Now found only in the lower Red River of the Cumberland system in Kentucky and Tennessee (1), in the Middle Fork of the Holston River in Virginia (54,59) (disappeared from the Upper South Fork (52)), possibly in the Stones River in Tennessee, in the Duck River, Tennessee, from Wilhoite Mill downstream to Columbia (1) and in 1972 was found in the Clinch River in the Cedar Bluff area above Norris Reservoir (1,49).

E. turgidula Former distribution:- Upper White River system, Missouri and Arkansas; Tennessee River system, Tennessee and Alabama; Cumberland River system (32).

Had disappeared from Mussel Shoals in the Tennessee River, Alabama by 1963 (55). The population in the headwaters of the Elk River was destroyed in 1967 by quarry washings (29). Subsequently restricted to the Duck River in the vicinity of Normandy, Bedford Co., Tennessee which is the site of the Normandy dam. May now be extinct (18,29).

E. lefevrei Known only from Black River, Missouri and may now be extinct (19,26).

E. personata Former distribution:- Tennessee River system, Tennessee and Alabama; Cumberland River system, Tennessee; Ohio River system, from lower Wabash drainage to Cincinnati, Hamilton Co., Ohio (32).

Has not been seen for 50 years and is probably extinct (22). Had disappeared from Mussel Shoals in the Tennessee River, Alabama by 1963 (55).

E. haysiana Former distribution:- Tennessee River system, Virginia, Tennessee and Alabama; Cumberland River system, Kentucky and Tennessee (3,32). Had disappeared from Mussel Shoals on the Tennessee River by 1963 (55). Last recorded population (1972) was in the Clinch River, Virginia, between St Paul and Dungannon, a distance of about 10 miles (16 km) (18,49). Probably now extinct (22,23).

E. flexuosa Former distribution:- Tennessee River system, Tennessee and Alabama; Cumberland River system, Kentucky; Ohio River system, from lower Wabash River, Indiana to Ohio River, Jefferson Co., Ohio (32).

It may now be extinct as it has not been collected since the construction of the Tennessee Valley Authority dam on the Tennessee. Although once locally abundant in the Ohio River, near Cincinnatti, Hamilton Co., Ohio, it has not been collected there since 1900 (18,22,32).

E. lewisi In 1964 it still occurred in the Cumberland River but may now be extinct as it has not been collected since the construction of Wolf Creek Dam (22). Used to occur in the lower section of the Clinch River which is now impounded (49).

E. stewardsoni Former distribution:- Tennessee and Cumberland River systems, Tennessee and Alabama (32).

Was never found in very great numbers and is now presumed extinct, not having been seen for 50 years (22,32). Used to occur in upper Clinch River (63).

E. metastriata Alabama and Black Warrior Rivers of the Tombigbee system in Alabama; Etowah and Connasauga Rivers in Georgia (3,22).

E. othcaloogensis Upper Coosa River system: Othcalooga Creek, Gordon Co., Georgia and the Cahaba River, Alabama (3,22).

E. sulcata Tennessee River system, Tennessee and Alabama; Cumberland River system, Kentucky and Tennessee; Ohio River system, Wabash River drainage and Ohio River to Scioto River drainage, Ohio; St Lawrence River system, Lake Erie drainage (15,32).

Thought to be extinct but found in Tennessee in the Cumberland River in 1976 (9) and in 1977-78 in the same river in the vicinity of Rome Landing, Smith Co. (29). Had disappeared from Mussel Shoals, Tennessee River, Alabama by 1963 (55). Has not been recorded from Michigan this century but may still persist in St Joseph of the Maumee, or in streams tributary to Lake Erie or St Clair and in the Detroit River (43). E. s. delicata has been described from streams tributary to western Lake Erie and Lake St Clair in Michigan (19) and also from Lake St Clair (21). E. s. sulcata, although originally present in the lower Tennessee, is now known only from the Cumberland River system where it does not appear to be reproducing (20). It was considered extinct in 1976 but is apparently occasionally found in the Green River, Kentucky, (22,31,32).

POPULATION Unknown for all species, but from specimens in collections it is clear that many species now extinct or rare, were once abundant. These include E. florentina, E. penita, E. capsaeformis (once abundant in Tennessee River system), and E. brevidens. E. triquetra is the most successful member of the genus in that it is widely distributed and the most generally abundant (32).

HABITAT AND ECOLOGY Like other unionids, Epioblasma is adapted to highly oxygenated riffle habitats, now largely eliminated by man (39), and the ecology of most species is generally similar to that of other freshwater mussels. E. brevidens is found in moderate-sized clear streams with rocky bottoms and is absent from smaller tributary streams (28,32); it is a winter breeder (28). E. torulosa is found in coarse sand and gravel at a depth of a few cm to 1-2 m. (32). E. perplexa has been described from riffle areas, buried in firmly packed sand or gravel, and is

135

relatively intolerant of silt and mud, favouring shallow, fast-flowing water (30). E. triquetra is found in riffle areas, usually deeply buried in gravel, stone and sand in fast-flowing, shallow waters (27,30,32,40). E. penita is found in gravel shoal areas with moderate to swift current (33,42) and E. sampsoni has been found in sand and gravel but never in mud (32). E. flexuosa was described from muddy, deep water. E. lewisi is found in shallow riffles (32).

SCIENTIFIC INTEREST AND POTENTIAL VALUE This genus is unique among freshwater mussels in the extent and variety of sexual dimorphism found among its members. Extensive studies have been made on it since the beginning of this century and the distribution patterns of the different species indicate former confluences between present day river systems (32). In the past the shells were used to make buttons (5).

THREATS TO SURVIVAL Factors threatening North American unionid mussels in general were discussed in the introduction to this section and apply to most members of the genus Epioblasma. More specific threats are outlined below:-

E. triquetra Urban pollution seems to have been a major factor in the decline of this species in the Clinton River, Michigan (24). In general it is not plentiful wherever it is found in Michigan (21). Pollution probably caused its disappearance from the Illinois River (36). It is reported to be very uncommon in Missouri (35) and uncommon in Kansas (27), Illinois (16) and Indiana (61).

E. brevidens This is widely distributed, and localized pollution or other habitat degradation is unlikely to cause its total extinction. However populations throughout Tennessee are being reduced by siltation, strip mine effluents and impoundments (28).

E. penita Adverse impacts from stream channelization and alteration of fish distribution in the section of the Tombigbee river where it occurs may have caused the extinction of this species. The construction of the Tennessee-Tombigbee waterway (an operation by the U.S. Army Corps of Engineers) involves dredging channels and constructing locks and dams to deepen the Tombigbee river. This causes physical and chemical alteration of the main channel of the river and lowering of water levels in parts of its tributaries, thus isolating populations which increases the likelihood of hybridization between species in the Tennessee and Tombigbee Rivers (33,42,46). Three of the four proposed impoundments between Gainsville, Alabama, and Amory, Mississippi had been completed by 1981 and a fourth was scheduled for completion in 1982. The river has thus been converted into a series of reservoirs (58). The extinction of E. penita in the Alabama River was probably due to the relatively recent modification of this free-flowing river to a series of impoundments (2).

E. torulosa Has been affected by impoundments, domestic sewage, treatment plant effluents, industrial outfalls, agricultural silt and pesticide run-off, dredging and channelization (37). The populations described as E. torulosa rangiana have been particularly affected. The Big Darby Creek population was threatened by impoundments in 1974; the small St Joseph River population may have been eliminated by pollution from cannery wastes, copper, zinc and cyanide outfalls and agricultural pesticides; the Elk River population was large but may have been lost as a result of the construction of the Sutton Dam, since the water is now cold and quite acid; the Green River population may have been eliminated as a result of impoundment and there have been no records of the Bass I., Lake Erie population for ten years (37).

E. florentina is endangered by pollution from several sources. In 1978, the population in the Cedar Bluff area of the Clinch River system Tennessee was threatened by a proposal to discharge sewage into Little River, a tributary of the

Clinch (34).

E. curtisi is threatened by a proposed plan for the damming of the Little Black River with the aim of retarding floodwaters and providing recreational amenities, and the channelization of a portion of it (13).

E. walkeri is threatened by pollution is different parts of its range. Mine acids and municipal wastes have resulted in low concentrations of dissolved oxygen below Adairville; untreated effluent from a meat packing plant flows into the Red River system; at Murfreesboro in the west fork of the Stones River the dissolved oxygen content is very low; lead, mercury and a history of accidental spills (such as the collapse of a retaining wall in 1967 at Carbo, Virginia, sending 130 million gallons (591 million litres) of toxic industrial waste into the river) have caused pollution in the Clinch River. The construction of a dam at Columbia, Tennessee threatens to inundate 50 miles (80 km) of the Duck River, and channelization of the Upper Clinch river will probably also destroy populations (1,4).

E. sulcata delicata is said to be endangered in Michigan on account of its rarity (21), and the remaining Tennessee population of E. sulcata in the Cumberland River may disappear if incidental collection continues (29).

CONSERVATION MEASURES TAKEN Although many surveys have been carried out to determine the status of these species, many of which are now officially listed as Endangered, with a few exceptions little conservation action has been taken for them. E. torulosa gubernaculum and E. walkeri have been studied, (with the seven other mussel species listed as Endangered in Virginia), by the Fish and Wildlife Service's Virginia Cooperative Fishery Research Unit and the Biology Department of the Virginia Polytechnic Institute and State University with the aim of devising management plans (59), and E. curtisi is being similarly studied in Missouri (6). The Environmental Impact Research Program of the U.S. Army Engineer Waterways Experiment Station includes a Mollusk Study, the objectives of which are to collect information on sampling methods, biological and ecological requirements and habitat creation for selected common and Federally listed Endangered molluscs (17).

The status of individual species, both Federally and in different states, is given below. In addition, a list of rare and endangered molluscs in the eastern U.S.A. was drawn up in 1971 on which Epioblasma species feature as 'presumed extinct' (= X) or 'rare or endangered' (= E) (19). Several species are listed on Appendix I or II of CITES (Convention on International Trade in Endangered Species of Wild Fauna and Flora). These probably do not occur in trade unless taken incidentally during fishing for commercially valuable naiads. Those listed on Appendix I may not be traded for commercial purposes between party states; those on Appendix II may be traded provided an export permit from the country of origin has been issued.

E = endangered   T = threatened   S = of special concern
X = extinct   R = rare   P = possibly extinct

(N.B. see references for definitions of categories, since these are not necessarily the same as the IUCN definitions)

E. triquetra E Virginia (23); T Michigan (24); E Alabama (22); R Missouri (26); E/T Indiana (47).

E. brevidens T Alabama (22); E Virginia (23); S Tennessee Heritage Program (28).

E. penita E Alabama (22); E U.S. (19). In 1980 the U.S. Fish and Wildlife Service published a notice of status review to determine whether this species should be proposed for listing under the U.S. Endangered Species Act (42); the necessary information has been obtained (58) but the species has not yet been listed.

E. arcaeformis S Tennessee Heritage Program (28); X Alabama (22); X U.S.A. (19).

E. lenior P Tennessee Heritage Program (49); X Alabama (28); X U.S.A. (19).

E. torulosa E Tennessee (28,49).

E. t. torulosa X Alabama (22); E U.S.A. (19); E U.S. Endangered Species List (38); Appendix I CITES.

E. t. rangiana R Michigan (24); E Ohio (45); E Canada (25); E U.S.A. (19); Appendix II CITES.

E. t. gubernaculum E Virginia (23); E Tennessee (28,49); E U.S.A. (19); E U.S. Endangered Species List (38); Appendix I CITES.

E. sampsoni X Alabama (22); X U.S.A. (19); E U.S. Endangered Species List (38); Appendix I CITES.

E. propinqua X Alabama (22); X U.S.A. (19).

E. biemarginata P Tennessee Heritage Program (28); X Alabama (22); X U.S.A. (19).

E. capsaeformis E Virginia (23).

E. florentina X Alabama (22).

E. f. florentina E Tennessee (28,49); X U.S.A. (19); E U.S. Endangered Species List (38); Appendix I CITES.

E. f. curtisi E Missouri (26); E U.S.A. (19); E U.S. Endangered Species List (38); Appendix I CITES.

E. f. walkeri E Tennessee (28).

E. walkeri E Tennessee (49); E Virginia (23); E U.S.A. (19); E U.S. Endangered Species List (7); Appendix I CITES.

E. turgidula X Alabama (22); E Missouri (26); E Tennessee (28,49); X U.S.A. (19); E U.S. Endangered Species List (38); Appendix I CITES.

E. lefevrei E Missouri (26); E U.S.A. (19).

E. personata X Alabama (22); X U.S.A. (19).

E. haysiana E Virginia (23); X Alabama (22); P Tennessee Heritage Program (28); E U.S.A. (19).

E. flexuosa X Alabama (22); X U.S.A. (19).

E. stewardsoni X Alabama (22); X U.S.A. (19).

E. lewisi X Alabama (22); P Tennessee (28); X U.S.A. (19).

E. metastriata T Alabama (22); E U.S.A. (19).

E. othcaloogensis E Alabama (22); E U.S.A. (19).

E. sulcata R Michigan (48); E Tennessee Heritage Program (28).

E. s. delicata E U.S. Endangered Species List (38); E Ohio (45, = E. s. perobliqua);
Appendix I CITES (= E. s perobliqua); E U.S.A. (19, = E. s perobliqua).

E. s. sulcata X Alabama (22); E U.S.A. (19).

CONSERVATION MEASURES PROPOSED Recovery plans should be set up by the
U.S. Office of Endangered Species for those species known still to exist, but which
are limited to single or very few populations. Surveys should be carried out to
confirm the status of those species thought to be extinct, and recovery plans set up
should they be rediscovered. E. turgidula, for example, was recently rediscovered
but subsequently lost because of the absence of such a rescue programme (31).
Conservation of the habitat will clearly be the best approach for this group of
species, and full encouragement should be given to the Mollusk Study of the U.S.
Army Engineer Waterways Experiment Station.

Certain rivers would seem to merit particular attention. The naiad fauna of the
Clinch River above Norris Reservoir (Virginia and Tennessee) appears to be one of
the best preserved of the entire area. In spite of rubbish and obvious pollution in a
number of places, this part of the river still retains most of the species recorded
half a century ago. Twenty eight naiads endemic to the Cumberland and Tennessee
Rivers are still found there, including six species of Epioblasma (49). The Green
River probably has the finest representative Ohioan naiad fauna in existence, and
this may reach its peak at Munfordville, Kentucky. In the early 1960s, 64 species or
forms, including E. triquetra and E. torulosa, were recorded from the entire system
and the Munfordville site has yielded a greater variety of naiad species than any
other (44). For its size, the Duck River in Tennessee is one of the richest, having
over 50 species of extant naiads (37). Improved water pollution and siltation
control efforts are required in these rivers (8,29) and particular attention should be
paid to the habitat of Epioblasma species. In the Powell River implementation of
reclamation measures would improve the habitat for several endangered mollusc
species and lead to possible recolonisation of headwater areas (12). Action in these
rivers would contribute to the conservation of eight forms of Epioblasma (E.
torulosa gubernaculum, E. haysiana, E. brevidens, E. triquetra, E. florentina
florentina, E. walkeri, E. turgidula and E. sulcata sulcata).

In Canada, the Sydenham River in south-western Ontario is the richest site for
freshwater mussels, 32 unionid species having been recorded from it. A portion of
the eastern branch near Alvinston is particularly important, and since most of the
river is still unpolluted and unrestricted by dams, consideration should be given to
creating a reserve to protect at least one population of E. torulosa (25). For other
populations a minimum requirement is to maintain some reasonably natural habitat,
in particular, free-flowing streams (37).

Immediate action needs to be taken to ensure the future survival of E. penita and
the four other naiad species found in the same area of the Tombigbee River. In the
Upper Tombigbee R., the river bendways, in the upper reaches of each
impoundment where cut-offs have been constructed, resemble pre-impoundment
conditions and may be the only viable areas in which to maintain naiad populations.
An effort to re-establish flow and a clean gravel substrate in these places should be
made immediately, precautions to prevent siltation should be taken and
channelization projects for the tributaries should be discouraged (58). The U.S.
Army Engineer Waterways Experiment has begun a project to design and monitor a
gravel bar habitat for aquatic insects and molluscs to be placed in the Tombigbee
River. This may benefit the remaining E. penita population (60).

The conservation of E. brevidens in Tennessee will depend on the maintenance of water quality in the medium-sized rivers where it still occurs and on the curtailment of siltation and stripmine effluents (28). If any individuals of E. sampsoni are found, an artificial propagation programme should be initiated and the species could be introduced into its former range (31). Similarly, the relict population of E. sulcata in the Cumberland River, Tennessee, could be transplanted and an attempt made to propagate the species (29). E. sulcata and E. torulosa have been proposed for upgrading to Endangered status for the State of Michigan (43). In 1975 it was suggested that additional survey work should be carried out to determine the status of E. sulcata delicata in the drainages flowing into western L. Erie (21). The upper Castor River in Missouri should be declared critical habitat for E. curtisi (57).

REFERENCES

1. Anon(1977). Tan Riffle Shell determined to be endangered. Dept. of Interior News Release, U.S. Fish and Wildlife Service, 9 September.
2. Bates, J.M. (1962). The impact of impoundment on the mussel fauna of Kentucky Reservoir, Tennessee River. Amer. Midl. Nat. 68(1): 232-236.
3. Burch, J.B. (1975). Freshwater Unionacean Clams (Mollusca: Pelecypoda) of North America. Revised edition, Malacological Publications, Michigan.
4. Clarke, A.H. (1970). Introduction. In Clarke, A.H. (Ed.), Papers on the Rare and Endangered Mollusks of North America. Malacologia 10(1): 7-8.
5. Coker, R.E. (1919). Freshwater mussels and mussel industries of the United States. Bur. Fish. Bull. 36: 13-89. Bur. Fish. Doc. 865 (quoted in 19).
6. Buchanan, A.C. (1981). The distribution and habitat of the Curtis' Pearly Mussel, Epioblasma florentina curtisi (Utterback, 1915) in southeastern Missouri. (Abstract). Bull. Am. Mal. Union, Inc. 1981: 43.
7. U.S.D.I. Fish and Wildlife Service (1977). Endangered and threatened wildlife and plants. Determination that the Tan Riffle Shell is an Endangered Species. Federal Register 42 (163): 42351-42353
8. Imlay, M.J. (1977). Competing for survival. Water Spectrum 9(2): 7-14.
9. Isom, B.G., Gooch, C. and Dennis, S.D. (1979). Rediscovery of a presumed extinct river mussel, Dysnomia sulcata (Unionidae). The Nautilus 93 (2-3): 84.
10. Isom, B.G. and Yokley, P. (1968). Mussels of Bear Creek watershed, Alabama and Mississippi, with a discussion of area geology. Amer. Midl. Nat. 79: 189-196.
11. Isom, B.G. and Yokley, P. (1968). The mussel fauna of Duck River in Tennessee, 1965. Amer. Midl. Nat. 80: 34-42.
12. Dennis, S.D. (1981). Mussel fauna of the Powell River, Tennessee and Virginia. Sterkiana 71: 1-7.
13. McMillan, W. (1979). Channelization threatens otters, mussels, Little Black. Ozark Guardian. Sept. 1979: 2-3.
14. Ortmann, A.E. (1924). The naiad-fauna of Duck River in Tennessee. Amer. Midl. Nat. 9: 3-47.
15. Ortmann, A.E. (1925). The naiad-fauna of the Tennessee River System below Walden Gorge. Amer. Midl. Natur 9: 321-373.
16. Parmalee, P.W. (1967). The fresh-water mussels of Illinois. Ill. State Mus. Popular Sci. Ser. 8: 101-108 (quoted in 18).
17. Miller, A.C. (1982). The mollusk study. In U.S. Army Engineer Waterways Experiment Station, CE Report of

Freshwater Molluscs Workshop. Vicksburg, Miss.

18. Stansbery, D.H. (1970). Eastern freshwater mollusks (1). The Mississippi and St Lawrence River systems. In Clark, A.H. (Ed.), Papers on the rare and endangered mollusks of North America. Malacologia 10 (1): 9-20.

19. Stansbery, D.H. (1971). Rare and endangered mollusks in the Eastern United States. In Jorgensen, S.E. and Sharp, R.W. (Eds), Proceeding of a Symposium on the rare and endangered mollusks (naiads) of the U.S. Dept. of Interior, Fish and Wildlife Service.

20. Jenkinson, J.J. (1981). Endangered or threatened aquatic molluscs of the Tennessee River system. Bull. Am. Mal. Union, Inc. 1981: 43-45.

21. Van der Schalie, H. (1975). An ecological approach to rare and endangered species in the Great Lakes region. Michigan Acad. 8(1): 7-22.

22. Stansbery, D.H. (1976). Naiad mollusks. In Boschung, H. (Ed.), Endangered and threatened plants and animals of Alabama. Bull. Alabama Mus. Nat. Hist. 2: 42-52.

23. Dennis, S. (1979). Freshwater and terrestrial molluscs. In Linzey, D.W. (Ed.) Endangered and Threatened Plants and Animals of Virginia. Proc. Symp. May 1978, Virginia Polytechnic Institute and State University, Blacksburg, Virginia.

24. Strayer, D. (1980). The freshwater mussels (Bivalvia: Unionidae) of the Clinton River, Michigan with comments on man's impact on the fauna, 1870-1978. The Nautilus 94(4): 142-149.

25. Clarke, A.H. (1976). The endangered molluscs of Canada. In Mosquin, T. and Suchal, C. (Eds), Canada's Threatened Species and Habitats. Proc. Symp. May 1976. Ottawa, Canada.

26. Nordstrom, G.R., Pflieger, W.L., Sadler, K.C. and Lewis, W.H. (1977) Rare and Endangered Species of Missouri. Missouri Dept. of Conservation and U.S. Dept. Agric. Soil Conservation Service.

27. Murray, H.D. and Leonard, A.B. (1962). Handbook of Unionid Mussels in Kansas. Mus. Nat. Hist. Univ. of Kansas, Lawrence, Kansas.

28. Bogan, A.E. and Parmalee, P.W. (in press). The mollusks, Vol. 2. In Tennessee's Rare Wildlife, Tennessee Heritage Programme, Nashville, Tennessee.

29. Bogan, A.E. and Parmalee, P.W. (1979). Endangered or Threatened Mollusks of Tennessee. University of Tennessee, Knoxville.

30. Dennis, S. (1969). Pennsylvania mussel studies. Final report. Eastern Michigan University, Ypislanti, Michigan.

31. Imlay, M.J. (1981). Memo to J.M. Engel, 22 January.

32. Johnson, R.I. (1978). Systematics and zoogeography of Plagiola (= Dysnomia = Epioblasma), an almost extinct genus of freshwater mussels (Bivalvia: Unionidae) from middle North America. Bull. Mus. Comp. Zool. 148(6): 239-320.

33. Paradiso, J.F. (1976). Status of mussels and fishes in the Tennessee-Tombigbee Waterway. Memo to Assistant Secretary for Fish and Wildlife and Parks, 17 March.

34. Stansbery, D.H. (1978). In litt. to State Water Control Board, Richmond, Virginia, 24 August.

35. Buchanan, A.C. (1980). Mussels (Naiades) of the Meramec River Basin, Missouri. Missouri Dept. of Conservation,

Aquatic Series 17, Jefferson City, Missouri.

36. Starrett, W.C. (1971). A survey of the mussels (Unionacea) of the Illinois River: a polluted stream. Illinois Nat. Hist. Survey Bull. 30(5):1-403.

37. Stansbery, D.H. (1974). In litt. to A. Clarke, 8 April.

38. U.S.D.I. Fish and Wildlife Service (1976). Endangered status for 159 taxa of animals. Federal Register 41(115): 24062-24067.

39. Fuller, S.L.H. (1980). Freshwater mussels (Mollusca: Bivalvia: Unionidae) of the upper Mississippi River. Final report 79-24F. Academy of Natural Sciences, Philadelphia.

40. Fuller, S.L.H. (1980). Historical and current distributions of fresh-water mussels (Mollusca: Bivalvia: Unionidae) in the Upper Mississippi River. In Rasmussen, J.L. (Ed.), Proc. UMRCC Symp. on Upper Mississippi River Bivalve Mollusks: 72-80, Upper Mississippi River Conservation Committee.

41. Clarke, A.H. (1981). Determination of the precise geographical areas occupied by four endangered species of freshwater molluscs. Final Report to USFS, Minnesota, Contract No 14-16-003-81-019.

42. U.S.D.I. Fish and Wildlife Service (1980). Endangered and threatened wildlife and plants; review of the status of five mollusc species. Federal Register 45(72): 24904-24905.

43. Master, L.L. (1982). In litt., 26 January.

44. Stansbery, D.H. (1965). The naiad fauna of the Green River at Munfordville, Kentucky. Ann. Rep. of Am. Mal. Union : 13-14.

45. Anon (1974). Endangered wild animals in Ohio. Ohio Dept. of Natural Resources, Div. of Wildlife.

46. U.S.D.I. Office of Endangered Species (1976). Status of mussels and fishes in the Tennessee-Tombigbee Waterway. Memorandum to Assistant Secretary for Fish and Wildlife and Parks.

47. Anon. (1980). Indiana: non-game and endangered species conservation. A preliminary report. In Ref. 56.

48. Anon. (1980). Michigan. In Ref. 56.

49. Stansbery, D.H. (1973). A preliminary report on the naiad fauna of the Clinch River in the Southern Appalachian mountains of Virginia and Tennessee (Mollusca: Bivalvia: Unionoida). Bull. Am. Mal. Union. Inc. 38th Ann. Meeting, 1972: 20-22.

50. Stansbery, D.H. (1976). The occurrence of endangered species of naiad molluscs in Lower Alum and Big Walnut Creeks. Report to Ohio Dept. of Transportation. OSUMZ Report 17.

51. Stansbery, D.H. (1961). A century of change in the naiad population of the Scioto River system in central Ohio. Ann. Rep. Am. Mal. Union: 20-22.

52. Stansbery, D.H. and Clench, W.J. (1977). The Pleuroceridae and Unionidae of the Upper South Fork Holston River in Virginia. Bull. Am. Malac. Union. :75-78.

53. Stansbery, D.H. (1965). The molluscan fauna. Ch.11 in Prufer, O.H et al. (Eds), The McGraw Site - a Study in Hopewellian Dynamics N.S. 4(1), Cleveland Museum of Natural History.

54. Stansbery, D.H. and Clench, W.J. (1974). The Pleuroceridae and Unionidae of the Middle Fork Holston River in Virginia. Bull. Am. Mal. Union: 51-54.

55. Stansbery, D.H. (1964). The Mussel (Muscle) Shoals of the

Tennessee River revisited. Ann. Rep. Am. Mal. Union. 1964: 25-28.

56. Anon. (1980). Eastern States Endangered Wildlife. U.S. Dept. Interior, Bureau of Land Management, Alexandria, Virginia.

57. Stansbery, D.H. (1978). In litt., 29 September, to Office of Endangered Species, U.S.D.I., Washington D.C.

58. Williams, J.D. (1982). Distribution and habitat observations of selected Mobile Basin unionid mollusks. In Miller, A.C. (Compiler), Report of Freshwater Mollusks Workshop, Vicksburg, Miss.

59. Anon. (1982). Virginia's endangered mussels studied by State's Co-op Fishery Research Unit. Endangered Species Technical Bulletin 7(3): 6-7.

60. Miller, A.C. (1982). Habitat development for freshwater mollusks in the Tombigbee River near Columbus, Mississippi. In Miller, A.C. (Compiler), Report of Freshwater Mollusks Workshop. Vicksburg, Miss.

61. Goodrich, C. and van der Schalie, H. (1944). A revision of the mollusca of Indiana. Am. Mid. Nat. 32(2): 257-326.

62. Clarke, A.H. (1973). On the distribution of Unionidae in the Sydenham River, Southern Ontario. Malacological Rev. 6: 63-64.

63. Ortmann, A.E. (1918). The nayades (freshwater mussels) of the upper Tennessee drainage. With notes on synonymy and distribution. Proc. Amer. Philos. Soc. 57(6): 521-626.

64. Stansbery, D.H. (1974). An annotated list of the naiad mollusks of the Muskingum River in Eastern Ohio (Mollusca: Bivalvia: Unionoida). Museum of Zoology Ohio State University. Unpubl. manuscript.

We are very grateful to S. Chambers, A.H. Clarke, M.J. Imlay, R. Johnson, L. Masters, D.G. Smith and D. Stansbery for providing information and comments on this data sheet.

SPENGLER'S FRESHWATER MUSSEL   INDETERMINATE

Margaritifera auricularia (Spengler, 1793)

Phylum   MOLLUSCA   Order   UNIONOIDA

Class   BIVALVIA   Family   MARGARITIFERIDAE

SUMMARY A large freshwater mussel which once occurred in many West European rivers but which now seems to be verging on extinction. Surveys are required to determine where it still exists and, when it is found, the necessary measures should be taken to ensure its future survival.

DESCRIPTION Shell 120-165 mm in length, very thick and heavy. Detailed description in (2).

DISTRIBUTION Originally occurred throughout much of western, central and southern Europe. Since about 1850 it has been reported living in the following rivers: the Tagus (Portugal), the Ebro and Guadalquivir (Spain); the Po (Italy); and the Adour, Allier, Arros, Aube, Charente, Dordogne, Doubs, Garonne, Loire, Lot, Oise, Seine, Saône, Somme, Tarn and Vesle (France). However, some of these records may be based on fossils (4). It has long been extinct in England, East and West Germany, Czechoslovakia, Luxembourg, Switzerland and central Italy and in France it was uncommon by 1930. It has not been recorded at any site recently (1,4,6,7). A subspecies (M. a. maroccana Pallary) occurs in Morocco, but no information is available on its distribution.

POPULATION Unknown.

HABITAT AND ECOLOGY Little is known of its mode of life but it is presumed to be very similar to M. margaritifera although not calcifuge (4). All reported occurrences are from large rivers (3,4,6,8). In the River Ebro in Spain it favours quiet pools at depths down to 6 m (3,8).

SCIENTIFIC INTEREST AND POTENTIAL VALUE This species has a long and interesting fossil record and some of the shells have been associated with stone axes and Neolithic implements (4,5,9). It has been valued for its nacre (mother-of-pearl) and pearls (4).

THREATS TO SURVIVAL Like other pearl mussels it probably has a very slow reproductive cycle coupled with a high longevity. This would cause the species to be very vulnerable to heavy exploitation and may account for its widespread disappearance. Pollution has also been cited as a factor (4).

CONSERVATION MEASURES TAKEN None.

CONSERVATION MEASURES PROPOSED Surveys should be carried out to determine its current distribution, and then action should be taken to provide adequate protection for extant populations.

REFERENCES 1. Bless, R. and Nowak, E. (1977). Rote Liste der Muscheln (Bivalvia). In Blab, J., Nowak, E., Trautmann, W. and Sukopp, H. (Eds), Rote Liste der gefährdeten Tiere und Pflanzen in der Bundesrepublik Deutschland. Kilda Verlag, Greven.
         2. Ellis, A.E. (1978). British Freshwater Bivalve Mollusca. Synopses of the British fauna (N.S.) 11, Lin. Soc., London and Academic Press.

3. Haas, F. (1916). Sobre una concha fluvial interesante (Margaritana auricularia Spglr.) y su existencia en Espana. Boln. Soc. aragon. Cienc. nat. 15: 33-44 (cited in 4).

4. Preece, R.C., Burleigh, R., Kerney, M.P. and Jarzembowski, E.A. (in press). Radiocarbon age determinations of fossil Margaritifera auricularia (Spengler) from the River Thames in west London. Journal of Archaeological Science.

5. Wüst, E. (1916). Sitzung am 28th Februar 1916 (Postglacial fossil finds of Margaritana sinuata in Germany). Schr. Nat. ver. Schleswig Holstein 16: 352-353.

6. Germain, L. (1930-31). Mollusques terrestres et fluviatiles. Faune de France 21,22. Lechevallier, Paris.

7. Reuter, M.J. (1974). Révision des Margaritanidae, Unionidae et Dreissenidae du Grand-Duché de Luxembourg. Unpub. thesis.

8. Haas, F (1917). Estudios sobre la Nayades del Ebro. Boln. Soc. aragon. Cienc. nat. 16: 71-82 (cited in 4).

9. Kennard, A.S. (1923). The Holocene non-marine Mollusca of England. Proc. malac. Soc., Lond. 15: 241-259.

We are very grateful to M. Kerney for supplying the information for this data sheet and to F. Giusti, D.G. Smith, K. Spitzer, H. Turner and the Musée d'Histoire Naturelle de Luxembourg for their additional comments.

FRESHWATER PEARL MUSSEL   VULNERABLE

Margaritifera margaritifera (Linnaeus, 1758)

Phylum   MOLLUSCA   Order   UNIONOIDA

Class   BIVALVIA   Family   MARGARITIFERIDAE

SUMMARY The Freshwater Pearl Mussel once had a wide distribution throughout
northern Europe, eastern North America and Eurasia. It has been extensively
exploited for its pearls which are highly valued, and in a number of countries such
as the U.K. it is threatened by over-collecting. In other countries its range is
being reduced through habitat destruction and pollution. Collection by SCUBA
divers should be discouraged and reserves to protect breeding stock should be
created.

DESCRIPTION Shell thick and elongate. Size variable but usually about 10 cm
long and 5 cm high. Umbones low and often unevenly eroded by the high levels of
carbonic acid in the soft waters where it lives. Periostracum brown to black;
inner surface nacreous (5,12). May produce quite sizeable pearls (5). Young shells
are light yellowish brown streaked with green rays (12). Can be confused with the
common lowland species of Unio and Anodonta but is distinguished by its lower
umbones (5).

DISTRIBUTION Found in eastern North America, northern Europe (including
Scandinavia) and Eurasia (32). The distribution map for this species in Europe is
given in (23), but information is very incomplete for eastern Europe.

It occurs in Belgium and probably in Holland (16). At the beginning of this century
it was common in Luxembourg but in 1973 shells were found only in five rivulets
(Troine, Clerve, Wiltz, Sûre and Our) in the north of the country. Living
specimens were collected from the Sûre and Our Rivers only (52), and it is not

known if the species is still extant.

In France in 1930 it was common in certain rivers in the Pyrénées, Auvergne and Vosges, and was found less frequently in the west of France (44). It has now disappeared from the Dordogne and the Loire but is still found in the Pyrénées and Brittany and in small rivers in the Morvan and Massif Central including tributaries of the Yonne, Allier, Dordogne and Vienne (42,43).

Found in the British Isles, except in the east and south-east of England, south of a line from Scarborough to Beer Head. In Wales, south-west and northern England only a few scattered rivers now contain mussels. A large proportion of Scottish rivers still contain Pearl Mussels, and in Ireland they are found in several rivers except those in the central limestone plain and Shannon basin (10,11,16,24,38,41).

Found in Scandinavia including the Lofoten Islands, Norway (16). The species has a wide distribution in mainland Norway from the southernmost parts to the province of Finnmarken, mainly in coastal areas (7,13,28,29). It has been reported from Iceland but the records have never been confirmed (45). In Sweden it occurs from Scania to Lapland (16). In Finland occurs in 20 watercourses mainly in the north between 64°N and 67°N, its southernmost locality being the River Mustionjoki on the south coast. In south-west Finland populations occur locally and in small numbers (35). In Denmark it is found in the River Varde Aa in west Jutland and has been introduced into others e.g. Skern Aa, Sneum Aa and Kongeaaen (7,14,16), although apparently these populations have died out (51).

In Germany it is restricted mainly to Bavaria with isolated populations occuring on the Lüneburger Heide (14,16,17), the subalpine mountains of Hessen (Vogelsberg) and in the Saxonian Vogtland and Elster regions of East Germany (21). On the Lüneburger Heide, the Pearl Mussel used to be found in rivers draining into the Elbe and Weser systems. By 1954 the species was extinct in most waters running into the Elbe except for an isolated population in one tributary, the Este; in the Weser system it still occurs in the Lachte and the Lutter, tributaries of the Aller (9). In the U.S.S.R. occurs in the Volga watershed and Rivers Don and Dnieper (16,21).

In Austria it is restricted to the north-west and is found exclusively in rivers running from the granitic highlands. Although not occurring in the Danube itself, it once occurred abundantly in tributaries of the upper Danube in which the water has a low calcium content such as the Doblbach, Kosslbach and Pfuda river systems. To-day these rivers have only scattered, adult specimens and a few small banks of 50-100 individuals (53,56). Used to be numerous in Lower Silesia, Poland but almost certainly extinct now (26). A search in 1956 revealed only dead shells in an area in the province of Wroclaw which was previously known to contain this species (48). In Czechoslovakia only six populations still survive, mainly in south Bohemia (46).

In North America it is found on the Atlantic coast from Newfoundland, Canada, down to Delaware Co., Pennsylvania, U.S.A., (40°N) and is confined to the region east of Appalachians (16,33,36).

POPULATION Abundance varies from country to country. In Europe the area with the highest density is probably Scotland (40), although parts of Germany, the north-west U.S.S.R. and northern Sweden also have healthy populations (21).

In Britain, the population numbers in rivers in the south-west, Wales and northern England are generally very low; in Scotland the mussel is still abundant in some areas such as the rivers Spey and Kerry, but many populations are very reduced, e.g. the River Ythan in Aberdeen; other populations are small because of the nature of the habitat (38). Forty seven new populations have been found since

1970, six of which are in England and Wales, and it is thought that there may be others as yet undiscovered (41).

In West Germany there are currently thought to be about 150 000 pearl mussels, 130 000 of these occurring in Bavaria (2); the population in the Fichtelgebirge numbers about 20 000 - 25 000 individuals divided among seven streams (3). The population in the Lachte and Lutter was estimated at 50 000 individuals in 1954 (9).

In Sweden the species used to be more common in most areas and densities were previously much higher (7). The population in Denmark is assumed to be very small (51).

HABITAT AND ECOLOGY Usually restricted to waters relatively poor in lime (although it may occur in hard water (11)) with not too slow a current and not too high a temperature e.g. in forest and mountain regions or in areas with crystalline rock (14,16). Upper limit of altitude distribution is 500-600 m (2). Generally not found in rivers on plains but watercourses on sandy heaths, such as the Lunëburger Heide, Germany and south-west Jutland, Denmark, can satisfy its ecological requirements (9,14). May also flourish in clear river lakes with only an imperceptible current if suitable rivers flow through them (10). Unable to withstand drought and found only in permanent rivers (16). Typically found in water 0.5-1.5 m deep in a mixture of boulders, stones and sand (10,14), half buried in the substrate (10,15) but not found in sediment or silt (14). Tends to be sedentary as an adult but may move short distances leaving tracks in substrate (14,16).

The host fish in central Europe are the Stream or Brown Trout, Salmo trutta fario (2,22) and the Salmon Salmo salar (40); other native fish may once have been hosts (22). In the U.S.A. the Brook (Salvelinus fontinalis), Brown (Salmo trutta) and Rainbow (Salmo gairdneri) Trout are common hosts (12,22), but the European Pearl Mussel is unable to develop in Rainbow Trout introduced from North America (22). In New England, U.S.A., the Pearl Mussel spawns in the autumn, probably to coincide with the activities and movements of the host fish. The Brook Trout breeds between September and December, and the Pearl Mussel releases its glochidia when the fish aggregate over the spawning grounds (31,34). Studies in the Fichtelgebirge, West Germany, showed that the parasitic stage was of variable length but the percentage of gravid females remained remarkably constant at 30 per cent. The proportion of infected fish appears to depend on the density of the mussels. Young mussels were found being released readily in May or June by host trout (1,19).

Juvenile mussels are rarely found and little is known of their behaviour and ecology for their first few years or until they are about 5 cm long (30). This is partly due to the fact that young individuals bury themselves for protection from predators such as trout and from the movement of stones over the river bed (11). Since the water in which Pearl Mussels occur is generally poor in carbonates there is little material available for formation of shells and growth is consequently slow. Sexual maturity is not reached until the age of 12-15 years (41) (in New England 7-9 years (31)) and most individuals achieve ages of 60 years or more; the maximum life expectancy is about 100 years. However there is some dispute as to whether it is valid to measure age by counting annual lines on the hinge ligament, the method generally employed (21).

SCIENTIFIC INTEREST AND POTENTIAL VALUE Has been highly prized in Britain since pre-Roman times for its pearls (11,38) and is still currently fished in a number of countries for this purpose (38). It is one of the largest freshwater bivalves and the longest lived. It is interesting in that it is generally only found in rivers with low levels of calcium, despite the fact that large quantities are required to make its shell (which can weigh over 100 g) and that such acid waters cause chemical erosion of the shell (38). <u>Margaritifera</u> is now a taxonomically isolated genus, belonging to a very ancient Cretaceous family with only a few living representatives.

<u>THREATS TO SURVIVAL</u> Now extinct in many areas where it was formerly common mainly as a result of its sensitivity to pollution, alteration of water courses, and, in some cases, pearl fishing (7,14,19). A comparison of records made before and after 1950 has revealed the decline of the species in a number of European countries: for example, in Norway the species has been recorded in the past from 55 fifty km grid squares but it is now confirmed living in only 34 (62 per cent) of these; in Finland and Britain it is confirmed from only 45 per cent of its recorded distribution and in Ireland from 19 per cent (23).

A survey in Finland showed that the species had undergone a catastrophic decline and was on the verge of extinction in south and central Finland and in areas of Lapland, partly because it had been collected extensively for pearls since 1750. It was afforded protection in 1955 and the population is recovering in some areas but it is still threatened throughout the country by water pollution, regulation of water levels and deepening and straightening of watercourses. Predation by the introduced Musk Rat (<u>Ondatra zibethicus</u>) has been cited as a threat but this is open to doubt (7,35). No individuals less than 7 cm long have been found in the southernmost population of the River Mustionjoki and it is thought that this population no longer breeds. It may soon disappear completely from rivers where pollution cannot be held in check (35).

In Sweden there have been a number of reports of decimation due to pearl fishing; for example by 1905 in parts of the River Ljusman in the province of Härjedalen, only isolated individuals remained. In southern Sweden the draining of fens and marshes has meant that brooks and rivers dry out more than they used to (7), and the species appears to have disappeared from some places as a result of acidification (29). Acid rain may also have caused a reduction of its range in Norway. In Denmark the population in the River Varde Aa in West Jutland decreased as a result of pollution from industrial waste (7). Isolated living specimens were found in 1981 and 1982 but the population is currently threatened by pollution from fish-farms and from mercury in the sediment from a drug factory (51).

Since the beginning of the 20th century the number of Freshwater Mussels in Central Europe has decreased by more than 90 per cent and it is thought that very few of the remaining populations are still able to reproduce (3). In Germany, in the Fichtelgebirge mountains of Bavaria, there were thought to be about 700 000 mussels in 1914, compared with the present population of 25 000. Young mussels were found in only one of five rivers investigated in this region, and in two of the streams the youngest specimens were over 60 years old (1). The beginning of industrialization in this area precipitated the species' decline (3) and current damaging factors include domestic and industrial waste, stream channelization, loss of bank vegetation and intensive agriculture (1). It has been found that, contrary to earlier opinion, glochidia are produced by all these populations and development proceeds normally on the host fish, of which there is no shortage (3). It now appears that there are a number of populations in Germany where the youngest specimens are 50-60 years old but which are still fertile (2). The critical stage appears to be when the young mussels leave the host fish; even a slight increase in pH, electrical conductivity, chromium, phosphate or calcium in the water or sediment can inhibit the growth of juveniles (3). It has been estimated that there is a 90 per cent mortality rate in young mussels, possibly associated with the fact that they bury themselves in the top layer of the substrate where pollutants accumulate (19). An increase in eutrophication, causing a higher organic content in the sediment, may also be deleterious and may be an even more important factor than direct pollution (4). In the last 20 years the increase in the use of streams for trout farming has also been damaging. For example in 1976 in one area a bank containing more than 100 mussels was left to dry out, the stream having been channelled into the trout farm. Commercial exploitation has not had a major effect on German populations since in the past generally only foresters were permitted to fish for pearls, and this practice has now been prohibited (1). On the Lüneburger Heide there has been a considerable decrease in the number of rivers occupied by the species during this century and no young have been recorded for 20 years (6,9,37). This is thought to be due mainly to alteration of the drainage and water systems, since the species now can only be found in parts of the Lachte and Lutter rivers which have not been altered (9). In the Vogelsberg in Hessen, it seems unlikely that the remaining populations will be able to survive or reproduce since they are so scattered; it was found that most individuals had reached a considerable age and young ones were virtually absent (21). Again this seems to be due to the long term effects of pollution from domestic waste. It is thought that a decline in numbers of host fish and alteration of watercourses may also play a role here; for example a large mussel population was destroyed by alteration of a brook at Altefeld (20).

Pollution is reported to have contributed to the decline of the Pearl Mussel in Lower Silesia, Poland (26,48), Czechoslovakia (47), U.S.S.R. (7), Luxembourg (52) and Austria (56). Austrian populations were further depleted by collection for the mother-of-pearl industry earlier this century (57).

In the U.K. juveniles constitute no more than 5 per cent of the population (7,38) and one population study showed that only 7 per cent of the population was less than 30 years old (41), suggesting that pollution is having an important effect in this country. In a few rivers the species has become extinct as a result of refuse dumping, and it is also threatened by siltation and eutrophication, destruction of the substrate (e.g. dredging pools to create improved salmon runs) and by obstruction of migratory movements of the host fish such as the building of dams for hydro-electric schemes (38). Gravel dredging is probably an important factor in the River Mourne, River Bann, and other larger rivers in Northern Ireland since it makes downstream water turbid (25). The mussel remains widespread and not in danger of extinction in the U.K. as a whole, but may eventually be lost in England and Wales and will almost certainly become extinct in some rivers (41). Over-exploitation has contributed to the decline of the Pearl Mussel in Britain, particularly in Scotland where it has declined most recently (41). There has been a pearl fishery in Britain since pre-Roman times, but the effect of human predation must have been negligible until the end of the 18th century. Over-fishing has been reported occasionally during the last 200 years (11) and now large numbers of Pearl Mussels are killed indiscriminately in Britain by amateur treasure seekers (11,38). Skin divers and sub-aqua clubs take mussels from deep pools which were previously remote and inaccessible and which provided breeding reservoirs (38,41). Professional pearl fishing in the U.K. has declined considerably. Some professionals have a responsible attitude, returning mussels alive to the river and only opening adults. Traditionally pearl fishermen would leave 12-15 years between exploitations to allow recovery, and the more experienced fishermen take only large specimens with scarred shells (41). Pearl fishing by itself is unlikely to lead to the extinction of the species since it will become uneconomic when the population is reduced to a low level (11), but amateur fishing will continue because of the enjoyment it gives.

Populations in eastern North America vary in their size and susceptibility. New England populations still exist in good numbers, but the increasingly acidic rainfall and accelerated siltation from land development could endanger them in the future. Outside New England M. margaritifera occurs only as isolated populations in small tributary streams of a few major drainages. These populations, because of their small size, face greater dangers. In northern New York M. margaritifera formerly occurred commonly in several streams flowing into Oneida Lake but now appears to be declining. The species was found in the early twentieth century in three tributary streams to the Schuykill River system in eastern Pennsylvania but recently was found in only one of the three. Reduction of these populations probably resulted from organic and inorganic pollution and increased siltation. The status of other populations in the lower Hudson River system of south eastern New York is unknown (42). In New England the species is restricted to small streams, which have limited resources to support a top predator such as the Brook Trout, the host fish; together with habitat deterioration, this factor is causing a decrease in host availability for the Pearl Mussel, affecting its future survival (31).

CONSERVATION MEASURES TAKEN Has been protected in Finland since 1955, and is listed as vulnerable (7,27,35). It was protected in Denmark earlier this century and between 1910-1912 was introduced into a number of rivers to expand its range (7), but by 1955 it was thought to be extinct in Denmark and protection was withdrawn (51). In Sweden pearl fishing was prohibited for a limited period in the province of Härjedalen because of population depletion (7). In Norway, although the mussel is not protected by law, pearl fishing rights belongs to the owner of the river (29).

The species is protected by law in Poland (48); attempts made to reintroduce it in some areas in 1965 were unsuccessful (26,48). In France pearl collection in the past was an important activity in the Vosges, but collection has been prohibited throughout the country since 1979 (43). It is protected by law in Czechoslovakia and occurs in one nature reserve in the Sumava mountains in South Bohemia (46) (and see habitat sheet). In Austria it is protected under fisheries legislation (57).

Listed in the German national Red Data Book as 'threatened with extinction' (8). Attempts have been made in Germany for several centuries to manage the Pearl Mussel populations. In Hessen the mussel was protected in the Middle Ages under a royal prerogative (17) and the pearl fishery on the Lüneburger Heide was controlled until 1705 by the Dukes of Brunswick-Lüneburg at Celle. However, when their control ended the mussel populations were heavily plundered. In the Fichtelgebirge in the late 19th and early 20th century the landed gentry were responsible for pearl fishing and mussel thieves were punished severely (3). Pearl fishermen in Bavaria were required to use a special tool which enabled them to search for pearls without killing the mussels. Pearl fishing is banned completely in this region now (38) as is the building of fish ponds near Pearl Mussel rivers (49). In the 18th century the mussel was introduced into the southern Odenwald in the Steinach to increase its distribution. Although specimens are still found in this region other introductions have proved unsuccessful (17). Attempts have also been made to infect suitable fish with glochidia in order to find new hosts (7). In 1968 mussels on the Lüneburger Heide were transplanted into a small stream in order to carry out detailed studies of their life cycle. It was hoped that increasing the density of the population would improve their reproductive capacity (18). In 1972 a similar project was set up in the Vogelsburg in Hessen when a large proportion of the population was put into a protected stretch of stream in the Oberwald at Grebenhain. There was a good supply of host fish which it was hoped would improve reproductive potential. However, the project has not been entirely successful as the majority of individuals in the population have been found to be over forty years old and in 1976 there was a setback when the water level dropped drastically (17,18,21). Every year 500-1000 Brown Trout, infected with glochidia are released into the Fichtelgebirge, but it is thought that the chances of new populations becoming established are small (49). The population in Bavaria has recently been mapped and on the Lüneburger Heide the biology of young mussels is being studied (19). Considerable efforts are being made by the Faculty of Animal Ecology at the University of Bayreuth to draw public attention to the seriousness of the problem in Germany, and in June 1982 a petition requesting action was sent to a variety of governmental agencies and national and international non-governmental organizations (50). No action has yet been forthcoming.

CONSERVATION MEASURES PROPOSED The plight of the Pearl Mussel emphasizes the need for much stricter pollution control measures throughout Europe and North America, and for greater control of channelization and alteration of watercourses. Until the need for such measures is more generally recognized, it is important that reserves should be created in areas which are as yet unaffected. For the Fichtelgebirge area in Bavaria it has been suggested that an area free from pollution by domestic waste should be protected. Black Alders could be planted along the river bank to provide shade which would lessen the rate of organic production in the water and thus provide better conditions for the development of juvenile mussels (3). It is also hoped that a channel could be built along the river to divert sewage presently entering upstream of the mussels. Three rivers have been proposed as reserves in Bavaria (49). In Czechoslovakia, the nature reserve where this species occurs should be extended to include the entire montane stream (47). In Poland efforts are required to determine whether living populations still exist and if their presence is confirmed strict protective measures will be required (48).

Methods of pearl fishing should be controlled and divers should be educated about the damage they can cause (7,39). For example, since young mussels never contain pearls, mussels under 9 cm in length should not be disturbed when pearl hunting. Only shells with scars on them, indicating the possibility of a pearl having formed, should be opened and divers should be made aware of the fact that some rivers have mussels with no pearls in them; in these cases it is useless to open the shells. With care it is possible to use tongs to open a shell to look for a pearl, and then to close it again unharmed, and this practice should be encouraged. SCUBA clubs should be discouraged from pearl fishing and consideration should be given to limiting fishing to professional collectors who would have to pay a licence fee and use tongs. A minimum size limit could be set but knowledge of the biology of the species is currently not well enough advanced to know what the size should be. Restocking of depleted areas is probably impractical at the moment (41). Care should be taken not to trample over river beds while collecting (39). It is unlikely that the pearl fishery would ever become economically viable again in Germany, but if the remaining populations on the Lunéberge Heide were protected, it might be possible to make use of them on a small scale (9). M. margaritifera is to be included in new legislation for protected animals in Luxembourg. It has been proposed for inclusion in the next edition of the national Red Data Book for the U.S.S.R. (54).

CAPTIVE BREEDING Freshwater pearl mussels culture has been carried out in Upper Austria since the last century, initially in the Perlbach, a branch of the Doblbach. In the 1950s the breeding station was transferred to Rutzenberg at Schärding (55). Successful breeding has been reported, but the programme has been discontinued (56).

REFERENCES 1. Bauer, G. (1979). Untersuchungen zur Fortpflanzungsbiologie der Flussperlmuschel (Margaritana margaritifera) im Fichtelgebirge. Arch. Hydrobiol. 85(2): 152-165.
2. Bauer, G. (1981). In litt., 12 January.
3. Bauer, G. and Thomas, W. (1980). Die Ursachen für den Rückgang der Flussperlmuschel um Fichtelgebirge und Massnahmen für ihren Schutz. Natur und Landschaft 55 (3): 100-103.
4. Bauer, G., Schrimpff, E., Thomas, W. and Reimer, H. (1980). Zusammenhänge zwischen dem Bestandsrückgang der Flussperlmuschel (Margaritifera margaritifera) im Fichtelgebirge und der Gewässerbelastung. Arch. Hydrobiol 88(4): 505-513.

5. Beedham, G.E. (1972). Identification of the British Mollusca. Hulton Group Keys, Hulton Educational Publications Ltd., Bucks.

6. Bischoff, W.D. and Utermark, W. (1976). Die Flussperlmuschel in der Lüneburger Heide, ein versuch ihrer Erhaltung. 30 Jahre Naturschutz und Landschaftspflege Niedersachen: 190-204.

7. Bjork, S. (1962). Investigations on Margaritifera margaritifera and Unio crassus. Acta Limnologica 4: 1-109.

8. Bless, R., Nowak, E., and Ziegelmeier, E. (1978). Rote Liste der Muscheln (Bivalvia). In Blab, J., Nowak, E. and Trautmann, W. (Eds), Rote Liste der gefährdeten Tiere und Planzen in der Bundesrepublik Deutschland. Natur-schutz Aktuell Nr. 1, Kilda Verlag, Greven.

9. Boettger, C.R. (1954). Flussperlmuschel und Perlenfischerei in der Lüneburger Heide. Abh. Braunschw. Wissensch. Ges. 6: 1-39.

10. Boycott, A.E. (1936). The habitats of freshwater Mollusca in Britain. J. Anim. Ecol. 5: 116-186.

11. The Earl of Cranbrook (1976). The commercial exploitation of the freshwater pearl mussel, Margaritifera margaritifera L. (Bivalvia; Margaritiferidae) in Great Britain. J. Conch. 3: 87-91.

12. Ellis, A.E. (1978). British freshwater bivalve Mollusca. Synopses of the British fauna (N.S.) 11, Lin. Soc., London and Academic Press.

13. Esmark, B. (1886). On the land and freshwater Mollusca of Norway. J. Conch. 5: 90-131.

14. Hendelberg, J. (1961). The fresh-water pearl-mussel, Margaritifera margaritifera L. Rep. Inst. Freshw. Res. Drottning 41: 149-171.

15. Hertel, R. (1959). Die Flussperlmuschel (Margaritana margaritifera L.) in Sachsen. Abh. Ber. Staatl. Mus. Tierkunde Dresden 24: 57-82.

16. Jackson, J.W. (1925). The distribution of Margaritana margaritifera in the British Isles. Part I and Part II. J. Conch. 17: 195-211, 270-278.

17. Jungbluth, J.H. (1976). Das Flussperlmuschel-Projekt im Vogelsberg-ein Beitrag zum Artenschutz. Natur und Mensch. Naturhistorische Gesellschaft Nürnberg e.v.

18. Jungbluth, J.H. (1980). Biotopschutz-Projekte zur Bestandssicherung gefährdeter Arten am Beispiel der Flussperlmuschel (Margaritifera margaritifera L.). Gesellschaft für Ökologie (Freising-Weihenstephan 1979) 8: 321-325.

19. Jungbluth, J.H. (1981). In litt., 24 February.

20. Jungbluth, J.H. and Kühnel, U. 1977 (1978). Wassergüte-Untersuchungen an Perlmuschelbächen. Gesellschaft fur Ökologie (7) Kiel: 317-322.

21. Jungbluth, J.H. and Lehmann, G. (1976). Untersuchungen zur Verbreitung, Morphologie und Okologie der Margaritifera - Populationen an den atypischen Standorten des jungtertiären Basaltes im Vogelsberg/Oberhessen (Mollusca: Bivalvia). Arch. Hydrobiol. 78(2): 165-212.

22. Jungbluth, J.H. and Utermark, W. (in press). Die Glochidiose der Salmoniden in Mitteleuropa: Infektion der der Bachforelle Salmo trutta fario L. durch die Glochidien der Flussperlmuschel Margaritifera margaritifera (L). Fisch und Umwelt.

23. Kerney, M.P. (1976 (1975)). European distribution maps of Pomatias elegans (Müller), Discus ruderatus (Ferussac) Eobania vermiculata (Müller) and Margaritifera margaritifera (Linne). Arch. Moll. 106 (4/6): 243-249.
24. Kerney, M.P., (1976). Atlas of the Non-marine Mollusca of the British Isles. European Invertebrate Survey. Conchological Society of Great Britain and Biological Records Centre, Institute of Terrestrial Ecology, Huntingdon.
25. Kerney, M. (1981). In litt. 3 February.
26. Krakowska, E. (1978). Skojka perlorodna Margaritifera margaritifera L. (Mollusca. Bivalvia) - gatunek chroniony. (Margaritifera margaritifera L. (Mollusca, Bivalvia) as a protected species). Przeglad Zoologiczny 22(1): 48-54.
27. Malmstrom, K.K. (1975). Suomen uhanalaiset elain-ja kasvilajit on luettebitu. Suomen Luonto 2: 81-84.
28. Okland, J. (1976). (Distribution of some freshwater mussels in Norway, with remarks on European Invertebrate Survey). Fauna, Oslo. 3: 3-40.
29. Okland, J. (1981). In litt., 30 January.
30. Smith, D.G. (1976). Notes on the biology of Margaritifera margaritifera (Lin.) in central Massachusetts. Am. Mid. Nat. 96: 252-256.
31. Smith, D.G. (1978). Biannual gametogenesis in Margaritifera margaritifera (L.) in Northeastern North America. Bull. Am. Mal. Union. 1978: 49-53.
32. Smith, D.G. (1980). Anatomical studies on Margaritifera margaritifera and Cumberlandia monodonta (Mollusca: Pelecypoda: Margaritiferidae). Zool. Journ. Linn. Soc. 69: 257-270.
33. Stober, Q.J. (1972). Distribution and age of Margaritifera margaritifera in a Madison River mussel bed. Malacologia 11: 343-350.
34. Taylor, D.W. and Uyeno, T. (1966). Evolution of host specificity of freshwater salmonid fishes and mussels in the North Pacific Region. Venus 24: 199-209.
35. Valovirta, I. (1977). Report of mapping of non-marine molluscs in Finland. Malacologia 16 (1): 267-270.
36. Walker, B. (1910). The distribution of Margaritana margaritifera in North America. Proc. Malac. Soc. Lond. 9: 126-145.
37. Wellmann, G. (1938). Untersuchungen über die Flussperlmuschel (Margaritana margaritifera L.) und ihren Lebensraum in der Lüneburger Heide. Z. Fisherei 36: 489-603.
38. Williams, J.C. (1980). In litt., 2 September.
39. Young, M. (1980). Pearl mussels. Letters to the editor. Treasure Hunting.
40. Young, M. 1982). In litt., 20 January.
41. Young, M. and Williams, J. (1983). The status and conservation of the freshwater pearl mussel (Margaritifera margaritifera Linn.) in Great Britain. Biol. Cons. 25(1): 35-52.
42. Mouton, J. (1982). In litt., 1 February.
43. Real, G. and Testud, A-M. (1980). Données preliminaires sur les mollusques continentaux proteges ou reglementés en France. Haliotis 10 (1): 75-86.
44. Germain, L. (1930-31). Mollusques terrestres et fluviatiles. Faune de France 21,22. Lechevallier, France.
45. Mandahl-Barth, G. (1938). Land and freshwater Mollusca.

The Zoology of Iceland 4 (65): 31 pp.
46. Pecina, P., Capicka, A. (1979). An atlas of endangered and protected animals (in Czechoslovakia). 220 pp. Prague.
47. Spitzer, K. (1982). In litt., 11 October.
48. Ferens, B. (1965). Animal species under protection in Poland. Publication for the Dept. of Interior by the Scientific Publications Foreign Cooperation Center of the Central Institute for Scientific, Technical and Economic Information, Warsaw, Poland.
49. Bauer, G. (1982). In litt., 20 August.
50. Jungbluth, J.H. (1982). In litt., undated.
51. Jensen, F. (1982). In litt., 1 November.
52. Reuter, M.J. (1974). Révision des Margaritanidae, Unionidae et Dreissenide du Grand-Duché de Luxembourg. Unpub. thesis.
53. Modell, H. (1965). Die Najadenfauna der oberen Donau. Veröff. Zool. Staatssamm lung München. 95: 159-304.
54. Ghilarov, M. (1982). In litt., 23 November.
55. Grohs, H. (1957). Die Flussperlmuschelzucht bei Schärding, Oberösterreich. Zeitschrift der Deutschen Gesellschaft für Edelstein kunde 19.
56. Grohs, H. (1983). Unpub. Supplement to ref. 55.
57. Grohs, H. (1983). In litt., 13 December 1982.

We are very grateful for assistance and helpful comments provided by G. Bauer, B. Bölscher, H. Grohs, F. Jensen, J.H. Jungbluth, M. Kerney, W. Kühnelt, J. Mouthon, J. Okland, A. Reidel, M. Richardot-Coulet, D.G. Smith, K. Spitzer, H. Walden, J.C. Williams and the Musée d'Histoire Naturelle de Luxembourg in the compilation of this data sheet.

INTRODUCTION About 24 000 species of land snail have been described all of which are in the class Gastropoda. The majority are pulmonates which means they have a true lung but 4000 are in the order Mesogastropoda and have no lung. The latter also have no gills and gas exchange takes place across the mantle cavity. Land snails are in general small, often inconspicuous and are difficult to collect and identify. In many cases dissection is essential to identify a species since details of the reproductive system are often important in the systematics of a group. Since many museum collections are based only on the shells, their taxonomy is often poorly understood (17). The majority of species are ground dwelling, with fairly dull coloured shells but a few are arboreal and have brightly coloured shells with variegated patterns, often showing remarkable polymorphism.

Several aspects of their biology make terrestrial molluscs highly susceptible to rapid environmental changes. Most species require a humid environment and usually live beneath logs or in leaf mould on the forest floor. Arboreal species are often nocturnal, attaching themselves to the undersides of leaves during the day and in dry periods. A few species can tolerate considerable desiccation and in very dry periods may retract into their shells, sealing the aperture with an epiphragm of mucus or calcified mucus. They generally feed on decaying vegetation or fungi and lichen, but some are carnivorous, feeding on worms, other snails and a variety of other invertebrates.

Land snails are generally hermaphroditic and undergo a lengthy courtship prior to copulation. Their reproductive systems tend to be highly complex, displaying endless variations which have presumably evolved to prevent interbreeding between the numerous apparently similar species. A small number of large, yolky eggs are produced, often with thin calcareous shells, and are laid in leaf mould or other damp, sheltered locations. Several species brood their eggs internally and give birth to live young. Reproductive rates in such species tend to be low but may be offset in the natural state by long reproductive lives. This strategy tends to evolve in situations relatively free of predation and snails are often highly vulnerable to increased predation pressure (1,15,17).

Land snails are especially diverse on islands. This is partly due to the fact that vertebrate predators are few or absent in such places, but also because many species require very small 'survival areas'. Such species tend to have a low natural vagility and may have extremely small ranges, for example, limited to a small portion of a mountain valley or to a grove of trees. A number of pulmonate families have undergone enormous speciation in the Pacific islands, and in some places land snail diversity is astonishing. For example, on Rapa Island, up to 21 species have been collected from one locality (2,18). The endemic land snail fauna of Hawaii is renowned for its diversity and includes species from several families including Hydrocenidae, Helicinidae, Pupillidae, Cochlicopidae, Achatinellidae, Endodontidae, Helicarionidae, Zonitidae, Succineidae and Amastridae (this whole family being endemic to Hawaii) (6,20). The isolation of Tasmania from the Australian mainland has resulted in a unique invertebrate fauna in which almost half the snail fauna is endemic, including a number of unusual forms such as Anoglypta, Caryodes and Beddomeia (38). New Zealand land snails also exhibit a very high level of endemicity and in climax forests over 40 species have been taken from a single square foot (0.09 m^2) of litter (18,21). In Madagascar, 359 of the 377 known species are endemic (22). Isolated patches of forest may also have high levels of endemicity. In the Usambara Mountains in Tanzania 45 per cent of the molluscs in the region are endemic and in the Streptaxidae, a family of predatory snails, endemicity is as high as 62.5 per cent (see Threatened Community account) (24).

SCIENTIFIC INTEREST AND POTENTIAL VALUE Land snails have long been favourite subjects of study for biologists, as they are slow moving and easy to catch, are easily marked and are generally tolerant of laboratory conditions. They have been used extensively for research into the basis of natural selection, on account of the high degree of diversity found both within and between populations. Some of the strongest evidence that natural selection acts on genetically determined visible characters has been produced from studies on land snails such as Partula (see review)(4,5). Land snails are also important subjects of study for biogeographers, providing evidence for theories on the distribution of land masses and species (16).

Terrestrial molluscs play an important ecological role. They dispose of large quantities of dead plant matter and provide food for a wide variety of animals (25). Compared with marine molluscs, their use as food by man is fairly small, but in several countries snails are considered a gastronomic delicacy. Helix pomatia is famous for its role in French cooking (see review) but other species such as Helix aspersa, and increasingly Achatina fulica, the Giant African Snail, are also eaten in large quantities. Taiwan is now a major exporter of the latter (34).

Shell collectors have recently become interested in land snails, particularly the colourful tree species. Liguus virgineus from Haiti, L. fasciatus from Florida, Polymita from Cuba and a wide variety of species from the Philippines and Papua New Guinea are involved in the international shell trade (26,27,28).

THREATS TO SURVIVAL Land snails include some of the most vulnerable terrestrial invertebrates, and many species have become extinct over the last century. This is particularly apparent on Pacific islands. The unbalanced nature of the original fauna arriving on such islands and the subsequent lack of immediate competition often results in the development of very delicately balanced communities which can be easily disturbed. The degree of land snail extinction in the outer islands of Polynesia (Rapa Island, Marquesas, Mangareva, the Hawaiian Chain and the Austral Group) has been described as catastrophic (15). By 1934, the land snail fauna of Mangareva was reduced to a single forest patch; Rapa Island probably lost 50 per cent of its endemic species between 1934 and the mid-1960s; and very few endemic species were found in the Marquesas or on Kauai in the 1970s. In the Society and Samoan Islands ground-dwelling taxa are fast becoming extinct (15). In Hawaii, of approximately 1061 endemic snail species, about 600 have probably become extinct recently and another 200-400 are probably endangered (10).

Many species in the family Achatinellidae are seriously endangered. The subfamily Achatinellinae of approximately 100 species is endemic to Hawaii and provides a good example of the rapid decline of this group. Recent searches have failed to find many species in the four genera, Achatinella (see review), Partulina, Perdicella and Newcombia. No trace of Hawaii Island's three Partulina species have been found recently; Oahu's only Partulina has not been found for decades; on Molokai only five of the eight species have been found recently, most of which were very rare and local; and Lanai has only two or three Partulina species in the small amount of remnant native forest. On Molokai only one living individual of the island's sole Perdicella species has been found recently (8,56).

All members of the Partulidae are probably threatened (see review for Moorean species) and the Short Samoan Tree Snail Samoana abbreviata, may now be extinct. In 1926 this species was abundant and widely distributed on Tutuila but despite strenuous efforts it was not collected in 1975 (16). In the Pacific, more than 100 species in the family Endodontidae have become extinct within the last 50 years (35). In the 1830s, 13 species were collected on Rarotonga in the Cook Islands but in 1965 only two were found. In the 1840s living endodontoids were recorded from Mangareva in the Gambier Islands but in 1934 only the dead remains of 25 species were found. In Hawaii, of 125 endodontid species recorded before 1850, there are probably fewer than a dozen left today (36) and on Oahu and Kauai, traces of endodontids have been found only in isolated high mountain patches of native plants (23).

In Europe, North America and Australasia, areas where the land snail faunas are comparatively well-studied, many species are also seriously threatened. In all areas habitat destruction and introductions are the main threats, although in a few cases over-exploitation has clearly played a role, and often all three factors may be important.

1. Habitat destruction Rapidly increasing human populations on most Pacific islands are causing large-scale deforestation for fuel, agriculture and development of the tourist industry. Land snails depend on native forests for food and shelter and are consequently declining rapidly. For example, with two exceptions (Libera fratercula and Rhysoconcha) no endodontids in the Pacific have been taken from disturbed primary forest or secondary vegetation zones. This may be a result of changes in litter composition and/or humidity levels when the forest is opened up to drying by sunlight, or to subtle alteration of the food source. Because they are restricted to the ground strata snails are among the first taxa to be destroyed or displaced by human disturbance.

In the prehuman era Hawaii was covered with forest from the sea shore to the timber line. The Polynesians caused considerable destruction through forest fires and the introduction of domestic animals which trampled the vegetation, and subsequently the process has been speeded up by the introduction of plants, nematodes and plant diseases (20). On Mangareva in the Gambier Islands where there used to be a rich snail fauna, an expedition in 1934 found nothing of the native forest left except a tiny plot on Mt Mokoto at about 500 m, grass having replaced most other vegetation. In 1934 Rapa Island still had about 40 per cent of its native forest, but since then it has been disappearing fast, the process hastened by the introduction of goats, cattle and horses (10,23,37). As the native human population increased, a large part of the island forest was cut down for fuel, the original vegetation being replaced by grasses, ferns and weeds. Although the population had dropped to 300 by the 1960s and cultivation had declined, deforestation is probably still taking place (37). No specimens of the 17 endodontids known from Rapa Island could be found during the most recent collecting trip in 1963 (23).

In New Zealand, loss of habitat is rapidly reducing the ranges of many endemic snails. The Flax Snail Placostylus hongii is now restricted to Poor Knights islands and P. ambagiosus is found only in a few remnants of coastal forest in the far north. Papusuccinea archeyi is a xerophytic species and survives only in plant associations on sand dunes. It is now restricted to a small area on North Island although fossils show that its former distribution was much wider. The Paryphantidae are carnivorous forest dwellers, dependent on high humidity and natural plant cover. Many species in the genus Paryphanta have a discontinuous distribution with high concentrations in small areas and their extinction seems inevitable as a result of the removal and opening up of lowland forest for agriculture (21,32). In Tasmania the native molluscan fauna is still present in surprisingly large numbers but the dominant species have been introduced by man. Anoglypta (see data sheet) and Caryodes have been able to survive in small areas and are well adapted to the remnants of the Tasmanian temperate rain forest with its high moisture content and deep litter. However, large areas of the forest have been cleared and drastically altered through the growth of agriculture, monocultures, forestry and urban development (38).

In Europe loss of woodlands and hedgerows and cultivation of calcareous downland, all of which provide suitable damp or chalky habitats, have caused a decrease in the range of several species (25). Loss of fenlands is responsible for the decline of many species in the genus Vertigo such as V. angustior, V. geyeri, V. genesii and V. moulinsiana which are widespread but very local and found only in calcareous fens and marshes. Woodland species and relic species with small ranges are particularly vulnerable, including Elona quimperiana, Catinella arenaria, Truncatellina arcyensis, Vallonia spp., Trichia spp. and Trochoidea geyeri (39). Elona quimperiana, for example, is found only in the west of Brittany, France and in north eastern Atlantic coastal areas of Spain in deciduous forest and damp heathlands. Its biology and ecology are poorly known and there is concern that it could become endangered through habitat loss (40,41).

In the U.S.A. surveys have revealed many species threatened through habitat loss. In the eastern U.S.A., 22 species are considered to be potentially endangered on account of their limited ranges or because they are found in areas undergoing rapid development (48). For example, most forms of the well-known Florida Tree Snail (Liguus spp.) are now extinct in the Keys, eliminated by the growth of holiday homes and tourist resorts. Fortunately several have been transplanted into the Everglades National Park where they stand a good chance of survival. The Painted Snake Coiled Forest Snail Anguispira picta is threatened by lumbering in its limited area of distribution in Tennessee. Succinea ovalis chittenangoensis, the Chittenango Amber Snail is known only from the spray zone talus under the waterfall after which it is named, and is vulnerable on account of its small range (33).

Little information is available for Africa and many species have only been recorded from their type localities. In South Africa, where the terrestrial molluscan fauna is comparatively well-studied the number of endemic species is high; for example, the proportion of endemics reaches 90 per cent in the genus Gulella. Many species are forest dwellers, and the area of coastal bush and forest along the east coast is particularly vulnerable. Other areas important for molluscs include the Chirinda Forest (Mt Selinda) and the Victoria Falls Rain Forest, both in Zimbabwe (47).

Terrestrial snails are even less well-studied in Asia, where many species may become extinct before they are known to science. In the Philippines, deforestation and fire had eliminated much of the endemic snail fauna by 1957, and even in Palawan, probably the least accessible and developed island in the country, much of the original fauna has probably gone (10).

2. Introduced species The introduction of carnivorous snails to many Pacific islands in an attempt to control the Giant African Snail Achatina fulica has caused enormous damage to endemic snail populations. A. fulica was confined to the African mainland and Madagascar until the beginning of the 19th century when it was introduced to Reunion Island in the Indian Ocean, purportedly for making soup for the Governor's mistress. Within a few years populations had reached pest proportions, the snail showing a preference for cultivated plants. It subsequently spread to India and colonized large areas of the Asiatic mainland, and has since spread eastwards across the Pacific, becoming established in Tahiti, the Hawaiian chain, Vanuatu and New Caledonia (12).

Attempts at biological control of A. fulica were made with the carnivorous snails, Gonaxis quadrilateralis from Africa and Euglandina rosea from Florida (10,19). Studies carried out to investigate the effectiveness of these species have produced no satisfactory results, although one study cited decreases in the abundance of A. fulica at several Oahu sites. However, no attempt was made to determine whether this was due to predation or to disease and other factors (7,11). Results from a second study suggested that the carnivorous snails preyed selectively on juvenile A. fulica (14) but this study did not consider reproductive seasonality as a factor determining presence or absence of young A. fulica, and at least one study has shown that in Oahu no egg production occurred in study populations between late December and early May (9). Furthermore, it was found that although E. rosea may attack adults of A. fulica, the latter usually escape and can regenerate damaged tissue. In the Society Islands it has been found that A. fulica populations are undergoing a decline in numbers even where E. rosea has not been introduced (57).

Despite the inconclusiveness of the experiments, E. rosea was selected for introduction into several countries including Mauritius, the Seychelles, Reunion, Vanuatu and New Caledonia (12). In Hawaii, where it was introduced in 1955, E. rosea is moving out of the lower limestone regions where A. fulica is usually found, to occupy arboreal habitats in the mountains. In its natural habitat, E. rosea preys on the Florida Tree Snail Liguus and where it is introduced it tends to prefer the endemic tree snails to A. fulica. There are numerous accounts of collecting trips on Oahu where large numbers of E. rosea were found but not the expected endemic Achatinella. The presence of E. rosea has been cited as one of the major factors in the decline of this genus (see review) (10,19). On Kauai E. rosea has probably assured the extinction of many endemic Amastridae, including the genus Carelia (16). In 1978 A. fulica arrived in American Samoa and in spite of requests not to do so, the Department of Agriculture introduced E. rosea in 1980. There are fears that this will destroy eight endemic land snail species. In Tahiti and Moorea on the Society Islands, species of the genus Partula have disappeared wherever E. rosea has been introduced (57) with particularly serious consequences on Moorea (see review for Moorean viviparous snails). In the long run E. rosea may prove more difficult to eradicate than A. fulica. A further serious concern is that E. rosea is implicated in the spread of rat lungworm (Angiostrongylus cantonensis) which causes eosinophylic meningocephalitis in humans (19).

Introduced snails have been found on other islands. On Kauai no endemic land snails were found at 14 out of 15 collecting stations in 1975 but the following introduced snails were found: Achatina, Bradybaena, Subulina, Euglandina, and Prosopeas. The carnivorous West African snail, Streptostele (Tomostele) musaecola was found on Tutuila in American Samoa, where it may cause the extinction of endemic land snails since no equivalent predators previously existed (16).

Introduced ants have probably had serious affects on many Pacific land snails. For example, eight introduced ants have been recorded from Rapa (45), and may have contributed to the disappearance of the endemic endodontids. The present distribution of the endodontid Thaumatodon hystrellicoides, found only on Upolu in Western Samoa, correlates with the absence of introduced ants, especially the rapacious, predatory Pheidole megacephala (36). This is native to Africa and has spread through commerce to almost all of the more humid parts of the tropics (45). The reason for the apparent endodontid absence wherever ants are common probably relates not to adult predation, since the apertural barriers of endodontids are thought to be effective against this, but to egg or juvenile predation. The habit of egg deposition in the shell umbilicus common to endodontids would provide no protection against ant mouth parts. Ants could easily prevent successful reproduction of endodontids by continually removing eggs from the umbilical cavities (23,36). Pheidole megacephala is now found on many islands; in Hawaii it occurs from the seashore to the beginnings of damp forest and is considered responsible for the decline of several endemic molluscs (20).

Introduced vertebrate predators are also having damaging effects in a number of places, particularly on small islands. For example, in New Zealand, rats, mice and hedgehogs provide a greater predatory pressure on Paryphanta than did the flightless Weka Gallirallus australis which was the only endemic predator (21).

3. Exploitation It is unlikely that many land snails are currently endangered by over-exploitation. Research on the Florida Tree Snail in the Everglades has shown that populations can withstand a considerable harvest (26). However, problems may arise when threats, such as habitat destruction, occur as well. Papustyla pulcherrima (see review) is an example of the many land snails now quite heavily involved in the ornamental shell trade, which have very small ranges in forests with considerable timber potential. Species collected heavily for food such as Helix pomatia (see review) may also be threatened.

4. Pollution Balea perversa and Clausilia bidentata, two European snails, are known to be susceptible to atmospheric pollution from sulphur dioxide (13). With the spread of the effects of acid rain, such species could come under increased pressure.

CONSERVATION In North America and Europe, awareness is growing of the need for land snail conservation. Several countries now include snails in their wildlife protection legislation. The U.K. lists Monacha cartusiana, Catinella arenaria and Myxas glutinosa under the Wildlife and Countryside Act of 1981 (31). France has legislation prohibiting the collection of Helix melanostoma, H. aperta, H. tristis, Tacheocampylaea raspaili, Otala apalolena, Elona quimperiana and Rumina decollata, and controlling the collection of H. aspersa, H. pomatia and Zonites algirus (42) but no provision is made for habitat protection. The Red Data Book for West Germany lists large numbers of terrestrial snails (30). The U.S.A. lists Discus macclintocki (Iowa), Papustyla pulcherrima (Papua New Guinea), Polygyriscus virginianus (Virginia) and the genus Achatinella (Hawaii) as Endangered and Succinea chittenangoensis (New York), Triodopsis platysayoides (West Virginia), Mesodon clarki nantahala (N. Carolina), Anguispira picta (Tennessee) and Orthalicus reses (Florida) as Threatened under the Endangered Species Act (29). A recovery plan has been drawn up for Succinea chittenangoensis (43).

The European Invertebrate Survey has initiated mapping schemes for molluscs in about 25 countries (55). Atlases have been produced for Britain (44), Hessen (West Germany)(49) and Hungary (46) and are in preparation for West Germany (50), Norway (51), Finland (52), northern Italy (53), and Spain (54). Mapping schemes facilitate the identification of threatened species at the national level. For example, mapping has shown that approximately 25 out of the 190 non-marine mollusca in the U.K. are threatened, two of which have become extinct in this country during this century, although still occurring elsewhere in Europe. Once the broad national patterns have been established, finer scale mapping is necessary to determine the exact status of rarities and to identify the small and often vulnerable habitats on which they depend (55).

Very little action has been taken in other countries. From work carried out in the Pacific it appears that the islands nearer the 'New Guinea' core region have lower extinction rates than do islands further away although why this should be so is not clear. For example in the Society and Samoan Islands, although many ground-dwelling species are fast becoming extinct, many arboreal or semi-arboreal taxa seem to be abundant even in semi-disturbed conditions. Work done in 1973 indicates that little or no extinction of endemics has occurred in Vanuatu. Of the areas in the Pacific not sampled since the 1940s, the Palau and Caroline groups have the greatest recorded number of land snails and may offer the best chance of identifying 'saveable species' (16).

The main forms of conservation action required are:

1. Further research There are serious gaps in the knowledge of the land snails of the southern hemisphere and its unique habitats that are rapidly being destroyed. At the Seventh International Malacological Congress in 1980 it was recommended

that governments, universities, museums and conservation agencies should be urged to encourage research on the taxonomy of these species. Many species have not been recorded since collecting trips earlier in the century and efforts should be made to document their current status. The National Academy of Sciences in the U.S.A. includes the southern hemisphere among its research priorities in tropical biology. Encouragement should be given to the European Invertebrate Survey mapping schemes which will provide the essential data on which effective conservation measures can be based. Ultimately it is hoped that maps can be constructed for the major continental areas. It has been proposed that a centre should be established with overall responsibility for co-ordination of data (55).

2. Control of introductions A malacologist should be employed to study potential control procedures for Achatina fulica. Recent introductions of Euglandina rosea, such as that in American Samoa, should be eradicated.

3. Creation of reserves. In some cases it may be possible to create reserves fairly easily and very effectively; for example it has been recommended that the upper reaches of Mt Matafao on Tutuila, American Samoa, should be made a reserve as the slopes are too steep to be used for any other purpose and they contain many of the endemic snail species of the island (16). On Hawaii about one quarter of the land area is now protected, primarily for water conservation, although enforcement of regulations is far from perfect. Isolated uninhabited islands have considerable potential as refuges, having escaped large scale environmental damage by man. Although endodontids are nearly, if not entirely, extinct in the main Hawaiian island, two species were reported in 1980 to be thriving on the small island of Nihoa in the North-western Hawaiian islands. Nihoa is included within a Federal Wildlife Refuge which may permit the survival of these two species (3).

4. Trade controls These may not be necessary at present but trade in the more popular food and ornamental shell trade species should be monitored. In the Big Cypress National Preserve in Florida, tree snails (L. fasciatus) may be collected from 1 October to 31 March, but a bag limit of ten snails of each colour form per person has been imposed (26).

REFERENCES

1. Barnes, R.D. (1980). Invertebrate Zoology, 4th Ed. Saunders College, Philadelphia, and Holt, Rinehart and Winston.
2. Carlquist, S. (1965). Island Life: a Natural History of the Islands of the World. The Natural History Press, Garden City, N.Y.
3. Christensen, C.C. (1981). In litt., 29 June.
4. Clarke, B. (1975). The causes of biological diversity. Scientific American 233: 50-60.
5. Clarke, B., Arthur, W., Horsley, D.T. and Parkin, D.T. (1978). Genetic variation and natural selection in pulmonate molluscs. In Fretter, V. and Peake, J. (Eds) Pulmonates Vol.2A Systematics, Evolution and Ecology. Academic Press, London.
6. Cook, C.M. and Kondo, Y. (1960). Revision of Tornatellinidae and Achatinellidae. BPBM Bull. 221: 1-303.
7. Davis, C.J. and Butler, G.D. (1964). Introduced enemies of the Giant African Snail, Achatina fulica Bowdich, in Hawaii. Proc. Hawaiian. Ent. Soc. 18: 377-389.
8. Hart, A.D. (1981). In litt., 23 January.
9. Kekauoha, W. (1966). Life history and population studies of Achatina fulica. Nautilus 80: 3-10, 39-46.
10. Kondo, Y. (1980). Endangered land snails, Pacific. Unpub. report to IUCN Species Conservation Monitoring Unit, Cambridge, U.K.
11. Mead, A.R. (1961). The Giant African Snail: a Problem in

Economic Malacology. Univ. Chicago.

12. Mead, A.R. (1979). Economic malacology with particular reference to Achatina fulica. In Fretter, V. and Peake, J. (Eds), The Pulmonates Vol. 2B, Academic Press, London.

13. Holyoak, D.T. (1978). Effects of atmospheric pollution on the distribution of Balea perversa (Linnaeus) (Pulmonata: Clausiliidae) in southern Britain. Journal of Conchology 29: 319-323.

14. Nichida, T. and Napompeth, B. (1975). Effect of age specific predation on age distribution and survival of the Giant African Snail, Achatina fulica. Proc. Hawaiian Ent. Soc. 22: 119-123.

15. Solem, G.A. (1974). The Shellmakers: Introducing Mollusks John Wiley and Sons, N.Y.

16. Solem, G.A. (1975). Final Report 14-16-0008-873. Office of Endangered Species, Washington D.C.

17. Solem, G.A. (1978). Classification of the land mollusca. In Fretter, V. and Peake, J. (Eds) Pulmonates Vol. 2A Systematics, Evolution and Ecology. Academic Press, London.

18. Solem, G.A, (1980). The reasons for land snail diversity. Haliotis 10(2): 131.

19. Van der Schalie, H. (1969). Man meddles with Nature - Hawaiian style. The Biologist 51: 136-146.

20. Zimmerman, E.C. (1948). Insects of Hawaii Vol.1. University of Hawaii, Honolulu.

21. Climo, F.M. (1975). The land snail fauna. Ch 11. In Kuschel, G. (Ed.), Biogeography and Ecology of New Zealand. Junk, The Hague. 89 pp.

22. Van Bruggen, A.C. (1980). Notes on the African element among the terrestrial molluscs of Madagascar. Haliotis 10(2): 32.

23. Solem, G.A. (1976). Endodontoid Land Snails from Pacific Islands. Part I. Family Endodontidae. Field Museum Press. 501 pp.

24. Rodgers, A. and Homewood, K.H. (1979). The Conservation of the East Usambara Mountains, Tanzania: a Review of Biological Values and Land Use Pressure. Zoology Dept., University of Dar es Salaam, Tanzania.

25. Kerney, M. and Stubbs, A. (1980). The Conservation of Snails, Slugs and Freshwater Mussels. Nature Conservancy Council Interpretive Branch, Shrewsbury, U.K.

26. Abbot, R.T. (1980). The shell trade in Florida: status, trade and legislation. Special Report 3., TRAFFIC (U.S.A.), Washington D.C.

27. Wells, S.M. (1982). Aspects of the shell trade in the Philippines. TRAFFIC Bulletin 4(1): 2-6.

28. Wells, S.M. (1982). Marine conservation in the Philippines and Papua New Guinea with special emphasis on the ornamental coral and shell trade. Report to Winston Churchill Memorial Trust, London.

29. U.S.D.I. (1980). Endangered and threatened wildlife and plants. Federal Register 45(99): 33768-33779.

30. Blab, J., Nowak, E. and Trautmann, W. (1977). Rote Liste der gefährdeten Tiere und Pflanzen in der Bundesrepublik Deutschland. Kilda Verlag, Greven.

31. Anon. (1981). Wildlife and Countryside Act 1981 Chapter 69. H.M. Stationery Office, London.

32. Watt, J.C. (1979). New Zealand invertebrates. Keynote Paper II. In A Vanishing Heritage: the Problem of Endangered Species and their Habitats. Nature Conservation Council, Wellington; 273 pp.

33. Solem, A. (1974). Endangered status of eastern United States land snails. Final Report Contract No. 14-16-0008-764,OES, USDI.
34. Mead, A.R. (1980). The Giant African Snails enter the commercial field. Paper given at the 7th International Malacological Congress, Perpignan, France, September 1980.
35. Solem, G.A. (1981). In litt., 23 January.
36. Solem, G.A. (1971). Wiped out by man and still unsung. In Jones, T.C. (Ed.), The Environment of America J.G. Ferguson Publishing Co. Pp.200-203.
37. Clarke, J.F.G. (1971). The Lepidoptera of Rapa Island. Smithsonian Contributions to Zoology 56. 282 pp.
38. Smith, B.J. and Kershaw, R.C. (1979). Field Guide to the Non-marine Molluscs of South Eastern Australia. ANU Press, Canberra.
39. Kerney, M.P. and Cameron, R.A.D. (1979). A Field Guide to the Land Snails of Britain and North-west Europe. Collins, London.
40. Daguzan, J. (1980). Contribution a l'étude de la croissance de Elona quimperiana (de Férussac) (Gasteropode Pulmone Stylommatophore) vivant en Bretagne occidentale. Haliotis 10(2): 41.
41. Gittenberger, G. (1979). On Elona (Pulmonata, Elonidae fam. nov.). Malacologia 18: 139-145.
42. Real, G. and Testud, A.M. (1980). Données préliminaires sur les mollusques continentaux protegés ou reglementés en France. Haliotis 10(1): 75-86.
43. Anon. (1982). Regional briefs. Endangered Species Technical Bulletin 7(1): 2.
44. Kerney, M.P. (1976). Atlas of the Non-marine Mollusca of the British Isles. European Invertebrate Survey/Conchological Society of Great Britain/Biological Records Centre, Institute of Terrestrial Ecology, Huntingdon.
45. Wilson, E.O. and Taylor, R.W. (1967). The ants of Polynesia. Pacific Insects Monog. 14: 1-109.
46. Pinter, L., Richnovsky, A. and Szigethy, A.S. (1979). Distribution of the recent Mollusca of Hungary. Soosiana, Suppl. 1.
47. Van Bruggen, A.C. (1981). In litt., 12 February.
48. Hubricht, L. (1981). The endangered land snails of the eastern United States. Proceedings of the Second American Malacological Union Symposium on the Endangered Mollusks of North America. Bull. Am. Mal. Union, Inc. : 53-54.
49. Jungbluth, J.H. (1978). Prodromus zu einem Atlas der Mollusken von Hessen. Fundortkataster der Bundesrepublik Deutschland, Teil 5. Universistät des Saarlandes, Saarbrücken, 165 pp.
50. Ant, H. and Jungbluth, J.H. (1979). EIS-Beiträge aus der Bundesrepublik Deutschland. Malacologia 18: 185-195.
51. Okland, J. (1979). Distribution of environmental factors and freshwater snails (Gastropoda) in Norway: use of European Invertebrate Survey principles. Malacologia 18: 211-222.
52. Valovirta, I. (1977). Report of mapping of non-marine molluscs in Finland. Malacologia 16: 267-270.
53. Bishop, M.J. (1976). I molluschi terrestri della provincia di Novara. Atti della Società italiana di scienze naturali, e del Museo civile di storia naturale, Milano. 117: 265-299.
54. Ibanez, M., Abuso, M.R. and Álvarez, J. (1976). El cartografiado de los seros vivas en Espana. Trabajos y

Monografias del Departamento de Zoologia, Universidad de Grenada 2: 1-10.

55. Kerney, M. (1982). The mapping of non-marine Mollusca. Malacologia 22 (1-2): 403-407.

56. Severns, M. (1981). A vanishing breed - Hawaii's Partulina. Hawaiian Shell News 29(10) N.S. 262: 3.

57. Pointier, J.-P. and Blanc, C. (1982). Achatina fulica dans les Iles de la Société. Rapport de Mission, Muséum National d'Histoire Naturelle Ecole Pratique des Hautes Etudes, Antenne de Tahiti, Papeete.

We are very grateful to S. Chambers, M. Kerney and P. Morden for comments on this section.

LITTLE AGATE SHELLS or
OAHU TREE SNAILS

Achatinella Swainson, 1828
A. apexfulva (Dixon, 1789)
A. bellula Smith, 1873
A. bulimoides Swainson, 1828
A. byronii Wood, 1828
A. concavospira Pfeiffer, 1859
A. curta Newcomb, 1853
A. decipiens Newcomb, 1854
A. fulgens Newcomb, 1853
A. fuscobasis Smith, 1873
A. leucorraphe (Gulick, 1873)

A. lila Pilsbry, 1914
A. lorata (Férussac, 1824)
A. mustelina Mighels, 1845
A. pulcherrima Swainson, 1828
A. pupukanioe Pilsbry & Cooke, 1914
A. sowerbyana Pfeiffer, 1855
A. swiftii Newcomb, 1853
A. taeniolata Pfeiffer, 1846
A. turgida Newcomb, 1853

The other species of the genus are probably extinct, although a few were recorded in the mid 1900s:

A. abbreviata Reeve, 1850
A. buddii Newcomb, 1853
A. caesia Gulick, 1858
A. casta Newcomb, 1853
A. cestus Newcomb, 1853
A. decora (Férussac, 1821)
A. dimorpha Gulick, 1858
A. elegans Newcomb, 1853
A. juddii Baldwin, 1895
A. juncea Gulick, 1856
A. lehuiensis Smith 1873

A. livida Swainson, 1828
A. papyracea Gulick, 1856
A. phaeozona Gulick, 1856
A. rosea Swainson, 1828
A. spaldingi Pilsbry & Cooke, 1914
A. stewartii (Green, 1827)
A. thaanumi Pilsbry & Cooke, 1914
A. valida Pfeiffer, 1855
A. viridans Mighels, 1845
A. vittata Reeve, 1850
A. vulpina (Férussac, 1824)

Phylum	MOLLUSCA	Order	STYLOMMATOPHORA
Class	GASTROPODA	Family	ACHATINELLIDAE

SUMMARY Of the 41 described species of Achatinella, a genus endemic to Oahu, Hawaii, U.S.A., 14 were declared extinct and 25 rare and endangered in 1970 (only two were considered common); by 1979, 22 species were presumed extinct and 19 species endangered. Overcollecting at the turn of the century caused depletion of populations in accessible areas. There is still some collection but the main threats now are habitat destruction (forest fires) and introduced plant and animal species, particularly the carnivorous snail Euglandina rosea. Slow growth, low reproductive rates and a low dispersal rate combine to make any perturbation in the habitat a serious threat to their existence. The genus has been listed as Endangered under the U.S. Endangered Species Act.

DESCRIPTION The shells are variously patterned and coloured shades of red, orange, yellow, brown, green, grey, blue, black and white. Each shell is unique in size, shape, colour and pattern. The colours are usually contained in the periostracum and may be an adaptation to reflect sunlight, as the snails are often difficult to see amongst the foliage. The taxonomy is still not finalized and there may only be 12 true species although 41 are named. Particular forms are unique to particular valleys and intermediate forms are found on the ridges between them; many colour forms are restricted to colonies measuring only a few square metres. The basic shell shapes are globular, conical and ovate, demarking three subgenera. Adult shells vary from 1.25 cm to slightly over 2.7 cm long depending on species; shells are usually smooth and glossy but a few have slightly sculpted surfaces. The number of shell whorls varies between five and seven, and may vary within a species. Some species are either dextral or sinistral; others are both. A small 'tooth' (lamella) protrudes from the central column and is a distinctive character of the genus (7,8,10).

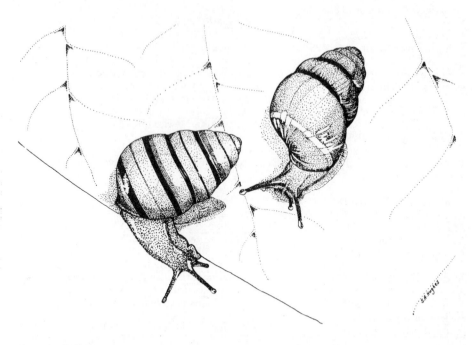

DISTRIBUTION Oahu, Hawaii, U.S.A. Surviving populations found from about 300 m (Koolau) to just over 1100 m (Waianae) in native and introduced forests of both mountain ranges (9,13). Fossil deposits and data from former collectors indicate that Achatinella once occurred in lowland valleys, along coastal plains and even near the sea at Kahuku (1,7,8). Thirty six species apparently evolved in the Koolau range; the five species in the Waianae range probably evolved from two species which had migrated from Koolau (4). Details in (8,9).

POPULATION Rough estimates of population sizes were made in 1979. A. turgida was thought to have a population size of less than 50; A. bulimoides, A. taeniolata and A. bellula less than 100; A. fuscobasis less than 200; A. lila and A. pulcherrima less than 300; A. apexfulva, A. byronii, A. curta, A. concavospira, A. decipiens, A. fulgens, A. pupukanioe and A. swiftii less than 1000; A. sowerbyana less than 5000 and A. mustelina less than 10 000 (8).

HABITAT AND ECOLOGY Arboreal, generally inhabiting native shrubs and trees. Nocturnal, although may be active in day during and after heavy rains. During the day the snails usually rest alone or in small clusters sealed to branches, trunks or undersides of leaves; at night they graze on leaf and bark surfaces for microscopic algae, fungi, liverworts and detritus (7,10). The radula of Achatinella is specialized for algal cell piercing. It has about 56 000 very tiny teeth which overlap closely and apparently puncture and partly scrape loose the cells of algal films found on the surfaces of leaves and twigs (11). The snails are sedentary, and individuals may never move from a single tree; some prefer certain types of tree but are not necessarily found on all trees of that species (7,10). A. lila is found on Ohia (Metrosiderus polymorphus) trees and the surrounding low shrubs and vines (17). The animals move about at night as they feed, but there is little movement between isolated colonies. They are hermaphrodites but not self-fertilizing, and they breed throughout the year, usually bearing one live young. A study of A.

mustelina showed that this species, unlike many gastropods, has a relatively constant but very slow growth rate of about 2 mm per year throughout the growing portion of its life. Maximum size was calculated to be achieved at 6.9 years, assumed to be the age of an individual at first reproduction. The potential life span was calculated to exceed 9.25 years (16). A study on A. lila showed a rapid growth rate in the juvenile stages slowing down as the animal matured. The largest snail in the sample, 17 mm in length, was estimated to be 5.27 years old (17). A. mustelina appears to have a very low annual fecundity of about 0.4 offspring per adult per year. In the native state this low fecundity is probably offset by a lengthy reproductive life (6).

SCIENTIFIC INTEREST AND POTENTIAL VALUE The genus Achatinella includes some of the most beautiful, varied and highly developed land snails in the world (7). The endemic Hawaiian land snail fauna is considered by some workers to be the most remarkable of all land snail faunas and Achatinella forms one of the most distinctive elements of the Hawaiian biota (14). Oahu's rugged topography of ridges, valleys and ravines is largely responsible for Achatinella's diversity by isolating tree snail populations geographically. The genus has provided a major source of material for the study of evolution, both historically and at present (7,14). Early Hawaiians used the shells for body ornaments and barter (8).

THREATS TO SURVIVAL Achatinella is highly vulnerable to human activities because the various species have a) small geographical ranges b) low reproductive rates c) virtually no defence mechanisms against some exotic predators and d) a general dependency on relatively intact native forest conditions (1,7). Some species of Achatinella at lower elevations were threatened with extinction as early as 1850 as a result of forest destruction and cattle trampling (2). The Polynesians who arrived between 500-600 A.D. presumably destroyed or altered much lowland forest. The island was rediscovered by Captain Cook in 1778, after which early Caucasian visitors and settlers introduced cattle, goats, sheep, horses and many other animals (the Hawaiians brought fowl, pigs, dogs and the Polynesian Rat). Shallow-rooted endemic plants were unable to survive heavy grazing. As the undergrowth was eaten away, the soil dried out rapidly and by 1935 serious destruction had taken place which was further hastened by the introduction of foreign nematodes and plant diseases as well as exotic plants (14). By 1978, approximately 85 per cent of the original forest cover had been destroyed or radically altered, most woodland below 400 m having been cleared for agriculture (1). Native forest now exists mainly in the upper valleys, at the heads of ravines and on upper mountain ridges (8).

The existence of feral pigs and goats, still officially sanctioned for hunters, is incompatible with the preservation of remnant native forest (8). Numerous forest fires, from military firing practice or from other causes, have denuded many lower mountain slopes and ravines since the 1930s. Fast growing exotic trees have been introduced to replace natural forest that has been lost, to conserve water and to prevent soil erosion, but less than half the Achatinella species have adapted to them. Principal exotics include Australian Eucalyptus, Ironwood, Norfolk Island pine, Silk Oak, Paper Bark and Guava(1,8). Uluhe, the native False Staghorn Fern (Dicranopteris linearis), is replacing native forest possibly as a result of fires and feral animals opening up the understorey. In healthy native wet forest ecosystems this fern is present but inconspicuous (1) but in Oahu some forest areas now covered with this species show little sign of regeneration (8). Clidemia hirta, an exotic ornamental, has become a problem in the Koolau range where it forms dense stands, occupying large areas of wet forest understorey and hinders native plant growth and regeneration; it is now spreading to the Waianae range (1). The Coffee Twig Borer Beetle (Xylosandrus compactus: Scolytidae) destroys the tips of Osmanthus, the native olive on which Achatinella is often found (4).

In the prehuman era, Achatinella probably had no active predators, although several species of native forest birds may have eaten a few (3). Predation by a number of introduced species is now heavy, and the long pre-reproductive period and low fecundity of the genus makes it extremely vulnerable. Mice and three species of rat have been introduced. The Brown or Arboreal Roof Rat (Rattus rattus) is a major problem as it climbs trees well and lives in the forest (15). Many rat-killed snails from at least ten species have been found in both mountain ranges (1,8,9). The introduced ant Pheidole megacephala is ubiquitous from the seashore to the edges of damp forest, and preys on Achatinella (14). Euglandina rosea, a carnivorous snail from Florida, was introduced to combat the Giant African Snail Achatina fulica which is a plant pest but has spread and now feeds on Achatinella. Dead Achatinella shells have been found often with Euglandina in both mountain ranges and where Euglandina has been long established, live Achatinella are usually very rare (8). The introduction of Euglandina has proved to be a most serious error (12). Furthermore, its increase may lead to an increase in populations of the predatory flatworm Geoplana sp., which could pose an additional threat to Achatinella (18).

Over-collecting had a major effect on Achatinella in the last century and the period 1850-1900 was known as the years of 'land shell fever' when hundreds of thousands or possibly millions of shells were collected. Many local populations were wiped out as a result. The 1930s and 1940s were the last years of widespread abundance (8,11). Some accessible populations, near hiking trails, are still susceptible to collecting, because of increasing pressure on remaining native forests for recreational use. Approximately 80 per cent of Hawaii's population lives on Oahu (4,8), and the island is undergoing rapid development (16).

A potential new threat to these animals arises from the possible aerial application of Malathion for suppression of pestiferous fruit fly populations in the proposed Tri-fly Eradication Program in Hawaii. It is possible that no studies of the effect of pesticides on Achatinella species will be carried out (5).

CONSERVATION MEASURES TAKEN The genus has been listed as Endangered under the U.S Endangered Species Act (18). The surviving species of the genus occur within State Forest Reserves or Conservation Districts and consequently should be protected (8) but currently little interest is being shown in the preservation of native forest, and on islands other than Oahu large tracts have been cleared recently for planting Macadamia nuts or other crops. The Army has stated that measures are being taken to minimise the threat of accidental forest fires caused by training activities to the native forest habitats of Achatinella. These include building firebreaks around military firing areas and curtailing exercises using live ammunition during droughts (18).

CONSERVATION MEASURES PROPOSED A malacological-botanical survey team should be formed to determine the feasibility of exotic plant (mainly Clidemia) control and eradication and to study the excessive growth of the Uluhe Fern. There should be increased hunting pressure on all introduced feral animals, including rodents, in forest areas occupied by endemic tree snails. The threat of forest fires and other potentially damaging activities by U.S. armed forces should be curtailed in the extensive areas of native forest that they lease from the State of Hawaii. Collection of live individuals should be permitted only for scientific research. Attempts should be made to stop the spread of Euglandina and to eradicate it from forests (8). The local inhabitants of Oahu should be educated about the threat to Achatinella and its uniqueness and effective reserves should be created in the remaining native forests (7).

CAPTIVE BREEDING Specimens have been kept for brief periods of time in captivity but copulation has not been observed (8). Specimens of A. mustelina, collected in 1976, survived well in captivity and seven of the ten produced offspring within a month of collection (6), and A. lila has also been kept in captivity successfully (17). Further knowledge of the biology of this species in the wild is necessary before experimentation in captive breeding is carried out.

REFERENCES   1.   Chambers, S.M. and Williams, L.K. (1980). Endangered and threatened wildlife and plants. Proposed Endangered status for Achatinella, a genus of Hawaiian tree snails. Federal Register 45(125): 43358-43360.

2.   Baldwin, D.D. (1886). The Landshells of the Hawaiian Islands. Hawaiian Almanac and Annual. Honolulu. Pp. 55-63.

3.   Carlquist, S. (1970). Hawaii, a Natural History. Natural History Press, Garden City, New York.

4.   Gagné, B.H., Kay, E.A. and Langford, P.S. (1975). A survey of Achatinella on Oahu, Hawaii Sept-Dec. 1974. Report to Office of Endangered Species, USDI.

5.   Gagné, W. (1980). CCH comments on Achatinella. Conservation Council for Hawaii Newsletter Oct. 1980.

6.   Hadfield, M.G. and Mountain, B.S. (in press). A field study of a vanishing species, Achatinella mustelina (Gastropoda: Pulmonata), in the Waianae Mountains of Oahu. Pacific Science.

7.   Hart, A. D. (1978). The onslaught against Hawaii's tree snails. Natural History 87 (10): 46-57.

8.   Hart, A. D. (1979). A survival status report on the endemic Hawaiian tree snail genus Achatinella (Swainson) from Oahu. Report to OES, USDI. Unpub. report.

9.   Pilsbry, H.A. and Cooke, C.M. (1914) Achatinellidae. Manual of Conchology. 2(22).

10. Kondo, Y. (1970). Extinct land molluscan species. Colloquium on Endangered Species of Hawaii. Unpub. report.

11. Solem, G.A. (1974). The Shell Makers: Introducing Mollusks. John Wiley and Sons, New York.

12. Van der Schalie, H. (1969). Man meddles with nature - Hawaiian style. The Biologist 51: 136-146.

13. Welch, D'alte A. (1938). Distribution and variation of Achatinella mustelina Mighels, in the Waianae Mountains, Oahu. Bernice P. Bishop Museum Bulletin 152 Honolulu, 164 pp.

14. Zimmerman, E.C. (1948). Insects of Hawaii Vol. 1. Univ. Hawaii Press, Honolulu. 206 pp.

15. Atkinson, I.A.E. (1977). A reassessment of factors, particularly Rattus rattus L., that influenced the decline of endemic forest birds in the Hawaiian Islands. Pacific Science 31(2): 109-133.

16. Young, G. (1979). Which way Oahu? National Geographic Magazine 156(5): 652-679.

17. Severns, R.M. (1981). Growth rate of Achatinella lila, a Hawaiian tree snail. Nautilus 95(3): 140-143.

18. U.S.D.I. (1981). Endangered and threatened wildlife and plants; listing the Hawaiian (Oahu) tree snails of the genus Achatinella as Endangered species. Federal Register 46(8): 3178-3182.

We are very grateful to A.D. Hart for assistance with this data sheet, and to C. Christensen, M. Hadfield, F. Howarth and Y. Kondo for additional information and helpful comments on the early drafts.

Partula aurantia Crampton, 1932   P. suturalis Pfeiffer, 1855
P. dendroica Crampton, 1925   P. taeniata Mörch, 1850
P. exigua Crampton, 1925   P. tohiveana Crampton, 1925
P. mirabilis Crampton, 1925
P. mooreana Hartman, 1880   Samoana diaphana (Crampton & Cooke, 1953)
P. olympia Crampton, 1925   S. solitaria (Crampton, 1932)

Phylum	MOLLUSCA	Order	STYLOMMATOPHORA
Class	GASTROPODA	Family	PARTULIDAE

SUMMARY This group of land snails, endemic to the island of Moorea in the Society Islands, is used extensively in research on genetics. Their numbers and range are diminishing rapidly as a result of the introduction of the carnivorous snail Euglandina rosea to combat the Giant African Snail Achatina fulica. Considerable effort is being put into the maintenance of captive breeding colonies, but it is feared that the majority of these species will be extinct in the wild by 1986.

DESCRIPTION There is rich polymorphism in the colour and form of the shells of these species. Colours vary from pale to dark brown with variable numbers of bands; for example individuals of Partula taeniata range from white to dark brown. Individuals also differ in size and in presence and degree of striping (spiral lines) and striations (fine vertical lines) (2,7). Detailed descriptions given in (7).

DISTRIBUTION Moorea, Society Islands. Several species are sympatric. P. suturalis and P. taeniata are the most widely distributed and both have a number of subspecies. P. s. vexillum is the most widespread subspecies, occurring everywhere except the north central area (the range of P. dendroica) and the eastern and south-eastern valleys which are occupied by P. s. strigosa (6). P. dendroica is an allopatric replacement of P. suturalis in the north central area of the island and probably only deserves the rank of geographical race (the two forms cross freely in the laboratory) (13). P. tohiveana inhabits a restricted area high on the slopes of Mt Tohiveana and P. olympia a restricted area on Mt Monaputa. These are probably geographical races of the same species as they are the terminal members of a ring species, the other end of which is P. suturalis (13). P. mooreana is found in the south-west and central parts of the island and high in the southern and south-eastern valleys, coexisting with P. suturalis over much of its range. P. exigua and P. aurantia occur in the north-east of the island and the latter may hybridize with P. suturalis. Samoana solitaria occurs high on the slopes at the southern end of Faatoai Valley. S. diaphana occurs at 600 m on a high plateau about 1.5 km to the north-east of Mt Tohiveana and has also been found on the slopes of Mt Teaharoa (3,16).

POPULATION Unknown. Densities in favourable habitats exceed 10 per m^2 but two per m^2 is usual in undisturbed habitats (10).

HABITAT AND ECOLOGY Live on bushes and trees in the mountains (3). They do not occur on the dry coastal strips although P. suturalis may be found as low as 40 m. The lower parts of the spurs between the valleys are covered mainly with grasses and xerophytic ferns which provide a barrier to Partula. They require a thick growth of shrubs and trees and the high level of moisture which makes such vegetation possible and are found on the higher parts of ridges where the vegetation is thicker. P. s. vexillum often lives in trees as high as 15 m from the ground; young individuals appear to live lower in the trees than older ones (6). They feed on decaying vegetation, the actual species of which is not crucial, but the most common food plants are succulent herbs, ferns, arums and Pandanus (7). Some of the sympatric species show ecological differences, especially P. suturalis and P. taeniata (5). They are hermaphroditic and ovoviviparous (9). Partula reared in captivity probably have a growth rate of 10-20 mm a year (11).

SCIENTIFIC INTEREST AND POTENTIAL VALUE Following the classic study of Crampton (6), these species have been extensively studied. Like other Partula species, they provide excellent material for discovering patterns of genetic variation and differentiation on account of their unique combination of ovoviviparity, low mobility, short generation time, ease of culture in the laboratory and extensive genetic polymorphism at both the morphological and molecular levels (2,4,8,12,16).

THREATS TO SURVIVAL Partula on Moorea could be extinct by 1986 as a result of the introduction of the carnivorous snail, Euglandina rosea, to combat a crop pest, the Giant African Snail, Achatina fulica. E. rosea was introduced to Moorea in 1977 at Paopao in the north-east of the island. It has since spread into the mountains and now occurs in undisturbed forest at distances of up to about 2.5 km from its centre of origin (16,17). By 1980 no Partula were found in the area of introduction of E. rosea although previously individuals had been abundant (4,16). About half of the range of P. aurantia was represented by the valley of Paraoro, which in 1980 seemed to be already devoid of snails, and in the next valley of Paparoa, Euglandina was already present (10). By July 1982 Euglandina had spread over much of the north-east of the island. P. aurantia must now be considered extinct. The subspecies P. taeniata striolata is probably also extinct, and P. suturalis and P. olympia have been entirely eliminated from the north-east of the island. With one exception, no Partula can be found in the areas in the north-east now occupied by E. rosea. The exception is P. exigua, of which a few scattered individuals were found in Faamaariri and Mouaputa valleys. It is not known why these survived and whether they will do so indefinitely. An incidental consequence of the spread of E. rosea is the raising of P. olympia to the status of a good biological species since the intergrading population of this and P. suturalis has been eliminated (16). If E. rosea has a comparable effect on the two Samoana species, these could disappear by 1987 (16).

CONSERVATION MEASURES TAKEN An attempt is being made to establish expanding breeding colonies in captivity, with the aim of passing colonies on to zoological gardens and reintroducing them to Moorea if E. rosea is eliminated or dies out. Six of the nine Moorean species are now held in laboratories in Nottingham (U.K.), Charlottesville (U.S.A.), Perth (Australia) (4) and in a 'snailarium' in Jersey, Channel Islands, and at London Zoo. Partula was proposed for inclusion on Appendix II of CITES (Convention on International Trade in Endangered Species of Wild Fauna and Flora) but the proposal was not implemented (1) as the snails are not involved in trade. All Partulidae are protected under Dutch wildlife legislation for non-native species (14).

CONSERVATION MEASURES PROPOSED Support should be given to captive breeding projects for these species. It is unlikely that any methods will be developed to eradicate E. rosea within the next few years, but if techniques become available or if this species dies out naturally, captive bred Partula should be re-introduced to their original ranges. Major efforts must be made to prevent the introduction of E. rosea to any other Pacific islands.

CAPTIVE BREEDING Laboratory colonies of Partula from Moorea and Tahiti have been maintained in the U.K. since 1962 (8), and colonies of P. suturalis, P. mooreana, P. aurantia, P. mirabilis and P. tohiveana are currently maintained in the U.S.A. (10) and Australia. A 'snailarium' was set up at the Jersey Wildlife Preservation Trust in 1981 for the subspecies Partula taeniata nucleola. The snails are kept in simple glass aquaria and fed on a mixture of porridge oats, cuttle bone and baby food. Four individuals have been bred so far (15). An attempt is being made to maintain colonies of four species at London Zoo (18). It is known that individuals can live in captivity for 10 years but it is not known if expanding colonies can be maintained (4).

REFERENCES
1. Anon. (1977). Proposals for amendments to CITES appendices. Special working session, Geneva 1977.
2. Carlquist, S. (1965). Island Life. A Natural History of the Islands of the World. The Natural History Press, Garden City, New York.
3. Clarke, B.C. (1980). In litt., 13 June.
4. Clarke, B.C. (1980). An attempt to rescue endangered species of Partula. Project proposal to IUCN.
5. Clarke, B. and Murray, J. (1969). Ecological genetics and speciation in land snails of the genus Partula. Biol J. Linn. Soc. 1: 31-42.
6. Clarke, B.C. and Murray, J. (1971). Polymorphism in the Polynesian land snail Partula suturalis vexillum. In Creed, R. (Ed.), Ecological Genetics and Evolution. Blackwell, Oxford. Pp. 51-64.
7. Crampton, H.E. (1932). Studies on the variation, distribution and evolution of the genus Partula. The species inhabiting Moorea. Carnegie Institute of Washington Publ. 410: 1-335.
8. Johnson, M.S., Clarke, B. and Murray, J. (1977). Genetic variation and reproductive isolation in Partula. Evolution 31: 116-126.
9. Lipton, C.S. and Murray, J. (1979). Courtship of land snails of the genus Partula. Malacologia 19 (1): 129-146.
10. Murray, J. (1981). In litt., 13 February.
11. Murray, J. and Clarke, B. (1966). The inheritance of polymorphic shell characters in Partula (Gastropoda). Genetics, 54: 1261-1277.
12. Murray, J. and Clarke, B. (1968). Partial reproductive isolation in the genus Partula (Gastropoda) on Moorea. Evolution 22: 684-699.
13. Murray, J. and Clarke, B. (1980). The genus Partula on Moorea: speciation in progress. Proc. Roy. Soc. B. 211: 83-117.
14. Anon. (1980). Aanhangsel a van het besluit ter uitvoering van Artikel 3 van de wet bedreigde uitheemse diersoorten. Staatsblad 454.
15. Tonge, S. (1982). In litt., 27 July.
16. Clarke, B.C., Murray, J. and Johnson, M. (in prep.). The extinction of endemic species by a programme of biological control.
17. Pointier, J.P. and Blanc, C. (1982). Achatina fulica dans les iles de la Société. Rapport de Mission, Museum National d'Histoire Naturelle, Antenne de Tahiti, B.P.562, Papeete, Tahiti.
18. Bertram, B. (1983). Pers. comm., 6 January.

We are very grateful to B. Clarke, M. Hadfield, F. Howarth, J. Murray and S. Tonge for help with this data sheet.

PLANT'S GULELLA SNAIL   VULNERABLE

Gulella planti (Pfeiffer, 1856)

| Phylum | MOLLUSCA | Order | STYLOMMATOPHORA |
| Class | GASTROPODA | Family | STREPTAXIDAE |

SUMMARY A medium-sized land snail, Gulella planti occurs in a restricted area
around Durban, South Africa, and is a member of a genus containing a number of
localized endemics. It is threatened by bush clearance and extensive urbanization.

DESCRIPTION The shell is pupiform in shape and is generally smooth and glossy with
7-8 whorls. Shell sculpture may vary from complete absence to being fairly well
marked with riblets or growth striae. Although small, averaging 17.8 x 9.00 mm, it
has the largest shell among southern African Gulella. Compared with other Gulella, it
has limited apertural dentition (2). Detailed description in (1).

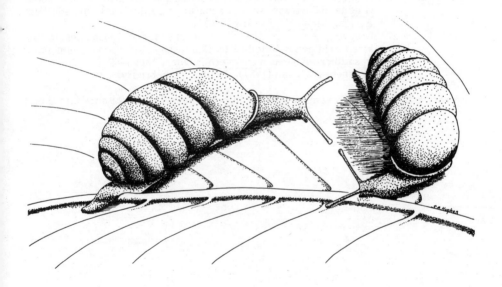

DISTRIBUTION Occurs in a restricted area, mainly centred in and around Durban, on
the coast of Natal and Zululand, from south of Durban to north of the Tugela River
and some distance inland. The restricted ranges reported for many species of this
genus may be due simply to lack of records, resulting from the small size of these
animals, their cryptic life style and their occurrence in remote localities. However,
this is unlikely to be the case with G. planti which is comparatively large and has a
well-studied range (2).

POPULATION Unknown.

HABITAT AND ECOLOGY Occurs in coastal bush and similar types of vegetation,
and is carnivorous. There are no detailed studies of its ecology. It shares its habitat
with a number of congeneric species, none of which have shells over 12 mm long,
which suggests that its large size may be an adaptation to some unique local situation

177

(2). Within the genus, species with large shells tend to be those with restricted distributions (1).

SCIENTIFIC INTEREST AND POTENTIAL VALUE Because of the difficulty of obtaining live material it has still never been dissected (apart from the radula), and virtually nothing is known of its biology and natural history (2). Like many other terrestrial molluscs, it is possible that it will become extinct before any scientific details have been obtained.

THREATS TO SURVIVAL It is threatened by bush clearance on the coast of Natal. The natural range of the snail may once have covered parts of what is now Durban (2,3), but this area has already been altered extensively through urbanization and the development of ports and holiday areas.

CONSERVATION MEASURES TAKEN None.

CONSERVATION MEASURES PROPOSED Its status should be investigated throughout its range (2), and a reserve created in a suitable area.

REFERENCES   1.   Bruggen, A.C. van (1973). Distribution patterns of the genus Gulella (Gastropoda: Pulmonata: Streptaxidae) in southern Africa. Malacologia 14: 419-425.
              2.   Bruggen, A.C. van (1980). Size clines and subspecies in the streptaxid genus Gulella Pfr. (Mollusca, Gastropoda, Pulmonata) in southern Africa. Zool. Verh. Leiden. 180: 1-62.
              3.   Bruggen, A.C. van (1981). In litt., 12 February.

We are very grateful to A.C. van Bruggen for information provided for this data sheet.

GRANULATED TASMANIAN SNAIL   ENDANGERED

Anoglypta launcestonensis (Reeve, 1853)

Phylum   MOLLUSCA   Order   STYLOMMATOPHORA

Class   GASTROPODA   Family   CARYODIDAE

SUMMARY This large land snail in a monotypic genus is endemic to Tasmania, Australia. It is found in an area of only 3000 ha and is very susceptible to land clearance. The creation of a reserve will be essential to guarantee its long-term survival.

DESCRIPTION Shell large (25-35 mm), with five whorls; conical with low spire having a series of pronounced granular ridges and spiral lines; base convex and smooth. Upper surface is yellowish green, brown and black with a narrow yellow line near periphery; base is chocolate brown with bright yellow broad band (1,2).

DISTRIBUTION Confined to 3000 hectares of temperate rainforest in north-east Tasmania, Australia (2,5).

POPULATION A rough estimate of under 5000 individuals has been made. Populations appear to occur in small pockets or patches (3). Only a few specimens have been taken live (2).

HABITAT AND ECOLOGY Confined to deep, wet leaf litter in temperate rainforest (2), particularly in fern gullies. Details of ecology not known (3).

SCIENTIFIC INTEREST AND POTENTIAL VALUE A monotypic genus endemic to Tasmania in the family Caryodidae which is endemic to eastern Australia. It is a large spectacular species, typifying the uniqueness of the Tasmanian fauna (2,5).

THREATS TO SURVIVAL The species is very susceptible to the extensive land clearing operations which are being carried out throughout south-east Australia, including Tasmania, for the introduction of pasture grasses and monocultures of pines, wheat and other crops (3,4).

CONSERVATION MEASURES TAKEN None.

CONSERVATION MEASURES PROPOSED This is an example of an invertebrate whose future survival could be ensured through the creation of a comparatively small but adequately protected reserve. If other suitable areas of habitat exist, translocation experiments could perhaps be attempted.

REFERENCES   1. Cox, J.C. (1868). A Monograph of Australian Land Shells. Sydney.
             2. Smith, B.J. and Kershaw, R.C. (1979). Field Guide to the Non-marine Molluscs of South-eastern Australia. ANU Press, Canberra.
             3. Smith B.J. (1980). In litt., 30 July.
             4. Smith, B.J. (1979). Survey of non-marine molluscs of south-eastern Australia. Malacologia 18: 103-105.
             5. Smith, B.J. and Kershaw, R.C. (1981). Tasmania Land and Freshwater Molluscs. Fauna of Tasmania, Univ. of Tasmania.

We are very grateful to B.J. Smith for information provided for this data sheet.

MANUS GREEN TREE SNAIL   RARE

Papustyla pulcherrima Rensch, 1931

Phylum	MOLLUSCA	Order	STYLOMMATOPHORA
Class	GASTROPODA	Family	CAMAENIDAE

SUMMARY This well known brilliant green tree snail is much sought after by collectors. It is endemic to the rain forest of Manus Island, northern Papua New Guinea, and although not immediately under threat is highly vulnerable to large scale deforestation and possible over-collection. It is listed on Appendix II of CITES, and as Endangered under the U.S. Endangered Species Act.

DESCRIPTION The shell is about 4 cm long, and is an intense pea-green colour with a yellow band along the suture, which effectively camouflages it amongst the foliage in which it lives. The colour is contained in the periostracum and wears off with age to reveal a yellow layer underneath. The animal is tan with a lateral brown stripe on either side (1,2). A detailed description is given in (2). It is frequently referred to under the name of Papuina pulcherrima.

DISTRIBUTION Manus Island, Admiralty Archipelago, northern Papua New Guinea (2). Few professional malacologists have collected specimens of Papustyla and most specimens in museums have been obtained by traders, explorers and incidentally by collectors of other animal material. As a result locality data tend to be poor (2) but the snail has recently been found several kilometres inland from Lorengau, the provincial capital (1,4).

POPULATION Unknown.

HABITAT AND ECOLOGY No scientific studies have been carried out, and the only information available is from anecdotal accounts. The snail is restricted to rain forest. The main host trees are Dillenia (Dillenaceae) and Astronia (Melastomaceae), although it may also be found on several other trees and on large climbing aroids. Only a few individuals are found in a tree. They are inactive during the day and are found attached to the undersides of leaves, five or more metres above the ground. Reproduction and development may possibly be as in the Hawaiian Achatinella (1).

SCIENTIFIC INTEREST AND POTENTIAL VALUE Its intense green coloration makes Papustyla pulcherrima unique among all known species of land molluscs, and as a result it is highly prized by shell collectors. It is traditionally used by Manus islanders for decorative purposes, and is currently used in modern jewellery (1). Its restriction to a single island in Melanesia is an example of the striking adaptive radiation and geographical isolation which species within the Camaenidae have undergone. Many islands in the Melanesian Archipelago have one or several endemic tree snail species which are proving to be of great interest in studies of evolutionary zoology (2).

THREATS TO SURVIVAL The Manus Green Tree Snail appears to be relatively host specific to trees with timber potential, and logging is almost certainly the main threat. However, the shells are still highly sought after by collectors. In the past large numbers were bought by tourists and exported by dealers (5). It has been reported that branches of trees were cut down to obtain single snails (1) but such claims are probably exaggerated (4,5). In 1977 the snail was said to have retreated into the wilder central forests of the island and populations were thought to have been much reduced or eliminated by collecting and agriculture within a several mile radius of Lorengau (3). In 1981 however, the snail was found relatively easily, and there was no evidence for particularly heavy collection. Between January and June 1981, 154 shells had been exported by the New Guinea Shell Agency, although 834 specimens had been collected and deposited at the Lorengau Department of Business Development that year. Shells were also on sale in Lorengau market (4).

CONSERVATION MEASURES TAKEN At one time Manus villagers attempted to 'farm' snails by collecting them from the wild and raising them in large cages, but this was unsuccessful. The species is listed on Appendix II of CITES (Convention on International Trade in Endangered Species of Wild Fauna and Flora), which means that trade between party states is only allowed if a valid export permit is provided by the country of origin. It is listed as endangered under the U.S. Endangered Species Act which means that imports into the U.S.A. are prohibited, although specimens are reportedly still entering the country (6).

CONSERVATION MEASURES PROPOSED The Manus Green Snail is only one of a number of land and tree snails being collected for export in Papua New Guinea (4). Since most of these are barely known to science the opportunity should be taken of obtaining live specimens for anatomical studies, and surveys and ecological studies should be carried out to determine more precisely their distribution patterns and population biology. Provided the current trade continues to be monitored, it is unlikely to have any noticeable effect on populations. Logging could pose a major threat, however, and if this potential is realised, reserves will be required in appropriate sites.

REFERENCES
1. Harrison Gagné, B. (1981). Up a tree with the Manus green snail. Hawaiian Shell News 24(5) N.S. 257:1, 8-9.
2. Clench, W.J., and Turner, R.D. (1962). Monographs of the genera Papustyla, Forcartia, and Meliobba (Papuininae: Camaenidae). The Malacological Society of Australia.
3. Pyle, R.M. (1980). In litt. to U.S. Fish and Wildlife Service, 4 February.
4. Wells, S.M. (1982). Marine Conservation in the Philippines and Papua New Guinea with special emphasis on the ornamental coral and shell trade. Report to Winston Churchill Memorial Trust, London.
5. Pitman, R.W. (1977). Manus Island's green tree snails at home. Hawaiian Shell News 25(4), N.S. 208: 9-10.
6. Harrison Gagné, B. (1982). In litt., 7 September.

We are very grateful to B. Harrison Gagné and J. van Goethem for reviewing this data sheet.

MADEIRAN LAND SNAILS

VULNERABLE

Family   PUPILLIDAE

Leiostyla abbreviata (Lowe, 1852)
L. cassida (Lowe, 1831)
L. corneocostata (Wollaston, 1878)
L. gibba (Lowe, 1852)
L. lamellosa (Lowe, 1852)

Family ENDODONTIDAE

Discus defloratus (Lowe, 1854)
D. guerinianus (Lowe, 1852)

Family   HELICIDAE

Caseolus calculus (Lowe, 1854)
C. commixta (Lowe, 1854)
C. sphaerula (Lowe, 1852)
Discula leacockiana (Wollaston, 1878)
D. tabllata (Lowe, 1852)
D. testudinalis (Lowe, 1852)
D. turricula (Lowe, 1831)
Geomitra moniziana (Paiva, 1867)
Helix subplicata (Sowerby, 1824)

Phylum   MOLLUSCA

Class   GASTROPODA

Order   STYLOMMATOPHORA

Family   See above

SUMMARY One hundred and seven endemic land snails have been described from the Madeiran Archipelago. Many of these have very restricted distributions, and are highly vulnerable to habitat alteration such as over-grazing by domestic animals and the development of tourism. The conservation plans which have been drawn up for the islands should be implemented as soon as possible and should take the ecological requirements of the endemic molluscan fauna into consideration.

DESCRIPTION The Madeiran archipelago consists of seven islands, all within 30 km of each other and divided into three groups: Madeira itself, Porto Santo and its offshore islets, and the three Deserta islands. Of the 176 land gastropods that have been described from the islands, 107 are endemic to the archipelago, only 30-40 species occur also in Europe and North Africa and only seven on any of the other Atlantic islands.

The majority (96) of the endemic group are helicids, and even within the islands these tend to be rare and have restricted distributions. Only three species (Discula polymorpha, Heterostoma paupercula and Clausilia deltostoma) are present on all three island groups and most of the rest occur on one or two groups only. Ponta do Garajau on Madeira probably supports the most diverse molluscan fauna (4). Each island group has its own characteristic species and the following table lists the number of species recorded from different islands (1):-

	No. of species	Size of island	
		sq. miles	km²
Madeira	96	740.0	(1924)
Porto Santo	57	39.0	(101)
Baixo	14	1.5	(3.9)
Cima	17	0.3	(.8)
Deserta Grande	23	10.4	(27.0)
Bugio	27	3.3	(8.6)
Chao	9	0.3	(.8)

DISTRIBUTION   The following species are limited to single or very few populations:

Leiostyla corneocostata a brown snail from Porto Santo, Pta Calheta (1).

Geomitra moniziana a snail with an opaque white shell from Gaula and Canico in south east Madeira (1) and from Ribeiro de Porto Nova and San Vicente (8).

Caseolus commixta a snail with a brownish-white, granular textured shell from Ilheu de Baixo, Porto Santo (1).

C. calculus described in 1848 as being very scarce and localized. It was found most abundantly on Ilheu de Cima but was also recorded from the Pico d'Anna Ferreira and the Pico Branco (1).

C. sphaerula Pico Branco, Porto Santo (1).

Discula leacockiana: Pico d'Anna Ferreira, Porto Santo (1).

D. tabellata the smallest Discula, it has a strongly sculptured but delicate shell, greyish-yellow in colour, mottled with brown. Found on the dry maritime slopes of Ponta Garajau in south Madeira and has also been recorded from Cabo Girao, west of Funchal (1,8).

D. testudinalis tortoiseshell coloured shell with fine granular markings. Limited range at Pedragal in the north of Porto Santo. It is also rare as a sub-fossil but has been found in this form at Campo de Baixo (1,8).

D. turricula a small, reddish-brown, granulose, turreted land snail, which is one of the most distinct and beautiful in the archipelago (1). Endemic to the Ilheu de Cima or Upper Island, an islet lying some 400 m off the south-east point of the island of Porto Santo (2). It abounds under the numerous large basaltic lumps littering the top of the island, 120 m above sea level (1). Studies of shells in collections have revealed that, by the turn of the century, this animal had become slightly larger. Furthermore, between about 1900 and 1910 there was a change in mean shell shape from a higher- to a lower-spired form (9,10).

Helix subplicata the largest Helix in the archipelago (slightly larger than the European H. aspersa), pale olive-yellow-brown in colour with a glossy surface and no markings. Living specimens recorded only from Ilheu de Baixo although sub-fossil forms have been found on Porto Santo (1). In the 1930s only sub-fossil forms were found on Ilheu de Baixo although the whole island was not searched (8). May now be extinct (5).

Other species have a wider distribution but live in threatened habitats such as original woodland. Several have not been seen this century, and some have not been recorded for at least a hundred years (5).

Leiostyla abbreviata a small, strongly ribbed snail, said to be rare on Madeira as early as 1878 (1).

L. cassida a snail with a comparatively large, solid shell with reddish-brown striations and irregular white longitudinal dashes. Was found abundantly as a sub-fossil at Canical, but in 1878 was said to be uncommon in the living form, although it was found at the head of the Ribeira de Santa Luzia in south Madeira among vegetable detritus near the waterfall. Has also been recorded from Ribeira de Sao Jorge in the north of the island. Its preferred habitat is probably damp wooded ravines at intermediate altitudes (1).

L. lamellosa one of the rarest of this group, even in 1878. Found only in south Madeira at intermediate altitudes in the Vasco Gil ravine and the Ribeira de Santa Luzia (1).

L. gibba although fairly common in the sub-fossil beds at Canical, was already very rare in 1878. Two specimens were found among loose plant detritus at the base of rocks near the head of the Ribeira de Santa Luzia, south Madeira (1).

Discus guerinianus one of the most 'elegant' snails of the archipelago with a polished, variegated shell with reddish-brown and pale yellow transverse bands. In 1878 it was said to be rare and confined to the damp wooded areas of Madeira at high and intermediate altitudes in the interior of the island, beneath stones and decaying vegetable refuse (1).

D. defloratus a single specimen has been found at Pico d' Arribentao in the mountains above Funchal, Madeira (1).

POPULATION Unknown.

HABITAT AND ECOLOGY Madeira is a high island rising to 1860 m with a moist forest fauna in the high north western part and a different dry habitat fauna on the southern coastal strip. There is a sharp transition in the molluscan fauna from the south coast of the main part of the island eastwards to Ponta de Sao Lourenco where the fauna more closely resembles that of the Desertas, and where there is similar short vegetation including Suaeda and Mesembryanthemum. The south-eastern stretch of coastline from Canical to Funchal is relatively dry and low lying, similar to Ponta de Sao Lourenco although the vegetation is very disturbed by cultivation (4). Snails are generally absent from the coniferous plantations.

The ecology of Porto Santo and the Desertas is similar to that of the dry coastal facies of Madeira. The three Desertas, Ilheu de Cima and Ilheu de Baixo are all relatively dry and stony with thin soil and vegetation cover. The top of the Ilheu de Cima is covered with quite dense thickets of sapling trees and bushes, but with a sparse ground cover of lichens, moss and Mesembryanthemum (2). Porto Santo is larger, with mountains rising to 517 m and an extensive sandy beach backed by low lying terrain (1,2,3,4).

SCIENTIFIC INTEREST AND POTENTIAL VALUE The archipelago offers a rich field for the study of evolutionary radiation of an island fauna, with many problems still to be investigated. The diverse molluscan fauna with its high frequency of endemism is partly accounted for by the isolated position of the islands in the Atlantic and effective barrier to dispersal between the islands provided by the sea. There are numerous instances of evolutionary divergence between similar species on different islands in the archipelago. The most interesting radiations have occurred in the drier areas rather than in the mountains. Sub-fossil deposits, particularly at Canical on Madeira, provide some indication of the evolutionary history of the species, and may enable a detailed picture of changes with time to be constructed. With the exception of a few widespread and some introduced species, most have only distant affinities with the recent palaearctic fauna and virtually none with the recent north African fauna. They are most closely related to the early tertiary molluscs of the western Palaearctic and are therefore of interest as a relict fauna containing several primitive genera, now extinct elsewhere (2,4,5,6,7).

D. turricula is particularly interesting since around the turn of the century it underwent a change in shell shape. Shell dimensions are in part inherited but may also be influenced strongly by environmental factors, and some alteration in the environment of Ilheu de Cima may have caused an ecological or genetic response in this species. It is possible that the building of the lighthouse in 1900 could have permanently changed the microclimate of the islet surface. Furthermore the new human population on the previously uninhabited island could have significantly altered the vegetation pattern. The total effect of these changes may have been to cause the selection of a broader, lower-spired animal, better suited to the new environment (2). A sub-fossil species, closely similar to D. turricula has a more widespread distribution in Porto Santo and it may be possible to construct a more detailed picture of changes with time using the fossil material (10).

THREATS TO SURVIVAL All the habitats of the endemic molluscs are threatened by development and/or erosion. The volcanic soils are very fragile and erode rapidly after mechanical disturbance or when the vegetation is removed. Such areas used to support an endemic low scrub cover, much of which has now gone because of over-grazing by introduced cattle, sheep, goats and rabbits (6,13), resulting in the disappearance of the fauna. Grazing by domestic cattle is particularly severe on Ponta de Sao Lourenço, and much damage is being done by feral goats and rabbits on Deserta Grande (6).

The unique fossil bed at Canical is being removed to provide sand for building purposes, and further development of holiday homes is a possibility in this area (6). It is particularly striking that the preferred habitats of snails and tourists, namely dry coastal areas, closely coincide. It was noticed during a survey at Garajau that where the land is undisturbed there is a greater diversity and higher density of molluscs than where it is in use (4).

Since so many endemics occur in single populations or in very small ranges, even small scale developments could result in extinctions. For example, D. turricula could easily be wiped out either by over-collecting or further human exploitation of the island, since Porto Santo is becoming increasingly popular as a tourist centre (11).

CONSERVATION MEASURES TAKEN Little effort has been made so far to prevent the extinction of this important molluscan fauna. Madeira is now an autonomous region governed by the Junta Autonoma de Madeira, and a project for the creation of a series of natural reserves on the island and for the establishment of a regional park has been prepared by the Regional Secretariat for Planning and Finance in conjunction with the Jardin Botanico do Funchal (12). A detailed conservation plan has been produced, aimed primarily at the conservation of Laurisilva forest and its many endemic plants. Some of the areas may provide protection for molluscs, particularly forest dwellers, and the proposed Sao Lourenço Park should improve the situation for some of the dry zone species. Feral goats have been removed from the Ilheu dos Desembarcadoures and some of the other small islands (Desertas, Cima, Baixo) are now partly protected. The Desertas are being considered for addition to the National Parks network (14,15).

CONSERVATION MEASURES PROPOSED Research urgently needs to be carried out to determine the areas in greatest need of protection, and the kind of protection required. Studies on the Madeiran molluscan fauna should be encouraged and should be closely integrated with conservation activities in the archipelago. The project drawn up for IUCN by the Jardin Botanico Canario 'Viera y Clavijo' (12), should be studied to determine to what extent the proposed reserves would provide protection for molluscs. International aid is required to launch this project. It will include the setting up of a conservation service and an education campaign which would then be maintained by the regional government.

Habitat protection is the highest priority. Since most species are small and difficult to find, over-collecting is unlikely to pose a threat (5). In particular there is urgent need to control grazing of domestic stock and feral animals and to review proposals for further development of the tourist industry, to ensure that it does not conflict with areas important for their molluscan faunas (5,6) particularly in the south coastal region of Madeira and on Porto Santo (15).

REFERENCES

1. Wollaston, T.V. (1878). Testacea Atlantica. Reeve, London.
2. Pettit, C.W.A. (1977). An investigation of variation in shell form in Discula (Hystricella) turricula (Lowe, 1831) (Pulmonata: Helicacea). J. Conch. 29: 147-150.
3. Mandahl-Barth, G. (1943). Systematische Untersuchungen über die Heliciden-Fauna von Madeira. Abh. Senckenberg Naturf. Ges. 469: 1-93.
4. Cook, L.M., Jack, T. and Pettit, C.W.A. (1972). The distribution of land molluscs in the Madeiran Archipelago. Boletim do Museu Municipal do Funchal. 26(112): 5-30.
5. Walden, H.W. (1982). In litt., 12 April.
6. Cook, L.M. (1982). In litt., 25 March.
7. Cockerell, T.D.A. (1922). Land snails of the Madeira islands. Nature (London) 109: 446.
8. Nobre, A. (1931). Moluscos terrestres, fluviais e das aguas salobras do arquipélago da Madeira. Barcelos.
9. Cockerell, T.D.A. (1922). Porto Santo and its snails. Nat. Hist. N.Y. 22: 268-270.
10. Cook, L.M. and Pettitt, C.W.A. (1979). Shell form in Discula polymorpha. J. Moll. Stud. 45: 45-51.
11. Pettitt, C. (1981). In litt., 10 March.
12. Bramwell, D., Montelongo, V., Navarro, B. and Ortega, J. (1981). Informe sobre la conservacion de los bosques y la flora de la isla de Madeira. Jardin Botanico Canario 'Viera y Clavijo' and IUCN.
13. le Grand, G. (1981). Rapport sur Madeira. Report to ICBP, Azores, September.
14. le Grand, G. (19820. In litt., 2 November.
15. Bramwell, D. (1982). In litt., 9 December.

We are very grateful to L. Cook and C. Pettitt for providing the first drafts of this data sheet and to D. Bramwell, G. le Grand and H. Walden for additional information.

ROMAN or EDIBLE SNAIL   RARE

Helix pomatia Linnaeus, 1758

Phylum	MOLLUSCA	Order	STYLOMMATOPHORA
Class	GASTROPODA	Family	HELICIDAE

SUMMARY The exploitation of Helix pomatia for food has almost certainly been
the cause of its overall decline over much of Europe, and of many of the local
extinctions that have been reported. Habitat loss may seriously affect remaining
populations. Legislation controlling exploitation exists in many countries but little
attempt has been made to improve habitat protection.

DESCRIPTION The Roman or Edible Snail (Escargot de Bourgogne) is the largest
north-west European snail (13). The shell varies in size (30-50 mm x 32-50 mm)
and is thick, coarsely striated and globular. It is coloured cream, pale yellow and
pale brown, with up to five rather faint brown spiral bands. The animal itself is
pale yellow grey with a yellowish border to its foot. The organic outer layer of
the shell may be lost as the animal ages and it appears chalky white; in calcareous
and limestone regions this may aid in concealment (22). The size and thickness of
the shell probably afford good protection and it is rare to find the remains of
shells on thrush anvils (3). Ridges on the shell lip indicate the age of the snail
since maturity, and individuals may live for more than nine years. Unlike most
invertebrates, the adults have a long life, but mortality of eggs and juveniles is
very high (22).

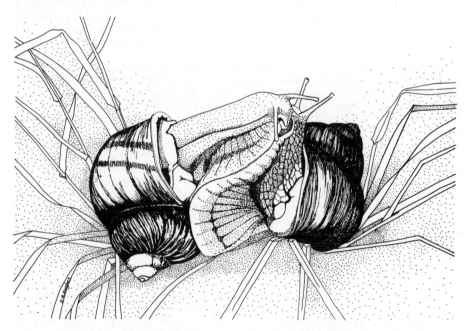

DISTRIBUTION Central and south-east Europe. Widespread in Central Europe,
extending westwards to central France and south-east England, and north to the
south Baltic coast. Extends as far south as Italy and northern Greece (13,32).
Occurs naturally in eastern France, West Germany, Austria, Belgium, Switzerland,
Luxembourg, Poland (where it is found throughout the country but with an

irregular distribution in the Carpathians (30) and it is probably only indigenous in the south (39)), Czechoslovakia, Bulgaria, Romania, Yugoslavia, Hungary and northern Italy (22). Dispersal seems to be poor and its distribution is characterized by small isolated colonies (15). Naturally has an alpine distribution (may occur as high as 2200 m in the Alps (13,43)) but has been introduced by man into lower regions (10). Introduced into countries including Norway, Sweden, Finland, Denmark, western France and the Netherlands (5,13,22), and Britain probably by the Romans (22). In the Netherlands imports during the 16th and 17th centuries, as well as more recently, have given rise to many small populations (5,6). Introduced into the U.S.A. in central Michigan (26). Distribution maps exist for the species for the U.K. (21), Hungary (35), France (8), and Austria (41) and one is in preparation for Switzerland (43).

POPULATION Unknown.

HABITAT AND ECOLOGY Usually requires limestone or calcareous soils, generally in open woodland and downland. It is found mainly in woods, hedges and tall herbage and can be a pest in vineyards (13). In Germany its distribution corresponds to calcareous regions (10). In Britain it is found on chalk in the North Downs and Hertfordshire, and on limestone in Gloucestershire. On the North Downs and in the Cotswolds it is usually found on steep scarp slopes and valley sides with a south or south-west aspect, often on grasslands at the margins of woods, on broad woodland rides, or in mixed scrub and grassland. Elsewhere found on artificial banks such as railway cuttings, hedges and chalk pits (23). In Romania, found at an altitude of 800-1000 m, in gardens, orchards, parks and forests, preferring moist places but not necessarily calcareous soils (16). In Austria it is common in broadleaf lowland forests especially along the Danube and other large rivers but it is scarce in open meadows, coniferous forests and on slopes above the treeline. It is not restricted to calcareous soils (42). In Switzerland it is most abundant in the limestone, dolomite and marl regions of the Alps, Jura and Swiss Plateau but also occurs in regions with siliceous bedrock (43).

In the Autumn Helix pomatia hibernates under twigs and leaves, having dug a hole in the earth which protects it from predators such as foxes, badgers, and rats, and also from severe climatic conditions. The shell aperture is sealed with an epiphragm, a layer of mucus with a high content of calcium carbonate and calcium phosphate which stiffens into a calcareous plate. It emerges in April or May and discards the epiphragm (14,22), probably in response to mild wet weather. Immature Helix pomatia tend to be nocturnal (3,20); adults have thicker shells which afford greater protection and they are active in the open during the day. Loss to predators is probably neglible, although some, particularly juveniles, may be eaten by rats, birds or carnivorous insects in the winter when hibernating (5). Active adults and juveniles froth vigorously when attacked. Activity is closely related to humidity; in very hot weather they often climb up the vegetation after rain possibly to escape the hottest layers of still air near the ground among the vegetation. They feed on a variety of plants (22,28), and individuals have small ranges of about 30 m in diameter (29).

Helix pomatia is hermaphrodite, and mutual fertilization occurs during copulation. Mating usually takes place after emergence from hibernation, from May to June, and only in wet weather. Courtship behaviour is lengthy; the animals rear up together and may remain with their soles adpressed for several hours, which may help to ensure cross-fertilization. More than two snails may be involved in the early stages of courtship. The final fertilization is preceded by the firing of calcareous 'love darts'. The eggs are laid in a flask-shaped cavity 6 cm deep in the soil. Laying occurs from May to August and may take up to 48 hours. An average of 40 eggs are laid which hatch after 3-5 weeks. On hatching the young snails eat the calcareous shell of the egg and may eat unhatched eggs or even other young snails, remaining in the egg cavity to emerge when substantial rain has fallen. Time taken to reach maturity varies from 2-5 years. When a snail is sexually mature there is no further increase in shell dimensions but the shell continues to thicken as new material is laid down on the inside (22,28). Mortality is highest among eggs and juveniles (5,15,24,29), largely as a result of desiccation. The reproductive potential of the species is high; each snail may lay over 100 eggs per season (24). Adults tend to be very long-lived, in some cases up to 10 years, although 5-6 years is more common (5,24,29) but reproductive success is low, which may be a major factor in the decline of the species. Since adult life is so long there is little recruitment of new adults to maintain population numbers and in England many populations have been found to contain a very low proportion of young snails (24).

SCIENTIFIC INTEREST AND POTENTIAL VALUE Helix pomatia is used for dissection and physiological experiments in educational establishments (32), for medical research and most importantly for food, for which it has been exploited on a large scale since the last century. At the beginning of this century about 70 million snails were consumed in France per year but a tenfold increase in consumption took place in less than 50 years (7). It is now eaten mainly in France, the U.K., Switzerland and the U.S.A. (32), and could perhaps provide an economic crop if managed rationally. It is currently collected for export in Austria, Germany, Hungary, Poland, Romania, Switzerland, Spain and Yugoslavia (11,32), although Austria and Switzerland have almost stopped export (10). Austria exports H. pomatia to France to the value of 3 million Austrian schillings a year, and exports are only limited by the supply of snails (19). In Poland H. pomatia has been collected by co-operatives in large numbers since the war, for export to West Europe and the U.S.A.. Exports increased from just over 91 tonnes in 1957 to over 205 tonnes in 1961, and between 1951 and 1961 40 million snails were exported (30). Switzerland is now primarily an importer and 236 609 kg of H. pomatia were imported in 1980, 90 per cent from Hungary and the rest from Poland and Spain (11). Romania exports 40-60 million per year; Hungary exported 40 million in 1974 and Germany exported 48 million between 1971 and 1973 (30,32).

THREATS TO SURVIVAL Mapping sites pre- and post-1950 suggested a marked decline in France (8,10) and fears of local extinctions in the U.K. were expressed as early as 1909 (28). Snails are reported to be declining in many places in Switzerland (4), the Netherlands (5,6) and may be threatened in Austria (42) but populations in Luxembourg, Czechoslovakia, Sweden and Bulgaria are reported to be healthy (37,38,40). The primary cause of the depletion is probably uncontrolled exploitation although when populations fall below a certain density, commercial collection ceases (39). As old snails are more easily found, declining populations with a high proportion of old snails are particularly vulnerable to collectors. However, such populations will become extinct anyway unless the cause of breeding failure is found and corrected (24). Numbers of snails collected in Yugoslavia each year since 1957 showed a decrease in spite of an increase in price paid per kg (25). In Poland H. pomatia has disappeared from some areas such as Cracow and Zamosi as a result of over-collecting, and many snails are wasted since the business is badly organized (30). Collection of snails by Mediterranean immigrants in Switzerland has been cited as one of the reasons for the decline in this country (43) Although collection for food is the main problem, populations in Limburg, Netherlands, are reported to have been depleted through collection for scientific research (6).

Pesticides (e.g. spraying of copper sulphate on vines in Bourgogne), fires, road-building, housing developments and loss of woodland have also had a serious effect (5,6,8,12,23). In Britain many sites are becoming less suitable as habitats as a result of scrub encroachment (22,24).

CONSERVATION MEASURES TAKEN Legislation concerning collection exists in a number of countries. Usually closed seasons vary from March to July or April to

191

June and shells with a diameter of less than 30 mm cannot be collected (32). The genus <u>Helix</u> (except for <u>H. aspersa,</u> a common pest species), was proposed for inclusion in Appendix II of CITES (Convention on International Trade in Endangered Species of Wild Fauna and Flora). This proposal was not accepted, mainly because the listing of <u>H. pomatia</u> under CITES would have little effect unless the countries of export had adequate regulations controlling collection and as long as no illegal trade was documented (11).

West Germany enacted its first snail collecting laws in 1936, and most states have passed laws to ensure that a site can only be exploited once in three years; the legislation has become more stringent in the last few years (32). In 1980 a decree was passed permitting collection only between 1 April and 15 June, and shells with a diameter of less than 30 mm may not be collected (34). In Bavaria the annual catch since then has been monitored and shows a slight increase, which has been interpreted as showing the efficacy of these management practices in improving production (12). In Austria sites may only be exploited once in three years, but regulations are poorly enforced (42). <u>H. pomatia</u> is listed in the regional Red Data Book for the province of Steiermark (46).

Hungary has a size restriction of 28-38 mm (27,31,32). The Netherlands imposed a total ban on collecting in 1973 (5). In Switzerland some cantons banned collecting for 32 years from 1973 (32) and it is now protected in 18 of the 26 cantons (4); in part of the country only snails larger than 32 mm may be collected (32). Poland has a minimum size limit of 28 mm for collection (30). In the province of Trento in Italy, collection of the genus <u>Helix</u> is forbidden between 1 April and 30 June; in the rest of the year collection is permitted provided the daily catch does not exceed 1 kg per person, and collection is forbidden from one hour after sunset to one hour before sunrise (2). In Luxembourg, <u>H. pomatia</u> is on the threatened species list (18) and since 1967 collection has been prohibited unless the written consent of the landowner is obtained (44). In France collection of adults is forbidden between 1 April and 30 June and individuals less than 30 mm in diameter may not be taken at all. Individual départements have the right to strengthen this law (1,9). Recently efforts have been made to publicise these regulations through the production of posters (45). In Bulgaria collection is forbidden from 20 April to 20 May (40).

Munich University has a research project in progress to study the effects of collecting on <u>H. pomatia</u> (12). In Britain, the Netherlands and probably other countries as well, <u>H. pomatia</u> occurs in a number of small nature reserves or protected areas (6,32); in Poland it occurs in the Ojcow National Park (30) and in Switzerland at the eastern edge of the Swiss National Park in the lower Engadine (43).

<u>CONSERVATION MEASURES PROPOSED</u> An increase in the minimum size limit for collection would afford the species greater protection and would give each snail an opportunity to mature and reproduce. With the present three year collecting cycle in force in some countries, a population could still be seriously reduced if unfavourable conditions such as dry weather occurred in the years with no collection; the collecting cycle should therefore be increased to four years or more. Similarly the closed season is probably too short since unfavourable weather conditions could displace the egg laying peak so that it falls outside ir. It is during the period of mating and egg laying that snails are most visible and hence most easily collected (32). It has been recommended that in Poland a closed season should be implemented between 1 April and 31 July and that the minimum size limit should be raised to 35 mm. In areas where snails are heavily exploited, collection should be stopped for 3-5 years to allow the population to regenerate. Special care should be extended to populations in national parks and nature reserves, where collection should be controlled and a quota fixed each year in agreement with the Board of the National Park or the provincial conservator of

nature (30,39). Management of areas which were once open habitat and thus suitable for this species, by cutting trees and shrubs periodically, could raise productivity. Since dispersal is so poor it might be necessary to introduce snails into areas that are thought to be suitable, even if these are close to already existing populations (24). There is urgent need for more detailed, intensive population ecology studies in areas of exploitation so that effective legislation can be implemented (32).

CAPTIVE BREEDING Farming has been attempted a number of times in different countries, but so far has not been entirely successful (30,32) partly because the slow growth rate and poor reproductive capacity do not render this species ideal for captive breeding (10). So-called 'farms' are usually places where snails from the wild are fattened, or are enclosed wild areas where collecting takes place. Snails appear to be particularly susceptible to diseases when intensively reared (17,19,32). Experimental captive breeding has been attempted in Poland (30), Holland (5), Austria (19) and Hungary (33). In France research into snail culture is carried out using a number of species at the Centre Universitaire d'Héliciculture de Besançon (36). It takes a minimum of two years for a snail to reach maturity; to achieve this the juveniles born in the summer must be maintained at 20°C throughout the winter, although in the second winter they can be put into hibernation (10). An experimental breeder at Delden in the Netherlands managed to raise a population of 1500 from 22 individuals in 1970, in an area of 60 m^2. However, rearing attempts starting with several thousand specimens have failed. Experimental snail farms were still in operation at Apeldoorn and Delden in the Netherlands in 1974 (5). Captive breeding in Austria has also not always been successful largely because of disease (19), and experiments undertaken in Hungary were not pursued although results initially looked promising (33). However, it should be possible to raise the productivity of wild populations by rearing and releasing young snails. Adult snails could be collected from the wild to lay their eggs in captivity; the young snails would be taken through their first winter in small containers and then released into the wild, having passed the period of highest natural mortality (24).

REFERENCES
1. Anon. (1979). Espèces protegées. Courrier de la Nature 61: 40-44.
2. Barletta, G. (1976). I molluschi e la legge. Conchiglie 12 (9-10): 173-182.
3. Boycott, A.E. (1936). Habits of young Helix pomatia. J. Conch. Lond. 20: 224.
4. Burckhardt, D., Gfeller, W. and Müller, H.U. (1980). Animaux protégés de Suisse. Ligue Suisse pour la Protection de la Nature (LSPN), Basle. 224 pp.
5. Butot, L.J.M. (1974). De geshiedenis en de verespreiding van de Wijngaardslak in de oostelijke en noordelijke provincies van Nederland. Levende Nat. 77: 166-80.
6. Butot, L.J.M. (1975). De Wijngaardslak in Limburg. Publicaties van het Natuurhistorish Genootshap in Limburg XXV (2&3): 6-23.
7. Cadart, J. (1955). Les escargots (Helix pomatia L. et Helix aspersa M.). Savoir en Histoire Naturelle 24 Paris, Paul Lechevalier.
8. Chevallier, H. (1973). Repartition en France et importance economique de l'escargot de Bourgogne, Helix pomatia Linné. Haliotis 3: 177-83.
9. Chevallier, H. (1979). Protection des escargots et reglementation du ramassage en France. L'Escargot Ecologiste 6: 11-14.
10. Chevallier, H. (1980). Les escargots du genre Helix commercialisés en France. Haliotis 10 (1): 11-23.

11. Dollinger, P. (1981). News from Switzerland. TRAFFIC Bulletin 3 (2):18.

12. Falkner, G. (1980). Protection et exploitation de l'Escargot de Bourgogne (Helix pomatia L.) en Bavière (Allemagne). L'Escargot Ecologiste 7: 41-45.

13. Kerney, M.P. and Cameron, R.A.D. (1979). A Field Guide to the Land Snails of Britain and Northwest Europe. Collins, London. 288 pp.

14. Lind, H. (1968). Hibernating behaviour of Helix pomatia L. (Gastropoda, Pulmonata). Videnski Medd. Dansl. Naturn. Foren. 131: 129-151.

15. Lomnicki, A. (1971). Struktura i regulacja wielkosci populacji slimaka winniczka, Helix pomatia L. a niektore zagadnienia jego ochrony (Population structure and regulation of population size of the Roman snail H. pomatia L. and some problems of its conservation). Ochr. Przyr. 36: 189-255.

16. Lupu, D. (1981). In litt., 8 March.

17. Meynadier, G., Bergoin, M. and Vago, C. (1964). Bacteriose epizootique chez les Helicidés (Mollusques). Anatomie van Leeuwenboek 30: 76-80 (quoted in 22).

18. Mousset, A. and Pelles, A (1975). Wirbellose Tiere. In In Luxembourg geschützte Tiere. Natura (Luxemburger Liga für Nature und Umweltschutz), Luxemburg.

19. Nawratil, O.J. (1969). Probleme der Massenvermehrung von Helix pomatia L. (Weinbergschnecken). Malacologia 9: 135-41

20. Pollard, E. (1973). Growth classes in the adult Roman snail (Helix pomatia L.). Oecologia 12: 209-212.

21. Pollard, E. (1974). Distribution maps of Helix pomatia L. J. Conchol. Lond. 28: 239-242.

22. Pollard, E. (1975). Aspects of the ecology of Helix pomatia L. J. Anim. Ecol. 44: 305-329.

23. Pollard, E. (1980). In litt., 14 April.

24. Pollard, E. and Welch, J. (1980). The ecology of Helix pomatia L. in England. Haliotis 10 (1): 25-28.

25. Popeskovic, D., Zivanovic, D. and Mucalica, Z. (1964). Vingradski puz (Helix pomatia L.) kao hrana i izvosni materijal. Vet. Glasn. 18: 1157-8. (quoted in 22).

26. Solem, A. (1981). In litt., 23 January.

27. Soos, L. (1936). Az ehetö csiga életmodja és tenyésztése Természettud. Közl., 68: 1-4 (quoted in 22).

28. Taylor, J.W. (1909). A monograph of the land and freshwater mollusca of the British Isles Part 16: 145-244. Taylor Bros., Leeds.

29. Tischler, W. (1975). Zur Biologie und Ökologie der Weinbergschnecke (Helix pomatia). Faun. Okol. Mitt. 4: 283-298.

30. Urbanski, J. (1963). Slimak winniczek Helix pomatia L. - jego systematyka, biologia, znaczenie gospodarcze i ochrona. Ochr. Przyr. 29: 215-254.

31. Vasarhelyi, I. (1965). Az eticsiga gyüjtéséröl. Allatt. Közl. 52(1-4): 143-146 (quoted in 22).

32. Welch, J.M. and Pollard, E. (1975). The exploitation of Helix pomatia L. Biol. Conserv. 8: 155-160.

33. Gyarfas, J. (1982). In litt., 28 April.

34. Anon (1980). Federal Republic of West Germany Species Protection Decree - B Art Sch V., 25 August 1980.

35. Pinter, L., Richnovsky, A. and Szigethy, A.S. (1979).

Distribution of the recent mollusca of Hungary. Soosiana, Suppl. I.

36. Deray, A. (1980). Structure et vocation du Centre Universitaire d'Héliciculture de Besançon. Haliotis 10(1): 53-58.
37. Spitzer, K. (1982). In litt., 11 October.
38. Walden, H.W. (1982). In litt., 11 October.
39. Stepczat, K. (1976). Wystepowanie, Zasoby, uzyskiwanie i ochrona slimaka winniczka (Helix pomatia L.) w Polsce. Uniwersytet im Adama Mickiewicza w Poznaniu, Seria Zoologica, Poznan. 3: 68 pp.
40. Deltshev, C. (1982). In litt., 5 November.
41. Klemm, W. (1973). Die Verbreitung der rezenten Land-Gehäuseschnecken in Österreich. Supplement 1. des Catologus Faunae Asutriae. Denkschriften der österreichischen Akademie der Wissenschaften (mathematisch naturwissenschaft liche Klasse) 117.
42. Kühnelt, W. (1982). In litt., 17 December.
43. Turner, H. (1982). In litt., 3 December.
44. Anon. (1967). Règlement grand-ducal du 28 avril 1967 portant restriction du droit de ramasser des escargots.
45. Anon. (1982). L'escargot est protegé. Le Courrier de la Nature 79: 10.
46. Gepp, J. (Ed.) (1981). Rote Listen Gefährdeter Tiere der Steiermark. Österreichischen Naturschutzbundes, Landesgruppe Steiermark.

We are very grateful to L. Butot, H. Chevallier, C. Deltshev, P. Dollinger, F. Giusti, M. Kerney, W. Kühnelt, D. Lupu, E. Pollard, A. Reidel, K. Spitzer, H. Turner, H.W. Walden, J.M. Welch and the Musée d'Histoire Naturelle de Luxembourg for information for this data sheet.

ANNELIDA

Polychaetes, earthworms and leeches

INTRODUCTION The Annelida comprise a variety of animals with a worm-like form, their most distinguishing characteristic being the division of the body into segments. They have no solid skeleton but hydraulic pressure within the body cavity provides some degree of rigidity. Movement takes place through alternate contraction of the circular and longitudinal muscles in the body wall. Detailed knowledge of this phylum is still very sparse and even estimates of the number of living specimens vary enormously, from 8700 to 14 000. There are three classes, Polychaeta, Oligochaeta and Hirudinea.

The polychaetes or bristle worms include the most primitive annelids and make up the biggest class. They are mainly marine although there a few estuarine forms. They are often strikingly beautiful and very colourful and range in size from one mm long to a metre or more with, in free-living forms, paired lateral appendages (parapodia) which are used for crawling or swimming. Burrowing and tubiculous polychaetes commonly occur in huge numbers on the ocean floor, forming a major part of the fauna of the sea bed. They are an important source of food for many fish, particularly bottom-living species. As in many invertebrates, reproduction involves the release of eggs and sperm into the sea where fertilization takes place. A particular phenomenon, called epitoky, is illustrated by many of the bottom-living forms. Special reproductive individuals, or epitokes, are formed which leave their bottom burrows and swim to the ocean surface where spawning takes place. A well known polychaete which produces epitokes is the Palolo Worm, Eunice viridis. The non-reproductive form occupies rocks and coral crevices below the low tide mark and releases its epitokes in October and November at the beginning of the last lunar quarter.

The Hirudinea (leeches) is the smallest annelid class, in which over 500 species are known. Three quarters of them are blood-sucking ectoparasites and the remainder are predatory on other invertebrates such as worms, snails and insect larvae. Although rarely restricted to a single host species, blood-sucking leeches are usually confined to one class of vertebrates or, in a few cases, invertebrates. Marine, terrestrial and freshwater species are known, but leeches are most abundant in northern temperate freshwater lakes and ponds. Much of the North American leech fauna is shared with that of Europe. Generally they prefer shallow vegetated water which offers protection from predation as well as a substrate for locomotion and for deposition of their cocoons. In favourable environments, huge numbers may be found. Relatively few tolerate rapid currents or acid waters but they can withstand considerable oxygen depletion, and areas of high organic pollution may support large populations of, for example, Helobdella stagnalis (1,16)

The Oligochaeta are probably the best known class since they include the earthworms. Since most of the following data sheets concern earthworms, the significance of this group is discussed in greater detail than the previous two. There are about 3000 species of earthworm, of which probably only 10 per cent are well known. Many species have very restricted distributions but some are found worldwide, particularly some of the lumbricids originally native to Europe. These are generally species that are able to occupy disturbed habitats, are at their most successful in cultivated land and can survive transportation by man to colonize alien environments. European species are now the dominant earthworms in temperate climates, probably because intensive agriculture was first developed in Europe and subsequently spread to other parts of the world with early colonizing nations. For example the earthworm fauna of the larger Chilean cities has been found to consist of European species only (2). Similarly members of the Pheretima group (Megascolecidae) have spread with agriculture through the tropics from South-East Asia.

Although earthworms are terrestrial species they still retain many characteristics of their aquatic ancestors, their adaptations to life on land being behavioral rather than physiological. In spite of this they are found in almost all habitats except deserts and the driest soils. They are burrowers and often occur in huge numbers, especially if the soil contains a high proportion of organic matter, or at least has a surface layer of humus. Pastures and orchards in Europe usually have an average of 200 per m^2 and often have up to 500 earthworms per m^2. Their distribution is affected by factors such as moisture, texture and acidity. Young worms and small species are generally restricted to a few centimetres of the upper humus; larger species may be limited to the upper soil level which contains some organic matter, whereas the larger species, such as the group including the common Lumbricus terrestris, range from the surface to several metres in depth. Earthworms can migrate to much deeper levels for protection against desiccation or low temperatures, and some species are capable of going into a period of quiescence or 'diapause' during which, in dry periods, they may lose as much as 70 per cent of their water content (1).

Earthworms range in size from a fraction of a centimetre to over 6 m in length. They feed on a wide variety of organic materials, most of which are extracted from the large quantities of soil that pass through the gut while burrowing, although some feed on leaves or other plant materials directly by pulling them into their burrows. Earthworms in laboratories have been known to live for 6-10 years but the life span of most species is probably much shorter in natural unprotected conditions. Lumbricid earthworms reach sexual maturity between six months and a year of age, depending on environmental conditions and the species (1,2).

The Oligochaeta also contains about 1000 aquatic species found in habitats as diverse as semi-terrestrial sites (including water in tree stumps and bromeliads) and the deep ocean trenches. The biomass of oligochaetes in intertidal stretches of some estuaries and sheltered coasts may exceed that of many other benthic organisms, and the true diversity of salt-water species is only now being appreciated. Many species, particularly in the Tubificidae, are able to tolerate high levels of dissolved organic matter (25,34).

SCIENTIFIC INTEREST AND POTENTIAL VALUE Earthworms have generated considerable interest and, compared with many invertebrates, have been intensively studied in spite of the fact that they are not important either as pests or for food. This is largely due to their convenience as experimental animals and to their effect on soil fertility, first perceived by Aristotle, further elaborated on by Gilbert White of Selbourne in 1770, and later made the subject of one of Darwin's major works (3).

The activities of earthworms that have most effect on the soil are firstly the ingestion and partial breakdown of organic matter and the ejection of material as casts, and secondly their burrowing which brings subsoil to the surface. These activities improve the mixing of soil fractions and are responsible for a large proportion of the fragmentation of litter in woodlands of the temperate zone. Since earthworms often make a large contribution to the total biomass of invertebrates in the soil, their excreta and mortality increase the amount of mineralized nitrogen available for plant growth. Burrowing also aerates the soil and improves its water holding capacity (2).

New methods of agriculture now being used render the activities of earthworms even more essential in maintaining soil fertility than in the past. Soil often receives minimal cultivation as seeds can be sown by direct drilling and weeds are removed with herbicides. Where earthworm populations are low, ley farming, or the inclusion of grass as part of crop rotation, may have a beneficial effect. Grassland has been shown to have a higher density of earthworms than arable land, probably because more organic matter is available for food in the former. Adding organic manure may improve populations, and the effect of healthy earthworm populations on crop yields is such that worms are being bred commercially for this purpose (2).

Research is being carried out into a number of other uses for earthworms. Experiments are in progress to determine ways in which they can be most efficiently used for disposal of organic waste, either producing a useful fertilizer, or the worm crop itself being used as a source of protein. Worm meal has a high protein content (58-71 per cent on a dry-weight basis), competing well with meat or fish meal in experimental trials, and there is good reason to believe that it will become an important component in commercial feeds for intensive livestock industries and fish culture (8). The bait market is another major consumer of earthworms. For example, huge quantities (all introduced species) are exported from Canada to the U.S.A. each year for this purpose, the worms having been collected at night from golf courses and sports fields (30). Vermiculture is big business in several countries. The Japanese are currently the foremost earthworm producers with the U.S.A. second and the Philippines third (8). Farms in the U.S.A produce as many as 0.5 million worms per day (2). Lumbricids are cultured in the more northerly states of the U.S.A. and the peregrine African earthworm Eudrilus eugeniae (Eudrilidae), the African Night-crawler, is cultured in the southern states. The potential of worms as human food is also being investigated (some people, such as the Australian aborigines (28) and the Maoris in New Zealand, have traditionally made use of earthworms in this way) but worms are unlikely to become important in this way for many years (8). By contrast the Palolo Worm has long been important in Samoa. The spawning period is awaited with great expectancy, and as the epitokes rise to the ocean surface they are scooped up in armfuls. The epitokes also provide food for fish, birds and other predators (1).

Many annelids contain compounds of value to man. Earthworms could prove to be a source of new drugs and new species should be studied with this in mind. Folk medicine in several countries has prescribed earthworms for a variety of ailments including jaundice, smallpox, haemorrhoids and rheumatism. More recently earthworm extracts have been shown to have an antipyretic activity that could be made use of in medicine. Some earthworms have a bioluminescence reaction which is dependent on peroxide and can be used as a clinical test for this substance (26). However, as yet there is no substantial evidence to suggest that earthworms could make any significant contribution to human or animal medicine, although the supposed rheumatism curing properties may be worth investigating (8). The use of the leech in medicine is well known, and current work on the Medicinal Leech Hirudo medicinalis is discussed in the data sheet. Haementeria ghilianii from French Guiana, the largest leech in the world, may prove to be the most important species in research on account of its easily dissected and simple nervous system, the extracts that can be obtained from it which have potential medicinal applications, and its large eggs which permit embryological studies, unlike those of the Medicinal Leech. It is now bred in large numbers in Berkeley, California (17). The polychaete Lumbrineris brevicirra produces a toxin in its integument which has been used in the development of an insectide, effective against pests such as Colorado Beetle, Mexican Bean Beetle and insect strains resistant to organophosphates and organochlorides. It is also reported to be essentially nontoxic to warm-blooded animals and decomposes fairly rapidly in biological tissues and the natural environment (29).

Aquatic oligochaetes play a major role in the retrieval of organic matter from the sediments of lakes and rivers. Leeches and aquatic oligochaetes have been considered as potential indicators of pollution, but many species that are an obvious feature of polluted waters also occur in clean habitats and there appears to be no good correlation between environmental variables and the species' observed distributions (16, 25, 31).

THREATS TO SURVIVAL   The conservation status of most annelid species is, not surprisingly, poorly known. For earthworms this is largely because most sampling is done in agricultural areas, and as a result species unable to survive in soils disturbed by cultivation are seldom recorded, and come to be regarded as being very rare. In many cases, if sampling is carried out in undisturbed habitats, such rare species are found (4). However, many have extremely limited geographical distributions and may be highly specialized, morphologically and physiologically, to fit a limited range of soil conditions. A large number probably are seriously at risk from deforestation, climatic changes, or a combination of both. For example, in Australia where there are about 400 described earthworms with probably many more undiscovered, several of those with small ranges have already been exterminated by large scale cane farming in north Queensland. Other species in south Australia, dependent on native vegetation, are now only found along road sides or even on traffic islands where the indigenous eucalyptus trees have been left (5,23,27).

In South Africa, earthworms in the families Acanthodrilidae and Microchaetidae are probably already extinct in Transvaal and Natal as a result of deforestation and soil erosion. Since the Europeans arrived in South Africa about 97 per cent of the indigenous forest has been destroyed and much of the grassland has been replaced with exotic trees and crops. As a result the organic matter content of the soil is a fraction of what it was originally (6).

The American Pacific Northwest is unusual in terms of its earthworm ecology since most of the areas moist enough for earthworms are still inhabited by native species. This is because forest land is still left with native tree species, an unusual practice still dominating the Northwest timber industry largely because of the value of Douglas Fir. However, this is changing as fertilizers, introductions of new species, wide thinning, scarification and other more extreme practices are used. The next twenty years may see a dramatic change in the number of endangered species since all heavily ploughed and grazed areas now mainly have introduced species (19). A second refugium for native species is in the southern states from the Carolinas to Texas and Oklahoma (20).

One of the consequences of the loss of native vegetation and its accompanying earthworm fauna is the further expansion of the ranges of peregrine earthworms, species such as the lumbricids that are able to colonize disturbed habitats. The latter are very rarely actually responsible for the disappearance of native species; in very few cases are they ever found together, and usually the peregrine species only move in when the native species have been eliminated through loss of habitat (2). For example, in New Zealand the rich native earthworm fauna (all in the family Megascolecidae) is disappearing wherever native vegetation is being cleared and replaced by pasture, and lumbricids from Europe subsequently colonize these areas (7). In South America the native Acanthodrilidae are disappearing from the Chilean-Patagonian subregion and the peregrine Lumbricidae are becoming more common. In Brazil, a similar situation exists with the native Glossoscolecidae and other peregrine species in the families Megascolecidae and Octochaetidae (24).

Changes in agricultural practices also affect earthworm populations. Large scale arable farming, with no livestock, can cause a lowering of the organic content of the soil and thus a reduction in food for earthworms. It is thought that this may have caused the decline in earthworm numbers noticed in eastern areas of the U.K., and contributed indirectly to a catastrophic decline in the rook Corvus frugilegus population in the same area (9). Several insecticides have been found to affect numbers and biomass of earthworms in the soil. In the long term this could have a serious effect, not only on earthworm populations, but also on soil aeration (21). Primitive agricultural practices may be affecting native earthworm populations in South America, since periodical burning of stubble leads to calcination of the superficial layers of the soil (24).

Earthworms with very restricted ranges are vulnerable to many of man's activities. The monotypic Lutodrilus multivesiculatus, (in the monogeneric family Lutodrilidae) is known only from river mudbanks in a few localities in south-east Louisiana and is probably a relic of the most recent North American glaciation (18). Komarekiona eatoni (also in a monospecific genus in the monogeneric family Komarekionidae) is another relic of the last glaciation. Although populations are known from Indiana, Illinois, Virginia and Tennessee, these are male-sterile, parthenogenetic forms. The only sexually reproducing forms known occur below or east of the Appalachia Mountains in North Carolina (20,22). The species has been considered for the preliminary list of North Carolina endangered species, since the population has been largely destroyed by construction work (10) and may even be too close to extinction to be rescued (18). Island species could also be at risk, such as the genus Microscolex which forms 2-3 superspecies on islands of the southern oceans, and the two lumbricid species which form the subfamily Diporodrilinae in Corsica (4).

There is even less information available on the status of aquatic annelids but several species have very limited ranges. A large part of the family Lumbriculidae, for example, is known only from Lake Baikal, where 90 per cent of the oligochaetes recorded from the open waters are endemic. In the coastal belt oligochaetes account for at least 20 to 30 per cent of the total mass of invertebrates and on silty or sandy substrates at greater depths they make up 40 to 100 per cent (25,36). This fauna has been under considerable threat from pollution resulting from rapidly increasing industrialization and urban growth on the lake shores although efforts are now being made to control this (37). Endemic tubificids are found in Yugoslavia and Albania in the lakes Ochrid, Prespa, Dojran and Skadar which could become threatened in a similar way to Lake Baikal. Endemic species in the Haplotaxidae, found in groundwater springs and caves in karst areas in the Pyrennées and Roumania, could be threatened by pollution (25). Haplotaxis ornamentus, described in 1980 from Great Lake and Arthur's Lake, Tasmania, may be vulnerable to changes in the lake level. Both lakes have had their levels raised for hydro-electric purposes but this species has only been found in the original levels of the lakes (35).

For leeches there is evidence that if surveys were carried out, a number of species would be found to be threatened. The data sheet on the Medicinal Leech Hirudo medicinalis, gives one such example. A survey of leeches in South Carolina revealed that one marine species, parasitic on skates and rays, might now be extinct within the state, although it still occurs elsewhere (15). Other species give cause for special concern. For example, Ozobranchus margoi is a marine leech found exclusively on the sea turtles Chelonia mydas and Caretta caretta which are listed as Endangered and Vulnerable respectively in the IUCN Amphibia-Reptilia Red Data Book (33). Clearly the future survival of this leech is linked with that of the sea turtles (15).

Marine polychaetes are less likely to be seriously threatened. In some places, however, there have been local depletions of populations of the lugworms Arenicola and Nereis, as well as destruction of habitat, through the activities of bait diggers (32). Oil pollution may affect some species. Worms living in sandy or muddy beaches may be killed in large numbers if toxic oil penetrates the substratum. Different species vary in their sensitivity to oil, and tolerant, hardy species, which are the first to recolonize an oil damaged sediment, dominate the early stages of restoration of the community. Lugworms that eat contaminated sediments play an important role by bringing oily residues to the surface and hastening their degradation (38).

It is possible, although scientific evidence is at present lacking, that the Palolo Worm of the Pacific is becoming rarer. It is extremely difficult to carry out any survey of its population size in view of its habitat, and it is well known that the annual harvest of the epitokes is extremely variable. However it has been noticed that the Palolo has apparently completely disappeared from almost the entire northern coast of Upolu in Samoa. In the 1940s and 1950s the worms disappeared from Palolo Deep, where they were very abundant in the 1930s. The few places where they can still be found seem to be in areas where there is little agricultural activity and practically no construction or earthmoving work on the neighbouring coast. The Palolo has also gone from Suva harbour in Fiji, possibly as a result of the general pollution to be found there now (14).

CONSERVATION Given the high densities at which earthworms can occur, the future of rare species could be safeguarded fairly easily by the preservation of small areas of native vegetation. This would be feasible for the endemic Australian and North American species. For example the Australian Giant Earthworm Digaster longmani which is found in rain forest areas between Tamborine and Kyogle along the Queensland–New South Wales border may be well provided for by existing National Parks (11). New Zealand also has a high number of endemic earthworms which do not seem to be under any major threat at present, but for which action may need to be taken in the future (4). If reserves are created for annelids, appropriate management techniques are needed as it has been found, for example, that trampling in some National Parks in Australia is reducing earthworm populations (23).

Protection may be necessary for the breeding areas of the Palolo Worm in the Pacific. In 1979 the Palolo Deep Marine Reserve was established at Upolu in Samoa but, as mentioned earlier, the Palolo Worm appears to have disappeared there. Reserves specifically for this species have been recommended for Fiji and Western Samoa, such as at the known breeding area in Salammau (12). In view of the lack of knowledge about the ecology, distribution and population biology of this species, it is essential that research should be initiated on it as soon as possible (13).

REFERENCES
1. Barnes, R.D. (1980). Invertebrate Zoology, 4th Ed. Saunders College, Philadelphia.
2. Edwards, C.A. and Lofty, J.R. (1979). Biology of Earthworms, 2nd Ed. Chapman and Hall, London.
3. Darwin, C. (1881). The Formation of Vegetable Mould through the Action of Worms. Murray, London. 326 pp.
4. Sims, R.W. (1980). In litt., 2 June.
5. Jamieson, B. (1980). In litt., 10 July.
6. Ljungstrom, P.O. (1972). Introduced earthworms of South Africa. On their taxonomy, distribution, history of introduction and on the extermination of endemic earthworms. Zool. Jb. Syst. Bd. 99:1-81.
7. Lee, K.E. (1961). Interactions between native and introduced earthworms. Proc. N.Z. Ecol. Soc. 8:60-62.
8. Sabine, J. (1982). Earthworms as a source of food and drugs. In Satchell, J.E. (Ed.), Proceedings of the Darwin Centennial Symposium on Earthworm Ecology. Grange-over-Sands, Cumbria, October 1981.
9. Tabor, R. (1981). Wildlife diary. Wildlife 23 (10): 21.
10. Cooper, J.E. and Shelley, R.W. (1977). Other invertebrates. In Cooper J.E., Robinson, S.S. and Funderburg, J.B. (Eds), Proc. Symp. on Endangered and Threatened Plants and Animals in N. Carolina. N.C. State Mus. Nat. Hist., Raleigh.

11. Marks, E.M. (1969). The invertebrates. In Webb, L.J., Whitelock, D. and Brereton, J.L.G. (Eds), The Last of the Lands. Milton, Jacaranda

12. Dahl, A.L. (1980). Regional Ecosystems Survey of the South Pacific Area. Technical Paper 179, South Pacific Commission, Noumea, New Caledonia.

13. Raj, U. (1982). In litt., 9 February.

14. Marschall, K.J. (1981). In litt., 25 October.

15. Sawyer, R.T. (1976). Leeches of special concern from South Carolina. In Forsythe, D.M. and Ezell, W.B. (Eds), Proceedings of the First South Carolina Endangered Species Symposium, Charleston, South Carolina.

16. Sawyer, R.T. (1974). Leeches (Annelida: Hirudinea). Chapter 4. in Hart, G.W. and Fuller, S.L.H. (Eds,) Pollution Ecology of Freshwater Invertebrates. Academic Press, London.

17. Branning, T. (1982). Learning from a giant leech. National Wildlife 20(3): 35-37.

18. McMahan, M.L. (1976). Preliminary notes on a new megadrile species, genus and family from the southeastern United States. Megadrilogica 2(11): 6-8.

19. Fender, W.H. (1982). In litt., 28 April.

20. Gates, G.E. (1974). On a new species of earthworm in a southern portion of the United States. Bull. Tall Timbers Res. Stn. 15: 1-13.

21. Thompson, A.R. (1971). Effects of 9 insecticides on the numbers and biomass of earthworms in pasture. Bull. Env. Contamination and Toxicology 5: 577-586.

22. Reynolds, J.W. (1977). The earthworms of Tennessee (Oligochaeta). III. Komarekionidae with notes on distribution and biology. Megadrilogica 3(4): 65-69.

23. Wood, T.G. (1974). The distribution of earthworms (Megascolecidae) in relation to soils, vegetation and altitude on the slopes of Mt. Kosciusko, Australia. J. Anim. Ecol. 43: 87-106.

24. Righi, G. (1982). In litt., 4th February.

25. Brinkhurst, R.O. and Jamieson, B.G.M. (1971). Aquatic Oligochaeta of the World. Oliver and Boyd, Edinburgh.

26. Jamieson, B.G.M. (1981). The Ultrastructure of the Oligochaeta. Academic Press, London and New York.

27. Jamieson, B.G.M. (1974). Earthworms (Oligochaeta: Megascolecidae) from south Australia. Trans. Roy. Soc. S. Aust. 98: 78-112.

28. Brough Smyth, R. (1878). The Aborigines of Victoria. Vol.1. John Ferres, Government Printer, Australia.

29. Ruggieri, G.D. (1976). Drugs from the sea. Science 194: 491-497.

30. Tomlin, A.D. (1982). The earthworm bait market in North America with special reference to Ontario, Canada. In Satchell, J.E. (Ed.), Proceedings of the Darwin Centennial Symposium on Earthworm Ecology, Grange-over-Sands, Cumbria, October 1981.

31. Brinkhurst, R.O. and Cook, D.G. (1974). Aquatic earthworms (Annelida: Oligochaeta). In Hart, C.W. and Fuller, S.L.H. (Eds), Pollution Ecology of Freshwater Invertebrates. Academic Press, New York. Pp.143-156.

32. NCC (1979). Nature Conservation in the Marine Environment. Report of the NCC/NERC Joint Working Party on Marine Wildlife Conservation, Interpretive Branch,

Nature Conservancy Council, Shropshire, U.K.

33. Groombridge, B. and Wright, L. (1982). The IUCN Amphibia-Reptilia Red Data Book, Part I. IUCN Conservation Monitoring Centre, Cambridge, U.K.

34. Brinkhurst, R.O. (1982). British and Other Marine and Estuarine Oligochaetes. Cambridge University Press, Cambridge, U.K.

35. Brinkhurst, R.O. and Fulton, W. (1980). On Haplotaxis ornamentus sp. nov. (Oligochaeta, Haplotaxidae) from Tasmania. Rec. Queen. Vict. Mus. 72.

36. Kozhov, M. (1963). Lake Baikal and its Life. W. Junk, The Hague.

37. Galaziy, G.I. (1980). Lake Baikal's ecosystem and the problem of its preservation. J. Mar. Tech. Soc. 14(7): 31-38.

38. R.C.E.P. (1981). Oil Pollution of the Sea. Eighth report of the Royal Commission on Environmental Pollution, October 1981. H.M.S.O., London.

We are very grateful to R.O. Brinkhurst, W.M. Fender, G.E. Gates, S. James, B.G.M. Jamieson, K. Marschall, U. Raj, G. Righi and R.W. Sims for providing information for this section.

MEDICINAL LEECH   INDETERMINATE

Hirudo medicinalis Linnaeus, 1758

Phylum   ANNELIDA   Order   ARHYNCHOBDELLAE

Class   HIRUDINEA   Family   HIRUDINIDAE

SUMMARY The range and abundance of the Medicinal Leech was much reduced in Europe as a result of over-collecting in the last century and it is now extinct in many countries. It is currently threatened by habitat loss, changing agricultural practices, and renewed collecting pressure for medical research. Loss of marsh habitat may be the current major threat, since this is also causing a decrease in the abundance of frogs, which are hosts of the leech.

DESCRIPTION The body is elongate and flattened and is widest at about the 16th somite, tapering anteriorly and posteriorly. In extreme extension it reaches 100-125 mm, with a width of 8-10 mm; in extreme contraction it is 30-35 mm long and 15-18 mm wide. The dorsal surface is usually olive green, richly variegated with reddish-brown, yellowish-green, orange and black. The pattern is very variable, generally based on three pairs of reddish-brown or yellowish longitudinal stripes often interrupted by black spots on the last ring of each somite. The ventral surface is usually yellowish-green and more or less spotted with black, and has a pair of black marginal stripes. The jaws have somewhat fewer than a hundred sharp teeth and make a Y-shaped incision in the host's tissue (9).

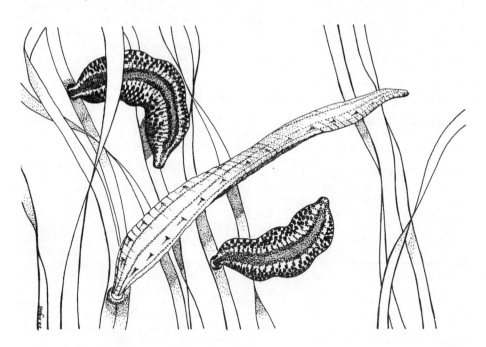

DISTRIBUTION Its range formerly extended from western and southern Europe to the Ural Mountains and the countries bordering the eastern Mediterranean (9,18).

It was introduced into several parts of Scandinavia but the extent to which it occurs there naturally is not fully known. For Finland many records exist up to the beginning of this century for the area as far north as 63°N, but it is now only known from a few localities on the south-west mainland coast and on Aland (52). It was reported as extinct in Norway in 1854 (7), but has been found at six localities in the south of the country since 1960 (49,50,51). In Sweden it appears to have been relatively common in the 19th century but to have declined during this century (1,53). It is now known from a small number of localities, mainly in Scania, the southernmost province, and on the Baltic islands Öland and Götland (54). It is currently known from 33 localities in Denmark (57).

Considered almost extinct in Germany (West and East) by 1922 (1) but a few localities probably still persist (56). Rediscovered in 1946 in Holland after decades of absence and a survey of its current distribution has been carried out (5). In Luxembourg it still occurs in small ponds and rivulets a few kilometres south of the city of Luxembourg (59,60,61,62).

It was once common in Britain but was becoming less frequent as early as 1816 (12); declared extinct in the U.K. in 1910 (9) but subsequent isolated records exist for the New Forest, Islay, Anglesey, Sutherland and Yorkshire (4,16,17,25,26) and a number of other sites including Cumbria, Lincolnshire, Norfolk, Argyll (28,29,30,43). The Anglesey site probably no longer contains Hirudo as the lake has been drained and refilled (20). The most recent U.K. record is from Kent in 1981; the survival of the leech in this locality may be related to the presence of an introduced population of Marsh Frogs, Rana ridibunda, which are still common (22). It has not been recorded in Ireland for over 100 years (14).

Occurs in France in the Camargue (41), probably still in the Marais de Carentan in Normandy (42) and possibly in other localities. Three localities were reported in Italy during the 1970s, but there are very few other recent records in this country (15). It probably still occurs in Switzerland (56). Still occurs in Hungary and said to be common in the Kisbalatan on the Hungarian plain (18,47). It occurs in Czechoslovakia but is said to be very rare and to be extinct in many parts of the country (55). Still abundant in some parts of Bulgaria (58). Thought to be present in Yugoslavia, Greece and Turkey (36). It seems likely that, as attention is drawn to this species, more locality records will become available.

POPULATION Unknown.

HABITAT AND ECOLOGY Occurs under stones in freshwater ponds, streams and marshes, often in areas where farm animals graze (25). In Denmark and Germany Medicinal Leeches have been found in intensely eutrophic village ponds and also in temporary pools, but never in lakes or marshes (3). The localities where they have been found in Finland since the 1960s are marshes with a pH of 6-7 (52). They appear to be able to live in areas too anoxic for most other leeches (10) and are often found in sluggish and stagnant waters. Young leeches feed on frogs and tadpoles (13). Adults are external blood sucking parasites on warm-blooded vertebrates and can eat 2-5 times their own weight at a single meal. Digestion is slow (they may require 200 days to digest a meal) and medicinal leeches have been reported to go without food for up to one and a half years (2). Little is known of their life cycle. It probably takes at least two years for the breeding stage to be reached in the wild and there may be a delay of one to nine months between copulation and cocoon deposition. The scelerotized cocoons are about 10 mm long and are laid in a damp place (14). Leeches are reported to bury themselves in cold weather and during hot periods when the marshes dry out (34,36).

SCIENTIFIC INTEREST AND POTENTIAL VALUE The Medicinal Leech was used intensively in the 19th century for blood-letting therapy, and an enormous trade developed. France was the biggest user and it has been estimated that in the 19th century over a thousand million leeches were imported into this country alone (18). Blood-letting is still used occasionally as a treatment in the remoter rural parts of Europe (34,36) but this practice gradually died out in most countries at the beginning of the century. Recently, however, interest in the species has been revived in the medical and scientific world on account of the compounds that can be obtained from it. It secretes a powerful anticoagulant called hirudin which is used in research into the mechanism of the human blood-clotting process. Hirudin is supplied by a number of pharmaceutical firms in Europe on a commercial basis, most of which is obtained from Hungarian leeches (18,32,33,34,35), although there are reports that they are also collected in Italy (34,46), France (35), Yugoslavia (36,46), Greece and Turkey (36). It is estimated that 12 000 kg of leeches are used a year (35). A firm in Italy collects about 20 000 specimens a year (46). At least five other medically useful substances have been isolated from Hirudo salivary glands, including a highly specific enzyme which dissolves blood clots (31). When the structure of these substances have been determined finally it is hoped that large quantities will be able to be synthesized (18,44). It is thought that the substances of greatest importance (hirudin, eglin and bdellin), could be produced by genetic engineering eventually, and that collection for medicinal purposes on a large scale could stop within 5-10 years (40). Recently Medicinal Leeches have been supplied in increasing numbers to hospitals for use in micro-surgery and skin-grafting operations (35,37); for example, a firm in Slough, U.K. imports about 5000 a year for use in dissection and microsurgery (37). Leeches have a highly visible and very simple central nervous system which can be manipulated in the laboratory (38,39), and it is estimated that neurophysiologists, especially in the U.S.A. probably use some 2000 specimens a year for dissection and experimentation (33). H. medicinalis is reported to be important in the diet of the Glossy Ibis in Hungary (47).

THREATS TO SURVIVAL   The Medicinal Leech is said to be endangered in the U.K. (18) and Czechoslovakia (55), vulnerable in Bulgaria (58), declining in Finland (52), Luxembourg (60) and France (35) although not currently threatened in Sweden (54) or Denmark (57). Its status in other European countries needs to be determined.

The enormous magnitude of the leech trade and the use of leeches as a form of blood-letting therapy in the last century contributed directly to its present status. By the beginning of the 19th century local sources were depleted in western Europe, especially in France, Britain and Germany. Leech dealers then started importing, primarily from the vast marshes of Hungary and the Balkan countries (18). In the 1940s the leech was still being collected in reservoirs in the northern Caucasus and Ukraine for use in research in the U.S.S.R. but there were already fears that these would become exhausted (24). In 1972 it was still being used in the Lithuanian S.S.R. but the local supply was depleted (27). The recent revival in scientific and medical interest in the species has renewed pressure on it since thousands of leeches are required to obtain small quantities of purified hirudin. Pharmaceutical firms in Europe are said to be unable to fulfil the needs of medical researchers in Europe and North America (18).

Changes in farming methods may also have contributed to the decline of the leech by destroying suitable habitats. Closer control of horses and cattle, and the habit of watering them at troughs rather than at natural ponds, has decreased host availability and opportunites for expansion of its range (3,13,52). Paradoxically, in Sweden, it is thought that changes in farming methods may have benefited the species since free-ranging cattle herds have recently been introduced (54).

General loss of marsh habitat has been cited as a factor in its decline in France, Bulgaria and Czechoslovakia (36,55,58). The gradual drying out of marshes has been said to make it difficult to obtain leeches in commercial quantities in France (35). The Camargue population may be suffering a decline on account of the recent two-year drought which has caused many marshes to dry out for long periods and others to have increased in salinity (41). The Marais de Carentan in

Normandy is currently threatened by drainage for agriculture in spite of the fact that it is nominally protected (42). The disappearance of the leech may also be associated with a decline in frog populations. Frogs are the main hosts for juveniles and have equally been affected by loss of marsh habitat (8,11). The abundance of the leech in Kent, U.K., is probably due to the large population of Marsh Frogs still to be found there (45).

CONSERVATION MEASURES TAKEN Several countries introduced legislation in the last century in an attempt to control the leech trade. In 1823 Hanover forbade the export of leeches. In 1827 the Austrian government leased to two dealers the exclusive rights for five years to use special reservoirs for leech collecting. In 1828 leech exportation was forbidden from Sardinia for two years. Leech exportation was stopped in 1844 from Wallachia and in 1850 from Spain and Portugal due to depletion of leeches. In 1848 the Russian government put a high tariff on the export of leeches and declared the months from May to July closed to collection. Hungary is reported to have had an embargo on leech exports but this no longer exists (33). Leech breeding programmes were started in a number of countries in attempts to meet demands in the face of declining supplies but were generally unsuccessful in economic terms(18).

CONSERVATION MEASURES PROPOSED In view of the sparse information available on the current distribution of H. medicinalis, the first step should be a thorough survey of its range. It is becoming apparent that small populations have frequently been overlooked and that there is a need for the collation of all available distribution data. A three-year study is to be made of the biology and conservation potential of the newly discovered Kent population where the owners of the site are being co-operative with regard to its protection (45). However, given the number of countries which have reported a decline, it seems that the species should be considered threatened until further information becomes available. It is being considered for listing under a new decree for protected animals in Luxembourg (62).

It is being proposed for listing on the Appendix II of CITES (Convention on International Trade in Endangered Species of Wild Fauna and Flora in Commerce). If the proposal is accepted, commercial trade in the species between countries which are parties to CITES would only be permitted provided a valid export permit from the country of origin had been obtained. This measure would afford some control over the trade and would provide a means of monitoring it.. The U.S. trade in leeches collected in the wild should be controlled. Wild populations should be protected, and breeding programmes set up to supply the demand for live Medicinal Leeches and for hirudin. Reintroduction into the wild could be attempted. The use of alternative species such as Hirudinaria spp., the buffalo leeches still common in Asia, should be investigated. Other species such as Haemopis sanguisuga and Haementeria ghilianii should be used for educational purposes in schools and Hirudo medicinalis only for medical work (18). H. sanguisuga is already increasingly being used in schools. In fact, H. ghilianii from French Guiana may be more suitable for dissection purposes (38) but great care would need to be taken if these species are imported live in large quantities to avoid their introduction into the wild.

A detailed project has been drawn up which would involve the development of a breeding facility to produce an annual standing crop of about 10 000 H. medicinalis, (patterned after a similar facility established in California for Haementeria ghilianii), long term ecological studies, genetic engineering and tissue culture studies and further research into the medical application of leech compounds (48).

CAPTIVE BREEDING Breeding programmes were started in the mid-nineteenth century in France, Germany, and the U.S.A. (6,7,10). Research into leech breeding has been undertaken in the U.S.S.R. and under laboratory conditions it was found that two generations a year could be produced and that leeches suitable for medicinal use could be reared in 6-12 months (27). However, the technical problems of breeding leeches coupled with the decline in their use led to the abandonment of leech farming in Europe and the U.S.A. Experiments have also been carried out on the captive rearing of leeches, to determine the hosts which must be provided in order to raise newly hatched leeches to a size suitable for medical use, as quickly as possible (24). In the 18th and 19th centuries many Medicinal Leeches were imported into Finland and pond culture was attempted but with no success (52). Recently the giant Amazonian leech Haementeria ghilianii has been bred successfully in captivity (21) and it is thought that the same should be possible for the more temperamental Medicinal Leech (18).

REFERENCES

1. Arndt, W. (1940). Als Heilmottel gebrauchte Stoffe Q. Blutegel pp. 524-573. Die Rohstoffe des Tierreiches Q. Berlin.

2. Barnes, R.D. (1980). Invertebrate Zoology 4th Ed. Saunders College, Philadelphia.

3. Bennike, S.A.B. (1943). Contributions to the ecology and biology of the Danish freshwater leeches (Hirudinea). Folia Limnol. Scand. 2: 1-109.

4. Blair, W.N. (1928). Notes on Hirudo medicinalis the medicinal leech, as a British species. Proc. Zool. Soc. London (1927): 999-1002.

5. Dresscher, T.G.N. and Higler, L.W.G. (1982). De Nederlandse Bloedzuigers. Hirundinea. Wetenschappelijke Mededelingan. Koninklijke Nederlandse Naturhistorische Vereniging 154: 64pp.

6. Egidy, H.E. (1844). Die Blutegel zucht, nächst aüsfuhrlicher Beschreibung der Blutegel, seiner Arten und Varietäten, 185 pp.

7. Fermond, C.H. (1854). Monographie des Sangsues Medicinales Paris 512 pp.

8. Groombridge, B. (1980). Amphibians and Reptiles. Part IV. In A Draft Community List of Threatened Species of Wild Flora and Vertebrate Fauna. Vol I. Nature Conservancy Council, London.

9. Harding, W.A. (1910). A revision of the British leeches. Parasitology 3: 130-201.

10. Hessel, R. (1884). Leech culture. Bull. U.S. Fish Comm. 4: 175-176.

11. Honegger, R.E. (1978). Threatened Amphibians and Reptiles in Europe. European Committee for the Conservation of Nature and Natural Resources, Council of Europe, Nature and Environment Series 15., Strasbourg.

12. Johnson, J.R. (1816). A Treatise on the Medicinal Leech. London 8vo.

13. Mann, K.H. (1955). The ecology of the British freshwater leeches. J. Anim. Ecol. 24: 98-117.

14. McCarthy, T.K. (1975). Observations on the distribution of the freshwater leeches (Hirudinea) of Ireland. Proc. R. Ir. Acad. 75B: 401-451.

15. Minelli, A. (1979). Sanguisughe d'Italia. Catalogo orientativo e considerazioni biogeografiche. Lavori Soc. Ital. Biogeogr. N.S., 4: 279-313.

16. Reynoldson, T.B. (1952). A record of the leech Hirudo medicinalis from Islay, with brief mention of other species. Scott. Nat. 64: 164-166.

17. Reynoldson, T.B. (1956). An Anglesey record of the

medicinal leech <u>Hirudo medicinalis</u> L. <u>Trans. of the</u>
<u>Anglesey Antiquarian Society and Field Club</u>: 54.

18. Sawyer, R.T. (1976). The medicinal leech, <u>Hirudo</u>
<u>medicinalis</u> L., an endangered species. In Forsythe, D.M. and
Ezell, W.B. (Eds), <u>Proc. 1st South Carolina Endangered</u>
<u>Species Symposium</u>.

19. Sawyer, R.T. (1980). In litt., 8 September.

20. Sawyer, R.T. (1980). In litt., 13 October.

21. Sawyer, R.T., Le Pont, F., Stuart, D.K. and Kramer, A.P.
(1981). Growth and reproduction of the giant glossiphoniid
leech <u>Haementeria</u> <u>ghilianii</u>. <u>Biological Bulletin</u> 160:
322-331.

22. Scofield, A.M. (1981). In litt., 14 August.

23. Scofield, A.M. (1981). <u>A Check List of the Helminth</u>
<u>Parasites of Domestic Animals of the United Kingdom.</u>
Animal Health Division, Hoechst U.K. Ltd., Milton Keynes.

24. Sineva, M.V. (1944). (Observations on breeding the medicinal
leech). <u>Zool. Zhurn.</u> 23 (6): 293-303.

25. Warwick, T. (1961). The vice county distribution of the
Scottish Freshwater leeches and notes on the ecology of
<u>Trocheta bykowskii</u> (Gedroyc) and <u>Hirudo medicinalis</u> in
Scotland. <u>Glasg. Nat.</u> 18: 130-135.

26. Whitehead, H. (1952). The medicinal leech (<u>Hirudo</u>
<u>medicinalis</u> L.) in Yorkshire - another record. <u>Naturalist</u>
Hull no. 843: 158.

27. Zapkuvene, D.V. (1972). (Breeding and growing of medicinal
leeches under laboratory conditions). <u>Lietuvos TSR Moksl.</u>
<u>Acad. Darb. Ser. C.</u> (Trudy Acad. Nank <u>Litovski CCP, Ser. B)</u>
3(59): 71-84.

28. Elliott, J.M. and Tullett, P.A. (1982). <u>Provisional Atlas of</u>
<u>the Freshwater Leeches of the British Isles.</u> Freshwater
Biological Association, Occasional Publication no. 14.

29. Hale, Dr. (1981). Pers. comm.

30. Waterhouse, Mr. (1981). Pers. comm.

31. Sawyer, R. (1981). In litt., 29 November.

32. Vargha, B. (1982). In litt., 29 January.

33. Sawyer, R. (1981). In litt., 12 September.

34. Feliciano, A. (1981). In litt., 7 October.

35. Desbarax, J. (1981). In litt., 30 June.

36. Nell, E. (1981). In litt., 6 October.

37. Lucas, J. (1982). In litt., 20 May.

38. Branning, T. (1982). Learning from a giant leech. <u>National</u>
<u>Wildlife</u> 20(3): 35-37.

39. Muller, K.J., Nicholls, J.G. and Stent, G.S. (Eds.) (1981).
<u>Neurobiology of the Leech.</u> Cold Spring Harbour Laboratory.

40. Fritz, H. (1981). In litt., 7 December.

41. Britten, R. (1982). In litt., 22 March.

42. Debout, G. and Provost, M. (1981). Le Marais de la
Sangsuriére. <u>Le Courrier de la Nature</u> 74: 10-18.

43. Ratcliffe, D.A. (Ed.) (1977). <u>A Nature Conservation Review</u>
<u>Vol. 2.</u> Cambridge University Press. p. 203.

44. Petersen, T.E., Roberts, H.R., Sottrup-Jensen, L. and
Magnusson, S. (1976). Primary structure of hirudin, a
thrombin-specific inhibitor. In <u>Protides of the Biological</u>
<u>Fluids.</u> Peters, H. (Ed.), Proc. 23rd Colloquuim, Brugge.
1975 Pergamon Press.

45. Scofield, A.M. (1982). In litt., 26 May.

46. Oriano, G. (1981). In litt., 10 October.

47. Keve, (1968). Über die Arealveränderungen von <u>Plegadis</u>

falcinellus (L.). Zoologische Abhandlungen, Staatliches Museum für Tierkunde in Dresden, 29(13): 169.

48. Sawyer, R. (1982). Project proposal for studies on Hirudo medicinalis.
49. Tvermyr, S. (1965). Legeiglen (Hirudo medicinalis L.) finnes enna frittlevende i Aust-Agder. Fauna (Oslo) 18: 136-139.
50. Nilssen, J.P. (1980). Acidification of a small watershed in southern Norway and some characteristics of acidic aquatic environments. Int. Revue Ges. Hydrobiol. 65: 177-207.
51. Okland, K.A. (1982). In litt., 5th October. (Data from "File of freshwater invertebrates of Norway", at the Section of Limnology, University of Oslo, Norway).
52. Terhivuo, J. (1982). In litt., 20 September.
53. Forselius, S. (1952). Blodigeln (Hirudo medicinalis L.) i Norden. Särtryck ur Sr. Faun. Rery 3: 67-79.
54. Dahm, A.G. (1982). In litt., 15 September.
55. Spitzer, K. (1982). In litt., 11 October.
56. Elliott, J.M. (1982). Pers. comm., 23 August.
57. Kirkegaard, J.B. (1982). Pers. comm. to H. Enghoff.
58. Deltshev, C. (1982). In litt., 5 November.
59. Hoffman, J. (1955). Faune hirudinéenne du Grand-Duché de Luxembourg. Institut Grand-Ducal, Section des Sciences des Sciences naturelles, physiques et mathématiques, Archives N.S. 22: 200-202.
60. Hoffman, J. (1955). Signalement d'une importante station de Hirudo medicinalis L. an Grand-Duché de Luxembourg. Institut Grand-Ducal, Section des Sciences naturelles, physiques et mathématiques, Archives N.S. 22: 213-222.
61. Hoffman, J. (1960). Notules hirudinologiques, II. Nouvelle station de Hirudo medicinalis au Grand-Duché. Instit. Grand-Ducal, Section des Sciences naturelles, physiques et mathématiques, Archives N.S. 27: 289.
62. Meyer, M. (1982). In litt., 9 November.

We are very grateful to R. Sawyer for information supplied for this data sheet, and to R. Britten, A.G. Dahm, C. Deltshev, J.M. Elliott, H. Enghoff, C. Holmquist, A. Minelli, K.A. Okland, M. Richards, A.M. Scofield, J.O. Solem, A. Soos, K. Spitzer, J. Terhivuo and staff of the Musée d' Histoire Naturelle and the Administration des Eaux et Forêts in Luxembourg for their helpful comments.

NORTH AMERICAN GIANT EARTHWORMS   ENDANGERED

Megascolides americanus Smith, 1897   Washington Giant Earthworm
M. macelfreshi Smith, 1937   Oregon Giant Earthworm

Phylum   ANNELIDA   Order   HAPLOTAXIDA

Class   OLIGOCHAETA   Family   MEGASCOLECIDAE

SUMMARY The earthworm fauna of the Pacific Northwest states, U.S.A., includes
two threatened endemic species. Both are restricted to sites of little or no soil
disturbance in areas where most of the suitable land has been converted to
agriculture. If these small remaining sites were to be altered substantially, both
species would become extinct. They are among the largest of North American
earthworms.

DESCRIPTION Both species are large, pale, thick-bodied worms with a clitellum
extending from segments 13 to about 22 and numerous nephridia per body segment.
Megascolides macelfreshi may attain a length of 60 cm. M. americanus is somewhat
smaller, although of the same order of magnitude (5,6). When disturbed, species of the
genus Megascolides contract, becoming short and turgid, unlike lumbricids which
wriggle on disturbance (3). A new genus is being described for these and several
related species, all endemic to the U.S.A. (7). M. macelfreshi was formally considered
synonymous with M. wellsi (3), but it is now thought to be distinct.

DISTRIBUTION M. americanus was described from Pullman, Washington, U.S.A. (5,6).
Its total distribution is unknown, but it is probably limited to the moister part of the
Palouse district of Washington State and perhaps Idaho (it has been collected from
Moscow, Idaho), an area in which very little natural or undisturbed habitat remains
(1,2,3). In 1978 specimens were collected one mile east of Pullman beneath a hawthorn
thicket and from the junction of highways 95 and 195 on the Washington-Idaho border
(7). M. macelfreshi was described from Salem, Oregon (2,6), and is known from a series
of highly localized sites in the Willamette Valley, another largely altered region
(1,2,3). Recent collections are from Aurora (Marion County) and Palmer Creek
(Yamhill Country), Helmick State Park (Polk County) and the Calapooya River near
Tangent (Linn County) (7).

HABITAT AND ECOLOGY From the little that is known of the ecology of M.
americanus, it appears to favour the native hawthorn shrubland which has nearly
disappeared from the Palouse country of south-east Washington, having been replaced
by wheat monoculture. One recent collection was from a site beneath Douglas Fir
trees (7).

M. macelfreshi is restricted to deep, little disturbed soils in moist forest, usually a
mixture of Douglas Fir, Grand Fir and Bigleaf Maple, but also pure Douglas Fir
woodlots and oak-ash woodland on occasion. Although a resident of the Willamette
Valley, it is no longer found in the open valley grasslands but is found instead in the
forests which intrude into or hem the valley (7).

The food supply of these worms consists of nearly pure organic residues with only a
slight admixture of soil. Activity generally takes place near the surface, although
aestivation may occur at 3-5 metres depth (5,7). Both species are indicators of
remnant ecological conditions that were once widespread but have since become
extremely restricted (7).

SCIENTIFIC INTEREST AND POTENTIAL VALUE Endemic to the American Pacific
Northwest, these worms have probably evolved convergently with the giant earthworms
of Australia. They are indicative of primary soil conditions in the Palouse region of

Washington and the Willamette Valley of Oregon, the former a loess situation and the latter a postglacial outwash trough.

Because of their great size, they could have considerable value as experimental animals if bred, or collected sparingly. Intrinsically, both species have value as spectacular members of the regional fauna and as probably the largest worms in North America.

THREATS TO SURVIVAL   Most of the original habitat for both species has been converted to croplands or pasture. These conversions result in drastically altered or lowered food supply for all worms, and favour the adventive lumbricids, especially Aporrectodea trapezoides. Part of the range is also being industrialized, urbanized and developed for suburban housing. Injury from ploughing is fatal to affected individuals, although overall this is probably not as great a mortality factor as altered food supply. Clear-felling, if followed quickly by natural regeneration of forest, does not in itself extirpate the worms, but conversion to agriculture or heavy chemical treatment for forest management is likely to do so. Scrub and woodland removal near streams is another activity detrimental to these animals (7).

Both the Oregon and Washington Giant Earthworms are very sensitive to mechanical and chemical damage. They have been unable to compete well with alien, adventive lumbricid earthworms in open grassland where the habitat has been significantly altered (3), but may be able to recover in areas in which a reservoir population has remained and forest is allowed to regenerate naturally (7).

CONSERVATION MEASURES TAKEN Several searches have been carried out recently, one of which led to the rediscovery of M. americanus in 1978 (see above), although the species was not found at most of the sites examined (7). This species was also looked for on Rose Creek Preserve, a Nature Conservancy reserve in Whitman Co., Washington, but was not found (8) although there are indications that it may be present (9). Both species may be difficult to find except under favourable conditions on account of the depth of their burrows. The Oregon and Washington Offices of the Nature Conservancy have shown interest in protecting one or more of these sites per species and a survey for additional sites of M. macelfreshi has been carried out (10,11,12), the results of which are not yet available.

CONSERVATION MEASURES PROPOSED The following suggestions have been made to the Nature Conservancy, North West Region (4): (1) determine ownership and owner attitudes for the Palmer Creek and Calapooya Creek sites in the Willamette Valley; (2) in early spring, check the previously recorded sites for present status, as well as other likely sites along the Calapooya drainage, in the Willamette Greenway, on the Rose Creek Preserve and adjacent Smoot Hill Preserve of Washington State University and Idler's Rest Preserve (Idaho) of the Nature Conservancy, as well as Magpie Hill, near Pullman and the Cogswell-Foster Preserve of the Nature Conservancy in Oregon; (3) encourage the Washington and Oregon Natural Heritage Programs and other field personnel to carry out field checks in other likely localities; (4) petition the Office of Endangered Species of the U.S. Fish and Wildlife Service for a full review of status of these taxa as potentially Endangered or Threatened Species under the Endangered Species Act. In addition, one or more habitat reserves should be established through acquisition in fee or conservation easement, to be managed for representative worm populations. The Oregon and Washington Offices of the Nature Conservancy are the most appropriate organizations to carry out these activities, and the non-game programmes of the two state wildlife agencies should be retained.

REFERENCES   1.   Altman, L.C. (1936). Oligochaeta of Washington. Wash. Univ. Pub. in Biol. 4 (1): 1-137.
             2.   Macnab, J.A. and McKey-Fender, D. (1947). An introduction to Oregon earthworms with additions to the Washington list. North Science 21 (2): 69-75.

3. Macnab, J.A. and McKey-Fender, D. (1948). North American Plutellus and Megascolids with synonymical notes (Annelida, Oligochaeta). Amer. Midland Naturalist 39 (1): 160-164.
4. Pyle, R.M. (1978). Memorandum: big worms. Unpubl; Northwest Office, The Nature Conservancy. 1 p.
5. Smith, F. (1897). Upon an undescribed species of Megascolides from the United States. Amer. Naturalist 31: 203.
6. Smith, F. (1937). New North American species of earthworms of the family Megascolecidae. Proc. U.S. Nat. Mus. 84: 157-181.
7. Fender, W. (1981). In litt.
8. Guthrie, D. (1978). In litt., 19 December.
9. Hudson, B. (1978). In litt., 28 December.
10. Hoffnagle, J. (1981). Pers. comm.
11. Krause, F. (1981). Pers. comm.
12. Margolis, K. (1982). Pers. comm., February.

We are very grateful to W. Fender and D. McKey-Fender for the information supplied for this data sheet.

GIANT GIPPSLAND EARTHWORM or KARMAI   VULNERABLE

Megascolides australis McCoy, 1878

Phylum   ANNELIDA   Order   HAPLOTAXIDA

Class   OLIGOCHAETA   Family   MEGASCOLECIDAE

SUMMARY One of the largest earthworms in the world, this species is endemic to a relatively small area in Victoria, Australia. Little is known of its ecology and population dynamics. Some anecdotal reports suggest that populations are declining, although others suggest the species is still locally common. In view of its small range, action should be taken to ensure that an adequately large area is fully protected.

DESCRIPTION One of the largest earthworms. Body large and cylindrical with 300-500 rings. May reach up to 4 m in length and 4 cm in diameter. An average worm is probably about 1.5 m long and about 2 cm in diameter. Detailed description is given in (2,10).

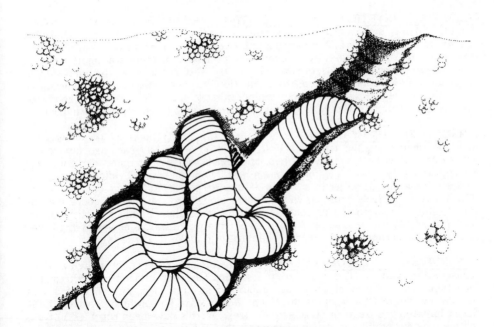

DISTRIBUTION Occurs in an area of approximately 100 000 ha in South Gippsland, Victoria, Australia, centred on Loch Korumburra and Warragul; the distribution is patchy (1,4,5). It is confined largely to the hilly areas of the western Strzelecki Ranges and the undulating country around Warragul (9).

POPULATION Unknown, although reports by local people indicate that locally it can be very common. One report quoted a density of one worm per 100 sq. feet (30 m^2) in parts of the Bass Valley (9).

HABITAT AND ECOLOGY Lives on sloping land in open forest and established pasture areas, where the main agricultural activity is dairy farming. This land receives light top dressings of superphosphate which do not seem to affect the worms adversely, and there is fairly regular light harrowing but little ploughing. Areas receive moderate rainfall in most seasons (5). It appears that the worm is confined to the deep grey soil formed from Cretaceous rocks in the western Strzelecki Ranges, and to the alluvial areas derived from this soil to the north and south-west (9). Worms are found close to the surface in winter when the ground is wet, as they favour damp soil. During the summer they retreat to greater depths. Unlike most worms, they have a permanent burrow system and may live in colonies, but each individual has its own hole (1,4). Burrows are said to be about 25 mm in diameter and contain masses of faecal soil material and occasionally the cocoons. Worms held in the hand contract and shoot out a milky fluid to a distance of several inches from dorsal pores. It is thought that the primary role of this fluid is to moisten and lubricate the wall of the burrows to make the passage of the worm easier. Worms move fast along the burrow system by alternately contracting and extending the body (9). A gurgly, sucking sound is reported to be audible as the worm moves through its burrow (10). They feed on organic matter in the soil (1,4). The worms are reported to emit an odour resembling creosote which may repel birds although the Kookaburra is known to eat them (9). Copulation has never been recorded but probably occurs underground. The cocoons are about 5 cm long, light yellow to dark brown in colour, depending on age, and are tough and leathery with a stalk at each end. They each contain one embryo (10).

SCIENTIFIC INTEREST AND POTENTIAL VALUE The Giant Gippsland Earthworm has been known as a scientific oddity for many decades. Visiting scientists and overseas biological supply houses have been interested in acquiring specimens for museum display and teaching purposes, sustaining a small trade in the preserved specimens over many years (9). The worm is of special value in experimental work on invertebrate giant fibre action potential neurophysiology (5). The secretion used for moistening burrow walls is reputedly good for rheumatism (10). It is the type species of a large genus currently under revision.

THREATS TO SURVIVAL Some local people say that reports of its occurrence have declined since the land was first opened up 50 or 60 years ago but others report that it is still widespread and common (9). There is some demand from biological supply houses and this is being met by one or two individuals locally but there is no information as to the size of the market or its effect on populations (5). Large individuals could be vulnerable to ploughing, and their cocoons may be susceptible to the more arid conditions found in cultivated soils (3). In 1931 its greatest enemy was said to be the plough but the worm was so abundant that even intensive cultivation was thought to be unlikely to make it scarce (8).

CONSERVATION MEASURES TAKEN As a result of a request for a review of its conservation status (11) a study of the species has been carried out (9), using a questionnaire to tap local knowledge about the species. The only protected area within the range of the worm is the Mount Worth State Park, which consists of 164 ha of forest in the Western Strzelecki Ranges. Worms have been reported in parts of the park but nothing is known of their numbers or status there (9).

CONSERVATION MEASURES PROPOSED More detailed field studies on specific populations should be carried out to determine their composition. Research is required on the breeding cycles and precise habitat requirements of the Giant Gippsland Earthworm. Proposals for creating reserves specifically for the species should consider an area close to its centre of distribution (9) and it may be necessary to place the worm on the Victoria State list of endangered animals (5). Ploughing and herbicide application should be stopped in appropriate areas (6,7,8).

CAPTIVE BREEDING Several people have claimed to have successfully bred the worm in captivity but none of these experiments have been recorded in the literature (9).

REFERENCES
1. Eve, J.E. (1974). Gippsland worms. Victorian Nat. 91: 36-8.
2. McCoy, F. (1878). Prodromus of Victoria. Vol.I - Decade 1; 91: 124-5.
3. Sims, R.W. (1980). In litt., 2 June.
4. Smith, B.J. (1974). More on Gippsland worms. Victorian Nat. 91: 124-5.
5. Smith, B.J. (1980). In litt., 12 August.
6. Jamieson, B. (1980). In litt., 10 July.
7. Anon (1978). Victoria. Wildlife in Australia 15 (3).
8. Marks, E.N. (1969). The invertebrates. In Webb, L.J., Whitelock, D., and Brereton, J.L.G. (Eds), The Last of the Lands, Milton, Jacaranda.
9. Smith, B.J. and Peterson, J.A. (1982). Studies of the Giant Gippsland Earthworm Megascolides australis McCoy, 1878. Vic. Nat. 99: 164-172.
10. Spencer, W.B. (1888). The anatomy of Megascolides australis (the giant earthworm of Gippsland). Trans. R. Soc. Vict. 1(1): 3-60.
11. Land Conservation Council, Victoria (1973). Report on the Melbourne Study Area. Melbourne L.C.C.

We are very grateful to B.J. Smith and R.W. Sims for providing information for this data sheet.

Diplotrema spp. Spencer, 1900
Chilota spp. Michaelsen, 1899
Microscolex spp. Rosa, 1887
Udeina spp. Michaelsen, 1910

Phylum   ANNELIDA   Order   HAPLOTAXIDA

Class   OLIGOCHAETA   Family   ACANTHODRILIDAE

SUMMARY South Africa has 90 endemic acanthodrilines including 59 species in the endemic genus Udeina. Most of these are limited to indigenous forests and consequently their range has been reduced enormously with the destruction of forests for agriculture or replantation with exotic trees. It is feared that many species may be extinct already.

DESCRIPTION The group of genera to which these species belong is recognized mainly by the structure of the excretory system (7). These acanthodrilines are unable to go into diapause, have no typhlosole and have either rudimentary dorsal pores or none (4) (the typhlosole may be an adaptation to life in soils poor in nutrients (5)). They probably undergo parthenogenesis (3). A detailed description appears in (6).

DISTRIBUTION Diplotrema 14 species are found in the western Cape and two in the eastern Cape of South Africa; Chilota 12 species in the western Cape, one in Natal; Microscolex three in the western Cape, three in the eastern Cape, one in Natal, two in Transvaal, Orange Free State and Lesotho; Udeina 48 species in the western Cape, six in the eastern Cape, seven in Natal and six in Transvaal, Orange Free State and Lesotho (7). They tend to have very restricted distributions; for example one species is known from only 4 sq. miles (10 km^2) (5).

POPULATION Unknown.

HABITAT AND ECOLOGY All South African acanthodrilines have evolved in forests. They require very wet cool soils, rich in organic matter; the continuous canopy of the forest and thick layer of litter insulating the soil from radiation and reducing evaporation means that there is little fluctuation in the temperature or humidity of the soil (4). Mainly found in regions which have a rainfall of over 20 inches (500 mm) per annum and tend to be restricted to permanently moist habitats (6). Some are found in upland valley bogs of the South West Cape and one species has been recorded from the sandy soils of the Cape Flats (6). The worms are restricted to the litter layer, or just below it, and depend on vegetable debris for food. Two species of Udeina (U. kinbergi and U. transvaalensis) are found in arable land, but they probably normally inhabit grasslands. One species of Microscolex (Eodrilus arundinis) is found in alkaline non-forest soils. These three species have probably invaded these habitats from forests quite recently, and still require wet soils (4).

SCIENTIFIC INTEREST AND POTENTIAL VALUE This group of poorly known species appears to be indicative of natural forest in South Africa. They are examples of species which may become extinct before they are fully known to science.

THREATS TO SURVIVAL The endemic acanthodrilines are gradually being exterminated as a result of the destruction of indigenous forests (2). By the 1950s, native forest had been reduced by 97 per cent since the arrival of Europeans in South Africa, and now covers only 0.2 per cent of the land (1). Most of the remaining native forest consists of stands of trees, a few metres in width, along brooks. Plantations of exotics raise forest cover to one per cent of the land but soils under these support no native earthworms. The most important limits to spread outside native forest are soil temperature and moisture. Deforestation is often carried out with, or is followed by, burning which destroys most soil fauna. Acanthodrilines appear to be less adaptable than many other soil organisms and are not even able to colonize original grasslands, where soils are more similar to forest soils than in cultivated land (4). Cultivation leads to a reduction in numbers of both species and individuals (6). Although there is no apparent competition between native and introduced species of earthworms, the latter rapidly colonize areas where the former have been eliminated (5,6). Acanthodrilines may also be affected by pollution and pesticides to which introduced species may be more resistant (5).

It is now probably too late to save or map the acanthodrilines in Natal since native forest is disappearing so fast, although those of Cape Province have not suffered as much. Since many species have narrow ranges, a large number have probably become extinct in the last few hundred years (4).

CONSERVATION MEASURES TAKEN None known.

CONSERVATION MEASURES PROPOSED Surveys should be carried out to determine the present status of these species, and suitable forest reserves should designated. The Nature Conservation Section of the Cape Provincial Administration should be approached to survey existing forest reserves.

REFERENCES 1. Acocks, J.P.H. (1953). Veld types of South Africa. Mem. Bot. Surv. S. Afr. 28: 192 pp. (quoted in 3).
2. Edwards, C.A. and Lofty, J.R. (1977). Biology of Earthworms. Chapman and Hall, London.
3. Ljungstrom, P.O. (1969). On the earthworm genus Udeina in South Africa. Zool. Anz. 182: 370-9.
4. Ljungstrom, P.O. (1972). Taxonomical and ecological note on the earthworm genus Udeina and a requiem for the South African acanthodrilines. Pedobiologia 12: 100-110.
5. Ljungstrom, P.O. (1972). Introduced earthworms of South Africa. On their taxonomy, distribution, history of introduction and on the extermination of endemic earthworms. Zool. Jb. Synt. 99: 1-81.
6. Pickford, G.E. (1937). A monograph of the acanthodriline earthworms of South Africa. Cambridge. 612 pp.
7. Sims, R.W. (1978). Megadrilacea (Oligochaeta). In Werger, M.J.A. (Ed.), Biogeography and Ecology of Southern Africa. W. Junk, The Hague.

We are very grateful to A.J. Reinecke for commenting on this data sheet.

Microchaetus spp. Rapp, 1849
Tritogenia spp. Kinberg, 1867

Phylum   ANNELIDA   Order   HAPLOTAXIDA

Class   OLIGOCHAETA   Family   MICROCHAETIDAE

SUMMARY A group which includes the largest earthworm in the world, the South African endemic giant earthworms are characteristic of primary grasslands and savannahs. They are under threat from increased desertification caused by over-grazing and intensive cultivation.

DESCRIPTION Microchaetus spp. range from 3 to 105 cm in length when contracted and most are longer than 20 cm. Stretched worms may reach 6-7 m. Microchaetus microchaetus is the largest earthworm in the world, reaching 7 m in length with a diameter of 2-3 cm. The small species are grey and the larger ones a conspicuous green. They are soft and slow moving but can break off the hind part of the body in order to escape (5). Five species of Tritogenia and 33 species of Microchaetus are endemic to South Africa (1). T. gisea, the only widespread endemic in the former genus, is parthenogenetic (3).

DISTRIBUTION South Africa, mainly south of Orange and Vaal Rivers. Distribution is correlated with underlying water table: they are found wherever the water rises to at least 40-50 cm below the surface regardless of the moisture content of the surface soil. Tritogenia occurs in Natal; Microchaetus occurs in the Western and Eastern Cape, Natal and Transvaal and one species is found in Lesotho (1,5). Microchaetus microchaetus is endemic to a restricted area of the Eastern Cape near King William's Town (4).

POPULATION Unknown. Populations of Microchaetus modestus are known to reach densities of 80-100 worms /m^2 (7).

HABITAT AND ECOLOGY Found in primary grasslands and savannahs, and in soils with an arid surface layer where they can avoid competition with other earthworms as a result of their lower moisture preference (1). Microchaetus is found in fields and meadows, usually in clayey/sandy soil (5). The worms are subsoil feeders (1) and only emerge on the surface after heavy rain. If the soil becomes too dry they burrow deeply and go into diapause (a state of suspended activity) (5).

SCIENTIFIC INTEREST AND POTENTIAL VALUE A species of Microchaetus has given rise to a unique terrain, the Kommetje Flats, near King William's Town. 'Kommetjies' are ring-like mounds of worm casts about 1 m in diameter with a central bowl 30-100 cm deep. The species responsible is not known but could be Microchaetus microchaetus which is present there. Some farmers regard kommetjies as aids to reducing erosion since rainwater collects in the bowls and seeps slowly into the ground causing less soil to be washed away (1,5).

THREATS TO SURVIVAL South African giant earthworms are being exterminated by the lowering of the water table and the formation of desert conditions in savannahs (2). Grasslands are being replaced by karroid type vegetation and karoo species, as over-grazing by domestic cattle and intensive cultivation has led to wind and rain erosion, lowering of the organic content of the soil and large diurnal soil temperature fluctuations (1). As the Karoo Desert has advanced, the western edge of the range of microchaetids has been pushed eastwards (3) and their range is decreasing (8). Casts of Microchaetus have been destroyed by bulldozers where they have made farming impossible (2,3,5). These species may also have been affected by climatic changes (6). No recent surveys of Tritogenia spp. have been carried out (8).

CONSERVATION MEASURES TAKEN None known. However, M. microchaetus occurs in the Mountain Zebra National Park which affords it some protection (8).

CONSERVATION MEASURES PROPOSED Further information is required on all species; surveys of their distribution should be undertaken and reserves set up in representative habitats. The South African National Parks Board's Research Station should be approached to assist in surveys of earthworms in the various Parks under their control.

REFERENCES   1.   Sims, R.W. (1978). Megadrilacea (Oligochaeta). In Werger, M.J.A. (Ed.), Biogeography and Ecology of Southern Africa. W. Junk, The Hague.
2.   Edwards, C.A. and Lofty, J.R. (1977). Biology of Earthworms. Chapman and Hall, London.
3.   Ljungstrom, P.O. (1972). Introduced earthworms of South Africa. On their taxonomy, distribution, history of introduction and on the extermination of endemic earthworms. Zool.Jb.Synt. 99: 1-81.
4.   Hey, D. (1978). The Fauna and Flora of Southern Africa and their conservation. Department of Nature and Environmental Conservation, Cape Town.
5.   Ljungstrom, P.O. and Reinecke, A.J. (1969). Ecology and natural history of the microchaetid earthworms of South Africa. Pedobiologia 9: 152-157.
6.   Sims, R.W. (1980). In litt., 2 June.
7.   Reinecke, A.J. and Ryke, P.A.J. (1972). The ecology of Microchaetus modestus. Wet. Bydraes, P.U. Series B. nr. 42 (In Afrikaans).
8.   Reinecke, A.J. (1982). In litt., 12 January.

We are very grateful to A.J. Reinecke and R.W. Sims for information supplied for this data sheet.

ARTHROPODA

The Arthropoda is a vast assemblage of invertebrates comprising about one million described species from a living total conservatively estimated at two million, several times the number of all other animal species combined (3). Their adaptive diversity has enabled representatives to survive in virtually every habitat, and in many respects they are the most succesful invaders of the land. The annelid ancestry of the arthropods is evident in their metameric segmentation which is generally apparent, although it may be obscured by fusion or loss of segments, or by specialized appendages (1).

The distinguishing feature of arthropods, and one to which many other evolutionary innovations are related, is the chitinous exoskeleton or cuticle. This covers the entire body and is divided into inter-connecting plates which permit movement. Growth is facilitated by periodic moulting. Each segment probably originally bore one pair of appendages, but these have become reduced and specialized as paddles or legs for locomotion, as reproductive intromittent organs, as gills, as mouthparts, or as sensory organs (2).

Living arthropods have been traditionally divided into two subphyla, the Mandibulata with antennae, and the Chelicerata without antennae. The Chelicerata still stand as a monophyletic group, although within this group the exact relations of the Pycnogonida are unclear. However, most zoologists agree that the Mandibulata is an artificial assemblage, probably of three distinct and separate evolutionary lines from the annelidan stock, which resulted in the extinct Trilobitomorpha and the extant Crustacea and Uniramia (4). Only one of the four arthropodan lines, the Uniramia, evolved on land. These animals are mandibulate and antennate, but the name uniramian refers to their unbranched appendages. The Uniramia comprises the Insecta, Chilopoda, Diplopoda, Symphyla and Pauropoda, the last four of which were formerly grouped into the Myriapoda, a name still used informally for convenience. There is increasing evidence that at least the uniramians, and perhaps all four arthropod groups, evolved independently from annelid stock, each developing a chitinous exoskeleton and jointed appendages. If this is so, then the Arthropoda should be considered as a superphylum, embracing the four lines as phyla.

Despite the strong evidence and fairly general agreement that the Arthropoda is polyphyletic, there is still considerable opposition to the loss of the traditional phylum. Hence, in this book, as in references (1) and (5), the four evolutionary lines are recognized as subphyla of the phylum Arthropoda.

REFERENCES
1. Barnes, R.D. (1980). Invertebrate Zoology. Saunders College, Philadelphia. 1089pp.
2. Barrington, E.J.W. (1967). Invertebrate Structure and Function. Nelson, London. 549 pp.
3. Levi, H.W. (1982). Arthropoda. In (5).
4. Manton, S.M. (1978). The Arthropoda: Habits, Functional Morphology and Evolution. Oxford University Press, Oxford. 527 pp.
5. Parker, S.P. (1982). Synopsis and Classification of Living Organisms. McGraw Hill, New York. 2 vols., 1232 pp.

CLASSIFICATION OF THE ARTHROPODA

Subphylum: Trilobitomorpha (extinct trilobites)

Subphylum: Chelicerata

    Class: Merostomata (horseshoe crabs, extinct eurypterids)
    Class: Arachnida (spiders and their allies)
    Class: Pycnogonida (sea-spiders)

Subphylum: Crustacea (crustaceans)

Subphylum: Uniramia

    Class: Insecta (insects)
    Class: Chilopoda (centipedes)
    Class: Diplopoda (millipedes)
    Class: Symphyla (symphylans)
    Class: Pauropoda (pauropodans)

CHELICERATA

Horseshoe crabs, arachnids and sea-spiders.

INTRODUCTION The arthropod subphylum Chelicerata comprises aquatic and terrestrial forms. The body is divided into two sections, the cephalothorax (prosoma), and the abdomen (opisthosoma). There are no antennae and the first pair of appendages are feeding structures called chelicerae. The second pair of appendages, the pedipalps, are modified for functions which differ between the three classes.

MEROSTOMATA The horseshoe crabs, or king crabs, comprise a very ancient class of chelicerates with only four living species, all marine. They are discussed in detail in the following review.

ARACHNIDA The arachnids are essentially terrestrial chelicerates without book gills, although a few species of mites are secondarily aquatic. Evidence of metamerism varies from the clear segmentation of scorpions, the most primitive arachnids, to the total loss of segmentation and the fused cephalothorax and abdomen of the specialized mites. The appendages common to all arachnids are a pair of chelicerae, a pair of pedipalps and four pairs of legs. Arachnids (with the exception of many mites) are largely predatory, other arthropods forming the principal prey. Many arachnids are equipped with pincers, poison fangs or poison stings to immobilize prey, and true spiders have spinnerets for production of silk.

The Arachnida are here divided into 13 orders (22). Five of these are minor groups each with less than 100 species worldwide. The Uropygi (whip scorpions, c. 85 spp.), Schizomida (schizomids, c. 80 spp.), Amblypygi (tailless whip scorpions, c. 70 spp.) and Ricinulei (ricinuleids, c. 33 spp.) are all tropical or subtropical, while the Palpigradi (palpigrades, c. 60 spp.) are found in southern U.S.A. and the Mediterranean region (29). The Scorpiones (scorpions, c. 1200 spp.) are the oldest known terrestrial arachnids (30). They are nocturnal, mainly tropical and subtropical and usually between 3 and 9 cm long, although the largest species reaches 18 cm. Their pedipalps are developed into claws and there is a poison sting at the tip of the long, flexible abdomen (30). The Pseudoscorpionida (pseudoscorpions, c. 2000 spp.) are tiny arachnids, rarely longer than 8 mm (31). They are found throughout the world, living under bark and stones, and in plant litter, soil, moss, nests, caves, coastal flotsam, and houses. The pedipalps are superficially similar to those of scorpions, but have poison glands at their tips. They feed mainly on small arthropods such as collembolans and mites. The Solpugida (solpugids, solifugids, sun or wind scorpions, over 900 spp.) are tropical or subtropical arachnids that earn their common names from the diurnal habits of many species, and the great speed with which they run (32). Many species prefer an arid climate. They may reach 7 cm in length, and have very large, heavy chelicerae used in catching any small animal, including vertebrates. The pedipalps resemble legs but have special adhesive organs at their tips, used in prey capture. The Araneae (spiders, c. 60 000 spp.) are probably the largest group of arachnids after the mites, with many species yet to be described. They are variable in size (0.5 mm to 9 cm in length) and are found throughout the world. The chelicerae have fangs which, with a few exceptions, transmit venom of varying potency; the pedipalps are short and leg-like in females, but adapted as copulatory organs in males. The abdomen is unsegmented in all except the most primitive species, and ends in modified appendages called spinnerets which produce silk. This silk is a liquid mixture of proteins which hardens when drawn out into a thread. The Opiliones (daddy longlegs or harvestmen, c. 4500-5000 spp.) are superficially similar to spiders but differ in having very slender legs, no poison nor silk glands, and no obvious division between cephalothorax and abdomen (33). They are most abundant in humid tropical and temperate climates. They usually have a body length of 5 to 10 mm, but they may be as short as 1 mm or as long as 20 mm, with a leg length of up to 160 mm. In general harvestmen are predatory, but scavenging is also important (4,13,21,22).

The last three orders, Opiliocariformes, Parasitiformes and Acariformes, were formerly joined into one polyphyletic order, the Acari or Acarina, which is now recognized as a subclass (34). These orders are the mites and ticks, which comprise 30 000 described species, and at least as many undescribed species. They are undoubtedly the most important orders with respect to their impact on man, and they occur from polar to equatorial regions. Most species are less than 1 mm in length, although many are larger. The Opiliocariformes consist of a single family of brightly coloured, omnivorous, primitive mites resembling harvestmen (4). The Parasitiformes are divided into three suborders, the Mesostigmata, Holothyrina and Ixodida. The Mesostigmata is a large group of parasitic and free-living mites which includes the Dermanyssidae, important parasites of birds (red mites) and mammals. The Holothyrina is a small suborder with two families of large predatory mites confined to Australia, New Zealand and the Indo-Pacific. The Ixodida (Metastigmata) are the large parasitic mites usually known as ticks. The head of hard ticks (Ixodidae) is directed forward and is covered by a hard plate. The soft ticks (Argasidae) have a leathery integument, and no plate above the head, which is on the underside. There are about 300 species of ticks worldwide, all with recurved teeth for piercing the skin and feeding on the blood of reptiles, birds or mammals. Some may reach a length of up to 3 cm when gorged. The Acariformes consist of three suborders, the Prostigmata, the Oribatida and the Astigmata. The Prostigmata, or Trombidiformes, are widely distributed mites which include parasites of plants (red spider mites, Tetranychidae, and gall mites, certain Eriophyidae), of invertebrates (e.g. velvet mites, Trombidiidae) and of vertebrates (e.g. harvest mites, Trombiculidae). Most of the free-living species, including the snout beetles (Bellidae), are predators. Also included are the follicle mites (Demodicidae), marine water mites (Halacaridae) and the often colourful freshwater mites (Hydrachnellidae and Unionicolidae). The Oribatida (beetle or moss mites) or Cryptostigmata is a large group of heavily sclerotized mites common in soil and decaying plant material. Finally, the Astigmata, or Sarcoptiformes, include the scabies or mange mites (Sarcoptidae, Psoroptidae) and storage mites such as cheese mites (Acaridae) (4,13,21,22).

PYCNOGONIDA The sea-spiders are exclusively marine, spider-like animals. There are about 500 species which are generally 1-10 mm in body length. However, some deep sea species may grow to a body length of 6 cm, and a leg span of almost 75 cm. The often narrow body has four, five or six pairs of very long legs. Most species are drab, but a few are green, and some deep sea forms are red. Pycnogonids occur throughout the world's seas, but are apparently more common in colder waters. Although some are able to swim, the majority are bottom-dwellers, ranging from the intertidal zone to depths of over 6000 m. Many sea spiders are carnivorous, feeding on hydroids, soft corals, anemones, bryozoans and sponges. Other species feed on algae, micro-organisms or detritus (4).

SCIENTIFIC INTEREST AND POTENTIAL VALUE The horseshoe crabs are of great interest as evolutionary relicts, and they are used for a variety of purposes ranging from spearheads to fertilizer and material for medical tests. They are discussed in detail in the following review.

Scorpion poisons have received attention since earliest times, although only recently has it been concluded that they are of two kinds (13). One has local effects and is comparatively harmless, while the other is neurotoxic, resembling certain kinds of snake venom, and can be extremely dangerous or even fatal. Amongst the species with dangerous venom are the North African species Leiurus quinquestriatus, Androctonus australis and Buthus occitanus, and the North

American Centruroides sculpturatus and C. gertschii (13).

Almost all spiders (Araneae) have poison glands in the cephalothorax, but very few species are dangerous to man. The widow spiders are web spinners whose sedentary females may bite if molested. The Black Widow or Button Spider (Latrodectus mactans) is found in warm climates around the world and its neurotoxic venom causes severe abdominal pain, nausea, muscular spasms and respiratory paralysis (4,21), which may lead to death, particularly in children (4). In Brazil the bite of the wolf spider Lycosa raptoria (Lycosidae) causes necrosis, while Phoneutria fera (Ctenidae) has a neurotoxic venom with very painful effects (21). Funnelweb mygalomorphs (Dipluridae) of the genus Atrax (in Australia) and Trechona (in South America) are dangerously poisonous (4,21).

Araneae are also of interest in their ability to produce silk, which is used in a variety of ways. All spiders make silk draglines which are trailed behind for safety, and silk egg cocoons. In some species silk is used for ballooning, in which small and young spiders release a line into the wind, and float off to a new area. Perhaps as a result of this habit, spiders are early colonizers of islands and disturbed areas. A small web-spinner was the first animal to be recorded on Krakatoa, nine months after its eruption in 1883 (12). Spiders have been captured at altitudes of almost three miles (5 km) above the earth's surface, and airmen have recorded passing through great numbers of silk filaments (12). Many spiders use silk to line burrows or to build nests, and silk snares are made in various forms, including cobwebs and sheet, funnel and orb webs. The bolas spiders (Mastophora spp.) capture passing moths by throwing a sticky globule dangling on a silk line (21). Species of the genus Nephila build webs one metre or more in diameter and their silk is the strongest natural fibre known. The webs, once twisted and matted, are used by South Sea Islanders for bags and fish nets (21). The production of silk by spiders is too slow to offer any commercial possibilities. However, one spider is itself the centre of a substantial level of trade. The Red-knee Tarantula (Brachypelma smithi) is an increasingly popular pet in Europe and North America and it is reviewed separately in this volume.

Undoubtedly the arachnids of most scientific interest are the economically important mites and ticks. Heavy infestations of ticks on domestic animals may cause anaemia or infection with various diseases. Ticks are common on many wild animals but may be very restricted in their distributions. Ornithodorus vansomereni is known only from the nests of swallows and the Cliff Chat on Ukazi Inselberg near the Kora Reserve, Kenya (3). Red spider mites such as Metatetranychus ulmi are probably the most economically important of the phytophagous mites, causing damage by sucking plant juices and spreading viral infections. Some mites, such as Acarus siro (Acaridae), are pests of stored foods, while others, like the follicle mites Demodex spp., cause skin diseases such as mange (4,13,21). Predatory mites of the family Phytoseiidae are among the most important natural enemies of phytophagous mite pests, primarily Tetranychus spp. (red spider mites). Natural populations of the predators, augmented by populations reared in laboratories, have been succesfully used in control of mite pests in crops of strawberries, apples, and citrus in the U.S.A. and in glasshouse crops such as tomatoes, cucumbers and peppers in western Europe and the U.S.S.R (1). The cryptostigmatid, or oribatid, mites are a major component of the soil fauna, and they are significant in soil formation and maintenance (25).

THREATS TO SURVIVAL Horseshoe crabs are harvested by clam diggers to protect their clam beds, by fishermen for eel bait, and by medical personnel for material for the lysate test. Some mortality may result from pollution and land reclamation. Details are given in the following review.

Threats to Arachnida are poorly documented, and will probably remain so without a substantial investment in taxonomic studies. However, the major threat is undoubtedly loss of habitat, a factor which is generally the result of increasing human population pressure and the overwhelming destructive force of modern technology. Several categories of habitat destruction may be recognized.

1. Damage to caves.
The threats to cave arachnids are comparatively well documented since arachnids often constitute an important part of the rich invertebrate fauna which survives in such protected environments. Changes in surface water courses or land use may have severe effects on subterranean cave communities. In the Mojave Desert, east of Death Valley, California, the endemic whip scorpion Trithyreus shoshonensis survives the cold winters in the upper Shoshone Cave, which is heated by thermal ground water (10). The cave is threatened by fire, visitor pressure, vandalism, and by-passing of the water source (26). The U.S. Office of Endangered Species has been petitioned to review the species and has prepared a ruling, but this has not been published (26). Malheur Cave, in Oregon, U.S.A., is an unusual 1100 m long lava tube containing a 400 m lake. The lake is an important source of humidity, supporting a diverse cave fauna including an endemic pseudoscorpion Apochthonius malheuri (5,8). At one time the cave was used as a meeting place for the local Masonic order, and the sawdust used on the floor enriched the cave community (7). It was feared that visitor pressure might cause harm to the pseudoscorpion and other endemic invertebrates, and the cave has now been closed to the public (5,7). Spiders (Araneae) are frequently endemic to isolated cave systems, as are the two North Carolina troglophilic species Nesticus sp. in Bat Cave (2), and Ivesia carolinensis in Linville Caverns (14). Other endemic spiders from the Kocevje Caves, Yugoslavia and from Kauai Island, Hawaii are discussed in the reviews that follow. Harvestmen (Opiliones) and mites (Acari) may also be endemic to caves; the harvestman Banksula melones is restricted to a series of caves in California (see separate review), and the mites Elliotta howarthi (6), Flabellorhagidia sp. nov. and other Rhagidiidae are confined to lava tubes in Washington, U.S.A. (17,27 and review of Deadhorse Cave). Cueva los Chorros in Puerto Rico contains 17 species of mites, some of which are new to science (23 and separate review).

2. Drainage of wetlands.
The Great Raft Spider Dolomedes plantarius is restricted to Redgrave and Lopham Fens Nature Reserve in Suffolk, U.K. It generally hunts by running across the surface of still waters, but it is able to submerge and attack small fish. It was formerly widespread in marshes and fens, but virtually all its habitats have now been drained for development (15). Many other spiders are restricted to wet environments and would be threatened by changes in the water table. Acanthophyma (Lasiargus) gowerensis is found only in salt marshes in Glamorgan, Wales, and south-west Ireland, while Glyphesis cottonae occurs only in Sphagnum bogs in the New Forest and Cumbria, U.K. (15). Wicken Fen in Cambridgeshire, U.K., is gradually drying out and certain spiders associated with waterlogged ground, including Dolomedes fimbriatus, Singa heri, Centromerus incultus and Pirata piscatorius, have not been recorded there for many years (15).

3. Forestry.
This category includes both the logging of virgin forests, which now mainly occurs in the tropics, and the development of plantation forestry. Felling of tropical rain forests is probably threatening many species of arachnids, as it is insects (see Insecta account). At least six of the 13 arachnid orders are virtually confined to the tropics and subtropics and many of the others have their greatest species diversity there. Unfortunately there are insufficient data to estimate the impact on Arachnida of development in the tropics.

Not all invertebrates characteristic of old trees depend directly upon wood or its breakdown products. The pseudoscorpion Dendrochernes cyrneus and the spider Lepthyphantes midas live as predators only within ancient trees in U.K. woodlands

(19). Thyreostenius parasiticus is the most common spider high up in deciduous trees in Wytham Wood, Oxford. U.K., but is otherwise found only in artificial habitats (15). Microhexura montivaga is a funnelweb tarantula (Dipluridae) endemic to spruce and fir forest in North Carolina, U.S.A., where it occurs in moss. It is of special concern since it has not been found in recent years (14). The trap-door spider Cyclocosmia torreya (Ctenizidae) is a species of special concern in Florida, since it is known to be sensitive to logging of its woodland habitat (16). Similarly, the Florida population of the orb-weaving spider Eustala eleuthra (Araneidae) may be threatened by destruction of tropical hardwood hammocks (16). Sphodros abboti and S. rufipes both build tubular webs on the sides of trees and bushes in woodlands in south-east U.S.A.; they cannot survive severe disturbance of their habitat (16).

4. Agriculture and commercial development.
A survey in Florida has revealed several endangered or threatened spiders. The Lake Placid Funnel Wolf Spider Sosippus placidus (Lycosidae) is possibly extinct because of the development of citrus plantations in its type locality (16). The same threat applies to McCrone's Burrowing Wolf Spider Geolycosa xera (Lycosidae) (16). The Rosemary Wolf Spider, Lycosa ericeticola, occurs only on a small area next to an expanding open-cast mine in Interlachen, Florida (16).

5. Residential development.
An unusual case of a victim of residential development is the Florida population of the Dusky-headed Tailless Whip Scorpion Paraphrynus raptator, which was formerly common in outside lavatories and cisterns in Key West. Although it is found in Honduras, Guatemala, and Mexico, the Florida population may be endangered by gradual modernization of residential facilities (16). Also in Florida, Cesonia irvingi (Gnaphosidae) is known only from Key West and Bob Allen Keys, which are subject to considerable development pressure (16). The best known site for the Ladybird Spider (Eresus niger) in Hampshire, U.K., was destroyed by building, and the spider was not found between 1906 and 1980 (15,28). The El Segundo and other relict sand dunes of Los Angeles County, California, are the last remnants of a once extensive dune system which has been reduced by development. At least two species of mite, Eremobates sp. nov. and Erythraeus tuberculatus, are endemic to the dune remnants, which are threatened by development of tennis courts, golf courses and introduced plants (see El Segundo review).

CONSERVATION Conservation of horseshoe crabs is described in the following review. There are no data on the conservation needs of sea-spiders.

Various species of arachnids have been listed as being in need of conservation or given legislative protection, sometimes from threats of an unspecified nature. The spiders Eresus niger (Eresidae), Misumena vatia (Thomisidae) and Dolomedes fimbriatus (Pisauridae) are protected in Switzerland (11). In Sweden the spiders Araneus nordmanni and Atypus affinis and the pseudoscorpions Allochernes wideri and Anthrenochernes stellae are considered to be under threat (18). In the Federal Republic of Germany several harvestmen are listed: Nelima silvatica is threatened with extinction, Ischyropsalis hellwigi and Odiellus spinosus are threatened, and Astrobunus laevipes and Nemastoma dentigerum are potentially threatened (9). In U.K. the Great Raft Spider Dolomedes plantarius and the Ladybird Spider Eresus niger are protected by law (20).

Rare species in heathlands, grasslands, freshwater marshes and ancient woodlands in industrial countries are particularly at risk from loss of their habitats. However, the problem is particularly insidious in the areas of highest diversity of arachnids, the tropics and sub-tropics, because of the extent of taxonomic ignorance of these areas (24). It seems likely that extinctions of undescribed species are occurring regularly. Without the necessary taxonomic research, attempts to prevent undocumented extinctions can only be made by protection of representative habitat types. Such an arbitrary system is far from satisfactory and, for example, would certainly fail to prevent losses in cave fauna. There is no substitute for an increased effort in the fields of arachnid taxonomy and basic ecology, with simultaneous efforts to protect the habitats of those species in greatest danger.

REFERENCES

1. Ables, J.R. and Ridgway, R.L. (1981). Augmentation of entomophagous arthropods to control pest insects and mites. In Papavizas, G.C. (Ed.), Biological Control in Crop Production. Beltsville Symposium in Agricultural Research 5: 273-304. Allenhead, Osman and Co., New Jersey.

2. Anon. (1981). TNC News 31 (6), Nov-Dec.

3. Anon. (1982). Review of Habitat Status of Some Important Biotic Communities in Kenya. Division of Natural Sciences, National Museum, Kenya.

4. Barnes, R.D. (1980). Invertebrate Zoology. Saunders College, Philadelphia. 1089 pp.

5. Benedict, E.M. (1973). Malheur Cave as a biospeleological resource. The Speleograph 9: 47-48.

6. Benedict, E.M. (1981). Editorial. North American Biospeleology Newsletter 25: 1.

7. Benedict, E.M. and Gruber E.H. (1981). Ecology of Malheur Cave, Harney County, Oregon. Proc. Int. Congress Speleology. Bowling Green, Kentucky. Pp. 480-482.

8. Benedict, E.M. and Malcolm, D.R. (1973). A new cavernicolous species of Apochthonius (Chelonethida: Chthonidae) from the western United States with special reference to troglobitic tendencies in the genus. Trans. Am. Micros. Soc. 92: 620-628.

9. Blab. J., Nowak, E. and Trautmann, W. (1981). Rote Liste der Gefährdeten Tiere und Pflanzen in der Bundesrepublik Deutschland. Naturschutz Aktuell Nr. 1. Kilda Verlag, Greven.

10. Briggs, T.S. and Hom, K. (1972). A cavernicolous whip-scorpion from the northern Mojave Desert, California (Schizomida: Schizomidae). Occ. Papers Calif. Acad. Sci. 98: 7 pp.

11. Burckhardt, D., Gfeller, W., and Müller, H.U. (1980). Animaux Protégés de Suisse. Ligue Suisse pour la Protection de la Nature, Bâle. 224 pp.

12. Carson, R.L. (1951). The Sea Around Us. Staples Press, London. Pp. 88-91.

13. Cloudsley-Thompson, J.L. (1968). Spiders, Scorpions, Centipedes and Mites. Pergamon, Oxford. 278 pp.

14. Cooper, J.E., Robinson, S.S., and Funderburg, J.B. (Eds) (1972). Endangered and Threatened Plants and Animals of North Carolina. North Carolina State Museum of Natural History, Raleigh, North Carolina.

15. Duffey, E. (1974). Changes in the British spider fauna. In Hawksworth, D. (Ed.), The Changing Flora and Fauna of Britain. Academic Press, N.Y. and London. Pp. 293-305.

16. Edwards, G.B. and Wallace, H.K. (1982). Araneae section. In Franz, R. (Ed.), Invertebrates. Vol. 6. Rare and Endangered Biota of Florida. University Presses of Florida, Gainesville. 131 pp.

17. Elliott, W.R. (1976). New cavernicolous Rhagidiidae from Idaho, Washington and Utah (Prostigmata: Acari: Arachnida). Occ. Papers Texas Tech. Univ. Mus. 43: 1-15.

18. Ehnström, B. (1982). Data given in letter from H.W. Walden in litt., 11 October.
19. Harding, P. (1978). Pasture-woodlands in lowland Britain and their importance for the conservation of epiphytes and invertebrates associated with old trees. CST Report No. 211, NCC Research Contract HF3/03/118. 51 pp.
20. Her Majesty's Government (1981). Wildlife and Countryside Act 1981, Chapter 69. Her Majesty's Stationery Office, London. 128 pp.
21. Levi, H.W., Levi, L.R. and Zim, H.S. (1968). Spiders and Their Kin. Golden Press, New York. 160 pp.
22. Parker, S.P. (Ed.) (1982). Synopsis and Classification of Living Organisms. McGraw Hill, New York. 2 vols, 1232 pp.
23. Peck, S.B. (1981). The invertebrate fauna of tropical American caves, Part III: Puerto Rico, an ecological and zoogeographical analysis. Biotropica 6: 14-31.
24. Pyle, R., Bentzien, M. and Opler, P. (1981). Insect conservation. Ann. Rev. Ent. 26: 233-258.
25. Wallwork, J.A. (1970). Ecology of Soil Animals. McGraw Hill, London. 283pp.
26. U.S.D.I. Fish and Wildlife Service (Unpubl. MS). Proposed endangered status for the Shoshone Cave Whip-scorpion (Trithyreus shoshonensis).
27. Zacharda, M. and Elliott, W.R. (1981). Holarctic cave mites of the family Rhagidiidae (Actinedida: Eupodoidea). Proc Eighth Int. Congress. Speleology, Bowling Green, Kentucky. Pp. 604-607.
28. Anon. (1980). Back from the dead. The Observer (U.K.), 10 February, p. 4.
29. Levi, H.W. (1982). Arthropoda. In (22), vol. 2, p. 71.
30. Francke, O.F. (1982). Scorpiones. In (22), vol. 2, pp. 73-75.
31. Muchmore, W.B. (1982). Pseudoscorpionida. In (22), vol. 2, pp. 96-102.
32. Muma, M.H. (1982). Solpugida. In (22), vol. 2, pp. 102-104.
33. Shear, W.A. (1982). Opiliones. In (22), vol. 2, pp. 104-110.
34. Johnston, D.E. (1982). Acari. In (22), vol. 2, p. 111.

We are very grateful to P.M. Hillyard, H.K. Hyatt and F.M. Wanless for their comments on this review.

Limulus polyphemus (Linnaeus, 1758)
Tachypleus tridentatus (Leach, 1819)
T. gigas (Müller, 1785)
Carcinoscorpius rotundicauda (Latreille, 1802)

Phylum	ARTHROPODA	Order	XIPHOSURA
Class	MEROSTOMATA	Family	LIMULIDAE

SUMMARY Once widespread and diverse in the Palaeozoic, horseshoe crabs are now 'living fossils' with only four species in three genera still surviving. Limulus polyphemus is found on the Atlantic coast of North America and the other three species occur in the coastal waters of Asia, from Japan south to Indonesia. Although still abundant in some areas, horseshoe crabs are exploited intensively by man for animal fodder, bait for eel fishing, fertilizer and most recently for biomedical purposes. Animals are usually harvested from breeding beaches, and some control and management of collection is probably required.

DESCRIPTION Horseshoe crabs are light greenish grey to dark brown with a heavily armoured body and may reach a length of 60 cm. Carcinoscorpius rotundicauda is the smallest species and is about 30 cm long (35) and Limulus polyphemus is usually about 50 cm long. Males are generally smaller than females. The carapace is horseshoe-shaped and is convex above and concave below, a shape which facilitates pushing through mud or sand and also protects the ventral appendages (26). The colour of the adult carapace, before it is discolored by erosion and the activity of bacteria and attached organisms, may be related to substrate colour and water turbidity. Northern populations tend to have dark brown or green carapaces wheras those along the east and west coasts of Florida, U.S.A., are a much lighter brown (32). The crabs have been described as 'walking museums' since the carapace provides a suitable habitat for large numbers of species including algae, coelenterates, flat worms, bryozoans and molluscs (26). There is a long mobile tail spine or telson which aids forward movement and helps to right the animal should it be accidentally overturned. Immature crabs have spines along the dorsal surface of the telson which proportionately diminish in size as the animal grows. These may help the animal to burrow or deter predators (26). Two pairs of small appendages, in front and behind the mouth, assist the pincer-tips and heavy bases of the walking legs in moving food into the mouth. The fifth pair of walking legs has a larger, heavier base which is used to crush thin-shelled molluscs. This pair of legs also has a distal whorl of spatulate spines that sweep silt away while burrowing (26).

DISTRIBUTION Each species has an intermittent distribution within its range, temperature probably being the limiting factor for the northern ranges of L. polyphemus and Tachypleus tridentatus (26).

L. polyphemus is found on the east coast of America from Nova Scotia to the Yucatan Peninsula. It consists of a number of geographically isolated races which can also be distinguished morphometrically and physiologically, and which are probably separated on the basis of salinity and temperature. Each of the major estuaries along the eastern coast of the U.S.A. has a Limulus population. The largest individuals are found in estuaries from Georgia to New Jersey. Those in the Gulf of Mexico and north of Cape Cod are appreciably smaller, suggesting that Limulus optimally is a temperate species (22,32). Occasionally L. polyphemus is found in European waters but it has never established a reproducing population there; the specimens that have been found have probably been brought over from the fishing grounds off eastern North America by fishermen and then released (27).

Limulus polyphemus

T. tridentatus has been reported from western and southern Japan, Taiwan, Philippines, North Borneo, Vietnam and Sumatra, Indonesia (15,33).

T. gigas and C. rotundicauda have been reported from the coast of India, Thailand, Malaysia, Singapore, Indonesia, the Philippines and Borneo and have essentially the same ranges (15,33).

All three Asian species are found sympatrically at Kota Kinabalu on the north-west coast of Borneo and throughout much of Indonesia and T. tridentatus and C. rotundicauda are found sympatrically along the south-west coast of the Philippines including Palawan (33,34).

POPULATION In general horseshoe crabs are abundant in the areas where they occur. In 1956 there were thought to be more than one hundred thousand adults in the Cape Cod Bay area (30) and currently L. polyphemus is often brought up in trawls and dredges off the east coast of the U.S.A. (32). The population in waters off Massachusetts, north of Cape Ann, is also probably large as 1640 crabs have been tagged there (32). The largest of the populations along the east coast of the U.S.A. is centred on Delaware Bay (13,32); in 1977 and 1979 peak spawning activity was estimated to involve 305 500 adult animals (13). In coastal waters off New Jersey, crabs are abundant in the inshore 3 miles (5 km) from Cape May to Atlantic City, but much less common further north (14).

HABITAT AND ECOLOGY Horseshoe crabs are marine bottom dwellers living in sandy and muddy bays and estuaries. Limulus polyphemus has been well studied and most of this account deals with this species. In Florida, Gulf of Mexico, adults are found to a depth of 30 m, but major concentrations are found at about 5-6 m, and the animals only return to the beach to breed (25,31). Limuli have been dredged from depths to 246 m and at distances as far as 56 km off the Atlantic coast (26,32). C. rotundicauda differs from the other species in that it is often

236

found in brackish water in rivers; it has been found about 4 km from the sea in Thailand, and has been reported 90 miles (145 km) inland from the sea in India (15).

Horseshoe crabs are omnivorous scavengers, feeding on a wide variety of prey including worms such as Nereis, molluscs, and algae, which they dig for, using their pincer-tipped appendages to pass it to the mouth (14,15,16,26). Limulus can be destructive of soft-shelled clam populations such as Mya arenaria, (17,19,30), consuming at least one hundred 10-20 mm clams per day (17). The large concentrations of crabs in the inshore regions of southern New Jersey may be responsible for the poor recruitment in this region, of juvenile Surf Clams, Spisula solidissima, as observations since 1976 have shown that juvenile Surf Clams are the most frequently occurring items in Limulus gut samples (14,16). The crabs appear to be able to locate clams up to 2-3 ft (60-90 cm) away (19). Adults exhibit a nocturnal activity pattern which is highest at full moon (23). In Florida waters there is little or no long range movement of individuals or populations (23). Along the Atlantic coast the migrations may be much greater, estimated as up to 16 km (32), 33.8 km (as shown by tagging results (12)) and 56 km (from dredging results (26)). Adults usually move along the bottom with a stiff-legged gait although they can swim in quiet water (32). During their first 2-3 years moulting may occur several times a year, the animals digging themselves into the sand to protect themselves from predators (16). Subsequently they moult once a year or less, in the northern Gulf of Mexico during the winter months and in Massachusetts in summer (23,25).

Breeding is characterized by the migration of huge numbers of crabs into shallow intertidal waters where they congregate along the shores of bays and estuaries. The mating season for L. polyphemus depends on latitude: from April to June in the south of its range and from May to July in the north (7,22,46). T. tridentatus in Japan breeds between July and August. T. gigas in Thailand breeds between April and August with a peak in June to July. C. rotundicauda in the Gulf of Siam, Thailand, lays its eggs all the year round (15). Breeding shows strong lunar and tidal rhythms, animals appearing at, and within a few days of, the new and full moon and within two hours of high tide (23,46) but in some northern populations breeding is less dependent on the lunar cycle (7). Although some populations have been shown to breed exclusively at night (23,49), a more recent study has shown daytime breeding (46).

Wave surge guides the crabs to and from the beach, although other mechanisms may be involved as well; crabs move faster and further under strong wave conditions and the greatest number of crabs emerge on spring high tides with high wind and maximum conditions of wave surge (20,21,31). The males move parallel to the shoreline in a metre or more of water until they intercept and attach themselves to the females heading directly for the beach (23). The couples then proceed to the beach and the male clasps the posterior end of the female and fertilizes the eggs as they are laid (9). A predominance of male crabs is common on breeding beaches and more than one male may be associated with each female. In L. polyphemus the breeding sex ratio may be as high as 14 males:1 female (22). The ratio in T. tridentatus seems to be 1:1. Solitary males of T. gigas and C. rotundicauda are seen in Thailand suggesting that in these species males out number females during breeding (15).

The female lays her eggs in an excavation about 15 cm deep, over a period of two hours in the Gulf of Mexico, but over three hours where the tidal amplitude is greater, as in Delaware (23,25,26,46). The siting of nests may depend on the balance of predation. Florida nests are usually at the high tide line, perhaps to avoid predation by fish. In other areas nests may be intertidally situated, perhaps in response to the presence of large numbers of shoreline-wading birds (46). Clutch size varies from 200-300 eggs (12) to 2000-30 000 (46). The Indo-Pacific species lay larger but considerably fewer eggs (15,26). After making one or more nests, the crabs leave the beach when the tide ebbs, stranded animals digging themselves in to await the next high water (16). The breeding habits of T. gigas appear to be similar to those of L. polyphemus. C. rotundicauda breeds on mud in mangroves along river shorelines and where there is 0.5-1.0 m of water at high tide (24).

Newly laid horseshoe crab eggs are sticky and occur as tightly clumped balls. They hatch in approximately 5 weeks, depending on the ambient temperature and other environmental factors (18,31). The 'trilobite' larvae remain in distinct aggregations, gradually moving to the surface. Circadian activity rhythms in the larvae synchronize activity with water levels on the beach so that the larvae are released into the water on an appropriately high tide. On emergence they exhibit a 'swimming frenzy', moving vigorously and continually for hours. They are strongly positively phototactic and orient themselves to any available light source (18,31). About 6-8 days after leaving the nest they moult into the first juvenile instar which resembles the adult except for telson length. At this stage their behavioural pattern changes abruptly and they cease the nocturnal swimming characteristic of the trilobite, and become a diurnal benthic animal, alternately crawling on the substrate surface and burying in the sand. They play an important role in the intertidal community as predators on bivalves and polychaetes (18,23,25). They undergo a series of moults through different juvenile instars on the intertidal flats and shallow water areas near the adult breeding beach, moving into increasingly deep water as they grow, and becoming subtidal by the time they are mature (23,25,31,32). After low tide juveniles bury themselves for protection from invading pelagic predators (25). Most activity is seen in the summer in both adults and juveniles; in the winter they may bury themselves as protection from the cold (25). Horseshoe crabs probably reach maturity at three years in Florida waters but throughout most of their range to the north the normal time is 9-11 years (22), which would mean that populations could take some time to recover from over-exploitation (26). The total life span may vary with latitude but they can live as long as 14-19 years (38).

Although all phases of the life cycle are highly adapted to avoid predation (25,31), a number of predators besides man are known. Eggs that are washed out of the nest are eaten by minnows and other fish (16,22,26) and the nests may be preyed on by wading birds (16). Waders, such as the Willet Catoptrophus semipalmatus, prey on larvae that become stranded on the beach after hatching (18). Adults may be attacked by Tiger Sharks (Galeocerdo curvieri) (25) and horseshoe crabs have been found to make up 95 per cent of the diet of Loggerhead Turtles in Mosquito Lagoon, Florida (10).

SCIENTIFIC INTEREST AND POTENTIAL VALUE   The massive emergence of crabs on beaches of the Atlantic and Gulf of Mexico is one of the most spectacular phenomena of the North American sandy shoreline (23). However, the main scientific interest in this group lies in the fact that they are 'living fossils', the sole surviving members of a larger group which flourished about three hundred million years ago. They are related to arachnids and give some idea of what the ancestral aquatic arachnid may have looked like. The horseshoe crab was widespread and diverse in the Paleozoic, the majority of species in the fossil record occurring in brackish to freshwater environments. Forms almost identical to Limulus date from the Triassic, some 230 million years ago, and very similar crabs appear in the Devonian, some 400 million years ago (23,31). It is the closest living relative of the trilobites (26), and the resemblance of the larval instar to certain trilobites has long been a subject for comment, and is sometimes cited as an indication of the phylogenetic antiquity of the species (18). As a living fossil it may also provide information on genetic variation and rates of evolution (5,6).

Horseshoe crabs are probably the best known of all marine animals used in the laboratory (23). As a result of some 40 years of research, its lateral compound eye is one of the best understood examples of sensory receptor function (29). It is a particularly good model for studying ecologically dictated modifications of behaviour as all phases and especially its breeding habits are accessible to field and laboratory observations (31).

Horseshoe crabs have been used by man in other ways. Coastal American Indians used the long serrated telsons as fish spear heads (32). The eggs are eaten in Thailand. Crabs are dried and varnished and sold in large numbers as souvenirs. They are used as animal fodder for poultry and pigs, and as fertilizer. Crab meal has a protein content of 46 per cent, and the roe is reported to increase the growth of broiler fowl and the rate of egg laying, although a fishy flavour is imparted to both flesh and eggs (16). The largest harvest in the U.S.A. is centred on the Delaware Bay population. In 1855 on the Cape May shore of Delaware Bay, some 750 000 horseshoe crabs were reported taken on about a half mile (1 km) of beach. Fishery data from the 1850s to the 1930s indicate that millions of adult crabs were harvested annually, but that subsequently catches decreased steadily. By 1960, 42 000 crabs were being taken and as a result of public outcry against the odours from the fertilizer plants this fishery was closed (13). Limulus is currently caught in large numbers in Chesapeake Bay as bait for the commercial eel fishery (23). FAO fishery statistics suggest that the number of Limulus caught annually in the U.S.A. is again increasing. No catch was recorded for the years 1970-73 but in 1976 an exceptional 927 tonnes were caught, most of which went to a poultry offal rendering plant. Recently the processing of Limulus for fertilizer and animal feed has declined, but the catch for eel bait and pharmaceutical laboratories has continued to increase. The U.S.A. and FAO recorded catches increasing from 23 tonnes in 1977 to 232 tonnes in 1980 (1,2,3).

More recently Limulus has become increasingly important in the field of biomedical research. An extract of the horseshoe crab affords at least partial amelioration of leukemia in mice (8), and most importantly the horseshoe crab provides the basis for a very sensitive in vitro test for bacterial endotoxins (11,39). With the discovery of the Limulus lysate test, U.S. Federal Drug Agency approved the use of this species as an acceptable replacement for the rabbit assay by pharmaceutical companies. Contamination of vaccines and intravenous fluids with endotoxins is a major problem and tests are carried out to detect them. Horseshoe crab blood contains only one type of cell, the amoebocytes, which can be easily separated from the whole blood. An extract of these lysed cells forms a gel in the presence of endotoxins, and this test is both more sensitive and more rapid than other tests, and may be cheaper. The Limulus lysate test has recently been proved useful in testing for gonorrhoea (4). T. tridentatus is also effective for the lysate test and large numbers have been collected for this purpose (15).

THREATS TO SURVIVAL L. polyphemus preys on clams, and clam diggers in the U.S.A. killed the crabs in large numbers in the past, when they moved into clam beds in the spring. This, combined with the fact that they were also collected for animal food, has led to a decline in their overall population since the last century. The highest catches reported in fishery statistics occurred in the late 1920s and early 1930s when between four and five million crabs were harvested annually. Since then the annual harvest has declined with catches dropping from some two million animals to one million during the 1940s. In the 1950s, 0.25-0.5 million were being caught annually (36). The taking of L. polyphemus for eel baiting in U.S.A. waters may be the most significant source of mortality and an estimated 60 000 crabs are killed annually for this purpose (39). The Delaware Bay population has endured massive mortalities at different times during the past 100 years (26,39) and it appears that although crabs are resilient to heavy harvests, once the population is reduced it may take at least a decade of little or no harvesting for population levels to build up again (13).

Currently there is concern as to whether the species can withstand exploitation for biomedical use. Its survival rate and population dynamics are still poorly understood and it is feared that the widespread and accelerated use and bleeding of crabs could threaten the species. Large quantities of both L. polyphemus and T. tridentatus are taken for the Limulus lysate test, generally from breeding beaches. The lysate bleeding of horseshoe crabs probably need not pose a threat if the animals are handled carefully and returned to the water. It has been estimated that a third of the plasma can be taken without causing damage (15) and a study in the U.S.A. of 10 000 animals indicated that there was only a 10 per cent increase in mortality due to bleeding (47). Massachusetts law, however, prohibits the return of the crabs to the sea despite a federal statute that requires it (39). Lysate producers who do not have bleeding facilities on the coast also contribute to mortality (39).

The T. tridentatus population in Japan is reported to have declined as a result of pollution and land reclamation (37). Studies on the effect of different types of pollution on Limulus have shown that juveniles tend to be quite tolerant of halogenated hydrocarbons (41,42), but that animals exposed to oil show increased respiratory rates and delayed moulting. The long term effects of pollution could have severe impacts on populations, particularly as the benthic habits of the crabs could cause them to encounter harmful concentrations of halogenated hydrocarbons absorbed onto sediment (40,41,42).

CONSERVATION MEASURES TAKEN The Association for the Conservation of Horseshoe Crabs was established in Japan in 1978 and is campaigning to protect these species (15,43).

CONSERVATION MEASURES PROPOSED The emerging importance of Limulus to biomedical research, as the source of Limulus lysate, requires a more complete knowledge of the biology of this species, so that it can be managed wisely as a biological resource (23,31). A key factor in its survival may be the continued existence of extensive unspoiled habitats where breeding and the development of young can occur with relatively little or no disturbance from man (26). If Limulus amoebocytes could be cultured this would eliminate the necessity for harvesting wild animals for biomedical purposes (44).

CAPTIVE BREEDING Several attempts have been made to breed horseshoe crabs in captivity. The early stages have been reared (28,39,45) but so far none of the four species has been raised successfully to the completion of the life cycle, except in one case where success was probably fortuitous (39).

REFERENCES   1. FAO, (1982). Statistics provided by E. Aykyuz.
          2. Anon. (1980). International Commission for the Northwest Atlantic Fisheries Statistical Bulletin for 1978. Vol. 28. Table 3, p. 69.
          3. Shuster, C. (1982). In litt., 25 August.
          4. Bernstein, A. (1982). Gonorrhea I.D. Health June 1982 : 17-18.
          5. Selander, R.K., Yang, S.Y., Lewontin, R.C. and Johnson, W.E. (1970). Genetic variation in the horseshoe crab (Limulus polyphemus), a phylogenetic 'relic'. Evolution 24: 402-414.
          6. Riska, B. (1981). Morphological variation in the horseshoe crab (Limulus polyphemus), a 'phylogenetic relic'. Evolution 35: 647-658.
          7. Cavanaugh, C. (1975). Observations on mating behaviour of Limulus polyphemus. Biol. Bul. 149: 422.
          8. Ruggieri, G.D. (1976). Drugs from the sea. Science 194: 491-497.
          9. Pomerat, C.H. (1933). Mating in Limulus polyphemus. Biol.

Bull 64(2): 243-252.

10. Mortimer, J.A. (1980). The feeding ecology of the sea turtles. Unpubl. manuscript, 17 pp.

11. Levin, J. and Bang, F.B. (1968). Clottable protein in Limulus: its localization and kinetics of its coagulation by endotoxin. Thromb. Diath. Haemorrh. 19: 186-197.

12. Shuster, C.N. (1950). Observations on the natural history of the American horseshoe crab, Limulus polyphemus. In 3rd Rept., Investigations of the methods of improving the shellfish resources of Massachusetts. Woods Hole Oceanographic Institution, Contribution No. 564: 18-23.

13. Shuster, C.N. and Botton, M.L. (in press). An estimate of the 1977 spawning population of the horseshoe crab, Limulus polyphemus, in Delaware Bay. Estuaries.

14. Botton, M.L. (1981). Predation by horseshoe crabs, Limulus polyphemus, on commercially important bivalve species in New Jersey. Paper presented at 1981 meeting of the National Shellfisheries Association.

15. Sekiguchi, K. and Nakamura, K. (1979). Ecology of the extant horseshoe crabs. In Cohen, E. (Ed.), Biomedical Applications of the Horseshoe Crab (Limulidae). Alan R. Liss, Publ. N.Y. Pp. 37-45.

16. Shuster, C.N. (1971). Xiphosurida. In Encyclopaedia of Science and Technology, 3rd Ed., Mc Graw-Hill Book Co., N.Y.

17. Botton, M.L. (1981). Food habits of breeding horseshoe crabs in Delaware Bay. Bull. of the New Jersey Academy of Science 26: 68.

18. Rudloe, A. (1979). Locomotor and light response patterns of larvae of the horseshoe crab, Limulus polyphemus (L). Biol. Bull. 157: 494-505.

19. Smith, O.R. (1953). Notes on the ability of the horseshoe crab, Limulus polyphemus to locate soft shell clams, Mya arenaria. Ecology 34: 636-637.

20. Rudloe, A. and Herrnkind, W.F. (1976). Orientation of Limulus polyphemus in the vicinity of breeding beaches. Mar. Beh. Physiol. 4: 75-89.

21. Rudloe, A. and Herrnkind, W.F. (1980). Orientation by horseshoe crabs, Limulus polyphemus, in a wave tank. Mar. Beh. Physiol. 7: 199-211.

22. Shuster, C.N. (1958). On morphometric and serological relationships within the Limulidae, with particular reference to Limulus polyphemus (L). Dissertation Abstracts. 18(2): 371-372.

23. Rudloe, A. (1980). The breeding behaviour and patterns of movements of horseshoe crabs Limulus polyphemus, in the vicinity of breeding beaches in Apalachee Bay, Florida. Estuaries 3: 177-183.

24. Sekiguchi, K., Nishiwaki, S., Makioka, T., Srithunya, S., Machjajib. S., Nakamura, K. and Yamasaki, T. (1977). A study of the egg-laying habits of the horseshoe crabs, Tachypleus gigas and Carcinoscorpius rotundicauda, in Chonburi area of Thailand. Proc. Jap. Soc. Sys. Zoo. 13: 39-45.

25. Rudloe, A. (1981). Aspects of the biology of juvenile horseshoe crabs, Limulus polyphemus. Bull. Mar. Sci. 31: 125-133.

26. Shuster, C.N. (1982). A pictorial review of the natural history and ecology of the horseshoe crab, Limulus

polyphemus, with reference to other Limulidae. In Bonaventura, J. and Bonaventura, C. (Eds.) Physiology and Biology of Horseshoe Crabs: Studies on Normal and Environmentally Stressed Animals. Alan R. Liss, Publ. N.Y. Pp. 1-52.

27. Wolff, T. (1977). The horseshoe crab, (Limulus polyphemus) in north European waters. Vdensk. Meddr. Dansk. Naturrh. Foren. 140: 39-52.

28. Kropach, C. (1979). Observations on the potential of Limulus Aquaculture in Israel. In Cohen, E. (Ed.), Biomedical Applications of the Horseshoe Crab (Limulidae). Alan R. Liss, Publ., N.Y. Pp. 103-106.

29. Adolph, A.R. (1971). Recording of the optic nerve spikes underwater from freely moving horseshoe crabs. Vision Research 11: 979.

30. Turner, H.J. (1956). The shellfish predators of Massachusetts. Mass. Dept. of Nat. Res.: 1-20.

31. Rudloe, A. (1979). Limulus polyphemus: a review of the ecologically significant literature. In Cohen, E. (Ed.), Biomedical Applications of the Horseshoe Crab (Limulidae). Alan R. Liss, Publ., N.Y. Pp. 27-35.

32. Shuster, C.N. (1979). Distribution of the American Horseshoe Crab, Limulus polyphemus (L). In Cohen, E. (Ed.), Biomedical Applications of the Horseshoe Crab (Limulidae). Alan R. Liss., Publ. N.Y. Pp. 3-26.

33. Sekiguchi, K. and Nakamura, K. (1980). Sympatric distribution pattern of three species of Asian Horseshoe Crabs. Proc. Jap. Soc. Syst. Zool. 18: 2-4.

34. Sekiguchi, K., Ayodhyoa, Nakamura, K., Yamasaki, T., Sugita, H. and Shishikura, F. (1981). Geographic distribution of horseshoe crabs in Indonesian seas. Proc. Jap. Soc. Syst. Zool. 21: 10-14.

35. Sekiguchi, K., Nakamura, K. and Seshimo, H. (1978). Morphological variation of a horseshoe crab, Carcinoscorpius rotundicauda, from the Bay of Bengal and the Gulf of Siam. Proc. Jap. Soc. Syst. Zool. 15: 24-30.

36. Shuster, C.N. (1960). Horseshoe 'Crabs'. Estuarine Bulletin 5(2): 3-8.

37. Sekiguchi, K. (1982). In litt., 14 January.

38. Ropes, J.W. (1961). Longevity of the horseshoe crab, Limulus polyphemus L. Transactions of the American Fisheries Society 90: 79-80.

39. Pearson, F.C. and Weary, M. (1980). The Limulus amebocyte lysate test for endotoxin. Bioscience 30(7): 461-464.

40. Strobel, C.J. and Brenowitz, A.H. (1981). Effects of Bunker C oil on juvenile horseshoe crabs (Limulus polyphemus). Estuaries 4(2): 157-159.

41. Laughlin, R.V. and Neff, J.M. (1977). Interactive effects of temperature, salinity shock and chronic exposure to No. 2 fuel oil on survival, development rate and respiration of the Horseshoe Crab, Limulus polyphemus. In Wolfe, D. (Ed.), Fate and Effects of Petroleum in Marine Organisms and Ecosystems. Plenum Press, New York.

42. Neff, J.M. and Giam, C.S. (1977). Effects of Aroclor 1016 and Halowax 1099 on juvenile horseshoe crabs, Limulus polyphemus. In Vernberg, F.J., Calabres, A., Thurberg, F.P., and Vernberg, W.B. (Eds.), Physiological Responses of Marine Biota to Pollutants. Academic Press, New York. 462 pp.

43. Nakamura, K. (1982). In litt., 19 February.

44. Pearson, F.C. and Woodland, E. (1979). In vitro cultivation of Limulus amoebocytes. In Cohen, E. (Ed.), Biomedical Applications of the Horseshoe Crab (Limulidae). Alan R. Liss, New York. Pp. 93-99.

45. French, K.A. (1979). Laboratory culture of embryonic and juvenile Limulus. In Cohen, E. (Ed.), Biomedical Applications of the Horseshoe Crab (Limulidae). Alan R. Liss, New York. Pp. 61-72.

46. Cohen, J.A. and Brockmann, H.J. (in press). Breeding activity and mate selection in the horseshoe crab, Limulus polyphemus. Bull. Mar. Sci.

47. Rudloe, A. (in press). The effect of heavy bleeding on mortality of the Horseshoe Crab, Limulus polyphemus. J. Invert. Path.

We are very grateful to M.L. Botton, J.A. Cohen, J. Levin, K. Nakamura, A. Rudloe, and K. Sekiguchi for providing information and commenting on this data sheet, and particular thanks go to C.N. Shuster for who has sorted out many of the problems encountered in dealing with such a large subject.

RED-KNEE TARANTULA SPIDER INSUFFICIENTLY KNOWN

Brachypelma smithi (F.O. Pickard–Cambridge, 1897)

Phylum	ARTHROPODA	Order	ARANEAE
Class	ARACHNIDA	Family	THERAPHOSIDAE

SUMMARY Brachypelma smithi is a large (14–17 cm span) black tarantula with attractive red and pale red markings. It is increasingly in demand in the U.S.A. and Europe as a household pet, but there are no scientific data on the status of the wild populations. The Red-knee Tarantulas are harvested from western Mexico, and it is recommended that trade should be carefully monitored.

DESCRIPTION The Red-knee Tarantula, Mygalomorph, or Bird-eating Spider, is a large ground-dwelling species with contrasting coloration which blends well with the harsh sun and shade conditions of the often exposed habitat. Adult spiders reach a diagonal span of c. 14 cm, though some exceptional males reach 17 cm. The colour is similar in both sexes and is predominantly black with a pale red margin to the carapace and a distinct pale red clypeal fringe along the front of the head. The abdomen is black, flecked with long pale hairs. The colouring of the legs is quite remarkable; the hairs on the femurs are black with a pale red margin on the dorsal extremity; the patellas are also pale red with distinct short rust-red hairs dorsally. The hairs on the tibias are black with a band of pale red hairs distally. The metatarsus is almost uniform black with some pale hairs distally on the dorsal surface. Long, light tipped hairs and trichobothria (fine sensitive setas) are distinct on all tibias and metatarsal joints. Adult males are armed with an articulate intromittent organ at the tip of each palpus (11).

DISTRIBUTION The Red-knee Tarantula is known from Mexico west of the Sierra Madre Occidental, and from a single indefinite record from Costa Rica (2). Pickard-Cambridge originally described the species from specimens taken in Guerrero State, south-west Mexico (12). Other published locations include Colima, Nayarit and Mexico States (1), with unpublished records from Sinaloa (9) and Sonora States (9). Since there are no records from Guatemala, El Salvador or Nicaragua, the single uncertain collection from Costa Rica requires further substantiation. Unless further evidence is forthcoming, the Red-knee Tarantula may be considered endemic to Mexico (2).

POPULATION No population estimates have been attempted, and at the time of writing it has not been possible to discover the difficulties involved in collecting the spider. Predation is known to be very high on the juvenile spiders (1), and parasitic wasps and flies may also take a toll, as they do with other theraphosid spiders (8). However, the adults may be very long-lived once they have reached maturity, which takes between four and seven years (11). An adult has been kept in captivity for over 13 years (2), giving a possible total lifespan of over 20 years.

HABITAT AND ECOLOGY Brachypelma smithi is predominantly a burrowing species of semi-desert habitats. The burrows are usually in sand or soil banks protected from rain. The predominant food is insects, though some small vertebrates and carrion may also be devoured. Mature males are nomadic and mate with mature females from mid- to late summer. Gestation varies from two to four months depending on climatic conditions. Two to four hundred eggs are deposited in the cocoon web lining the burrow, or in a bowl-shaped cocoon web constructed in hollows between or beneath rocks or natural debris. The web is finally torn around its top and folded over the egg mass to form the ball-like cocoon. Ambient conditions are controlled to some extent by the mother, who moves the cocoon to a suitable temperature regime within the burrow. Nymphs emerge simultaneously approximately nine weeks after laying and remain with the mother for 12 to 16 days before dispersing (11). Predation and other natural causes may account for up to 98 per cent mortality in young spiders (1). Males mature at about the twentieth instar two to three years before females, preventing cross-breeding of close relatives. Females reach sexual maturity after about seven years (11). Like many of the New World Theraphosidae, B. smithi is armed with urticating hairs which are intensely irritating to man, and may be fatal to small rodents (1,3,4). Although the Red-knee Tarantula is favoured as a pet because of its fairly docile demeanour, it is capable of inflicting a painful bite with its poisonous jaws.

SCIENTIFIC INTEREST AND POTENTIAL VALUE The ecology and taxonomy of the spider family Theraphosidae remain in a state of considerable neglect. They are of interest to science in that they have evolved in parallel with the araneomorphs, or true spiders, but have remained relatively primitive in lifestyle (11). Because tarantulas are generally large and easy to maintain, they are favoured research animals for arthropod physiologists. Selected studies range from measurement of blood pressure (15) and work on nerve action potentials (13), to metabolic investigations (14).

THREATS TO SURVIVAL Demand for unusual and exotic pets has increased dramatically over recent years and the Red-knee Tarantula can be bought from dealers in London for about ₤20. B. smithi is a colourful representative of a generally drab family and therefore a highly prized species. The spiders are quite stout, have a relatively strong cuticle, travel well, and fetch high prices. In consequence many small colonies may have been completely destroyed with no apparent monitoring to control over-collecting. Indiscriminate collecting must eventually jeopardise the recovery of colonies and reduce the populations and range of the species. Unfortunately there is no documentation of the numbers of Tarantulas exported from Mexico or imported into the U.S.A. and Europe.

Ploughing and other disturbance to marginal agricultural land in Mexico may be restricting the range of the species (6).

CONSERVATION MEASURES TAKEN No conservation measures have been taken specifically for the Red-knee Tarantula. Mexican legislation is very general and prohibits all hunting and export of wild animals except under licence (5,16). The regulating authority is the Ministry of Agriculture and Livestock. The issuing of licences is subject to the scrutiny of experts, who determine the quantities which may be collected, and from where (17). At present the export and import of any type of animal, with the exception of those bred in captivity and those on special official programmes, is prohibited (17). In practice, it is not clear to what extent the laws are applied to invertebrates, and an authoritative statement is being sought. There is no apparent restriction on import of the Tarantula into the U.S.A. or Europe and world trade seems to be channelled via American dealers.

CONSERVATION MEASURES PROPOSED Attempts at captive breeding have been successful though environmental conditions need careful control (10). A breeding programme would be very valuable for the pet trade but would require a careful study of the species' natural environment (11). The number of Tarantulas collected is unknown and monitoring should be started as soon as possible. Initial studies would concentrate on the past and present range of the Tarantula, its density and its population dynamics. Only then can trade statistics be examined in the light of suitable biological information. If trade appears to be causing a permanent decline in populations, then the Red-knee Tarantula should be considered for listing under the Convention on International Trade in Endangered Species of Wild Fauna and Flora (CITES). International trade may then be carefully controlled.

REFERENCES
1. Baerg, W.J. (1958). The Tarantula. University of Kansas Press, Lawrence. 88pp.
2. Baerg, W.J. and Peck, W.B. (1970). A note on the longevity and molt cycle of two tropical theraphosids. Bull. Br. Arachnol. Soc. 1: 107-108.
3. Cooke, J.A.L. (1972). Stinging hairs: a tarantula's defense. Fauna 4: 5-8.
4. Cooke, J.A.L., Roth, V.D. and Miller, F.H. (1972). Urticating hairs of Theraphosid spiders. Am. Mus. Novit. 2498: 43pp.
5. Cooke, J.A.L. (1982). In litt., 9 July.
6. Levi, H.W. (1982). In litt., 16 August.
7. Levi, H.W., Levi, R.L. and Zim, H.S. (1968). A Guide to Spiders and their Kin. Golden Press, New York. 160 pp.
8. Minch, E.W. (1980). Notocyphus dorsalis arizonicus Townes (Hymenoptera: Pompilidae), a new host record of theraphosid spiders. Wasmann J. Biol. 37: 24-26.
9. Minch, E.W. (1982). In litt., 5 November
10. Murphy, F. (1980). Keeping spiders, insects and other land invertebrates in captivity. Bartholomew, Edinburgh. 96 pp.
11. Nussel, P.M. (1982). In litt. 14 July.
12. Pickard-Cambridge, F.O. (1897). Arachnida Araneida. In Biologia Centrali-Americana. Godman and Salvin, London. Pp. 1-40.
13. Rathmayer, W. (1965). Neuromuscular transmission in a spider and the effects of calcium. Comp. Biochem. Physiol. 14: 673-687.
14. Ross, R.H. and Monroe, R.E. (1970). Utilization of acetate-1-^{14}C by the tarantula Aphonopelma sp. and the scorpion Centruroides sculpturatus in lipid synthesis. Comp. Biochem. Physiol. 36: 765-773.

15. Stewart, D.M. and Martin A.W. (1974). Blood pressure in the tarantula _Dugesiella_ hentzi. J. Comp. Physiol. 88: 141-172.
16. Secretaria De Agricultura Y Ganaderia, Subsecretaria Forestal Y De La Fauna, Departamento De Conservacion De La Fauna. (Undated). _Ley Federal de Caza, Direccion General de la Fauna Silvestre._ Ediciones del Departamento de Divulgacion Forestal, Mexico D.F.
17. Fuentes, J.V. (1982). Letter from the Permanent Assistant, Direccion General de la Fauna Silvestre, in litt., 3 December.

We are grateful to P.M. Nussle for preliminary information for this review, and to J.A.L. Cooke, P.M. Hillyard, H.W. Levi, P. Liedo, N. Platnick, D.J. Stradling, F.M. Wanless and particularly E.W. Minch for further comment.

Troglohyphantes gracilis Fage, 1919
T. similis Fage, 1919
T. spinipes Fage, 1919

Phylum	ARTHROPODA	Order	ARANEAE
Class	ARACHNIDA	Family	LINYPHIIDAE

SUMMARY These small, obligate subterranean spiders from Yugoslavia are of considerable biogeographic interest. Their very limited range makes them susceptible to human impact.

DESCRIPTION These three species of Troglohyphantes are very small (body length 2-4 mm, legs 6-11 mm), pale, delicate spiders with reduced eyes. The three species are very similar (2).

DISTRIBUTION Troglohyphantes similis is known only from a single cave south of Kocevje, Slovenia, Yugoslavia. The other species are known from a few caves clustered around Kocevje Polje (a large karst depression some 41 km long). Each species has a maximal known extent of about 10 km (2).

POPULATION Unknown.

HABITAT AND ECOLOGY These species are true troglobites, and occur only in cave and other subterranean habitats of karst limestone. The known localities have saturated humidity, stable microclimates, and total or near total darkness. The spiders make sheet webs about 5 cm in diameter in which they catch Collembola and small flies (2).

SCIENTIFIC INTEREST AND POTENTIAL VALUE Of the 54 Troglohyphantes species known from Yugoslavia, these are among the most limited in distribution. They provide unusual insights into subterranean speciation. Although the three species are closely similar and have contiguous, allopatric ranges, no intermediate forms have been found, indicating the presence of subterranean barriers to dispersal, possibly associated with karst watersheds (2).

THREATS TO SURVIVAL These species are presently under no known threat but their limited distribution makes them susceptible to industrial development or other surface modification that may occur (1).

CONSERVATION MEASURES TAKEN None known.

CONSERVATION MEASURES PROPOSED The species should be monitored periodically. Land management authorities for the vicinity of the caves should be notified of their biological importance, and protective agreements or reserve designation sought (1).

REFERENCES 1. Crawford, R.L. (1982). In litt., 20 February.
2. Deeleman-Reinhold, C.L. (1978). Revision of the cave-dwelling and related spiders of the genus Troglohyphantes Joseph (Linyphiidae), with special reference to the Yugoslav species. Slovenska Akademija Znanosti in Umetnosti, Razred za Prirodoslovne Vede, Dela 23: 1-223.

We thank R.L. Crawford for the information in this review, and B. Vogel for further comment.

Adelocosa anops Gertsch, 1973

Phylum	ARTHROPODA	Order	ARANEAE
Class	ARACHNIDA	Family	LYCOSIDAE

SUMMARY This obligate subterranean blind spider is found in an extremely limited cave ecosystem on the Hawaiian island of Kauai, U.S.A. The larger caves are under increasing pressure from visitors and surface development.

DESCRIPTION Adelocosa anops has a total body length of 10-20 mm; the cephalothorax is light brown or orange with no eyes whatsoever and long translucent orange bristly legs; the abdomen is dull white (5,7). Adelocosa is a monotypic genus (7) and the species is known in Hawaii as pe'e pe'e maka'ole.

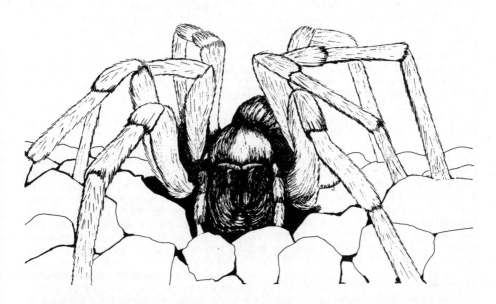

DISTRIBUTION Adelocosa anops is known only from the deep zone of its type cave (Koloa Cave) (2) and one additional cave on the south-east coast of Kauai Island, in the Hawaiian group (U.S.A.). These caves are lava tubes resulting from an eruption of Koloa. The spider inhabits small cavities impenetrable to man, as well as the caves themselves (3), but its distribution is limited to a single lava flow series (7).

POPULATION Unknown.

HABITAT AND ECOLOGY The spider does not spin webs but actively stalks and overwhelms other subterranean invertebrates, even without the benefit of eyes (7,8,10). It requires a permanent moisture source (10) and is suited to the cave temperature of 25-26°C and 100% humidity. The No-eyed Big-eyed Wolf Spider has a low reproductive capacity (15-30 eggs per clutch), and long developmental period (one year) (7). The adult lifespan is at least six months (7).

SCIENTIFIC INTEREST AND POTENTIAL VALUE The No-eyed Big-eyed Wolf Spider is one of only two obligatory cave lycosids known. The other species, Lycosa howarthi, is from Hawaii itself and has weakly developed eyes (7). All other members of the family have large, well-developed eyes, hence their common name, Big-eyed Wolf Spiders. The Kauai Caves also contain a subterranean detritivorous amphipod Spelaeorchestia koloana (Crustacea: Amphipoda: Talitridae) (2). This blind, long-legged white amphipod is part of the diet of Adelocosa and is also an endemic monotypic genus.

THREATS TO SURVIVAL The greatest threat to these invertebrate troglobites is the withdrawal and pollution of groundwater due to surface development (3). There is a proposal to develop tourist facilities in the region, to include enlargement of existing hotels and building of roads and a sewage plant (7). Water is already deficient, and natural water sources have been disrupted. Moisture in the caves now comes from surface irrigation of sugar cane (7). The spider and other invertebrates are extremely sensitive to desiccation and a failure in water supply could be disastrous (7). Agriculture has already ruined the largest lava cave in the area when it was covered with 5 m of sugar cane bagasse (waste residue) (7). The elimination of tree roots by surface deforestation may present a further problem since roots are an important source of food to the cave fauna (3,6). Human visitors may affect the cave habitats by trampling, littering, smoking, vandalism, destroying tree roots or altering microclimatic conditions. The Koloa caves are Civil Defence shelters, and well known to local people. Although the spiders would probably survive in remote crevices, human interference could destroy those spiders accessible to scientists (3,7).

CONSERVATION MEASURES TAKEN A notice of review from the Office of Endangered Species of the U.S. Fish and Wildlife Service was followed by a proposal in 1978 (4) to list the spider as Endangered and the amphipod as Threatened. Since the two-year time limit of the listings then expired, these proposals were withdrawn in 1980 (1). Further proposals may be made in the future. Some beneficial co-operation has been obtained from the landowners, including the McBryde Sugar Company, the Moana-Kauai Corporation and Grove Farm. Survey and research funds have been extended by the National Science Foundation, the U.S. Fish and Wildlife Service, the Moana-Kauai Corporation, and the Xerces Society. The Xerces Society grant (8) involved operations recommended as a result of a biospeleological survey of 180 ha designated for golf course and resort development. This area includes many very small caves, only two of which were found to harbour small populations of the spider and amphipod. Moana-Kauai Corporation plans to establish reserves for these two caves within their proposed golf course (8). Efforts are being made to attract the spider and amphipod into other caves (8).

CONSERVATION MEASURES PROPOSED The largest and most stable populations of both the spider and the amphipod occur in Koloa Cave No. 2, and formal reserve status for this cave should be established. Long term monitoring of the populations within the two reserve caves on the Moana-Kauai Corporation resort property is necessary to ensure that the size of these reserves is adequate for the survival of the species. Management regulations need to be developed and should include provisions for limiting access to the caves and restrictions on the type of modifications allowed on the surface near the caves. Endangered species status should be re-proposed and all the provisions of the U.S. and Hawaii Endangered Species Acts brought to bear upon the issue (7).

REFERENCES 1. Bentzien, M.M. (1980). Endangered and threatened wildlife and plants: notice of withdrawal of expired proposals, for listing eight arthropod species. Fed. Register 45: 58171.
2. Bousfield, E.L. and Howarth, F.G. (1976). The cavernicolous fauna of Hawaiian lava tubes, 8. Terrestrial Amphipoda

(Talitridae), including a new genus and species with notes on its biology. Pacific Insects 17: 144-154.

3. Crawford, R.L. (1982). In litt., 20 February.
4. Ekis, G. and Opler, P. (1978). Endangered and threatened wildlife and plants. Proposed listing and critical habitat determination for two Hawaiian cave arthropods. Fed. Register 43: 26084-26087.
5. Gertsch, W.J. (1973). The cavernicolous fauna of Hawaiian lava tubes, 3. Araneae (Spiders). Pacific Insects 15: 163-180.
6. Howarth, F.G. (1973). The cavernicolous fauna of Hawaiian lava tubes, 1. Introduction. Pacific Insects 15: 139-151.
7. Howarth, F.G. (1979). The no-eyed big-eyed wolf spider (Adelocosa anops) Gertsch; proposal to U.S. Fish and Wildlife Service. In litt.
8. Howarth, F.G. (1979). Grant proposal for a salvage survey of lava tubes near Koloa, Kauai. Xerces Soc. 2 pp.
9. Howarth, F.G. (1979). In litt., 20 July.
10. Howarth, F.G. (1980). The zoogeography of specialized cave animals: a bioclimatic model. Evolution 34: 394-406.

We thank F.G. Howarth for information supplied on this topic, and R.L. Crawford, P.A. Opler and B. Vogel for further comments.

GLACIER BAY WOLF SPIDER   RARE

Pardosa diuturna Fox, 1937

Phylum   ARTHROPODA   Order   ARANEAE

Class   ARACHNIDA   Family   LYCOSIDAE

SUMMARY This moderate-sized wolf spider is found on glaciers and barren terrain over a very limited area in south-east Alaska, U.S.A. Although it is under no known threat, its restricted distribution should be monitored.

DESCRIPTION Pardosa diuturna has a body form typical of a wolf spider and is 7-8 mm in length exclusive of legs. Its colour is uniform black with no trace of a pattern (1,4).

DISTRIBUTION Pardosa diuturna is known only from the vicinity of Glacier Bay in south-east Alaska, U.S.A. The six known localities (1,2,5) all fall within a very narrow range of latitude, 58° 36'N to 58° 51'N, and are within 285 km of each other in an east-west direction. It is most unusual for an active, surface spider to have such a restricted distribution (3).

POPULATION Unknown.

HABITAT AND ECOLOGY This species has so far only been found in habitats barren of vegetation, including high alpine fellfield, glacial outwash, and the surface of glaciers (1,2,5). A major food source may be wind-blown insects. Like most Pardosa spp., P. diuturna is diurnal and its activity depends on incident solar radiation. Most Pardosa are dark in colour but unlike most of its congeners, P. diuturna is pure black, perhaps an adaptation for making maximum use of solar energy (3).

SCIENTIFIC INTEREST AND POTENTIAL VALUE P. diuturna is one of a small assemblage of spider species in several families that forage on snowfields, glaciers, and glacial outwash. Most of these species are far more widely distributed. None has been studied thoroughly. P. diuturna would make an interesting case study and it would be valuable to discover the reasons for its restricted distribution (3).

CONSERVATION MEASURES TAKEN Much of the rugged and inaccessible habitat of P. diuturna is in Glacier Bay National Monument. The species is not known to be in jeopardy (3).

CONSERVATION MEASURES PROPOSED No specific measures are contemplated. The National Park Service should be aware of the presence of these organisms and monitor their presence periodically.

REFERENCES   1.   Chamberlin, R.V. and Ivie, W. (1947). The spiders of Alaska. Bull. Univ. Utah, Biol. Series 10 (3): 1-103.
              2.   Crawford, R.L. and Mann, D.H. (In prep.). Arachnid fauna of the Lituya Bay region, coastal southeast Alaska.
              3.   Crawford, R.L. (1982). In litt., 20 February.
              4.   Fox, I. (1937). Notes on North American lycosid spiders. Proc. Ent. Soc. Wash. 39: 112-115.
              5.   Vogel, B.R. (1982). Pers. comm. to R.V. Crawford.

We thank R.L. Crawford for information on this topic, and B. Vogel for further comment.

MELONES CAVE HARVESTMAN   VULNERABLE

Banksula melones Briggs, 1974

Phylum   ARTHROPODA   Order   OPILIONES

Class   ARACHNIDA   Family   PHALANGODIDAE

SUMMARY   The Melones Cave Harvestman occurs in a small number of limestone caves in central California, U.S.A. A new dam on the Stanislaus River and quarrying threaten four major populations. Newly discovered colonies may be similarly affected in the future, or damaged by cavers.

DESCRIPTION   Banksula melones is a small (less than 10 mm including legs), light orange cave harvestman with large, well-developed eyes. Detailed descriptions are given in the literature (1,3).

DISTRIBUTION   Of the 18 localities now known, 16 are in caves within a 2 km radius of the confluence of the Stanislaus River and South Fork of the Stanislaus River, Tuolumne County, California, U.S.A. The other two are 5 km south-west of that confluence, in the drainage of Coyote Creek in Calaveras County (7). Extensive surveys have been undertaken and the known distribution is not expected to be greatly enlarged in future.

POPULATION   Total populations are unknown. These harvestmen are very small and may inhabit tiny subterranean crevices as well as the caves available for study. Banksula melones has been found in so many caves in the limited area of its distribution that the population is probably continuous between, as well as within, caves. The population transferred to Transplant Mine (see below) has apparently grown to exceed densities of natural populations. However, this may be a temporary condition in a population as yet unequilibrated (2,4,7).

HABITAT AND ECOLOGY   These harvestmen and their congeners live in relatively shallow, generally well drained limestone caves (1,6). Banksula california lacks corneas and retinae, but four other species, including B. melones, have functional eyes. All Banksula are confined to caves and only B. melones has been collected near or at the cave twilight zone (3). They prey on Collembola (springtails) and other microfauna dwelling among organic cave detritus. Excess moisture and deficiency of organic material are limiting factors (3). Length of instars (time between moults) is several months and individuals have been observed to survive in the absence of prey for 43 days. This species shares the caves with a number of other cave endemics, and serves as an indicator for the system as a whole. Other caves may contain other species of Banksula, some newly discovered and equally threatened (1,3,7).

SCIENTIFIC INTEREST AND POTENTIAL VALUE   The genus Banksula, and the species B. melones in particular, represent an instance of extreme endemism and evolutionary radiation in a tightly circumscribed, limited cave system. As such, they present an outstanding collective opportunity for studies in evolutionary biology, especially in the area of autochthony (in situ evolution) and near-sympatric speciation. In addition, the so-called Transplant Mine population is the only colony of arthropods which has been rescued in this way, and has great importance as an experiment in conservation methodology (7).

THREATS TO SURVIVAL   Populations in McLean's Cave, Scorpion Cave and Vulture Cave will probably be inundated when the Army Corps of Engineers finishes the New Melones Dam on the Stanislaus River (5). The McNamee's Cave colony may be destroyed if the owners decide to quarry there. The Transplant

Mine population faces some danger from excess moisture and lack of organic input. The other 14 naturally occurring populations will encounter increasing threat from human impact as caving increases in popularity. Trampling could have substantial impact in these caves, especially for species common on well-developed soil surfaces, such as species of Banksula (7).

CONSERVATION MEASURES TAKEN   The U.S. Fish and Wildlife Service, the U.S. Army Corps of Engineers and the World Wildlife Fund (U.S.A.) have given contracts or grants for the study and survey of Banksula melones and other cave fauna of the Calaveras Formation (6,7). Representatives of WWF (U.S.A.) and IUCN/SSC have visited the area and descended McLean's Cave and the Transplant Mine with local cavers and conservationists. The U.S. Army Corps of Engineers, seeking to avoid the species' extinction from the area to be inundated by their dam, carried out a transfer operation. The site chosen for the rescue, known as the Transplant Mine, was an abandoned mineshaft above the high-water contour for the impoundment. This action and subsequent monitoring has been discussed in detail (2,4-7). Reservations have been expressed (4) concerning apparent excess moisture in the mine, the need for a reliable supply of organic debris and potential adverse effects of inbreeding in the small foundation population. The Corps intends to deal with these problems and has already taken action to reduce moisture in the mine. Along with B. melones, a portion of the overall fauna of McLean's cave was transplanted to the mine, and additional surveys over a period of years will be required to determine whether the displaced section of a cave fauna can adapt and survive under the new conditions. The Army Corps also gave a contract to explore the overall impact of the New Melones Reservoir project on the cave resources of the area (6). Furthermore, the Corps plans (7) to acquire the major cave areas and to institute a management plan which will include limiting human visitors. These actions are largely on behalf of the invertebrate fauna of the cave system, including B. melones and other endemic harvestmen. Finally, the Friends of the River and other conservation organizations are campaigning to prevent construction or completion of the dam, or to limit the height to which filling is permitted. The U.S. Fish and Wildlife Service, Office of Endangered Species, was petitioned to list B. melones as a federally Threatened or Endangered species but did not do so. Federal conservation action has come from the Army Corps of Engineers rather than the Office of Endangered Species.

CONSERVATION MEASURES PROPOSED   The Army Corps of Engineers should be urged to fulfil the intent of its conservation proposals for the areas. Government ownership and management of the cave areas seems to be the best way to ensure safety of the unique Calaveras Formation cave fauna. The Transplant Mine operation should be monitored and sustained because of its importance. Additional survey and taxonomic work is needed to catalogue and record the distributions of cave endemics of the area. Ecological research should be encouraged and supported. The opportunity exists for this to become a model cave fauna conservation effort and study system, in spite of the losses that will accrue from the New Melones impoundment. Private and state efforts to limit the height of impoundment should be encouraged in order to minimize damage both above and below ground. Both surface and subterranean components are essential to the continued existence of this and other species endemic to the area.

REFERENCES   1.   Briggs, T.S. (1974). Phalangodidae from caves in the Sierra Nevada (California) with a redescription of the type genus (Opiliones: Phalangodidae). Occ. Papers Calif. Acad. Sci. No. 198, 15 pp.

2.   Briggs, T.S. (1975). Biological transplant project, New Melones Lake, California - final report. U.S. Army Corps of Engineers, Sacramento Dist., Calif. 5 pp.

3.   Briggs, T.S. and Ubick, D. (1981). Studies on cave harvestmen of the central Sierra Nevada with descriptions of

new species of <u>Banksula</u>. <u>Proc. Calif. Acad. Sci</u>. 42: 315-322.

4.   Elliott, W.R. (1978). The new Melones Cave harvestman transplant - final report. U.S. Army Corps of Engineers, Sacramento Dist., Calif. 62 pp.

5.   Elliott, W.R. (1981). Damming up the caves. <u>Caving Int. Mag</u>. 10.

6.   McEachern, J.M. and Grady, M.A. (1978). An inventory and evaluation of the cave resources to be impacted by the New Melones Reservoir Project, Calaveras and Tuolumne Counties, California - final report. U.S. Army Corps of Engineers, Sacramento Dist., Calif.

7.   Rudolph, D.C. (1979). Status of the Melones Cave harvestman in the Stanislaus River Drainage - final report. Contract #14-16-0009-79-009, U.S. Fish and Wildl. Serv., Wash. D.C. 20 pp.

We thank R.L. Crawford, W.R. Elliott, F.G. Howarth and D.C. Rudolph for information on this subject.

CRUSTACEA

INTRODUCTION Over 38 000 crustacean species are known, most of which are marine, although some are found in fresh water and some are terrestrial. The majority of planktonic animals throughout the world are crustaceans, but the group is also well represented in bottom faunas from coastal waters to abyssal depths. Individuals range in size from microscopic planktonic forms to large bottom-living spider crabs whose leg span may reach over 3.5 m and lobsters which may weigh over 10 kg (1,2,68).

Crustaceans are extremely diverse in detailed structure and life style, but most are recognizable by their general body plan. The segmented body is usually partly or entirely covered by a carapace, and the chitinous jointed external skeleton is often strengthened with calcium salts. The segments carry appendages which are adapted for a variety of functions including feeding and locomotion, and crustaceans are unique among arthropods in having two pairs of antennae. Gills, usually associated with the limbs, are present in many species but the small forms tend to respire through the general body surface. Many species are scavengers, in which case the appendages may be adapted for picking up food; others, with suitably adapted appendages, are filter-feeders and predators. Growth, as in all arthropods, takes place through a series of moults, when the old exoskeleton is discarded. Many crustaceans tend to be at their most vulnerable in the period before the new skeleton hardens, a factor that may be important in their conservation. With a few exceptions, sexes are separate and after fertilisation the females lay eggs, which in many species are carried in brood pouches. In most aquatic species a series of free-swimming planktonic larval stages is a common feature of the life cycle but sometimes the larval development is passed in the egg and the animals hatch in the post-larval stage (1,2,68).

Crustacean taxonomy is still a matter of some dispute. It is based largely on segmentation and the adaptations of appendages for different uses. For example Anaspides, a fairly primitive malacostracan (see data sheet), is fully segmented, whereas in lobsters and shrimps several segments are fused. Traditionally eight subclasses have been recognized but two new subclasses have been described since 1981.

The smaller, more primitive species were once grouped together as the Entomostraca. The marine Cephalocarida is a primitive group of crustaceans, discovered in 1955, of which only nine species are so far known. The Branchiura and a new subclass described in 1983 are entirely parasitic, and the marine Mystacocarida are minute forms found living between sand grains. The primitive Branchiopoda is a widely distributed and often very abundant group found mainly in fresh water. It includes the fairy and tadpole shrimps, which have adaptations, such as drought-resistant eggs, to life in temporary water bodies and the Cladocera or water fleas which play an important role in most freshwater ecosystems. There are also several marine species, including many planktonic forms, and the brine shrimp Artemia, which occurs in salt pans and ponds throughout the world, sometimes in high densities. The Ostracoda, characterized by a bivalved carapace, resembling a shell, which completely encloses the head, body and limbs, is a group widespread and abundant in sea and fresh water. The Copepoda which, with 8500 species, is the second largest crustacean subclass, contain marine, freshwater and parasitic animals and the tiny, usually transparent, herbivores with no carapace, which dominate the animal plankton. The subclass Remipedia was described in 1981 on the basis of a single species found in a marine cave in the Bahamas. The Cirripedia or barnacles are probably the most familiar of the small subclasses. Most members of this group are marine and live attached to rocks and other firm substrates by means of a cement-like secretion. A mantle encloses the body and limbs and is usually strengthened by calcareous plates which make the animal highly resistant to exposure to air and extremes of temperature.

The Malacostraca, with approximately 23 000 species, makes up 60 per cent of all crustaceans and is the largest and most varied subclass. It includes the more highly developed, larger crustaceans, the bodies of which often have heavily calcified exoskeletons and carapaces. The most highly developed and important order within this group are the Decapoda with 8500 species. These include some of the crustaceans most familiar to man, such as shrimps, prawns, lobsters, crayfish and crabs. The shrimps and prawns in the suborder Natantia are swimming animals with light external skeletons frequently laterally flattened. Lobsters and crabs in the suborder Reptantia are bottom-dwelling, walking animals. Examples in the data sheets include the Coconut Crab, which is a hermit crab in the infraorder Anomura (unlike true crabs in the infraorder Brachyura, hermit crabs live in mollusc shells which protect their soft abdomens) and several species of freshwater crayfish in the infraorder Astacura. About 450 species of freshwater crayfish are known, mostly from temperate regions, where several are threatened. They live in all types of waters, (except those exhibiting extremes of temperature and salinity), seeking shelter beneath stones, within thick vegetation, or among plant debris. Many species are found in caves and subterranean water systems (69).

The order Euphausiacea includes the marine shrimp-like krill. Because of its large size and widespread abundance, Euphausia superba is the most well-known species, and is the one most commonly referred to as krill. It has a circumpolar Antarctic distribution, and is pelagic, occurring in huge numbers in the open ocean where it tends to aggregate in large swarms (67). The Amphipoda comprise mainly marine species, characterized by their curved, laterally flattened bodies lacking a carapace. The Isopoda are flattened dorso-ventrally and include marine and freshwater species as well as the largest group of truly terrestrial Crustacea, the woodlice or pill bugs. Reviews for several isopod species follow, as examples of the many crustaceans which are highly vulnerable on account of their specialized habitats and restricted ranges in caves and subterranean waters.

There are several other minor groups within the Malacostraca, including the Syncarida, a primitive group of freshwater crustaceans including the Tasmanian anaspids (see review).

SCIENTIFIC INTEREST AND POTENTIAL VALUE Crustaceans are a vital link in the food chains of aquatic ecosystems. For example, the Copepoda provide an important link between phyto-plankton and fish. Some species, including many of the water fleas (Cladocera) are the basic food for commercial fish, particularly juveniles (1), and the isopod Asellus aquaticus is a major food for freshwater fish in Europe (6). Krill is an important food for most vertebrates in the southern oceans, including five whale species, four seal species, many fish and birds and several squid (67).

Freshwater crayfish are often major croppers of aquatic vegetation and their disappearance can cause premature ageing of lakes (8), since without crayfish the dense root systems of aquatic plants and their annual deposition can fill shallow lakes with great quantities of organic matter. The scavenging habits of crayfish ensure that they contribute to the recycling of organic matter (both living and dead) in the ecosystem. In turn, because of their relatively high fecundity and wide range in sizes, they serve as prey for a large number of secondary consumers, including trout, perch, reptiles, amphibians, birds, and mammals such as otters and racoons (2,9). Other scavenging crustaceans which may play an important ecological role are the hermit crabs which recycle dead organic matter on beaches.

Many freshwater and terrestrial crustaceans are of interest to biologists on account of their relict distributions and interesting evolutionary history. Both freshwater and terrestrial isopods have evolved directly from marine ancestors and their current distributions provide useful data on early geological events. The suborder Phreatoicidea, for example, is a very ancient group, fossil phreatoicids from Middle Pennsylvanian deposits in Illinois dating back some 120 million years (5). The phreatoicid fauna of Great Lake in Tasmania is also of interest with two endemic genera (Mesacanthotelson and Onchotelson) and nine endemic species (4). The Ostracoda have the most extensive fossil record of any crustacean group, stretching back over 500 million years. Certain oil-bearing strata are rich in certain ostracod fossils which can be used as indicator species when searching for new oilfields (1).

The main value of crustaceans to man is as a food resource and they are a major element in both commercial and subsistence fisheries. Of greatest economic importance are the prawn and shrimp fisheries, the world catch of which totalled over 1.5 million tonnes in 1980 (12,13). FAO has produced a catalogue of almost 350 species of shrimps and prawns that are used by man or are considered to be of potential commercial value throughout the world (58). Over 20 major types of crab fisheries are listed by FAO, and the 1980 world production totalled over 800 000 tonnes (12).

The lobster is probably the most valuable and popular crustacean and the world lobster catch (including both clawed and spiny lobsters) totalled over 150 000 tonnes in 1980 (12). Spiny lobsters of the family Palinuridae are found in all major tropical and temperate oceans and support subsistence, large-scale commercial or recreational fisheries in most areas. The 1980 catch totalled nearly 70 000 tonnes, 70 per cent coming from the Caribbean, eastern Indian Ocean and south-east Atlantic (12,15). The recreational harvest of spiny lobster used to be insignificant but, with the growth in popularity of SCUBA and skin diving, has become important in several areas (17). Although lobsters are now considered a luxury food in most countries, in the past they were taken for use as fertilizer, bait and as a subsistence food (16). The clawed lobsters, Homarus americanus in the west Atlantic and H. gammarus and Nephrops norvegicus from the eastern Atlantic and Mediterranean, have been exploited intensively for centuries (16). A wide range of other marine crustaceans are used as food by man, including barnacles which are eaten in some countries including the Philippines (49) and Spain (50).

The potential value of the krill stocks of the Southern Oceans has recently become a major issue. Krill fishing is still considered a pilot-scale activity, but the swarms that appear regularly in the Antarctic summer make this species easy to catch. Krill catches in 1980 totalled over 420 000 tonnes, of which the U.S.S.R. was responsible for 388 000 tonnes (12). Estimates of sustainable yield are extremely variable, ranging from a few million to tens of millions of tonnes. Krill is now considered unlikely to have any major value as a human food resource, but krill protein concentrate could have potential as animal feedstock although other species currently provide a more economical source of fish meal (10,13,67).

Freshwater crustaceans accounted for 82 787 tonnes of the 1980 world crustacean catch (12). In Europe crayfish have been consumed in large quantities since Roman times. They are now considered a luxury food in most of the western world but in regions such as New Guinea and Madagascar they are an important source of quality protein (2). The U.S.A. is probably the major crayfish producer. The industry is centred on southern Louisiana where in recent years 30-60 million pounds (13-27 thousand tonnes) of Procambarus spp. (mainly P. clarkii) have been taken annually from natural swamps and marshes and about 50 000 acres (20 235 ha) of culture ponds. The wild harvest accounts for about 60 per cent of the total, and varies with the amount of sustained spring flooding (2,18,19). Pacifastacus leniusculus is exploited commercially in California, Oregon and Washington and Orconectes rusticus and O. virilis in Wisconsin. Sport fisheries for food crayfish also exist in these states (18,19). In other parts of the world, the wild harvest of crayfish is on a much smaller scale, although consumption may be high. In Europe interest in freshwater crayfish fisheries increased in the 1970s due to increased consumption and the availability of crayfish plague-resistant North American species (especially P. leniusculus), for stocking purposes. The European crayfish catch may now exceed 100 million individuals annually (59). The commercial potential of crayfish has probably not been fulfilled and greater use could be made of native species under efficient management techniques.

Apart from food, crustaceans are used by man in several other ways. The brine shrimp Artemia is used as fish food (2). Hermit crabs (usually the tree-climbing species Coenobita clypeatus) have become popular as pets in the U.S.A and more recently in Europe (11). Spiny lobsters are dried, lacquered and mounted for the souvenir trade in many parts of South East Asia (54). Crustaceans of various kinds are used as bait, including crayfish in the U.S.A. (18) and crabs in Thailand (52). Freshwater crayfish can be used as environmental indicators of water quality and recently have been used to help remove nutrients from effluent (21,22).

THREATS TO SURVIVAL Much of the information available on threats to crustacean populations is of an anecdotal nature. For marine species there is currently little evidence that any species are threatened with extinction but there are ample data indicating overfishing of food species. Freshwater and terrestrial crustaceans often have restricted ranges and are subject to habitat loss and pollution.

1. Habitat alteration Although this has greatest impact on freshwater and terrestrial species, marine species with limited coastal ranges may be vulnerable. For example, in California, a species of barnacle (Balanus aquila) appears to be endemic to the area between San Francisco and San Diego and is one of the last surviving members of a group known mainly from fossils (41). It is vulnerable to alteration of its coastal habitat. Another example is the tiny pea crab Parapinnixa affinis, which lives symbiotically with polychaete worms and which has only been recorded from a few bays in southern California. Although it was abundant in the first half of this century, all the areas from which it has been recorded have undergone extensive modification, and recent collecting efforts have not revealed any individuals (42).

Freshwater crayfish which are inhabitants of flowing waters are vulnerable to impoundments which significantly reduce faunal composition and variety and cause siltation. Channelization and dredging reduce the variety of habitats since riffle areas are changed or reduced, deeper pools are filled, and undercut banks and sheltered areas removed. A uniform stream is created, spreading water over a larger, shallower area, elevating summer temperatures, and increasing the effects of shifting of substrates and scouring during periods of flooding. Dredging removes sheltered portions and shallow areas that could support plant growth and collect detrital material, thereby reducing food resources and/or habitat. Rock cover may be removed or embedded into the substrate by heavy earth-moving equipment, while the remaining rock habitat is silted in as a result disturbances to the substrate or stream banks, or from spoil. Silt, from mining, deforestation, farming or other alterations and disturbances of land contours without appropriate re-vegetation, also reduces crayfish populations (27).

Studies have shown that silt in suspension is not in itself detrimental but heavy deposits on stream beds smother burrows and obliterate retreats under rocks and debris (27). In east central Kentucky, crayfish in Leatherwood Creek were reduced by 90 per cent when the stream bed became covered to a depth of 5-15 cm with clay silt as a result of strip mining (28). Sawmill operations that use streams for discarding sawdust have a particularly serious effect as the sawdust fills the interstices between rocks and affects the flow of water through them so that the oxygen becomes depleted; in such areas the crayfish fauna tends to be non-existent. Ash wastes smothering the substrate have also been reported to be detrimental (27).

Some species have extremely localized distributions and could be eradicated by a single major alteration of their habitats. For example the crayfish Cherax crassimanus and Engaewa subcoerulea are known only from single localities in the Karri Forest in south-west Australia (29). The isopod Caecidotea dimorpha is known only from two localities, a spring seep in Wayne Co., Missouri and a spring in Jackson Co., Arkansas, U.S.A. Lirceus megapodus is known only from a few springs in the St Francois Mountains of south-eastern Missouri (34). If these springs were to be capped or the drainage patterns altered these species could become as threatened as the Socorro Isopod (see review). In the Cuatro Ciénegas basin, Mexico, a number of endemic isopods are now threatened by drainage and canalization of the water courses in which they occur (see review of Cuatro Ciénegas snails) (7,35). The Californian freshwater shrimp Syncaris pacifica, once fairly widespread throughout three counties in the state, is now only found in restricted portions of three streams. It is threatened through further habitat alteration and introduction of exotic fish species and is now the only living species in the genus since S. pasadenae, from southern California, has already been extirpated (30). Many as yet undescribed species may already be extinct as a result of damming and hydro-electric power projects. Two undescribed phreatocid isopods from the sand of the shores of Lake Pedder, Tasmania, are probably now extinct, the lake having been flooded (5).

Burrowing and cave species are highly specialized and limited in their ability to compete outside their subterranean habitats, possessing behavioural, physiological, and morphological characteristics that reflect the low energy dynamics of their environments. They tend to have lower population densities and, in the case of cave crayfish, display a lower fecundity and longer period of maturation than their epigean (surface) relatives. Troglobitic crustaceans are vulnerable if they occur in populated areas where caves are likely to be subject to human visitation or to the effects of development of the surrounding environment. In Alabama, U.S.A., of the 14 species listed as threatened or of special concern, over half are cave species (31). The State of Virginia, U.S.A. has designated seven cave isopods as threatened or of special concern (32), including the Madison Cave Isopod (see review). A cavernicole isopod Miktoniscus linearis from Linville Caverns has been included in the preliminary list of endangered plant and animal species for North Carolina (33). Some of the primitive anostracans, restricted to ephemeral water bodies, may also be vulnerable to habitat alteration. Many species live in fresh or slightly saline pools in ranchland or farmland which farmers tend to fill in (56).

2. Pollution Pollution from toxic chemicals, alterations in hydrogen ion concentrations and depletion of oxygen levels can affect crustaceans. A certain amount of organic enrichment (eutrophication) of the water does not appear to be detrimental and may be beneficial to some species. Such enrichment increases plant growth and associated micro-organisms, thus supplying additional shelter and food for crayfish (27). Asellus aquaticus, a European freshwater isopod, has been found to thrive in waters subject to organic pollution as it tolerates low oxygen concentrations (6). However, some crayfish appear to be vulnerable to low oxygen concentrations as illustrated by the absence of crayfish in the polluted Connasauga River, Georgia, U.S.A. (27). In the south-eastern U.S.A. the insecticide Mirex, used to control fire ants (Solenopsis saevissima, a South American introduction), has been shown to be toxic to crayfish (36). The resistance of this chlorinated hydrocarbon to degradation and its additive effect in the environment make it especially harmful to lentic and burrowing species (31). Surface coal mining, besides increasing siltation, results in periodic acid poisonings from the breakdown of pyrites associated with coal deposits (27). Crayfish populations in Kentucky were reduced by 90 per cent as a result of acid poisoning and increased siltation in

the rivers caused by strip mining(63). There is evidence that the increasing acidification of lakes in Scandinavia affects freshwater crustaceans (6). Caustic alkaline effluents are also known to eradicate crayfish (27), and have caused the demise of the Pacifastacus leniusculus fisheries in the Willamette and Columbia Rivers (18). In Tasmania, a recent survey of salt lakes has shown that the isopod Haloniscus searlei now only occurs in two lakes, and that the largest population, in Township Lagoon, Tunbridge, is threatened since part of the lake is used as the local rubbish dump (40).

Pollution may affect marine species but there are fewer data available. Attached crustaceans such as barnacles are likely to be more vulnerable to oil pollution than mobile, subtidal species such as crabs and lobsters. However, heavy mortalities of the latter have been observed during large spillages where the oil has become dispersed in the water column in shallow waters. There are also reports of sublethal effects; for example, crude oil in sea water at a concentration of 0.9 ppm was observed to depress the appetite and alertness of adult lobsters and thus to interfere with feeding (62). As with other marine invertebrates, eggs and larvae are particularly sensitive to oil and the lethal threshold for lobster larvae appears to range between 2 ppm and 30 ppm. Crabs, both adult and juvenile, seem to be more hardy (61). However, populations of ghost crabs on the east coast of South Africa have shown an apparent decline in numbers due probably to the effects of oil pollution. Up to 95 per cent of crabs examined on the Zululand coast had oil on their bodies. The long term effects of a substantial decline in this important decomposer element could have major ecological consequences (55). Fresh oil, or oil that has been treated with dispersants, can cause high mortality among sedentary crustaceans, notably barnacles, although the oceanic goosebarnacle, Lepas fascicularis, will attach itself to floating tar balls (61).

3. Introductions Starting in the 1850s and periodically since, a crayfish plague has eliminated large numbers of freshwater crayfish in Europe. The plague is caused by a fungus, Aphanomyces astaci, originally endemic to North America, which attacks the exterior uncalcified parts of the cuticle. Outbreaks in old and new areas still occur frequently. Although most European species are attacked, North American species seem to be resistant and have been introduced into Europe. Orconectes limosus was introduced into Germany in the 1890s and is now found throughout much of western and central Europe where it is extensively cultivated as well as Procambarus clarkii and Pacifastacus leniusculus. However the two latter species appear to act as carriers for crayfish plague and there is little control to ensure that introductions do not carry disease (9,20). Furthermore there is the possibility that introduced crayfish may be vectors for viruses specific to fish (37,43). Introductions may also affect native species through competition. Unlike the native species, O. limosus appears to thrive in polluted waters. In Britain Austropotamobius pallipes (see review) could be affected by introductions of P. leniusculus. Within North America, P. leniusculus, Procambarus clarkii, Orconectes neglectus, O. rusticus and O. virilis have been extensively introduced to areas outside their ranges, and in some places appear to have replaced native crayfish (31,70).

There are other dangers if introductions are not carried out with care. For example, P. clarkii was introduced into Lake Naivasha in Kenya where there had previously been no crayfish. The introduced animals interfered with the established shallow water fishery by consuming netted fish and becoming entangled in the nets and their burrowing activities caused bank erosion. On the other hand they provided a large part of the diet of the commercially important predatory bass, Micropterous salmoides (39). In the U.S.A., P. clarkii has become a nuisance because of its burrowing activities which damage irrigation ditches and earth dams. The resulting burrow chimneys interfere with the operation of farm machinery; furthermore, this species feeds upon cultivated crops (38).

4. Exploitation Overfishing of marine crustacean stocks has taken place in many areas, and in others stocks tend to be fully exploited. Overfishing of lobster stocks is one of the most serious problems. The fisheries for the clawed lobsters Nephrops norvegicus (Norwegian Lobster), Homarus gammarus (Common Lobster) and H. americanus (American Lobster) are considered to be in biological and economic trouble. The N. norvegicus fishery has undergone a major expansion since the 1940s and there is concern that some stocks are now being overfished; increased fishing effort and declining catches have indicated over-exploitation of the U.S. fishery for H. americanus; and catches of H. gammarus in European waters are falling while prices continue to rise (16,23). The spiny lobster fisheries are not under such immediate threat, but in many parts of the world they have declined (24). For example the Jasus edwardsii fishery in the Chatham Islands, New Zealand, had declined dramatically by the early 1970s (25). The continual decline in stocks of Palinurus interruptus in southern California is said to be due to the taking of undersized lobsters (24). Reductions in spiny lobster Panulirus argus landings in Florida are probably associated with reduced post-larval recruitment and fishing activity in the nursery areas, and the recreational fishery in Biscayne Bay has significantly altered the density and age structure of the population (17). The spiny lobster fishery is reported to have declined in Thailand (especially Panulirus longipes and P. penicillatus) (54) and in Ceylon (57). Data available from most of the major spiny lobster fisheries show that annual total production is likely to fall very dramatically unless controls are placed on fishing effort (24).

There are many other examples of valuable crustacean stocks that are being over-exploited as a result of poor management of the fishery or lack of controls. The barnacle Pollicipes pollicipes is thought to be declining in northern Spain where it is collected as a luxury food (50). The sand crab Emerita emeritus (Hippidae) is classed as threatened in Thailand where it has disappeared from heavily collected beaches in Phuket (52). The Blue Crab (Callinetes sapidus) fisheries of Chesapeake Bay and Delaware Bay on the east coast of the U.S.A. have come under threat from over-exploitation and pollution (65).

Potential large scale harvesting of krill has become a major issue, not so much because of its effect on the krill itself, but because it could become a threat to the recovery of whale stocks. The key role played by krill in Antarctic food chains indicate that, as far as management decisions are concerned, krill cannot be considered in isolation. It is very difficult to predict the effect of intensive krill fisheries but the small scale harvesting carried out at present is likely to have little effect on the total krill population and will provide data essential for the formulation of a long-term management plan (67).

The exploitation of freshwater crustaceans is probably not currently a major threat. Crayfish are now largely a luxury food, and consumption has declined in many countries. The increasing development of aquaculture techniques is also taking pressure off wild populations. A working party on crayfish has been set up by the European Inland Fisheries Advisory Commission (EIFAC) to improve and develop crayfish fisheries in Europe (59). It must be emphasized however, that heavy collection of wild populations, combined with factors such as pollution, could rapidly cause a decline in the numbers of some species (18,26).

CONSERVATION Research is in progress to determine satisfactory methods of managing commercial crustacean species as sustainable resources, the two major aspects of which are management of fisheries and establishment of aquaculture operations.

1. Fishery Management An increasing awareness of the need to manage crustacean fisheries has stimulated a large body of work on the subject. Most crustacean fisheries are now probably subject to some form of legislation and control, but the success of different strategies is still open to much debate. For

lobsters, there are still very few fisheries that can provide data of sufficient quality and continuity to make a precise estimate of maximum sustainable yield for confident management (14). A variety of control methods are in use. Limited entry is used in many areas and has been successful in the Western Australia spiny lobster fishery (24) but in many places there are difficulties in controlling the numbers of licences issued. Quota systems have also met with mixed success.

A lobster management plan for the U.S. Homarus americanus fishery was initiated in 1972 (23). Attempts were made to regulate the H. gammarus fishery as early as 1849 in Norway and a management plan is currently being discussed for the European fishery (16). Current controls include minimum landing sizes (usually 74-83 mm carapace length), protection of ovigerous females (only in Spain, Denmark and Portugal), closed seasons, licences and limited numbers of traps. Size limits need to be fully researched, new ones set where necessary, and fishing effort must be controlled. This is likely to mean making a choice between recreational and professional fishing. The Nephrops norvegicus fishery will also need overall management (16), and it is recognized that a sound management strategy is important for the spiny lobster fisheries, with most emphasis placed on control of fishing effort (15). The main requirements for effective lobster fishery management are: provision of precise data to understand the effects of increased fishing pressure; determination of minimum size limits and provision of adequate inspection and legislative backing ; restriction of fishing effort and possibly the introduction of catch quotas; and an improvement in the relationship between professional and amateur fishermen (44).

Crab fisheries are also regulated by size limits and prohibitions on collecting ovigerous females (51). In northern Spain attempts have been made to control the recreational fishery for the barnacle Pollicipes pollicipes by banning collection using SCUBA equipment and setting a closed season (50).

Consideration is already being given to the type of management required should a major krill fishery develop (68). This needs to be regulated to prevent irreversible changes in krill populations, over-capitalization of krill fishing fleets and irreversible changes in populations dependent on krill. For the time being, all harvesting should be on an experimental basis as part of a scientific research programme, and designated areas where no krill may be taken should be protected. It is hoped that the Convention for Conservation of Antarctic Living Marine Resources will provide a mechanism for a broadly based approach to managing these resources (13).

Freshwater crustacean fisheries also require further control and management. In a number of countries crayfish collection is regulated under inland fisheries legislation. In Sweden and Finland the import of live crayfish is regulated and a permit is necessary for stocking crayfish. Harvesting of wild crayfish is controlled in Sweden, Finland, Spain and France with closed seasons and minimum size limits (16). In the Netherlands the native crayfish Astacus astacus is fully protected (see review). In the U.S.A. most of the west coast states have well-established regulations with size limits and closed seasons for crayfishing including bait fishing; Louisiana has very few controls but all states restrict the harvest of ovigerous females (18).

Management requirements for freshwater crayfish exploitation include regulations controlling the transfer of live crayfish for food or bait, protection of wild populations from over-exploitation, and management of native species to meet local demand for food. For example, Procambarus clarki was introduced for aquaculture into the Dominican Republic although the native Macrobrachium carcinus is already found in suitable freshwater habitats and is used for protein by some of the people. The latter can be grown in large numbers using aquaculture techniques and is as productive as the exotic species (46).

2. Aquaculture Commercial culture of various crustacean species is in progress but there is need for further research (63). Pond culture of freshwater crayfish was carried out in Europe for centuries but was then largely ignored until more recent times when it has become a major operation in several countries. Crayfish are popular for aquaculture since their ability to convert vegetation, dead fish and surplus fish feed into edible protein allow them to be combined with other animals such as fish. Furthermore the percentage survival to harvestable stages is relatively high (unlike lobsters and prawns) since there is no free-living planktonic larval stage, and hatcheries are not needed (9). Several species are currently farmed (20,64). The largest commercial enterprise is the farming of Procambarus clarkii in the southern U.S.A. but this American species is also cultured in Europe and many other countries. The yabby, Cherax destructor, is cultured in Australia, and the American species Pacifastacus leniusculus in Europe in at least 17 countries (20,64).

There are still no commercially viable lobster farming operations on account of the difficulties of rearing lobsters to marketable size but many advances in this direction have been made recently. Homarus americanus appears to have particular potential. Lobster hatcheries aimed at restocking fishing grounds with hatchery-reared juveniles, or extending fishing grounds by stocking new areas that have suitable habitat, have been in existence since the last century. However the success of such operations has never been satisfactorily monitored and the cost of rearing individuals through metamorphosis tends to make the process uneconomic (45). In France, fishermen are now supporting a scheme to restock over-collected grounds with hatchery-reared juveniles (66).

Although no marine crustaceans are cultured on a commercial basis in Europe or North America, certain species have been cultured for many years in the Indo-Pacific, where warm water and cheap labour abounds. Over 45 species of prawn have been cultured experimentally in nearly 50 countries, some of which have moved into commercial production. Prominent among these is Japan which produces over 1000 tonnes of cultured prawns a year.(64). The crab Scylla serrata has long been an incidental product of brackish water pond culture. Overfishing of this species resulted in the development of culture techniques and crabs are now raised as a subsidiary crop to milkfish. The ponds are seeded with small crabs which reach marketable size in six months (3,65). Research has shown that crab culture in general is not a commercial proposition on account of the slow growth rates of the larger species (3). However culture techniques have been developed for the Blue Crab (Callinectes sapidus) of the east coast of the U.S.A. and juveniles are being reared to restock depleted bays (65).

As yet, very little interest has been shown in the conservation of non-commercial crustaceans although some countries include vulnerable species in their lists of protected species. Crustaceans are included in the national Red Data Books for Switzerland and Germany (47,48). In the U.S.A., threatened freshwater crustaceans have been identified in many states and have been recommended for state and federal listing (53,60). For freshwater crustaceans the main requirements are increased pollution controls and protection of critical habitat. The following data sheets include examples of several threatened non-commercial species.

REFERENCES 1. George, J.D. and George, J.J. (1979). Marine life. Harrap, London.
2. Schmitt, W.L. (1973). Crustaceans. David and Charles (Holdings) Ltd., Newton Abbot, Devon.
3. Warner, G.F. (1977). The Biology of Crabs. Elek Science, London.
4. Williams, W.D. (1974). The Crustacea of Australian inland

waters. In Keast, A. (Ed.), Ecological Biogeography of Australia. W. Junk, The Hague.

5. Bayly, I.A.E., Lake, P.S., Swain, R. and Tyler, P.A. (1972). Lake Pedder: its importance to biogeographical science. In The Pedder Papers. Australian Conservation Foundation.

6. Okland, K.A. (1980). Ecology and distribution of Asellus aquaticus (L.) in Norway, including relation to acidification in lakes. Intern. rapport, IR 52/80. Sur Nedbors Virkning Po Skog og Fisk, SWSF - prosjektet, 1432 AS-NLH. Norway.

7. Cole, G.A. and Minckley, W.L. (1970). Sphaerolana, a new genus of cirolanid isopod from northern Mexico, with description of two new species. Southwest Natur. 9. 15: 71-81.

8. Abrahamsson, S. (1966). Dynamics of an isolated population of the crayfish Astacus astacus Linné. Oikos 17: 96-107.

9. Holdich, D.M., Jay, D. and Goddard, J.S. (1978). Crayfish in the British Isles. Aquaculture 15: 91-97.

10. Grantham, G.J. (1977). The Utilization of Krill. FAO Report GLO/SO/77/3, FAO Rome, 61 pp.

11. Gardner, R.G. (1977). Gardner's Treecrab Book. R.G. Gardner, Sacramento, California.

12. FAO (1981). 1980 Yearbook of Fishery Statistics. Catches and Landings Vol. 50. FAO, Rome.

13. FAO (1981). Review of the State of the World Fishery Resources. FAO Fisheries Circular 710, Revision 2, Rome.

14. Cobb, J.S. and Phillips, B.F. (Eds) (1980). The Biology and Management of Lobsters. Vol. 2. Ecology and Management. Academic Press, New York.

15. Morgan, G.R. (1980). Population dynamics of spiny lobsters. Ch. 5 in reference 14, Pp. 189-217.

16. Dow, R.L. (1980). The clawed lobster fisheries. Ch. 8 in reference 14 Pp. 265-316.

17. Davis, G.E. (1979). Management recommendations for juvenile spiny lobsters, Panulirus argus in Biscayne National Monument, Florida. South Florida Research Center Report M-530: 320 pp.

18. Huner, J.V. (1978). Exploitation of freshwater crayfishes in North America. Fisheries 3 (6): 2-5.

19. Huner, J.V. and Barr, J.E. (1980). Red Swamp Crayfish: Biology and Exploitation. Sea. Grant. Publ. No. L.S.U.-T-80-001, Louisiana State University Centre for Wetland Resources, Baton Rouge. Louisiana, U.S.A.

20. Karlsson, S. (1977). The freshwater crayfish. Fish Farming International 4(2): 8-12.

21. Goldman, C., Rundquist, J.C. and Flint, R.W. (1975). Ecological studies of the California crayfish Pacifastacus leniusculus, with emphasis on their growth from recycling waste products. In Avault, J. W. (Ed.), Freshwater Crayfish Papers from the 2nd Int. Symp. on Freshwater Crayfish, Baton Rouge, Louisiana, U.S.A. Pp. 481-487.

22. Rundquist, J., Gall, G. and Goldman C.R. (1977). Watercress-crayfish polyculture as an economic means of stripping nutrients. In Lindquist, O. (Ed.), Freshwater Crayfish, Papers from the 3rd Int. Symp. on Freshwater Crayfish, Kuopio, Finland 1976: 33-50.

23. Saila, S.B. and Marchesseault, G. (1980). Population dynamics of clawed lobsters. Ch. 6 in reference 14 Pp. 219-241.

24. Bowen, B.K. (1980). Spiny lobster fisheries management.

Ch. 7 in reference 14. Pp. 243-264.

25. Devine, W.T. (1982). Nature conservation and land-use history of the Chatham Islands, New Zealand. Biol. Cons. 23(2): 127-142.

26. Momot, N.T. and Gowing, H. (1977). Results of an experimental fishery on the crayfish Orconectes virilis. J. Fish. Res. Board. Can. 34: 2056-2066.

27. Hobbs, H.H. and Hall, E.T. (1974). Crayfishes (Decapoda Astacidae). In Hart, C.W. and Fuller, S.L.H. (Eds), Pollution Ecology of Freshwater Invertebrates. Academic Press, N.Y., London. Pp. 195-214.

28. Branson, B.A. and Batch, D.L. (1972). Effects of strip mining on small-stream fishes in east-central Kentucky. Proc. Biol. Soc. Washington 84 (59): 507-518.

29. Tingay, A. and Tingay, S.R. (1980). Rare, Restricted and Endangered Fauna of the Jarrah and Karri Forests. Research Publication, Conservation Council of Western Australia Inc.

30. Eng, L.L. (1981). Distribution, Life History, and Status of the California Freshwater Shrimp Syncaris pacifica (Holmes). Inland Fisheries Endangered Species Program, Special Publication 81-1. State of California, Department of Fish and Game.

31. Bouchard, R.W. (1976). Crayfishes and shrimps. Bull. Ala. Mus. Nat. Hist. 2: 13-20.

32. Holsinger, J.R. (1979). Freshwater and terrestrial isopod crustaceans (Order Isopoda). In Linzey, D.W. (Ed.), Proceedings of the Symposium on Endangered and Threatened Plants and Animals of Virginia. 1978, Virginia Polytechnic Institute and State University.

33. Endangered Species Committee (1973). Preliminary list of Endangered Plant and Animal Species in North Carolina. North Carolina Dept. of Natural and Economic Resources.

34. Pflieger, W. (1974). In Rare and Endangered Species of Missouri. Missouri Dept. of Conservation and U.S. Dept. of Agriculture and Soil Conservation Service.

35. Cole, G.A. and Minckley, W.L. (1966). Speocirolana thermydronis, a new species of cirolanid isopod crustacean from Central Coahuila, Mexico. Tuland Stud. Zool. 13: 17-22.

36. Ludke, J.L., Finley, M.T. and Lask, C. (1971). Toxicity of Mirex to crayfish, Procambarus blandingi. Bull. Environ. Contam. and Toxicol. 6(1): 89-96.

37. Unestam, T. (1976). The dangers of introducing new crayfish species. In Avault, J.W. (Ed.), Freshwater Crayfish. Papers from the 2nd Int. Symp. on Freshwater Crayfish, Baton Rouge, Louisiana, U.S.A. Pp. 522-561.

38. Bouchard, R.W. (1979). Taxonomy, distribution and general ecology of the genera of North American Crayfishes. Bull. Amer. Fisheries. Soc. 3(6): 11-19.

39. Lowery, R.S. and Mendes, A.J. (1977). Procambarus clarkii in Lake Naivasha, Kenya and its effects on established and potential fisheries. Aquaculture 11: 111-121.

40. Williams, W.D. (in press). Conservation. Ch. 11 in Life in Inland Waters. Blackwell Scientific Publications, Melbourne.

41. Newman, W.A. and Abbott, D.P. (1980). Cirripedia. Ch. 20 in Morris, R.H., Abbott, D.P. and Haderlie, E.C. (Eds), Intertidal Invertebrates of California. Stanford University Press, Stanford, California.

42. Wicksten, M.K. (1980). On the status of Parapinnixa affinis

Holmes, the California Bay pea crab. Unpub. brief to Office of Endangered Species, USDI, Washington D.C.

43. Huner, J.V. and Avault, J.W. (1978). Introductions of Procambarus spp. - a report to the Introductions Committee of the International Association of Astacology, 4th Biennial Meeting, Thonon-les-Bains, France. Freshwater Crayfish 4, Papers from the 4th Int. Symp. on Freshwater Crayfish. Pp. 191-194.

44. Bennett, D.B. (1980). Perspectives on European lobster management. Ch. 9 in reference 14. Pp. 317-331.

45. Van Olst, J.C., Carlberg, J.M. and Hughes, J.T. (1980). Aquaculture. Ch. 10 in reference 14. Pp. 333-384.

46. King, F.W. (1978). In litt., 5 December.

47. Burckhardt, D., Gfeller, W. and Müller, H.U. (1980). Animaux protégés de Suisse. Ligue Suisse pour la Protection de la Nature (LSPN), Bâle.

48. Türkay, M. (1977). Rote Liste der Zehnfüssigen Krebse (Decapoda). In Blab, J., Nowak, E., Trautmann, W. and Sukopp, H. (Eds). Rote Liste der gefährdeten Tiere und Pflanzen in der Bundesrepublik Deutschland. Kilda Verlag, Greven.

49. Anon. (1977). Inventory of endemic, endangered, rare, vanishing and most economically important species of the Philippine flora and fauna. Dept. of Nat. Resources and Nat. Sciences Resource Center, University of the Philippines.

50. Cendero, O. (1980). In litt., 30 May.

51. Anon. (undated). The Edible Crab. Zoology Leaflet 6, British Museum (Natural History), London.

52. Bain, J.R. and Humphrey, S.R. (1982). A profile of the endangered species of Thailand. Vol I. Florida State Museum, Office of Ecological Services.

53. Bouchard, R.W. (1976). Investigations on the status of fourteen species of freshwater decapod crustaceans in the United States. Part I. Troglobitic shrimps and western North American crayfishes. Report to U.S.D.I. Office of Endangered Species.

54. Bhatia, U. (1974). Distribution of spiny lobsters along the west coast of Thailand with observation on their fishing grounds. Research Bulletin 5, Phuket Marine Biological Center, Thailand.

55. Berry, P.F. (1976). Natal's ghost crabs. Afr. Wildl. 30: 35-37.

56. Hartland-Rowe, R. (1981). In litt., 19 February.

57. Hoffman, T. (1977). Lobsters will not be for ever. Wildlife and Nature Protection Society of Ceylon Newsletter 40: 7.

58. Holthuis, L.B. (1980). FAO Species catalogue. Vol. 1. Shrimps and prawns of the world. An annotated catalogue of species of interest to fisheries. FAO Fish. Synop. (125) Vol. 1: 261 pp.

59. Westman, K. and Pursiainen, M. (1982). Status of crayfish stocks and fisheries in Europe. Paper prepared for European Inland Fisheries Advisory Commission (EIFAC), 12th Session, Budapest, Hungary.

60. Bouchard, R.W. (1978). Report of the First Meeting of the IUCN/SSC Freshwater Crustacean Specialist Group. Tuscaloosa, Alabama, 13 April 1978.

61. RCEP (1981). Oil pollution of the sea. Eighth report of the Royal Commission on Environmental Pollution, October 1981, HMSO, London.

62. GESAMP (1977). The impact of oil on the marine environment. Joint Group of Experts on the Scientific Aspects of Marine Pollution. Reports and Studies No. 6, FAO, Rome.
63. Branson, B.A. and Batch, D.L. (1972). Effects of strip mining on small-stream fishes in east central Kentucky. Proc. Biol. Soc. Washington 84 (59): 507-518.
64. Wickins, J.F. (1982). Opportunities for farming crustaceans in western temperate regions. In Muir, J.F. and Roberts, R.J. (Eds), Recent Advances in Aquaculture. Croom Helm, London and Canberra.
65. Bardach, J.E., Ryther, J.H. and McLarney, W.O. (1972). Aquaculture. Wiley-Interscience, New York, London.
66. Anon. (1982). Support for lobster programme. Fish Farming International 9(9): 7.
67. Everson, I. (1977). The living resources of the Southern Ocean. FAO Report GLO/SO/77/1, FAO, Rome, 156 pp.
68. Abele, L. (Ed.) (1982). The Biology of Crustacea. Academic Press, London and New York.
69. Hobbs, H.H. 1974). Synopsis of the families and genera of crayfishes (Crustacea; Decapoda). Smithsonian Contributions to Zoology 164: 32 pp.
70. Schwartz, F.J., Rubelmann, R. and Allison, J. (1963). Ecological population expansion of the introduced crayfish Orconectes virilis. Ohio Jour. Sci. 63(6): 266-273.

We are very grateful to R.W. Bouchard, K. Bowler, T.E. Bowman, G. Boxshall, N.D. Bruce, L. Eng, I. Everson, J.S. Goddard, J.R. Holsinger, L.B. Holtuis, F. Howarth, J.V. Huner, R.W. Ingle, S. Karlsson, R. Lincoln, R.S. Lowery, G. Magniez, J. Reynolds, J.H. Stock and W.D. Williams for assisting with this section.

TASMANIAN   VULNERABLE
ANASPID CRUSTACEANS

Anaspides tasmaniae Thompson, 1893
A. spinulae Williams, 1965
Paranaspides lacustris Smith, 1908
Allanaspides helonomus Swain, Wilson, Hickman and Ong, 1970
A. hickmani Swain, Wilson and Ong, 1971

Phylum	ARTHROPODA	Order	ANASPIDACEA
Class	CRUSTACEA	Family	ANASPIDIDAE

SUMMARY All five species of this family of primitive freshwater malacostracans or syncarids are endemic to Tasmania, Australia. Anaspides tasmaniae is the most widely distributed and currently is not threatened. A. spinulae is known from one lake only and is vulnerable to any habitat change which might take place there. Paranaspides lacustris is vulnerable to changes in water level in the three lakes where it occurs. Allanaspides helonomus and A. hickmani occur in an area where a major hydroelectric scheme has been implemented, and the latter species is particularly vulnerable in view of its localized distribution. All five occur in national parks but there is concern that this does not necessarily ensure their future survival.

DESCRIPTION Anaspides tasmaniae is the largest extant syncarid, generally attaining a length of about 55 mm. It has a soft, chitinous integument which is straw yellow in colour with numerous black chromatophores in the skin beneath. The overall colour varies from olive green to pale brown, although cave specimens have greatly reduced pigmentation or none (10,14,16,20,21,26,27). In specimens from a cave in southern Tasmania a reduction in the pigment of the eyes was noted (28). The body is normally held straight and unflexed and is probably close to the ancestral crustacean form (19,21,27). Its only defence mechanism is an evasive flexion reaction which it makes if touched, and unlike crustaceans with carapaces, it is able to repeatedly flex to continual approaches of a predator (18). In captivity, it is more active at night than during the day. It can swim and jump but usually walks or runs on stones at the sides or on the bottom of streams or pools (14,21). A. spinulae is distinguished from A. tasmaniae only by the spination of the abdomen and telson, and may be a geographic race or subspecies (28). No fully adult males are known; the largest female recorded was 25 mm long (27).

Paranaspides lacustris has a 'shrimp-like' appearance with a pronounced dorsal flexure around the junction of the thorax and abdomen. It is almost translucent; immature animals are pale green while mature ones are brown and sparsely powdered with black dots, attaining a length of up to 25 mm. The animal has large antennal scales and a large tail fan. The differences in morphology between this species and Anaspides tasmaniae are probably related to the fact that the former tends to live among weeds, and swims, while the latter crawls among stones (14,21,21).

Allanaspides helonomus is small, reaching about 15 mm in length, and has a slight dorsal flexure. It is brown with a conspicuous transparent, oval dorsal window, the fenestra dorsalis (12,23,24). A. hickmani is distinguished by its larger, rectangular dorsal window, the tissue beneath which contains a bright red pigment (12,24). It attains a length of up to 11.5 mm. There is a detailed description in (24). Allanaspides occupies an intermediate position between Anaspides and Paranaspides (23).

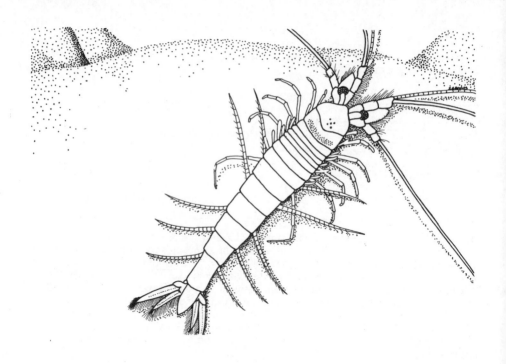

Anaspides tasmaniae

DISTRIBUTION All endemic to Tasmania, Australia. Anaspides tasmaniae is widely distributed and common in alpine and subalpine waters of Tasmania. Early records included North West Bay River and other localities on Mt Wellington, Mt Hartz, Mt Field and Mt Read (14,15,19,20,25). Does not occur in the north or east of Tasmania; the northern limit appears to be the edge of the Western Tiers and Black Bluff. In the south found from Mt Wellington near Hobart on the east coast to the west coast mountains including Frenchman's Cap and the Eldon Range (17,27,28). A. spinulae is known only from 54 specimens collected in Lake St Clair (27).

Paranaspides lacustris is restricted to Great Lake, Shannon Lagoon, Penstock Lagoon, Arthur's Lake and Woods Lake in the Central Plateau of Tasmania. Its distribution parallels the distribution of the native fish Paragalaxias (4,16,19,28). There are also unpublished reports of its occurrence in Lake River beneath Wood Lakes (29).

Allanaspides helonomus seems to be restricted to the Pedder Valley in south-west Tasmania. It is apparently reasonably common behind the eastern shores of Lake Pedder, but has also been recorded from the McPartlan Pass (2,23,24). A. hickmani is known only from its type locality, an area of about 2 acres (1 ha) at the McPartlan Pass in south-west Tasmania, about 5-6 km to the north of Lake Pedder; it may occur sympatrically with A. helonomus (24,28).

POPULATION Unknown.

HABITAT AND ECOLOGY Anaspides tasmaniae usually occurs in the highlands, although its total altitudinal range is from about 15 m to 1200 m above sea level. Most collections have been taken at altitudes in excess of 750 m (the altitude preference is 915-1066 m) and the species is therefore subject to near freezing temperatures (27,28). May even occur in pools with thick ice cover (25) or in areas under snow for 2-3 months of the year (14). Typical habitats are small upland streams, moorland pools, button grass pools, and in the south-west, alpine lakes (14,16,26,28). Its greater abundance in lakes in the south-west may be due to the fact that these are remote and do not contain the introduced Brown Trout <u>Salmo trutta</u> which is a known predator (28). Also occurs in caves in north central Tasmania near Mole Creek and in southern Tasmania near Ida Bay (10). Its ability to survive in caves may have been of value during the Tertiary periods of high temperature when caves may have provided oligothermal refuges (26). The species' absence from the Ben Lomond plateau in the north-east would be explained if it only tolerates water with very low amounts of dissolved matter; its absence from areas south of the Ben Lomond plateau is probably due to the drier and warmer climate found at lower altitudes (27).

The animal tends to crawl among stones on stream bottoms. Diet is omnivorous: submerged mosses and liverworts, periphytic algae, small invertebrates, worms, tadpoles and detritus have all been recorded (13,14,19). It filter feeds and also uses the mouth parts to collect small particles of algal and diatom growth which it scrapes off weeds and stones (3,13,14). The usual life span is three years but some specimens may survive for four years. For about the first year and a half of life, growth is not seasonally restricted. Breeding probably occurs after approximately 15 months, when males and females are about 18 mm long. Eggs are about 1 mm in diameter and purple, and are deposited singly on submerged vegetation. Most are laid in spring and hatch between June and October, taking about 8 months to develop, although those laid in autumn and winter may require up to 14 months (6). Most individuals have a two year interval between the time they are laid as eggs and the time they first breed, and most breed twice (19,21,27,28).

A. spinulae was collected at shallow depths, from 3.0 m to 4.5 m on a sandy bottom at least partly covered by flocculent algae. No specimens were obtained from the adjacent shore. Nothing is known of its ecology (27).

<u>Paranaspides</u> <u>lacustris</u> Found at depths in the range 0.2-10 m throughout the lake in suitable situations, mainly in the littoral zone where weeds are found, and often associated with beds of <u>Chara</u> sp. (4,13,16,20,27). Food consists mainly of fine detritus, which it filters, and algal material which is scraped off weeds with its mouthparts (3,13,14,15,27). Spends most of its time swimming amongst the weed or clinging to it (14). Egg laying probably occurs during late spring or summer and the life cycle only lasts about 18 months (4).

<u>Allanaspides</u> <u>helonomus</u> and <u>A. hickmani</u> both occur in areas referred to locally as button grass plains, where they live in the burrows of the burrowing crayfish <u>Parastacoides</u> <u>tasmanicus</u> and in surface pools. The burrows, which usually open into these pools, probably provide a favourable microhabitat (12,24). No other ecological data is available.

SCIENTIFIC INTEREST AND POTENTIAL VALUE Anaspides tasmaniae was first discovered in 1892 by George Thomson who remarked "I believe this form to be the most interesting crustacean which has been discovered for many years" (32). It is now well established that these species are Syncarida, a group which had previously only been described as fossils from the Palaeozoic, and which had been thought extinct for 200 million years (28). They therefore occupy a very important position in crustacean phylogeny as primitive freshwater Crustacea, the Anaspididae having arisen in the Palaeozoic, 100 million years earlier than the now dominant advanced crustaceans, the Decapoda. Primitive characters include the lack of carapace, poor differentiation between thorax and abdomen, the presence of exopodites on most or all thoracic limbs and the failure of females to brood eggs. They are believed to be morphologically similar to the ancestral Eumalacostraca (18).

Anaspides tasmaniae has been used for investigating the lateral giant fibre system of crustaceans (18). The genus Allanaspides is particularly important for its apparently unique ion transporting site located on the dorsal surface of the cephalothorax and known as the fenestra dorsalis. This may be related to these species' survival in acidic ion-deficient waters, since their environment is characterized by a low level of inorganic ions and a high level of organic matter giving rise to a pH of less than 5.0. However this is not a unique situation, and the differences between the fenestrae dorsalis of the two species suggests that selective pressures other than those associated with ionic regulation have also been involved (12,24).

THREATS TO SURVIVAL It is not clear to what extent the introduced Brown Trout (Salmo trutta) has affected populations of Anaspides tasmaniae. A known predator of Anaspides, the trout was introduced to many lakes in the 1800s and was multiplying rapidly at the beginning of this century (19). There is a good correlation between the presence of trout in lakes and the absence of A. tasmaniae (7,11) suggesting that the latter survives only in those places where predation by trout is not unduly heavy, or which are inaccessible to trout (28). Populations of A. tasmaniae in the Gordon River area may be threatened by the recent proposals to exploit this area's potential for timber and hydro-electric power (34). A. spinulae has not been collected since 1963, and could also have been affected by Brown Trout which has been introduced to Lake St Clair (22).

Paranaspides lacustris populations have fluctuated in size considerably in the past. In the early 1900s the species was apparently abundant in the Great Lake despite that fact that a known predator, the introduced Brown Trout, had been in the lake some forty years (14,15,20,21). In the 1920s and 1930s P. lacustris almost disappeared (13,14) but numbers had recovered by the 1940s (16). It was clearly common in 1963 (27) and 1966, but no further specimens were obtained in 1969, 1970 and 1972 (28). Both periods of apparently reduced population size (i.e. 1920s and 1930s, and after 1966) followed artificial elevations of the water level of the lake (7.5 m in 1922, and about 4.0 m in 1967) (27,28). Higher water levels destroy the weed beds on which the species is dependent and growth of new weed is slow so that populations would take some time to recover (13,14). In 1974 there were still no collections from the Great Lake, although it was common in Shannon Lagoon (28) and since 1975 it has been locally common in Great Lake (22). However, it is proposed to add a further nine metres depth to the Miena Dam on the Great Lake which could affect the species (4).

Much of the restricted geographical area in which Allanaspides helonomus occurs has been flooded for a major hydro-electric development (1,31). The limited distribution of A. hickmani makes it vulnerable to any activity which might destroy its habitat.

CONSERVATION MEASURES TAKEN Most species occur within a World Heritage Site called the Western Tasmania Wilderness National Parks. This comprises three national parks:- Southwest National Park, Franklin-Lower Gordon Wild Rivers National Park and Cradle Mountain-Lake St Clair National Park (30). Anaspides tasmaniae, A. helonomus and Allanaspides hickmanni occur in the Southwest National Park; A. spinulae and Anaspides tasmaniae within the Cradle Mountain Lake St Clair National Park; and A. tasmaniae within the Franklin-Lower Gordon Wild Rivers National Park. In addition A. tasmaniae occurs in the Hartz Mountains and Mount Field National Parks and several state reserves (22,33).

CONSERVATION MEASURES PROPOSED Although some of these species occur within national parks, these areas are by no means inviolate in Tasmania (1,31,34), and there is still room for improvement in the standard and procedures of environmental impact assessment in Australia (9). All necessary efforts should be

278

made to conserve the habitats of these unique species. For Paranaspides lacustris it may be necessary to reserve a lagoon not exposed to rapid changes in water level (8). In general, it would be advisable to carry out a thorough investigation whenever it is planned to introduce trout into any new system (e.g. lake or stream), and the planned introduction should not go ahead if anaspid crustaceans, especially Anaspides tasmaniae, are present.

REFERENCES

1. Australian Conservation Foundation (1972). Pedder Papers: Anatomy of a Decision. Australian Conservation Foundation, Victoria, Australia.

2. Bayly, I.A.E., Lake, P.S., Swain, R. and Tyler, P.A., (1972). Lake Pedder: its importance to biological science. In reference (1).

3. Cannon, H.G. and Manton, S.M. (1929). On the feeding mechanisms of syncarid Crustacea. Trans. R. Soc. Edinb. 56: 175-189.

4. Fulton, W. (in press). Distribution and other notes on Paranaspides lacustris. Aust. Soc. Limnol. Bull. 8.

5. Gordon, I. (1961). On the mandible of Paranaspides lacustris Smith - a correction. Crustaceana 2: 213-222.

6. Hickman, V.V. (1937). The embryology of the syncarid crustacean, Anaspides tasmaniae. Pap. Proc. R. Soc. Tasm. 1936: 1-35.

7. Knott, B., Suter, P.J. and Richardson, A.M.M. (1978). A preliminary observation on the littoral rock fauna of Hartz Lake and Hartz Creek, southern Tasmania, with notes on the water chemistry of some neighbouring lakes. Aust. J. Mar. Freshwater Res. 29: 703-715.

8. Lake, P.S. (1974). Conservation. In Williams, W.D. (Ed.), Biogeography and Ecology in Tasmania. W. Junk, The Hague.

9. Lake, P.S. (1980). Chapter 16. Conservation. In Williams, W.D. (Ed.), An Ecological Basis for Water Resource Management. Australian National University Press, Canberra. Pp. 163-173.

10. Lake, P.S. and Coleman, D.J. (1977). On the subterranean syncarids of Tasmania. Helictite 15: 12-17.

11. Lake, P.S. and Knott, B. (1972). On the freshwater crustaceans of the Central Plateau. In Banks, M.R. (Ed.), The Lake Country of Tasmania. Roy. Soc. Tasm., Hobart. Pp. 95-99.

12. Lake, P.S., Swain, R. and Ong, J.E. (1974). The ultrastructure of the fenestra dorsalis of the syncarid crustaceans Allanaspides helonomus and Allanaspides hickmani. Z. Zellforsch. 147: 335-351.

13. Manton, S.M. (1929). Observations on the habits of some Tasmanian Crustacea. Victorian Nat. 45: 298-300.

14. Manton, S.M. (1930). Notes on the habits and feeding mechanisms of Anaspides and Paranaspides (Crustacea, Syncarida). Proc. Zool. Soc. Lond. 791-800.

15. Nicholls, G.E. (1929). Notes on freshwater Crustacea of Australia. Victorian Nat. 45: 285-295.

16. Nicholls, G.E. (1947). On the Tasmanian Syncarida. Rec. Queen Vic. Mus. 2: 9-16.

17. Riek, E.F. (1959). The Australian Freshwater Crustacea. In Keast, A., Crocker, R.L. and Christian, C.S. (Eds), Biogeography and Ecology in Australia, W. Junk, The Hague.

18. Silvey, G.E. and Wilson, I.S. (1979). Structure and function of the lateral giant neurone of the primitive crustacean Anaspides tasmaniae. J. Exp. Biol. 78: 121-136.

19. Smith, G.W. (1908). Preliminary account of the habits and structure of the Anaspidae with remarks on some other freshwater crustacea from Tasmania. Proc. R. Soc. 80: 465-473.
20. Smith, G.W. (1909). The fresh water Crustacea of Tasmania with remarks on their geographical distribution. Trans. Linn. Soc. Lond. 11: 61-92.
21. Smith, G.W. (1909). On the Anaspidacea, living and fossil. Q. J. Microsc. Sci. 53: 489-578.
22. Swain, R. (1980). In litt., 26 August.
23. Swain, R., Wilson, I.S., Hickman, J.L. and Ong, J.E. (1970). Allanaspides helonomus gen. et sp. nov. (Crustacea: Syncarida) from Tasmania. Rec. Queen Vict. Mus. 35: 1-13.
24. Swain, R., Wilson, I.S., and Ong, J.E. (1971). A new species of Allanaspides (Syncarida, Anaspididae) from south-western Tasmania. Crustaceana 21: 196-202.
25. Thomson, G.M. (1894). On a freshwater schizopod from Tasmania. Trans. Linn. Soc. Lond. 6: 285-303.
26. Williams, W.D. (1965). Subterranean occurrence of Anaspides tasmaniae (Thomson) (Crustacea, Syncarida). Int. J. Speleology 1: 333-337.
27. Williams, W.D. (1965). Ecological notes on Tasmanian Syncarida (Crustacea: Malacostraca) with a description of a new species of Anaspides. Int. Revue Ges. Hydrobiol. Hydrogr. 50: 95-126.
28. Williams, W.D. (1974). Freshwater Crustacea. In Williams, W.D. (Ed.), Biogeography and Ecology in Tasmania. W. Junk, The Hague.
29. Williams, W.D. (1981). The Crustacea of Australian inland waters. In Keast, A. (Ed.), Ecological Biogeography of Australia. W. Junk, The Hague.
30. IUCN (1982). Protected Areas Data Unit (PADU) draft data sheets on Tasmanian National Parks.
31. Lake Pedder Committee of Enquiry (1974). The flooding of Lake Pedder: final report of Lake Pedder Committee of Enquiry. Australian Government Publishing Service, Canberra.
32. Thomson, G.M. (1893). Notes on the Tasmanian Crustacea with description of new species. Pap. Proc. R. Soc. Tasm. 1892: 45-76.
33. Swain, R. (1982). In litt., 13 August.
34. Brown, R. (1982). The Tasmanian wilderness campaign. Wildlife 24(12): 446-453.

We are very grateful to R. Swain for the information provided for this data sheet and to P.S. Lake for helpful comments.

MADISON CAVE ISOPOD   VULNERABLE

Antrolana lira Bowman, 1964

Phylum   ARTHROPODA   Order   ISOPODA

Class   CRUSTACEA   Family   CIROLANIDAE

SUMMARY Antrolana lira is an isopod known only from a single restricted subterranean water system at Madison's Cave in the Appalachian Valley, Augusta Co., Virginia, U.S.A. It is vulnerable to groundwater pollution and illegal visiting of the cave and is listed as Threatened under the U.S. Endangered Species Act.

DESCRIPTION Antrolana lira is eyeless and has an unpigmented, compact, flattened body. Females are larger (length up to 21 mm) than males (length up to 16 mm) and more numerous (1,6). A detailed description is given in (1).

DISTRIBUTION Found only in two lakes in Madison's Saltpetre Cave and one pool in nearby Steger's Fissure, Ridge Province, Augusta Co., Virginia, U.S.A. Madison Cave is situated just west of the South Fork of the Shenandoah River and 0.3 km south of Grottoes, and consists of two caves developed in the eastern side of Cave Hill. Steger's Fissure is located approximately 500 feet (152 m) north of the cave entrance and consists of a single limestone crevice about 115 feet (35 m) deep (1,2).

POPULATION Large numbers are seldom encountered (usually fewer than a hundred per month during field studies (6)) but given the subterranean nature of the Isopod's habitat it is difficult to estimate the population size (2,6). Individuals are more abundant in the east lake and the fissure lake than in the west lake (5).

HABITAT AND ECOLOGY A. lira is found in phreatic water (i.e. below the level of the water table) in deep lakes (2,5). Individuals are found in shallow portions of the lakes, resting on silt and talus (1), and also at depths of 10.7 m and 15.2 m (6) but nothing is known of the animal's habitat and ecology in the rest of the water system. Water level changes in the lakes are correlated with those in the South River, suggesting a subterranean connection between the pools and the river (2,5). Organic matter (usually wood) is present in all three lakes but is most abundant in the fissure (5). The Isopods are probably scavengers and are thought to be mainly carnivorous since insect parts have been found in their guts and individuals are attracted to shrimp bait (6). In both the cave and the fissure A. lira is found in the company of the endemic troglobitic amphipod Stygobromus stegerorum. Large, presumably sexually mature, females do not have brood plates which may indicate that the species is ovoviviparous as are species in at least one other genus in the family (2). No reproductive females have been found and the reproductive cycle is not known. However juveniles occur year round suggesting that there may not be a definite annual reproductive cycle. The age structure is skewed towards adults, indicating that the species has a low reproductive rate and that individuals are long-lived (6).

SCIENTIFIC INTEREST AND POTENTIAL VALUE Of the 17 or 18 troglobitic cirolanids recorded from the western hemisphere, all except A. lira occur in an arc surrounding the present day Gulf of Mexico. A. lira, the only member of its genus, is the only North American, freshwater cirolanid found north of Texas, Mexico and the West Indies (5). Evidence suggests that these subterranean freshwater isopods were derived from marine ancestors left as relics during periods of marine embayments. If this is so, A. lira may be the sole survivor of a lineage that dates back to marine invasions of the Appalachian region in the late Paleozoic (1).

THREATS TO SURVIVAL The cave has been operated as a commercial attraction at various times in the past. It is now in private ownership but unauthorized visitation has resulted in rubbish accumulation and siltation in the lakes. Visitors standing on the steep banks cause the clay talus to creep into the pools, reducing the size and quality of the already limited available habitat (2). Mercury pollution of the South River, caused by an E.I. Du Pont de Nemours and Co. factory at Waynesboro', Virginia, is also a threat since it could easily reach the cave through the underground water system (3).

CONSERVATION MEASURES TAKEN The Cave Conservancy of the Virginias, in conjunction with the current owner of the cave, has gated the entrance to discourage visitors (4). A cave management plan designed to protect both the cave and its unique organisms is being drafted (7). The species is designated as Threatened in Virginia (2) and is listed as Threatened under the U.S. Endangered Species Act (4).

CONSERVATION MEASURES PROPOSED It is strongly recommended that Madison's Saltpetre Cave and the associated karst ground-water aquifer be protected on a permanent basis, with limited access to the cave, for research and education only (2).

REFERENCES   1.   Bowman, T.E. (1964). Antrolana lira, a new genus and species of troglobitic cirolanid isopod from Madison Cave, Virginia. Int. J. Speleol. 1: 229-236.
2.   Holsinger, J.R. (1979). Freshwater and terrestrial isopod crustaceans (Order Isopoda). Proceedings of the Endangered and Threatened Plants and Animals of Virginia Conference. 1978.
3.   Bolgiano, R.W. (1980). Mercury contamination of the South, South Fork Shenandoah, and Shenandoah Rivers. Virginia State Water Control Board Basic Data Bulletin 46.
4.   U.S.D.I. (1982). Endangered and threatened wildlife and plants; listing the Madison Cave isopod as a threatened species. Federal Register 47(192): 43699-43701.
5.   Collins, T.L., and Holsinger, J.R. (1981). Population ecology of the troglobitic isopod crustacean Antrolana lira Bowman (Cirolanidae). Proc. 8th Int. Congress of Speleology, Bowling Green, Kentucky, July 1981. Vol 1: 129-132.
6.   Collins, T.L. (1982). An ecological study of the troglobitic cirolanid isopod, Antrolana lira Bowman, from Madisons Saltpetre Cave and Stegers Fissure, Augusta Co., Virginia. M.Sc. Thesis, Old Dominion University.
7.   Holsinger, J.R. (1982). In litt.

We are very grateful to T.E. Bowman, N.D. Bruce, J.R. Holsinger and F. Howarth, for reviewing this data sheet.

SOCORRO ISOPOD   ENDANGERED

Thermosphaeroma thermophilum (Richardson, 1897)

Phylum   ARTHROPODA   Order   ISOPODA

Class   CRUSTACEA   Family   SPHAEROMATIDAE

SUMMARY The Socorro Isopod is endemic to a single thermal outflow in Socorro
County, New Mexico, U.S.A. Having lost its natural habitat it has managed to
survive in the artificial environment of an abandoned bath house. Because of its
small numbers and limited distribution and habitat it is listed as Endangered by the
U.S. Fish and Wildlife Service, and a recovery plan has been drawn up in order to
implement action for its protection.

DESCRIPTION Thermosphaeroma thermophilum has a flattened body with seven
pairs of legs, two pairs of antennae on the head, and oar-like extensions (uropods)
on the last segment. The average length is 7.8 mm (range 4-13 mm) in males and
5.1 mm (range 4.5-6.0 mm) in females (3,9). Sexual dimorphism is marked, and
males may reach a much larger size (up to 60 mm^2) than females (up to 21 mm^2)
(10). Both sexes are greyish brown in colour with small black spots and lines which
run together to form a broad black band in the centre of each of the thoracic
segments. All the exposed edges of the body are tinged with bright orange (2,3,9).

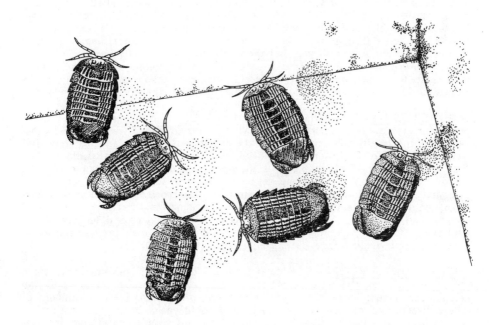

DISTRIBUTION Restricted to a thermal outflow from Sedillo Spring
approximately 3 km west of the Socorro City in the Socorro Mountains in
south-central Socorro County, New Mexico, U.S.A. (1). During the late
Pleistocene and early Holocene this and two other springs (Cook and Socorro) fed a
marsh in which T. thermophilum may well have lived, extending 0.5 mile (1 km)
east of Cook Spring (4). Both Cook and Socorro Springs are capped now and the
water is diverted to Socorro City (1).

POPULATION Population sizes probably vary seasonally and perhaps annually but the two published counts of the species are similar. In 1976 the population was estimated to consist of about 2 400 individuals (5), and in 1977 2 449 individuals were estimated (6). Densities as high as 210 individuals per 100 cm^2 may be reached. Males are much more abundant than females (10).

HABITAT AND ECOLOGY Although it has lost its natural habitat, the species has managed to survive in an artificial environment. The entire population is now found in the thermal water system of an abandoned bath house which is supplied with water from Sedillo Spring. This consists of a small (1 m x 2 m x 0.3 m) cement-lined animal watering tank, a smaller pool and approximately 40 m of open irrigation pipe. The majority of water from the spring is diverted to Socorro City for municipal use. The impoundments were constructed in the early 1900s when the spring was used by local residents as a recreation site (10). Water temperature ranges from 27 to 34°C (10) and the Isopods do not inhabit a nearby pool where the temperature is lower, suggesting that the species is adapted to thermally stable conditions and that its range was restricted long before the habitat was modified by man (10). The Isopod is found in two small pools and two runs. Much of the population is confined to the larger pool where the flat bottom is covered with 1-4 cm of finely divided substrate into which the animals burrow during the day, emerging at dusk (10). The primary food source is the blue-green algae which cover most surfaces, but detritus and dragonfly nymphs are also eaten (6,10), as well as injured conspecifics (7,10).

The reproductive cycle is still not fully known, but has been studied in captivity and investigated in the field (1,6,10). Throughout the year there are large fluctuations in abundance of females in various reproductive conditions. Gravid females are found through the year but are most abundant in the spring, suggesting that reproduction occurs all year but there is a slight seasonally-based cycle (10). Individuals probably live for eight months to a year. Predation pressure is very low, since there are no fish in the spring and the Isopods are unlikely to be accessible to birds. This may account for the high densities that are reached (10).

SCIENTIFIC INTEREST AND POTENTIAL VALUE The Socorro Isopod is of particular interest and importance in that it is one of only five fully freshwater isopods in the family Sphaeromatidae. The problem of how this species arrived at its present state of evolutionary adaptation is of interest to isopod specialists and the concept of land-lock fauna is of interest to biologists as a whole (9). Its reproductive behaviour is proving to be of interest to behavioural ecologists. Inter-male competition is intense and large males are more successful at obtaining 'high quality' females (that is, in terms of fecundity and proximity to sexual moult) than small males. Sphaeromatid sexual dimorphisms are largely if not exclusively sexually selected and the family as a whole may provide numerous tests of present evolutionary and sexual selection theory (10,11).

THREATS TO SURVIVAL Municipal and private water developments have completely altered the natural habitat of this species through capping of the original spring source and piping of the water to other areas (2). The amount of water diverted to the system is limited and could easily be stopped through a readily available cut-off valve. Protection of habitat from harmful contaminants and other negative impacts cannot be guaranteed as the habitat is on private land (1).

CONSERVATION MEASURES TAKEN The species is listed as Endangered under the U.S. Endangered Species Act and a recovery plan has been drawn up (1).

CONSERVATION MEASURES PROPOSED The recovery plan should be implemented. It recommends the following actions: Sedillo Spring should be fully protected to ensure that the existing private landowner avoids using land management practices which could alter the site detrimentally through contamination, erosion and the introduction of predatory or competitive species, and to ensure that a permanent flow of water is maintained. The area around the existing pools should be fenced and the status of the Isopod population and its habitat should continue to be monitored. Additional flows of water should be acquired, preferably from Socorro and Cook springs, and the possibility of expanding the present habitat in other ways should be investigated. Suggestions have also been made for introducing the species into other water systems such as Fort Harmony Spring although this is outside the historic range of the species. Such an operation should obviously be attempted only with extreme caution. The captive population should be maintained and further data collected on the species' biology and ecological requirements (1).

CAPTIVE BREEDING Captive populations have been established at New Mexico Department of Game and Fish, Santa Fe, the Rio Grande Zoo in Albuquerque, and at Dexter National Fish Hatchery in order to insure the genome against possible catastrophic extinction and to assure diversity among captive populations. As yet, breeding in captivity has not been successful (1).

REFERENCES   1.   U.S. Fish and Wildlife Service (1982). Socorro isopod (Thermosphaeroma thermophilum) recovery plan. U.S. Fish and Wildlife Service, Albuquerque, New Mexico. 16 pp.
2.   Hatch, M.D. (1979). Handbook of Species Endangered in New Mexico (Invertebrate Section). Pp. F1-F2.
3.   Richardson, H. (1897). Description of a new crustacean of the genus Sphaeroma from a warm spring in New Mexico. Proc. U.S. Nat. Mus. 20 (1128): 465-466.
4.   Hatch, M.D. (1977). Proposed endangered status for the species Exosphaeroma thermophilum. Unpub. rep., New Mexico Dept. Game and Fish.
5.   Hatch, M.D. (1976). The status of Exosphaeroma thermophilium. Unpub. rep., New Mexico Dept. Game and Fish.
6.   Shuster, S.M. (1977). The Socorro isopod. Unpub. rep., New Mex. Dept. Game and Fish, 25 pp.
7.   Richardson, H. (1905). Contributions to the natural history of the isopod. Proc. U.S. Nat. Mus. 27 (1350): 1-32.
8.   Anon. (1978). Socorro isopod. End. Sp. Tech. Bull. 3(1): 5.
9.   Bowman, T.E. (1981). Thermosphaeroma milleri and T. smithi, new sphaeromatid isopod crustaceans from hot springs in Chihuahua, Mexico, with a review of the genus. J. Crust. Biol. 1(1): 105-122.
10. Shuster, S.M. (1981). Life history characteristics of Thermosphaeroma thermophilum, the Socorro isopod. Biol. Bull. 161: 291-302.
11. Shuster, S.M. (1981). Sexual selection in the Socorro isopod Thermosphaeroma thermophilum (Cole) (Crustacea, Peracarida). Anim. Behav. 29: 698-707.

We are very grateful to T.E. Bowman, N.D. Bruce, F. Howarth and J.E. Johnson, for reviewing this data sheet.

Mexistenasellus wilkensi Magniez, 1972   Wilken's Stenasellid
Mexistenasellus parzefalli Magniez, 1972   Parzefall's Stenasellid

Phylum   ARTHROPODA   Order   ISOPODA

Class   CRUSTACEA   Family   STENASELLIDAE

SUMMARY These two troglobitic isopods are endemic to a single cave in Ciudad Valles, Mexico. Their restricted habitat is vulnerable to any alteration of the cave ecosystem, which should therefore be protected.

DESCRIPTION Mexistenasellus wilkensi is unusually large for a stenasellid, reaching 18 mm in length. The carapace is very thick and sometimes has calcareous deposits on it (2). Only female specimens have been found so far. M. parzefalli reaches 13 mm in length and is comparable in size to the largest European stenasellid (2,3). Detailed descriptions in (2,3).

DISTRIBUTION Known only from Cueva del Huisache, 27 km north-west of Ciudad Valles, San-Luis Potosi, Mexico (2), in the Sierra de la Colmena.

POPULATION Unknown. Although only small numbers have been found, larger populations probably exist within the karst among the fissures and subterranean rivers, as the aquifer is probably similar to that found in the neighbouring Sierra de El Abra (2). A greater number of M. wilkensi than M. parzefalli have been caught but it is not known if this is significant.

HABITAT AND ECOLOGY Both species are found in a small pool in the dark zone of the cave. The temperature of the water is 24°C and therefore comparable with the habitat of European species known from thermal springs, suggesting that the group as a whole is thermophilic. The Mexican species are often found near bat excreta in the water and may feed off this material, although stenasellids are generally predatory carnivores. M. wilkenseni has the adaptations of a very active predator and could even feed on M. parzefalli. It is unusual to have two similar species occurring together but they may avoid competition by specializing on different foods. It is thought that they may live for even longer than European species (for example, although only 12 mm long Stenasellus virei lives for fifteen years) as the intermoult period may be as long as a year (2).

SCIENTIFIC INTEREST AND POTENTIAL VALUE The Stenasellidae were long considered to be indigenous to the Old World, and have only recently been described from the western hemisphere, the first species discovered being Mexistenasellus coahuila from the Cuatro Ciénegas basin in Mexico (4). Many more forms are probably yet to be discovered in subterranean waters (2). M. wilkensi and M. parzefalli are therefore of interest for their paleogeographical relationships. They are almost certainly endemic to Cueva del Huisache which harbours a remarkable assemblage of troglobitic isopods, including a third endemic species Mexilana saluposi and an apparently undescribed anthurid (1).

THREATS TO SURVIVAL The cave is not currently under threat from any known human activity, but human visitation to the cave or any alteration to the subterranean water system could lead to the immediate extinction of these two species.

CONSERVATION MEASURES TAKEN None.

CONSERVATION MEASURES PROPOSED The cave and the subterranean water system should be protected and carbide lamps should not be permitted in the cave. Collection should be limited to adults (after reproduction) for research purposes only. Further biological research on these species as well as additional biological surveys of Mexican caves is necessary.

REFERENCES
1. Bowman, Th.E. (1975). A new genus and species of troglobitic cirolanid isopod from San Luis Potosi, Mexico. Occas. Pap. Mus. Texas. Tech. Univ. 27: 1-6.
2. Magniez, G. (1972). Deux Stenasellidae cavernicoles nouveaux de l'Amerique centrale: Mexistenasellus parzefalli n.sp. et Mexistenasellus wilkensi n.sp. (Crustacea: Isopoda: Asellota). Int. J. Speleol. 4: 19-31.
3. Magniez, G. (1973). Description du mâle de Mexistenasellus parzefalli (Crustacea Isopoda Asellota) cavernicole du Mexique et observations sur cette espèce. Int. J. Speleol. 5: 163-170.
4. Cole, G.A. and Minckley, W.L. (1971). Stenasellid isopod crustaceans in the western hempishere - a new genus and species from Mexico - with a review of other North American freshwater isopod genera. Proc. Biol. Soc. Washington 84: 313-326.

This data sheet is based on a draft by G.J. Magniez for which we are very grateful, and we would like to thank R.W. Bouchard, T.E. Bowman and F. Howarth for reviewing the sheet.

SHOUP'S CRAYFISH   VULNERABLE

Orconectes shoupi Hobbs, 1948

Phylum   ARTHROPODA   Order   DECAPODA

Class   CRUSTACEA   Family   CAMBARIDAE

SUMMARY Shoup's Crayfish is currently known only from Mill Creek in the vicinity of Nashville, Tennessee, U.S.A. It could be seriously affected by increased pollution or siltation.

DESCRIPTION This species is similar in appearance to many other North American crayfish, and the characteristics that identify and separate it from its congeners are of a specialized nature including thickened ridges on the rostrum and long-fingered chelae. Further details are given in (2,3).

DISTRIBUTION Known only from Mill Creek, a tributary of the Cumberland River in the vicinity of Nashville in Davidson and Williamson counties, Tennessee, U.S.A. Records exist for three other localities in Tennessee: 1) Big Creek (Elk River system), Giles County, 2) South Harpeth River (Harpeth River system), Davidson County and 3) Richland Creek (Cumberland River system), Davidson County, but attempts to collect additional individuals have been fruitless. The Big Creek and South Harpeth River localities probably represent introductions that did not become established. The Richland Creek record, dating from 1895, may be in error. Alternatively Orconectes shoupi has been displaced by O. placidus which dominates this locality, but is unknown from the Mill Creek system (1,2).

POPULATION Unknown.

HABITAT AND ECOLOGY Has been collected primarily in pool areas from under flattened, limestone slabs and rocks of varying sizes on a predominantly gravel and limestone bedrock substrate. Mill Creek, flowing over limestone deposits of Ordovician age, is silty with minor amounts of sand and mud. Crayfish were not common under rocks that rest on the bedrock portions of the creek. Adequate cover is necessary to support large populations of O. shoupi, since most adults are solitary animals except during the mating season. The feeding habits of this species are unknown, but it is probably omnivorous and opportunistic (1,3). Reproductive males have been collected during the months of March, April, May, July, October and November, and females bearing eggs in April. Additional collections made during the months of May and August did not contain any reproductive males. It seems from this information that male O. shoupi are primarily in reproductive form from late summer to early spring depending upon winter and early spring temperatures, and that egg laying occurs at least in early spring and is probably related to water temperatures (1).

SCIENTIFIC INTEREST AND POTENTIAL VALUE This species is a member of a primitive group within the genus Orconectes and is of zoogeographic and phylogenetic importance in the study of the origin of the genus Orconectes and relationships within the Cambaridae (1). It is probably used as a food resource by a number of vertebrates, especially fish, reptiles, amphibians and the Raccoon (Procyon lotor) (1,3).

THREATS TO SURVIVAL O. shoupi appears to be affected by the level of siltation, availability of critical habitat, and polluting agents other than those that lead to a certain amount of eutrophication and subsequent increase in algae which may serve as habitat and a food resource. Its limited distribution lies mainly within the city and suburbs of Nashville, Tennessee, and the water quality is therefore likely to deteriorate further (1).

CONSERVATION MEASURES TAKEN Has been proposed for listing under the U.S. Endangered Species Act.

CONSERVATION MEASURES PROPOSED It is essential to protect the Mill Creek watershed from siltation, alterations in stream physiography and further degradation of water quality. Research is needed on the life history and ecological requirements of this species (1).

REFERENCES   1.   Bouchard, R.W. (1978). Conservation status report for O. shoupi. In Report of First Meeting of the IUCN/SSC Freshwater Crustacean Specialist Group. Tuscaloosa, Alabama.
             2.   Bouchard, R.W. (1974). Crayfishes of the Nashville Basin, Tennessee, Alabama and Kentucky (Decapoda, Astacidae). Abstract. The ASB Bulletin 21(2): 41.
             3.   Hobbs, H.H. (1948). On the crayfish of the Limosus section of the genus Orconectes (Decapoda, Astacidae). Journal of the Washington Academy of Science 38(1): 14-21.

This data sheet is based on the conservation status report for O. shoupi by R.W. Bouchard and we are very grateful for his assistance.

NOBLE CRAYFISH   VULNERABLE

Astacus astacus (Linnaeus, 1758)

Phylum   ARTHROPODA   Order   DECAPODA

Class   CRUSTACEA   Family   ASTACIDAE

SUMMARY Once one of the most abundant crayfish in Europe and the most highly valued as a food item, Astacus astacus has been severely depleted as a result of the crayfish plague which has swept Europe since the 1860s. The species may now be vulnerable to habitat deterioration and competition with introduced species. Several countries control its exploitation, and a number of research projects on artificial breeding are in progress.

DESCRIPTION The Noble Crayfish varies in colour from green or blue to brown and sometimes almost black. The lower surface of its claws or chelae are reddish (4). The female reaches a total length of 12 cm, while the male may reach up to 16 cm (5). (Other common names for this species are given in (28)).

DISTRIBUTION Astacus astacus has a scattered but wide distribution throughout northern Europe (6,7), ranging from France (where all except the north-eastern populations are introduced), central Netherlands, Switzerland (introduced) (5), West Germany (13), Austria (32), northern Italy (introduced), Yugoslavia, Czechoslovakia, Poland, Hungary, Bulgaria, Romania and the U.S.S.R. northward to Sweden (south and Baltic coast area), Denmark, Norway and Finland in which countries it is the only native crayfish (7,28). It has reportedly been introduced into the British Isles (6) but there are no recent records (12). Its occurrence in Norway is a result of immigration or introduction from Sweden many centuries ago and it is found in the south and south-west of the country (28). In Finland it was introduced into rivers flowing into the Bothnian Bay at the turn of the century (18), and occurs naturally in other lakes throughout the western part of the country up to 65°N (19,28). In 1979 it was reported to occur in Lithuania, U.S.S.R. (14). It was introduced to Spain but did not survive, probably on account of plague (24). In Hungary it is found in the highland regions of the north-west (28). It is not known when it was introduced into Switzerland, where it is found mainly in the north-east and may be abundant in some lakes (28). It used to occur in many small rivers in Luxembourg, but is now thought to be extinct (30).

POPULATION Unknown. It was once abundant in Europe and is now generally scarce (10). Indications of abundance in some countries are given in (28).

HABITAT AND ECOLOGY Found along the banks of well oxygenated ponds, streams, lakes and rivers. Feeds on worms, aquatic insects, molluscs, vertebrates and plants and is nocturnal, spending the day in its burrow. It undergoes periodic moults and sexual maturity is reached during the fourth year (5,28). The breeding season in Germany and Switzerland is from October to November (4,5) and the eggs are carried by the female over the winter; hatching takes place the following May. In the Baltic countries egg laying takes place in the second half of November and the eggs hatch from the second half of June to the beginning of July (31). In some regions, mink may be a heavy predator (15); other predators in Baltic countries include eel, perch, burbot, pike, otter and muskrat (31). Details of parasites and diseases are given in (28).

SCIENTIFIC INTEREST AND POTENTIAL VALUE It is one of the most popular edible crayfish and is collected for food in many countries including Denmark, Finland, Hungary, Norway, Poland and Sweden. In most countries fishing is a recreational sport, and the commercial fishery is small. In Finland, about

1000-5000 semi-professional fishermen and 50 000 recreational fishermen take an estimated 2.5-4.0 million individuals a year. At the beginning of the century, before the crayfish plague struck the population, up to 20 million individuals were being taken a year. In 1900 15.5 million crayfish were exported but as a result of the plague, exports have declined and are now exceeded by imports (28). In Hungary, numbers of semi-professional fishermen have dropped from about 100 in 1960 to 10-15 in 1980. Fishing in Norway is mainly for recreation, although there are 10-15 semi-professional fishermen. The total annual catch is estimated at 20-40 tonnes, of which about 10-15 tonnes are exported. In Poland collection is mainly carried out by recreational fishermen. There are some semi-professionals but there have been no commercial catches since 1980. Annual catch between 1969 and 1978 has been estimated at 15 tonnes. In Sweden the current annual catch is probably less than 100 tonnes, although prior to 1907 when the plague arrived, it is estimated to have been at least 1000 tonnes (28).

THREATS TO SURVIVAL Has been eliminated throughout much of Europe as a result of the crayfish plague (9,28) and probably pollution. A. astacus is more sensitive to DDT than are other crayfish, and it is possible that its current distribution may partly reflect its sensitivity to existing pollution levels. Furthermore, females accumulate more DDT than males, which could be serious in the long-term if this affects reproductive success (17).

In the 1960s populations in Finland declined drastically through plague, pollution, dam construction and dredging. Dredging causes turbidity, which affects water quality and destroys the habitat of the crayfish. Only 20 of the 67 major watercourses in which A. astacus occurred remain uninfected by plague, and populations in the good crayfish rivers flowing into the Gulf of Bothnia have been devastated (28). Populations are increasing again but are still vulnerable to these factors (18,28) as well as to competition with introduced species, although A. astacus coexists in some lakes with introduced Pacifastacus leniusculus (19). In Lithanian the plague is the greatest threat (31).

In Norway it has probably been affected by acid rain and overfishing. For example, unrestricted and illegal fishing in a small lake and stream in Oslo rendered the local population extinct in only a few years (8).

Crayfish populations in Sweden have been affected most seriously by acidification of lakes. This causes a decrease in the number of fry and also a prolonged period when the shell is soft. Moulting, egg laying and hatching have been shown to be the most sensitive periods in the lifecycle of this species. In addition, eel (Anguilla anguilla) predation and the plague have reduced some populations (27,28) and the species has been found to decline in areas where another crayfish, Pacifastacus leniusculus has been introduced (20).

Its distribution in Denmark is reported to have been affected by pollution in rivers and lakes (28).

Populations have declined in the Netherlands, although it is not currently known how seriously the species is threatened or what will be the effect of the introduced American species Orconectes limosus (22,23). The decline is thought to be mainly due to deteriorating environmental conditions and the plague (28).

The crayfish plague is thought to have been responsible for its recent disappearance in Luxembourg (30).

In Switzerland it has disappeared from some localities (5). For example, a very dense population in Lake Bret (Canton of Vaud) suddenly disappeared for no known reason (28).

In the province of Steiermark in Austria, it is reported to be threatened by pollution, the crayfish plague and the introduction of the exotic crayfish Orconectes limosus (52).

In Yugoslavia it used to be abundant but by the 1960s efforts had to be made to restock populations (21).

In Hungary it was present throughout most of the country until the 1860s, after which the crayfish plague decimated most stocks. Populations have been recovering gradually, but the last twenty years have seen a dramatic decrease in abundance, due to river drainage, pollution and the plague (28).

Populations are said to be declining in Bulgaria as a result of pollution and habitat disturbance (29).

It is reported to be very rare in Czechoslovakia (26).

CONSERVATION MEASURES TAKEN In Finland crayfish trapping is permitted only between 21 July and 31 October and there is a minimum size limit of 10 cm. As from 1983 a state fishing licence will be required (18,28). In Lithuania fishing is strictly controlled with a minimum size limit of 10-11 cm and efforts are being made to control disease and water pollution (31).

In Norway fishing is permitted between 7 August and 7 April and there is a minimum size limit of 9.5 cm. In some areas fishermen may only put out a limited number of traps.

In Sweden there is a minimum size limit of 9 cm and there are local regulations concerning numbers of traps used and numbers of nights open for fishing. Attempts have been made to prevent acidification by liming the lakes and forty crayfish lakes have been treated. Efforts are also being made to control other forms of water pollution (28).

In Denmark fishing is allowed from 1 July to 30 September for females and from 1 April to 30 September for males; the minimum size limit is 9 cm.

The species has been protected in the Netherlands, since 1973 (2).

In France the minimum size limit for collection is 9 cm (from the top of the rostrum to the end of the telson). A single fisherman may use only six catching nets ('balances'); the open season is between 14 August and 15 September, and in regions where the species is considered to be near extinction, fishing is completely prohibited (11).

In Luxembourg collection is prohibited all the year round and the animal is to be included under new legislation for protected species (30).

Fishing is controlled throughout Switzerland, and all Astacidae are protected in eight of the 26 cantons (5). There is a minimum size limit of 12 cm, and Federal legislation requires that all Astacidae have a closed season of at least forty weeks each year (28).

In Hungary the fishing season is from 1 June to 15 October and there is a minimum catch size of 10 cm.

In Poland the fishing season is from 16 March to 14 October for males and 1 August to 14 October for females. There is a minimum size of 9 cm, and recreational fishermen may only use five traps.

Listed as seriously threatened in the West German Red Data Book (13).

Fishing is regulated in the province of Bolzano, Italy (1).

CONSERVATION MEASURES PROPOSED The Finnish Game and Fisheries Research Institute has conducted extensive research on crayfish populations (19) but further information is required on the species' distribution and the reasons for its decline. Research is needed to develop a plague resistant strain. In addition, protection of habitat and control of exploitation of wild populations should be considered. In Norway recommendations have been made for more stringent control measures and protection, and information campaigns supported by the Ministry of the Environment should be initiated (8). All three species of Astacus occurring in Czechoslovakia have been proposed for complete protection (26).

CAPTIVE BREEDING The species was reared in France in the 1860s in one of the first commercial crayfish culture enterprises (3) but commercial rearing is not currently widespread. Pond culture is still carried out in a few places in Europe (10). Artificial breeding experiments have been carried out in Lithuania, U.S.S.R. with the aim of increasing stocks in natural waters (16), and also in Finland (28) and Yugoslavia (21). Specimens from Denmark were introduced to Cyprus in 1976 and 1978 for aquaculture purposes; they are kept at the Experimental Freshwater Culture Station at Kalopanayiotis but it is not yet known if the population is firmly established (28). Aquaculture for restocking purposes is carried out in Norway where about 250 000 juveniles are produced annually (28).

REFERENCES 1. Anon. (1973). Norme per la protezione della fauna. Estratto dal Bolletino Ufficiale della Regione Trentino, 11 September.
2. Anon. (1973). Staatsblad 488. 6 August, Netherlands.
3. Arrignon, J. (1975). Crayfish farming in France. In Avault, J.W. (Ed.), Freshwater Crayfish. Papers from the 2nd Int. Symp. on Freshwater Crayfish 1974, Baton Rouge, Louisiana, U.S.A.
4. Bott, R. (1950). Die Flusskrebse Europas (Decapoda, Astacidae). Abh. Sen. Natur. Ges. 483: 36 pp.
5. Burckhardt, D., Gfeller, W. and Müller, H.U. (1980). Animaux protégés de Suisse. Ligue Suisse pour la Protection de la Nature (LSPN), Bâle.
6. Davies, A.W. (1965). Crayfish in Suffolk rivers. Trans. Suffolk Nat. Soc. 13: 11-13.
7. Gledhill, T., Sutcliffe, D.W. and Williams, W.D. (1976). Key to British Freshwater Crustacea: Malacostraca. Freshwater Biol. Assoc. Sci. Publ. 32. 71 pp.
8. Guldbrandsen, K.S. (1976). Rovfiske kan odelegge krepsebestanden. Fauna, Oslo 29 (3): 122-126.
9. Holdich, D.M., Jay, D. and Goddard, J.S. (1978). Crayfish in the British Isles. Aquaculture 15: 91-97.
10. Karlsson, S. (1977). The freshwater crayfish. Fish Farming International 4(2): 8-12.
11. Laurent, P.J. (1973). Astacus and Cambarus in France. In Abrahamsson, S. (Ed.), Freshwater Crayfish. Papers from the 1st Int. Symp. on Freshwater Crayfish, Austria 1972: Pp. 69-78.
12. Thomas, W. and Ingle, R. (1971). The nomenclature, bionomics, and distribution of the crayfish, Austropotamobius pallipes (Lereboullet) (Crustacea, Astacidae) in the British Isles. Essex Naturalist 32: 349-360.
13. Türkay, M. (1977). Rote Liste der Zehnfüssigen Krebse (Decapoda). In Blab, J. Nowak, E., Trautmann, W. and Sukopp, H. (Eds), Rote Liste der gefährdeten Tiere und

Pflanzen in der Bundesrepublik Deutschland. Kilda Verlag, Greven.

14. Matskyavichene, G.I. (1979). The biology of the crayfish of the Lithanian inland waters. Mokslas - Vilnyus, U.S.S.R.

15. Gulbrandsen, K.S. (1978). Astacus astacus food for Mustela vison. Fauna (Blindern) 31(I): 11-16.

16. Cukerzis, J.M., Sheshtokas, J. and Terentyev, A.L. (1978). Method for accelerated artificial breeding of crayfish juveniles. In Laurent, P.J. (Ed.), Freshwater Crayfish 4. Papers from the Fourth International Symposium on Freshwater Crayfish, Thonon-les-Bains, France 1978. Pp. 451-458.

17. Airaksinen, M., Valkama, E.-L. and Lindquist, O.V. (1977). Distribution of DDT in the Crayfish Astacus astacus L. in acute test. Freshwater Crayfish 3. Papers from the Third International Symposium on Freshwater Crayfish at the University of Kuopio, Finland. 1976: Pp. 349-356.

18. Niemi, A. (1977). Population studies on the crayfish Astacus astacus L. in the River Pyhaejoki, Finland. Freshwater Crayfish 3. Papers from the Third International Symposium on Freshwater Crayfish at the University of Kuopio, Finland 1976 : Pp. 81-90.

19. Westman, K. and Pursiainen, M. (1978). Development of the European crayfish Astacus astacus (L.) and the American crayfish Pacifastacus leniusculus (Dana) populations in a small Finnish lake. In Laurent, P.J. (Ed.), Freshwater Crayfish 4. Papers from the Fourth International Symposium on Freshwater Crayfish, Thonon-les-Bains, France 1978. Pp. 243-250.

20. Fürst, M. (1977). Flodkräftan och signalkräften i Sverige 1976. Information fran Sötvaltenslaboriet, Drottningholm (10): 32 pp. (quoted in 19).

21. Herfort-Michieli, T. (1978). L'écrevisse à pieds rouges en Slovenie depuis 1972. In Laurent, P.J. (Ed.), Freshwater Crayfish 4. Papers from the Fourth International Symposium on Freshwater Crayfish. Thonon-les-Bains, France 1978. Pp. 185 189.

22. Holtuis, L.B. (1981). In litt., 2 October.

23. Geelen, J.F.M. (1978). The distribution of the crayfishes Orconectes limosus (Rafinesque) and Astacus astacus (L.) (Crustacea, Decapoda) in the Netherlands. Zoologische Bijdr. 23: 4-19.

24. Hapsburgo-Lorena, A.S. (1978). Present situation of exotic species of crayfish introduced into Spanish continental waters. In Laurent, P.J. (Ed.), Freshwater Crayfish 4. Papers from the Fourth International Symposium on Freshwater Crayfish. Thonon-les-Bains, France 1978. Pp. 175-184.

25. Gtowacinski, Z., Bieniek, M., Dyduch, A., Gertychowa, R., Jakubiec, Z., Kosior, A. and Zananek, M. (1980). Situation of all vertebrates and selected invertebrates in Poland - list of 81 species, their occurrence, endangerment and status of protection. Warszawa-Krakow.

26. Pecina, P. and Capicka, A. (1979). An Atlas of Endangered and Protected Animals in Czechoslovakia. Prague, 220 pp.

27. Waldén, H. (1982). In litt., 11 October.

28. Westman, K. and Pursiainen, M. (Eds), (1982). Status of crayfish stocks and fisheries in Europe. Paper prepared for European Inland Fisheries Advisory Commission (EIFAC)

Twelfth Session, Budapest, Hungary 1982.

29. Deltshev, C. (1982). In litt., 5 November.

30. Musée d'Histoire Naturelle de Luxembourg (1982). In litt., 19 November.

31. Cukerzis, J.M. (1970). The Biology of Crayfish (Astacus astacus L.). 207 pp.

32. Gepp, J. (Ed.) (1981). Rote Listen Gefährdeter Tiere der Steiermark. Österreichischen Naturschutzbundes, Landesgruppe Steiermark.

We are very grateful to R.W. Bouchard, J.M. Cukerzis, C. Deltshev, L.B. Holthuis, R. Ingle, Musée d'Histoire Naturelle de Luxembourg, J. Okland, K. Spitzer, H.W. Walden and K. Westman for information provided for this data sheet and for helpful comments.

WHITE-CLAWED or   RARE
ATLANTIC-STREAM CRAYFISH

Austropotamobius pallipes (Lereboullet, 1858)

Phylum   ARTHROPODA   Order   DECAPODA

Class   CRUSTACEA   Family   ASTACIDAE

SUMMARY   The White-clawed Crayfish still has a fairly wide distribution
throughout the waterways of Europe, but is vulnerable in several ways. On the
continent, populations have declined in many areas as a result of crayfish
'plague'. There are fears that this could be introduced to Ireland and the U.K. as a
result of the current interest in aquaculture of imported exotic crayfish species.
North American crayfish are now farmed in many countries but it is not known
what the effects of competition between these and Austropotamobius pallipes
would be if large numbers of the former were to escape into the wild. Like
salmonid fish, whose requirements it shares, A. pallipes appears to be vulnerable
to degradation of freshwaters through pollution or an increase in acidification.
The species has considerable commercial potential in its own right, which should
provide a strong incentive for its conservation.

DESCRIPTION   The White-clawed Crayfish is reddish- to greenish-brown in colour,
the underside of the claws usually being white. It has a short keeled rostrum and a
single spine behind each eye. May reach 22 cm in body length (9,34). Other
common names for it are given in (33).

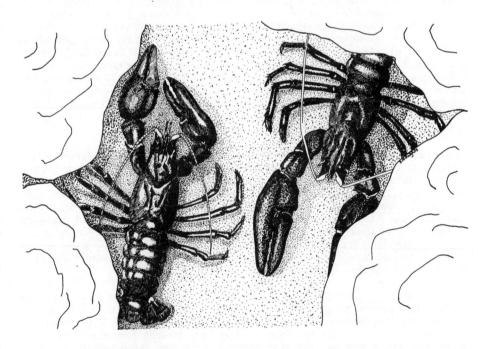

DISTRIBUTION   Found throughout much of Europe including West Germany,
Switzerland, France (1,18), the U.K. (9,29), Portugal (where it has a restricted
distribution)(31), north and central Spain, northern Italy, Yugoslavia (2,7,10), and it

has recently been found in a tributary of the River Gail in southern Carinthia, Austria (35). It appears to be restricted to mountain streams on the continent, but is more widespread in the U.K. and has a broad distribution in Ireland (21,23) where it is the only crayfish. Its distribution in the U.K., where it is the only naturally occurring crayfish, may be wider than is generally realized (3,12,16), but it does not occur in Scotland.

POPULATION Its population in the U.K. varies from year to year and during any one year (8,22,29). Short cycle variations are usually due to natural causes such as death from old age, the breeding cycle, or migration into deeper waters with the onset of winter. Longer lasting variations may be due to the migrations of a population from one locality to another, to overfishing or to disease or drought. In the U.K. large populations are found in some reservoirs and quarries (12,16) and there are indications that the species is more abundant than was previously thought (33). It is perhaps most abundant in Co. Westmeath in Ireland (the headwaters of the Boyne and of the Shannon tributaries) and in Co. Fermanagh (in the tributaries of the Erne in Northern Ireland) (25). It is said to be the commonest indigenous crayfish in France (19). In Switzerland, Austropatamobius pallipes is particularly abundant in certain lakes in Graubünden and some canals in Valais (33).

HABITAT AND ECOLOGY In France it occurs mainly in mountain streams; for example it has been found in Lozere at fairly high altitudes, is common in hill streams in Limousin and has been recorded from a subalpine lake near Chamonix (25). In Spain it is found mainly in the slow-flowing streams of the highland plains and the marshy areas of the Mancha (34). In the U.K., it is common in many alkaline waters such as chalk and limestone trout streams, but is also found in reservoirs and quarries filled with relatively still water, often to a depth of 7 m. Although it is absent from waters with a low pH value, water at about neutrality is suitable (16). In Ireland the distribution pattern corresponds broadly with that of carboniferous limestone or with superficial calcareous glacial drift (23). It is widespread and often abundant in the lakes and rivers of the lowlands (23) but is absent from lakes larger than 1000 ha (21), from many acid coastal streams and, with one exception, from the sandstone rivers of the south (24). Juvenile crayfish are known to be eaten by eels (Anguilla anguilla) (28), which may have restricted crayfish distribution in rivers such as the Bann and large lakes in Ireland although this has not been proven (24). Crayfish are also eaten by coarse fish and trout (20) but trout and crayfish populations commonly coexist (24).

Growth rate seems to depend on the environment. Irish lake populations reach maturity in their third year, when individuals are about 65-76 mm long. Growth and maturation of individuals in stream populations may be slower (hence the continental minimum size limit for harvesting of 90 mm) (24). In laboratory tests Austropotamobius pallipes is omnivorous and shows a preference for animal (e.g. Gammarus and Asellus spp.) over plant food (28). Studies in Ireland and the south of England have shown that Austropotamobius pallipes breeds in September and October, the eggs are laid soon afterwards and hatching takes place in the following May or June (12,13,21). In the north of England fertilization occurs in November and the eggs are carried until August (6).

Mortality among juveniles is high in the first few months of life away from the mother, due mainly to predation and problems associated with moulting. At release the young average 10 mm in length (12). They moult five or six times and overwinter at 16 mm. The following summer they undergo four moults and then overwinter at 24 mm and the next summer there are three moults and they overwinter at 37 mm. Subsequently there are one or two moults a year, and the animals may live for up to ten to twelve years (5). Although they can breed each year (12) they may not do so in poor environments or if temperatures are low (24).

SCIENTIFIC INTEREST AND POTENTIAL VALUE The White-clawed Crayfish used to be eaten extensively in the U.K. and is still a delicacy in some areas (4). Natural large populations in quarries and reservoirs could be cropped to meet current demand if subjected to a rational policy of fishery management and the development of artificial rearing methods has considerable potential (12,27). In other countries the species has considerable commercial value. Between 20 and 30 million individuals are caught annually in Spain by 80 professional, 10 000 semi-professional and about 900 000 recreational fishermen (33). The species may be a useful biological indicator of water quality (16).

THREATS TO SURVIVAL A. pallipes has declined in many parts of its range. In Spain and France it has been decimated by the crayfish plague (Aphanomyces) (8,28). In France its range is said to be decreasing yearly, as a result of increased industrialization and pollution (1,18). Populations have decreased in most water courses in the Haute-Loire or have totally disappeared (32). In Spain the decline is said to be due to dredging, pollution and overfishing (7,33). In Ireland populations have disappeared from several localities in the past hundred years (21), although crayfish diseases have not yet been identified in this country. Many cases of crayfish disappearance (e.g. from ponds around Dublin) are best explained by habitat alteration and urban growth (23). Populations appear to have gone from polluted stretches of the River Suir below Thurles, although they remain in unpolluted tributaries and in the river upstream from this point. The pollution is organic and relatively mild (agricultural wastes and town sewage) but it has resulted in siltation, weed growth and marked fluctuations in dissolved oxygen content (25). Drainage and changes in agricultural practices may also affect the species. In the U.K., Nanpantan Reservoir, Loughborough, contained an enormous population of A. pallipes until recently. In 1979 the reservoir was completely drained and many crayfish died. About 2500 crayfish were rescued and transferred to other water bodies (11). The large population in Markfield Quarry, Leicester, is threateded by plans for the quarry to be filled with rubble. Attempts are being made to prevent this, or to set up a rescue and transfer operation (30). There are a number of records of local populations of river crayfish being wiped out by the activities of farmers and industry, although immigration from upstream could eventually restock these areas (11). Crayfish populations in the U.K. were badly affected by the drought in 1976, especially in Kent where the species has virtually disappeared from the River Darwent (16). Such natural catastrophes combined with the factors mentioned above could have a long-lasting damaging effect on populations.

In the U.K. the increasing interest in commercial culture of introduced crayfish such as the American Pacifastacus leniusculus threatens the native crayfish which would probably be out-competed if the former were allowed to escape. Attempts to establish self-regenerating populations of P. leniusculus will probably necessitate large scale introductions, spread over several years. Increase in traffic from countries where crayfish plague is widespread increases the likelihood of its introduction into the U.K. Over the last three years juveniles of P. leniusculus have been introduced into a number of privately owned ponds, lakes and fish farms. They are imported from Sweden where crayfish plague occurs, although exports of artifically bred individuals are said to be plague-free (17). So far there are no import controls or requirements for disease-free certification, and the native population is vulnerable to any introductions, which could include wild-caught crayfish (12). Since the native crayfish has a wide distribution in the U.K., introductions will almost certainly spread to its waters. The effects of competition are unknown, but if the plague reaches the U.K. it seems likely that the resistant aliens would survive and multiply (12). Evidence is scant, but what there is suggests that A. pallipes is inferior to other crayfish species under such competitive circumstances. For example, it is restricted to hill streams in continental Europe but in Ireland, where it is the only crayfish, it has a broad distribution (24). Male crayfish of many species have been found to be more

299

sexually aggressive than <u>A. pallipes</u> and if cross-breeding occurred, it could result in the death of females of <u>A. pallipes</u> (13,14).

CONSERVATION MEASURES TAKEN In Spain and France there are size (usually a minimum length of 90 mm) and seasonal restrictions on the capture of the species and some areas are set aside as reserves (25). In Spain the minimum size for capture is 80 mm from eye to end of tail; only 80 animals may be caught on each licence granted, and fishing is limited to the period from 21 June to 31 August (daily) and during the rest of the year on Thursdays, Saturdays and holidays only (7). The species is listed as threatened in Portugal (31). In Switzerland fishing of all Astacidae is regulated, there is a minimum size limit of 10 cm (33) and <u>A. pallipes</u> is protected in eight of the 26 cantons (6). In Ireland crayfish cannot be taken by any type of net or trap without a licence, but they can be caught by hand with the consent of the owner of the water. There is a blanket prohibition on the import of exotic crayfish into either northern or southern Ireland (25). Transplantation of threatened populations is being attempted in the U.K. (30).

CONSERVATION MEASURES PROPOSED Cultivation of <u>A. pallipes</u> rather than of introduced species should be encouraged. The suitability of wild harvests of <u>A. pallipes</u> in areas where it is abundant should be assessed. Field trials for the best conditions for its aquaculture, and studies of the economic implications of using this species for export, should be assessed (24). The import of exotic crayfish should be controlled and there should be some form of checking to guarantee that imports are free of disease. Crayfish introductions should not be made until the ecological roles of <u>A. pallipes</u> and exotic species are better known (24).

CAPTIVE BREEDING In France <u>A. pallipes</u> is being cultured within its natural range above the Gorges du Tarn, Lozere, and in the Camargue, a hatchery has been set up for restocking purposes (1,24). In Spain one privately-owned and three state-owned establishments, including El Chaparillo, Cuidad Real, have aquaculture programmes (33). Several research institutes including the University of Dublin, Ireland, and Durham University, U.K., are carrying out captive breeding studies (12).

REFERENCES   1. Arrignon, J. (1975). Crayfish farming in France. In Avault, J.W. (Ed.), <u>Freshwater Crayfish</u>. Papers from the 2nd Int. Symp. on Freshwater Crayfish, Baton Rouge, Louisiana, 1974. Pp. 105-116.

2. Bott, R. (1950). Die Fluss Krebse Europas (Decapoda, Astacidae). <u>Abhandlungen der Senckenberischen Naturforschenden Gesellschaft</u> 483: 36 pp.

3. Bowler, K. (1979). Plague that has ravaged Europe. <u>Fish Farmer</u> 2.

4. Bowler, K. (1981). In litt., 8 April.

5. Brewis, J.M. and Bowler, K. (1982). The growth of the freshwater crayfish, <u>Austropotamobius pallipes</u>, in Northumbria. <u>Freshwater Biology</u> 12(2): 187-200.

6. Burckhardt, D., Gfeller, W. and Müller, H.U. (1980). <u>Animaux Protégés de Suisse</u>. Ligue Suisse pour la Protection de la Nature (LSPN), Bâle.

7. Cuellar, L. and Coll, M. (1978). First essays of controlled breeding of <u>Astacus pallipes</u>. In Laurent, P.J. (Ed.), <u>Freshwater Crayfish 4</u>, Papers from the 4th Int. Symp. on Freshwater Crayfish, INRA, Thonon-les-Bains, France 1978. Pp. 273-276.

8. Duffield, J.E. (1933). Fluctuations in numbers among freshwater crayfish, <u>Potamobius pallipes</u> Lereboullet. <u>J. Anim. Ecol.</u> 2: 184-196.

9. Gledhill, T., Sutcliffe, D.W. and Williams, W.D. (1976). Key to British Freshwater Crustacea: Malacostraca. Freshwater Biol. Assoc. Sci. Publ. 32. 71 pp.

10. Hapsburgo-Lorena, A.S. (1978). Present situation of exotic species of crayfish introduced into Spanish continental waters. In Laurent, P.J. (Ed.), Freshwater Crayfish 4, Papers from the 4th Int. Symp. on Freshwater Crayfish, INRA, Thonon-les-Bains, France, 1978. Pp. 175-184.

11. Holdich, D.H. (1981). In litt., 26 January.

12. Holdich, D.M., Jay, D. and Goddard, J.S. (1978). Crayfish in the British Isles. Aquaculture 15: 91-97.

13. Ingle, R.W. (1977). Laboratory and SCUBA studies on the behaviour of the female crayfish Austropotamobius pallipes. Rep. Underwat. Ass. 2(N.S.): 1-15.

14. Ingle, R.W. (1981). Pers comm., 9 April.

15. Ingle, R.W. and Thomas, W. (1974). Mating and spawning of the crayfish Austropotamobius pallipes. J. Zool. Lond. 123: 525-538.

16. Jay, D. and Holdich, D.M. (1981). The distribution of the crayfish, Austropotamobius pallipes, in British waters. Freshwater Biology 11: 121-130.

17. Karlsson, S. (1977). The freshwater crayfish. Fish Farming International 4(2): 8-12.

18. Laurent, P.J. (1973). Astacus and Cambarus in France. In Abrahamsson, S. (Ed.), Freshwater Crayfish, Papers from the 1st Int. Symp. on Freshwater Crayfish, Lund, Sweden 1972. Pp. 69-78.

19. Laurent, P.J. and Suscillon, M. (1962). Les écrevisses en France. Anals. Stn. Cent. Hydrobiol. Appl. 9: 333-397.

20. Moriarty, C. (1963). Food of perch (Perca fluviatilis, L.) and trout (Salmo trutta L.) in an Irish reservoir. Proc. R. Ir. Acad. 63B: 1-31.

21. Moriarty, C. (1973). A study of Austropotamobius pallipes in Ireland. In Abrahamsson, S. (Ed.), Freshwater Crayfish, Papers from the 1st Int. Symp. on Freshwater Crayfish, Lund, Sweden. Pp. 57-67.

22. Pixell-Goodrich, H. (1956). Crayfish epidemics. Parasitology 46: 480-483.

23. Reynolds, J.D. (1978). Ecology of Austropotamobius pallipes in Ireland. In Laurent, P.J. (Ed.), Freshwater Crayfish 4, Papers from the 4th Int. Symp. on Freshwater Crayfish, INRA, Thonon-les-Bains, France, 1978. Pp. 216-220.

24. Reynolds, J.D. (1979). The introduction of freshwater crayfish species for aquaculture in Ireland. In Kernan, R.P., Mooney, O.V. and Went, A.E.J. (Eds), The Introduction of Exotic Species: Advantages and Problems. Proceedings of a Symposium 4-5 January 1979, Royal Irish Academy, Dublin.

25. Reynolds, J.D. (1981). In litt. (undated).

26. Richards, K. and Fuke, P. (1977). Freshwater crayfish: the first centre in Britain. Fish Farming International 4(2): 12-15.

27. Rhodes, C.P. and Holdich, D.M. (1979). On size and sexual dimorphism in Austropotamobius pallipes (Lereboullet). A step in assessing the commercial exploitation potential of the native British freshwater crayfish. Aquaculture 17: 345-358.

28. Svardson, G. (1972). The predatory impact of eel (Anguilla anguilla L.) on populations of crayfish (Astacus astacus L.). Rep. Inst. Freshwater Res. Drottningholm 52: 149-191.

29. Thomas, W. and Ingle, R. (1971). The nomenclature, bionomics, and distribution of the crayfish, Austropotamobius pallipes (Lereboullet) (Crustacea, Astacidae) in the British Isles. Essex Naturalist 32: 349-360.

30. White, G. (1981). Markfield Quarry - threat of infilling. The Underwater Conservation Society Newsletter February 1981: 2.

31. Baeta Neves, C.M. (1959). Protection des animaux rares et menacés au Portugal. In Animaux et Végétaux Rares de la Region Méditerranéenne. La Terre et la Vie. Suppl.

32. Demars, J.J. (1978). Premières données sur les populations d'écrevisses de quelques cours d'eau du haut bassin Loire-Allier. Freshwater Crayfish 4. Papers from the 4th Int. Symp. on Freshwater Crayfish INRA, Thonon-les-Bains, France, 1978. Pp. 166-174.

33. Westman, K. and Pursiainen, M. (1982). Status of crayfish stocks and fisheries in Europe. Paper presented at European Inland Fisheries Advisory Commission (EIFAC) 12th Session, Budapest, Hungary 1982.

34. Hapsburgo-Lorena, A.S. (1982). In litt., 3 November.

35. Kühnelt, W. (1982). In litt., 17 December.

We are very grateful to K. Bowler, A.S. Hapsburgo-Lorena, D.M. Holdich, R.W. Ingle, W. Kühnelt and J.D. Reynolds for providing information and helpful comments on this data sheet.

Pacifastacus fortis (Faxon, 1914)

Phylum   ARTHROPODA   Order   DECAPODA

Class   CRUSTACEA   Family   ASTACIDAE

SUMMARY This crayfish is known only from tributaries of the Pit River in Shasta County, California, U.S.A. It is a slow-growing, relatively long-lived species with low fecundity and is adapted to living in cool, clear, spring-fed habitats. Threats to its survival include habitat alteration, exotic predators and competitors, an increasing human population in the area, and harvest for human consumption.

DESCRIPTION It is a small to medium sized crayfish (adults are 25-50 mm in total carapace length) and has the darkest overall colouring of any North American crayfish. Depending upon how closely and carefully it is examined and upon light conditions, the species appears to be generally black to dark green or brown dorsally and bright orange ventrally. A few individuals from some populations exhibit a light blue to blue-green colour dorsally and light orange to yellow ventrally, except the population in Sucker Spring, where most individuals are blue to blue-green. All colour forms except the blue are cryptic among the volcanic rubble of their habitat (1,5).

DISTRIBUTION Pacifastacus fortis inhabits the Pit River drainage in Shasta County, north-eastern California, U.S.A., where it is known from two tributary systems, Fall River and Hat Creek. In the Hat Creek subdrainage, populations have been found in Crystal, Baum and Rising River Lakes. In the Fall River subdrainage, populations occur in Fall River, Big Lake, Spring, Squaw and Lava Creeks, and in the Crystal and Rainbow Springs. An additional population occurs in a spring tributary of the Pit River at Pit Power House III, known as Sucker Spring, which lies between the two subdrainages (2,5).

POPULATION Uncertain. Estimates given for Hat Creek populations (6.89 crayfish per m^2 in Crystal Lake and 0.9 crayfish per m^2 in Baum Lake) relate only to prime habitat and over-estimate the actual density (3). Since population size in part reflects the amount of critical habitat (i.e. rocky cover), which is quite patchy in these lakes, these figures cannot be extrapolated to estimate the entire population. It should also be noted that this species was at times gregarious (sensu lato) with occasionally nearly a dozen or more individuals tolerating each other's presence under a single large rock.

HABITAT AND ECOLOGY Occurs in cool, clear, spring-fed lakes and streams, usually at or near a spring source, in waters which show relatively little annual fluctuation in temperature and remain cool during the summer. Prefers lentic and slowly to moderately flowing waters. Found only under rocks larger than 7.5 cm in diameter, usually on clean, firm sand or gravel substrate (1,5), although in Crystal Lake a fine, probably organic material 1-3 cm thick covered most of the bottom. It is most abundant where plants are absent and abundance is positively correlated with depth, distance from shore and mean stream width (3). The most important habitat requirement appears to be the presence of adequate rock rubble for cover. Some adaptability is possible, since at Sucker Spring no natural habitat remains and crayfish are living within the rock wall of a fish raceway (5). All known populations occur below 1036 m elevation (1). Very little is known about the diet of P. fortis. Specimens kept in aquaria have fed on both freshwater limpets and tubifex worms. The morphology of the mouthparts suggests that the species relies on predation, browsing of encrusting organisms, or detritus for its food. P. fortis, like most crayfish, is found solitarily; apparent gregariousness may be due to toleration of the proximity of other crayfish if space is limited. Individuals do not appear to be as aggressive as many other crayfish and only infrequently show the classic crayfish defensive posture of raised chelae.

Predators are unknown specifically, but a number of vertebrates could be expected to utilize this species for food. Local people in the Fall River Mills area in the past have reported finding trout with crayfish, probably P. fortis, in their stomachs (5). The available life history information from field collections indicates that the species is similar to its congeners, with copulation occurring in late September and October after the final moult of the season, egg-laying during the fall, and hatching the subsequent spring. Females appear to be about 28 mm in total carapace length when they reach sexual maturity. Data suggest that relatively few eggs (10 to 70) are laid and that fecundity, low at first, increases with age. By the third instar young crayfish are free-living miniatures of the adults (2,5).

SCIENTIFIC INTEREST AND POTENTIAL VALUE This species is a member of a primitive group dating back at least to the Miocene, and is of zoogeographic and phylogenetic importance in the study of the relationships within the genus Pacifastacus and the family Astacidae. Pacifastacus fortis and P. nigrescens (from the San Francisco Bay area, California) represent species of a group that once occupied a larger range which has been reduced through geological and related climatic changes, which raised stream gradients in some areas and created dry areas with seasonal streams in others (5).

THREATS TO SURVIVAL P. fortis probably no longer occurs at Fall River Mills, its type locality. This and other sections of the Pit River drainage have changed markedly through the development of the area for agricultural purposes, particularly by dyking and diversion of water. The construction of several power plants and reservoirs has changed much of the Pit River system from wild and free-flowing streams to a series of interconnected impoundments subject to periodic and unseasonal fluctuations in water flow. The increase in human population and use of the area has contributed to the fact that the special environmental conditions found historically in the middle reaches of the Pit River System have now been greatly reduced (5).

The presence of the introduced crayfish, Orconectes virilis and Pacifastacus leniusculus, in the Pit River drainage is a potential threat. Introduction of both exotics probably resulted from angling activities, since anglers were using crayfish for bait at least in the early 1960s in the Pit River. Both species are known to have displaced native species in other areas (1,2,7,8). O. virilis presently occurs in the Pit River below Pit Power House III and in Lake Britton, but although excellent habitat is available, as yet there is no evidence that the species exists in Hat Creek or Fall River. Since the Lake Britton population of O. virilis was probably introduced by fishermen, the likelihood of similar introductions into Hat Creek and Fall River is very high. P. leniusculus has recently been found in Crystal and Baum Lakes as well as in the Fall River, the source of these populations probably being the adjoining Crater Lake (Lassen County) where it was introduced for trout. The Crystal and Baum Lake populations have appeared since 1975, when a survey revealed no trace of this species. Both exotics appear to be faster growing, faster maturing, more fecund and more aggressive than P. fortis. They have been introduced to the area relatively recently and are currently expanding their range within the middle reaches of the Pit River System. Although it is not known if this expansion is at the expense of the native crayfish, the outlook at present is bad (5).

In 1979 P. fortis in Crystal Lake was collected by skin divers operating at night. Because of its nocturnal habit and the clear, shallow waters of its habitat, P. fortis is very vulnerable to this type of collection. The largest individuals, including ovigerous females, were taken. There was a noticeable reduction in the adult crayfish population (5) which has been slow to recover (4).

CONSERVATION MEASURES TAKEN In 1980 Pacifastacus fortis was listed as a rare species by the California Fish and Game Commission and is protected from take, possession, or sale within the State. Other regulations prohibit the take, possession or use for bait of any crayfish species at any time of year within the range of P. fortis. These regulations were enacted to protect P. fortis and prevent the spread of exotics by unintentional introductions.

CONSERVATION MEASURES PROPOSED A study of the ecology, life history, behaviour, and interaction with O. virilis and P. leniusculus would supply practical information to answer some of the questions concerning the survival potential of P. fortis. Because knowledge of food habits will be important in developing a viable management plan, studies to identify the components of the diet of P. fortis need to be initiated (5). Attempts should be made to transplant the species into isolated waters and legal protection is required to prevent over-exploitation by collectors (4).

REFERENCES 1. Bouchard, R.W. (1977). Distribution, ecology and systematic status of five poorly known western North American crayfishes (Decapoda: Astacidae and Cambaridae). In Lindqvist, O.V. (Ed.), Freshwater Crayfish. Papers from the Third International Symposium on Freshwater Crayfish, 1976, Kuopio, Finland. University of Kuopio. Pp. 409-423.

2. Bouchard, R.W. (1978/1979). Taxonomy, distribution and general ecology of the genera of Northern American crayfishes. Bull. Amer. Fisheries Soc. 3(6): 11-19.

3. Daniels, R.A. (1980). Distribution and status of crayfishes in the Pit River drainage, California. Crustaceana 38(2): 131-138.

4. Daniels, R.A. (1981). In litt., 4 March.

5. Eng, L.L. and Daniels, R.A. (1982). Life history, distribution and status of Pacifastacus fortis (Decapoda: Astacidae). Calif. Fish Game 68(4): 197-212.

6. Hobbs H.H. (1974). A checklist of the North and Middle American crayfishes (Decapoda: Astacidae and Cambaridae). Smithsonian Contributions to Zoology 166. 161 pp.

7. Riegel, J.A. (1959). The systematics and distribution of crayfishes in California. Calif. Fish Game 45: 29-50.

8. Schwartz, F.J., Rubelmann, R. and Allison, J. (1963). Ecological population expansion of the introduced crayfish, Orconectes virilis. Ohio J. Sci. 63: 266-273.

We are very grateful to R.W. Bouchard, R.A. Daniels and L.L. Eng for information supplied for this data sheet.

GIANT FRESHWATER CRAYFISH   VULNERABLE

Astacopsis gouldi Clark, 1936

Phylum   ARTHROPODA   Order   DECAPODA

Class   CRUSTACEA   Family   PARASTACIDAE

SUMMARY The largest freshwater crayfish in the world, Astacopsis gouldii has a limited distribution in north-west Tasmania, Australia. It has been depleted through over-exploitation and habitat alteration. Collection is controlled under Tasmanian fisheries legislation and it occurs in one small reserve but additional reserves are required to provide fuller protection.

DESCRIPTION Adults may attain a length of up to 40.5 cm (1), and individuals weighing 3 kg are not uncommon (2,4). The species is characterized by the conspicuous longitudinal carina in the centre of its rostrum (1,6). The body has spines or tubercles and a telson that is entirely calcified. There is one other species in this genus (10).

DISTRIBUTION Ranges over the western half of the north coast of Tasmania, Australia (6) in rivers and streams entering Bass Strait and in the Arthur River system in the north-west (9). It is absent from the Central Highlands and rivers of the south-east coast (3,5). The type locality is at Circular Head (1,6).

POPULATION Unknown.

HABITAT AND ECOLOGY A cold water species, it prefers deep, still pools with cover available under logs or vegetation, but may also be found in small, swiftly running streams and as far downstream as tidal waters (3). It can survive in a temperature range of at least 3-21°C (10). Animal material is probably used as food when available, although stream vegetation may also be eaten (10). Little is known of its breeding ecology. In rivers of north-western Tasmania, males and females occur in pairs between July and September (10), and spawning takes place in spring following moulting (11). Newly hatched young are found attached to females in late spring (1).

SCIENTIFIC INTEREST AND POTENTIAL VALUE Since its carapace is quite large in relationship to its tail (the most valuable part of the meat) the suitability of this animal for aquaculture seems to be limited (2), but it is occasionally taken for food. It is the largest freshwater crayfish in the world (10).

THREATS TO SURVIVAL Population numbers have decreased, probably as a result of habitat alteration, particularly the removal of vegetation cover along streams (7,10). Like other Australian crayfish species, it is very susceptible to the crayfish plague fungus Aphanomyces astaci (12). If this were to be introduced into Tasmania it could have very serious consequences for Astacopsis gouldi, given this species's limited range.

CONSERVATION MEASURES TAKEN Collection is regulated by the Inland Fisheries Commission and export is prohibited, except under permit, by the Tasmanian authorities (2). There is a closed season between May and July and during the fishing season only twelve Giant Freshwater Crayfish may be taken on any one day by one person. The legal minimum size is 4.5 inches (10 cm) carapace length, and specimens may not be offered for sale; fishermen must hold an inland angling licence. Female Giant Freshwater Crayfish in berry may not be taken (10). A reserve was established in 1968 at Caroline Creek, Mersey River drainage, North Tasmania, for the purpose of studying the habitat of the species in order to propose effective conservation measures. The area reserved consists of two parts, one surrounded by an exotic pine plantation and the other by eucalypts and ti-tree (10).

CONSERVATION MEASURES PROPOSED Additional reserves are required in suitable areas (7). It has been suggested that one could be established in north-west Tasmania, for example on the headwaters of the Duck River and the Montago River. This could be managed jointly by the Tasmanian National Parks and Wildlife Service and the Inland Fisheries Commission. In the reserve, habitat alteration and pollution, especially from pesticides, should be controlled and sufficient surveillance carried out to prevent poaching(8).

CAPTIVE BREEDING Attempts were made to maintain captive populations at Salmon Ponds, Plenty, but these failed mainly because of the species's aggressiveness which led to individuals killing each other (5). Specimens obtained from the Flowerdale River in 1960 died. In 1962 two adult females in berried condition were transferred from Caroline Creek to Plenty. One produced young but subsequently all died. In 1967 three crayfish were brought from the Blythe River to Salmon Ponds but these also died (10).

REFERENCES
1. Clark, E. (1938). Tasmanian Parastacidae. Papers and Proceedings of the Royal Society of Tasmania. Pp. 117-127.
2. Frost, J.V. (1975). Australian Crayfish. In Avault, J.W. (Ed.), Freshwater Crayfish. Papers from the 2nd Int. Symp. on Freshwater Crayfish, Baton Rouge, Louisiana, U.S.A. 1974. Pp. 87-95.
3. Gould, C. (1870). On the distribution and habits of the large freshwater crayfish (Astacus sp.) of the northern rivers of Tasmania. Proc. Roy. Soc. Tasmania August: 42-44.
4. Lake, P.S. (1974). Conservation. In Williams, W.D. (Ed.), Biogeography and Ecology in Tasmania. W. Junk, The Hague.
5. Lynch, D.D. (1969). The giant freshwater crayfish of Tasmania Astacopsis gouldi. Austr. Soc. Limnol. Bull. 2: 20-21.
6. Riek, E.F. (1969). The Australian freshwater crayfish (Crustacea: Decapoda: Parastacidae) with descriptions of new species. Aus. Journ. Zool. 17: 855-918.
7. Riek, E.F. (1981). In litt., 18 August.
8. Lake, P.S. (1981). In litt., 24 August.
9. Swain, R., Richardson, A.M.M. and Hortle, M. (in press). A revision of the Tasmanian genus of freshwater crayfish, Astacopsis Huxley (Decapoda: Parastacidae). Australian Journal of Marine and Freshwater Research 33.
10. Lynch, D.D. (1967). Synopsis of biological data of the giant freshwater crayfish Astacopsis gouldi Clark 1936. Inland Fisheries Commission, Tasmania.
11. Clark, E. (1936). The freshwater and land crayfishes of Australia. Mem. Nat. Mus. Vict. 10: 5-58.
12. Unestam, T. (1975). Defense reactions in and susceptibility of Australian and New-Guinean freshwater crayfish to European crayfish plague fungus. Aust. J. Exp. Biol. Med. Sci. 53(5): 349-359.

We are very grateful to P.S. Lake, E.F. Riek, and R. Swain for comments on this data sheet.

COCONUT or ROBBER CRAB   RARE

Birgus latro (Linnaeus, 1766)

Phylum   ARTHROPODA   Order   DECAPODA

Class   CRUSTACEA   Family   COENOBITIDAE

SUMMARY Probably the largest terrestrial arthropod in the world, this hermit crab is found throughout many of the islands of the Indo-Pacific. It has been intensively hunted by the local inhabitants in many places, and is reported to be extinct on a number of islands. Although detailed information is lacking there is a strong suggestion that it is declining in areas with heavy human populations. As well as being collected for food Coconut Crabs are taken for sale as curios. Protection should be afforded to this species in vulnerable areas.

DESCRIPTION This monospecific genus of land hermit crab varies in colour from purplish-blue to orange-red. For example, on Aldabra red individuals are more common than blue ones (42), but on South Sentinel all individuals are reported to be blue (41). The carapace is sharp-fronted and swollen posteriorly. The animal has a bulbous tail, relatively long and strong legs and large red eyes. Unlike most land or marine hermit crabs the adult carries no gastropod shell, although juveniles do so until they reach a certain size (8,10,13). The body is firm, symmetrical (except for unequal claws) and linearly arranged. By abandoning the shell-carrying habit it can grow to relatively gigantic proportions - crabs weighing 7 lbs (3 kg) and measuring 3 feet (1 m) from leg tip to leg tip are not unusual (13). Specimens collected from islands near Cebu, in the Philippines, reached weights of up to 3.5 kg (16), and weights of 15 kg have also been reported (29). An experiment showed that the animals can lift at least 28 kg (1). Sexual dimorphism is shown in the maximum size attained; in a study on Guam the largest female seen had a thoracic length of 4.7 cm whereas the largest male had a thoracic length of 7.6 cm (2); similar size differences were found on Aldabra (44).

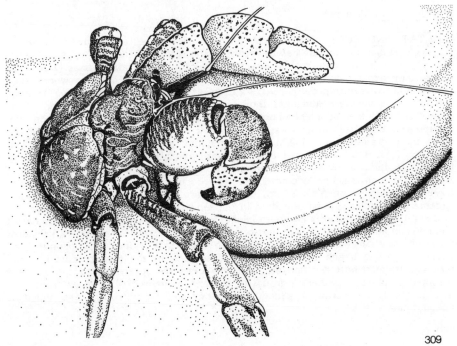

DISTRIBUTION Birgus latro is widely distributed throughout the Western Pacific and eastern Indian Oceans, occurring almost exclusively on oceanic islands or on small offshore islets adjacent to large continental islands (2). However, although it has a wide geographic spread, it probably does not occur on all atolls since those that are dry do not provide enough food (31).

In the Indian Ocean, it is found on islands south of the equator (21). Still occurs abundantly on Aldabra (43) but is extinct on almost all the islands in the Seychelles group (23); occurs on some of the numerous small islands off the Tanzanian coast (5); in the Andaman and Nicobar Islands now found only on South Sentinel Island (21,22) and at Galathea Bay where it was said to be common in 1967 (35,40); formerly occurred on Mauritius but exterminated in the second half of the 19th century; in 1939 almost exterminated in the Keeling Islands (except for uninhabited North Keeling I.) (21); in the Chagos Archipelago, found on both of the northern atolls and on most of the islands of the Great Chagos Bank (25); widely distributed on Christmas Island (4).

Has not been recorded from the East African mainland coast, the Maldives and Laccadives, the islands near Peninsula Malaysia and Sumatra, the western part of the Malay Archipelago nor the islands in the South China Sea (21). In 1981, found on Sipadan Island, about 30 miles (48 km) off the east coast of Sabah, Malaysia (33). In 1939 was still present in parts of Indonesia, the Philippines and Taiwan (21). Currently is found on Olango Island, off Mactan (Cebu) in the Philippines (28). In Indonesia it has recently been reported from the Togian Islands, Central Sulawesi (24) and in 1963 from the north coast of Irian Jaya (15). A detailed account of its distribution in the 1950s on the small islands off the north coast of Irian Jaya is given in (46). In Papua New Guinea has been reported from Rantan and Sae Islands (32) and Los Negros Island (30) in Manus Province.

In the Pacific found in Fiji and the Marshall Islands but not in the Hawaiian group, Wake or Midway (13,31). Recorded from Gardner Island in the Phoenix Islands (7), the Ellice Islands and Tuamotus (31), Fanning Island (14), the Marianas (2) and Vanuatu (36,37).

POPULATION Unknown.

HABITAT AND ECOLOGY Almost entirely terrestrial and drowns in water (11,44), although the female has to return to the sea to release her eggs and crabs may go to the beach and drink seawater to maintain their salt balance (2,26). On Guam they are most typically found in coastal limestone forests where they establish burrows within the porous, solution-pitted limestone substrate. On some islands in the northern Marianas they occupy burrows dug in the soil or in the interstices of the boulder-cobble shorelines (2). On Olango Island in the Philippines, they live in burrows in coral rock in thick undergrowth. These are 0.5-1 m in diameter and 1-2 m deep, and may extend 4-6 m in a horizontal direction (28). On barrier reef islands crabs have been found to live in shallow burrows in the substrate or hidden among Pandanus roots and fallen coconut fronds. On such islands Coconut Crabs may occur throughout the island, but on larger high islands, such as Guam, they are rarely found in the interior (2). On Aldabra the species is most abundant in the sandy coconut grove at Anse Mais and in damp Pandanus thickets on South Island, but is also found in the barren coastal champignon and throughout the platin. Normally the crabs inhabit rock crevices, although they can burrow in sand (42). On Christmas Island, the Coconut Crab occurs from the beach through the shore terrace to the highest parts of the plateau. Its distribution is uneven and some areas are apparently avoided. Crabs frequently collect together in groups of a dozen or more (10). Larger crabs may inhabit the best areas of an island, with sandy soil for burrowing and abundant coconut palms (13).

Burrows, which are inhabited during the day, provide protection from desiccation and can be defended against conspecific intruders. At night crabs bring large pieces of food, such as coconuts and pandanus fruits, to the burrow where they may remain for several days feeding, unlike other hermit crabs which feed communally. It is not clear whether individual crabs maintain the same burrow for long periods of time, or whether they just enter any available burrow (2,13). Before native rats disappeared from Christmas Island, Coconut Crabs tended to climb into low bushes or up tree trunks at night, presumably to escape predation (3). As they increase in size they inhabit larger burrows. When threatened they retreat into their burrows and may seal the entrance with a claw (13). In competition over food the largest crab always wins, and compared with other hermit crabs they have a simplified behavioural repertoire and are basically asocial.

The crabs feed largely on rotting coconuts on the ground, inserting their pincers into the hole at the top of the seed and scooping the flesh out (1,29,44). However they are also successful on islands with no coconuts and will apparently scavenge anything organic (13), eating fallen leaves and fruit (9,40) particularly Pandanus (42). In the Togian Islands, Sulawesi, they feed on vegetation and probably on the nuts of Terminalia catappa (34). Although they feed mainly on plant material (21), they also eat the moulted exoskeletons of other crustaceans which probably provide them with calcium for growth and thickening of the carapace (2,13). On Aldabra there are a few records of crabs feeding on dead tortoises (42) and in turtle nesting areas they are predators of hatchlings (43). They may also be cannibalistic, a factor which could help to keep population numbers down, although in Guam no cannibalism was observed (2,13). Although under natural conditions Coconut Crabs usually seem to be nocturnal, on uninhabited islands in the northern Marianas they are also active during the day which may be related to human predation by man (2,28); in New Guinea they have been reported to emerge on days with heavy rain or cloud (15); on South Sentinel island in the Andamans they were seen during the day only within gloomy vegetation (1).

Most hermit crabs engage in lengthy courtship but mating in Coconut Crabs is quick, simple and infrequent (13). On the basis of spermatophore morphology, it has been suggested that copulation occurs in the water (17), but field observations suggest that it occurs on land (2,13). Subsequently the fertilized eggs are extruded from the female's body and carried beneath her abdomen, held in place by three specialized abdominal appendages (2). When the eggs are ready for hatching the female walks down to the edge of the sea and releases the larvae (2,19), usually at night and at high tide. Larval release may be related to lunar and tidal rhythms (13,44). It may well be selectively advantageous for an animal such as the Coconut Crab, which releases its larvae in inshore waters, to do so during spring tide to ensure the greatest opportunity for the eggs to be flushed off the reef flat where egg predation may be quite severe, out to open ocean waters where predation may be reduced and a more constant supply of edible phyto- and zooplankton may be available for the developing larvae. By releasing the larvae at night the risk of predation is probably lower. The larval stages consist of a pelagic phase lasting 17-28 days and an amphibious phase of 21-28 days during which the young crabs migrate on to land having occupied gastropod shells (13,18,19). They compete with the young of other hermit crabs for food and shells, increasing in size and changing shells with each moult (2,13). It has been reported that juveniles will use a half coconut as a shell on reaching a size when they are too large for any mollusc shell (26).

Usually when they reach a size where the carapace measures about one inch (2.5 cm) across, the shell-carrying habit is given up (2,13). On Enewetok the smallest crab found without a shell was 2.2 cm in carapace length and on Guam approximately 0.84 cm (2). Young Coconut Crabs in captivity have been reported carrying shells for about 2.5 years (19). The shell-living habit is probably retained in the young to protect it from desiccation, and also possibly from predation, during this vulnerable stage in its life history (18).

Coconut Crabs are reported to be slow growing (9). Extrapolation of the growth figures obtained suggests that crabs in excess of 10 cm are about 5 years old (12). A study of moulting in captivity revealed that the crab remains hidden in its burrow, which it plugs with soil, for about 30 days during this period. The abdomen becomes swollen prior to this, presumably with a stored source of energy; the exuviae are eaten in the burrow before the crab emerges (12). On Guam it appeared that crabs had specific areas for moulting where they buried themselves in the soil (2); similar behaviour is reported from Aldabra (44).

SCIENTIFIC INTEREST AND POTENTIAL VALUE Probably the largest extant terrestrial arthropod (13), this unique and fascinating component of many Indo-Pacific islands is among the largest of terrestrial animals on islands uninhabited by man, and is relatively free of predators. On coral islands it is an important scavenger and its burrowing activity is probably helpful in moving surface organic matter underground (31). It has been used extensively by man for food and as a fishing bait, and is an important element in the Chamorro culture in the Marianas (2,34). Dried and mounted specimens are popular as tourist souvenirs (2,4,10); for example, in the 1960s Coconut Crab carapaces were exported to Hong Kong from Irian Jaya (15). Its abdominal fat is alleged to be an aphrodisiac (1).

THREATS TO SURVIVAL The Coconut Crab has been exterminated from parts of its range as a result of intensive hunting by coral island inhabitants, who consider it a delicacy and an aphrodisiac. It is now uncommon throughout much of its present range although it may be abundant on smaller and less densely inhabited islands (13,24,31,15). Breeding populations on small islands could be eliminated easily through intensive collection (34). Introduced pigs, rats, monitor lizards and monkeys have been implicated in Coconut Crab predation but their effects must be felt most strongly by the younger crabs (13).

It is thought that the decline of its range along the Indian coast is due to over-exploitation and disturbance from expanding human settlements (22,35). In the Chagos Archipelago in 1975, the species was most abundant and individuals were larger on the islands of the Great Chagos Bank which were evacuated in the 1930s; on the northern atolls of Peros Banhos and Salomon, which were cultivated until the 1970s, individuals were smaller and fewer in number. At present only the southern atoll of Diego Garcia is inhabited and the future of the species on the northern part may be assured (25), but it should probably be protected, since it is still eagerly collected by visitors and residents of the Archipelago (27). On Christmas Island preserved specimens injected with formalin are sold and lacquered, or given to island residents or the crews of visiting ships and have also been exported to Singapore for sale. The curio market leads to the selection of the larger specimens (in contrast within Guam where small crabs are taken: see below) and is often wasteful as crabs are discarded if damaged during preparation (4).

In the 1950s it was feared that the Coconut Crab might be exterminated through over-collecting in the part of Indonesia known as the Dutch East Indies; however a survey showed that it was still present in almost all the localities from which it had been recorded two and a half centuries before (20). On Olango Island in the Philippines, fishermen used to catch an average of ten crabs a night but this is no longer possible (28). In Papua New Guinea it may be threatened by over-collection for food (32). In Vanuatu Coconut Crabs command high prices in hotels and restaurants and are caught in large numbers (37); populations are showing a marked decline in some areas (39).

On Guam, which supports a large human population and which has been subject to considerable environmental modification, Coconut Crabs have become scarce. The development of the coastal zone during recent years has modified or destroyed much of its preferred habitat. Coastline alteration reduces the total availability of Coconut Crab habitats, reduces access to the ocean for gravid females and inhibits recolonization by juvenile crabs. Crabs are collected by setting out coconuts as bait and waiting until night when the crabs come out. Harvesting pressure is probably heaviest on larger crabs since these are most in demand for food and also because large crabs tend to drive smaller crabs away from the bait. A few small crabs are collected to be dried, lacquered and sold as tourist souvenirs. There is a continuous but rather specialized demand for crab as a luxury food item, and since the Guam population cannot satisfy demand, crabs are imported from the northern Mariana Islands. Coconut Crabs on the islands of Asuncion and Guguan in the northern Marianas appear to be flourishing presumably because these islands are uninhabited (2).

More than a decade after the nuclear bomb tests of the 1950s on Bikini and Eniwetok, Coconut Crabs were still 'hot' because they eat their own exoskeletons following a moult in order to recycle the calcium; radioactive strontium was therefore being recycled long after radiation on other parts of the island had returned to acceptable levels (13).

CONSERVATION MEASURES TAKEN Export for the tourist curio trade from Christmas Island has been banned (10). In Papua New Guinea villagers in some areas have been asked not to collect crabs for food in view of their rarity (32). On Saipan in the northern Mariana Islands, Municipal Ordinance No. 25-22-1974 prohibits sale of Coconut Crabs for any purpose except human consumption, the collection of crabs with carapace widths less than 3 inches (7.5 cm), and collection of crabs of any size between 1 June and 30 September, the period of greatest reproductive activity. Public Law No 1-18 prohibited collection on Aguiguan Island for three years from 1979; after that if stocks are thought to be large enough to justify harvesting, they could be collected as authorized by the mayor of Tinian. Similarly, on the islands north of Saipan, a one year moratorium was introduced in 1977 after which harvesting was allowed on a regulated basis (2). Coconut Crabs on Aldabra are protected (24,45).

CONSERVATION MEASURES PROPOSED Surveys must be carried out to determine the current distribution of the Coconut Crab and to ascertain the extent to which it is being collected for use as food or curios. Protection is probably required on a number of islands (11) and could be most easily implemented on uninhabited islands such as South Sentinel Island in the Andamans (1). A proposal for a reserve in the Togian Islands, Sulawesi, is being drawn up (34). Legislation is now being prepared in Vanuatu which will protect females fully and prohibit the taking of males with a thorax length of less than 14 cm (38).

Detailed recommendations have been made for the conservation and management of Coconut Crab populations on the Mariana Islands which may also be appropriate for other areas. Sanctuaries should be established where the harvesting of crabs is prohibited. These should be reasonably large and have broad access to the ocean, and should include limestone forest and coconut dominated habitat. Various of the uninhabited northern islands in the Commonwealth of the Northern Mariana Islands, such as Guguan and Ascuncion, would be appropriate. Sanctuaries already exist on Guam in the form of large military installations which control tracts of relatively undisturbed natural crab habitat. Few military personnnel are interested in hunting the crabs and civilian access is limited on such bases. Three proposed conservation areas to be designated under the Guam Comprehensive Development Plan would also be appropriate as sanctuaries: the Seashore Park and Wildlife Reserve near Umatac, the Open Space areas on the south-east coast and the Open Space and Wildlife Reserve near Yigo. Harvesting should be discouraged in such areas and annual surveys should be made to assess stock densities (2). The collection of crabs smaller than the median size of female reproductive activity (carapace length of 90 mm) should be prohibited, as should the sale of mounted crabs as souvenirs. This latter use competes with the use of crabs as food. Collecting Coconut Crabs on beaches should be prohibited within two days prior to and six days after the new moon; this would permit female crabs to release their eggs into the sea undisturbed (2).

CAPTIVE BREEDING   Studies in the Andaman and Nicobar islands have indicated that it may be possible to farm Coconut Crabs for commercial purposes (6), and tentative experiments on their culture have been carried out at the Marine Station in Maribago, Philippines (16). However, to date this has not been successful, and further research is required on growth rates and techniques for mass culture of the larval stages. If this is found to be feasible, young crabs raised from larvae could be used to restock natural areas or transferred to 'grow-out' facilities for further rearing. Rearing of terrestial stages would probably be most successful in a large area with a variety of natural conditions available (2).

REFERENCES   1.   Altevogt, R. and Davis, T.A. (1975). Birgus latro, India's monstrous crab. A study and an appeal. Bull. Dep. Mar. Sci. Cochin. 7: 11-23.
2.   Amesbury, S.S. (1980). Biological studies on the coconut crab (Birgus latro) in the Mariana Islands. University of Guam, Marine Laboratory Technical Report no.66.
3.   Andrews, C.W. (1900). A Monograph of Christmas Island (Indian Ocean). British Museum (Nat. Hist.), London. 337 pp.
4.   Anon. (1974). Conservation of endangered species on Christmas Island. Report from the House of Representatives Standing Committee on Environment and Conservation. Australian Government Publishing Service, Canberra. 45 pp.
5.   Bryceson, I. (1981). A review of some problems of tropical marine conservation with particular reference to the Tanzanian coast. Biological Conservation 20: 163-171.
6.   Chatterjee, S.K. (1977). Wildlife in the Andaman and Nicobar Islands. Tigerpaper 4(1): 2-5.
7.   Ellis, A. (1936). Adventuring in Coral Seas. Angus and Robertson, Sydney, Australia.
8.   Gibson-Hill, C.A. (1947). Field notes on the terrestrial crabs. Bull. Raffles Mus. 18: 43-52.
9.   Gillison, A.N. (1975-76). Report on the conservation of vegetation on Christmas Island - Indian Ocean. Unpublished report.
10.   George, R.W. (1978). The land and freshwater crabs of Christmas Island. Unpublished report.
11.   Harms, J.W. (1932). Die Realisation von Genen und die konsekutive Adaption II Birgus latro L., als Landkrebs und seine Beziehungen zu den Coenobitiden. Z. Wiss. Zool. 140: 167-290.
12.   Held, E.E. (1963). Moulting behaviour of Birgus latro. Nature 200: 799-800.
13.   Helfman, G.S. (1979). Coconut crabs and cannibalism. Natural History 88 (9): 77-83.
14.   Herms, N.B. (1926). Diocaladra taitensis (Guerin) and other coconut pests of Fanning and Washington Islands. Philippine J. Sci. 30: 243-71.
15.   Holthuis, L.B. (1963). Contributions to New Guinea Carcinology IV. Further data on the occurrence of Birgus latro (L.) in west New Guinea (Crustacea, Decapoda, Paguridae). Nova Guinea, Zoology 18; 355-359.
16.   Horstmann, U. (1976). Some aspects on the culture of the

coconut crab (Birgus latro). Unpub. paper, University of San Carlos, Philippines. 10 pp.

17. Matthews, D.C. (1956). The probable method of fertilization in terrestrial hermit crabs based on a comparative study of spermatophores. Pacif. Sci. 10: 303-309.

18. Reese, E.S. (1968). Shell use; an adaption for emigration from the sea by the coconut crab. Science 161: 385-386.

19. Reese, E.S. and Kinzie, R.A. (1968). The larval development of the coconut or robber crab Birgus latro (L.) in the laboratory. Crustaceana Supplement 2. Studies on Decapod Larval Development. Pp. 117-144.

20. Reyne, A. (1939). On the distribution of Birgus latro L. in the Dutch East Indies. Archives Neerlandaises de Zoologie 3: 239-247.

21. Reyne, A. (1939). On the food habits of the Coconut Crab with notes on its distribution. Archives Neerlandaises de Zoologie 3: 283-320.

22. Saharia, V.B. (Ed.), (1980). Wildlife in India. Dept. of Agriculture and Co-operation, Min. of Agriculture, Government of India.

23. Salm, R.V. (1978). Conservation of Marine Resources in Seychelles. Report on current status and future management for the Government of Seychelles. IUCN, Morges, Switzerland.

24. Salm, R.V. (1981). In litt., 4 April.

25. Sheppard, C.R.C. (1979). Status of three rare animals on Chagos. Envir. Conserv. 6(4): 310.

26. Sheppard, C.R.C. (1980). A shell for rent. Underwater World Sept. 1980: 30-31.

27. Sheppard, C.R.C. (1981). In litt., 8 April.

28. Storch, V., Cases, E. and Rosito, R. (1979). Recent findings on the coconut crab, Birgus latro (L.). Philippine Scientist 16: 57-67.

29. Vogel, H.H. and Kent, J.R. (1970). Life history, behaviour and ecology of the coconut crab, Birgus latro. Bull. Ecol. Soc. Amer. 51:40.

30. Wells, S.M. (1981). Pers. obs., July.

31. Wiens, H.J. (1962). Atoll environment and ecology. Yale, 532 pp.

32. Whitaker, R. (1979). A preliminary survey of the Crocodile resources in the island provinces of Papua New Guinea (Appendix) UNDP Project.

33. Philips, A. (1981). Pers. comm.

34. Salm, R. (1982). In litt., 13 March.

35. Anon. (1981). Rare and Endangered Animals of India. Zool Surv. Ind., Government of India.

36. Baker, J.R. (1929). Man and Animals in the New Hebrides, Routledge.

37. Dickinson, D. (1981). Conservation for whose benefit? Naika 1: 4.

38. Dickinson, D. (1981). Vanuatu's Laws on Conservation. Naika 2: 2.

39. Dickinson, D. (1981). In litt., 8 November.

40. Daniel, A. and Premkumar, V.K. (1967). The coconut crab Birgus latro (L.) (Crustacea: Paguridae) in the Great Nicobar Island. J. Bombay Nat. Hist. Soc. 64: 574-579.

41. Davis, T.A. and Altevogt, R. (1976). Giant turtles and robber crabs of the South Sentinel. Yojana (Planning Commision, India) 15 August : 77-78.

42. Grubb, P. (1971). Ecology of terrestrial decapod crustaceans on Aldabra. Phil. Trans. Roy. Soc. Lond. B. 260: 411-416.
43. Alexander, H.G.L. (1979). A preliminary assessment of the role of the terrestrial decapod crustaceans in the Aldabran ecosystem. Phil. Trans. Roy. Soc. Lond. B. 286 : 241-246.
44. Alexander, H.G.L. (1976). An ecological study of the terrestrial decapod crustaceans of Aldabra. Ph.D. Thesis, University of London.
45. Stoddart, D.R. and Morris, M.G. (1980). A Management Plan for Aldabra. Royal Society, London.
46. Holthuis, L.B. (1959). The occurrence of Birgus latro (L.) in Netherlands New Guinea (Crustacea: Decapoda Paguridea). Contributions to New Guinea Carcinology III. Nova Guinea N.S. 10: 303-310.

We are very grateful to H.G.L. Alexander, A.K. Ghosh, C. Huxley, R.V. Salm and C.R.C. Sheppard for information supplied for this data sheet.

INSECTA

Insects

INTRODUCTION The Insecta is one of five classes in the subphylum Uniramia.
The Uniramia comprises those arthropods which have mandibles, one pair of
antennae, and other appendages which are primitively unbranched. The other four
classes in the subphylum are the Chilopoda, Diplopoda, Pauropoda and Symphyla,
which are collectively known as the myriapoda, although they are quite distinct
from each other. Uniramians occur worldwide and are predominantly terrestrial,
but with many freshwater and a few marine representatives (4,54).

The class Insecta includes about 0.75-1 million described species, which are
commonly believed to represent less than half of all living insects, the rest being
undescribed. However, opinion on this total number varies very widely because of
constant new discoveries and extrapolations, particularly from work in the
tropics. The Insecta is not only the biggest single class in the animal kingdom, but
it also exceeds the number of all other animal species combined. The variety of
form almost defeats general description, but insects may be defined as arthropods
with a tracheate respiratory system, and a body divided into a head with one pair
of antennae, a thorax with three pairs of legs and up to two pairs of wings, and an
abdomen (11). Insects occur almost everywhere in the terrestrial and freshwater
world, but are poorly represented in marine habitats and absent from the ocean
depths (4). Their success is attributed largely to the evolution of flight, which has
improved dispersal, escape from predators, and access to food and optimal
environmental conditions (4). Insects feed on every imaginable source of
nourishment, including the leaves, galls, stems, sap, wood, roots, pollen and fruits
of plants, the blood and tissues of invertebrate and vertebrate animals, fungi,
protozoa, yeasts and bacteria, and the processed or decomposing dead remains of
plants and animals, including timber, books, humus, leather, bones, keratin,
corpses and excreta (11,35,63,72). The beetle Niptus hololeucus can survive on
cayenne pepper or sal ammoniac and the fly Psilopa petrolei inhabits pools of
crude petroleum (15).

SCIENTIFIC INTEREST AND POTENTIAL VALUE The substantial benefits to
mankind from insects generally receive less publicity than the economic losses
that result from their roles as vectors of diseases (63) and as pests of agriculture,
fibres and stored goods. The main benefits are ecological, economic, scientific
and aesthetic (57).

Because of their vast numbers and variety, insects are dominant components of
many food webs in both the production and decomposition divisions of ecosystems.
The biomass of insects in most freely-draining soils far exceeds that of the more
evident birds and mammals above the surface. Because insects are small in size,
yet so diverse and ecologically important, they may be affected by relatively
minor perturbations of their habitats. Since environmental quality has become of
major concern, terrestrial and particularly aquatic insects have therefore been
utilized as valuable indicators of ecological conditions (6). Conspicuous insects
such as butterflies and dragonflies are particularly useful in monitoring changes;
the decline of many European butterflies, although mainly due to loss of suitable
habitat, has been linked to acid rain resulting from industrial pollution (33).

In both ecological and economic terms insects are useful as pollinators. Many wild
plants depend upon insects, and those plants are part of a genetic pool which may
be of known or as yet undiscovered value. Bees are the most important
pollinators, and are vital for the reproduction of many crops (21,27). Several other
orders of insects also include pollinators, and have evolved as a result of mutually

317

beneficial developments in insects and flowering plants since the Triassic period.

The natural control which predatory or parasitoid insects have over phytophagous insects, and which phytophagous insects have over their hostplants, has often been under-estimated in the past. Many potential insect and plant pests are held in check by natural enemies, a balance that may not be evident until those controls are removed. This benefit is probably the greatest economic contribution of the insects, although it is virtually impossible to evaluate. One of the disadvantages of monocultural methods of agriculture, particularly of exotic crops, is that natural enemies may be absent or unable to control their host insects and weeds, which consequently reproduce unchecked to pest proportions. Under these circumstances biological control by introduction or enhancement of predatory, parasitoid or phytophagous species may be preferable to insecticidal or herbicidal treatment. Sometimes integrated control may be required, using a combination of natural, biological and chemical methods. Biological and integrated control methods have become increasingly important since the unselective effects and ecological drawbacks of chemical insecticides have been realized. Insects and other arthropods are the most common agents in biological control programmes. The Commonwealth Institute of Biological Control has over the past 25 years developed and utilized biological pest control programmes worldwide. The economic benefits far exceed the cost of development, and the methods are cheaper than chemical application both in terms of money and energy. Some of the more spectacular programmes are estimated to save about £1 million per year. These include control of noctuid moths (Mythimna) in New Zealand vegetables and pastures, control of coconut leaf-mining beetles (Promecotheca) by eulophid wasps, and control of winter-moth (Operophtera) in deciduous trees in Canada (16,17). Important successes against weeds include the elimination of prickly pears (Opuntia) from Queensland, the West Indies and other islands by the moth Cactoblastis cactorum; control of lantanas (Lantana spp.) in Uganda and elsewhere by the lacebug Teleonemia scrupulosa and control of waterweeds in Florida, on the Nile and in some of the East African lakes by a variety of beetles, mites and grasshoppers (16,17).

Many cultures derive part of their nourishment from insects or their products. Beetle grubs, grasshoppers, locusts, termites and some lepidopterous larvae and pupae are consumed. Scale insects (Homoptera: Coccoidea) produce excremental manna, which is still collected from the desert floor in Sinai (72). Apiculture is of value both in the production of honey and in the pollination of crops (21,27). Insects have been important in the development of cloth and dyes. Substantial economic benefits accrue from the ancient skill of sericulture (2): silkworms, which are the caterpillars of various saturniid moths, notably Bombyx mori, have been in domestic use longer than any other insect. They were originally used in ancient China and the industry is said to have been founded in 2640 BC (2). Scale insects were the source of the cochineal (carmine) dye formerly used in Mexico by the Aztecs and in ancient Greece and Rome (72), but they have now been replaced by synthetic dyes.

Scientifically and aesthetically, insects make a substantial contribution in terms of their fascination, their beautiful or bizarre appearance, their improvements to man's environment, their usefulness in teaching and in biogeographical (10,38), environmental (6), synecological and autecological investigations. Economic returns indirectly accrue from the essential basic research on genetics, evolutionary biology and medicine in which Drosophila and various Lepidoptera are often used as biological tools.

In this short review it is impossible to cover fully the wide range of man's interest in insects, but it is evident that their impact is considerable, and that the known or potential value stored in their diversity should be preserved for the future.

THREATS TO SURVIVAL The causes of the decline and extinction of insects may be broadly divided into two categories, natural threats and man-induced threats. Natural changes in the populations of insect species, the evolution of new species and the extinction of unsuccessful species through natural selection, immigration and emigration are beyond the scope of this review. Man-induced changes are currently by far the most important factors affecting insects worldwide; they operate on a time-scale drastically foreshortened from the ponderous natural changes over geological time. Through technological innovation, man is able to create or destroy landscapes and biotopes within a geological instant, permitting no opportunity for natural selection of adaptable genotypes. Man-induced changes that affect insects are here divided into five categories: 1) changes to land, 2) changes to water, 3) atmospheric pollution, 4) changes to closely associated fauna and flora, and 5) specific threats to the species in question. The first four categories, divided into ten sections, deal with the insects's habitat or biotope, the components which make up the integrated network of physical and biological parameters of which the species is a part, and which are essential to the species' survival. Impact in these categories is generally created as a direct result of, or enhanced by, increasing human population pressure. The fifth category, divided into two sections, includes threats peculiar to the species itself, but not necessarily to its environment. In this document only a few examples of threatened insects can be given, and the lists are by no means comprehensive.

1. Changes to the terrestrial environment.
1a. Forestry.
Deforestation of tropical equatorial regions is widely considered to be the greatest single threat to insect species. Tropical rain forest covers an area of only 9.35×10^8 ha (64), only 6.3 per cent of the 14.9×10^9 ha which comprise the total of all land on Earth (5), yet at least 50 per cent of the world insect fauna is believed to live within the tropical moist forest boundaries (49). Despite this great density of species, there is evidence that population densities are often lower than is found in temperate species (22,23), and ranges may be very small. It has been estimated that tropical moist forests are being converted at a rate of between 73×10^5 (24) and 20×10^6 ha per year (50), depending on the criteria used. This represents between 0.8 and 2.1 per cent of the total available area. The loss of insect species is inevitable in the wake of conversion on this scale, but the impact is largely undocumented. The invertebrate faunas of the tropical forest regions are very poorly known and a large proportion of species remain undescribed (see review of the Rain Forests of Gunung Mulu). In the species reviews which follow, the emphasis on temperate species is an artefact which reflects the lack of data from the tropics, and not the actual world distribution of the extent of threats to insects.

Mixed reforestation of logged tropical forest is greatly to be encouraged, particularly when logging patterns maintain corridors of access to undisturbed areas. In this way disrupted insect populations may recover. However, the northern temperate practice of reforestation with coniferous monocultures generally reduces the diversity of insect species. In mountainous regions of Europe the ranges of the Apollo Butterfly (Parnassius apollo, see review) and Erebia species (brown and ringlet butterflies) have been severely restricted in this way (33). Plantation forestry often includes drainage of forest bogs, and Carterocephalus palaemon (the Chequered Skipper) seems to be susceptible to this change, at least over parts of its range (33).

1b. Agricultural conversion.
Agriculture is one of the most extensive methods of land use and has probably resulted in the greatest losses of native insect populations, particularly when it follows forest clearance. However, few extinctions have so far resulted from agriculture in temperate latitudes because of a low incidence of restricted endemicity in the recent alluvial soils best suited to farming. Many such areas

have only developed since the last Ice Age ended 10 000 years ago, a short evolutionary time compared to, for example, the 60 million years of relatively stable conditions over Africa from 20°N to the Cape. In most of lowland U.K. butterflies are now unable to survive the intensification of farming, and populations have been severely reduced (47). Intensification includes a range of factors detrimental to insects, including use of insecticides and herbicides, destruction of hedgerows, drainage, short fallow periods, and use of fertilizers. In North America the conversion of prairie to agriculture has had disturbing effects, but has caused no documented extinctions. The populations of two sphinx moths, Euproserpinus euterpe and E. wiesti, have been greatly reduced by planting of cereal crops (69) and more recently by insecticidal sprays (see review of Wiest's Sphinx Moth). In Nepal, agricultural encroachment in the Kathmandu Valley is threatening the Relict Himalayan Dragonfly (Epiophlebia laidlawi, see review). The greatest threats are in the tropics where largely undocumented reductions in populations and in species diversity may be occurring daily as a result of conversion of tropical forests and woodlands to agriculture. The conversion of lowland and montane forest in Hawaii to pineapple and sugarcane plantations has probably resulted in the loss of several hundred species of native insects (52, see review of picture-winged flies). In Jamaica timber operations and coffee plantations have divided and reduced the populations of the Homerus Swallowtail (Papilio homerus, see review).

1c. Pastoralism.
In Europe permanent grassland, especially calcareous grassland, is a very important habitat supporting many species of insects that cannot survive the reduction in plant species diversity which results from application of fertilizers (46). Over-grazing or the prevention of natural grazing can also reduce insect diversity. Improvement of grassland by application of fertilizer and herbicides has resulted in great reductions in the ranges of the large blue butterflies and the ants that host their caterpillars (see review of large blues). In the 1950s the introduction to Britain of myxomatosis in order to reduce rabbit grazing seriously depleted the populations of the butterflies Lysandra bellargus (Adonis Blue), Maculinea arion (Large Blue) and Hesperia comma (Silver-spotted Skipper). In Australia heavy grazing by sheep threatens an undescribed wingless grasshopper Achurimima sp., which is known from only six small populations inhabiting lightly grazed native pastures (37). Another grasshopper, Keyacris scurra, has been eliminated from vast areas of Australia since intensive sheep grazing has eliminated the Kangaroo Grass Themeda australis which it uses for shelter (37,41). Browsing by sheep may pose a threat to the immature stages of butterflies with very limited distributions, such as Boloria acrocnema (Uncompahgre Fritillary, see review) in U.S.A. Grazing is incompatible with Hesperia dacotae (Dakota Skipper, see review) because of the consequent loss of plants that are important sources of nectar. There are no data on the effects of pastoralism on the insects of the African or South American grasslands, but it may be surmised that escalating population pressure will lead to over-grazing and intensification of grazing, with possible deleterious effects on the insect fauna.

1d. Urbanization and industrialization.
The expansion of towns and cities as a result of population growth inevitably causes the destruction of natural habitat. Major threats to insects have resulted from the spread of cities like San Francisco and Los Angeles in California, U.S.A. Three endemic dune butterflies, Cercyonis sthenele sthenele, Glaucopsyche xerces (Xerces Blue) and Icaricia icarioides pheres are now extinct (67). The Antioch Sand Dunes in Contra Costa County, California, have been designated as Critical Habitat for the Antioch Dunes Evening-primrose (Oenothera deltoides howellii) and the Endangered Lange's Metalmark Butterfly (Apodemia mormo langei) (68). Furthermore, the well-being of the following insects is a matter of concern and has been the subject of enquiries by the U.S. Office of Endangered Species: Middlekauf's Katydid (Idiostatus middlekauffi), the Antioch Weevil (Dysticheus

rotundicollis), the Antioch Robber Fly (Cophura hurdi), the Valley Mydas Fly (Raphiomydas trochilus), the Antioch Vespid Wasp (Leptochilus arenicolus), the Antioch Tiphiid Wasp (Myrmosa pacifica), the Antioch Sphecid Wasp (Philanthrus nasalis), the Antioch Andrenid Bee (Perdita scitula antiochensis) and the Yellow-banded Andrenid Bee (Perdita hirticeps luteocincta). The Antioch Dunes have been almost completely destroyed by industrialization and two species are feared to be already extinct. The Antioch Katydid (Neduba extincta) is known only from a single specimen collected in 1937 (60), and the Antioch Anthicid Beetle Anthicus antiochensis has not been seen since 1953.

Recreational areas close to large cities are particularly susceptible to soil erosion, denudation and compaction as a result of trampling by people, animals and vehicles. An increasing problem is the abuse of hilly terrain and coastal dunes by users of off-road vehicles ranging from motor-cycles to four-wheel drive trucks. The larval burrows of the tiger beetle Cicindela dorsalis have been virtually eliminated from beaches of the western U.S.A. (74). The dune systems at Antioch and El Segundo (see review) have suffered damage, as has the coastal habitat of Belkin's Dune Fly (Brennania belkini, see review) and the Wandering Skipper (Panoquina errans, see review).

2. Changes to the aquatic environment.
2a. Drainage and channelization of wetlands.
Wetland biotopes occur at all altitudes and include estuaries, marshes, fens, bogs and mires. They may be found in open country, as in the English fens, or in forested country, as in Fennoscandia and montane areas of central Europe (see review of the mires of the Sumava Mountains in Czechoslovakia), and all are extremely vulnerable to change. As long ago as 1847/48 the draining of the English fens resulted in the loss of the Large Copper Butterfly (Lycaena dispar dispar), the first documented extinction of a butterfly (20). Fen remnants are now only preserved by careful management. Eight of the 15 endangered butterflies in Europe are wetland species (33). The British distribution of the Mole Cricket Gryllotalpa gryllotalpa has been reduced by draining of fenlands and meadows, as well as the use of insecticides, and the species is listed in the U.K. Wildlife and Countryside Act. Drainage and development of subcoastal swamps is threatening the Brazilian Fluminense Swallowtail (Parides ascanius, see review), but information from tropical areas is generally lacking. Threats to insects and other invertebrates may be expected to result from major tropical developments such as the channelization into the Jonglei Canal of the Sudd, a marsh into which the White Nile flows in Sudan. Minor improvements to land on a large scale may lower the water table and have disastrous results for insects. In Britain the extinctions of the damselflies Coenagrion armatum and Lestes dryas were almost certainly caused by such changes (42,43).

In the U.S.A., capping of springs led to the loss of the fritillary butterfly Speyeria nokomis caerulescens (71). Cultivation of prairie marshes has severely restricted the ranges of the Dakota Skipper (Hesperia dacotae, see review), the satyrid Lethe eurydice fumosa (69) and the Delta Green Ground Beetle (Elaphrus viridis, see review). In Australia the tapping of streams threatens at least five species of Odonata, and two more are threatened by sand-mining of their dune lake habitat (37).

2b. Water impoundment.
The containment of rivers kills insects whose larvae are adapted to fast-flowing water, such as the net-veined midges (Diptera: Blepharoceridae) (see reviews of Edwardsina gigantea and E. tasmaniensis). The larvae of Trichoptera (caddis-flies) are also at risk and in Tasmania three Lake Pedder endemics Taskiria mccubbini, Taskiropsyche lacustris (Kokiriidae) and Archeophylax vernalis (Limnephilidae) may already be extinct as a result of the new impoundment of the lake. Another endemic to Lake Pedder, Westriplectes pedderensis, may have

survived the inundation but is still regarded as endangered (73). Further extinctions are expected if the impoundment of the Lower Gordon River, also in Tasmania, proceeds as planned. Adult Blepharoceridae, Trichoptera, Plecoptera (stoneflies) and Ephemeroptera (mayflies) are poorly dispersed and often have very restricted ranges (see reviews). Inundation of river banks and beaches destroys specialized habitats whose fauna may be unable to adapt to the new conditions (see review of Colombia Tiger Beetle, Cicindela columbica). In some cases inundation of valleys may threaten the whole, or a major part, of an insect's range. The ground beetle Carabus olympiae (Carabidae) is one of the rarest insects in Italy. It was believed to be extinct in 1928, but in 1942 was rediscovered in a montane site in Piedmont, which remains its only known locality (78). The site will be flooded if the proposed dam in the Valsessera Valley proceeds (59).

2c. Water pollution.
Pollution of streams by acid mine drainage, silage pit effluent and siltation has probably had profound effects on aquatic insect communities, and industrial pollution has at least transient ill-effects. Blepharoceridae are unable to tolerate siltation, and dragonfly and damselfly larvae are sensitive ecological indicators of pollution (see review of Shining Macromia Dragonfly, Macromia splendens). With the exception of Hydropsyche contubernalis, all caddis-flies in the River Rhine are severely reduced in numbers because of heavy urban and industrial pollution; Tobias Caddis-fly (H. tobiasi) is probably extinct (see review). The unusual Lake Tahoe Stonefly (Capnia lacustris), which spends its entire life history at depths of 200-264 feet (60-81 m) in Lake Tahoe on the California/Nevada border (36), is now threatened by urban pollution, particularly from the Nevada side.

3. Atmospheric pollution.
3a. Acid rain.
Although only circumstantially correlated, it seems that acid rain resulting from the solution of atmospheric pollutants in rain is responsible for declines and losses of insect, particularly butterfly, populations in Europe (33) and North America (48). Several widespread species of butterfly, including Parnassius apollo, are reported to have suffered severe declines in areas to the north and east of the principal industrial areas of western Europe from which atmospheric pollution is carried on the prevailing south-westerly winds (33).

4. Changes to associated fauna and flora.
4a. Loss of host.
Insects that rely on specific animals or plants require healthy populations of their hosts for their own survival. Loss of the American Chestnut (Castanea dentata) from most of North America after accidental introduction of Chestnut Blight (Endothia parasitica) has pushed at least five microlepidopterans to the brink of extinction (53). Parasites of threatened vertebrates are probably becoming extinct at least as fast as their hosts, although there are few data available (see review of Pygmy Hog Sucking Louse, Haematopinus oliveri). Invertebrates are often parasitized too; the Dusky Large Blue (Maculinea nausithous, see Large Blue review) is parasitized by the hymenopteran Neotypus sp. which is threatened by the decline of its host's range in Europe (66).

4b. Exotic introductions.
Intentional or accidental introductions of exotic animals and plants will invariably upset the balance of arthropod species, sometimes irreparably. Introduced plants may dominate native species, indirectly causing loss of native insects. This has occurred in Hawaiian and Californian coastal dunes after the intentional introduction of European Beach Grass (Ammophila arenaria) and ice-plants (Mesembryanthemum spp.) (57,62 and see El Segundo review). Introduction of the European Hare (Lepus capensis) to Laysan in the leeward Hawaiian Islands caused the extinction of several noctuid moths (52). The accidental introduction of the

322

Black Rat (Rattus rattus) to Lord Howe Island caused the probable extinction of the stick-insect Dryococelus australis (see review). In New Zealand the giant wetas (Deinacrida spp., see review) have been severely threatened by rodent introductions throughout their range. The once numerous damselfly Megalagrion pacificum has probably been virtually exterminated by the mosquito fish (Gambusia affinis) which was introduced into Hawaii to control mosquitoes (45).

5. Threats to individual species.
5a. Over-collecting.
Although this is often cited as a cause of insect decline (e.g. Carabus olympiae (78), Parides hahneli, see review) and may indeed aggravate critical situations (as it possibly did in the loss of the Large Copper from the U.K.), there are no cases of extinction or loss of populations directly attributable to indiscriminate collecting (57). The vast majority of insects produce far more offspring than will survive the natural forces of predation, parasitism and starvation. Nevertheless, accusations of over-enthusiastic collecting frequently arise, generally related to ethically dubious areas of trade in wild-caught specimens (see reviews of Boloria acrocnema and Euproserpinus wiesti). Most entomologists and entomological traders take great care to ensure that their activities are not harmful and, if possible, are beneficial in terms of scientific knowledge of species' needs, or of enhancing wild populations by captive breeding and avoidance of collecting in the wild. Guidelines are published by all major entomological societies (32).

5b. Pesticides.
Widespread use of artificial chemicals is often cited as a major factor in the loss of insect populations, but has not resulted in any documented extinctions of species. The decline of insects in rural areas of Europe is due largely to the general intensification of agriculture, only one aspect of which is the use of insecticides and herbicides. Pesticide use in native habitats, particularly on islands, should be viewed with great caution (57, see review of picture-winged flies). Severely reduced populations may be susceptible to aerial application of insecticides, and particular care is needed to avoid drift away from agricultural areas. Wiest's Sphinx Moth (Euproserpinus wiesti, see review) was almost exterminated when the area around its last known locality was sprayed for grasshopper control. The final extinction of the English population of Aporia crataegi (Black-veined White) has been attributed to orchard spraying (33).

CONSERVATION Concern for declining insect populations has been voiced since the first half of the 19th century, and the historical aspects of insect conservation have been reviewed elsewhere (55,57). The present momentum began in the 1960s; legislative measures were taken in the 1970s and are being strengthened in the 1980s.

1. Legislation.
International agreement is contained in the Convention on International Trade in Endangered Species of Wild Fauna and Flora (CITES), which controls and monitors import and export of listed species, and to which 79 countries have now acceded. Appendix I, which contains no insects, is a list of species in which trade is subject to strict regulation, and is virtually prohibited except to fulfil scientific needs. Appendix II lists species in which trade is regulated or monitored. All birdwing butterflies (Papilionidae: Ornithoptera, Trogonoptera and Troides spp., see review of Queen Alexandra's Birdwing) and the Apollo Butterfly (Parnassius apollo, see review) are listed in Appendix II. These species may be commercially traded, but an export permit from the country of origin is required before specimens may be removed from that country, or imported into another state which is a party to CITES.

A complete discussion of national laws pertaining to insects is beyond the scope of this review, and only a few points will be mentioned. All north European countries

323

give greater emphasis to protecting habitats and biotopes than to passing legislation concerning individual species. Nevertheless, in many cases individual species are protected by law, often where they may not be adequately protected within reserve areas or where a threat from trade or over-collecting is perceived. In Great Britain the Wildlife and Countryside Act 1981 gives total protection to only 14 insects, but there is great emphasis on the protection and management of reserves. The governmental Nature Conservancy Council notifies landowners of Sites of Special Scientific Interest (SSSIs) that they may own, and requires to be consulted about all proposed developments in these sites. Some SSSIs have been selected on entomological grounds and an increasing number have been established as National Nature Reserves (NNRs). Entomological information is used in selecting NNRs, which have an important role in preserving the habitat of many insect species. For example, 32 out of the 37 British species of dragonflies breed in NNRs, and the other reserves protect significant populations of rare butterflies such as the Swallowtail (Papilio machaon), the Black Hairstreak (Strymonidia pruni) and the Chequered Skipper (Carterocephalus palaemon). Other important insect localities are protected in the reserves of voluntary bodies such as the National Trust, the many County Trusts which support the Royal Society for Nature Conservation, and the Royal Society for the Protection of Birds (47).

Several other north European countries, including Belgium (39), West Germany (8,9), France (13,75), Switzerland (12) and Poland (51,76), give legislative protection to named insects, and all European countries protect habitats in national reserves. Legislation to conserve insects is generally absent in southern Europe and the Mediterranean region, although Spain has published its own Lepidoptera Red Data Book (19) and conservationists there (29) and in Italy (78), are pressing for legislation. In many other areas of the world legislation relating to insects is confined to bans on collecting or trade of one or a few species (e.g. Brazil, see review of Fluminense Swallowtail), or a ban on all collecting (e.g. parts of Australia (37,55), Mexico (see review of Red-knee Tarantula) and most recently Kenya. Exceptional legislation has been passed in Papua New Guinea, protecting their seven rarest species of birdwing butterflies and providing for prevention of collecting and trade, and for habitat preservation and management (see Queen Alexandra's Birdwing review). At the same time butterfly farming has become a significant local industry, as it has been for some time in Taiwan (61). In P.N.G. two unprotected birdwings, Ornithoptera priamus and Troides oblongomaculatus, and a huge saturniid moth, Coscinocera hercules, contribute most of the income (70).

The 1973 U.S. Endangered Species Act requires the Secretary of the Interior to protect endangered species through the Office of Endangered Species in the Fish and Wildlife Service. In September 1982 the Office of Endangered Species had listed 762 organisms as Threatened or Endangered, of which 49 were invertebrates and 13 insects. Regrettably, the present administration accords low priority to invertebrates (14) and the last insect listing was on 8 August 1980 (the Delta Green Ground Beetle). This is not for lack of candidates, or for lack of public interest in the protection of threatened invertebrates. Many individual states in the U.S.A. have passed their own local legislation to protect insects or their habitats (7,55,56).

2. Research and management.
The management of threatened insects began in the U.K., where it is now the responsibility of the Institute of Terrestrial Ecology. The centres at Furzebrook and Monks Wood have been particularly active in management and conservation of rare insects. Through good management populations of the Swallowtail Papilio machaon britannicus survive well in the Norfolk Broads, U.K., although its reintroduction to Wicken Fen was unsuccessful despite excellent studies (47). Research has achieved some success in protecting the Black Hairstreak (Strymonidia pruni), various fritillaries and other butterflies. The Large Blue

Maculinea arion has become extinct in the U.K. despite extensive research efforts (see separate review), but recent studies promise greater success with Lysandra bellargus, the Adonis Blue (47). Insect conservation has always been of prime importance to the Royal Entomological Society of London, whose Joint Committee for the Conservation of British Insects directs and stimulates research and management programmes.

The 1973 U.S.A. Endangered Species Act has prompted detailed investigations and reviews to accompany the procedures involved in listing species (notice of review, proposed rulemaking, final rulemaking). The final rulemaking and its supporting documents often constitute the definitive statement on the autecology of the species concerned. The U.S. Office of Endangered Species, in co-operation with appropriate state agencies, has funded several projects relating to insect conservation, including work on the Karner Blue (Lycaeides melissa samuelis) in New York, six butterflies in California (3, and see San Bruno Mountain review) and recovery plans for Schaus' Swallowtail in Florida (25, and see review) and the Delta Green Ground Beetle in Jepson Prairie in California (34). Section 7 of the U.S. Endangered Species Act requires all federal agencies to co-operate with the Office of Endangered Species in carrying out conservation programmes. In the past this has included the U.S. Army Corps of Engineers, the Forest Service, the Bureau of Land Management and others. Through political lobbying and fund-raising, private organizations such as the U.S. Nature Conservancy (not to be confused with the Nature Conservancy Council in U.K.) and the Xerces Society have made great progress in furthering the cause of insect conservation in the U.S.A. (56).

The insect Specialist Groups now formed within the Species Survival Commission of IUCN cover Odonata, Lepidoptera and Formicidae as well as cave faunas. Each has drawn up a list of world priorities; a discussion of those of the Odonata Specialist Group is in press (44). The work of the insect Specialist Groups is evident in the lack of balance of the species reviewed in this section, which is biased towards the Lepidoptera, Odonata and Formicidae. The Specialist Groups are of great value in centralizing worldwide conservation data; more invertebrate Specialist Groups will be formed to examine the many orders of insects which are under-represented in the following pages.

3. Recording and mapping.
Distribution studies are fundamental to the conservation of every kind of animal and plant species. Without such information there is no way of assessing species abundance or vulnerability (47). The Biological Records Centre (BRC) at Monks Wood Experimental Station initiated insect recording schemes in 1967 and many detailed maps have been produced. In 1969 the BRC joined with other European recording centres to begin the European Invertebrate Survey. This became funded by Unesco's International Union for Biological Sciences and expanded to become one of the constituent commissions of IUBS, known as the International Commission for Invertebrate Survey (ICIS). This has resulted in detailed maps being produced by many countries, including Austria, Belgium (39), France, West Germany, Great Britain (31), Ireland, Luxembourg, the Netherlands, Norway, Poland, Rumania and Spain (30). Similar work in the U.S.A. has been initiated by the North American Invertebrate Survey which is also part of ICIS, and exemplary volumes on the Lepidoptera and Coleoptera have been produced in Korea (82).

4. Red data books and other lists of threatened insects.
Distribution studies in Europe have given the background for a review of the threatened butterflies of Europe (33) and for lists of some of the threatened insects of several European countries or regions, including Austria (Steiermark only, 83), Belgium (39), Finland (81), West Germany (8,9), Luxembourg (1), Switzerland (12,80), Poland (51,76,79) and Spain (19). Several other red data books are in preparation, including one for Great Britain. A preliminary list of

threatened insects of the U.S.S.R. is available (65) and will be expanded for inclusion in the forthcoming second edition of the U.S.S.R. Red Data Book (28). Lists of threatened insect species have been prepared by several states in the U.S.A., including New York, California, Kansas, Utah (57), Florida (26), Virginia (40) and North Carolina (18).

IUCN, which began the first series of Red Data Books in 1966, plans to continue with further volumes on insects. The priority is a review of the swallowtail and birdwing butterflies (Papilionidae), and a volume on the insects and other occupants of threatened cave systems is planned.

5. The future.
It has been emphasized that documentation of the conservation needs of invertebrates has been virtually restricted to the Nearctic, Palearctic and Australasian regions. The biogeographical index to this volume is swamped by examples from north temperate countries despite positive efforts to seek examples from the tropics. The greatest concentrations of species currently at risk probably occur in the tropical countries of Africa, South America and South East Asia, where equatorial strips or island-like areas have remained climatically favourable for millions of years. It is in these regions where conservation studies and research on insects must be greatly increased, and where the taxonomy of most invertebrate groups is very poorly known.

By far the greatest threat to insects is destruction of their natural habitat, and many countries are taking strong measures to gazette reserves and national parks. However, national parks and reserves must be adequately protected and must not be eroded by financial interests, even during times of recession. Elements remote from reserves but essential to their integrity must also be preserved, particularly watersheds. Insects can be of value in choosing sites for reserves; in South America data on the density, distribution and biogeography of insects are proving to be valuable in this regard (10,38). The needs of insects do not always comply with those of vertebrates and it is not safe to assume that protection of large areas for vertebrates will automatically safeguard the insect diversity. Because of their small size and relatively modest needs, insects are able to occupy ecological niches which are more numerous and far smaller in all dimensions (space, time etc.) than those of vertebrates. For this reason impact assessments may operate on too large a scale, and harmful effects may be overlooked. In most cases there is no substitute for surveys specifically to assess the insects that need to be considered. The important invertebrate surveys being carried out in Europe and North America should themselves be extended, and the same approach should be used in tropical centres throughout the world. To strengthen the case for habitat protection it is essential to document further the contents of tropical biomes, to list and describe the species and to investigate their ecological requirements. Both intensive and extensive surveys are needed in order to identify the conservation requirements of insects in those habitats where they are most abundant, using new or existing local taxonomic and ecological centres wherever possible. These aims are fully endorsed by the World Conservation Strategy, which emphasizes the need to protect living resources for ecological reasons, for their genetic diversity and for sustainable utilization (75).

REFERENCES
1. Anon. (1975). In Luxemburg geschützte Tiere. Natura, Luxembourg.
2. Anon. (1978). Silk. Entomology Leaflet No. 7. British Museum (Natural History), London.
3. Arnold, R.A. (1978). Survey and status of six endangered butterflies in California. Calif. Dept. Fish and Game, Non-game Wildlife Investigations E-1-1. Study V, Job 2.20. Final report, 95 pp.
4. Barnes, R.D. (1980). Invertebrate Zoology. Saunders

College, Philadelphia. 1089 pp.

5. Bartholomew, J. and Son Ltd. (1980). The Times Atlas, Comprehensive Edition. Times Books, London. 123 maps + 227 pp.

6. Baumann, R.W. (1979). Rare aquatic insects, or how valuable are bugs? Great Basin Naturalist Mem. 3: 65-67.

7. Berger, T.J. and Neuner, A.M. (1979). Directory of State Protected Species. Assoc. Syst. Collect. Lawrence, Kansas. 158 pp.

8. Blab, J. and Kudrna, O. (1982). Hilfsprogramm für Schmetterlinge. Naturschutz Aktuell. Kilda-Verlag, Greven. 135 pp.

9. Blab, J., Nowak, E. and Trautmann, W. (1981). Rote Liste der gefährdeten Tiere und Pflanzen in der Bundesrepublik Deutschland. Naturschutz Aktuell No. 1. Kilda-Verlag, Greven. 66 pp.

10. Brown, K.S., Jr. (1982). Paleoecology and regional patterns of evolution in neotropical forest butterflies. In Prance, G.T. (Ed.), Biological Diversification in the Tropics. Columbia University Press, New York. Pp. 255-308.

11. Brown, W.L. (1982). Insecta. In (54), pp. 326-328.

12. Burckhardt, D., Gfeller, W. and Müller, H.U. (1980). Animaux Protégés de Suisse. Ligue Suisse pour la Protection de la Nature, Bâle. 224 pp.

13. Burton, G.N. (1980). French insects now protected. AES Bull. 39: 58-59.

14. Chambers, S.M. (1981). Protection of mollusks under the Endangered Species Act of 1973. Bull. Am. Malacol. Union Inc. 1981: 55-59.

15. Cloudsley-Thompson, J.L. (1968). Spiders, Scorpions, Centipedes and Mites. Pergamon Press, Oxford. 278 pp.

16. Commonwealth Agricultural Bureaux (1982). Biological Control Service, 25 Years of Achievement. C.A.B., Farnham Royal, U.K. 24 pp.

17. Commonwealth Agricultural Bureaux (1982). Biological Control Service, Commonwealth Institute of Biological Control. C.A.B., Farnham Royal, U.K. 16 pp.

18. Cooper, J.E., Robinson, S.S. and Funderburg, J.B. (1977). Endangered and Threatened Plants and Animals of North Carolina. North Carolina State Museum of Natural History, Raleigh, North Carolina.

19. De Viedma, M.G. and Gomez Bustillo, M.R. (1976). Libro Rojo de los Lepidopteros Ibericos. Instituto Nacional para la Conservacion de la Naturaleza, Madrid. 117 pp.

20. Duffey, E. (1968). Ecological studies on the large copper butterfly Lycaena dispar batavus Obth. at Woodwalton Fen National Nature Reserve, Huntingdonshire. J. Appl. Ecol. 5: 69-96.

21. Else, G., Felto, J. and Stubbs, A. (undated). The Conservation of Bees and Wasps. Nature Conservancy Council, London. 13 pp.

22. Elton, C.S. (1973). The structure of invertebrate populations inside tropical rainforest. J. Anim. Ecol. 42: 55-104.

23. Elton, C.S. (1975). Conservation and the low population density of invertebrates inside neotropical rain forest. Biol. Conserv. 7: 3-15.

24. FAO/UNEP (1981). Tropical Forest Resources Assessment Project. (3 vols). FAO, Rome.

25. Florida Game and Fresh Water Fish Commission (1982).

Recovery Plan for Schaus' Swallowtail Butterfly (Heraclides (Papilio) andraemon bonhotei Sharpe). U.S. Fish and Wildlife Service, Atlanta. 57 pp.

26. Franz, R. (Ed.) (1982). Invertebrates. In Vol. 6, Rare and Endangered Biota of Florida. University Presses of Florida, Gainesville. 131 pp.

27. Free, J.B. (1982). Bees and Mankind. Allen and Unwin, Hemel Hempstead. 176 pp.

28. Ghilarov, M. (1982). In litt., 5 October.

29. Gomez-Bustillo, M.R. (1981). Protection of Lepidoptera in Spain. Beih. Veröff Naturschutz Landschaftspflege Bad.-Württ. 21: 67-72.

30. Heath, J. (1971). The European invertebrate survey. Acta Entomol. Fennica 28: 27-29.

31. Heath, J. (Ed.) (1973-1979). Provisional Atlas of the Insects of the British Isles. Biological Records Centre, Monks Wood, Huntingdon.

32. Heath, J. (1981). Insect conservation in Great Britain (including a code for insect collecting). Beih. Veröff. Naturschutz Landschaftspflege Bad.-Württ. 21: 219-223.

33. Heath, J. (1981). Threatened Rhopalocera (Butterflies) of Europe. Council of Europe, Strasbourg. 157 pp.

34. Holland, R. and Arnold, R.A. (1982). Jepson Prairie Recovery Plan. U.S. Fish and Wildlife Service, Sacramento, California. 82 pp.

35. Imms, A.D. (1973). Insect Natural History. Collins, Fontana, London. 348 pp.

36. Jewett, S.G., Jr. (1963). Stonefly aquatic in the adult stage. Science 3554: 484-485.

37. Key, K.H.L. (1978). The conservation status of Australia's insect fauna. Australian Nat. Parks and Wildlife Service Occ. Paper 1. 24 pp.

38. Lamas, G. (1982). A preliminary zoogeographical division of Peru based on butterfly distributions (Lepidoptera, Papilionoidea). In Prance, G.T. (Ed.), Biological Diversification in the Tropics. Columbia Univ. Press, New York. Pp. 336-357.

39. Leclercq, J., Gaspar, C., Marchal, J.-L., Verstraeten, C. and Wonville, C. (1980). Analyse des 1600 premières cartes de l'atlas provisoire des insectes de Belgique, et première liste rouge d'insectes menacés dans la faune Belge. Note Fauniques de Gembloux No. 4. Faculté des Sciences Agronomiques de l'Etat, Gembloux. 103 pp.

40. Linzey, D.W. (Ed.) (1979). Endangered and Threatened Plants and Animals of Virginia. Center for Environmental Studies, Blacksburg, Virginia.

41. Marks, E.N. (1969). The Invertebrates. In Webb, L.J., Whitelock, D. and Brereton, J.L.G. (Eds), The Last of the Lands. Milton, Jacaranda.

42. Moore, N.W. (1976). The conservation of Odonata in Great Britain. Odonatologica 5: 37-44.

43. Moore, N.W. (1980). Lestes dryas Kirby - a declining species of dragonfly (Odonata) in need of conservation: notes on its status and habitat in England and Ireland. Biol. Conserv. 17: 143-148.

44. Moore, N.W. (In press). Conservation of Odonata - first stage towards a world strategy. In Gambles, R. (Ed.), Advances in Odonatology.

45. Moore, N.W. and Gagné, W.C. (1982). Megalagrion pacificum

(McLachlan) - a preliminary study of the conservation requirements of an endangered species. Rep. Odon. Specialist Group Int. Un. Conserv. Nat. 3. 5 pp.

46. Morris, M.G. (1978). Grassland management and invertebrate animals - a selective review. Sci. Proc. Royal Dublin Soc. (A) 6: 247-257.

47. Morris, M.G. (1981). Conservation of butterflies in the United Kingdom. Beih. Veröff. Naturschutz Landschaftsplege Bad.-Württ. 21: 35-47.

48. Muller, J. (1972). Is air pollution responsible for melanism in Lepidoptera and for scarcity of all insect orders in New Jersey? J. Res. Lepid. 10: 189-190.

49. Myers, N. (1979). The Sinking Ark. Pergamon Press, Oxford. 307 pp.

50. Myers, N. (1980). Conversion of Tropical Moist Forests. Report to the National Academy of Sciences. National Research Council, Washington D.C.

51. Novak, I. and Spitzer, K. (1982). Ohrozeny svet hmyzu (Endangered World of Insects). Ceskoslovenska Akademie Ved, Prague. 138 pp.

52. Opler, P.A. (1976). The parade of passing species: extinctions past and present. Sci. Teach. 43: 30-34.

53. Opler, P.A. (1979). Insects of American chestnut: possible importance and conservation concern. In McDonald, W. (Ed.), The American Chestnut Symposium. West Virginia Univ. Press, Morgantown. Pp. 83-85.

54. Parker, S.P. (1982). Synopsis and Classification of Living Organisms. McGraw Hill, New York. 2 vols., 1232 pp.

55. Pyle, R.M. (1976). The eco-geographic basis for Lepidoptera conservation. Unpubl. Ph. D. thesis, Yale University. 369 pp.

56. Pyle, R.M. (1976). Conservation of Lepidoptera in the United States. Biol. Conserv. 9: 55-75.

57. Pyle, R., Bentzien, M. and Opler, P. (1981). Insect conservation. A. Rev. Ent. 26: 233-258.

58. Ratcliffe, D.A. (Ed.) (1977). A Nature Conservation Review. Cambridge Univ. Press, Cambridge. 721 pp.

59. Raviglione, M. (1982). In litt., 18 December.

60. Rentz, D.C.F. (1977). A new and apparently extinct katydid from the Antioch Sand Dunes (Orthoptera: Tettigoniidae). Ent. News 88: 241-245.

61. Severinghaus, S.R. (1977). The butterfly industry and butterfly conservation in Taiwan. Atala 5: 20-23.

62. Slobodchikoff, C.N. and Doyen, J.T. (1977). Effects of Ammophila arenaria on sand dune arthropod communities. Ecology 58: 1171-1175.

63. Smith, K.G.V. (Ed.) (1973). Insects and other arthropods of medical importance. British Museum (Natural History), London. 561 pp. + 12 pls.

64. Sommer, A. (1976). An attempt at an assessment of the world's tropical moist forests. Unasylva 28: 5-24.

65. Tanasijtshuk, V.N. (1981). Materials for the 'Red Book' of the U.S.S.R. on insects. Entomologicheskoye Obozreniye 60: 699-711. In Russian, English translation available.

66. Thomas, J.A. (1982). In litt., 3 December.

67. Tilden, J.W. (1956). San Francisco's vanishing butterflies. Lepid. News 10: 113-115.

68. U.S.D.I. Fish and Wildlife Service (1978). Determination that 11 plant taxa are endangered species and 2 plant taxa are threatened species. Federal Register 43: 17910-17916.

69. U.S.D.I. Fish and Wildlife Service (1978). Proposed endangered or threatened status or critical habitat for ten butterflies or moths. Federal Register 43: 28938-28945.
70. Vietmeyer, N.D. (1979). Butterfly ranching is taking wing in Papua New Guinea. Smithsonian 10: 119-135.
71. Wielgus, R.S. (1974). A search for Speyeria nokomis caerulescens (Holland) in southern Arizona. J. Res. Lepid. 11: 187-194.
72. Wigglesworth, V.B. (1964). The Life of Insects. Weidenfeld and Nicolson, London. 360 pp.
73. Williams, W.D. (1982). In litt., 19 January.
74. Wilson, D.A. (1970). Three subspecies of cicindelid threatened with extermination. Cicindela 2: 18-20.
75. D'Ornano, M. and Mehaignerie, P. (1979). Liste des insectes protégés en France. J. Officiel de la Rep. Française, 22 August.
76. Ferens, B. (1957). Animal Species Under Protection in Poland. Nat. Prot. Res. Centre of Polish Acad. Sci., Krakow, transl. and publ. in 1965 by U.S. Dept. of the Interior.
77. IUCN/UNEP/WWF (1980). World Conservation Strategy: Living Resource Conservation for Sustainable Development. IUCN, Gland.
78. Tassi, F. (1972). Gli insetti nella protezione della natura. (Insects in nature conservation). Atti del 9 Congresso Nazionale Italiano di Entomologia, Siena, 21-25 June 1972. 17 pp. + 5 pls.
79. Dabrowski, J.S. and Krzywicki, M. (1982). Ginace I Zagrozone Gatunki Motyli (Lepidoptera) W Faunie Polski. Czesc 1: Papilionoidea, Hesperioidea and Zygaenoidea. Panstwowe Wydawnictwo Naukowe, Warszawa-Krakow. 171 pp.
80. Gfeller, W. (1975). Geschütze Insekten in der Schweiz. Bull. Soc. Ent. Suisse 48: 217-223.
81. Mikkola, K. (1979). Vanishing and declining species of Finnish Lepidoptera. Notul. Entomol. 59: 1-9.
82. Kim, Chan-Whan (1976-1979). The Distribution Atlas of Insects of Korea. Series 1 Rhopalocera, Series 2 Coleoptera, Series 3 Hymenoptera and Diptera. Korea University Press, Seoul.
83. Gepp, J. (1981). Rote Listen Gefährdeter Tiere der Steiermark. Sonderheft Nr. 3 des Steirischen Naturschutzbriefes. Herausgegeben im Auftrag und Verlag des Österreichischen Gesellschaft für Natur- und Umweltschutz, Graz.

We are grateful to J. Heath, N.W. Moore, M.G. Morris and J.A. Thomas for their comments on this review.

LARGE BLUE LAKE MAYFLY   RARE

Tasmanophlebia lacus-coerulei Tillyard, 1933

| Phylum | ARTHROPODA | Order | EPHEMEROPTERA |
| Class | INSECTA | Family | SIPHLONURIDAE |

SUMMARY The largest species in the genus Tasmanophlebia, the Large Blue Lake Mayfly is known to occur only in five small lakes near the summit of Mt Kosciusko, New South Wales, Australia.

DESCRIPTION Both adult and nymphal stages of the species were described by Tillyard (2). The full-grown nymph is 35-40 mm in length, and the adult wingspan exceeds 30 mm (2).

DISTRIBUTION The species has only been collected in five lakes: Lake Coothapatamba, Lake Albina, Club Lake, Blue Lake and Hedley Tarn, all near the summit of Mt Kosciusko, Australia. It is restricted to permanent lenthic waters and there are no other similar habitats in which it could occur (1).

POPULATION Unknown.

HABITAT AND ECOLOGY Tasmanophlebia nymphs appear to be detritus feeders, consuming dead organic material on the lake bottom. The nymphs are common in all five lakes but nothing is known of their growth rate, life history, or predators (1). The habitat on Mt Kosciusko around 7000 feet (2135 m), where the lakes lie, is alpine meadow (1).

SCIENTIFIC INTEREST AND POTENTIAL VALUE The genus Tasmanophlebia is one of only two siphlonurid genera occurring on both mainland Australia and Tasmania; thus it has zoogeographic significance. This is the only mayfly in its family to occur in all five lakes on Mt Kosciusko (1). A useful model system for local autochthony (evolution in situ) is therefore available.

THREATS TO SURVIVAL Lake Albina is presently receiving septic tank effluent from a hikers' hut located nearby, and Blue Lake is a popular campsite for walkers. Concern has been expressed about water quality in both lakes. Siphlonurid mayflies are sensitive to changes in water quality (1).

CONSERVATION MEASURES TAKEN All five lakes are located within the Kosciusko National Park and some water quality investigations have been carried out on Lake Albina and Blue Lake (1).

CONSERVATION MEASURES PROPOSED Water quality monitoring programmes should be continued and broadened to include biological monitoring. Periodic surveys of the status of Tasmanophlebia populations by experienced limnologists should also be carried out (1). National Park management schemes should take into account the necessity of protecting these alpine tarns and their endemic fauna.

REFERENCES
1. Campbell, I.C. (1982). In litt., 9 February.
2. Tillyard, R.J. (1933). Mayflies of the Mt. Kosciusko region, NSW: Introduction and Siphlonuridae. Proc. Linn. Soc. NSW. 58 1-32.

I.C. Campbell kindly prepared the draft for this account and W.D. Williams provided further comment.

SMALL HEMIPHLEBIA DAMSELFLY   ENDANGERED

Hemiphlebia mirabilis Selys, 1868

Phylum   ARTHROPODA   Order   ODONATA

Class   INSECTA   Family   HEMIPHLEBIIDAE

SUMMARY A minute and extremely primitive damselfly, Hemiphlebia mirabilis seems to have disappeared from its known habitats in the flood-plain lagoons in Victoria, Australia, and may now be extinct. Most of the species' localities have been destroyed by damming and drainage.

DESCRIPTION H. mirabilis is a tiny, bright metallic green damselfly (wingspan 20-25 mm) with white anal appendages. The base of the discoidal cell is open in the forewing, a character found otherwise in only one living species and in fossil species from the Permian to early Mesozoic. The postnodal crossveins do not coincide with the crossveins below them, another archaic feature (4,7). The larva has been described in (6).

DISTRIBUTION Originally described, apparently in error, from "Port Denison" (= Bowen) in northern Queensland, Australia, the species was subsequently rediscovered on flood-plain lagoons of the Goulburn River at Alexandra, Victoria, and on the middle to upper course of the Yarra River, Victoria (7). Recent searches in the vicinity of Bowen have failed to disclose the species, but there is one recent unconfirmed report from Wilson's Promontoy Victoria. Another species (Synlestes weyersi) described by Selys from "Port Denison" and probably sent to him by Weyers, who also collected the original material of H. mirabilis (3,4), is now known to be confined to southern Victoria. Hence the originally designated type locality has been intepreted as erroneous in favour of a Victoria site (7).

POPULATION Unknown.

HABITAT AND ECOLOGY The only details of habitat are those of Tillyard (6) and Dobson (1). It appears that Hemiphlebia frequents reedy lagoons on flood plains, and depends on the seasonal flooding of those lagoons. The record from Wilson's Promontory can only be interpreted as coming from a vegetated dune lake or swamp, which could possess at least some features in common with flood-plain lagoons. Nothing is known of potential limiting factors in feeding or reproductive behaviour (7).

SCIENTIFIC INTEREST AND POTENTIAL VALUE Hemiphlebia mirabilis is considered to be of great interest by scientists and by the IUCN/SSC Odonata Specialist Group (5,7), which placed this species as its highest priority in world odonate conservation (5). The species is taxonomically isolated and is monotypic (the only member) within the family Hemiphlebiidae. It is usually even placed in its own superfamily because of its uninterpretable archaic traits (7).

THREATS TO SURVIVAL The major threat to Hemiphlebia is the destruction of its specialized habitat. The flood plain lagoons on the Goulburn River have been altered by the damming of the river upstream, which has eliminated flooding below. The flood plains themselves are affected by market gardening, other forms of agriculture, and cattle raising (7). The habitats on the Yarra River have been altered in similar ways (7). Surveys of the Goulburn in 1954 disclosed abundant Hemiphlebia (4); in 1972, a few individuals persisted (2), but in 1978 none could be found (7). Recent surveys of the Yarra by A. Neboiss, principally in search of caddis-flies which have similar habitats, failed to reveal any of the damselflies (7). The flood plains were found to have been converted to agriculture, and the lagoons and swamps have disappeared.

CONSERVATION MEASURES TAKEN Funds are being sought to underwrite an intensive search for additional populations.

CONSERVATION MEASURES PROPOSED Should a search succeed in finding further colonies, suitable conservation measures will be proposed (7). A minimal reserve should seek to maintain the natural flooding regime and vegetational components in one or more suitable lagoon settings. If no site appears to be suitable for long-term maintenance as a naturally flooding lagoon, the possibility of intensive site management to stimulate necessary conditions (as has been done for lagoon-nesting bird species) should be investigated.

REFERENCES
1. Dobson, R. (undated). Notebooks held by the Australian National Insect Collection, CSIRO, Canberra.
2. Donnelly, T.W. (1974). Odonata collecting "Down Under". Selysia 6: 1-7.
3. Fraser, F.C. (1948). The identity of Synlestes weyersii Selys and its confusion with a new species. Bull. Mus. roy. Hist. nat. Belg. 24(16): 1-8.
4. Fraser, F.C. (1955). A study of Hemiphlebia mirabilis Selys (Odonata), a survival from the Permian. Ent. mon. Mag. 91: 110-13.
5. Moore, N.W. (1981). Minutes of the 2nd meeting of the Odonata Specialist Group, Chur, Switzerland. IUCN, 3 pp.
6. Tillyard, R.J. (1928). The larva of Hemiphlebia mirabilis Selys (Odonata). Proc. Linn. Soc. NSW. 53: 193-206.
7. Watson, J.A.L. (1982). In litt. 8 March.

J.A.L. Watson kindly provided the data for this review and N.W. Moore gave further comments.

FREYA'S DAMSELFLY   ENDANGERED

Coenagrion freyi Bilek, 1954

Phylum   ARTHROPODA   Order   ODONATA

Class   INSECTA   Family   COENAGRIONIDAE

SUMMARY Coenagrion freyi, extinct from its type locality and now known from only one or two alpine populations, represents an extreme postglacial relict. The nearest occurrence of its near relative C. hylas is 4000 km away in Siberia. Any kind of aquatic alterations may threaten its precarious existence in Austria and Switzerland. Surveys and protective measures for its habitat are required in the near future.

DESCRIPTION A fairly robust damselfly, the female more so than the male, Coenagrion freyi was only described in 1954 (1). The male is azure blue, heavily marked with black on the thorax and abdomen. A conspicuous black baseline runs along the sides of the abdominal segments, and a black peak points forward on the back of segments 2-7. The female has both a blue and a green form and is more copiously marked with black. Distinctive diagnostic features are found on the male terminalia and on the female pronotum and lamina mesostigmalis (1-3,7,8,11).

DISTRIBUTION While the related C. hylas occurs in Siberia and Manchuria, C. freyi has a very limited distribution in the central European Alps. The type population was found at the Zwingsee at Inzell, Germany (1), but became extinct there during the past 10-15 years (4,10). Currently just one or possibly two populations are known on small lakes in the Austrian (6) and Swiss Alps (5).

POPULATION No data on population sizes are available. The type population at Inzell must have been sizeable in the 1960s, when large numbers of the insect could be collected in a short time (8).

HABITAT AND ECOLOGY The few populations known are found at moderate altitude in the littoral zone of alpine lakes. The species breeds in stagnant waters, apparently favouring Equisetum beds in shallow, offshore waters (6,7,8). Adults fly in summer. The larval stage has not been studied in detail, but is considered to be of the cold stenothermal type (4). Further knowledge of the ecological requirements of the species is needed in order for management efforts to be reliable and informed.

SCIENTIFIC INTEREST AND POTENTIAL VALUE C. freyi is closely related to the Siberian and Manchurian C. hylas, to which it is assigned as a subspecies by some authors (3,4,6). Dumont has argued the reasons for considering it a distinctive taxon, although it is not very different in appearance from C. hylas (4). The isolated occurrence of a hylas group member of the genus in Europe is considered remarkable (4), particularly in that it occupies a sub-mountainous environment. The nearest Siberian station, on the lower Yenesei River, lies 4000 km or more from known C. freyi sites, and C. hylas frequently breeds on the plains (4). C. freyi is therefore a significant post-glacial relict, having arisen from a superspecies with a boreo-alpine distribution (5). Its disjunction from C. hylas is believed to have resulted from the cessation of the Würm glaciation (4). Better understanding of its relationships could be enlightening from both recent evolutionary and biogeographical standpoints.

THREATS TO SURVIVAL Overcollecting has been suggested as a cause for the disappearance of C. freyi from the Zwingsee, but this is considered unlikely (5). A much more tenable explanation would hold development and eutrophication of the lake responsible. By 1964 the damselfly was already considered seriously endangered (7). In 1970 half the lake was transformed into a swimming pool for an adjacent hotel, the other half being managed in part as a nursery pond for trout rearing. The necessary Equisetum beds still existed, but the only odonate to be found was a common, ubiquitous species (4). Any significant alteration of the aquatic habitats still occupied by the species could represent important risks, and collecting cannot be entirely ruled out as a threat in the insect's depleted condition.

CONSERVATION MEASURES TAKEN No conservation measures have been taken to date, although warnings were published in 1964 and 1971 (4,7). The IUCN/SSC Odonata Specialist Group has given C. freyi early priority on its list of species requiring conservation attention (9). The possible Swiss site (documented only by photographs, not specimens) has been kept secret as a security measure against possible over-collecting (5).

CONSERVATION MEASURES PROPOSED A thorough search of the Carinthian Alps should be conducted in order to survey likely localities for possible additonal colonies of C. freyi (4). Dragonfly collectors should be petitioned to take specimens of C. freyi sparingly, if at all, or sites kept secret until such time as population levels are considered secure. More importantly, protective steps should be taken to prevent eutrophication and disturbance of the littoral zone in lakes found to harbour the insect (5). This may require the purchase of land or the jurisdictional establishment of nature reserves, as well as active management and research on the species' needs.

REFERENCES
1. Bilek, A. (1954). Eine neue Agrionide aus Bayern (Odonata) Nachr. Bl. Bayer. Ent. 3: 97-9.
2. Bilek, A. (1955). Das bisher unbekannte Männchen von Agrion (= Coenagrion) freyi Bilek 1955 (Odonata). Nachr. Bl. Bayer. Ent. 4: 89-91.
3. Bilek, A. (1957). Agrion freyi eigene Art oder Subspecies von Agrion hylas Trybom. Nachr. Bl. Bayer. Ent. 6: 2 pp.
4. Dumont, H.J. (1971). Need for protection of some European dragonflies. Biol. Conserv. 3: 223-228.
5. Dumont, H.J. (1981). In litt., 12 February.
6. Heideman, H. (1974). Ein neuer Fund von Coenagrion hylas. Odonatologica 3: 181-185.
7. Lieftinck, M.A. (1964). Aantekeningen over Coenagrion hylas (Trybom) in Midden Europa (Odonata, Coenagrionidae). Tijdschr. Ent. 107: 159-66.
8. Lohmann, H. (1967). Notizen über Odonatenfunde in Chiemgau. Deutsch. Ent. Z., N.F. 14: 363-9.
9. Moore, N.W. (1980). Minutes of inaugural meeting, Odonata Specialist Group, IUCN.
10. Schmidt, Eb., (1977). Ausgestorbene und bedrohte Libellenarten in der Bundesrepublik Deutschland. Odonatologica 6: 97-103.
11. Schmidt, Er., (1956). Über das neue Agrion aus Bayern (Odonata). Ent. Z. 66: 233-6.

The data for this review were kindly furnished by H. Dumont, and N.W. Moore added further comment.

SAN FRANCISCO FORKTAIL DAMSELFLY   ENDANGERED

Ischnura gemina (Kennedy, 1917)

Phylum	ARTHROPODA	Order	ODONATA
Class	INSECTA	Family	COENAGRIONIDAE

SUMMARY This small damselfly (suborder Zygoptera), confined to disturbed urban sites in the area of San Francisco Bay, California, U.S.A., is threatened by habitat modifications, pollution, and other human activities.

DESCRIPTION The anterior surface of the thorax is completely black and there are no antehumeral stripes. These features are shared with Ischnura denticollis from which I. gemina differs only in details of the abdominal appendages (3,5). The male final-stage nymph of I. gemina is primarily distinguished from related species by the shape of the caudal appendages (cerci) (2). The general colour of the nymph is buff, green or dark brown with pale markings (2).

DISTRIBUTION The known distribution of the species is restricted to coastal California, U.S.A., from Point Reyes, Marin County in the north to Santa Cruz County and Monterey County in the south (3). If these records represent the entire distribution of the species, it has a surprisingly restricted range for a temperate zone pond dragonfly.

POPULATION The species was described from only two males and one female (5) and until 1978 it was considered extinct. At the time there were only 21 males and 16 females in the world's collections. Three further specimens from Monterey County have since been found in the University of Florida collections. The species has been found at eight additional localities in the San Francisco Bay area, all but one in urban areas liable to habitat alteration (3).

HABITAT AND ECOLOGY The very restricted geographic range suggests some ecological requirement which will be difficult to determine from a study of the existing remnant populations. Adults have been reared from eggs or field-captured nymphs fed on various infusoria and mosquito larvae (2). The adults mature 5-10 days after transformation and thereafter live for 6-23 days (4). A relatively long life span and a long flight season (March-November) is thought to be an adaptation to the foggy San Francisco climate (4). No details of adult sexual behaviour are known.

SCIENTIFIC INTEREST AND POTENTIAL VALUE Small, localized populations of I. gemina may owe their origin to founders from other populations nearby (4). Such discontinuous distribution may enhance differential selection pressure, a possible advantage in overcoming man-made disturbances. The species certainly seems to be very adaptable and in the San Francisco Bay area it is active in conditions of climate unsuitable for most other diurnal insects (4).

THREATS TO SURVIVAL Of the seven urban San Francisco populations one has now become extinct due to new construction, another in a city park has heavy foot and motorbike traffic, and a third was recently dredged, although some of the nymphs miraculously survived (3). Each of two areas supporting the largest populations do not exceed one hectare, and two other sites are moderately to heavily polluted (3).

CONSERVATION MEASURES TAKEN Point Reyes National Seashore contains some habitat for the species. A grant has been given by the Xerces Society for conservation studies by R.W. Garrison, who has published substantial new information (2,3,4).

CONSERVATION MEASURES PROPOSED The cost of small preserves in the San Francisco area would be prohibitive. No doubt the best hope for this species is in the protection of the population at the Point Reyes National Seashore. The importance of protecting this population must be quickly brought to the attention of the appropriate officials of the U.S. National Park Service (1). There is a need for further life history studies, for determinations of precise ecological requirements, and for a description of adult sexual behaviour (1).

REFERENCES   1. Bick, G. (1982). In litt., 3 February.
              2. Garrison, R.W. (1981). Description of the larva of Ischnura gemina with a key and new characters for the separation of sympatric Ischnura larvae. Ann. Ent. Soc. Am. 74: 525-530.
              3. Garrison, R.W. and Hafernik, J.E. (1981). The distribution of Ischnura gemina (Kennedy) and a description of the andromorph female (Zygoptera: Coenagrionidae). Odonatologica 10: 83-91.
              4. Garrison, R.W. and Hafernik, J.E. (1981). Population structure of the rare damselfly, Ischnura gemina (Kennedy) (Odonata: Coenagrionidae). Oecologia (Berl.) 48: 377-384.
              5. Kennedy, C.H. (1917). Notes on the life history and ecology of the dragonflies (Odonata) of central California and Nevada. Proc. U.S. Natn. Mus. 52: 483-635.

G.H. Bick and R.W. Garrison kindly provided information on this species, and N.W. Moore and D. Paulson gave further comments.

RELICT HIMALAYAN DRAGONFLY   VULNERABLE

Epiophlebia laidlawi (Tillyard, 1921)

Phylum   ARTHROPODA   Order   ODONATA

Class   INSECTA   Family   EPIOPHLEBIIDAE

SUMMARY A relict species of Himalayan dragonfly, Epiophlebia laidlawi is one of two surviving species of the suborder Anisozygoptera which flourished in the Mesozoic Era. The species is limited to the eastern Himalaya Mountains in India and Nepal, where it inhabits steep valleys under threat from land use change and environmental alteration. Although scientific investigation is incomplete, the species is certainly vulnerable and may be endangered in the Kathmandu Valley where it is best known.

DESCRIPTION Epiophlebia laidlawi has both damselfly and dragonfly characteristics. The adult wings are stalked like those of damselflies, but the body features are more like those of dragonflies. Further descriptive details appear in the literature (2,3,4,5,8). The larva was discovered in 1918, the adult not until 1963 (8,4).

DISTRIBUTION According to available data the species has been found only in the eastern Himalaya Mountains, between Darjeeling, India, and the Kathmandu Valley, Nepal (1,5,7).

POPULATION Unknown. Numbers are substantial where populations survive, with larvae appearing quite numerous in suitable streams (5).

HABITAT AND ECOLOGY In the only related species, the Japanese E. superstes, the adult dragonflies seem to be restricted to steep valleys at altitudes from 2100 to 2800 m. The larvae are found in clear, cold, rapidly running streams, and are thought to take about seven years to reach the adult stage (2,5,8). Similar conditions probably apply to the Himalayan species. Larvae have been found at altitudes between 2000 and 2700 m, adults from 2400 to 2732 m (5). While it is believed that the potential habitat may extend down to 2000 m, none could be found below 2300 m in a recent survey, due to landscape alteration (5). The flight period of adults is thought to be May to July, and oviposition probably takes place on vegetation beside running water. Elastostema hookerianum (Urticaceae) may be used in the Siwapuri Valley since it grows abundantly in and around waterfalls (5). The Japanese E. superstes uses Elastostema involcratum and even liverworts as oviposition sites (5).

SCIENTIFIC INTEREST AND POTENTIAL VALUE The Anisozygoptera is one of three surviving suborders of the Odonata. Only two living species are known, E. laidlawi in the Himalaya and E. superstes in Japan. These relicts are of special phylogenetic interest since they combine characters of the damselflies (Zygoptera) with those of the true dragonflies (Anisoptera). They are considered to be near the origin of the latter (5). Few groups of animals have survived virtually unchanged since the Triassic, the time before the evolution of the mammals and birds, when dinosaurs were begining to develop. Equivalent examples among vertebrates are the coelacanth fish Latimeria and the rhyncocephalian reptile Sphenodon. The phylogenetic importance of E. laidlawi is therefore outstanding. It shares its general habitat with numerous other species of great interest and so can be thought of as an indicator of a habitat of much wider interest. Furthermore, there is evidence to suggest that the changes in land use which threaten the survival of E. laidlawi also threaten the economic prospects of farmers in the region, as well as public health (5). Thus conservation of the dragonfly's habitat could have long-term benefits for the people of the region.

THREATS TO SURVIVAL A number of land-use changes in the eastern Himalaya threaten the continued existence of E. laidlawi. Deforestation, reclamation, cattle grazing and tourism at high altitudes are all taking place; all of these activities can be deleterious to the natural communities in the region (5). Due to an expanding human population, cultivation is reaching a high altitude. Watercourses below 2000 m tend towards chronic pollution (5). Cutting of tree branches ·for cattle food, fertilizer, fuel and market goods exacerbates forest décline and erosion. Perhaps the greatest threat is the grazing of cattle right up to the summits (5). Future threats are likely to involve construction of a highway on the south-western slopes of Mt Siwapuri and the intensive tourism it would open up on the mountain itself (5). These factors particularly jeopardize the population in the Siwapuri Mountains around the Kathmandu Valley, but similar impacts are considered likely throughout the limited range of the dragonfly (5).

CONSERVATION MEASURES TAKEN To date no measures have been taken in terms of actual management in situ. In 1981 WWF approved and funded a project of the IUCN/SSC Odonata Specialist Group pertaining to this species. Under the direction of an authority on the Japanese species of Epiophlebia, a survey was undertaken to determine the present status of E. laidlawi and to make conservation recommendations. Much of the information reported here devolved from that survey (5).

CONSERVATION MEASURES PROPOSED Strict maintenance of the habitats known to harbour the Relict Dragonfly will be necessary to ensure its survival. Additional colonies should be sought, and one or more reserves established within which activities will be limited to those compatible with the natural site components (5). In particular, the headwaters of the Bagmati River require protection. The IUCN/SSC Odonata Specialist Group recommended that the Nepal Research Centre be approached in order to determine the most effective way in which to seek this goal in concert with Nepalese authorities (6).

REFERENCES 1. Asahina, S. (1958). On a rediscovery of the larva of Epiophlebia laidlawi Tillyard from the Himalayas (Odonata, Anisozygoptera). Tombo 1(1): 1-2.

2. Asahina, S. (1961). Taxonomic characteristics of the Himalayan Epiophlebia larva (Insecta, Odonata). Proc. Japan Acad. 37: 42.

3. Asahina, S. (1961). Is Epiophlebia laidlawi Tillyard (Odonata, Anisozygoptera) a good species? Rev. Gesamt. Hydrobiol. 46: 441-446.

4. Asahina, S. (1963). Description of the possible adult dragonfly of Epiophlebia laidlawi from the Himalayas. Tombo 6(3/4): 18-20.

5. Asahina, S. (1982). Survey of the relict dragonfly Epiophlebia laidlawi Tillyard in Nepal, May 1981. Rep. Odon. Specialist Group Int. Un. Conserv. Nat. 1: 6pp. + 6 figs.

6. Moore, N.W. (1981). Minutes of the second meeting of the Odonata Specialist Group, Chur, Switzerland. Report to IUCN. 3 pp.

7. Tani, K. and Miyatake, Y. (1979). The discovery of Epiophlebia laidlawi Tillyard 1921 in the Kathmandu Valley, Nepal (Anisozygoptera, Epiophlebidae). Odonatologica 8: 329-332.

8. Tillyard, R.J. (1921). On an anisozygopterous larva from Himalayas. Rec. Indian Mus. 22: 93-107, 1 pl.

We are grateful to S. Asahina for providing information for this review, and to N.W. Moore and J.A.L. Watson for further comment.

FLORIDA SPIKETAIL DRAGONFLY   VULNERABLE

Cordulegaster sayi Selys, 1854

Phylum   ARTHROPODA   Order   ODONATA

Class   INSECTA   Family   CORDULEGASTRIDAE

SUMMARY This large, conspicuously coloured dragonfly (suborder Anisoptera), now present only in northern Florida, U.S.A., is represented by two populations, one of which is seriously threatened by a housing development.

DESCRIPTION The adult is large (wingspread 10 cm) and very handsome, with a black abdomen ringed with bright yellow, thorax vertically striped with magenta, yellow and white, eyes a peculiar grey-yellow green and face with a black moustache marking (3). A more detailed description is available in the literature (6). Cordulegaster sayi is also known as Say's Spiketail (3,4).

DISTRIBUTION Although there are several old records of this species from various states in the U.S.A. (5,6), its present occurrence in any state other than Florida is extremely unlikely (1). There are three males in the Cornell University collection from Lake City (Columbia County), Florida, and one specimen has been collected from each of the following Florida localities: Torreya State Park (Liberty County), San Felasco State Preserve (Alachua County), and near Keystone Heights (Clay County). There appears to be a small population at Gold Head Branch State Park, and there is a larger one within the city limits of Gainesville (Alachua County) (3).

POPULATION Unknown.

HABITAT AND ECOLOGY Small, shallow woodland streams seem to be important for oviposition. Nymphal development may occur in creeks but nymphs may also occur in seepage areas adjacent to the upper limit of silt deposits. Nymphs hide in the silt with only the head and tip of the abdomen exposed (4). Nearby open weedy fields are an asset for mating, for feeding (adults prey primarily on small wasps) and for sexual maturation of the newly emerged winged insect. Thus, there seems to be a three-fold habitat requirement: small woodland streams, adjacent seepage areas, and nearby open fields. The adults fly only briefly, from the first week of March through to the first week of April, and nymphal development is thought to require 3-4 years (2,3).

SCIENTIFIC INTEREST AND POTENTIAL VALUE The principal interest in the genus Cordulegaster is in the unique method of oviposition, the phylogenetic origin of the ovipositor, and the relationship of the holarctic genus Cordulegaster to the only other members of the family, the oriental genera Anotogaster and Neallogaster. The behaviour of adults and the long period of nymphal development are of great interest, but poorly understood (1).

THREATS TO SURVIVAL The largest population is in a rapidly expanding section of the city of Gainesville where it is being acutely threatened by a housing development. Here in the 1970s a sewer line and a road were constructed through the seepage area and houses took over most of the formerly open fields used for maturation and feeding (3). Tree cutting promoted water run-off and scouring of some of the seeps. Use of pesticides for mosquito abatement is another threat to the endangered population (3).

CONSERVATION MEASURES TAKEN Florida Spiketails have been found in Goldhead Branch and Torreya State Parks, and in San Felasco Hammock State

Preserve, but these are not known to represent viable populations (1,4). No specific measures have been taken other than a survey of existing records (3).

CONSERVATION MEASURES PROPOSED Only about one hectare of the Gainesville seep area remains intact. The possibility of purchasing, fencing and managing this area as a small preserve should be pursued. If this metropolitan population becomes extinct, there is only the hope that the population at Goldhead Branch State Park may be viable or that other populations may be discovered in northern Florida. An extensive survey should be undertaken to check these possibilities (1).

REFERENCES 1. Bick, G. (1982). In litt., 3 February.
2. Dunkle, S.W. (1981). The ecology and behaviour of Tachopteryx thoreyi (Hagen) (Anisoptera: Petaluridae). Odonatologica 10: 189-199.
3. Dunkle, S.W. (1981). What is happening to Say's Spiketail? Unpublished manuscript.
4. Dunkle, S.W. and Westfall, M.J., Jr. (1982). Say's Spiketail. In Franz, R. (Ed.), Rare and Endangered Biota of Florida Vol. 6. Invertebrates. University Presses of Florida, Gainesville. Pp. 35-36.
5. Needham, J.G. (1903). Aquatic insects of New York State. Part III. Bull. N.Y. St. Mus. 68: 218-276.
6. Needham, J.G. and Westfall, M.J. (1955). A manual of the dragonflies of North America (Anisoptera). Univ. Calif. Press, Berkeley.

We are grateful to G.H. Bick for the preliminary draft of this review, and to N.W. Moore and D. Paulson for additional comment.

SHINING MACROMIA DRAGONFLY   RARE

Macromia splendens Pictet, 1843

Phylum	ARTHROPODA	Order	ODONATA
Class	INSECTA	Family	CORDULIIDAE

SUMMARY The Shining Macromia is restricted to south-west France, and is isolated from other macromiine dragonflies. This large spectacular insect requires protection from pollution and other disturbance of the streams that it occupies.

DESCRIPTION As the specific name suggests, Macromia splendens is a most attractive dragonfly (L. splendere: to shine), described by Baron E. de Selys Longchamps as "la Macromie éclatante" - the Shining Macromia (5,6). In his diagnosis of the species de Selys describes the general colour as bronzed black, with a yellow head bearing a black tranverse mark. The first eight segments of the abdomen bear yellow dorsal marks (6). Useful notes on the rather confused original description of the species are given by Morton (4).

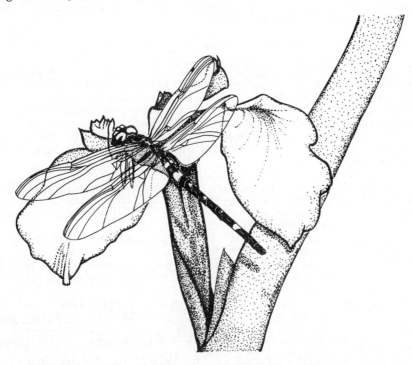

DISTRIBUTION Today, M. splendens is restricted to the Départements Lot, Dordogne, Charente, Gard, Var and Hérault of south-west France. Earlier sparse records include Portugal and Spain, but these have not been recently confirmed (2,3).

POPULATION Early writings give an impression of the great rarity of the species (4), but more recently it has been described as not rare in its present localities, and even occasionally common (3). However, its emergence on a biennial cycle makes accurate assessments difficult.

HABITAT AND ECOLOGY   The Shining Macromia breeds in slow-running, montane streams at middle altitudes. Records suggest biennialism, with regular appearance of adults only every two years (2). The natural history of the adult and aquatic larva is discussed in detail in (1,2,3).

SCIENTIFIC INTEREST AND POTENTIAL VALUE   This particularly conspicuous and beautiful dragonfly is the only European component of a large worldwide, mainly tropical and subtropical genus of Corduliidae (3). The species takes rather an isolated position both in regard to its morphology and its occurrence (3). It may be a Eurasian remnant of a much richer preglacial fauna which largely became extinct as a result of the cold climate during the Pleistocene glaciation (3). The various subspecies of its nearest relative, M. amphigena, occur in north-east and eastern Asia, from Siberia to Japan (2). The aquatic larval instars of dragonflies have specific habitat requirements and it has been suggested that they may be of value as ecological indicators of unpolluted fresh waters (2).

THREATS TO SURVIVAL   Although other western Palearctic macromiines apparently became extinct due to climatic changes in the Pleistocene, M. splendens seems to be well adapted to its present conditions, barring major human-induced change. The present threats and factors responsible for its decline are water pollution and stream channelization. Overcollecting has not yet proved damaging but has the potential to do so (2).

CONSERVATION MEASURES TAKEN   None.

CONSERVATION MEASURES PROPOSED   It has been suggested that M. splendens warrants an effort to protect at least part of its range (2). Sufficient safeguards could be achieved by preventing pollution on a number of tributaries of the Garonne, notably the River Lot and its chief tributaries, the Cele, the Aveyron and the Crieulon (2). Additional prime habitats should be investigated for local conservation attention. A detailed survey should be undertaken and a plan prepared for the protection of a number of populations in different parts of the range of the species.

REFERENCES   1.   Bilek, A. (1969). Erganzende Beobachtungen zur Lebensweise von Macromia splendens (Pictet 1843) und einigen anderen in der Guyenne vorkommenden Odonata-Arten. Ent. Z. 79: 117-124.
2.   Dumont, H. J. (1971). Need for protection of some European dragonflies. Biol. Conserv. 3: 223-228.
3.   Lieftink, M.A. (1965). Macromia splendens (Pictet, 1843) in Europe, with notes on its habitat, larva and distribution (Odonata). Tijdschr. Ent. 108: 41-59.
4.   Morton, K.J. (1924). Macromia splendens at last: an account of dragon-fly hunting in France. Entomologist's Mon. Mag. 61: 1-5.
5.   Selys Longchamps, E. de (1871). Synopsis des Cordulines. Bull. Acad. Belg. (2) 31: 238-565.
6.   Selys Longchamps, E. de and Hagen, H.A. (1850). Revue des Odonates ou Libellules d'Europe. Mém. Soc. Sci. Liège 6: xxii + 408 pp., 6 tables, 11 plates.

We thank H. Dumont for his contributions to this review, and N.W. Moore for further comment.

OHIO EMERALD DRAGONFLY   ENDANGERED

Somatochlora hineana Williamson, 1931

Phylum   ARTHROPODA   Order   ODONATA

Class   INSECTA   Family   CORDULIIDAE

SUMMARY This dragonfly, known from only four localities in Ohio and Indiana in mid-west U.S.A., has not been collected for 26 years. The species may already be extinct, but intensive searches for relict populations are needed.

DESCRIPTION This is a rather large Somatochlora with yellow labrum, metallic green frons, and black tibiae. On the dark thorax are two yellow stripes, the second being slightly wider and shorter than the first. The adult is well-described, beautifully illustrated (6) and clearly differentiated from all other North American species (3). The nymph is undescribed.

DISTRIBUTION The species is recorded only from Logan, Lucas and Williams Counties in north-west Ohio and from Lake County in north-west Indiana, U.S.A. (2,4,6). Only one specimen is known from the Indiana site and this is believed to have been a stray (7).

POPULATION In 1931, three males and four females were recorded from Logan County, Ohio, collected in 1929 and 1930 (6). There were no additional reports until 1953 when a single (probably stray) male was recorded from Gary, Lake County, Indiana, collected in 1945, and 17 males and 4 females were recorded (4) from Lucas and Williams Counties, Ohio, collected between 1949 and 1956. Thus the total number of individuals ever collected is 21 males and 8 females, and none have been taken in the last 26 years. S. hineana is almost certainly absent from the Gary area, Indiana, where there is heavy pollution from steel mills and associated industry (1). Indeed, because only one individual was collected from this area, it is doubtful if a viable population was ever present (7). With regard to the three north-western Ohio counties, the species has not been collected in Logan County since 1930 in spite of intensive collecting. It is probably extinct there. To judge from the number of individuals collected in Williams County, Ohio (one male and one female) the population there, if present, must be very small. The largest number of specimens (16 males, 3 females) were from Oak Openings State Park in Lucas County at the western end of Lake Erie, in the urban and heavily industrialized Toledo area. Because of the habitat requirements of this species, its presence today in the Toledo area is questionable (1).

HABITAT AND ECOLOGY The species has been found along a dredged channel of a small stream flowing through heavy swamp woods (6), but all other collections are from bogs (4). The adults are easy to capture but the flight season is short (4). All collections were between 7 June and 4 July. A description of breeding places for the genus Somatochlora (5) probably applies to S. hineana: rare, local and found only where original conditions have been little disturbed; found only in wilder districts and even there occupies remote places.

SCIENTIFIC INTEREST AND POTENTIAL VALUE Unless it is already too late, it is clearly of great scientific and ethical value to prevent the extinction of this attractive dragonfly.

THREATS TO SURVIVAL The species may already be extinct. If populations can be found an assessment of threats will be needed.

CONSERVATION MEASURES TAKEN No conservation measures have been taken specifically for this dragonfly, but the creation of Oak Openings State Park has provided incidental habitat protection if a population does survive there (1).

CONSERVATION MEASURES PROPOSED An immediate intensive search for viable populations of this possibly extinct species should be mounted. The three Ohio counties are the only possible localities, with the most likely being at Oak Openings State Park in Lucas County. If a population is still present there, the park authorities must be informed of the presence of this very rare species and their co-operation sought to maintain relatively undisturbed conditions (1).

REFERENCES   1.   Bick, G. (1982). In litt., 3 February.
             2.   Montgomery, B.E. (1953). Notes and records of Indiana Odonata, 1951-1952. Proc. Indiana Acad. Sci. 62: 200-202.
             3.   Needham, J.G. and Westfall, M.J. (1955). A manual of the dragonflies of North America (Anisoptera). Univ. Calif. Press, Berkeley.
             4.   Price, H.F. (1958). Additional notes on the dragonflies of north-western Ohio. Ohio J. Science 58: 50-62.
             5.   Walker, E.M. (1925). The North American dragonflies of the genus Somatochlora. Univ. Toronto Studies, Biol. Ser. 26: 1-202.
             6.   Williamson, E.B. (1931). A new North American Somatochlora (Odonata-Cordulinae). Occ. Pap. Mus. Zool. Univ. Mich. 225: 1-8.
             7.   Dunkle, S.W. (1982). Pers. comm. to G.E. Drewry, 25 August.

G.H. Bick kindly supplied the original draft of this review, and N.W. Moore and D. Paulson gave further comments.

MOUNT ST HELENS GRYLLOBLATTID   VULNERABLE

Grylloblatta chirurgica Gurney, 1961

Phylum   ARTHROPODA   Order   GRYLLOBLATTARIA

Class   INSECTA   Family   GRYLLOBLATTIDAE

SUMMARY Grylloblatta chirurgica is of great scientific interest because it is found only in a single lava flow and its caves on Mt St Helens, U.S.A. The type population may be extinct and the largest remaining population is threatened by indirect effects of Mt St Helens' volcanic activity. The other populations may be threatened in the future by human impacts.

DESCRIPTION The Mt St Helens Grylloblattid is an 18 mm long wingless insect, shaped like an earwig or rove beetle, with long antennae and cerci; its colour is yellow-brown with small but distinct black compound eyes. For a technical description see (4). Grylloblattids are also known as rock crawlers or ice crawlers.

DISTRIBUTION Grylloblatta chirurgica has been found only at 280-950 m in lava tube caves and associated habitats in a single lava flow, the 'Cave Basalt', on the south slope of Mt St Helens, Washington, U.S.A. There are records from nine individual caves (1,4). The lava flow is about 10 km long and 1-3 km wide. The flow is only about 1900 years old (3) and this species could not have evolved in such a short time. Ancestral populations must have been present in the previously caveless terrain surrounding the lava flow. Such populations may still exist but are undiscovered. An undescribed grylloblattid has recently been found on the south slope of Mt St Helens, above Butte Camp (1380 m). This alpine population does not appear to belong to the species G. chirurgica (2).

POPULATION Unknown.

HABITAT AND ECOLOGY Individuals are generally found under rocks, but sometimes in the open on the walls or floors of the caves. Presumably they also occur in the rock-fissure habitat which permeates the lava flow surrounding and between the caves. Conditions in these subterranean habitats include constant high humidity, reduced temperature variation compared to the surface (recorded temperatures in deep parts of these caves range from 2.5 to 9.7°C), total or near total darkness, and limited food resources (1). G. chirurgica is not, however, restricted to subterranean habitats. It has been found under rocks just outside a cave entrance, and rarely, usually at night during or just after snowfall, has been seen outside on the snow surface (1). To some extent this parallels observations on Grylloblatta spp. living in ice caves in central Oregon (10). These insects were supposed to migrate out of caves to the hypolithion in winter when cave temperatures drop below freezing with consequent reduction of humidity. However, the caves in whch G. chirurgica is generally found contain no ice, and never drop below freezing point. Dates of collection suggest that the movement to the surface follows no strict seasonal pattern (1).

The feeding and reproductive biology of G. chirurgica is totally unknown. Presumably, like its congeners, it is a predator and scavenger on dead arthropods. Juveniles have been found in all seasons of the year (1).

SCIENTIFIC INTEREST AND POTENTIAL VALUE Grylloblattids have fascinated entomologists since their discovery because of their comparative rarity, supposed primitive phylogenetic position, relictual distribution (9,10) and unusual temperature preference; the better-known alpine species are most active at temperatures near 0°C (7,10). Few genera of insects have evoked wider interest

347

or more speculation. None of the studies to date have involved G. chirurgica, but it would make an excellent subject for study in the future, as its habitat is relatively accessible and specimens may be found at any time of year. The biology and behaviour of the species is unknown, its geographic distribution is anomalous, its taxonomic affinities remain obscure, and the temperature of its habitat is significantly higher than preferred temperatures reported for other Grylloblatta spp. It would be valuable to investigate all these points. The special interest is further enhanced by its apparently endemic occurrence in a geologically young habitat, and its ecological relationship to the current vulcanism of Mt St Helens (1).

THREATS TO SURVIVAL Ape Cave, the type locality of G. chirurgica, has been developed by the U.S. Forest Service as a visitor attraction. Intensive and continuous human visitation has resulted in the virtual disappearance of the cave's fauna, including Grylloblatta. None have been collected there since 1965 and the Ape Cave population may be extinct. Some of the cave's fauna returned during the period May 1980 to March 1982, when the area was closed due to eruptive activity of Mt St Helens, but no Grylloblatta were noted (1). The largest population of G. chirurgica now known is in and around Spider Cave. There is some danger that an alluvial mudflow (a slow, water-transported flow of mud and debris caused by recent volcanic activity, and diverted by man-made roads) will engulf the cave (5,6).

No specimens of G. chirurgica have been seen since the eruption of Mt St Helens in May 1980 (1). However, little biological research has been done in the relevant caves during that period (1). There is no real reason to expect that the populations in the lower caves have been or will be affected by present or future volcanic activity. Mt St Helens has been active periodically for thousands of years, and the species has managed to survive until now. However, if human visitation to the other caves increases in the future, their populations could suffer in the same manner as that of Ape Cave (1).

CONSERVATION MEASURES TAKEN Most of the caves in which G. chirurgica occurs are on lands of the Gifford Pinchot National Forest. At least one cave is on private land within the national forest and one is outside the forest boundary. Prior to the 1980 activity of Mt St Helens, no specific action was taken for conservation of G. chirurgica. When it became clear that the Spider Cave population was threatened with inundation this, and the problem of visitor impact on other cave populations, was pointed out in comments on the U.S. Forest Service draft land management plan and environmental impact statement for the area (13), as well as in formal meetings between speleologists and U.S. Forest Service personnel (12). U.S. Forest Service officials agreed, both verbally and in the final management plan (14) to "determine the magnitude of threat to the area from flood-washed debris and what, if any, measures to reduce future damage are appropriate", and also to limit public access to caves as appropriate to protect their fauna and features. These management measures remain to be carried out (1).

CONSERVATION MEASURES PROPOSED Ideally, the habitat of Grylloblatta chirurgica should be included in a Mt St Helens National Monument. Legislation to this effect is pending along with an array of less protective land use designations (8). Of all proposals, the national monument would best protect natural diversity on Mt St Helens, including this species and other invertebrates (11). In the absence of this, pressure should be brought to bear on the U.S. Forest Service to fulfil their commitment to protect the Cave Basalt and its features and fauna. A search for possible non-cave populations of this species is needed. Of more immediate importance, steps should be taken to protect the Spider Cave population from engulfment by alluvial mudflow (1).

REFERENCES   1. Crawford, R.L. (1982). In litt., 3 April.
2. Crawford, R.L. (1982). In litt., 30 April.
3. Greeley, R. and Hyde J.H. (1972). Lava tubes of the Cave Basalt, Mount St Helens, Washington. Bull. Geol. Soc. Amer. 83: 2397-2418.
4. Gurney, A.B. (1961). Further advances in the taxonomy and distribution of the Grylloblattidae (Orthoptera). Proc. Biol. Soc. Wash. 74: 67-76.
5. Halliday, W.R. (1981). Further observations of the effects of post-eruption phenomena on caves of Mount St Helens, Washington. Western Speleological Survey, Bulletin. 65: 1-5.
6. Halliday, W.R. (1981). St Helens: observations January 10-11. Speleograph. 17(2): 34-37.
7. Henson, W.R. (1957). Temperature preference of Grylloblatta campodeiformis Walker. Nature 179(4560): 637.
8. Hooper, D. (1982). The spoils of St. Helens. Pacific North West 16(2): 36-46.
9. Kamp, J.W. (1963). Descriptions of two new species of Grylloblattidae and of the adult of Grylloblatta barberi, with an interpretation of their geographic distribution. Ann. Ent. Soc. Am. 56: 53-68.
10. Kamp, J.W. (1973). Biosystematics of the Grylloblattodea. Thesis, University of British Columbia. 276 pp.
11. Pyle, R.M. (In press). The impact of recent vulcanism on Lepidoptera. In Vane-Wright, R.I. and Ackery P. (Eds), The Biology of Butterflies.
12. Seesholtz, D.M. (1981). Reports to the district ranger, St Helens Ranger District, on meetings between speleologists and Gifford Pinchot National Forest on June 6 and September 26, 1981. Unpublished, private distribution.
13. U.S. Forest Service (1981). Draft environmental impact statement: Mount St Helens land management plan. Vancouver, Wash., 2 April 1981. 162 pp. + 12 map sheets.
14. U.S. Forest Service (1981). Final environmental impact statement: Mount St Helens land management plan. Vancouver, Washington, October 1981. 288 pp. + 12 map sheets.

We thank R.L. Crawford for the original draft used in this account, and R.I. Gara for further comment.

GIANT WETAS   VULNERABLE

Deinacrida carinata Salmon, 1950   Herekopare I. Weta
D. fallai Salmon, 1950   Poor Knights Weta
D. heteracantha White, 1842   Wetapunga
D. rugosa Buller, 1871   Stephens I. Weta

Phylum   ARTHROPODA   Order   ORTHOPTERA

Class   INSECTA   Family   STENOPELMATIDAE

SUMMARY Among the largest insects in the world, the giant wetas are now
confined to small offshore islands in New Zealand. They are threatened by
predation from introduced animals and modification of the habitat.

DESCRIPTION Deinacrida heteracantha is New Zealand's largest and heaviest
insect and is one of the largest insects in the world. An adult female 'Wetapunga'
(20) may measure 85 mm x 32 mm excluding the ovipositor (16), and an egg-laden
female may weigh up to 71 g (13,14). The adult is mid-brown, wingless and has a
squat body. The tibias of the hind legs are armed with long spines which in
defence are raised high above the succulent abdomen and suddenly drawn down and
back so that the spines inflict a wound on the attacker (6,15). The vigorous
movements of the legs produce a loud rasping sound from file-like ridges on the
body. This is quite effective in frightening birds but apparently does not deter
rats (16). D. fallai, the 'Poor Knights Weta' (20), was not described until 1950.
Adults are similar to D. heteracantha and have a deep chocolate-brown body
mottled with darker brown, and a dark brown stripe down the abdominal tergites
2-8. The ovipositor is a pale red-brown (13). The adult D. rugosa, the 'Stephens
Island Weta' (20), is light brown with a transverse row of raised dots on the
posterior margin of the dorsal abdominal segments. The hind tibia has at least
four pairs of sharp spines. As in all four species, females are larger than males
(13,14). D. carinata is a much smaller species, light brown in colour with the
transverse dark and light notch-like markings on the hind edge of the tergites
which are characteristic of the first instars of the other species (17). An
annotated bibliography of New Zealand wetas has been published (18).

DISTRIBUTION D. heteracantha is now certainly known only from Little Barrier
Island, New Zealand (13), although some individuals may persist on the mainland
(17). A population was discovered there during the 1960s but the single locality
has since been burned and cleared (15,19). Being an arboreal species, D.
heteracantha can co-exist with the Maori or Polynesian Rat (Rattus exulans) and
the Norway Rat (Rattus norvegicus) and still occurred in the northern part of the
North Island last century (2,19). In the first half of the last century D.
heteracantha occurred throughout the forests of the northern part of North Island,
and on Great Barrier Island (8). The accidental introduction of the Ship or Black
Rat (Rattus rattus) was the likely cause of its extermination there, and it is not
understood why the small recently-discovered population was able to persist
(2,17). D. fallai probably developed in isolation on the Poor Knight's Islands and
has never occurred on the mainland (13,17). It has been described as reasonably
plentiful on both Aorangi and Tawhiti Rahi, and it may also occur on some of the
smaller islands in the group (13). The occurrence of the other two species, D.
rugosa and D. carinata, on mainland New Zealand has never been authenticated
although they must once have been there. It is possible that they were eliminated
by the Maori Rat before the advent of Europeans. Now, D. rugosa is confined to
Stephens and Mana Islands (11), and possibly Middle Trio Island (12,19). D.
carinata is only found on Herekopare Island in Foveaux Strait (19).

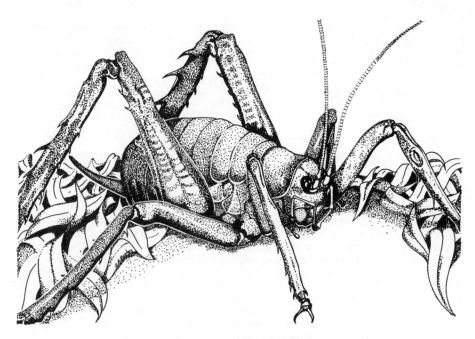

Stephens Island Weta (<u>Deinacrida</u> <u>rugosa</u>)

POPULATION In 1950 <u>D</u>. <u>heteracantha</u> was described as being fairly common (14) and in 1978 the population was estimated at 3000 by a park ranger (6). The population of <u>D</u>. <u>fallai</u> is unknown, but is believed to be secure (17). Surveys in 1976 and 1977 revealed 34 <u>D</u>. <u>rugosa</u> on Stephens Island, an estimated several hundred on Mana Island, but no live wetas, nor sign over the past decade on Middle Trio Island (9,10,11). The populations of <u>D</u>. <u>fallai</u> and <u>D</u>. <u>carinata</u> are unknown.

HABITAT AND ECOLOGY Prior to human settlement terrestrial mammals and mammalian predators were absent from the New Zealand ecosystem. Seals and two species of bat were the only exceptions. Consequently the well-known development of fat, heavy, ground-dwelling birds occurred. The situation was similar with the terrestrial invertebrates, although it has not been so well documented. It is likely that a number of species were eliminated by the Maori Rat in pre-European times. In New Zealand the rodent ecological niche was filled by wetas, which are sometimes known as 'invertebrate mice' (16,19). They are nocturnal, hiding during the day beneath rocks, logs or bark, in low vegetation, crevices, or holes in trees. They are mainly herbivorous, feeding on the leaves of shrubs, grasses and herbs, and possibly on ripe fruits, as well as occasional invertebrate carrion. <u>D</u>. <u>heteracantha</u> and <u>D</u>. <u>fallai</u> are mainly arboreal, with much longer antennae and relatively longer legs than <u>D</u>. <u>rugosa</u> and <u>D</u>. <u>carinata</u>, which are primarily ground dwellers with shorter antennae and legs, and heavier build. The latter are therefore much more vulnerable to rodent predation (19). <u>D</u>. <u>heteracantha</u> has been found in Kanuka (<u>Leptospermum</u> <u>ericoides</u>) forest, on Kohekohe (<u>Dysoxylum</u> <u>spectabile</u>), and weta faeces have been found at the base of Nikau Palms (<u>Rhopalostylis</u> <u>sapida</u>). At night wetas occur on the trunks and fronds of a number of European fan palms (e.g. <u>Chamaerops</u> <u>humilis</u>), on the branches of Mahoe (<u>Melicytus</u> <u>ramiflorus</u>) and on Pohutukawa (<u>Metrosideros</u> <u>excelsa</u>) in the vicinity of fan palms. <u>D</u>. <u>rugosa</u> on Stephens Island occurs in low entanglements of <u>Muehlenbeckia</u>, on tussocks and in grassland (9). Giant wetas are usually solitary although they may occur in pairs in the summer. Wetas are preyed upon by a large number of vertebrates including cats, rats, a harrier (Circus approximans), a kingfisher (Halcyon sancta), a morepork (Ninox novaeseelandiae), the Tuatara (Sphenodon punctatus) and lizards (13,19). The life span of giant wetas is at least two to three years. Mating occurs over several days during spring and summer. Eggs are laid in moist soil at night during the summer. Up to 600 eggs may be laid, singly or in groups of up to five, by a single female.

SCIENTIFIC INTEREST AND POTENTIAL VALUE Wetas belong to a primitive group of Orthoptera and their study provides valuable information on comparative aspects of insect biology. Currently scientists are studying their neurobiology, sound production, communication and social behaviour, ecology, diet, and photobiology (1,3,4,5,7). Their evolution on an isolated island system provides an excellent opportunity for the study of adaptive radiation (3,6).

THREATS TO SURVIVAL The main threats to the survival of these wetas are the invasion of their sanctuaries by introduced rodents (especially the Ship Rat) and habitat destruction (2,8,13,19). D. heteracantha can survive in the presence of the Maori and Ship Rats, D. rugosa with mice, D. carinata with cats, and D. fallai with pigs (long since eliminated), even though all of these prey upon them (19). Predation by Tuataras on D. rugosa on two of its islands is not a serious threat. The low numbers of this species on Middle Trio Island (if indeed it still exists there) may be due to the Wood Hens (Gallirallus australis) which were illegally introduced there but have since been removed (9,10,17).

CONSERVATION MEASURES TAKEN All four species are protected by the Wildlife Amendment Act 1980 and occur in legally protected nature reserves (13). Both Stephens Island and Middle Trio Island are flora and fauna reserves. The Poor Knight's Islands were made a sanctuary in the early 1920s (13), and neither Aorangi nor Tawhiti Rahi have rats or other introduced mammals. Wild cats have recently been eliminated from Great Barrier Island and Herekopare Island (17). Surveys have been carried out to ascertain the status of D. rugosa, and an attempt has been made to establish a population on another suitable rodent-free island sanctuary (12).

CONSERVATION MEASURES PROPOSED The populations of these four wetas should be monitored regularly and their habitats managed to provide the most favourable conditions, once these are known. Further information is needed as a basis for effective management plans. Every precaution against invasion by rats should be taken, and the protecting laws rigidly enforced. Consideration should be given to establishing additional colonies on suitable islands and to obtaining data necessary for the establishment of captive breeding colonies (12,17).

Two species, D. fallai and D. rugosa, have been successfully bred in captivity, and this would probably be possible for all species. Captive specimens can be fed on lettuce, but cannibalism tends to occur if they are overcrowded (13). D. fallai has been reared successfully through two successive generations on Buddleia davidii. Top priority could be given to the establishment of a breeding programme since surplus individuals could be sold in order to pay for the research needed.

REFERENCES 1. Ball, E.E. and Field, L.H. (1981). Structure of the auditory system of the New Zealand weta Hemideina crassidens (Blanchard, 1851) (Orthoptera, Ensifera, Gryllacridoidea, Stenopelmatidae). I. Morphology. Cell and Tiss. Res. 217: 321-343.

2. Buller, W.L. (1895). On the wetas: a group of orthopterous insects inhabiting New Zealand, with descriptions of two new species. Trans. N.Z. Inst. 29: 143-147.

3. Field, L.H. (1978). The stridulatory apparatus of the New Zealand wetas in the genus Hemideina (Insecta: Orthoptera:

Stenopelmatidae). J. Roy. Soc. N.Z. 8: 359-375.
4. Field, L.H. (1980). Observations on the biology of Deinacrida connectens (Orthoptera: Stenopelmatidae), an alpine weta. N.Z. J. Zool. 7: 211-220.
5. Field, L.H. (1980). The tick sound of a giant weta, Deinacrida rugosa (Orthoptera: Stenopelmatidae: Deinacridinae). N.Z. Ent. 7: 176-183.
6. Field, L. (1980). In litt., 14 August.
7. Field, L.H., Hill, K.G. and Ball, E.E. (1980). Physiological and biophysical properties of the auditory system of the New Zealand weta Hemideina crassidens (Blanchard, 1851) (Ensifera: Stenopelmatidae). J. Comp. Physiol. 141: 31-37.
8. Hutton, F.W. (1980). The Stenopelmatidae of New Zealand. Trans. Proc. N.Z. Inst. 29: 208-242.
9. Meads, M.J. (1976). Visit to Stephens Island, Cook Strait, from 22nd April to 3 May 1976: a preliminary report. Unpubl. report, Ecology Division, DSIR, File 4/15/9.
10. Meads, M.J. (1976). Visit to Chetwode Islands, Middle Trio Is., Cook Strait; Maud Island, Marlborough Sounds, from 23 September to 7 October 1976: a report. Unpubl. report, Ecology Division; DSIR, File 4/15/9.
11. Meads, M.J. and Moller, H. (1977). Report of a visit to Mana Island in September 1977. Unpubl. report., Ecology Division, DSIR, File 4/15/8.
12. Meads, M.J. and Moller, H. (1978). Introduction of giant wetas (Deinacrida rugosa) to Maud Island and observations on tree wetas, paryphantids and other invertebrates. Unpubl. report, Ecology Division, DSIR, File 4/15/9.
13. Richards, Aola M. (1973). A comparative study of the biology of the giant wetas Deinacrida heteracantha and D. fallai (Orthoptera: Henicidae) from New Zealand. J. Zool. 169: 195-236.
14. Salmon, J.T. (1950). A revision of the New Zealand wetas Anostostominae (Orthoptera: Stenopelmatidae). Dominion Museum Records in Entomology, Wellington 1: 1-177.
15. Watt, J.C. (1963). The rediscovery of a giant weta Deinacrida heteracantha on the North Island mainland. N.Z. Ent. 3: 9-13.
16. Watt, J.C. (1975). The terrestrial insects. In Kuschel, G. (Ed.), Biogeography and Ecology in New Zealand. Junk Publishers, The Hague.
17. Ramsay, G.W. (1982). In litt., 1 September.
18. Ramsay, G.W. (1979). Annotated bibliography and index to the New Zealand wetas (Orthoptera: Stenopelmatidae, Rhaphidophoridae). DSIR Information Series No. 144. New Zealand Department of Scientific and Industrial Research, Wellington.
19. Ramsay, G.W. (1978). A review of the effect of rodents on the New Zealand invertebrate fauna. In Dingwall, P.R., Atkinson, I.A.E. and Hay, C. (eds), The Ecology and Control of Rodents in New Zealand Nature Reserves. Information Series No. 4. Dept. Lands and Surveys, Wellington.
20. Ramsay, G.W. and Singh, P. (1982). Conservation and collecting. In Guide to New Zealand entomology. Bull. Ent. Soc. N. Z. 7: 28-35.

We are grateful to L.H. Field, M.J. Meads and G.W. Ramsay for information provided for this account.

LORD HOWE ISLAND STICK-INSECT

ENDANGERED

Dryococelus australis (Montrouzier, 1885)

Phylum	ARTHROPODA	Order	PHASMATODEA
Class	INSECTA	Family	PHASMATIDAE

SUMMARY Formerly common on Lord Howe Island, but thought to have been exterminated by rats since 1918, this stick-insect may still be precariously confined to a small island known as Ball's Pyramid, 700 km east of the Australian mainland.

DESCRIPTION The Phasmatodea (stick-insects or walkingsticks) are remarkable in closely resembling the twigs or foliage upon which they rest and feed. Phasmatodea may also be known as Phasmatoptera, Phasmida or Cheleutoptera (11). The earliest description of Dryococelus australis was by MacGillivray in 1853, who found Lord Howe Island "alive with colonies of a singular cricket-looking wingless insect between four and five inches which the local people on the island have designated the land lobster" (5). D. australis is a large shiny, heavy bodied, wingless species of the mainly Papuan subfamily Eurycanthinae (3,4). The general colour is dark red-brown with a black tinge on head and thorax; the mesosternum and membranous coxal areas are pale (2). The total body length is about 120 mm, the hind femur 23 mm, ovipositor sheath, 24 mm (2). The male has greatly thickened hind femora with a few heavy spines beneath (3,4).

DISTRIBUTION D. australis is endemic to the Lord Howe Island group, 700 km north-east of Sydney, New South Wales, Australia (7).

POPULATION D. australis was formerly abundant on Lord Howe Island (5,7,8), but in 1918 the accidental introduction of the omnivorous and predatory Black Rat (Rattus rattus) from a grounded ship is believed to have caused its extinction there (6). However, in 1964 a recently dead female was found on the nearby island of Ball's Pyramid (7). A search there in 1969 revealed more dead remains, in "woody bush" at 1600 ft (488 m), and in a bird's nest (7), but a search in 1981 was entirely unsuccesful (11). Ball's Pyramid, a spectacular 650 m spine of volcanic rock, is only about 0.5 km^2 in area with little vegetation (10). If the stick-insect still survives there, the population may be assumed to be precariously low.

HABITAT AND ECOLOGY Lord Howe and adjacent volcanic islands were formed in the Miocene and have since been eroded to 10 per cent of their original size (9). The island is mountainous with a sub-tropical climate, and vegetation ranging from closed forest (including rain forest) to open grassland (1). Ball's Pyramid, 23 km south of Lord Howe Island, is a narrow pinnacle apparently supporting little vegetation (9,10). D. australis is probably nocturnal, hiding by day in cavities in the trunks of living trees (3) or rotten stumps (2), and feeding on leaves by night. Since Ball's Pyramid is far more exposed and less vegetated than Lord Howe Island, it would appear to present a less stable and less suitable habitat for the stick-insect.

SCIENTIFIC INTEREST AND POTENTIAL VALUE The isolation of the Lord Howe Island group has resulted in the evolution of many endemic species. However, many of these have become extinct since the arrival of settlers on the island in 1788, because of hunting and competition with introduced species (6,7,8,9). D. australis is hopefully one of the surviving endemic species, and is probably representative of a much wider and little-known invertebrate fauna peculiar to this island group.

THREATS TO SURVIVAL Introduction of rats or other non-native mammals to Ball's Pyramid would almost certainly cause extinction of the stick-insect. In addition to man-made threats, the exposed position and small size of the island make the habitat and vegetation susceptible to storm damage.

CONSERVATION MEASURES TAKEN No specific measures have been taken to protect D. australis. However, most of Lord Howe Island and all immediately surrounding islands, including Ball's Pyramid, are covered by provisions of the Lord Howe Island (Amendment) Act 1981, which invokes Part V of the New South Wales National Parks and Wildlife Act, 1974. These areas now form the Lord Howe Island Permanent Park Preserve. The legislation is designed to protect and preserve habitat types, and requires the N.S.W. National Parks and Wildlife Service and the Minister of Lands to prepare a management plan (10). This plan has not yet been written. The island group was proposed in 1982 for the Unesco/MAB World Heritage List of protected places (10). A decision on this proposal is expected soon.

CONSERVATION MEASURES PROPOSED The Lord Howe Island group certainly warrants inclusion in the World Heritage List. It remains to be seen whether the proposed management plan will consider Dryococelus australis. As a preliminary action, Ball's Pyramid should be thoroughly surveyed for the stick-insect. Since the habitat there is vulnerable it is important to return a protected population to Lord Howe Island. Such a move would necessitate rat control. Many stick-insects will breed in captivity and the establishment of a successful laboratory colony of D. australis would ensure its future survival, if living insects can be found.

REFERENCES
1. Clark, S.S. and Pickard, J. (1974). Vegetation and environment. Aust. Nat. Hist. 18(2): 56-60.
2. Gurney, A.B. (1947). Notes on some remarkable Australasian walkingsticks, including a synopsis of the genus Extatosoma. Ann. Ent. Soc. Am. 40: 373-396.
3. Key, K.H.L. (1970). Phasmatodea (stick-insects). In The Insects of Australia. A Textbook for Students and Research Workers. CSIRO. Melbourne University Press, Melbourne. Pp. 348-359.
4. Key, K.H.L. (1978). The conservation status of Australia's insect fauna. Aust. Nat. Parks Wildl. Serv. Occ. Paper No. 1, 24 pp.
5. MacGillivray, J. (1853). The private journal of John MacGillivray, concerning his voyages in H.M.S. Herald. Two unpublished vols. Admiralty Library: London.
6. Recher, H.F. (1974). Colonisation and extinction. Aust. Nat. Hist. 18 (2): 64-68.
7. Smithers, C.N. (1970). On some remains of the Lord Howe Island phasmid Dryococelus australis (Montrouzier) (Phasmida) from Ball's Pyramid. Ent. Mon. Mag. 105: 252.
8. Smithers, C., McAlpine, D., Colman, P. and Gray, M. (1974). Island invertebrates. Aust. Nat. Hist. 18(2): 60-63.
9. Sutherland, L. and Ritchie, A. (1974). Defunct volcanoes and extinct horned turtles. Aust. Nat. Hist. 18 (2): 44-49.
10. Watson, J.M.C. (1981). Convention Concerning the Protection of the World Cultural and Natural Heritage. Nomination for the World Heritage List. In litt.
11. Key, K.H.L. (1982). In litt., 1 November, 3 December.

We are grateful to K.H.L. Key, T.J. Kingston, J. Marshall, J.D. Ovington and C.N. Smithers for comments on this review.

ST HELENA EARWIG   VULNERABLE

Labidura herculeana (Fabricius, 1798)

Phylum   ARTHROPODA   Order   DERMAPTERA

Class   INSECTA   Family   LABIDURIDAE

SUMMARY Labidura herculeana is the world's largest earwig. Confined to the
extreme north-east of the island of St Helena in mid-Atlantic, it was described in
1798, but only rediscovered in 1965. More recent searches have failed to find it,
perhaps because of unsuitable collecting methods. However, dead remains outside
its last known locality indicate a shrinking range.

DESCRIPTION This stunning insect is the largest earwig in the world, with a body
length of 36-54 mm and additional forceps of 15-24 mm. The largest specimen is a
male of 78 mm total length preserved in the collections of the Musée Royal de
l'Afrique Centrale, in Tervuren, Belgium (1). Females tend to be smaller, with
relatively shorter forceps. Labidura herculeana is wingless, and distinguished by
its entirely black body, reddish legs and short elytra. The forceps of the male are
cylindrical, smooth and exceptionally long (2). A detailed taxonomic description is
available in the literature (2).

DISTRIBUTION Labidura herculeana is only known from the island of St Helena,
situated in mid-Atlantic almost 8° south of the Equator. The earwig is very
localized, being confined to Horse Point Plain in the north-east, near to
Prosperous Bay. Remains have been found on the south and south-east flanks of
Flagstaff, several kilometres away, but no living specimens have ever been found
there (3).

POPULATION Unknown.

HABITAT AND ECOLOGY Horse Point Plain is dry and barren with stony soils
supporting small bushes and tufts of grass (2). Earwig remains, principally forceps,
tend to accumulate under large, partially embedded stones and at the bases of the
vegetation (2). Most living specimens have been found under stones, or near
burrows in the soil under stones. These burrows, used by the earwigs as escape
routes, extend for some distance into the soil but become indistinguishable from
soil fissures. The insects are behaviourally adapted to dry conditions, being
nocturnal and active during the summer rains, seeking underground shelter at the
onset of the dry season. Mating seems to occur between December and February
and females with eggs have been observed in March (2).

SCIENTIFIC INTEREST AND POTENTIAL VALUE Although the earwig fauna of
St Helena consists only of four species, it is of interest in the inclusion of two
totally wingless species in addition to the endemic described here, which has
elytra only. Reduction or loss of wings is a feature throughout the Dermaptera,
but is possibly more prevalent on islands (2). The St Helena Earwig is clearly of
great interest, being the largest species in the world. The only other earwig which
approaches L. herculeana in size is Titanolabis colossea from Australia, which
reaches a total length of up to 65 mm (2).

THREATS TO SURVIVAL Although confined to the north-east of St Helena, this
earwig was described as not uncommon during the Belgian Zoological Expedition
from Tervuren in 1965-67 (2), when 40 specimens were collected (1). However, its
limited distribution even then caused some concern about its future status (2).
More recently, reports have been received that searches by a local resident have
failed to find the earwig (3). Confirmation of these reports is being sought.

In view of the fact that remains of the earwigs, but not the earwigs themselves, have been found near Flagstaff, it may be assumed that its range is indeed decreasing. The chitinous forceps seem to survive desiccation and decomposition for some time (2), bearing witness to this reduction. The cause of the decline is unknown.

CONSERVATION MEASURES TAKEN None. There are no protected areas on St Helena at present.

CONSERVATION MEASURES PROPOSED A preliminary survey is needed, to assess the exact status of the species. A study of ecological requirements would identify any threats and provide information needed to set up suitable habitat protection and management measures.

REFERENCES   1.   Basilewsky, P. (1982). In litt., 9 November.
             2.   Brindle, A. (1970). Dermaptera. La faune terrestre de l'Ile de Sainte-Helene (Pt.1). Ann. Mus. Roy. Afr. Centr., Zool. 181: 213-227.
             3.   Brindle, A. (1982). In litt., 3 August.

We are grateful to A. Brindle for information on this species, and to P. Basilewsky and J. Marshall for further comment.

OTWAY STONEFLY   ENDANGERED

Eusthenia nothofagi Zwick, 1979

Phylum   ARTHROPODA   Order   PLECOPTERA

Class   INSECTA   Family   EUSTHENIIDAE

SUMMARY   This stonefly is a species from the very primitive subfamily Eustheniinae, which is of zoogeographic importance. Only 14 specimens are known, all collected in 1932 from the Otway Ranges, Victoria, Australia. Since then it has not been definitely identified.

DESCRIPTION   Eusthenia nothofagi is a large (38 mm), brachypterous species of stonefly with reddish-purple wings, distinguished from the related E. venosa by the male genitalia (6).

DISTRIBUTION   The exact collecting localities are unknown, but all 14 specimens were taken in the Otway Ranges in Victoria, Australia. Abundant eustheniid material collected elsewhere in Victoria indicates that E. nothofagi is restricted to the Otway Ranges (2).

POPULATION   Unknown.

HABITAT AND ECOLOGY   The habitat and ecology of this particular species are unknown, but probably resemble those of E. venosa described in (4). The nymphs are stream dwellers and the adults, being flightless or virtually so, are presumably to be found near streams.

SCIENTIFIC INTEREST AND POTENTIAL VALUE   The eustheniid stoneflies are considered to be the most archaic and least evolved Plecoptera (5). The circum-antarctic distribution of the family and the unusual distributions of some of the sub-families are of great zoogeographic interest (1).

THREATS TO SURVIVAL   Streams in the Otway Ranges have already been considerably affected by clearing of native forest for agriculture and plantations of exotic trees, mostly Pinus radiata (3). Future threats to the area include proposals to establish a wood chip industry in a substantial part of the remaining native forest (3).

CONSERVATION MEASURES TAKEN   None.

CONSERVATION MEASURES PROPOSED   A survey of streams in the Otway Ranges is required to establish the present distribution and abundance of the species. Once it has been rediscovered, studies on its life history and general ecology will be needed (2). Enquiries are under way into the possible existence of more recent specimens in collections.

REFERENCES   1.   Campbell, I.C. (1981). Biogeography of some rheophilous aquatic insects in the Australasian region. Aquatic Insects 3: 33-43.
             2.   Campbell, I.C. (1982). In litt., 24 May.
             3.   Campbell, I.C. (1982). In litt., 3 August.
             4.   Hynes, H.B.N. and Hynes, M.E. (1975). The life histories of many of the stoneflies (Plecoptera) of south-eastern mainland Australia. Aust. J. Mar. Freshwat. Res. 26: 113-153.
             5.   Illies, J. (1965). Phylogeny and zoogeography of the

Plecoptera. Ann. Rev. Ent. 10: 117-140.

6. Zwick, P. (1979). Revision of the stonefly family Eustheniidae (Plecoptera), with emphasis on the fauna of the Australian region. Aquatic Insects 1: 16-50.

We are grateful to I.C. Campbell for a draft of this reiew, and to H.B.N. Hynes, S. Jewett, A. Neboiss and W.D. Williams for further comment.

MOUNT KOSCIUSKO WINGLESS STONEFLY

RARE

Leptoperla cacuminis Hynes, 1974

Phylum	ARTHROPODA	Order	PLECOPTERA
Class	INSECTA	Family	GRIPOPTERYGIDAE

SUMMARY One of only two apterous Australian stonefly species; known only from a small stream in the head waters of the Snowy River near the summit of Mt Kosciusko, Australia.

DESCRIPTION The nymphs are pale yellow, without wing stubs (5). The mature adult is wingless and has a head and body length of 8 mm, an amber head and thorax, short brown legs, a light brown abdomen, short cerci and long (4-4.5 mm) antennae. The male genitalia are inconspicuous, which is unusual in flightless species (2,3).

DISTRIBUTION This stonefly is known only from a single small stream in the headwaters of the Snowy River at 2135 m, immediately below the summit of Mt Kosciusko (2230 m), New South Wales, Australia (2).

POPULATION Unknown.

HABITAT AND ECOLOGY Nymphs live under stones in the stream and in summer adults may be found amongst herbage on the banks (1). It has been suggested that *Leptoperla cacuminis* has a simple annual cycle: adults emerge in February and lay eggs which hatch in May to grow to emergence in February (6).

SCIENTIFIC INTEREST AND POTENTIAL VALUE There are very few apterous stoneflies worldwide, and *L. cacuminis* is one of only two examples in Australia (2). This species is unusual in spending a large proportion of its life cycle under snow (1). Another stonefly, *Austrocercella hynesi*, is also recorded only from the headwaters of the Snowy River on Mt Kosciusko, where it seems to be abundant (4).

THREATS TO SURVIVAL Possible threats to the very limited range of this species and *Austrocercella hynesi* have not been assessed.

CONSERVATION MEASURES TAKEN The stonefly is known only from the Mt Kosciusko National Park, where it is protected from development pressures and collectors (1).

CONSERVATION MEASURES PROPOSED *L. cacuminis* is certainly restricted to the Snowy Mountains, but other possible localities within the range need to be investigated. National Park authorities should be aware of this unique insect, in order that they may protect its habitat from disturbance.

REFERENCES
1. Campbell, I.C. (1981). In litt., 9 February.
2. Hynes, H.B.N. (1974). Comments on the taxonomy of Australian Austroperlidae and Gripopterygidae (Plecoptera). Aust. J. Zool. Suppl. 29: 1-36.
3. Hynes, H.B.N. (1974). Observations on the adults and eggs of Australian Plecoptera. Aust. J. Zool. Suppl. 29: 37-52.
4. Hynes, H.B.N. (1981). Taxonomical notes on Australian Notonemouridae (Plecoptera) and a new species from Tasmania. Aquatic Insects 3: 147-166.
5. Hynes, H.B.N. (1982). New and poorly known Gripopterygidae (Plecoptera) from Australia, especially

Tasmania. <u>Aust. J. Zool.</u> 30: 115-158.

6. Hynes, H.B.N. and Hynes, M.E. (1975). The life histories of many of the stoneflies (Plecoptera) of south-eastern mainland Australia. <u>Aust. J. Mar. Freshwat. Res.</u> 26: 113-153.

We are grateful to I.C.Campbell for information in this account, and to H.B.N. Hynes, S. Jewett, A. Neboiss and W.O. Williams for further comment.

MOUNT DONNA BUANG WINGLESS STONEFLY   RARE

Riekoperla darlingtoni (Illies, 1968)

Phylum	ARTHROPODA	Order	PLECOPTERA
Class	INSECTA	Family	GRIPOPTERYGIDAE

SUMMARY One of only two wingless Australian stoneflies, this species is apparently restricted to small temporary streams within a kilometre of the summit of Mt Donna Buang, Victoria, Australia.

DESCRIPTION This is a small, brown, wingless stonefly ranging from 6 mm (males) to 12 mm (females) in body length. Both sexes have conspicuous antennae up to 12 mm in length (5).

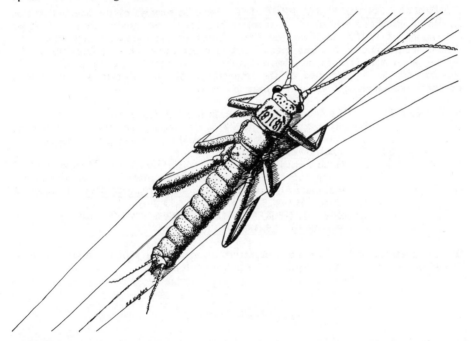

DISTRIBUTION The insect has been collected only within a kilometre of the summit of Mt Donna Buang in Victoria, Australia, in spite of efforts to find it elsewhere (2,5).

POPULATION Unknown.

HABITAT AND ECOLOGY Mt Donna Buang is completely forested with Eucalyptus trees and a dense understorey of Nothofagus. The stonefly adult is reported to occur commonly in rolled pieces of bark near the small temporary streams near the summit of the mountain (3). Nymphs occur under stones and in silty gravel of tiny streams, surviving dry periods by descending into the gravel of the stream bed (3,4). The life span is believed to be rather less than three years (3). Hatching occurs from February to May (autumn) and emergence occurs in September (spring) two years later (3).

SCIENTIFIC INTEREST AND POTENTIAL VALUE The species is of interest both because of its apterous (wingless) condition and its ability to survive the drying of its habitat. Only one other species of Australian stonefly is known to have this ability. _Riekoperla darlingtoni_ also has a longer life history than any other member of the genus so far studied (2). The Mt Donna Buang habitat is considered especially propitious for the evolution and persistence of such species (5).

THREATS TO SURVIVAL It is unclear to what extent the species is threatened. Mt Donna Buang is a popular tourist attraction for the people of Melbourne, especially in winter. Several large car parks and various other facilities have been constructed to serve visitors (1). The stream where the nymphs are most abundant skirts one of the car parks and any extensions of these facilities would be deleterious to the stoneflies (4).

CONSERVATION MEASURES TAKEN The summit of Mt Donna Buang is a Scenic Reserve. However the provisions of such a designation are not wholly protective. No specific measures have been taken on behalf of this insect (1).

CONSERVATION MEASURES PROPOSED Periodic surveys of the species' status should be undertaken in order to monitor population changes or trends. Further extensions of visitor facilities on the summit of Mt Donna Buang should be discouraged, not only for this species but also for the continued viability of the unique ecosystem (1). Management authorities of the Mt Donna Buang Scenic Reserve should be informed of the importance of the habitats present, and a co-operative agreement sought for their protection.

REFERENCES
1. Campbell, I.C. (1982). In litt., 9 February.
2. Hynes, H.B.N. (1974). Comments on the taxonomy of Australian Austroperlidae and Gripopterygidae (Plecoptera). Aust. J. Zool. Suppl. 29: 1-36.
3. Hynes, H.B.N. and Hynes, M.E. (1975). The life histories of many of the stoneflies (Plecoptera) of south-eastern mainland Australia. Aust. J. Mar. Freshwat. Res. 26: 113-153.
4. Hynes, H.B.N. (1982). In litt., 13 July.
5. Illies, J. (1968). The first wingless stonefly from Australia. Psyche 75: 329-333.

We are grateful to I.C.Campbell for information on this subject, and to H.B.N. Hynes, S. Jewett, and W.D. Williams for helpful comments.

PYGMY HOG SUCKING LOUSE   ENDANGERED

Haematopinus oliveri Mishra and Singh, 1978

Phylum	ARTHROPODA	Order	ANOPLURA
Class	INSECTA	Family	HAEMATOPINIDAE

SUMMARY The Pygmy Hog Sucking Louse is a blood-sucking ectoparasite known only from the Pygmy Hog (Sus salvanius), an endangered species probably restricted to north-west Assam in northern India. Both the hog and its attendant louse are threatened by destruction of thatch-scrub savanna.

DESCRIPTION The Anoplura are blood-sucking ectoparasites of mammals; they are wingless and cannot live independently of the host at any stage of the life-cycle. The eyes are reduced or absent, and the mouthparts highly modified for piercing and sucking. The body is dorsoventrally flattened with a leathery integument, and the legs are strongly developed with powerful claws for attachment to the host's hairs (12). Haematopinus oliveri is a parasite of the Pygmy Hog (Sus salvanius). The female louse is 3.9 to 4.2 mm in total length; the male is unknown. The species has been described in some detail and differs from the other six species of Haematopinus by a combination of the absence of paratergite II, the shape of the thoracic sternal plate and the relatively short head (6,15,16).

DISTRIBUTION The seven species in the genus Haematopinus infest wild pigs from Africa and Europe right across the Orient to the Philippines (15,16). The Pygmy Hog was formerly found in southern Nepal, Sikkim, north-west Assam, the foothills of Meghalaya, northern Bangladesh and south-east Assam. It is now extinct in most of this range and probably restricted to small, endangered populations in north-west Assam, India (4,8,9,10,11).

POPULATION Unknown. The only specimens collected are three females (holotype and two paratypes) and one associated nymph, all taken from a Pygmy Hog in Darrang, north-west Assam in 1977 (6). In louse populations, as in many other parasites, most individuals are on a few animals, whilst most hosts have few or no parasites. Elimination of very few hosts from a population may, therefore, lead to virtual or complete elimination of lice from that host population (3). The Pygmy Hog was at one time feared extinct but reappeared in 1971 near Barnadi. That hog population is now virtually extirpated and the only certainly known surviving population is in Manas National Park and certain of its buffer Reserve Forests (8,9,10,11). The population there is believed to be only a few hundred individuals, but may be considerably less.

HABITAT AND ECOLOGY Anoplurans are generally very host-specific. Close physiological relationships between the parasite and the host's blood characters are important (1), but this is only one of a variety of interacting components, including host hair type, skin microclimate and host ecology (3,5). The majority of Anoplura carry symbiotic bacteria which are essential in providing B group vitamins (13,17). The lethal effect of abnormal host-blood on Anoplura may result from the death of these bacteria (1,12), or from the accumulation of crystalline blood chemicals which rupture the midgut (2,7). Details of the biology of the louse are not known, but in common with other Anoplura it certainly feeds solely on blood and cannot survive without the host (17). Mating occurs on the host, and the eggs are cemented to host hairs, probably hatching within two weeks (12). The habitat of the Pygmy Hog is primarily unburnt thatch-scrub savanna, composed of mixed scrub and tall grasses (4,8,9,10,11).

SCIENTIFIC INTEREST AND POTENTIAL VALUE The conservation needs of the Pygmy Hog, its louse, and even the probably specific bacterial symbionts of the louse, pose difficult questions. The louse could prejudice the survival of the Pygmy Hog, since Anoplura are known to be vectors of rickettsial and other diseases. However, the parasite is only likely to affect feeble hogs, of little value to the population as a whole. The bacteria of the louse produce B vitamins, and possibly could be of value in the rapidly evolving disciplines of genetic engineering and biotechnology.

THREATS TO SURVIVAL The Pygmy Hog Louse is threatened by loss of its host the Pygmy Hog, which is in turn threatened by loss and despoilment of habitat (9,11). The upland savannas of north India are fertile and ideal for conversion to agriculture. Illegal immigrants from Nepal and Bangladesh reinforce the already high demands on the area. The remaining habitat is subjected to reafforestation and thatch grass harvesting, both of which entail very destructive annual fires (10,14). In some areas hunting also substantially reduced populations that survived until recently (10). Unless the Pygmy Hog and its habitat are more effectively protected, the hog and its louse will probably become extinct (14).

CONSERVATION MEASURES TAKEN No measures have been taken specifically to protect the louse, although its host has been the subject of certain measures. The Pygmy Hog is listed in Appendix I of the 1973 Convention on International Trade in Endangered Species of Wild Fauna and Flora (CITES) so trade in this species is strictly regulated. In India the hog is included in Schedule I of the 1972 Wildlife (Protection) Act, which affords maximal protection to the animal itself. Unfortunately this is ineffective since the animal's habitat is being destroyed, often in the normal course of duties of the same Forest Department that is responsible for enforcing the Wildlife (Protection) Act (10).

CONSERVATION MEASURES PROPOSED The louse is likely to survive if the hog does. Habitat protection with full legal backing is required, particularly in the Manas National Park and its buffer forests, the only places where the hog is now definitely known to survive (11).

REFERENCES
1. Kellogg, V.L. (1913). Ectoparasites of the monkeys, apes and man. Science 38: 601-602.
2. Krynski, S., Kuchta, A. and Bela, E. (1952). Research on the nature of the noxious action of guinea-pig blood on the body-louse. (In Polish). Bull. Inst. Mar. Med., Gdansk 4: 104-107.
3. Lyal, C.H.C. (1982). In litt., 12 October.
4. Mallinson, J.C. (1971). The pigmy hog Sus salvanius (Hodgson) in northern Assam. J. Bombay Nat. Hist. Soc. 68: 424-433.
5. Marshall, A.G. (1981). The Ecology of Ectoparasitic Insects. Academic Press, New York and London.
6. Mishra, A.C. and Singh, K.N. (1978). Description of Haematopinus oliveri sp. nov. (Anoplura: Haematopinidae) parasitizing Sus salvanius in India. Bull. Zool. Surv. India 1: 167-169.
7. Nelson, W.A., Keirans, J.E., Bell, J.F. and Clifford, C.M. (1975). Host-ectoparasite relationships. J. Med. Ent. 12: 143-166.
8. Oliver, W.L.R. (1979). The doubtful future of the pigmy hog and the hispid hare; pigmy hog survey report, Part 1. J. Bomb. Nat. Hist. Soc. 75: 341-372.
9. Oliver, W.L.R. (1980). The pigmy hog. The biology and conservation of the pigmy hog Sus (Porcula) salvanius and the hispid hare Caprolagus hispidus. Jersey Wildlife

Preservation Trust Sp. Scien. Rep. 1 : 80 pp.

10. Oliver, W.L.R. (1981). Unpublished Red Data Sheet on the pygmy hog, Sus salvanius (Hodgson, 1847). In litt. to J. Thornback, 19 June.

11. Oliver, W.L.R. (1981). Pigmy hog and hispid hare: further observations of the continuing decline (or, a lament for Barnadi, and a good cause for scepticism). Dodo, J. Jersey Wildl. Preserv. Trust 18: 10-20.

12. Richards, O.W. and Davies, R.G. (1977). Imms' General Textbook of Entomology 10th Edition, Volume 2, Classification and Biology. Chapman and Hall, London. Pp. viii + 420-1354.

13. Ries, E. (1931). Die symbiose der Läuse und Federlinge. Z. Morph. Ökol Tiere, 20: 233-267.

14. Sakya, K. (1979). Pygmy hog conservation - a challenge. Tigerpaper 6: 8-11 (Publication of FAO/UNEP Regional Office for Asia and Far East).

15. Stimie, M. and Merwe, S. (1967). A revision of the genus Haematopinus Leach (Phthiraptera; Anoplura) Zool. Ans. Jena 180: 182-200.

16. Weisser, C.F. (1974). Haematopinus ludwigi nov. spec. from Sus verrucosus Philippines, and neotype designation for Haematopinus breviculus Fuhrenlialz from Taurotragus oryx pattersonianus Uganda (Haematopinidae, Anoplura) Zool. Ans. Jena 193: 127-142.

17. Wigglesworth, V.B. (1964). The Life of Insects. Weidenfeld and Nicholson, London. 360 pp.

We are grateful to J. Thornback, M. Jenkins and W.L.R. Oliver for access to unpublished material on the Pygmy Hog, and to C.H.C. Lyal for valuable comments on the draft.

Magicicada cassini (Fisher, 1851)
M. septendecim (Linnaeus, 1758)
M. septendecula Alexander & Moore, 1962

Phylum	ARTHROPODA	Order	HOMOPTERA
Class	INSECTA	Family	CICADIDAE

SUMMARY The nymphs of the periodical cicadas of the U.S.A. and Canada take 13 to 17 years to develop into adults, longer than any other insect. When they finally emerge from the ground and moult on vegetation, all three species are perfectly synchronized. Hundreds of thousands may emerge at one time. Nevertheless, two of the 'broods', or year classes, of periodical cicadas are already extinct, and others may be threatened.

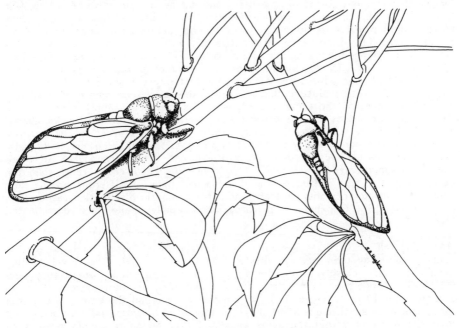

DESCRIPTION There are three morphologically distinct species of periodical cicadas, Magicicada septendecim, M. cassini and M. septendecula. Each has morphologically similar 17- and 13-year forms, the former named above, the latter known as M. tredecim, M. tredecassini, and M. tredecula respectively. Each pair of forms is commonly abbreviated to Decim, Cassini, and Decula respectively. Although the 13- and 17-year forms have different specific names, their specific status is in dispute. The two forms may be simply populations of the same species initially separated in time and subsequently virtually separated in space (14). The periodical cicadas are homopteran bugs with the mouthparts adapted for piercing and sucking plant tissues. Adult insects have large membranous wings held in a characteristic roof-like posture over the abdomen. M. septendecim is the largest species (30 mm long) with an orange or orange-striped abdomen, black tarsi and an orange patch behind and below the eye (15). Until 1962, when they were found to be separate species, M. cassini and M. septendecula were collectively known as the dwarf form of the periodical cicada. M. cassini has a black abdomen, black tarsi

369

and no orange patch, while M. septendecula has a black abdomen with orange stripes, orange tarsi and no orange patch (15). The three species can be distinguished in the field and have quite distinct songs (1). Periodical cicadas have red eyes which distinguishes them from the annually-breeding North American cicadas.

DISTRIBUTION Magicicada cassini and M. septendecula occur only in the U.S.A. east of the Great Plains, while M. septendecim has also been recorded from Ontario and Quebec in Canada. The three morphologically distinct species of periodical cicadas are sympatric over almost their whole range. However, the 13-year forms are confined to the Mississippi Valley and south-eastern states, while the 17-year forms occur in the northern, eastern and western sections of the range. A few 13- and 17-year broods overlap geographically, but they are rarely, if ever, found in the same patch of woods (5,6,13,14).

POPULATION The periodical cicadas are so-called because the imagos emerge synchronously only after a developmental period of 13 or 17 years. This is the longest development period of any insect and has been claimed to be the longest life-span (6,13). However, the wood-boring jewel beetle Buprestis aurulenta (Buprestidae) may live for up to 42 years (21), and certain termites are believed to be long-lived (3). Emergence is spectacular, with hundreds of thousands of fifth-instar nymphs burrowing out of the soil and climbing any vertical object on which to eclose (5,6,7). Although a given population of cicadas emerges only once every 13 or 17 years, the 13th or 17th year differs from place to place in the eastern U.S.A. All populations that emerge in the same year, wherever they may be, are members of the same 'brood', which is therefore simply a year class. Seventeen-year broods are numbered I-XVII, while 13-year broods are numbered XVIII-XXX. There are currently thought to be 13 of the 17-year broods and four of the 13-year broods. With only one or two exceptions, all three morphologically distinct species can be found in every brood, and the emergence of all three species is synchronized. However, M. septendecula is generally rarer than the other species (1,5,9). Although the distribution maps of different broods may appear to overlap (12), this generally indicates a mosaic distribution rather than coexistence (10). Coexistence is rare, but examples have been documented (2,4,10,18,19).

HABITAT AND ECOLOGY Periodical cicadas are found in woodlands, where the adult females lay their eggs in slits in tree branches (1,9). The tiny hatchling nymphs launch themselves into the air and crawl into cracks where they land (13,14,15). There they pierce tree rootlets and feed on dilute xylem fluid, to emerge after 13 or 17 years (7,8). Within the same brood the emergence of the three species is synchronized, but they are separated microspatially by occupying different habitats within the same woodland (5). M. septendecula prefers to oviposit in hickories and walnuts, and emerges in higher proportions under those trees than under oaks (5). M. septendecim and M. septendecula occur together in upland woods, but M. septendecim is less host-specific than M. septendecula. M. cassini is found mainly in floodplain woods, and on wooded slopes gradually replaces the other two species with increasing proximity to a stream (5). The three species survive quite well in secondary growth, but since trees characteristic of floodplains (e.g. American Elm) are often a component of upland secondary growth, the cicadas become intermixed, while remaining reproductively isolated (5).

SCIENTIFIC INTEREST AND POTENTIAL VALUE Most taxonomists agree that M. septendecula, M. septendecim, and M. cassini are separate species. However, there is some debate as to the status of the 13- and 17-year forms of each type. The primarily northern and southern distributions of the 17- and 13-year forms respectively suggested to early authors that these groups were evolutionarily distinct entities (1). More recent views envisage Pleistocene refugia occupied by

17-year cicadas, which gave rise to all present-day broods and forms (14). Seventeen-year nymphs of M. cassini differ from 13-year nymphs by having inhibited growth rates for the first four years (10). It seems probable that 13-year cicadas evolved from 17-year ones by virtue of a four-year acceleration. Similar accelerations have given rise to a succession of 17-year broods four years apart (8). Four-year accelerations are still taking place, and it is likely that the entire evolutionary process, including generation of the 13-year forms, may have taken place since the last glaciation (9). One-year accelerations also occur, but on a smaller scale, and probably as a result of climatic shocks (1,8). The result of these one- and four-year accelerations has been the evolution of the various broods and two forms of the three original species (8,14,20). Studies on the enzymes (14) and wing-vein morphology (17) of various broods and all species support and extend these ideas. It seems that some broods are polyphyletic, consisting of geographically separate populations which by chance now emerge in the same year (18). As a tool for evolutionary biologists the periodical cicadas are perhaps unsurpassed, giving a fascinating insight into the complexities of insect evolution.

Finally, the spectacle of the synchronous emergence of the periodical cicadas every 17 or 13 years should not be under-estimated. The immense pulse of life represented by the emergence of hundreds of thousands of cicadas, and the attendant gorged predators, is a superb sight (13). The egg-laying process may cause some damage to woody vegetation, but orchard growers avoid problems by the simple expedient of not planting new trees for 3-4 years prior to the emergence of a brood (13). This pest is relatively easy to predict.

THREATS TO SURVIVAL The conservation problems are unusual in being related to the sub-taxonomic broods rather than to the species as a whole. Two broods are already extinct, brood XI, which occupied the Connecticut River Valley (11), and brood XXI, which occurred along the Mississippi-Alabama border and in the Florida Panhandle (13). Brood XI, first recorded in 1767, was in decline as early as 1907. Brood XV was investigated in 1975 and was found only in one area of New Jersey; adults disappeared prematurely and no reproduction was noted. It is possible that XV was never a real brood, but only a four-year acceleration of Brood XI (18). Broods other than XI, XV, and XXI have not decreased significantly in range, but they are made up of a decreasing number of local populations.

The greatest threat to the periodical cicadas is habitat destruction. In 1902 an article in the New York Times quoted local entomologists as being afraid that "man would cement over or plow under these fascinating insects" (16). Vast areas of virgin forests in the eastern U.S.A. have been cleared for farmland, dams or suburban gardens (5,9). Floodplain forests have been cleared first because alluvial soil is more fertile and easier to farm than adjacent upland soil (9). The floodplain species M. cassini has thus been particularly at risk, although it is adaptable and has moved into secondary upland woods (5).

CONSERVATION MEASURES TAKEN No specific measures have been taken to conserve periodical cicadas. However, many populations are now located in national forests and national and state parks.

CONSERVATION MEASURES PROPOSED The smaller broods should be surveyed and monitored to ensure that not all their habitats are destroyed. A representative sample should be maintained within national parks and reserves. Under man's influence enormous environmental changes may occur during the developmental period of these cicadas. It is all too easy to forget an insect that only appears every 13 or 17 years.

REFERENCES

1. Alexander, R.D. and Moore, T.E. (1962). The evolutionary relationships of 17-year and 13-year cicadas, and three new species (Hompotera, Cicadidae, Magicicada). Misc. Publ. Mus. Zool., Univ. Mich. 121: 1-59.

2. Bryce, D. and Aspinwall, N. (1975). Sympatry of two broods of the periodical cicada (Magicicada) in Missouri. Amer. Midl. Nat. 93: 450-454.

3. Collins, N.M. (1981). Populations, age structure and survivorship of colonies of Macrotermes bellicosus (Smeathman) (Isoptera: Macrotermitinae). J. Anim. Ecol. 50: 293-311.

4. Craig, F.W. (1941). Observations on the periodical cicada. J. Econ. Ent. 34: 122-123.

5. Dybas, H.S. and Lloyd, M. (1974). The habitats of 17-year periodical cicadas (Homoptera: Cicadidae: Magicicada spp). Ecol. Monogr. 44: 279-324.

6. Gerhard, W.J. (1923). The periodical cicada. Field Mus. Nat. Hist. Leaflet 4: 41-54.

7. Lloyd, M. and Dybas, H.S. (1966). The periodical cicada problem. 1. Population ecology. Evolution 20: 133-149.

8. Lloyd, M. and Dybas, H.S. (1966). The periodical cicada problem. II. Evolution. Evolution 20: 466-505.

9. Lloyd, M. and White, J.A. (1976). On the oviposition habits of 13-year versus 17-year periodical cicadas of the same species. J. N.Y. Ent. Soc. 84: 148-155.

10. Lloyd, M. and White, J.A. (1976). Sympatry of periodical cicada broods and the hypothetical four-year acceleration. Evolution 30: 786-801.

11. Manter, J.A. (1974). Brood XI of the periodical cicada seems doomed. Mem. Connecticut Ent. Soc. 1974: 99-100.

12. Marlatt, C.L. (1907). The periodical cicadas. U.S. Dept. Agric. Bur. Ent., Bull. No. 71: 1-181.

13. Simon, C. (1979). Debut of the seventeen-year-old cicada. Nat. Hist. 88: 38-44.

14. Simon, C.M. (1979). Evolution of Periodical Cicadas: phylogenetic inferences based on allozymic data. Syst. Zool. 28: 22-39.

15. Simon, C.M. (1979). Brood 11 of the 17-year cicada on Staten Island: timing and distribution. Proceedings of the Staten Island Inst. of Arts and Sciences 30: 35-46.

16. Simon, C.M. (1980). In litt., 5 May.

17. Simon, C.M. (in press). Morphological differentiation in wing venation among broods of 13 and 17 year periodical cicadas. Evolution.

18. Simon, C.M. and Lloyd, M. (in review). Disjunct populations of periodical cicadas: relicts or evidence of polyphyly? J. N.Y. Ent. Soc.

19. Simon, C.M., Karban, R. and Lloyd, M. (1981). Patchiness, density and aggregative behavior in sympatric allochronic populations of 17-year cicadas. Ecology 62: 1525-1535.

20. White, J.A. and Lloyd, M. (1979). 17-year cicadas emerging after 18 years: a new brood? Evolution 33: 1193-1199.

21. Wood, G.L. (1982). The Guinness Book of Animal Facts and Feats. (3rd ed.). Guinness Superlatives, Enfield, U.K. 252 pp.

We are grateful to C.M. Simon for her comments on this sheet.

DELTA GREEN GROUND BEETLE   VULNERABLE

Elaphrus viridis Horn, 1878

Phylum   ARTHROPODA   Order   COLEOPTERA

Class   INSECTA   Family   CARABIDAE

SUMMARY The attractive and once probably widespread green and gold beetle Elaphrus viridis is now known only from one vernal pool in Jepson Prairie, California, U.S.A.. Its range has been eroded by agricultural changes, drainage and pipeline construction. It is listed as Threatened in the U.S.A. and draft recovery plans for the species and for Jepson Prairie have been prepared.

DESCRIPTION A distinctive metallic green and gold predatory ground beetle, E. viridis was collected in 1876 and described in 1878, but was not recorded again until 1974 (4). The adults are distinguished by their size, colour, spots, lack or reduction in size of the pits on the elytra and density of hairs (6).

DISTRIBUTION The species may have formerly ranged through Central Valley, California, but in 1974 was found only on the margins of two vernal pools (Olcott Pool and an unnamed pool) near Dozier, about 19 km south of Dixon, Solano County, California. The unnamed pool was recently diked and ploughed and at present the beetle is confined to Olcott Pool in Jepson Prairie (6). Preliminary searches of other potential habitat in this area have been unsuccessful, but it is thought that relict populations may be discovered eventually at various points within the Central Valley, including the San Joaquin Valley to the south (4).

POPULATION Densities were considered low in 1974 and 1975 when specimens were collected (4).

HABITAT AND ECOLOGY The beetle has not been studied in detail and its life history and ecology are poorly known (6). Most of the 800 species of ground beetles known from California are nocturnal, but the Delta Green Ground Beetle is active during the day (6). Both larvae and adult beetles probably take soft-bodied prey, such as chironomid larvae (6). Larvae undergo complete metamorphosis, with pupation usually taking place in chambers in wood or soil, with no cocoon (3). There is one generation per year, with adults present between February and May, after which they enter an obligate diapause (6). Oviposition is assumed to be during the wet winter and early spring, after the adult diapause has been broken (6). Early coleopterists assumed that the species inhabited muddy river borders (1). The 1974 and 1975 discoveries were at the damp margins of vernal pools, which are temporary shallow ponds that form over hardpan soils during winter rains, then gradually dry up during late spring and early summer (2). Pupation may occur in the damp soil around such pools, with the adults emerging as the soil cracks and dries. Individuals have also been collected more than 100 m from the pools and it seems likely that the beetles require both the residual moisture of the pools and the adjacent grassland. Formerly their life cycle may have been intimately tied to the annual flash flooding cycle in the Central Valley, their distribution being restricted to areas with dependable soil moisture for at least part of the year (4).

SCIENTIFIC INTEREST AND POTENTIAL VALUE The federally Endangered Solano Grass (Orcuttia mucronata) is also limited to an area around Olcott Pool (5,6). Elaphrus viridis is one of the rarest beetles in the U.S.A. and is an unusual carabid in being diurnal.

THREATS TO SURVIVAL Historical changes have probably slowly reduced the distribution and population size of this beetle. Wetland habitats including vernal pools, marshes, rivers and sloughs have been greatly modified in the past 150 years, and only remnants remain. A mapping study of Solano County showed that the vernal pool area has been continually reduced and fragmented. Control and restriction of the spring flooding cycle in the Central Valley may have disrupted the distribution of the beetle irreparably. The essential habitat requirements are not known, but it is assumed that the main limiting factor is availability of moist areas. The Olcott Pool population may also be threatened by grazing animals, which destroy the vegetation and compact the soil at the pool edges (4).

Many vernal pools have been lost to river channelization (loss of overflow), dam construction and the agricultural conversion of natural habitats. Phase II of the North Bay Aqueduct and waste water disposal for the city of Vacaville could affect the Critical Habitat of the beetle adversely if the needs of the species are not considered. Its current range could be affected by agricultural conversion, drainage or pipeline construction (5).

CONSERVATION MEASURES TAKEN E. viridis was listed as Threatened in September 1980, under the U.S. Endangered Species Act, and its known range was designated Critical Habitat (5). In December 1980 the U.S. Nature Conservancy bought the surface rights to approximately 1600 acres (648 ha) including Olcott Pool, to ensure that it will be managed to preserve its unique fauna and flora (4). A draft recovery plan for the area, known as the Jepson Prairie Preserve, has been prepared (6).

CONSERVATION MEASURES PROPOSED The recovery plan for Jepson Prairie (6) has as a primary objective the protection and management of the Delta Green Ground Beetle. This will include estimation of population sizes, and protection of three additional sites if other populations can be found. Other proposed measures include grazing restrictions, erection of fences, minimization of insecticide use, and further studies on the beetle's biology, ecology, distribution and habitat requirements (4,6).

REFERENCES   1. Andrews, F.G. (1978). Status report on Elaphrus viridis Horn, 1876 (Coleoptera: Carabidae). Unpublished report, California Dept. Food and Agriculture, Sacramento.
            2. Holland, R., and Jain, S. (1977). Vernal pools. In Barbour, M.G. and Major, J. (Eds). Terrestrial Vegetation of California. Wiley, New York. Pp. 515-533.
            3. Powell, J.A., and Hogue, C.L. (1979). California Insects. California Natural History Guide 44. University of California Press, Berkeley and Los Angeles. 388pp.
            4. Wilbur, S.R. (1981). Delta Green Ground Beetle recovery plan. Technical Review Draft. U.S. Fish and Wildlife Service, Oregon.
            5. U.S.D.I. Fish and Wildlife Service (1980). Endangered and Threatened Wildlife and Plants; Listing the Delta Green Ground Beetle as a Threatened Species with Critical Habitat. Federal Register 45(155): 52807-52810.
            6. Holland, R. and Arnold, R.A. (1982). Jepson Prairie Recovery Plan. U.S. Fish and Wildlife Service, Sacramento, California. 82pp.

We are grateful to R.B. Madge, R.T. Thompson and S.R. Wilbur for their comments on this review.

COLUMBIA TIGER BEETLE

ENDANGERED

Cicindela columbica Hatch, 1938

Phylum	ARTHROPODA	Order	COLEOPTERA
Class	INSECTA	Family	CICINDELIDAE

SUMMARY Formerly known from three rivers in the Pacific Northwest of the U.S.A., the Columbia Tiger Beetle has been destroyed over most of its historical range. A denizen of riverine sand-bars, the beetle has been susceptible to inundation by damming. Sandy shores of reservoirs are not suitable for recolonization. The last known colonies, in Idaho, are imperilled by proposed dams.

DESCRIPTION *Cicindela columbica* is a 12-14 mm tiger beetle with an iridescent black body and elytra, the latter with cream-coloured markings. It was originally described as a subspecies of *Cicindela bellissima* (3), but the author later elevated it to a full species (4), a decision subsequently supported by another specialist (6). Detailed descriptions have been published (3,6).

DISTRIBUTION Historical records (6) show a former range in the U.S.A. along the Columbia River as far west as Hood River, Oregon, and Cook, Washington; the Snake River east to Lewiston, Idaho and Clarkston, Washington, and in the lower Salmon River of Idaho. Currently the species is known only from Idaho, near Slate Creek and between Rice Creek and Eagle Creek on the lower Salmon River in Idaho County, and on the Clearwater River above Lewiston(11).

POPULATION The population has been drastically reduced by hydroelectric dams on the Snake and Columbia Rivers (1,3,6,8,9), probably to the point of extinction in Oregon and Washington (2). Three searches failed to find any populations on the Columbia River and none on the Snake River below the Lower Granite Dam (11). A survey conducted in 1979 (8,9) located the beetle on 14 of 49 sand-bars inspected on the lower Salmon River. The two largest populations were estimated at about 200 and 400 adults, and the overall population on the Salmon was considered healthy (8).

HABITAT AND ECOLOGY C. columbica occupies well-established riverine sand-bars which are not completely flooded by normal spring run-off (8). Vegetation on the bars is dominated by Salix, Xanthium and Glycyrrhiza. Surrounding terrain consists of basalt canyons with slopes generally exceeding 30 per cent, partially vegetated with Celtis, Cercocarpus and Pinus (2,8). Egg-laying and larval development are dependent upon a suitable undisturbed substrate. The larvae capture prey at the mouth of shallow burrows. Adults are active diurnal predators on small arthropods (2), and prefer sands with a surface temperature of 88-110°F (31-43°C), avoiding hotter substrates (8). When disturbed, they often fly out over the river or moist areas of the sand-bar (8). Although they are gregarious with individuals of their own species, they seem to isolate themselves from other tiger beetles (8). Four other species of Cicindela are found on the lower Salmon, and a total of six species have established themselves along new river shores resulting from dams downstream. The Columbia Tiger Beetle cannot do so, requiring old sand-bars that are immersed only rarely, during large floods (8). The largest colonies inhabit sand-bars approximately 0.5 km long and extending 100 m back from the water. The maintenance of suitable sand-bar habitat is probably dependent upon sufficient water flow in the affected rivers (2)

SCIENTIFIC INTEREST AND POTENTIAL VALUE The Columbia Tiger Beetle is a narrow endemic, closely attuned to pre-development conditions. It serves as a precise indicator of aboriginal sand-bar ecosystems in an area highly altered by major river impoundments. Whereas six other species of tiger beetles have been able to recolonize reservoir shores, this species cannot do so (8). The positive ecological qualities of cicindelid beetles, and the threats to them, have been stressed by biologists (e.g. 7).

THREATS TO SURVIVAL The primary threat to this species is the construction of dams and resultant impoundments. For example, the beetle was found downstream from Lewiston, Idaho, on the Snake River in 1970; the next year it was lost from that stretch when the Lower Granite Dam was completed (1). The deletion of the Lower Salmon from the River of No Return Wilderness Bill was in response to the interest of the U.S. Army Corps of Engineers in damming that portion of the river (9). Any threat to the Clearwater River site (11), or dam construction on the Lower Salmon, could precipitate the final extinction of this beetle (10). There has been concern that Idaho sites should be kept secret to prevent over-collecting by coleopterists, but most of the sites are quite inaccessible (8). Off-road vehicles, which damage cicindelid populations elsewhere, probably cause little harm to this particular species (7).

CONSERVATION MEASURES TAKEN Several searches for the beetle have been undertaken in recent years, notably a survey of the Lower Salmon River facilitated by the Bureau of Land Management and conducted by G. Shook (8). The U.S. Fish and Wildlife Service carried out a review of status of the species (10) following a public petition to do so, but has not published a proposed listing in the Federal Register. Regional entomologists hoped that the species might be preserved in the Hells Canyon National Recreation Area, but the 1979 survey failed to locate any populations in the Snake River Canyon.

CONSERVATION MEASURES PROPOSED Further efforts should be made to locate remaining populations in the historical range, particularly in Oregon and Washington (2). A search should also be conducted on the Salmon River above and below Riggins, Idaho (9). The Lower Salmon River should be formally protected as a scenic river, wilderness area, or by some other designation that would prevent dam construction (9). Any remnant should be assiduously protected by the agencies concerned. Additional ecological research may be necessary to provide a management basis for critical habitats. The Columbia Tiger Beetle should be considered for listing under the Endangered Species Act.

REFERENCES 1. Beer, F.M. (1971). Note on Cicindela columbica Hatch. Cicindela 3: 32.

2. Bentzien, M.M. (1979). Columbia tiger beetle Cicindela columbica Hatch. Unpubl. status summary, U.S. Office of Endangered Species. 5 pp.

3. Hatch, M.H. (1983). The Coleoptera of Washington: Carabidae: Cicindelinae. Univ. Washington Publ. Biol., Seattle 5: 225-240.

4. Hatch, M.H. (1949). Studies of the Coleoptera of the Pacific Northwest 1. Pan-Pacif. Ent. 25: 113-118.

5. Leffler, S.R, (1976). Tiger beetles of Washington. Cicindela 8: 21.

6. Leffler, S.R. (1979). Tiger beetles of the Pacific Northwest (Coleoptera: Cicindelidae). PhD Diss., Univ. Washington, Seattle. 731 pp.

7. Nagano, C.D. (1980(82)). Population status of tiger beetles of the genus Cicindela (Coleoptera: Cicindelidae) inhabiting the marine shoreline of southern California. Atala 8: 33-42.

8. Shook, G. (1979). Status report on the Columbia tiger beetle (Cicindela columbica Hatch) in Idaho (Cicindelidae: Coleoptera). Unpubl. report to U.S. Fish and Wildlife Service. 9 pp.

9.   Shook, G. (1979). Unpublished petition to list <u>Cicindela</u> <u>columbica</u> as a Federally Endangered Species. U.S. Office of Endangered Species. 6 pp.

10. U.S. Dept. of the Interior, Fish and Wildlife Service. (1980). Review of status of the Columbia tiger beetle. <u>Federal</u> <u>Register</u> 45 (43): 13786-13787.

11. Beer, F.M. (1982). In litt., 17 November.

We are grateful to F.M. Beer, G.D. Drewry, G.A. Dunn, C.D. Nagano and N. Stork for their comments on this review.

GIANT CARRION BEETLE   ENDANGERED

Nicrophorus americanus Olivier, 1790

Phylum   ARTHROPODA   Order   COLEOPTERA

Class   INSECTA   Family   SILPHIDAE

SUMMARY The largest of the North American carrion beetles, Nicrophorus
americanus was formerly widespread in the forest regions of the eastern North
American continent. Until recently it was thought that the last collection was in
1974, and the species was feared extinct. However, new records indicate an
extremely limited distribution in Rhode Island State, and further surveys in
primary forests are urgently needed.

DESCRIPTION Nicrophorus americanus is the largest North American silphid
(sexton, burying or carrion beetles), reaching a length of 25-36 mm (3,4,11). It is
easily distinguished by its large size, but also by the red frons and red pronotal
disc on a black ground colour (9). The antennal club is orange and the black elytra
have two pairs of scalloped red spots (1,4). Detailed modern descriptions of the
species have been made (1,4,9). N. americanus is closely related to the European
species N. germanicus, which is similar in its large size, the orbicular pronotum,
and the shape of the hind femora (9).

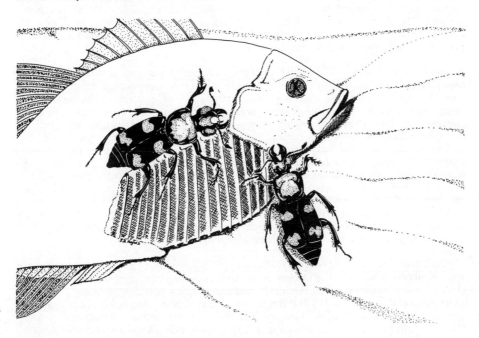

DISTRIBUTION The Giant Carrion Beetle was formerly widespread in the forest
regions of the eastern North American continent. Prior to 1960 the extensive
records include Alabama, Arkansas, Connecticut, Florida, Georgia, Illinois,
Indiana, Iowa, Kansas, Kentucky, Louisiana, Maryland, Massachusetts, Michigan,
Minnesota, Mississippi, Missouri, Nebraska (10), New Hampshire, New Jersey, New
York, North Carolina, Ohio, Pennsylvania, Rhode Island, South Carolina, South
Dakota, Tennessee, Texas, Virginia, Washington D.C. and Wisconsin in the U.S.A.,
and Nova Scotia, Ontario and Quebec in Canada. All these records are listed in

379

the report of a survey of American museums (14), unless otherwise indicated.

From 1960 onwards, the only records are from Michigan (1961 (4)), Illinois (1961 (3)), Indiana (1965 (14)), Missouri (1966 (14)), Nebraska (1969 (15)), Ontario (1972 (14)), Arkansas (1973/4 (3)), and Kentucky (1974 (5)). With no records after 1974, the species has been feared extinct (5,13), but a series dated 1974 to 1981 from the State of Rhode Island has recently come to light (7). Nevertheless, this must represent one of the most disastrous declines of an insect's range ever to be recorded.

POPULATION The size of the population in Rhode Island State is unknown, but evidently represents only a tiny proportion of the Giant Carrion Beetle's former numbers.

HABITAT AND ECOLOGY Although the Giant Carrion Beetle is one of the most distinctive members of the North American beetle fauna, virtually nothing has been published on its natural history (4,5). Fortunately the closely related European species N. germanicus has been studied, and it is believed that the two species may have similar basic habits (3). Like all Nicrophorus, a pair of adults buries small vertebrate carcasses in the soil. The male and female work together, lying on their backs beneath the carcass and using their legs to lever the body to soft ground up to a metre away (12). It is interred in a chamber probably 20 cm or deeper in the soil, thus preventing other scavengers, particularly flies, from finding the booty (3). As the corpse decomposes, it is fed upon by the adults and worked into a compact ball, with a conical depression which collects nutritious liquids. The female lays her eggs in the walls of a passage directly above the carcass, and the hatched larvae are fed on the liquids (3,12). Parental care usually continues right through to pupation (3,12).

In common with other large Nicrophorus, N. americanus was probably originally associated solely with mature mesic forests (3). Only there would the soil be of a suitable texture to allow the deep burial necessary to protect carcasses (3). In some instances, however, the Giant Carrion Beetle was able to utilize man-made habitats. In the 1920s the beetles were attracted to waste fish used as fertilizer on agricultural land in New York State, but when legislation prevented this practice, the beetles disappeared (8). Partitioning of resources between Nicrophorus species is achieved by different seasonal patterns and particularly by habitat preference (2). There is no evidence that certain species prefer a particular type of carcass, although a certain type of carcass may be more common in the preferred habitat. Hence, although N. americanus has been noted feeding on dead fish (8), this would not be its natural food in forests. There are few data on this aspect, perhaps because most specimens have been caught at lights at night, rather than on carcasses (4).

SCIENTIFIC INTEREST AND POTENTIAL VALUE The genus Nicrophorus is unique among beetles in the extent of parental co-operation and care of the young (12). The adults can produce a clearly audible buzzing sound by rubbing the elytra across the abdomen (12,13). The mechanism is used when the beetles are alarmed, and also in communicating with the larvae, a most unusual behaviour pattern (12,13). Carrion beetles are also important in their role as decomposers of organic matter. Only about one third of the carcass is consumed by the adults and young, the remainder being left to decompose into a nutritious contribution to soil fertility (11).

THREATS TO SURVIVAL A recent appraisal of the biology of Nicrophorus species concludes that the Giant Carrion Beetle is mainly dependent upon primary deciduous forest (3), a vegetation type now reduced to less than one per cent of its former area in the U.S.A. (6). Two other large species of Nicrophorus, N. germanicus in Europe and N. concolor in Japan and China, are also associated with temperate forests. N. concolor is common in the mature, undisturbed temperate forests which are still quite widespread in Japan. Conversely N. germanicus is suffering localization and reduction of its abundance throughout its range (3). So little is known about the present distribution of the Giant Carrion Beetle that it is impossible to assess any further threats.

CONSERVATION MEASURES TAKEN None.

CONSERVATION MEASURES PROPOSED Surveys are urgently needed in the major areas of primary forest remaining within the historical range of the Giant Carrion Beetle. If new populations are found, ecological studies will be necessary to determine the species' precise requirements. Suitable habitat, including any currently known localities, should be protected and managed in accordance with the findings. The U.S. Fish and Wildlife Service Office of Endangered Species should be encouraged to declare federal Endangered status to this beetle.

REFERENCES 1. Anderson, R.S. (1981). The biology and distribution of the Silphidae and Agyrtidae of Canada and Alaska (Insecta: Coleoptera). M.Sc. Thesis, Carleton University.
2. Anderson, R.S. (1982). Resource partitioning in the Carrion Beetle (Coleoptera: Silphidae) fauna of southern Ontario: ecological and evolutionary considerations. Can. J. Zool. 60: 1314-1325.
3. Anderson, R.S. (in press). On the decreasing abundance of Nicrophorus americanus Olivier (Coleoptera: Silphidae) in eastern North America. Coleop. Bull. 36.
4. Anderson, R.S. and Peck, S.B. (subm.). The Silphidae and Agyrtidae of Canada and Alaska. Coleoptera: Silphidae, Agyrtidae.
5. Davis, L.R., Jr. (1980). Notes on beetle distributions, with a discussion of Nicrophorus americanus Olivier and its abundance in collections (Coleoptera: Scarabaeidae, Lampyridae and Silphidae). Coleop. Bull. 34: 245-251.
6. Delcourt, H.R. and Harris, W.F. (1980). Carbon budget of the southeastern U.S. biota: analysis of historical change in trend from source to sink. Science 210: 321-323.
7. Drewry, G. (1982). In litt., 19 August.
8. Latham, R. (1977). In litt. to L.J. Milne, 7 January.
9. Madge, R.B. (1958). A taxonomic study of the genus Necrophorus in America north of Mexico (Coleoptera, Silphidae). M.Sc. thesis, University of Illinois.
10. Meserve, F.G. (1936). The Silphidae of Nebraska. Ent. News 47: 132-134.
11. Milne, L.J. (1982). In litt., 27 August.
12. Milne, L.J. and Milne, M. (1976). The social behaviour of burying beetles. Sci. Amer. 235: 84-89.
13. Milne, L.J. and Milne, M. (1980). The Audubon Society Field Guide to North American Insects and Spiders. A.A. Knopf, New York. 989 pp.
14. Perkins, P.D.P. (1980). North American Insect Status Review. Contract 14-16-0009-79-052 final report to U.S.D.I. Office of Endangered Species. 354 pp.
15. Ratcliffe, B.C. (1982). In litt., 26 August.

We are grateful to R.S. Anderson, G.E. Drewry, R.B. Madge, L.J. Milne, S.B. Peck, B.C. Ratcliffe, T.J. Spilman, R.T. Thompson and G.B. Wiggins for information provided for this account.

HERCULES BEETLE   VULNERABLE

Dynastes hercules hercules (Linnaeus, 1758)
Dynastes hercules reidi Chalumeau, 1977

Phylum   ARTHROPODA   Order   COLEOPTERA

Class   INSECTA   Family   SCARABAEIDAE

SUMMARY   Two subspecies of the Hercules Beetle, one of the largest insects in
the world, are threatened in the Lesser Antilles by destruction of rain forest for
plantation forestry, and by indiscriminate use of pesticides.

DESCRIPTION   The male Dynastes hercules is among the largest insects in the
world, reaching a length of 150-175 mm (4,16,17). Half of this length consists of a
pair of large claw-like toothed horns of formidable appearance (9). The upper
(pronotal) horn is an extension of the pronotum and is lined with a ridge of
bristles. The eyes and mouthparts are hidden beneath the lower (cephalic) horn,
which is an extension of the head. In the male the head and pronotum are shiny
brown/black and the elytra shiny olive green dotted with black (1,15). The female
has no enlarged horns and, apart from a light olive anterior third of the elytra, is
entirely dull brown and covered with bristles, particularly on the pronotum and
elytra (1,15). The two subspecies under threat, D. h. hercules and D. h. reidi, are
distinguished by their distribution and by features of the male cephalic horn (15).

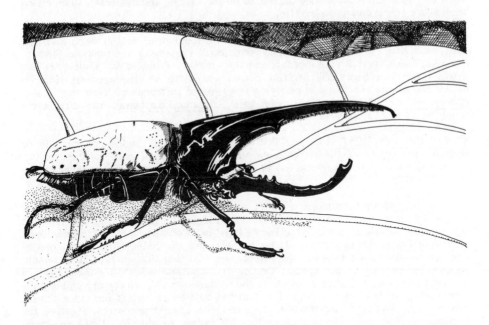

DISTRIBUTION   In a forthcoming revision of Dynastes hercules, eight races are
recognised (15):
Dynastes hercules hercules: montane areas of Basse-Terre, Guadeloupe, windward
and central mountain slopes of Dominica.
D. h. reidi: montane areas of Martinique and St Lucia.
D. h. alcides: Chiapas State, Mexico, through central and western Central
America to Costa Rica and Panama.

D. h. amazonicus: northern Para east to Amazonas State, Brazil, southern Amazonas, Colombia, south to central Bolivia; no data from Brazil interior, but occurs in Esprito Santo, Brazil.
D. h. bolivari: northern Venezuela west through Santander State, Colombia, south-west to Cauca and Chaqete into Oriente, Ecuador, north-east and central Peru.
D. h. chocoensis: Choco Mountains, Colombia, to Domingo Occidente, Ecuador.
D. h. ecuatorianus: montane localities in Napo Province, Ecuador, south to Iquitos, Peru.
D. h. glaseri: Trinidad and Tobago.

POPULATION Quantitatively unknown, but very variable throughout the range. The mainland subspecies are certainly numerous, and there are no reports of dropping numbers of D. h. glaseri on Trinidad and Tobago. D. h. reidi is particularly rare in Martinique where its habitat is disappearing (15), but it is also scarce in St Lucia (5). Only 11 specimens of this subspecies are known to be in the world's collections (6). The populations of D. h. hercules on Guadeloupe and Dominica are unknown, but are believed to be diminishing.

HABITAT AND ECOLOGY Dynastes hercules is a nocturnal and crepuscular inhabitant of humid neotropical forests. D. h. hercules and D. h. reidi are virtually confined to montane areas of the Lesser Antilles, but this may be more through necessity than choice. Observations on the beetles in nature indicate that monthly rainfall has a two-month delayed effect in increasing the density of beetles. They are most commonly seen in July and December (10). Flights are more numerous when the moon is full, and they appear to be for mating and dispersal, since flying beetles are young and unmated (10).

Larvae have been reared in the laboratory on a medium of rotten wood and cow dung, emerging after 15 months. The young adult remains in the medium for 4 to 6 weeks, feeds for 1.5 to 2 months, and then mates. Females lay their eggs two weeks later; it appears that mating occurs very late in vitellogenesis (10). In nature the larvae are found in decaying trees, often in trunks of 'Gommier blanc' (Dacryodes excelsa) and 'Paletuvier gris' (Amanoa caribaea), the life cycle requiring 13 to 24 months (4).

SCIENTIFIC INTEREST AND POTENTIAL VALUE Adult males are capable of changing the colour of the elytra from olive green to black in a few minutes (4). This is reported to be brought about by injecting water into the spongy layer of the cuticle (11,12).

The function of the enormous pronotal and cephalic horns in the male Hercules Beetle is a matter of some dispute. A common legend in Dominica is that at night the Hercules Beetle males (the 'Scieur de Bois' or 'Wood-sawyer') grasp tree limbs with their horns and saw them off by flying round and round (13). Local people find the neatly tooled twigs in the forest, not realising that in fact it is female cerambycid beetles of the species Oncideres amputator which lay their eggs in branches prior to sawing them off with their mandibles (4). Horns are found on a wide variety of beetles and their function has baffled biologists for more than a century (7). Suggested functions include defence against predators, fighting for females, digging, and even as a depository for excess weight (7). It has also been suggested that the horns have no function, and are somehow free of the constraints of natural selection (2). Although there have been no specific studies on D. hercules, work on many other horned beetles has established the role of the horns as weapons (7). They are mainly used in contests between members of the same species over critical resources (7). The often bizarre contests rarely result in physical damage, but serve to allow one of the contestants to remove his opponent from the desired resource, possibly a branch or a burrow (7).

384

Over geological time the insects have evolved a vast range of shapes and sizes. The smallest insect, the Fairy Fly Alaptus magnanimus, develops in the eggs of other insects and is a mere 0.21 mm long (16,17). Perhaps the largest insect was the extinct proto-dragonfly Meganeura from the Carboniferous period, with a 750 mm wingspan. In general a reduced size has proved to be advantageous to insects. Dynastes hercules is one of the largest living insects and as such is of interest not only as a spectacle, but also as a potential subject of enquiry into the physiological and ecological limitations to evolutionary success in insects.

THREATS TO SURVIVAL The most important threat to these two subspecies (and probably to many undocumented insects) is the destruction of the native forest on the Lesser Antilles. On Guadeloupe the forest is threatened with virtually complete destruction, and replacement with plantations of Mahogany (Swetenia macrophylla) and other trees (1,6). Chemical treatment of banana plantations on forest verges may adversely affect the beetles, particularly while they are feeding at night (1). Finally, collecting has been said to reduce populations of D. h. hercules (1), although this is in dispute (6) and requires substantiation. It is claimed that illegal trade is widespread, specimens over 14 cm long commanding high prices in the international market (1). Local inhabitants catch the beetle under street or house lights and sell them to dealers (1).

CONSERVATION MEASURES TAKEN Dynastes h. hercules is protected by French law on Guadeloupe and Martinique (1,3). Under this notice, the destruction or removal of any stage of development, the destruction, capture, or removal, the keeping for the purposes of collections and (either dead or alive) the transport, peddling, use, advertisement for sale, sale, or purchase of Dynastes hercules is forbidden (1,3). In the central part of Basse-Terre, Guadeloupe, is situated the 17 500 ha forested area of the proposed Parc Naturel de Guadeloupe (14). Theoretically this park should serve to protect the forest and its species, but the legislation regarding the park is not definitive and the area is threatened with plantations of Mahogany (6). There is no information on conservation measures taken in Dominica, Martinique or St Lucia.

CONSERVATION MEASURES PROPOSED Dynastes hercules was proposed for inclusion in the Convention on International Trade in Endangered Species of Wild Fauna and Flora (CITES) (1), but was subsequently withdrawn. Of greater importance seems to be the effective protection of the beetle's habitat on the Lesser Antillean Islands. Suitable sections of natural forest should be protected on as many individual islands as possible, in order to maintain the natural diversity of this beetle and of the other animals and plants. Any outstanding legislative measures for the Parc Naturel de Guadeloupe should be completed as soon as possible.

REFERENCES   1.   Anon. (1981). Convention on International Trade in Endangered Species of Wild Fauna and Flora (CITES). Proposition d'amendement: Inscription à l'annexe I de Dynastes hercules. (Withdrawn at 1981 CITES meeting New Delhi). In litt.
2.   Arrow, G.J. (1951). Horned beetles: a study of the fantastic in nature (Hincks, W.D., Ed.). Junk, The Hague.
3.   Burton, G.N. (1980). French insects now protected. AES Bull. 39: 58-59.
4.   Cartwright, O.L. and Chalumeau, F.E. (1978). Bredin-Archbold-Smithsonian Biological Survey of Dominica: The superfamily Scarabaeoidea (Coleoptera). Smithson. Conts. Zool. 279: iv + 32 pp.
5.   Chalumeau, F.E. (1977). Contribution à l'étude des Scarabaeoidea des Antilles (corrigenda et addenda aux Scarabaeoidea des Antilles Françaises). Bull. Soc. Linn. de

Lyon 7: 231-240.
6. Chalumeau, F.E. (1982). In litt., 16 August.
7. Eberhard, W.G. (1980). Horned beetles. Scient. Am. 242: 166-182.
8. Endrödi, S. (1976). Monographie der Dynastinae (Coleoptera) 6. Tribus: Dynastini. Acta Zool. Acad. Sci. Hung. 22: 217-269.
9. Evans, G. (1977). The life of beetles. George Allen and Unwin, London. 232 pp.
10. Gruner, L. and Chalumeau, F.E. (1977). Biologie et élevage de Dynastes h. hercules en Guadeloupe (Coleoptera, Dynastinae). Annls Soc. Ent. Fr. (N.S.) 13: 613-624.
11. Hinton, H.E. and Jarman, G.M. (1972). Physiological colour change in the Hercules Beetle. Nature 238: 160-161.
12. Hinton, H.E. and Jarman, G.M. (1973). Physiological colour change in the elytra of the Hercules Beetle, Dynastes hercules. J. Insect Physiol. 19: 533-549.
13. Honychurch, P.N. (1978). Dominica's National Park Wildlife. Dominica National Park Service: Roseau, 14 pp.
14. IUCN (1982). Directory of Neotropical protected areas. IUCN, Switzerland.
15. Reid, W.P. (1982). (Coleoptera: Scarabaeidae). A review of the known races of Dynastes hercules, and descriptions of four new forms. Unpublished manuscript.
16. Richards, O.W. and Davies, R.G. (1977). Imms' General Textbook of Entomology. 10th edition Chapman and Hall, London. Vol. 2, 1354 pp.
17. Wigglesworth, V.B. (1964). The life of insects. Weidenfeld and Nicholson, London. 360 pp.

We are grateful to M.E. Bacchus, F.E. Chalumeau, R. Gordan and W.P. Reid for comments on this sheet.

FRIGATE ISLAND GIANT TENEBRIONID BEETLE   RARE

Polposipus herculeanus Solier, 1848

Phylum	ARTHROPODA	Order	COLEOPTERA
Class	INSECTA	Family	TENEBRIONIDAE

SUMMARY Polposipus herculeanus is an unusually large tenebrionid beetle confined to Frigate (Frégate) Island in the Seychelles. Although under no immediate threat, the very localized world population should be carefully monitored.

DESCRIPTION Polposipus herculeanus is a large (length 25-30 mm), flightless tenebrionid of unusual appearance. The adults are pale grey to dark brown with a large (width 15 mm), rounded abdomen and relatively long legs (5). The chitinous forewings (elytra) are fused along the mid-line and covered with small, dark, shiny tubercles. The hindwings are absent (6). The well sclerotized larva reaches a length of 29-33 mm and is ochreous brown with the ninth segment castaneous brown (5). Photographs of adults are available in the literature (5,6).

DISTRIBUTION The species is confined to Frigate (Frégate) Island, 24 km south-east of La Digue and 45 km east of Mahé in the Seychelles group. It was originally described in 1848 from material reputedly from Bengal (7), but this locality was refuted in a redescription in 1922 (3). To confuse the issue, the species was described under the name Dysceladus tuberculatus in 1875, supposedly from material originating in Round Island, off Mauritius (9). This locality has also been contested and authorities in the British Museum (Natural History), where the specimens are held, suspect that an error has indeed occurred (14). In addition to Round Island, Mauritius, there are two Round Islands in the Seychelles, one between Cerf Island and St. Anne Island off Port Victoria, Mahé (1), and the other between Praslin and La Digue Islands (2). The collector of the Round Island specimen, Lt-Col. Pike, did visit the Seychelles as well as Mauritius, but not until 1871, the year after the specimen was accessed into the collection of the British Museum (Natural History) (6). Further intensive searches on Round Island, Mauritius have not produced a single specimen in the past 100 years (11). There is no affinity between the native fauna of the Seychelles and the Mascarenes and, on balance, it seems that the Round Island record was either a chance passenger on a piece of driftwood, or the result of human error.

POPULATION The tenebrionid is firmly established on Frigate Island and is locally fairly common (4,6). Nevertheless, the flightless adults are restricted in their ability to disperse and this confined species is rare on a world basis. Random population counts have been made in 30 trees in six patches in January 1978 and October 1982 (10). One patch showed a population decrease of more than 25 per cent, possibly as a result of storm damage, but there was no overall statistical difference between the results of the surveys (10).

HABITAT AND ECOLOGY Frigate Island has an area of about 202 ha and rises to a height of 125 m above sea level. The native vegetation has been completely cleared and much of the island is put down to coconuts and other crops to support the few inhabitants (8). However, secondary native vegetation occurs on substantial areas of steep, rocky terrain (12).

The larva of P. herculeanus lives in, and feeds upon, dead and decaying wood and bark (6). Adults have been collected under logs and palm leaves and from cracks in citrus and mango trees (4,6), but they are most commonly associated with the legume Pterocarpus indicus (Sangdragon), an introduced oriental tree (10,12). The tree has been on the island for over 200 years and was originally used for self-regenerating stakes in vanilla plantations (12). The plantations are all abandoned now, and the stakes have grown into trees (12). During the day the beetles hide 2-5 m from the ground under the loose layers of bark found particularly on large, old trees (4,6,10,12). They may occur individually or in clusters of up to 12, and as many as 22 have been found on one tree (10). At night the adults feed in the tree canopy or on the ground (10).

SCIENTIFIC INTEREST AND POTENTIAL VALUE Polposipus herculeanus is an exceptionally large and oddly shaped tenebrionid with interesting phylogenetic relationships and biology. The confined distribution is of great zoogeographic interest.

THREATS TO SURVIVAL There is no information on whether P. herculeanus has been reduced to its present distribution from a larger former range, and little is known of possible threats to its survival on Frigate Island. Cats formerly threatened the island bird life, particularly the threatened Seychelles Magpie Robin (Copsychus sechellarum), but they have now been eradicated (13). Rodents would be more dangerous and fortunately the island is free of rats, although the House Mouse (Mus musculus) is present (8,13). Any felling of the Pterocarpus indicus stands would undoubtedly have a severe impact on the populations of the tenebrionid.

CONSERVATION MEASURES TAKEN None specifically for the tenebrionid, although feral cats have been destroyed in order to protect the Magpie Robin (8,13).

CONSERVATION MEASURES PROPOSED The Frigate Island Giant Tenebrionid is not known to be under immediate threat. Nevertheless, the locally restricted population is susceptible to extinction. The status of the beetle should be monitored at intervals and any decline in the population investigated immediately. The owner of the island should be requested by the Seychelles Government to preserve stands of Pterocarpus indicus (12). Recommendations for future protection of the Magpie Robin include restricting the use of pesticides and improving measures to prevent accidental introduction of rats (8,13). These measures are equally important for the protection of the tenebrionid.

REFERENCES 1. British Admiralty (1978). Indian Ocean - Seychelles; Port Victoria and approaches. Chart No 722, scale 1: 12,500. M.o.D. Hydrographic Dept., Taunton.
2. British Admiralty (1980). Indian Ocean - Seychelles; Mahé, Praslin and adjacent islands. Chart No. 742, scale 1: 125,000. M.o.D. Hydrographic Dept., Taunton.
3. Gebien, H. (1922). V. Coleoptera, Heteromera: Tenebrionidae. The Percy Sladen Trust Expedition to the Indian Ocean in 1905, Vol. VII. Trans. Linn. Soc. Lond., 2nd Ser. Zoology 18: 261-324.
4. Lionnet, G. (1971). Frigate Island's giant beetle. Animals 13: 777.
5. Lionnet, G. (1972). The Seychelles. David and Charles, Newton Abbot. 200 pp.
6. Marshall, J.E. (1982). The larva of Polposipus herculeanus, with observations on its biology and phylogeny (Coleoptera: Tenebrionidae: Tenebrioninae) Syst. Ent. 7: 333-346.
7. Solier, M. (1848). Essai sur les Collaptérides. Blapsites. In Baudi, F. and Truqui, E. (Eds), Studi Entomologici. Torino. Pp. 149-370.

8. Todd, D.M. (1982). Seychelles Magpie Robin: Cat eradication on Frégate Island. Report to International Council for Bird Preservation. 34 pp.
9. Waterhouse, C.O. (1875). Description of some new genera and species of Coleoptera from South Africa, Madagascar, Mauritius, and the Seychelle Islands. Annls. Mag. Nat. Hist. 15: 403-414.
10. Watson, J. (1982). In litt., 15 November.
11. Williams, J.R. (1982). In litt., 10 November.
12. Robertson, I.A.D. (1982). In litt., 16 November.
13. Collar, N.J. (1982). Extracts from the Red Data Book for the Birds of Africa and Associated Islands. International Council for Bird Preservation, Cambridge. 76 pp.
14. Pope, R.D. (1982). Pers. comm., 15 October.

We are grateful to J.E. Marshall, on whose studies this sheet is based, and to N.J. Collar, R.D. Pope, I.A.D. Robertson, J. Watson and J.R. Williams for valuable background information.

LICHEN WEEVIL   VULNERABLE

Gymnopholus lichenifer Gressitt, 1966

Phylum   ARTHROPODA   Order   COLEOPTERA

Class   INSECTA   Family   CURCULIONIDAE

SUMMARY Gymnopholus lichenifer is a large weevil in the subfamily Leptopinae known only from Mount Kaindi and its environs in Papua New Guinea. It is characterized by a mutualistic relationship with a collection of plants and animals which live on its back, and is an example of the rich and diverse invertebrate fauna found on Mt Kaindi. The forests in which it lives are threatened by a variety of human activities and there is urgent need for their protection.

DESCRIPTION The Lichen Weevil is a large species, reaching a length of 23 mm (males), or 33 mm (females). The adult is slow moving, heavy-bodied and flightless. A detailed description is given in (3). On its back, the adult supports an association of plants which may include lichens, liverworts, moss, fungi and algae (3,4,9,10,11). Native lichens may cover the entire upper surface. Within this growth live a variety of invertebrates including protozoans, rotifers, nematodes, phytophagous mites and wingless insects such as springtails (Collembola), and bark lice (Psocoptera) (3,10). Isotomid Collembola have been observed feeding on the plants, and tardigrades probably also occur there. Several types of mite (Acarina) have been found, including a family of oval, eyeless mites (Symbioribatidae) known only from these weevils (1). The outer surface of the weevil is well adapted as a substrate for plants. Newly emerged adults have branched or rosette-like hairs and many scales resembling flattened metallic spines. These structures occur in special pits between tubercles and provide the attachment area for plants (6). A sticky secretion apparently traps spores that come into contact with the weevils. Plant spores may be wind-borne or transported by mites or other cohabitants of the weevil, and the plant community may take three to five years to develop (3).

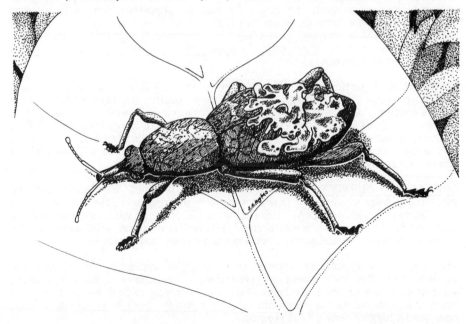

DISTRIBUTION Known only from Mt Kaindi and its environs on the south side of the Wau Valley, Papua New Guinea (3).

POPULATION Unknown.

HABITAT AND ECOLOGY Like other species of the genus, G. lichenifer adults feed on the leaves of various woody plants, vines or large grasses growing in the Nothofagus forest or at the upper forest edge. The larvae are subterranean root-feeders. In general Gymnopholus is not abundant within undisturbed forest but congregates at the interface of stunted forest and alpine grassland (3,6). It has been recorded from an altitude of 1700-2300 m (3). In sunny weather the weevils usually hang under leaves, providing shade for the plant community on their backs. In foggy, cloudy and wet weather they move about, but do not travel great distances and may remain on the same branch or sprig of leaves for several days (6). Males ride on the backs of females for long periods even when not actually mating, and individuals tend to be very long lived.

SCIENTIFIC INTEREST AND POTENTIAL VALUE The weevil genus Gymnopholus seems to be confined to the mountains of New Guinea (7). It includes some of the largest Papuan weevils, many of which are very beautiful, and the genus is particularly interesting from the point of view of its mutualistic relationships. G. lichenifer provides a particularly good example, supporting the most prolific fungal and lichen growth of all species so far observed. This mutualism is probably the result of a long evolutionary process in the constant humid environment provided by New Guinea montane forests. The process may have been helped by the fact that these weevils live comparatively long lives and suffer very little predation (9).

Flight loss is general in the genus but may not be of very ancient derivation, as most individuals have wings of about half to two thirds normal size and the elytra are often only weakly fused. Loss of flight would probably have been a prerequisite to the evolution of the mutualistic relationships, and undoubtedly contributed to local speciation on different mountain ranges. Species of the genus appear to be very actively evolving, with isolated populations on mountain tops producing many new races or subspecies (3,6). Plant growth on the weevils' backs probably provides protective camouflage (3). Besides G. lichenifer, four other species of this genus (G. urticivorax, G. marquardti, G. interpres and G. acarifer) are found in the Mt Kaindi region.

Mt Kaindi is of particular interest because of its long history of scientific, and especially entomological, study. It is the type locality of many endemic New Guinea species and is the most accessible mid-montane environment in the country (7). As a result of this and the establishment of the Wau Ecology Institute (WEI) at its foot, Mt Kaindi has become a prime research and educational area (8). It is still largely covered wth rain forest and is the only mountain in New Guinea with a road to the summit, making it of great value for education.

THREATS TO SURVIVAL The west and south sides of Mt Kaindi and the surrounding lower streams and rivers form the most important gold-mining area of New Guinea. Exploitation has increased with the price of gold, and the growing local population requires more land for vegetable farming and wood cutting. The existence of a road to the summit encourages poachers and squatters.

The weevils are reportedly scarcer as a result of collecting by visitors and disturbance to the environment by gold-mines, wood-gatherers, hunters, shifting agriculturalists, tourists and others. This is particularly serious for G. lichenifer since it is endemic to the mountain (5). Fortunately there is no pressure from commercial logging interests at present.

CONSERVATION MEASURES TAKEN No specific action to protect G. lichenifer has been taken. However, the Wau Ecology Institute is improving public education and protection of the area and has played a major role in the research carried out at Mount Kaindi. Trees have been planted on damaged slopes and in the WEI grounds, which is becoming a well developed nature reserve and teaching site. The Institute works closely with the traditional landowners and the Papua New Guinea Government, both of whom favour protection of the area. It is currently involved with marking boundaries clearly and preparing signs.

Part of the surrounding area already has some form of protection. The road is under consideration as a national walking track (The Bulldog National Walking Track) and the Ekuti range is under consideration as a wilderness area. Mt Missim, a special study area on the opposite side of the Wau valley, has been nominated as a national park. The McAdam National Park, a few kilometres down the Bulolo River from Wau valley, was the first to be designated in Papua New Guinea and protects a sample of a lower altitude vegetation zone, Araucaria (klinki and hoop pine) lower montane forest (2).

CONSERVATION MEASURES PROPOSED Although the Papua New Guinea government has approved measures to protect the Mt Kaindi area, it is unable to finance enforcement personnel or a visitor centre and is seeking international support. A reserve is being developed within the WEI grounds, but this is insufficient in size and altitudinal range to fulfil all the objectives of conservation and education. The establishment of Mt Kaindi as a reserve will enable this objective to be fulfilled, will ensure the future survival of endemic species such as G. lichenifer, and will complement the neighbouring reserves.

REFERENCES
1. Aoki, J. (1966). Epizoic symbiosis: an oribatid mite, Symbioribates papuensis, representing a new family, from cryptogamic plants growing on backs of Papuan weevils. (Acari: Cryptostigmata). Pacif. Insects 8: 281-289.
2. Gagné, W.C. and Gressitt, J. L. (1982). Conservation. Pt.7. In Gressitt, J. L. (Ed.), Biogeography and Ecology of New Guinea. Vol 2. Junk, The Hague.
3. Gressitt, J. L. (1966). Epizoic symbiosis: the Papuan weevil genus Gymnopholus (Leptopiinae) symbiotic with cryptogamic plants, oribatid mites, rotifers and nematodes. Pacif. Insects 8: 221-280.
4. Gressitt, J.L. (1969). Epizoic symbiosis. Entomol. News 80: 1-5.
5. Gressitt, J. L. (1980). In litt., 18 May.
6. Gressitt, J. L. (1982). Ecology and biogeography of New Guinea Coleoptera (Beetles). Pt. 4. Invertebrate Fauna. In Gressitt, J. L. (Ed.), Biogeography and Ecology of New Guinea Vol. 2. Junk, The Hague.
7. Gressitt, J. L. and Hornabrook, R.W. (1977). Handbook of Common New Guinea Beetles. Wau Ecology Institute Handbook No. 2. Papua New Guinea.
8. Gressitt, J.L. and Nadkarni, N. (1978). Guide to Mt. Kaindi: Background to Montane New Guinea Ecology. Wau Ecology Institute, Handbook No. 5. Papua New Guinea.
9. Gressitt, J.L. Samuelson, G.A. and Vitt, D.H. (1968). Moss growing on living Papuan moss-forest weevils. Nature, Lond. 217: 765-767.
10. Gressitt, J.L. and Sedlacek, J. (1970). Papuan weevil genus Gymnopholus. Second supplement with studies on epizoic symbiosis. Pacif. Insects 12: 753-762.

11. Gressitt, J.L., Sedlacek, J. and Szent-Ivany, J.J.H. (1965). Flora and fauna on backs of large Papuan moss-forest weevils. Science 150: 1833-1835.

We are grateful to the late J.L. Gressitt for the account on which this sheet is based. Further comments were kindly made by A. Allison, W.C. Gagné, M.G. Morris, S. Sakulas and R.T. Thompson.

Nemapalpus nearcticus Young, 1974

| Phylum | ARTHROPODA | Order | DIPTERA |
| Class | INSECTA | Family | PSYCHODIDAE |

SUMMARY The Sugarfoot Moth Fly is the only species of the primitive subfamily Bruchomyiinae known to occur in the Neartic region. There are only two known populations, both in Florida, U.S.A., one of which is imminently threatened by an extensive housing development.

DESCRIPTION The Psychodidae (moth flies and sand flies) are small to minute insects with hairy wings, short mouthparts, long legs and slender bodies (6). Most subfamilies, including this one, are harmless to man, but the Phlebotominae (sand flies) are vectors of various diseases. In general appearance Nemapalpus adults resemble dark mosquitoes, but the mouthparts are short and not adapted for bloodsucking (6). In some respects they also resemble microlepidopterans on the wing, and perch with their wings folded in a moth-like manner. N. nearcticus is dark brown with dark, fine hairs on unspotted wings. Wing length is about 3-4 mm, and the legs are long, about 5-6 mm (5). The type description (5) is given in the name Nemopalpus, now disused in favour of Nemapalpus (8).

DISTRIBUTION The type locality of the Sugarfoot Moth Fly is an area of mesic hardwood forest known as Sugarfoot Hammock, just north of S.W. 20th Avenue between Hogtown Creek and Highway I-75 in Gainesville, Alachua County, Florida, U.S.A. (5). The surrounding area is known as Sugarfoot Prairie (1). In 1981 a further population was found in Gulf Hammock, Levy County, Florida (6). Distribution of the bruchomyine flies, discussed in (2,3), is restricted to the New and Old World tropics, and this station for N. nearcticus marks their northerly limit.

POPULATION Common where it occurs, but restricted to only two known localities.

HABITAT AND ECOLOGY The species occurs only in mesic hardwood forest hammock dominated by Sweetgum (Liquidambar styraciflua), Hophornbean (Ostrya virginiana), Sugarberry (Celtis leavigata), Oak (Quercus prinus) and Red Bay (Persea borbonia). Records show adults to be on the wing from August to November, with the greatest level of activity corresponding with the highest rainfall in early September. The fly is most often found on tree trunks, in crevices or in stump-holes. Its biology is otherwise little-known, but preliminary laboratory rearings indicate a life-cycle (egg to adult) of 60 days (5). Moth flies in Europe tend to feed as larvae on decomposing excrement or vegetation on the forest floor. The adults may not feed at all.

SCIENTIFIC INTEREST AND POTENTIAL VALUE Discovered in 1971 (5), it is the only species of Nemapalpus known from North America. Most species occur in the older tropics. It is the only Nearctic member of the Bruchomyiinae, the most primitive subfamily of psychodids (5). The few species of this group are thought to be 'remnants of a group that has long passed its most successful period' (3). Thus N. nearcticus is of great interest to entomologists, both evolutionarily and biogeographically.

THREATS TO SURVIVAL A large development by the Sugarfoot Trust Inc., involving 800 apartments, with space for 2000 people and 1200 cars, has been approved in principle for 80 acres (32 ha) of the Sugarfoot Prairies east and south of Terwilliger Pond. The proposed alteration of the hammock area could result in the extermination of the Sugarfoot Hammock population of N. nearcticus (1).

CONSERVATION MEASURES TAKEN Following the original discovery of the species in 1971 several searches in other forests near Gainesville were unsuccessful (5). In October 1980 the Alachua County Commission voted to delay rezoning of Sugarfoot Prairie for development until January 1981, in order that further searches could be made (1). Those searches were also unsuccessful and revealed no other populations in the County (4). In summer 1981 specimens were found on tree trunks in Gulf Hammock, Levy County, Florida. (7). There is no information on whether the Sugarfoot Prairie development is going ahead, or whether any measures to protect the fly have been taken.

CONSERVATION MEASURES PROPOSED The Sugarfoot Trust Inc., cannot develop 181 acres (73 ha) of its 223 acres (90 ha) in the Terwilliger Pond area because they are low-lying and flood-prone. They intend to donate half of the lowlands to the University of Florida Foundation for environmental studies (1). However, the habitat of primary concern largely lies on the forest upland. Ideally the hammock should be acquired and managed by the state of Florida as an ecological reserve. In the event that the development goes ahead, as seems likely, the donated parcel of land should be demarcated in such a way as to protect at least a portion of the hammock habitat, although this might entail reducing the scope of the development. A founder population of N. nearcticus could be transplanted to San Felasco Hammock, a state reserve of several hundred acres situated approximately 18 km north of the existing colony (4). This area has suitable habitat, although the fly has not been found there . Adults may not be available, but a colony could probably be started from eggs and larvae collected in samples from the large tree holes where adults have previously been observed (4). World Wildlife Fund (U.S.A.) and the Xerces Society have expressed interest in such a project.

REFERENCES
1. Drummond, M.R. (1980). Fly prompts decision delay on Sugarfoot. Gainesville Sun 22 October, p. 1B.
2. Fairchild, G.B. (1952). Notes on Bruchomyia and Nemopalpus (Diptera, Psychodidae). Ann. Ent. Soc. Amer. 45: 259-280.
3. Quoate, L.W. (1961). Zoogeography of the Psychodidae (Diptera). Eleventh Int. Congr. Ent. Wien 1960. 1: 168-178.
4. Rogers, T.E. (1980). In litt., 10 December.
5. Young, D.G. (1974). Bruchomyiinae in North America with a description of Nemopalpus nearcticus n. sp. (Diptera: Psychodidae). Flor. Agric. Exper. Sta. J. Series No. 5223: 109-113.
6. Young, D.G. (1982) Sugarfoot Fly. In Franz, R. (Ed.), Rare and Endangered Biota of Florida Vol.6. Invertebrates. University Presses of Florida, Gainesville. Pp 82-83.
7. Young, D.G. (1982). In litt., 26 February.
8. Young, D.G. (1982). In litt., 5 August.

We are grateful to T.E. Rogers and D.G. Young for providing information for this sheet.

GIANT TORRENT MIDGE   ENDANGERED

Edwardsina gigantea Zwick, 1977

Phylum   ARTHROPODA   Order   DIPTERA

Class   INSECTA   Family   BLEPHAROCERIDAE

SUMMARY   The Giant Torrent Midge is a blepharocerid fly known only from Australia's Snowy Mountains region. The family is of some zoogeographic importance and this species is threatened by dam construction and river pollution.

DESCRIPTION   The Blepharoceridae (torrent or net-veined midges) are elongate, long-legged flies with a fine network of lines on their wings and the general appearance of very large mosquitoes. The larvae are less than 13 mm long, flattened, and have six ventral suckers. Both larvae and pupae are aquatic, the latter dark coloured, obovoid and ventrally flattened (3,4). The adults and pupae of E. gigantea have been described (5) but the larvae are unknown. This species is one of the largest members of the Australian Blepharoceridae, with a wingspan of up to 25 mm (6).

DISTRIBUTION   The species occurs only in the Snowy Mountains region of Australia where it formerly occupied a number of streams. Since the development of the Snowy Mountains hydroelectric scheme it is only known from two streams, Spencer's Creek and the Thredbo River below Thredbo village, New South Wales (2).

POPULATION   Unknown.

HABITAT AND ECOLOGY   The Blepharoceridae are confined to hilly and mountainous districts, where the larvae occur in rapidly flowing stony streams, attached to stones by their ventral suckers. The larvae are usually algal feeders. Adult Blepharoceridae fly weakly along the borders of streams, where the females prey on small Diptera and the males take nectar (3). Little is known of the ecology of this particular species (2).

SCIENTIFIC INTEREST AND POTENTIAL VALUE   The blepharocerids have a very wide but discontinuous range and are of considerable zoogeographic importance. The significance of their distributions in elucidating the possible origins of the southern continents and New Zealand, Tasmania and mainland Australia have recently been given much attention (1,5).

THREATS TO SURVIVAL   The Blepharoceridae are very intolerant of pollution, changes in stream level, and streams carrying sand or silt. This species has apparently been eliminated from a number of its former sites as a result of changes in river flow during and after dam construction (6). Both of the streams known to harbour the Giant Torrent Midge receive sewage effluent and any further decrease in water quality in these streams would constitute a serious threat to the species (2).

CONSERVATION MEASURES TAKEN   Both the known localities of the species lie within the Kosciusko National Park. Studies on water quality are being undertaken on the Thredbo River with the aim of at least preventing further deterioration in water quality, and preferably improving it (2).

CONSERVATION MEASURES PROPOSED A survey of the other streams in the Kosciusko National Park would be useful in evaluating the status of the species more precisely. Water quality monitoring should continue on Spencer's Creek, to ensure against further deterioration (2).

REFERENCES   1.   Campbell, I.C. (1981). Biogeography of some rheophilous aquatic insects in the Australian region. Aquatic Insects 3: 33-43.
2.   Campbell, I.C. (1982). In litt., 9 February.
3.   Richards, O.W. and Davies, R.G. (1977). Imms' General Textbook of Entomology. 10th Ed. 1354 pp. Vol. 2 Classification and Biology. Chapman and Hall, London.
4.   Williams, W.D. (1980). Australian Freshwater Life. Macmillan, Melbourne.
5.   Zwick, P.(1975). Edwardsininae (Diptera, Blepharoceridae): transantarctic dispersal or relict distribution? Verh. Internat. Verein. Limnol. 19: 3164-7.
6.   Zwick, P. (1977). Australian Blepharoceridae. Aust. J. Zool. Suppl. 46: 1-121.

We are grateful to I.C. Campbell for the original draft on this species, and to W.D. Williams for further comment.

TASMANIAN TORRENT MIDGE   ENDANGERED

Edwardsina tasmaniensis Tonnoir, 1924

Phylum   ARTHROPODA   Order   DIPTERA

Class   INSECTA   Family   BLEPHAROCERIDAE

SUMMARY The Tasmanian Torrent Midge was believed to be extinct when its type locality in Tasmania, Australia, was destroyed by a hydroelectric scheme in 1956. It was rediscovered in 1976 in the Dennison River but unfortunately this site is also threatened, by the Gordon hydroelectric scheme.

DESCRIPTION The Blepharoceridae (torrent or net-veined midges) are elongate, long-legged flies with a fine network of lines on their wings and the general appearance of very large mosquitoes. The larvae are less than 13 mm long, flattened, and have six ventral suckers. Both larvae and pupae are aquatic, the latter dark coloured, obovoid and ventrally flattened (4). Adult males and females of Edwardsina tasmaniensis have been described and redescribed (6). The larvae cannot be distinguished reliably from E. similis, and the pupae also lack distinct characters (6).

DISTRIBUTION E. tasmaniensis was originally described from the Cataract Gorge of the Esk River, Launceston, Tasmania, Australia in 1924 (3). The height and force of the water in the gorge were considered unique, and when the flow was diverted for hydroelectricity in 1956, the species was feared to be extinct (6). In 1976 it was rediscovered at a single site on the Denison River in south-west Tasmania (7). However, this site will be flooded if the proposed Lower Gordon hydroelectric scheme proceeds.

POPULATION Unknown.

HABITAT AND ECOLOGY The Blepharoceridae are confined to hilly and mountainous districts, where the larvae occur in rapidly flowing streams firmly attached to stones and rocks by their ventral suckers. The larvae are usually algal feeders. Adult Blepharoceridae fly weakly along the borders of streams, where the females prey on small Diptera and the males take nectar (3). E. tasmaniensis is similar to other blepharocerids in these respects, but requires extremely fast and turbulent water. This physical habitat seems to be the requirement which limits the species' distribution (2).

SCIENTIFIC INTEREST AND POTENTIAL VALUE The Blepharoceridae have been of considerable zoogeographic interest both in elucidating possible origins of the fauna of the southern hemisphere continents (5) as well as the faunal relationships between mainland Australia, Tasmania and New Zealand (1).

THREATS TO SURVIVAL Of the two known localities for the Tasmanian Torrent Midge, one has been modified by hydroelectric developments, thus eliminating the species from that site (6). The other site is similarily threatened, in this instance by the proposed Lower Gordon hydroelectric scheme (2). E. tasmaniensis was discovered in the 1920s, but in 1956 a hydroelectric power station began to divert most of the water in Cataract Gorge from the stream bed to turbine tunnels. Normal flow was drastically reduced, with the water being returned to the Esk downstream. Since the conditions required by the insect were thereby removed, it became extinct at the type locality (6). Impoundment of the Denison River is expected to bring about its total extinction (2).

399

CONSERVATION MEASURES TAKEN None.

CONSERVATION MEASURES PROPOSED A search for the species should be undertaken in south-west Tasmania to establish whether or not the species is present in areas not proposed for inundation (2). As this endemic species indicates, conditions in the Denison River gorge are probably unique in Australia since the impoundment of the Esk River's Cataract Gorge. The values of an unaltered, fast-flowing stream should be reconsidered with respect to the benefits of the proposed project.

REFERENCES 1. Campbell, I.C. (1981). Biogeography of some rheophilous aquatic insects in the Australasian region. Aquatic Insects 3: 33-43.
2. Campbell, I.C. (1982). In litt., 9 February.
3. Tonnoir, A.L. (1924). Les Blepharoceridae de la Tasmanie. Ann. Biol. Lacustre. 11: 279-291.
4. Williams, W.D. (1980). Australian Freshwater Life. Macmillan, Melbourne.
5. Zwick, P. (1975). Edwardsininae (Diptera, Blephariceridae) - transantarctic dispersal or relict distribution? Verh. Internat. Verein. Limnol. 19: 3164-3167.
6. Zwick, P. (1977). Australian Edwardsina (Diptera). Aust. J. Zool. Suppl. Ser. 46: 1-121.
7. Zwick, P. (1981). Australian Edwardsina (Diptera: Blephariceridae), new and rediscovered species. Aquatic Insects 3: 75-78.

We are grateful to I.C. Campbell for the original draft on this species and to W.D. Williams for further comments.

BELKIN'S DUNE TABANID FLY   ENDANGERED

Brennania belkini (Phillip, 1966)

Phylum   ARTHROPODA   Order   DIPTERA

Class   INSECTA   Family   TABANIDAE

SUMMARY Belkin's Dune Tabanid Fly is restricted to coastal sand dunes in southern California, U.S.A., and Baja California Norte, Mexico. Only one breeding colony is currently known, at Ballona Creek, California. Because its known habitat is being destroyed, Belkin's Dune Fly may be in danger of extinction.

DESCRIPTION Tabanidae are stoutly built flies with large, laterally extended, iridescent eyes. The proboscis is projecting, and adapted for piercing in the female. In most Tabanidae the females are haematophagous (blood-feeding), although this is apparently not the case with Belkin's Tabanid (2,3). Male tabanids feed on honeydew, nectar and plant juices. The adults of this species resemble bees and have been described in detail (1,4). The larvae are arenicolous (burrowing), but remain undescribed (3).

DISTRIBUTION Found on coastal sand dunes in southern California from Playa del Rey, Los Angeles County, California, south to Ensenada, Baja California Norte, Mexico. Only seven adult specimens are known from collections (three males and four females). Six are from California and a single specimen is from Mexico. Five of these specimens were collected before 1960. The only known breeding colony inhabits a small (1 ha) remnant sand dune at Ballona Creek, Los Angeles County, California (2,3).

POPULATION Unknown in Mexico, but clearly very low in California, U.S.A.

HABITAT AND ECOLOGY Very little is known about the ecological requirements of this extremely rare fly. It breeds only on coastal sand dunes, and a single larva has been found at a depth of 50 cm in the soil at Ballona Creek (3). The feeding requirements of the adults and juveniles are unknown. Adults are on the wing in late May to early July and may gain protection through their resemblance to bees (3).

SCIENTIFIC INTEREST AND POTENTIAL VALUE The early stages of B. belkini are of scientific interest, because juveniles of only eight pangoniine tabanids have previously been found (out of hundreds of known species). As stated, the females are unusual in not feeding on blood. Belkin's Dune Tabanid is a useful indicator of undisturbed coastal sand dune habitats (2).

THREATS TO SURVIVAL In at least the U.S.A. part of its range, Belkin's Dune Tabanid is severely endangered by human activities. The California coastal sand dunes are rapidly being destroyed by the introduction of exotic plants, off-road vehicle use and increased urban development (5, and see El Segundo Sand Dunes review). Ballona Creek, the location of the only known breeding colony of B. belkini, may soon become a site for condominiums, a hotel and shopping centre (3). Belkin's Dune Tabanid is also known from the 1 ha El Segundo Blue Butterfly Preserve owned by the Standard Oil Corporation in El Segundo, Los Angeles County (6), and undoubtedly inhabits the much larger Los Angeles International Airport site (Airport Dunes) which is largely destined for development as a golf course.

CONSERVATION MEASURES TAKEN No specific conservation measures for Belkin's Dune Tabanid have been taken, although the El Segundo Blue/Standard Oil

Reserve incidentally protects a small population (6). Careful regulation of coastal development by the California Coastal Commission and other planning agencies may indirectly provide protection. Public efforts to protect the Airport Dunes as a nature reserve, if successful, would be of great benefit to this insect (6, and see El Segundo Sand Dunes review).

CONSERVATION MEASURES PROPOSED The few remaining intact sand dune habitats in southern California should be surveyed for B. belkini and agreements sought to ban incompatible activities where it is found. The population biology and ecology of B. belkini require study to develop management plans. Proper management measures should be undertaken on all possible sites, including the Standard Oil and Airport sections of the El Segundo Dunes. The largest possible biological preserve should be set aside at the Airport Dunes. Listing as an endangered species by U.S. Federal and State agencies is recommended (3).

Only one specimen has been collected in Mexico and surveys are needed to establish the status of the species in that country.

REFERENCES   1.   Middlekauff, W.W. and Lane, R.S. (1980). Adult and immature Tabanidae (Diptera) of California. Bull. Calif. Insect Surv. 22: 1-99.
2.   Nagano, C.D., Hogue, C.L., Snelling, R.R. and Donahue, J.P. (1981). The insects and related terrestrial arthropods of Ballona. In Schreiber, R. (Ed.), The biota of the Ballona Region, Los Angeles County, California. Report to the Los Angeles County Department of Regional Planning. Pp. E1-E89.
3.   Nagano, C.D. (1982). In litt., 20 January.
4.   Phillip, C.B. (1966). New North American Tabanidae. XVIII. New species and addenda to a nearctic catalog. Ann. Ent. Soc. Am. 59: 519-527.
5.   Powell, J.A. (1981). Endangered habitats for insects: California coastal sand dunes. Atala. (1978) 6: 41-55.
6.   Pyle, R.M. (in press). Urbanization and endangered insect populations. In Frankie, G.W. and Kohler, C.S. (Eds), Perspectives in Urban Entomology, Vol. II.

We are grateful to C.D. Nagano and R.S. Lane for the original draft on this species, and to J. Donahue and G.E. Drewry for further information and comment.

PICTURE-WINGED FLIES

Drosophila spp. of Hawaii

VULNERABLE

Phylum	ARTHROPODA	Order	DIPTERA
Class	INSECTA	Family	DROSOPHILIDAE

SUMMARY An extraordinary group of fly species of great scientific value, the large picture-wings of the genus Drosophila, are endemic to the high-altitude forests of the high Hawaiian Islands. There they have radiated into an unusually diverse assemblage of about 100 species. The ecosystems of which they are a part are threatened by habitat destruction and proposed aerial pesticide application.

DESCRIPTION First brought to the attention of the scientific community 20 years ago (1), the remarkable drosophilid fauna of Hawaii has since become the focus of a continuing interdisciplinary study involving systematics, genetics and population biology (2,3,4). They represent a unique biological situation. Not only is there an exuberance of endemic species (possibly as many as 500 in the genus Drosophila alone) but their size range is also extraordinary: "from unusually small species to absolute giants up to about a centimeter across" (1). Among the giants is included a morphologically diverse group of about 100 species with dark wing-spots which, together with body spots and stripes, form bizarre, species-specific patterns. These 'picture-winged' species include forms with strong development of secondary sexual characters and behaviour (5). Elaborate courtship displays make these flies miniature analogues of birds of paradise. Males establish territories on branches of trees and tree ferns as mating arenas; these are stoutly defended and serve as attractions for females (6).

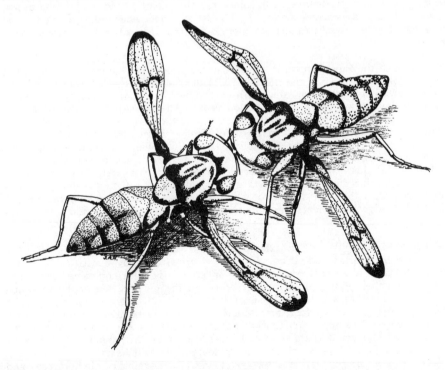

DISTRIBUTION The picture-wings and related smaller Drosophilidae are found only in the Hawaiian Archipelago, and the majority of species are also geographically confined to single islands or even single volcanoes within islands. Eighty-seven picture-wings have restricted distributions. The number of species endemic to single islands listed from north to south are as follows: Kauai 10, Oahu 27, Maui 19, Molokai 6 and Hawaii 25. Of those species which exist on two or more islands, the majority are shared by the two closely adjacent islands of Maui and Molokai. Single-island endemism is therefore a very strong tendency (11).

POPULATION Population sizes are generally small and as a particular host plant declines due to destruction of native forest, the flies also become rarer; very few are able to utilize exotic plant materials. Therefore populations are virtually absent in the lowlands (below 300 m elevation) where alien plants dominate almost to the exclusion of native vegetation (11).

HABITAT AND ECOLOGY Characteristically these Drosophila flies occur in the high-altitude forests (300 to 2000 m) where they breed in various native plant materials (7). Like the host plants that support them, the fly populations are very rare and patchy in their distributions, especially on the lava flows of young volcanoes where the plants have colonized only in recent geological times. The extensive secondary landscape of Hawaii, consisting of exotic plants on disturbed ground, is virtually devoid of the picture-winged flies and other native drosophilids. Additional ecological and evolutionary details are given in several papers (2,3,4).

SCIENTIFIC INTEREST AND POTENTIAL VALUE Drosophila fly species are the organisms of choice for fundamental genetic studies in evolutionary biology. Questions on genetic changes during speciation, the role of natural selection, and the development of new, true-breeding species are approached by scientists using Drosophila as an experimental tool. Since the flies can be bred in the laboratory, studies of genes, chromosomes, hybridization, behaviour, development and physiology can be brought to bear on these questions. Already a few answers have become clear (4).

The islands of the Hawaiian chain are volcanic and oceanic, having been formed in a serial manner. Kauai to the north-west is the oldest of the islands (5.6 million years); Hawaii (Big Island) at the south-east end of the chain is relatively young, having no lava flows older than 700 000 years. The formation of the islands in series is due to the movement of the Pacific tectonic plate over a hot spot in the earth's fluid mantle. These circumstances make it possible to designate the endemic species confined to the Big Island of Hawaii as relatively recent. Evidence indicates that some of these new species are in a state of genetic and environmental flux providing an unparalleled opportunity to catch the process of species formation in a nascent state. Combining these facts about natural populations with the fact that Drosophila are the accepted world paradigm for genetic investigation renders the scientific value of the Hawaiian fauna enormous (11).

THREATS TO SURVIVAL The existence of the picture-winged flies is seriously threatened by habitat alteration. Introduced predators may constitute a lesser threat. In Hawaii, the native Lobeliaceae have a woody habit and some form large trees. A number of the picture-wings breed on the curious yeasty decay which follows the death of a branch of one of these trees (7). Many similarly specific breeding sites are known for other species, involving a number of kinds of rare endemic trees. Indeed, a substantial number of species has been found in the remnants of the dry forests on the leeward sides of the islands. Clearing of forests to create pasture, sugar cane and pineapple plantations has had devastating effects on most of the native arthropod fauna (9). Logging of timber trees and tree ferns for orchid-bedding material furthers the damage. Furthermore, island ecosystems tend to be eroded by certain aggressive introduced weedy elements (8) so that the opening of the forest in any manner often initiates an unfavourable succession, whereby exotic species come to dominate or even to exclude native plants and animals. Among the most trenchant and invasive elements in Hawaiian forests are feral pigs, which open the forest, and Passiflora mollissima (Banana Poka Vine) which then invades. Although precise information is lacking it is considered likely that several introduced arthropod predators have affected picture-winged flies and other native insects. Among these are the voracious ant Pheidole megacephala (9,10) and more recently, a new and invasive strain of the Western Yellowjacket (Vespula pennsylvanica). Another, potentially devastating threat to these valuable arthropod ecosystems has recently been proposed by the U.S. Department of Agriculture. Known as the Tri-Fly Project, this is a scheme to attempt eradication of three species of exotic, pestiferous fruit flies (family Tephritidae) from the island chain. The proposed plans call for a knockdown phase involving the extensive use of insecticide-laden baits. Plans have been outlined for heavy and continued air drops over all forest areas and all islands in a manner recently (1981) employed in an attempt to control outbreaks of the Mediterranean Fruit Fly in California. Trials on the island of Lanai have already given conservationists grave cause for alarm. Such activities represent a distinct departure from the enlightened methods of biological control and integrated pest management which have characterized Hawaiian pest control since territorial days, and many biologists consider the risks unacceptable even if the prospects for pest eradication were great, which they are not. Although generalist exotic fruit flies may survive the chemical attack, highly specialized native drosophilidis would be highly at risk (11).

CONSERVATION MEASURES TAKEN All measures designed to conserve and perpetuate native Hawaiian forest conditions are potentially beneficial to these animals. The several national parks, the Maulua tract of the Nature Conservancy on Mauna Kea, and recovery plan activities on behalf of Endangered species of Hawaiian forest birds undertaken by the U.S. Fish and Wildlife Service, may be said to help with this problem. No measures specifically on behalf of the picture-wings have been enacted. Research and publicity by H.L. Carson, of the University of Hawaii Department of Genetics, and others is helping to raise positive awareness of these insects, their great value, and their jeopardized status (11).

CONSERVATION MEASURES PROPOSED Preservation of the picture-wings depends on protection of substantial areas on all islands from any agent that serves as an eroding force on native forests. Watersheds are of great importance to Hawaii's people, who live mostly in the lowlands. A simple way of protecting the rain forests would be to manage these watersheds so as to exclude introduced hoofed animals (feral cattle, pigs, goats, and deer) by tight fencing following elimination. The substantial cost of such an undertaking is well warranted, not only for these flies but also for the Endangered birds and many other organisms as well, and should be allocated by state or federal sources. In every instance, native vegetation should be retained and encouraged, and invasive exotics deterred, where possible through biological control. In addition to state and federal land use practices of this sort, commercial ranches and plantations should be encouraged and compensated to fence and protect substantial tracts in order to preserve the biological integrity of remnant rain forest ecosystems in their control. One of the problems is that the biological riches are spread out in such a way that virtually every area left contains species of unique interest. The compilation of inventories should be encouraged to identify as many of these as possible. Conservation methods should be designed most particularly to avoid the more drastic land uses such as destruction of native forests for timber or biomass conversion energy projects. State-wide pest eradication programmes based on heavy applications of insecticides could destroy many of the remaining endemic insects, and should be rejected as too risky. The state of Hawaii should consider up-grading the management classification of some of the land under its ownership or control in order to offer fuller protection to scarce resources. In particular, the Kilauea Forest Reserve on Hawaii, which harbours an array of Drosophila species, should be made a reserve in fact as well as in name, and the harvesting of tree ferns stopped. Hawaii does not possess large, conspicuous animals like the ungulates of Africa. Nevertheless, an utterly fascinating set of miniature ecosystems exist here; they deserve close attention and protection (11).

REFERENCES:
1. Zimmerman, E.C. (1958). Three hundred species of Drosophila in Hawaii? A challenge to geneticists and evolutionists. Evolution 12: 557-558.
2. Carson, H.L., Hardy, D.E., Spieth, H.T., and Stone, W.S. (1970). The evolutionary biology of the Hawaiian Drosophilidae. In Hecht M.K. and Steere W.C. (Eds), Essays in Evolution and Genetics in Honor of Theodosius Dobzhansky. Appleton-Century-Crofts, New York. Pp. 437-543.
3. Carson, H.L. and Kaneshiro, K.Y. (1976). Drosophila of Hawaii: systematics and ecological genetics. Ann. Rev. Ecol. Sys 7: 311-346.
4. Carson, H.L. (1982). Evolution of Drosophila on the newer Hawaiian volcanoes. Heredity 48: 3-25.
5. Hardy, D.E. (1965). Insects of Hawaii. Vol. 12. Diptera: Cyclorrhapha II, Series Schizophora, Section Acalypterae I. Family Drosophilidae. University of Hawaii Press: Honolulu. 814 pp.
6. Spieth, H.T. (1974). Mating behavior and evolution of the Hawaiian Drosophilia. In White, M.J.D. (Ed.), Genetic Mechanisms of Speciation in Insects. Australian and New Zealand Book Co., Sydney. Pp. 94-101.
7. Montgomery, S.L. (1975). Comparative breeding site ecology and the adaptive radiation of picture-winged Drosophila. Proc. Haw. Ent. Soc. 22: 65-102.
8. Mueller-Dombois, D., Bridges, K.W., and Carson, H.L. (1981). Island ecosystems: biological organization in selected Hawaiian communities. US/IBP Synthesis Series Vol. 15. Hutchinson Ross: Stroudsburg, Pa. 583 pp.
9. Zimmerman, E.C. (1959). Insects of Hawaii. Vol. I. Univ. Hawaii Press, Honolulu. 206 pp.
10. Zimmerman, E.C. (1970). Adaptive radiation in Hawaii with special reference to insects. Biotropica 2: 32-38.
11. Carson, H.L. (1981). In litt., 6 November.

We are very grateful to H.L. Carson for the original draft on this subject, and to P.F. Brussard, B. Wilcox and W. Gagne for further comment.

TOBIAS' CADDIS-FLY   EXTINCT

Hydropsyche tobiasi Malicky, 1977

Phylum	ARTHROPODA	Order	TRICHOPTERA
Class	INSECTA	Family	HYDROPSYCHIDAE

SUMMARY A former resident of the River Rhine in West Germany, Tobias' Caddis-fly has not been found for more than half a century despite surveys. It is the only species of caddis-fly documented as becoming extinct in recent times.

DESCRIPTION The species was described from old specimens in 1977 (3). Earlier it was mistakenly thought to belong to the partially sympatric Hydropsyche exocellata whose males, in common with H. tobiasi, have unusually large eyes. Early literature about H. exocellata is unreliable because the species has been confused with several others. The first good description and figure of H. tobiasi appeared in 1972 (5), under the name H. exocellata. H. exocellata was also described and figured in the same paper under the name H. borealis, but H. borealis is actually a regional form of H. contubernalis in north-western Russia. H. tobiasi differs from these other species in a number of morphological characters (3).

DISTRIBUTION All known collecting localities are near the River Rhine in West Germany, between Köln (Cologne) and Mainz. The species is well represented in early collections made without light traps, suggesting former abundance (5). Verified records are available for 1906-1914 (5) and the 1920s (6). Surveys near Bonn, Bad Säckingen, Wiesbaden-Schierstein, Oppenheim and Ingelheim in 1980 failed to find the species (1,2,4). It is believed that a formerly abundant species such as this could not have been overlooked by the surveys.

POPULATION Apparently nil.

HABITAT AND ECOLOGY As H. tobiasi was only described in 1977 and was previously mistaken for H. exocellata, its biology and ecological requirements are unknown. However, it may be assumed that, in common with all caddis-flies, the larvae develop in the river and the adults fly weakly nearby.

SCIENTIFIC INTEREST AND POTENTIAL VALUE Tobias' Caddis-fly is of interest as the first documented extinction of an aquatic caddis-fly. However, if a population survived, it could be a very sensitive and instructive indicator of the water quality in the River Rhine (4,5).

THREATS TO SURVIVAL (REASONS FOR DECLINE) The River Rhine has been subject to heavy urban and industrial pollution for several decades. All caddis-fly species with the exception of H. contubernalis survive only in very low numbers (4). H. tobiasi probably became extinct between 1920 and 1980.

CONSERVATION MEASURES TAKEN None.

CONSERVATION MEASURES PROPOSED Additional surveys should be conducted over a wide seasonal period at sites within the historic range of the species that have not recently been sampled. If the species is rediscovered, immediate steps should be taken to conserve and maintain its immediate and upstream environment. For other sensitive aquatic insects of the River Rhine, the message is clear: control pollution. It would be worthwhile to compare the non-endemic Rhine aquatic insect fauna with that of other, less polluted rivers, in order to determine the extent of local extinctions. In any case, it is now clear that polluted rivers can cause extinction at the species level.

REFERENCES: 1. Caspers, N. (1980). Die Macrozoobenthos-Gesellschaften des Rheins bei Bonn. Decheniana (Bonn) 133: 93-106.
2. Caspers, N. (1980). Die Macrozoobenthos-Gesellschaften des Hochrheins bei Bad Säckingen. Beitr. Naturk. Forsch. Süw Dtl. 39: 115-142.
3. Malicky, H. (1977). Ein beitrag zur kenntnis der Hydropsyche guttata - gruppe (Trichoptera, Hydropsychidae). Z. Arbgem. Öst. Ent. 29: 1-28.
4. Malicky, H. (1980). Lichtfallenuntersuchungen über die Köcherfliegen (Insecta, Trichoptera) des Rheins. Mainzer Naturw.Archiv 18: 71-76.
5. Malicky, H. (1981). In litt., 14 November.
6. Tobias, W. (1972). Zur Kenntnis europäischer Hydropsychidae (Insecta: Trichoptera), 1. Senckenbergiana biol. 53: 59-89.

We thank H. Malicky for providing information for this review, and L. Botosaneanu and J. Harrison for further comment.

EIGHT-SPOTTED SKIPPER   RARE

Dalla octomaculata (Godman, 1900)

Phylum   ARTHROPODA   Order   LEPIDOPTERA

Class   INSECTA   Family   HESPERIIDAE

SUMMARY Dalla octomaculata is a rare member of the cloud forest Lepidoptera on the Caribbean slopes of the Central Cordillera of Costa Rica and western Panama. This community is particularly rich in endemic species. Some deforestation has occurred recently, although this is not yet as severe as that on the Pacific slopes of the range.

DESCRIPTION Dalla octomaculata is a medium-sized pyrgine skipper with a dark chocolate ground colour, a large central cream spot on each hindwing, and three smaller cream spots on each forewing (4). As in all skippers, the antennae are hooked and all wing veins are unbranched and emanate from the cell.

DISTRIBUTION D. octomaculata is known only from the Central Cordillera of Costa Rica and western Panama. Only three specimens have been traced. The earliest (4) was captured at 6000-7000 ft (1850-2150 m) on Volcan Irazu, about 30 km east of San José in Costa Rica. The volcano has erupted since, and it is not known whether the skipper survives there. A second specimen was caught in Costa Rica this century before 1980, but no further information is available (7). The most recent capture (7 August 1979) was at 2010 m on the slopes of Cerro Pando in Bocas del Toro Province in Panama, 1.5 km from the Costa Rica border (6).

POPULATION Unknown.

HABITAT AND ECOLOGY Little information is available. The Panamanian specimen was caught at 2000 m in a very wet montane forest rich in epiphytes, especially mosses, ferns and bromeliads, large climbing aroids such as Monstera and Philodendron, and tree ferns. The skipper was caught in one of many small clearings, with 50 per cent canopy cover and a good herbaceous layer, beside a stream tributary to the Rio Changuinola. The foodplant is unknown. Other species in the area include the endemic hesperiid, Buzyges idothea (which was common in these clearings), and the butterflies Itaballia caesia tenuicornis (Pieridae), Oxeoschistus (Dioriste) euriphyle and O. puerta submaculatus (Satyrinae). Although the density and diversity of butterflies in this community is not high, the degree of endemism was the greatest encountered during a study in the highlands of western Panama (6). Buzyges idothea and Mesosemia grandis (Riodinidae), which occurred here at slightly lower altitude, are both found in a very similar habitat in the lower montane forest at 1600 m in the Monteverde Reserve, Cordillera de Tileran, Costa Rica. Two notable birds, the Black Guan (Chamaepetes unicolor) and the Collared Redstart (Myioborus torquatus) were not uncommon members of this community in both western Panama and at Monteverde.

SCIENTIFIC INTEREST AND POTENTIAL VALUE Although the Eight-spotted Skipper was discovered at the end of the 19th century (4), only two specimens have been captured since. The species is allied to D. cuparia (4) which occurs in the Andes of Bolivia, and may indicate relationships between the faunas of isolated mountain massifs of South and Central America (2). The biology of low density species is generally poorly known, particularly with regard to the critical area needed to support their populations. Further studies in this respect would be of considerable academic significance and could have direct conservation applications. This species is an aesthetically attractive insect, and indicative of a diverse community.

THREATS TO SURVIVAL There are still large tracts of forest at these altitudes on the Caribbean-facing slope of the mountain chain. However, forest clearance in the more accessible areas has already been substantial, and even in the more remote locality of the Panamanian specimen, a few clearings have been made recently (6). Very extensive clearance has already occurred at comparable altitudes and especially at slightly lower altitudes on the Pacific facing slope. Over half of the forest at this altitude has been cleared from the slopes of Volcan Irazu. As so little is known of the precise distributions of many such endemic species, the absolute threat to them is unknown. Unless areas are completely protected in each section of the mountain range, there are risks of extinctions.

CONSERVATION MEASURES TAKEN Several National Parks containing cloud forest have been created in Costa Rica, including Volcan Irazu (3), and one has been created at Cerro Campana (5) in Panama. Similar habitat occurs in Cerro Chirripo N.P., Volcan Poas N.P. and at Monteverde Reserve (Tropical Science Centre) in Costa Rica. In Panama, a forest preserve area has been proposed as a joint National Park with a comparable adjacent area in Costa Rica. Thus, both known localities are in reserves. In both types of protected land there is legislation against forest clearance and hunting.

CONSERVATION MEASURES PROPOSED The conservation measures taken to date are encouraging. However, forest is still being cleared within both types of reserve (5,6). Hunting, including that of the endemic Black Guan, was occurring in 1979 inside the western Panama preserve. The legislation must be enforced if the National Parks are to have conservation value. This will require the training and employment of additional forest guards or park rangers. More reserves should be established in sections of the mountain range with few or none at present, notably in eastern Chiriqui and in Veraguas Provinces of Panama. Parks must be sufficiently large to maintain viable populations of low density species, and should contain a range of altitudes, since many species are associated with a narrow altitudinal band (1). There is a tendency for only the high altitude, least economically productive areas to receive protection, and it is vital that intermediate altitudes should also be considered. Finance to support these measures may be met partly from tourists, and from scientists carrying out research. To encourage visitors, checklists and nature guides are needed for places other than Monteverde Reserve, for which good ones already exist. Interpretation of the forests' interest and value is needed, especially through schools. Public sympathy and awareness are necessary for long term conservation to succeed.

REFERENCES
1. Adams, M.J. (1973). Ecological zonation and the butterflies of the Sierra Nevada de Santa Marta, Colombia. J. Nat. Hist. 7: 699-718.
2. Adams, M.J. (1977). Trapped in a Colombian sierra. Geogrl Mag. 49: 250-254.
3. Boza, M.A. and Mendoza, R. (1981). The National Parks of Costa Rica. Incafao S.A., Madrid. 310 pp.
4. Godman, F.G. and Salvin O. (1879-1901). Lepidoptera - Rhopalocera. Vols. I, II, & III. Biologia Centrali - Americana. R.H. Porter, London with Dulau and Co., London.
5. Ridgley, R.R. (1976). A Guide to the Birds of Panama. Princeton University Press, Princeton. 394 pp.
6. Thomas, C.D. and Cheverton, M.R. Cambridge butterfly expedition to Central America. Unpublished report.
7. Museo de Insectos, University of Costa Rica, collection.

We are grateful to C.D. Thomas for the original draft of this review, and to K.S. Brown Jr., P.J. DeVries and P.A. Opler for additional comments. We apologise that, for technical reasons, we are unable to include the accents on Spanish words.

DAKOTA SKIPPER   VULNERABLE

Hesperia dacotae (Skinner, 1911)

Phylum   ARTHROPODA   Order   LEPIDOPTERA

Class   INSECTA   Family   HESPERIIDAE

SUMMARY The Dakota Skipper is a North American prairie endemic that now occurs only on scattered remnant tracts of native vegetation in four states of the U.S.A. Intensified agricultural land use is reducing the remaining habitat.

DESCRIPTION Hesperia dacotae is a relatively small skipper whose wariness, rapid flight, and cryptic colouring make observation against the vegetation of its habitat difficult. There is a pronounced sexual dimorphism. Adult males have a wingspan of 25-32 mm, females, 30-35 mm (5). Males are tawny orange above, with grey-brown borders strongly suffused with orange overscaling. There is a narrow black stigma, a raised structure of specialized sex scales on the forewing, just below the cell. Beneath, the hindwings and the apical region of the forewings are green-ochre with paler spots, the most prominent of which form a crescent on the hindwing which also shows dorsally. Females are darker above than males, with the tawny orange usually reduced. An irregular chain of pale spots is present on the forewing, the largest of which is glassy. Variation is discussed in (5,7). The larva is similar to other species in the genus, but can be distinguished from all other Hesperia larvae by the presence of pits on the ventral part of the head capsule (9). Early instars are grey-green, closely covered with fine tubercles each bearing a tiny seta; the head is black to red-brown, covered with shallow pits. The last instar is light brown or flesh-coloured with a pitted black head (9).

DISTRIBUTION Contemporary records of substantial populations are confined to 32 counties in four states: eastern North Dakota, eastern South Dakota, western Minnesota, and north-western Iowa. There are antiquarian records from south Manitoba (Canada), east Iowa, and Cook Co., Illinois (4,8,9). All records are within the region generally identified as mid-continent prairie, though the range includes both midgrass and tallgrass subdivisions. Within this area the skipper is confined to widely scattered remnant tracts of native vegetation (2).

POPULATION Unknown. The population is now divided into a number of probably effectively isolated populations as there is little in the biology or behaviour of skippers to suggest significant dispersal tendencies (2,6). Contemporary collection records exist for about 30 sites; four of which (one in Iowa, two in Minnesota, one in North Dakota) reportedly support substantial populations (4,7,8,11). Mark-recapture studies and visual estimates indicate an annual adult population of up to several thousand at one Minnesota site and the North Dakota site (8,9). No other estimates are available. Some sites are known to support only small populations, but most sites are poorly investigated. Considerable annual fluctuation in population size is possible, though stability has been reported in unspecified populations (8) and little fluctuation has been detected during four years of study at a Minnesota site (2).

HABITAT AND ECOLOGY H. dacotae is found only where native prairie vegetation remains relatively undisturbed, but its absence from many such tracts suggests more specific requirements (2). Edaphic factors (gravelly, calcareous, alkaline soil) are now known to be important (8,9). Earlier speculation that it is a species of wet prairie (5) is not correct (2). There is some evidence that in the preferred larval habitat midgrasses (Andropogon scoparius, Bouteloua curtipendula) are dominant, or edaphic conditions reduce the height of the vegetation (6). In North Dakota sites Alkali Grass (Zigadenus elegans) is invariably present, while various calciphile orchids and a lily are also typical (9). General agreement has been found between the records of the skipper and the area circumscribed by precipitation-evaporation ratios of 60 to 105 (8,9).

Females oviposit on any broad surface with some preference given to broad-leaved plants, especially Astragalus species. Grasses have not been observed as oviposition sites (9). Larvae feed on several grasses (Poaceae) and possibly on small sedges (Carex spp.). A preference for 'bunch grasses' has been reported (8). Andropogon scoparius, Bouteloua curtipendula, and Sporobolus heterolepis were the most utilized species in plots at a Minnesota site, with Stipa spartea and Koehleria cristata hardly fed upon (1). Larvae live in silk-lined, tubular shelters. In early instars these are vertical and mostly subterranean while those of late instars are constructed of silk and debris on the soil surface (2). Confined first instar larvae accept Poa pratensis, Koehleria cristata, Andropogon gerardi, Stipa spartea, Phleum pratense and Carex spp. for building purposes (9). Blades of grass nearby are cut off and pulled back into the shelter for feeding. Larvae seldom completely leave this tube, and feed mostly at night (9). Larval development is interrupted by winter diapause, usually in the fourth instar. Feeding and growth resume in the spring and pupation ordinarily occurs after the sixth instar in a shelter similar to the larval feeding shelter (2). The adult flight period is usually late June to late July, with only one generation per year throughout the range. Adults appear to live two to three weeks and require an ample nectar supply. The adults take nectar from several plants (oligolectic), preferring members of the Asteraceae, particularly Ratibida columnifera and Erigeron strigosus (9). Certain flowers widely used by other skippers are unacceptable to H. dacotae (9).

The interactions between the Dakota Skipper and two formerly major elements of the prairie ecosystem, fire and grazing by the North American Bison, are not well understood. Historically, grazing and occasional prairie fires may have maintained the habitat, and the adult skipper would probably have been forced to seek new locations (9). These movements were probably feasible because wide expanses of habitat were available at that time, but this is no longer so (9). Accumulation of dead grass in the absence of periodic burning or grazing is probably detrimental to the skipper. Intensive grazing by domestic cattle is not tolerated (8,9), but moderate or light grazing may be compatible (2). Many tracts where the skipper has survived have been regularly mowed for hay (8). Dakota Skipper prairies are ideally maintained by late season mowing (9). The effect of fire on the skipper probably also depends upon the time it occurs, with late summer/autumn and early spring fires least likely to be damaging. Three consecutive years of early spring controlled burns at a Minnesota site have not measurably depressed adult numbers (2).

SCIENTIFIC INTEREST AND POTENTIAL VALUE The Dakota Skipper is one of a small number of species endemic to the North American prairie biome (1,2). Its relationship to other members of the genus is obscure (4), but clarification of the phylogeny may shed light on the formation and history of the prairie (9,10,12,17). The recent nearly complete destruction of its habitat has presumably altered population structure, and this may afford opportunities for investigation of the genetic consequences as predicted by population genetics theory (2). This insect may be a useful indicator species in monitoring the effects of management on preserved tracts of prairie (2).

THREATS TO SURVIVAL Destruction of habitat for agricultural use or ranching has been the main cause of the Dakota Skipper's decline (2). The changing economy of agriculture is forcing the ploughing of tracts previously not thought to be worth cultivating. In addition, some of the flowers used for nectaring by the Dakota Skipper succumb to light grazing pressure, while the preferred species are lost to over-grazing (9). These factors may prevent the occupancy of grazing lands by the skipper (9). Gravel mining, housing developments, and irrigation projects threaten some sites. Though the skipper occurs on several nature preserves, some of these are small, making the long term prospects of their populations uncertain. Inappropriate management of preserves may pose an ironic danger (2).

CONSERVATION MEASURES TAKEN Several tracts where the skipper occurs are wholly or partially included in nature reserves, including three where substantial populations occur, and one which was acquired solely to protect the skipper (1). A study of the insect's biology and of the effects of habitat management by controlled burning is being conducted with grants from WWF-U.S., the Nature Conservancy, and the Xerces Society (2). A number of lepidopterists participated in a prairie skipper workshop in June 1980 at Lake West Okoboji (3). The purpose was to initiate a series of immediate and long-term studies dealing with H. dacotae and other rare and threatened prairie skippers.

A notice of review including the Dakota Skipper was published in the Federal Register in 1975 (13). A proposed ruling including Critical Habitat was published in 1978 (14), but later withdrawn because of amendments to the Endangered Species Act (15), and expiry of the two year limit on proposals (16). There is sufficient evidence to warrant another proposal of Threatened status and Critical Habitat under the Act, and this is under consideration by the Office of Endangered Species.

CONSERVATION MEASURES PROPOSED Habitat protection of the large population in north-central North Dakota is desirable. More information concerning populations on existing nature reserves should be obtained as a first step in incorporating the skipper in management planning. Areas used for larval development should be identified and protected from burning at critical times. Public agencies that administer sites where the skipper occurs (state parks, wildlife refuges, public grazing land) should be encouraged to consider protection of the skipper in their management plans, as the Nature Conservancy already does. Captive breeding involves difficulties in overwintering, mating, and obtaining ova. A captive breeding programme would impose undesirable selection pressures and is not immediately needed or recommended (2).

REFERENCES   1.   Dana, R.P. (1979). Investigation of life-history and response to prairie management techniques of three uncommon to rare prairie butterflies, the Dakota, Ottoe, and Pawnee Skippers. Unpubl. research proposal to World Wildlife Fund-U.S., and associated progress reports.
2.   Dana, R.P. (1981) In litt., 22 November.
3.   Downey, J.C. (1981). Studies on endangered prairie skippers. Atala 7: 27.
4.   Huber, R.L. (1975). Investigation into the proposal that Hesperia dacotae (Skinner) be considered for placement on either the threatened or endangered species lists: a preliminary survey of Minnesota populations. Report submitted to Minnesota Dept. of Natural Resources. 7 pp.
5.   Lindsey, A.W. (1942). A preliminary revision of Hesperia. Denison Univ. Bull., J. Sci. Lab., 37: 1-50.
6.   MacNeill, C.D. (1964). The skippers of the genus Hesperia in western North America, with special reference to California (Lepidoptera: Hesperiidae). Univ. Calif. Publ. Ent. 35. Univ. California Press, Berkeley and Los Angeles. 230 pp.
7.   McCabe, T.L. (1977). Skippers (Hesperioidea) of North Dakota. Fargo, N.D.: Department of Entomology and Agricultural Experiment Station, North Dakota State University. 70 pp.

8. McCabe, T.L. (1979). Report on the status of the Dakota Skipper (Lepidoptera: Hesperiidae; Hesperia dacotae (Skinner)) within the Garrison Diversion Unit, North Dakota, Contract Report submitted to U.S. Bureau of Reclamation. 46 pp.

9. McCabe, T.L. (1981). The Dakota skipper, Hesperia dacotae (Skinner): range and biology with special reference to North Dakota. J. Lepid. Soc. 35: 179-193.

10. Martin, P.S. (1975). Vanishings, and future, of the prairie. In Kesel, R.H. (Ed.), Grasslands ecology, geoscience and man 10: 39-49. School of Geoscience, Louisiana State University, Baton Rouge.

11. Opler, P.A. (1981). Management of prairie habitat for insect conservation. J. Nat. Areas Assoc. 1: 3-6.

12. Ross, H.H. (1970). The ecological history of the Great Plains: evidence from grassland insects. In Dort, W., Jr. and Jones, J.K., Jr. (Eds), Pleistocene and recent environments of the central Great Plains Pp. 225-240. University of Kansas Press, Lawrence.

13. U.S.D.I. Fish and Wildlife Service (1975). United States butterflies review of status. Federal Register 40 (55): 12691.

14. U.S.D.I. Fish and Wildlife Service (1978). Proposed Endangered or Threatened status or Critical Habitat for 10 butterflies or moths. Federal Register 43 (128): 28938-28945.

15. U.S.D.I. Fish and Wildlife Service (1979). Requirement to withdraw or supplement proposals to determine various U.S. taxa of plants and wildlife as Endangered or Threatened or to determine Critical Habitat for such species. Federal Register 44(45): 12382-12383.

16. U.S.D.I. Fish and Wildlife Service (1980). Notice of withdrawal of expired proposals for listing eight arthropod species. Federal Register 45 (171): 58171.

17. Wells, P.V. (1970). Historical factors controlling vegetation patterns and floristic distributions in the central plains region of North America. In Dort, W., Jr. and Jones, J.K., Jr. (Eds), Pleistocene and recent environments of the central Great Plains. Pp. 211-221. Univ. Kansas Press: Lawrence.

We are grateful to R.P. Dana for information used in compiling this sheet, and to J.C. Downey and P.A. Opler for further comment.

WANDERING SKIPPER   VULNERABLE

Panoquina errans (Skinner, 1892)

Phylum   ARTHROPODA   Order   LEPIDOPTERA

Class   INSECTA   Family   HESPERIIDAE

SUMMARY Restricted to the marine shoreline of California, U.S.A., and Baja California, Mexico, this species has been used as an indicator of coastal salt marsh habitat. Various leisure and commercial developments threaten Wandering Skipper populations.

DESCRIPTION This is a medium-sized, olive-brown skipper (2.5-3.0 cm) with pointed forewings. There are pale spots on both sides of the forewings. The undersides of the hindwings have veins lighter than the ground colour, with a row of small spots (10). The Wandering Skipper was at first considered to be a subspecies of Panoquina panoquinoides (Skinner), but it is now given specific status (1,6). The early stages have been described (2,4).

DISTRIBUTION The Wandering Skipper is found from Santa Barbara, California, south along the Pacific coast to Baja California, Mexico. P. errans exhibits clinal variation, and specimens from the Cape region of Baja California are larger and darker than those in the northern portion of the range (6).

POPULATION No information on populations is available.

HABITAT AND ECOLOGY The Wandering Skipper is always found in close association with the larval food plant, Salt Grass (Distichlis spicata) (2,4). P. errans seems able to live only on D. spicata that grows in moist soil which is at least wetted by high tides (8). High humidity levels are considered necessary for larval development. It was long thought (2) that the caterpillars feed only during

417

the late autumn and winter months, but it is now believed that they feed throughout the year as there are three flight periods (9). The adult takes nectar from a variety of flowering plants (9). Only one other skipper, Polites sonora siris, is restricted to the coast in the south-west United States (5).

SCIENTIFIC INTEREST AND POTENTIAL VALUE Investigation of the physiology of larvae of the Wandering Skipper may reveal how, unlike most other insects, it can tolerate large amounts of dietary salts. P. errans is an indicator of undisturbed coastal salt marsh habitat and has been used for this purpose at Ballona Creek, Los Angeles County, southern California (8,9).

THREATS TO SURVIVAL Several coastal populations of the Wandering Skipper have been destroyed in California. The causes may include insecticidal control of mosquitoes (9), damage to coastal habitats by off-road vehicles, and human and equine foot traffic (8). However, the primary threat to populations of P. errans is the filling or dredging of coastal wetlands. A colony among the near-shore canals in Venice, California is considered at risk due to commercial development (10).

CONSERVATION MEASURES TAKEN Urban development of the California sea coast is regulated by the California Coastal Commission, which is charged with preserving and protecting natural resources. This provides indirect protection for P. errans, which inhabits several areas of significant natural value (9). Under the name Panoquina panoquinoides errans, the Wandering Skipper was considered in 1975 for protection under the U.S. Endangered Species Act (3,11), but no further action was taken.

CONSERVATION MEASURES PROPOSED Southern and Baja California coastal salt marshes should be surveyed for populations of the Wandering Skipper. Use of insecticides and off-road vehicles and human and equine foot traffic should be controlled or prohibited. The welfare of P. errans colonies should be considered carefully in areas of urban or marina development. An investigation of the population dynamics of the skipper would be valuable for management purposes. This species should be reconsidered for protection under the U.S. Endangered Species Act (9).

REFERENCES 1. Brown, F.M. and Turner, T.W. (1966). Panoquina panoquinoides (Skinner) and Panoquina errans (Skinner) (Lepidoptera: Hesperiidae). Ent. News 27: 17-19.
2. Comstock, J.A. (1930). Studies in Pacific coast Lepidoptera (continued). Bull. Soc. Cal. Acad. Sci. 29: 15-19.
3. Donahue, J.P. (1975). A report on the 24 species of California butterflies being considered for placement on the federal lists of Endangered and Threatened Species. Report to the California Department of Food and Agriculture. 58 pp.
4. Emmel, T.C. and Emmel, J.F. (1973). The butterflies of southern California. Nat. Hist. Mus. L.A. Co., Sci. Ser. 26: 1-148.
5. Langston, R.L. (1974). Extended flight periods of coastal and dune butterflies in California. J. Res. Lepid. 13: 83-98.
6. MacNeill, C.D. (1962). A preliminary report on the Hesperiidae of Baja California (Lepidoptera). Proc. Calif. Acad. Sci. 4th Series. 30: 91-116.
7. MacNeill, C.D. (1975). Family Hesperiidae, The Skippers. Pp. 423-578 in W.H. Howe, (Ed.), The Butterflies of North America. Doubleday and Co., Garden City, N.Y. xiii + 633 pp.
8. Nagano, C.D., Hogue, C.L., Snelling, R.R. and Donahue, J.P. (1981). The insects and related terrestrial arthropods of

Ballona. In R. Schreiber (Ed.), The biota of the Ballona Region, Los Angeles County, California. Report to the Los Angeles County Department of Regional Planning.

9. Nagano, C.D. (1982). In litt., 20 January.
10. Pyle, R.M. (1981). The Audubon Society Field Guide to North American Butterflies. Knopf, New York. 916 pp.
11. U.S.D.I. Fish and Wildlife Service (1975). United States butterflies, review of status. Federal Register 40(55): 12691.

This review is based on a draft by C.D. Nagano, assisted by J.P. Donahue. Their contributions are gratefully acknowleged.

HARRIS' MIMIC SWALLOWTAIL BUTTERFLY   ENDANGERED

Graphium (= Eurytides) lysithous harrisianus (Swainson, 1822)

Phylum	ARTHROPODA	Order	LEPIDOPTERA
Class	INSECTA	Family	PAPILIONIDAE

SUMMARY Graphium lysithous is a species of kite swallowtail which mimics various species of Parides swallowtails in different parts of its range. The strikingly patterned subspecies G. l. harrisianus resembles the vulnerable species P. ascanius (also in this volume). Common in the eighteenth century, it was still found regularly around Rio de Janeiro, Brazil, until the 1940s. Nearly all known colonies have been destroyed by development and only a single known locality remains.

DESCRIPTION Harris' Mimic Swallowtail looks and behaves very much like Parides ascanius. It is a medium-sized black swallowtail with narrow tails and both wings crossed by a broad white band. The hindwing is variably marked with large red spots which appear similar to the rose-red patch on P. ascanius. Unlike the model, the mimic possesses a red streak at the base of the wings underneath. The antennae, tongue and tails are short for a swallowtail. The immature stages are identical to those of other subspecies of G. lysithous (1,3). The taxonomic position of this species is in dispute, and some authorities would place lysithous in the genus Eurytides. In addition, harrisianus has been considered to be a form, a subspecies and even a full species. However, the name Graphium lysithous harrisianus has been used here on the advice of authorities fully conversant with the taxonomic situation.

DISTRIBUTION G. l. harrisianus formerly occurred in southern Espirito Santo and along the whole coast of the State of Rio de Janeiro, Brazil, but is now known only from Barra de Sao Joao, on the eastern coast of Rio de Janeiro. Potential habitat and strong colonies of the model species (P. ascanius) occur widely in lowland Rio de Janeiro, but no G. l. harrisianus have been seen in these sites since 1945 (2,4). In the single known locality, four specimens were collected between 1977 and 1982 (2).

POPULATION Unknown, but certainly very low (2).

HABITAT AND ECOLOGY Little is known of the biology of the taxon. It flies in habitats adjacent to the lowland swamps occupied by its model, rarely in the swamps themselves, and sometimes on adjacent hillsides. Males do not obviously frequent hilltops in search of mates, as many other Graphium do, but are attracted to damp earth, where they were formerly often captured in Rio de Janeiro. Females are very rarely encountered, generally near their hostplants (various genera in the Annonaceae). The flight period is from September to November, with the pupal diapause lasting up to nine months (2,3,4).

SCIENTIFIC INTEREST AND POTENTIAL VALUE As the only clear Batesian mimic of the vulnerable Parides ascanius, this species occupies a unique position for mimicry research. Abundance ratios, association in the field (especially in light of the unusual situation of the model and mimic occupying slightly displaced microhabitats), synchrony of generations and other lines of enquiry suggest themselves for this model/mimic pair (5).

THREATS TO SURVIVAL The region in which the only known colony exists is undergoing development as a recreational area. There is to be some emphasis on wildlife conservation, but without special care the butterfly is likely to be lost in

421

the general alteration of habitat (2,4). At the current level of disruption, extinction is likely by 1990 (5).

CONSERVATION MEASURES TAKEN In June 1982 G. l. harrisianus was proposed to the Brazilian Government agency responsible for inclusion on the official list of species threatened with extinction.

CONSERVATION MEASURES PROPOSED Intensive searching is needed for any additional colonies which might occur in remote regions or in the Reserva Biologica Poco das Antas, Rio de Janeiro. Officials responsible for the planning and management of the recreation area under construction at Barro de Sao Joao should be informed of the species' presence and petitioned to include measures for its conservation in plans for the area. If further populations can be found, urgent studies on the swallowtail's management ecology should be undertaken.

REFERENCES   1.   Burmeister, A.M. (1879). Description Physique de la République Argentine Vol. V. (Lépidoptères), Atlas p. 9, No. 23.
2.   Callaghan, G., Laranja, J., Otero, L., Brown, K.S., Jr. (1982). Observations in 1940-1982 in Rio de Janeiro coastal lowlands. K.S. Brown Jr. In litt., 18 April et seq.
3.   D'Almeida, R.F. (1922). Mélanges Lépidoptérologiques, I. Etudes sur les Lépidoptères du Brésil. R. Friedländer und Sohn, Berlin.
4.   D'Almeida, R.F. (1966). Catalogo dos Papilionidae Americanos. Sociedade Brasileira de Entomologia, Sao Paolo, p. 278.
5.   Otero, L.S. and Brown, K.S., Jr. (in press). Biology and ecology of Parides ascanius (Cramer, 1775) (Lep., Papilionidae), a primitive butterfly threatened with extinction. Atala.

We are grateful to K.S. Brown Jr. for kindly providing this information, and to M.G. Morris for further comments. We apologise that, for technical reasons, we are unable to include the accents on Spanish words.

QUEEN ALEXANDRA'S BIRDWING BUTTERFLY   ENDANGERED

Ornithoptera alexandrae Rothschild, 1907

Phylum	ARTHROPODA	Order	LEPIDOPTERA
Class	INSECTA	Family	PAPILIONIDAE

SUMMARY Ornithoptera alexandrae is the world's largest butterfly. It is restricted to primary and advanced secondary lowland rain forest in or near the Popondetta Plain, a small area in the Northern Province of Papua New Guinea. Protected by law since 1966, the species is not often collected, but its habitat is now seriously threatened by the expanding oil palm and logging industries.

DESCRIPTION Queen Alexandra's Birdwing has an average head and body length of 7.5 cm. Large females may have a wingspan of more than 25 cm. The butterfly is sexually dimorphic, males being smaller (c. 18 cm wingspan) and brighter than females, with iridescent yellow, pale blue and pale green wing markings on a black ground colour. Females have cream markings on a dark chocolate-brown ground colour. The abdomen of both sexes is bright yellow and the ventral wingbases are bright red (3).

DISTRIBUTION The first female was collected in 1906 from the type locality high on the upper reaches of the Mambare River, well outside its present range (8). To date Ornithoptera alexandrae has only been recorded from nine 10 km grid squares on the Popondetta Plain in Northern Province, Papua New Guinea, and is known from only one other locality as a separate, high altitude population not far from the larger lowland population (12). It is reported that the 1951 eruption of Mt Lamington destroyed 250 km^2 of prime habitat, further fragmenting the patchy distribution already resulting from agriculture and logging (7).

POPULATION Conventional mark-recapture methods cannot be used to estimate numbers of O. alexandrae as the species flies high and is too infrequently seen. Larval counts are also low (only one or two may be located during a day's survey) and the leaves of the foodplant vine are often 40 m high in the upper canopy, effectively precluding observation of larvae.

HABITAT AND ECOLOGY O. alexandrae occurs with its larval food plant Aristolochia schlechteri (Aristolochiaceae) in secondary and primary lowland rain forest up to 400 m altitude on the volcanic ash soils of the Popondetta Plain, and in secondary hill forest on clay soils from 550 m to 800 m altitude in its other locality (6,10,11). It is strictly monophagous, although this is due to the oviposition specificity of the females as the larvae can mature equally well (and apparently better) on the softer-leaved Aristolochia tagala, a vine which is common and far more widespread throughout P.N.G. (14,15). O. alexandrae competes for its food plant with the more common birdwing Ornithoptera priamus. The effect of this competition is uncertain. The duration of the early stages (from egg to adult emergence) is in excess of four months and adults can live up to a further three months in the wild. Adults are subject to little predation but eggs are taken by ants, and larvae are preyed upon by toads, snakes, lizards and birds such as cuckoos, drongos and crow pheasants. However, it is believed that the aposematically (warning) coloured larvae and adults can, like other pharmacophagous butterflies, probably sequester the toxins that their food plants are known to contain, for their own protection against more general predators (9). Adults are strong fliers but appear to remain in home ranges, ignoring other available habitat. It has recently been documented that male butterflies swarm around a large timber tree Intsia bijuga (Leguminosae, known locally as Kwila), when it is in flower (7). Observations indicate that flying females will not accept males unless they have visited the flowers (7). Experimental confirmation of the behaviour pattern is needed, but the distribution of the tree may account for the absence of the butterfly from certain apparently suitable areas (7). The eggs are extremely large (4 mm diameter) and it has been calculated that females, if their ovaries are continuously productive, have the potential to lay about 240 eggs during their lifetime (10).

SCIENTIFIC INTEREST AND POTENTIAL VALUE O. alexandrae is the largest butterfly in the world and is aesthetically very attractive. The birdwings have long been held in high esteem by insect collectors and are in great demand worldwide. Species such as O. alexandrae, which are not only impressive but restricted in their range and hard to obtain, realize extremely high prices. Within the Division of Wildlife in P.N.G., there is already a marketing agency which supplies insect dealers with the unprotected insects of the country. If the long term future of O. alexandrae is safeguarded it could provide an extremely valuable income to the people of P.N.G. (5). Eventually the butterfly may become an added attraction to the growing tourist industry (13).

THREATS TO SURVIVAL The greatest current danger is the expanding oil palm industry in the Popondetta region, although cocoa and rubber plantations have also been a problem in the past. These have already claimed large tracts of forest known to have been habitat for O. alexandrae (1). Negotiations to exploit the reserves of wood in the Kumusi Timber Area are also in progress. Localized extinctions are occurring because of the growing human population in the area, and the clearing of forest to make food gardens. During the Second World War Popondetta was an important base, and at one time contained 26 airstrips (7).

CONSERVATION MEASURES TAKEN In 1966, the Fauna Protection Ordinance gave legal protection from collection to O. alexandrae and six other birdwings (4). The law has been stringently enforced on several occasions, resulting in fines for nationals and deportation of expatriates. Surveys by the Division of Wildlife are establishing the presence or absence of O. alexandrae in defined areas. A large

Wildlife Management Area (WMA), comprising approximately 11 000 ha of grassland and forest, has been established north of Popondetta. Several thousand cuttings of A. schlechteri are being prepared and an area of 4 ha of primary forest at the Lejo Agricultural Station, which is government land, is being planted as a future reserve and study area for O. alexandrae. The Wildlife Division has applied for a total of about 40 ha of government land that has been rejected for use as oil palm plantations because of the deeply dissected topography. The aim is to create reserves for O. alexandrae which, being government land, can be protected by law in perpetuity. A trial planting of A. schlechteri cuttings under tall, shady, mature (c. 14 years old) oil palms has been undertaken at the Popondetta Agricultural Training Institute to study the growth of the vines in this artificial habitat and to see whether O. alexandrae will eventually utilize them. Provincial wildlife officers regularly hold educational meetings with people in the Northern Province, to explain why the butterfly needs to be conserved. Representations for conservation of the species have been made to the Government of P.N.G. by several international bodies, including the IUCN/SSC Lepidoptera Specialist Group.

CONSERVATION MEASURES PROPOSED Negotiations to establish new WMAs are in progress between the Wildlife Division and interested landowners. Proposals for three reserve areas within the Kumusi Timber Area have been supported by the landowners and the timber company involved (Fletcher Forests, New Zealand). Implementation of the recent Conservation Areas Act (1978), which gives special protection to areas of "sites and areas having particular biological, topographical, geographical, historic, scientific or social importance", is being considered for certain sites. The Act also provides for the active management of such areas (2). It is hoped that portions of prime O. alexandrae habitat will be considered for inclusion under this Act.

It may be possible to breed O. alexandrae in captivity so that its biology and the reasons for its monophagy can be more closely studied. However, extremely large flight cages are required if the species is to behave normally in captivity, and the cost is prohibitive. Experiments to breed selectively for a culture of O. alexandrae which oviposits on Aristolochia tagala may prove rewarding (13).

REFERENCES   1.   Anon. (1976). Appraisal of the Popondetta Smallholder Oil Palm Development. Report No. 1160 Sept. 24th 1976. Dept. Primary Industry, P.N.G.
2.   Conservation Areas Act. (1978). No. 52 Independent State of Papua New Guinea. 12 September.
3.   D'Abrera, B. (1975). Birdwing Butterflies of the World. Lansdowne, Melbourne. 260 pp.
4.   Fauna Protection Ordinance. (1966). No. 19. Independent State of Papua New Guinea.
5.   Fenner, T.L. (1975). Proposal for experimental farming of protected birdwing butterflies with particular reference to Ornithoptera alexandrae. Unpubl. MS, Dept. Primary Industry. 5pp.
6.   Haatjens, H.A. (Ed.) (1964). General report on the lands of the Buna-Kokoda Area, Territory of Papua and New Guinea. C.S.I.R.O. Land Res. Ser. No. 10, 113 pp.
7.   Hutton, A.F. (1982). In litt., 20 June.
8.   Meek, A.S. (1913). A Naturalist in Cannibal Land. T. Fisher Unwin, London. 238 pp.
9.   Owen, D. (1971). Tropical Butterflies. Oxford University Press, Oxford.
10.   Parsons, M.J. (1980a). A conservation study of Ornithoptera alexandrae Rothschild (Lepidoptera: Papilionidae). First report. Wildlife Division, P.N.G. 89 pp.

11. Parsons, M.J. (1980b). A conservation study of <u>Ornithoptera</u> <u>alexandrae</u> (Rothschild) (Lepidoptera: Papilionidae). Second report. Aug. 1980. Wildlife Division, P.N.G. 16 pp.
12. Parsons, M.J. (1980c). A conservation study of <u>Ornithoptera</u> <u>alexandrae</u> Rothschild. Kumusi timber area survey. Third report, October 1980. Wildlife Division, P.N.G. 15 pp.
13. Pyle, R.M. and Hughes, S.A. (1978). Conservation and utilisation of the insect resources of Papua New Guinea. Consultancy Report to the Division of Wildlife. 157 pp.
14. Straatman, R. (1979). Summary of Survey on ecology of <u>Ornithoptera</u> <u>alexandrae</u> Rothschild. Consultancy report to the Dept. Ag. Stock & Fisheries, July 1970. 5 pp.
15. Straatman, R. (1971). The life history of <u>Ornithoptera</u> <u>alexandrae</u> (Rothschild). <u>J. Lepid. Soc.</u> 25: 58-64.

We are grateful to M.J. Parsons for assistance in the compilation of this review, and to B. D'Abrera, M.J.E. Coode, A.F. Hutton and M.G. Morris for further comment.

SCHAUS' SWALLOWTAIL BUTTERFLY   ENDANGERED

Papilio aristodemus ponceanus Schaus, 1911

Phylum	ARTHROPODA	Order	LEPIDOPTERA
Class	INSECTA	Family	PAPILIONIDAE

SUMMARY Schaus' Swallowtail was formerly locally abundant in Dade and Monroe Counties, Florida, U.S.A., but is now restricted to North Key Largo, Elliott Key and several smaller keys between them. The Key Largo population is severely threatened by development and insecticide spraying, despite being federally protected in the U.S.A. since 1976. The Elliott Key population is protected by the Biscayne National Park, but suffers dangerously wide fluctuations in numbers.

DESCRIPTION An attractive medium-sized swallowtail (wing-spread 86-95 mm, forewing expanse 45-55 mm), with brown wings dorsally yellow-spotted within a brown border and with a bold yellow band across the middle. The tails are long, brown, and yellow-edged, and the hindwings beneath have a chestnut brown median band inside a postmedian row of blue lunular spots. P. a. ponceanus is distinguished from the nominate subspecies by the reduction in width of the dorsal oblique yellow band, and more extensive yellow beneath. Spherical green eggs are deposited singly on the underside of young leaves. The cryptic larvae are brown mottled with tan, white and yellow, resembling bird droppings. When they are disturbed, a white bifurcated organ (osmeterium) protrudes from the nape and exudes a strong odour (1,5,10,14,16).

DISTRIBUTION This subspecies is only recorded from Florida, U.S.A. It was formerly locally common in areas of Dade County (including Miami) and Monroe County, including Elliott, Sands, Largo, Old Rhodes, Totten, Porgy, Adams, Upper and Lower Matecumbe and possibly Lignumvitae Keys (19). It has been destroyed in most of its range and is now restricted to localized colonies on North Key Largo in the south, Elliott Key to the north, and probably Old Rhodes, Totten, Adams, and Porgy Keys in between (1). Other subspecies are found on Cuba and the

Cayman Islands (P. a. temenes), Hispaniola (P. a. aristodemus), the Bahamas (P. a. bjorndalae) and possibly Puerto Rico (7,24)).

POPULATION   The first colony of P. a. ponceanus was found in the Brickell Hammock of Miami (17), but was apparently destroyed by the city's growth (14). It was subsequently rediscovered in the keys, but after a hurricane hit Lower Matecumbe in September 1935 P. a. ponceanus was feared extinct (10). Surveys after World War II showed that this was not so (11,12,13) but the most recent report of distribution only lists breeding populations on North Key Largo and the larger keys of the Biscayne National Park, such as Elliott Key (1,8). The Schaus' Swallowtail's normal population size seems to be small at all stages, although numbers may follow a cyclical pattern (15). Large numbers of adults (up to 100 per day) were recorded in the keys of the Biscayne National Park in 1972 (3,4,6,8) but populations were small during surveys there from 1973 to 1981 (6,15). However, a survey in May 1982 on Elliott Key recorded about 15 individuals, a cause for guarded optimism (23). Possible reasons for the population decline are given below. Recorded numbers on Key Largo have always been low (8), and most authorities consider that population to be doomed to extinction.

HABITAT AND ECOLOGY   Only tropical hardwood hammocks (patches of forest) containing the host plant Torchwood (Amyris elemifera) will support populations of Schaus' Swallowtail, although another rutaceous plant, the Wild Lime (Zanthoxylum fagara) has been observed as an oviposition site (1,16). Torchwood and Wild Lime are pioneering shrubs or small trees in whose shade sprout the hardwood seedlings which eventually form the hammock. They are therefore abundant at the edge of hammocks, but scarcer within the understorey of the mature trees (16). Schaus' Swallowtail probably continually colonizes regrowth areas partially destroyed by storm or fire (16). Females oviposit single eggs on young leaves at the tips of branches (16). Caterpillars hatch in four to seven days and moult four times at intervals of approximately 12 days. The final instar fastens itself vertically to a twig with silk and moults into a rusty brown or grey pupa (5). Emergence is normally slightly less than one year later, but some pupae in captivity have remained in diapause for two years (5,16). Annual population fluctuations suggest that this behaviour occurs in the wild and may be an adaptation to avoid unfavourable conditions while making best use of good rainfall years (1). Reproduction is correlated with the beginning of the rainy season (April to June) which, perhaps with light intensity or day length, seems to trigger a synchronous emergence of adults. There is some evidence to suggest a partial second brood in some years (7), for example, specimens were discovered during September in 1969 (22). Adult life span does not seem to exceed one month, but during that time the adults are quite capable of flying across open water to adjacent islands (5). Adults take nectar from blossoms of Guava (Psidium guajava) and Wild Tamarind (Lysiloma latisiliqua) in the hammocks, or Cheese Scrub (Morinda roioc) on their edge (16).

SCIENTIFIC INTEREST AND POTENTIAL VALUE   The ability of pupae to stay in diapause opportunistically for two seasons may be unusual in tropical or subtropical butterflies. The Schaus' Swallowtail is a rare example of an essentially tropical butterfly resident in a peripheral habitat in the only suitable area of the U.S.A. However, this is only one example of a number of less conspicuous tropical invertebrate species threatened in Florida by environmental degradation (9). Although only given subspecific rank, Schaus' Swallowtail is widely separated spatially from its conspecifics and natural reintroduction from conspecific stock is an unlikely event.

THREATS TO SURVIVAL   From the 1940s until today the range of Schaus' Swallowtail has been progressively and irrevocably eroded by the destruction of hardwood hammocks by private development and the leisure industry. Schaus' Swallowtail is no longer present on Upper or Lower Matecumbe (1), and it is nearly lost from North Key Largo. In the latter locality prime habitat has been subjected to fires, development, and aerial spraying against mosquitoes (6). Despite its protection under federal law (19), extinction of Schaus' Swallowtail from North Key Largo seems inevitable. Inadvertent injury to or destruction of deposited eggs, larvae or pupae of Schaus' Swallowtail is not an offence (19). Such activities are inevitable on building sites and the law does little to protect the species' range. The extent of threat to Elliott Key, and to other smaller keys to the north of Key Largo, is presumably lessened by their inclusion in the Biscayne National Park. However, restriction to such a small range would inevitably increase the threat of extinction from natural disasters such as hurricanes, frost or disease (18). Natural disasters such as the 1935 hurricane, which struck only four months after the species was rediscovered on Lower Matecumbe Key (2), have threatened the swallowtail populations in the past (6). Successive droughts since the early 1970s, combined with hard winter conditions in 1977/78, have also been detrimental (1). Throughout its chequered history Schaus' Swallowtail has suffered at the hands of collectors (14). Such a rare species fetches high prices. Even though the Schaus' Swallowtail is designated by the U.S. Fish and Wildlife Service as Threatened, adult specimens may be taken in unprotected areas without federal permits (19).

CONSERVATION MEASURES TAKEN On 17 April 1975 the U.S. Fish and Wildlife Service published a proposed ruling of Threatened status for Schaus' Swallowtail (18). It was recognised that the range of the butterfly was greatly depleted, but its status was limited to Threatened because the current protection of the population on Elliott and other keys in the Biscayne National Park was considered substantial (18). In the Final Rulemaking of 28 April 1976, the eggs and immature stages were protected from collectors, although not from inadvertent damage, and collection of adults was permitted outside protected areas (19). However, this legislation was reinforced by Florida state law, which banned collection of adults or immature stages of any federally listed species, except by special permit (21). In 1978 an authority considered the Biscayne National Park large enough to maintain Schaus' Swallowtail, unless natural catastrophe critically reduced the population (7). The staff of the National Park are aware of the butterfly, and co-operative in its conservation (7). Nevertheless, since the Final Rulemaking in 1976, the populations there have declined inexplicably. Recently, the Fish and Wildlife Service proposed a new Crocodile Lake National Wildlife Refuge on Key Largo (20). The report for the proposal notes the presence of Schaus' Swallowtail on North Key Largo, but it is not clear whether populations have been located within the proposed Refuge (20). Schaus' Swallowtail is classified as endangered in the most recent assessment by authorities in Florida (1,9) and a new draft recovery plan by the Florida Game and Fresh Water Fish Commission, recommends that Schaus' Swallowtail be federally listed as Endangered (24).

CONSERVATION MEASURES PROPOSED The recently published recovery plan for Schaus' Swallowtail gives details of the objectives which must be achieved to protect the butterfly from extinction (24). The priorities include protection of extant colonies and re-establishment of colonies within the species' historic range. The Biscayne National Park and Crocodile Lake National Wildlife Refuge will partially fulfil these aims, but should be expanded to include various important hammocks (24). Increased public awareness could result in voluntary conservation by informed developers and landowners in North Key Largo. Local authorities should inform planners of the butterfly's vulnerability. Collecting of adults should be further discouraged. The federal protection of eggs and immature stages, but not of adults, is a half-measure which invites abuse since it implies that the butterfly's situation is not very serious. As the new recovery plan suggests, the present federal Threatened status should be raised to Endangered as soon as possible (24). Collecting should be by permit only and should be restricted to males unless there is provision for rearing and release of progeny (1). Scientific research should be encouraged, to facilitate large-scale artificial rearing and re-introductions (24). Suitable sites include Lignumvitae Key, a preserve of the Nature Conservancy (23), or the proposed Crocodile Lake Refuge on Key Largo (20). There is a great need for detailed scientific study of the population dynamics of Schaus' Swallowtail. Although habitat destruction and excessive collecting will harm the species, very little is known of the effects of parasites on young stages, or of fire and hurricane on maintenance of suitable habitat.

REFERENCES

1. Baggett, H.D. (1982). Schaus' Swallowtail. In Franz, R. (Ed.), The Rare and Endangered Biota of Florida, Vol. 6, Invertebrates. University Presses of Florida, Gainesville. Pp. 73-74.

2. Behler, J.L. (1972). Schaus' Swallowtail Papilio aristodemus ponceanus. Animal Kingdom 75(2).

3. Brown, L.N. (1973). Populations of a new swallowtail butterfly found in the Florida Keys. Flor. Nat. April 1973: 25.

4. Brown, L.N. (1973). Populations of Papilio andraemon bonhotei Sharpe and Papilio aristodemus ponceanus Schaus (Papilionidae) in Biscayne National Monument, Florida. J. Lepid. Soc. 27: 136-140.

5. Brown, L.N. (1974). Haven for rare butterflies. Nat. Parks and Cons. Mag. July 1974: 10-13.

6. Covell, C.V. Jr. (1976). The Schaus swallowtail: a threatened subspecies? Insect World Dig. 3: 21-26.

7. Covell, C.V., Jr. (1978). Project Ponceanus and the status of the Schaus swallowtail (Papilio aristodemus ponceanus) in the Florida Keys. Atala 5 (1977): 4-6.

8. Covell, C.V., Jr. and Rawson, G.W. (1973). Project Ponceanus: a report on first efforts to survey and preserve the Schaus swallowtail (Papilionidae) in southern Florida. J. Lepid. Soc. 27: 206-210.

9. Franz, R. (Ed.), (1982). The Rare and Endangered Biota of Florida, Vol. 6, Invertebrates. University Presses of Florida, Gainesville. 131 pp.

10. Grimshawe, F.M. (1940). Place of sorrow: the world's rarest butterfly and Matecumbe Key. Nature Mag. 33: 565-567, 611.

11. Henderson, W.F. (1945). Papilio aristodemus ponceana Schaus (Lepidoptera: Papilionidae) Entomol. News 56: 29-32.

12. Henderson, W.F. (1945). Additional notes on Papilio aristodemus ponceana Schaus (Lepidoptera: Papilionidae). Entomol. News 56: 187-188.

13. Henderson, W.F. (1946). Papilio aristodemus ponceana Schaus (Lepidoptera: Papilionidae). Entomol. News 57: 100-101.

14. Klots, A.B. (1951). Field Guide to the Butterflies. Houghton Mifflin, Boston. 349 pp.

15. Loftus, W.F. and Kushlan, J.A. (1982). The status of Schaus' swallowtail and the Bahama swallowtail butterflies in Biscayne National Park. Report M-649, National Park Service, Everglades N.P., Homestead, Florida.

16. Rutkowski, F. (1971). Observations on Papilio aristodemus ponceanus (Papilionidae). J. Lepid. Soc. 25: 126-136.

17. Schaus, W. (1911). A new Papilio from Florida, and one from Mexico (Lepid.) Entomol. News 22: 438-439.

18. U.S.D.I. Fish and Wildlife Service (1975). Proposed threatened status for two species of butterflies. Federal Register 40(78): 17757.

19. U.S.D.I. Fish and Wildlife Service (1976). Determination that two species of butterflies are threatened species and two species of mammals are endangered species. Federal Register 41(83): 17736-17740.
20. U.S.D.I. Fish and Wildlife Service (1980). Ascertainment report: Crocodile Lake National Wildlife Refuge, Monroe County, Florida. 25 pp.
21. Baggett, H.D. (1982). Statement from 39-27.02 of the Rules of the Florida Game and Fresh Water Fish Commission. In litt., 24 October.
22. Brown, C.H. (1976). A colony of Papilio aristodemus ponceanus (Lepidoptera: Papilionidae) in the upper Florida Keys. J. Georgia Entomol. Soc. 11: 117-118.
23. Covell, C.V., Jr. (1982). In litt., 18 May.
24. U.S.D.I. Fish and Wildlife Service. (1982). Schaus' swallowtail butterfly recovery plan. U.S.D.I. Fish and Wildlife Service, Atlanta, Georgia. 57pp.

We are grateful to H.D. Baggett, L.N. Brown, C.V. Covell, Jr., M.G. Morris and P.A. Opler for comments on this review.

HOMERUS SWALLOWTAIL BUTTERFLY   VULNERABLE

Papilio homerus Fabricius, 1793

Phylum	ARTHROPODA	Order	LEPIDOPTERA
Class	INSECTA	Family	PAPILIONIDAE

SUMMARY Papilio homerus is the largest species in the genus, rivalling in size some of the birdwings (Troidini) of South East Asia. This, and the fact that it is uncommon and restricted to the island of Jamaica, have made it particularly prized by collectors. However, the main threat seems to be destruction of habitat. The once continuous population is already divided into two isolated pockets.

DESCRIPTION Homerus Swallowtail is a very large (forewing about 75 mm) black and yellow swallowtail with broad wings and conspicuous tails. The ground colour of the upperside forewing is black or very dark brown, usually with four or five subapical yellow spots and a broad yellow discal band extending across the entire wing, including a yellow bar across the cell. The costa is conspicuously serrate. The underside of the forewing is similar, but the subapical spots are usually reduced to one. The upperside of the hindwing has a ground colour of black to dark brown, with a broad yellow discal band whose margins are distinct anteriorly but more diffuse posteriorly, and with large, powder-blue post-discal spots or circular markings and reddish submarginal lunules. The underside of the hindwing has the yellow band much reduced. The sexes are similar but the male has hair-scales on the abdominal margin of the hindwings. The larva has an osmeterium, a protrusible repugnatorial sac characteristic of the Papilionidae (1), but is reluctant to use it. The fully grown caterpillar is dark green on back and sides, and brown beneath (4). The sides have an eye-spot, a white band, and brown markings (4).

DISTRIBUTION P. homerus is confined to the Caribbean island of Jamaica, where it has been recorded from seven of the 13 parishes. It is now restricted to two strongholds, the Blue Mountains of St Thomas and Portland (eastern population) and the 'Cockpit Country' of Trelawny Parish (western population) (1,4,6). There is no record on any other Caribbean island, despite suggestions to the contrary (2). The nearest relatives are apparently P. gammarus and P. abderus in Mexico (6).

POPULATION Little information is available. P. homerus is said to be quite common in a few favoured localities, but the total population is likely to be small. The population range is rapidly shrinking due largely to reductions of suitable habitat and limited distribution of food plants (6). Predation of larval instars is high (6). The once continuous eastern and western populations are now apparently disjunct (6). P. homerus populations seem to be free of the marked fluctuations in numbers characteristic of some other Caribbean swallowtails (4).

HABITAT AND ECOLOGY P. homerus frequents mountain slopes and gullies at fairly low elevations (150-600 m, occasionally higher). Flight is slow but powerful, and occurs throughout the day (0900-1800 hrs). The adults bask on high trees and bushes, visiting flowers daily, but only for short periods of time. Preferred nectar sources include species of Blechum, Bidens, Asclepias, Lantana, a malvaceous plant, and possibly a local Ipomea (6). The larval food plants are Hernandia cataepaefolia (eastern population) and H. troyiana (western population) (6). Oviposition has also been observed on Octotea in both locations (6). Thespesia populnea (Seaside Mahoe) has been recorded as a food plant (4) but does not occur in the known distribution of the insect (6). It may have been confused with Hernandia which is locally known as Water Mahoe (6). Seasonality is not marked, but the adults fly during February to April, September and October (4).

SCIENTIFIC INTEREST AND POTENTIAL VALUE As the largest species of Papilio, P. homerus is of considerable scientific importance (3,5). Its endemic occurrence in Jamaica is of zoogeographical interest. The zoogeography of the Papilionoidea of the Caribbean islands is not fully understood, and this species is not in a distinct group or subgenus (3,5). P. homerus is of great aesthetic value and a rational farming programme would help conservation efforts, and be an appealing commercial venture. The species has been reared with some difficulty in captivity (6) and further research is needed.

THREATS TO SURVIVAL The destruction and alteration of habitat for timber and coffee plantations are the main threats to this species. The number of sightings is declining and both populations appear to be limited to a few square kilometres (6). The rate of habitat destruction in the east has slowed since 1952, but that population is still on the decline (6). The western population is also suffering a reduction in range, but further surveys are needed. Collecting is very difficult in such mountainous country, but commercial collecting by expatriates may be a minor threat. The extent of collecting by Jamaicans is unknown. Up to $900 has been paid for a single specimen (P.A. Opler, pers. comm.).

CONSERVATION MEASURES TAKEN There have been no conservation measures specifically for this species.

CONSERVATION MEASURES PROPOSED A detailed survey of the species and its ecological requirements is needed, in order that consideration may be given to conservation. Some form of habitat protection seems essential, probably in the form of a nature reserve or national park. Commercial farming would lessen pressure on wild populations, and funds should be made available to provide suitable facilities (6). Collecting of wild specimens should be limited to those used for scientific purposes.

REFERENCES   1.   Brown, F.M. and Heinemann, B. (1972). Jamaica and its Butterflies, E.W. Classey Ltd., London.
              2.   D'Abrera, B. (1981). Butterflies of the Neotropical Region,

Part. 1, Papilionidae and Pieridae. Landsdowne Editions, Melbourne.

3. Munroe, E. (1961). The classification of the Papilionidae (Lepidoptera). Can. Ent. Suppl. 17, 51 pp.

4. Riley, N.D. (1975). A Field Guide to the Butterflies of the West Indies. Collins, London.

5. Rothschild, W. and Jordan, K. (1906). A revision of the American Papilios. Nov. Zool. 13: 411-752.

6. Turner, T.W. (1982). In litt., 12 August.

We thank M.G. Morris for the original draft of this account, and F.M. Brown, O.H.H. Mielke and T.W. Turner for further comment.

Parides ascanius (Cramer, 1775)

Phylum	ARTHROPODA	Order	LEPIDOPTERA
Class	INSECTA	Family	PAPILIONIDAE

SUMMARY Parides ascanius is a strikingly beautiful butterfly which occupies only a small fraction of its potential 'restinga' habitat (subcoastal swamps and thickets) in the state of Rio de Janeiro, south-east Brazil. It is a primitive species, lacking vigour, and most of its habitat is being drained and subdivided for residential, industrial, and recreational purposes. Many colonies active before 1970 are now extinct. Because of habitat pressure it is on the Brazilian list of animals threatened with extinction, the only insect so designated. Only about ten self-supporting colonies are now known, many in areas under pressure for development. Large colonies need to be located in regions suitable for establishment of conservation units.

DESCRIPTION P. ascanius is a large swallowtail with a black ground-colour broken by a broad white transverse band from the forewing costa to the hindwing inner margin, strongly coloured with rose on the hindwing, especially anally, and small hourglass-shaped submarginal red spots on the hindwing. The body is black, with a short abdomen showing red intersegmental marks (2). The larva is dark red-brown with creamy yellow tubercles and side stripes (1). Adult and juvenile morphology place P. ascanius very close to the widespread south Brazilian P. bunichus and almost equally close to the more restricted south Brazilian P. proneus, indicating that it is one of the most primitive members of the genus Parides (2). The name 'Fluminense' is taken from the Portuguese word meaning an inhabitant or associate of the state of Rio de Janeiro. Synonyms of this species are Battus orophobus D'Almeida and Parides orophobus (D'Almeida).

DISTRIBUTION Parides ascanius is presently known only from widely scattered points near the coast of the state of Rio de Janeiro, south-east Brazil, between the mouth of the Rio Paraiba do Sul (to the north) and Itaguai (to the south-west); formerly it may have occurred in favourable habitats on the coast of Sao Paulo state, but it is not surely known there today despite substantial searches (2).

POPULATION Unknown.

HABITAT AND ECOLOGY Parides ascanius inhabits only subcoastal and lowland swamps and thickets where its larval foodplant Aristolochia macroura (Aristolochiaceae) is abundant, and suitable flowers (mostly Compositae and Verbenaceae) are available to adults throughout the year (2). Very patchy distribution and wide-ranging male promenading indicate dispersed, possibly unstable, colonies with extensive competition from the sympatric P. zacynthus and P. anchises nephalion, the two most advanced members of the genus, both strong and aggressive species (2). P. ascanius occurs mostly on turf-sandy soils in vegetation of low to medium height and flies only in the sun during the morning and late afternoon (2). It can be reared quite easily, and kept in captivity in proper conditions (2). The habitat is very rich in endemic plants and animals, almost all of which have far wider and less patchy distributions than P. ascanius, which seems to be a relict (2).

SCIENTIFIC INTEREST AND POTENTIAL VALUE As a primitive and relict species with unusual affinities, P. ascanius has value for scientific study. Its beauty is widely recognised and specimens sell for a high price when available. Its close systematic relationships with two other species suggest that evolutionary studies would be of broad value and interest (2).

THREATS TO SURVIVAL   Habitat destruction throughout its range, but particularly near the city of Rio de Janeiro, has rendered P. ascanius increasingly scarce (2). Many colonies known before 1970 no longer exist because of swamp drainage for development into recreational areas, banana plantations, pasture or building (2). In addition, there is some danger from competition with other Parides species (2).

CONSERVATION MEASURES TAKEN   In an executive action (3481-DN) on 31 May, 1973, the Brazilian National Parks agency (through the IBDF, Brazilian Institute for Forest Development) placed P. ascanius on the official list of Brazilian animals threatened with extinction (as the synonym Battus orophobus D'Almeida). It is the only insect presently given this status, and is thereby protected from commerce (2).

The recently-established (1974) Federal Biological Reserve of Poço das Antas, north-east of Rio de Janeiro, is a 5000 ha area which includes at least 1000 ha of ideal P. ascanius habitat. Suitable areas may be extended by river management. P. ascanius is known to live in the reserve in adequately large colonies (2). A small but permanent colony of P. ascanius is also present in a swampy forest of less than 2 ha in the Parque Reserva Marapendi, Restinga de Jacarepagua in Rio de Janeiro. However, the presumed source of this colony is a patch of the food plant with its own colony in an unprotected nearby swamp, which is due to be drained (2).

CONSERVATION MEASURES PROPOSED   The Parque Reserva Marapendi is too small and inadequately protected. Attempts to transplant the food plant into suitable parts of the reserve have failed. Key parts of the nearby swamp should be protected from draining and building in order to ensure the future of this small colony (2). Should this colony fail, however, the population left in the Reserve of Poço das Antas cannot be assigned full and exclusive responsibility for preservation of P. ascanius. The main hope for the species lies in 1000 km^2 of almost impenetrable swamps in lowland Rio de Janeiro, between Itaguai and Campos. Colonies may well be flourishing there, this is not certain (2). When possible, surveys in this region should be carried out. Commerce in wild specimens should be discouraged internationally as well as nationally (2). Official captive breeding of the species in its habitat could be encouraged (2). Populations in reserves need to be extended and carefully monitored, to guarantee the well-being of this butterfly in coastal Rio de Janeiro (2).

REFERENCES   1. D'Almeida, R.F. (1966). Catalogo dos Papilionidae Americanos. Soc. Brasil. de Entomologia, Sao Paulo. 366 pp.
              2. Otero, L.S. and K.S. Brown, Jr. (In press). Biology and ecology of Parides ascanius (Cramer, 1775) (Lepidoptera, Papilionidae), a primitive butterfly threatened with extinction. Atala.

We thank K.S. Brown Jr. for providing information on this subject, and O. Mielke and M.G. Morris for further comment. We apologize that, for technical reasons, we are unable to include accents in Spanish words.

HAHNEL'S AMAZONIAN SWALLOWTAIL BUTTERFLY   RARE

Parides hahneli (Staudinger, 1882)

Phylum	ARTHROPODA	Order	LEPIDOPTERA
Class	INSECTA	Family	PAPILIONIDAE

SUMMARY Parides hahneli has been known for nearly a century, but only in the past decade have any reasonably dense colonies been discovered. It occupies a very specialized habitat in the lower middle Amazon Basin of Brazil and only three localities have ever been found for it. Until the most recent locality was found, just 20 collected specimens were known. A very primitive species, it may hold the key to the evolution of this swallowtail genus.

DESCRIPTION Parides hahneli is a medium-large swallowtail butterfly (80-100 mm) with a remarkable appearance. The forewings are long, narrow and rounded, the hindwings tailed unlike other members of its genus in the Amazon Basin (2,3). The wings are translucent yellow with black borders and cross-bars, yellower on the hindwing, with a red spot near the tail below. The species resembles sympatric ithomiine butterflies (Methona, Thyridia), which it apparently mimics, perhaps gaining protection thereby (2). The larva, dark brown with yellow rings, resembles that of P. pizarro steinbachi and P. vertumnus (4,5).

DISTRIBUTION Until 1970 Parides hahneli was known only from the region of Maues, south of the middle Amazon Basin in Brazil, but it was then discovered in the Rio Arapiuns area more to the east. There are sight and unconfirmed capture records from the region of Manaus (and Manacapuru) to the north-west of Maues, but no other records have been located for this distinctive species in spite of extensive searching by lepidopterists and commercial collectors over many years (2).

439

POPULATION Unknown. The butterfly is very rare and has a patchy distribution.

HABITAT AND ECOLOGY P. hahneli is apparently restricted to ancient sandy beaches now covered by scrubby or dense forest vegetation (2). Like many other troidine swallowtails using Aristolochia as a host plant, P. hahneli is probably extremely localized in occurrence, limited by the distribution and density of the plant. In addition, the adults need to feed on a continuous supply of nectar plants. The particular host species of Aristolochia for P. hahneli is not identified but it is also used at least by the sympatric P. chabrias ygdrasilla (1,5). Sympatric butterflies include Heliconius egeria, also partial to sandy areas and very local in conjunction with its foodplant, Passiflora glandulosa. Many plants are endemic to the Maues region. In general, the butterfly seems to occupy a poor but very specialized habitat which is not very diverse but has a high degree of endemism, demonstrated by this insect and other organisms (2). P. hahneli may suffer from food plant competition with sympatric troidine swallowtails (2).

SCIENTIFIC INTEREST AND POTENTIAL VALUE P. hahneli was probably the first member of its genus to invade the Amazon Basin. It retains a primitive pattern, morphology and tails, and is related to species of southern Brazil and Central America. It seems to represent a bridge between these archaic Parides, now pushed to the margins of the Neotropics, and the modern species in the P. aeneas and P. achises groups. It persists in a marginal biotope in the central Amazon, where it is extremely rare. This situation offers many opportunities for research on tropical ecological and evolutionary processes, which can only be solved in the context of the living organism (2). The species is beautiful (and rather bizarre), participates in mimicry rings not usual in other Parides, and currently sells for a very high price (2). It may not be possible to rear this species in captivity, and its collecting sites are a closely guarded secret in view of its commercial value. Scientific work on the species has hardly begun.

THREATS TO SURVIVAL Over-collecting for commercial purposes represents the only current threat. However, there is potential for deleterious habitat changes prior to its ecological requirements being fully understood (2).

CONSERVATION MEASURES TAKEN None to date.

CONSERVATION MEASURES PROPOSED The butterfly may occur in the Amazon National Park between Itaituba and Maues, but this is not likely. A search should be made in the small Ducke reserve near Manaus (2). A general study of the habitat and colonies of P. hahneli should be undertaken, with a view toward proposing a reserve near Maues or Arapiuns. Such a reservation would permit the survival of this swallowtail as well as the many other endemic organisms of the region, and would be a significant component of the eastern Brazilian land conservation system (2). Repeatedly scarred by rivercourse changes and affected by Pleistocene drying, this area is unique in its biological properties.

REFERENCES 1. Brown, K.S. Jr., Damman, A.J. and Feeny, P. (1981). Troidine swallowtails (Lepidoptera: Papilionidae) in southeastern Brazil: natural history and foodplant relationships. J. Res. Lepid. 19: 199-226.
2. Brown, K.S. Jr. (1982). In litt. 19 January.
3. D'Almeida, R.F. (1966). Catalogo dos Papilionidae Americanos. Soc. Brasil. de Entomologia, Sao Paulo. 366 pp.
4. Moss, A.M. (1919). The papilios of Para. Nov. Zool. 26: 295-319.
5. Manuscript material preserved with the Moss collection in the British Museum (Natural History).

We are most grateful to K.S. Brown Jr. for information on this species, and to O. Mielke and M.G. Morris for further comment.

APOLLO BUTTERFLY   VULNERABLE

Parnassius apollo (Linnaeus, 1758)

Phylum   ARTHROPODA   Order   LEPIDOPTERA

Class   INSECTA   Family   PAPILIONIDAE

SUMMARY The Apollo Butterfly, the first insect to be included under the CITES
treaty, is a large parnassian of montane and northern Eurasia. Prone to local
subspeciation, the Apollo has many named regional populations. A number of
these are extinct or endangered although others are still numerous, and the future
of the species is viewed with alarm by lepidopterists. Inclusion of suitable habitat
within national parks has not always proved adequate protection.

DESCRIPTION The Apollo is a large (50-80 mm wingspan) butterfly of the
primitive subfamily Parnassiinae. The rounded wings are chalky white with black
spots, grey markings and transparent areas lacking scales. The pattern, density
and intensity of marking varies according to locality, but the hindwings always
have striking scarlet spots. The antennal shaft is grey with darker rings (16).

DISTRIBUTION Parnassius apollo was formerly widely distributed in Europe and
Asia, although its range is now somewhat depleted. In northern Europe it has been
recorded from Norway, Denmark, Sweden and Finland across to Latvia and central
Siberia in the U.S.S.R. In central and southern Europe it has been recorded in
Spain, France, Switzerland, Italy, Liechtenstein, Austria, West Germany, Holland,
East Germany, Poland, Czechoslovakia, Hungary, Yugoslavia, Greece, Bulgaria and
Romania eastwards into the Ukraine and Bol'shoy Kavkaz (Caucasus) in the
U.S.S.R. (15,16,33-35). Populations are often isolated, disjunct and local (15,16).

POPULATION Population sizes vary dramatically from colony to colony, but density seldom seems to be very high. No specific estimates are known.

HABITAT AND ECOLOGY This butterfly, a relict of the glacial epoch, occurs in subalpine situations between 750 and 2000 m in the Alps and associated ranges, but near sea level in the northern parts of its range. The larvae feed exclusively on stonecrops (Sedum spp.). There is one generation per year, with over-wintering in the egg stage. Its population biology differs from one colony to another and individuals may be numerous or extremely rare. The Apollo seems to be able to withstand a certain degree of grazing of its habitats, which are commonly rocky and relatively xeric (35). Major alterations affecting the host plant are not tolerated (15,16). Additional details for specific races are available in the literature (10,18,20,22,29,31,32).

SCIENTIFIC INTEREST AND POTENTIAL VALUE All of the parnassians are of interest as primitive members of the swallowtail family, whose post-coital females possess an elaborate sphragis (pairing prevention device). P. apollo in particular has long been prized by collectors, who aim to possess as many of the different variants as possible. Traffic in Apollo therefore comprises a substantial value and volume of trade (26). Rare or very limited subspecies may command impressive prices. As a large, conspicuous and very attractive butterfly, the Apollo has a certain value as a tourist attraction in some alpine areas, and this could be developed as an appreciative resource. The high degree of differentiation among separated populations offers evolutionary biologists an array of adaptive strategies for study. The genetics of the species are of great interest (30).

THREATS TO SURVIVAL P. apollo is reported to have declined and become rare or endangered in all or part of the following countries: Bulgaria, Czechoslovakia, Finland (18,19), France, West Germany (1,3,35), East Germany (extinct), Greece, Italy, Liechtenstein (34), Netherlands, Norway, Poland (7), Romania, Spain (30), Sweden (17), Switzerland (5,13) and U.S.S.R. (Ukraine, Carpathians and Crimea) (15,33). There is disagreement over the importance of over-collecting in bringing about these declines. Collecting has been considered a potential threat in Finland under the current depleted conditions of populations (18,19). Over-collecting is reported to have occurred in Spain (26), Silesia (27) and Italy, where an effort was made to extirpate purposefully a rare, local race of Apollo in order to enhance the value of specimens already in hand (4). Market collecting has also been held partly responsible for the Polish decline of the species (8). None of these assertions contain numerical data. One lepidopterist, finding seven Apollos killed on 1 km of road, suggested that vehicles were a greater mortality factor than collectors (28), and a motorway system near Bozen, South Tyrol (Italy) nearly wiped out a local race of Apollo (26). However, most observers agree that it is not collectors or cars, but 'Tannensoldaten' (ranks of conifers in plantations) that threaten Europe's parnassians. Examples of afforestation interfering with Apollo survival have been cited in Bavaria (6,25), Poland (8), Spain (14) and Switzerland (15). Succession of suitable habitat to scrubland is another recurring threat, for instance in Poland and Switzerland (15). Climatic change (Finland) and acid rain (Norway) are among purported causes of the widespread withdrawal of P. apollo from much of Fennoscandia (15), although agriculture has also been blamed in Finland (19) and the causes in Sweden are not properly understood (17). Urbanization has been given as another reason for decline in Finland and tourist development is implicated in Bulgaria (15). On the whole, habitat change seems to be a far more important threat to Apollo than collecting. Whatever the major cause or causes, the effects are dramatic at least locally: in France, two subspecies are extinct, one endangered and seven vulnerable (12,15); in East Germany, the insect is extinct (15); and in Poland the decline is widespread and continuing - between 1950 and 1965 three colonies disappeared from Pieniny National Park and many other Polish populations are considered endangered (7-12,22,23,24,32).

CONSERVATION MEASURES TAKEN P. apollo was the first invertebrate to be included in Appendix II of the Convention on International Trade in Endangered Species of Wild Fauna and Flora (CITES), requiring monitoring of imports and exports to and from signatory states (26). Laws exist (15) at the national, regional or local level to protect P. apollo in the following countries: Austria, Czechoslovakia, Finland, France, Switzerland, West Germany (since the 19th century (26)) and Poland. However, these laws and regulations usually address protection of individuals (i.e., collecting restrictions) instead of habitats, and may have little effect. For example, despite legal protection accorded in 1952, the Apollo has continued to decline dramatically in Poland (7,8,22,24). P. apollo does exist in a number of national parks and other protected areas, such as Tatra and Pieniny National Parks in Poland, but this does not always lead to its conservation (11) since specific management measures seldom have been taken. An attempt has been made to reintroduce P. apollo into the Pieniny Mountains in Poland (12,23); the results are not yet known. Many publicity items have been produced stressing the need for conservation of P. apollo, such as a poster issued by the Fédération Française des Sociétés de Sciences Naturelles, depicting four of the threatened French races of the butterfly.

CONSERVATION MEASURES PROPOSED Many authors have called for specific measures on behalf of P. apollo. In Poland the status of the butterfly inside and outside the national parks should be assessed in order to identify the factors detrimental to its survival, and the government should implement the aims of the 1952 law protecting the butterfly (11,23). Reserves for Apollo are needed in the mountains of Crimea (31). The Spanish Lepidoptera Red Data Book (30) offers suggestions for conserving the butterfly in the Pyrenees, and strong measures (including a new national park for this and other wildlife) are recommended in a later Spanish appraisal (14). Artificial breeding of two generations per year (instead of the usual one) has been suggested as a means of reinforcing populations or enhancing reintroduction attempts in Scandinavia (20). Reintroductions should be attempted only with larvae or pupae, to avoid emigration of the adults (21). Past attempts to release butterflies have not succeeded. All countries where the species occurs should sponsor detailed surveys of Apollo colonies, and take steps to establish reserves and implement management measures where possible. Reforestation with conifers should be avoided and scrub succession should be arrested in important Apollo habitats. The provisions of CITES should be enforced, to determine levels of commercial trade in the butterfly and its possible impact on populations. Attention should also be given to Parnassius phoebus and P. mnemosyne, which may be at least as threatened as P. apollo in Europe (15,18,19,26).

REFERENCES

1. Anon. (1976). Rote Liste bedrohter Tiere in Bayern (Wirbeltiere und Insekten). Schrift. Natur. Lans. Bayern 3: 1-12.
2. Bernardi, G., Nguyen, T. and Nguyen, T.H. (1981). Inventaire, cartographie et protection des Lépidoptères en France. Beih. Veröff. Naturschutz Landschaftspflege Bad.-Württ. 21: 59-66.
3. Blab, J., Nowak, E. and Trautmann, W. (1977). Rote Liste der Gefährdeten Tiere und Pflanzen in der Bundesrepublik Deutschland. Naturschutz Aktuell 1: 1-67.
4. Bourgogne, J. (1971). Un témoignage de plus sur la destruction de la Nature (Papilionidae). Alexanor 7: 50.
5. Burckhardt, D., Gfeller, W. and Muller, H.U. (1980). Geschützte Tiere der Schweiz. Schweiz. Bund. für Naturschutz, Basel. 223 pp.
6. Christensen, G. (1975). Wer rottet aus? Entomol. Zeitschrift 85: 246-48.

7. Dabrowski, J.S. (1975). Some problems in the preservation of butterflies in Poland. Atala 3: 4-5.
8. Dabrowski, J.S. (1980). The protection of the Lepidopterofauna - the latest trends and problems. Nota Lepid. 3: 114-118.
9. Dabrowski, J.S. (1980). O stanie zagrozenia lepidopterofauny w niektorych parkach narodowych Polski. Wiad. Entomol. 1: 143-149.
10. Dabrowski, J.S. (1980). Mizeni biotopu jasone cervenookeho - Parnassius apollo (L.) v Polsku a nutnost jeho aktivni ochrany (Lepidoptera, Papilionidae). Cas. Slez. Muz. Opava (A) 29: 181-185.
11. Dabrowski, J.S. (1981). Remarks on the state of menacing the lepidopteran fauna in National Parks. Part 11 (general): the National Park in Tatra. Zesz. Nauk. Uniw. Jaqiellonskiego, Prace Zool 27: 77-100.
12. Dabrowski, J.S. and Palik, E. (1979). Uwagi o stanie zagrozenia w parkach narodowych, Czesc 1: Zmiany zachodzace we wspolczesnej lepidopterofaunie Pieninskiego Parku Narodowego, ze szczegolnym uwzglednieniem zanikania gatunku Parnassius apollo (L.), (Lepidoptera: Papilionidae), Dokumentacja n/t na zlec. Kom. Nauk.: 'Czlowiek i Srodowisko' PAN, 1-38 (maszynopis).
13. Gfeller, W. (1975). Geschützte Insekten in der Schweiz. Mitt. Schweiz. Ent. Ges. 48: 217-213.
14. Gomez-Bustillo, M.R. (1981). Protection of Lepidoptera in Spain. Beih. Veröff. Naturschutz Landschaftsflege Bad.-Württ. 21: 67-72.
15. Heath, J. (1981). Threatened rhopalocera (butterflies) in Europe. Council of Europe, Nature and Environment Series No. 23, 157 pp.
16. Higgins, L.G. and Riley, N.D. (1980). A Field Guide to the Butterflies of Britain and Europe. 4th Ed. Collins, London. 384 pp.
17. Janzon, L-A. and Bignert, A. (1979). Apollofjärilen i Sverige, Fauna Flora, Upps. 74: 57-66.
18. Mikkola, K. (1979). Vanishing and declining species of Finnish Lepidoptera Notul. Ent. 59: 1-9.
19. Mikkola, K. (1981). Extinct and vanishing Lepidoptera in Finland. Beih. Veröff. Naturschutz. Landschaftspflege Bad.-Württ. 21: 19-22.
20. Nikusch, I. (1981). Die Zucht von Parnassius apollo Linnaeus mit jährlich zwei Generationen als Möglichkeit zur Erhaltung bedrohter Populationen. Beih. Veröff. Naturschutz Landschaftspflege Bad.-Württ. 21: 175-176.
21. Nikusch, I. (1982). First trials to save threatened populations of Parnassius apollo by transplantation to new suitable biotypes. Third Europ. Cong. Lepidop., Cambridge. Unpublished.
22. Palik, E. (1966). On the process of dying out of Parnassius apollo Linné. Tohoku Koncho Kenkyu 2: 45-47.
23. Palik, E. (1980). The protection and reintroduction in Poland of Parnassius apollo (Linnaeus) (Papilionidae). Nota Lepid. 2: 163-164.
24. Palik, E. (1981). The conditions of increasing menace for the existence of certain Lepidoptera in Poland. Beih. Veröff. Naturschutz Landschaftspflege Bad.-Württ. 21: 31-33.
25. Pfaff, G. (1935). Wer rottet aus? Entomol. Zeitschrift 49: 105-107.

26. Pyle, R.M. (1976). The eco-geographic basis for Lepidoptera conservation. Part iii. A review of world Lepidoptera conservation. Yale Univ. PhD. Thesis, Publ. Univ. Microfilm Int. 369 pp.

27. Rowland-Brown, H. (1913). Parnassius apollo in Germany. Entomologist 46: 289-290.

28. Schmiedel, R. (1934). Tragt der sammler die schuld am Ruckgang unserer Insektenfauna? Entomol. Rundschau 51: 174-177.

29. Svenson, I. (1981). Changes in the Lepidoptera fauna of Sweden after Linnaeus. Beih. Veröff. Naturschutz. Landschaftspflege Bad.-Württ. 21: 23-30.

30. Viedma, M.G. de and Gomez-Bustillo, M.R. (1976). Libro Rojo de los Lepidopteros Ibericos. Instituto Nacional para la Conservacion de la Naturaleza, Madrid. 120 pp.

31. Yermolenko, V.M. (1973). On protection of useful, relict and endemic insects of the Ukrainian Carpathian mountains and mountains of Crimea. In On Insect Protection. Armenian Acad. Sci. Symposium, Yerevan. Pp. 29-35.

32. Zukowski, R. (1959). Extinction and decrease of the butterfly Parnassius apollo on Polish territories. Sylwan 103: 15-30.

33. Heath, J. and Leclerq, J. (Eds) (1981). European Invertebrate Survey. Provisional Atlas of the Invertebrates of Europe, Maps 1-27. Institute of Terrestrial Ecology, Cambridge and Faculté des Sciences Agronomiques, Gembloux.

34. Biedermann, J. (1982). Lebensraum für Insekten. Liechtensteiner Umweltbericht June 1982, 4-5.

35. Blab, J. and Kudrna, O. (1982). Naturschutz Aktuell, Hilfsprogramm für Schmetterlinge. Kilda-Verlag, Greven. 135 pp.

We are grateful to J. Heath, O. Kudrna, K. Mikkola, M.G. Morris, M. Morton, I. Nikusch and J. Shepard for help in compiling this review.

FLORIDA ATALA HAIRSTREAK BUTTERFLY   VULNERABLE

Eumaeus atala florida (Röber, 1926)

Phylum   ARTHROPODA   Order   LEPIDOPTERA

Class   INSECTA   Family   LYCAENIDAE

SUMMARY The brilliant Florida Atala Hairstreak was believed to be extinct in
the Florida, U.S.A., part of its range until small colonies were discovered in 1959.
Currently threatened by habitat alterations, it may also be susceptible to
over-collecting because of its small numbers, sluggish behaviour and limited
distribution. Determined habitat protection and management measures are
required.

DESCRIPTION Although it is certainly a member of the subfamily Theclinae, the
Florida Atala lacks tails and is larger than most other hairstreak butterflies. The
forewing expanse ranges from 40 to 50 mm, and the wings are velvety black with
iridescent blue-green scaling. Males are heavily green-dusted across the forewing
and along the dorsal hindwing margins; females are blue-dusted across the cell
only. Below, both sexes have three rows of metallic blue dots and a coral or
scarlet patch near the bright red abdomen. The similar Cycad Butterfly (E.
minijas), which is a rarity in southern Florida and Texas, is a neotropical butterfly
with white-fringed wings; E. atala wings have black fringes (12). The mature
larvae are vermilion with two parallel dorsolateral rows of yellow spots, making
them highly conspicuous on the food plants; pupae are similarly coloured (1).

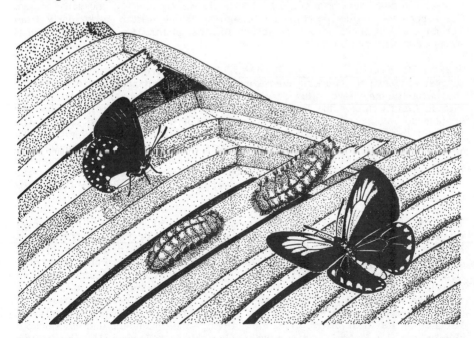

DISTRIBUTION Eumaeus atala is a Caribbean butterfly known from the Bahamas
(Andros, Abaco (4,6) and Grand Bahama), Cuba and Florida, U.S.A. (7). E. atala
florida has been claimed as strictly Floridian, but the hairstreak specialist H.K.
Clench considered Bahamian specimens as probably of the same subspecies (5).
Whatever the taxonomic position may be, the Florida population is under serious

447

threat and its destruction would be a great loss to the species as a whole. Colonies were formerly numerous in the southern part of Florida (Broward and Dade Counties) but are now known only from a small number of isolated colonies in Dade County (1). The butterfly was considered extinct in Florida for a number of years (13,15). The nominate subspecies E. a. atala is found in Cuba and the Bahamas (4,5,6).

POPULATION No precise total population estimate has been made, but the number of individuals in Florida is unlikely to exceed a thousand (2). These are confined to fewer than five breeding sites. Females are thought to produce about 40 eggs, only 10-20 per cent of which will reach adulthood in the wild (2). Brood size is therefore lower than average for butterflies by an order of magnitude, and replacement in nature is possibly very slow in comparison to that of most insects.

HABITAT AND ECOLOGY The Florida Atala inhabits brushy remnants of tropical hammocks and southern slash pine forest where the hostplant cycad Florida Coontie (Zamia pumila, Zamiaceae) occurs (11). It is possible that other species of Zamia may also be used. Cultivated and exotic cycads are reported as hosts of E. a. atala in Cuba (3), and E. a. florida may use these in Florida. However, cycads of the genus Zamia are still relatively widespread and abundant, while colonies of the Atala are few and highly localized. This suggests that there are microhabitat requirements not yet discerned by researchers (1). Females deposit the pale brown eggs in clusters of 5-15 at the tips of young Zamia shoots. Larvae, at first gregarious, later disperse as the initial hostplant becomes defoliated. Limiting factors may include host competition with the larvae of Seirarctia echo (Arctiidae) (9), and cannibalism, which is known to occur in captive final instars. Generation timing depends upon seasonal influences, and individuals have been recorded in nearly every month (1). Adults fly weakly and are unusually benign. They spend much time perching on foliage or taking nectar, particularly from Verbenaceae but also from a wide variety of other plants (16). The life history is further elaborated in (1).

SCIENTIFIC INTEREST AND POTENTIAL VALUE The bright colours of all stages suggest aposematism (warning coloration). The hostplants possess toxins (1), so the butterfly may have value for investigations in ecological chemistry. The Florida Atala's chief value for humans is as an aesthetic resource; its behaviour enables close observation and its appearance is superb on the wing or at rest. The rarity of the subspecies has caused collectors to offer substantial sums for the purchase of specimens from Florida. With its specific habitat requirements, E. a. florida is probably an indicator of conditions that have become greatly diminished in southern Florida. A journal devoted to insect conservation, published by the Xerces Society, is entitled Atala in recognition of this insect's rarity and conservation importance.

THREATS TO SURVIVAL The rapid rate of development in southern Florida over the past quarter-century seems to be responsible for the decline of this butterfly. Residential and industrial development, agriculture and forestry have brought about drastic reduction and fragmentation of the prime habitat. Forest fires have reduced the areas of secondary growth in which necessary ecological conditions prevail (1). The largest mainland colony was destroyed by bulldozers in December and January 1981/82 (16). The hostplant suffers commercial collecting in spite of being regarded as threatened, and this probably further imperils the Atala (1). Because of its local distribution, slow flight, small numbers and conspicuous colours, the Florida Atala is particularly vulnerable to collectors (1). Finally, as a patchy resident of southern Florida, the insect is always at risk from hurricanes. At least one entire colony is believed to have been destroyed in this way (13).

CONSERVATION MEASURES TAKEN Lepidopterists have always been aware of the precarious existence of E. atala florida, and have searched for new populations. This has resulted in discoveries after long periods during which it was considered to be extinct (1,8,13). In 1961, G.W. Rawson carried out a transplant of Florida Atala individuals from the only known colony (which was unprotected) to a site in Everglades National Park where conditions seemed suitable. This effort failed, probably because the founder population was destroyed by Hurricane Donna soon afterwards (13). Additional colonies were found in 1978, 1979 and 1980 (1,10). These populations have been studied and monitored by H.D. Baggett, J. Weintraub and others. The Xerces Society gave a grant to aid E. a. florida conservation and management ecology studies in 1981. The U.S. Office of Endangered Species published a notification of review of the status of the Florida Atala in 1975, but no further action has been taken (14).

CONSERVATION MEASURES PROPOSED Biologists and conservationists have recommended that taxonomic studies be carried out to determine the correct status of the Florida population, that steps be taken to preserve known habitat, and that collecting be controlled by state permit, with full protection accorded to immatures, females and host plants, with scientifically justified exceptions (1). A new status review should be conducted by the Fish and Wildlife Service to determine whether the butterfly should be protected under the U.S. Endangered Species Act. The feasibility of another introduction into Everglades National Park should be investigated, and studies supported to ascertain the whereabouts and condition of all critical habitats, and the management measures necessary for their perpetuation.

REFERENCES 1. Baggett, H.D. (1982). Florida Atala. In Franz, R. (Ed.), Invertebrates, Rare and Endangered Biota of Florida Vol. 6. University Presses of Florida, Gainesville. Pp. 75-76.

2. Baggett, H.D. (1980). In litt., 15 February et seq.

3. Bruner, S.C., Scaramuzza, L.C. and Otero, A.R. (1975). Catalogo de los insectos que atacen a las plantas economicas de Cuba. 2nd Ed. Academia de Ciencias de Cuba, Instituto de Zoologia, La Habana.

4. Clench, H.K. (1942). The Lycaenidae of the Bahama Islands (Lepidoptera: Rhopalocera). Psyche 49: 52-60.

5. Clench, H.K. (1977). A list of the butterflies of Andros, Bahamas. Ann. Carnegie Mus. 46: 173-194.

6. Clench, H.K. (1977). Butterflies of the Carnegie Museum Bahamas Expedition, 1976. Ann. Carnegie Mus. 46: 265-283.

7. Comstock, W.P. and Huntingdon, E.I. (1943). Lycaenidae of the Antilles (Lepidoptera, Rhopalocera). Ann. N.Y. Acad. Sci. 45: 49-130.

8. Funk, R.S. (1966). Record of Eumaeus atala (Lycaenidae) from the Florida Keys. J. Lepid. Soc. 20: 216.

9. Klots, A.B. (1951). A Field Guide to the Butterflies, East of the Great Plains. Houghton-Mifflin Co., Boston. 349 pp.

10. Lenczewski, B. (1980). Butterflies of Everglades National Park, S. Fl. Res. Cent. Rep. T-588. National Park Service, Homestead, FL. 110 pp.

11. Long, R.W. and Lakela, O. (1971). A Flora of Tropical Florida. University of Miami Press, Coral Gables, Florida. 962 pp.

12. Pyle, R.M. (1981). Audubon Society Field Guide to North American Butterflies. Knopf, N.Y. 916 pp.

13. Rawson, G.W. (1961). The recent rediscovery of Eumaeus atala (Lycaenidae) in Florida. J. Lepid. Soc. 15: 237-244.

14. U.S.D.I. Fish and Wildlife Service (1975). United States butterflies review of status. Federal Register 40(55): 12691.

15. Young, F.W. (1950). Notes on the disappearance of the Florida blue butterfly. (Eumaeus atala florida Röber, Lepid.,

Lycaenidae). <u>Lepid. Medlemsblad Lepid. Forening.</u> 1950-52: 1-2.

16. Weintraub, J. (1982). In litt., 18 August.

We are grateful to H.D. Baggett, M. Bentzien, C.V. Covell Jr., P.A. Opler, and G.W. Rawson for helpful comments, and particularly to J. Weintraub for valuable discussions.

LARGE BLUE BUTTERFLIES

Maculinea alcon Denis and Schiff., 1775	Alcon Large Blue	V
M. arion (L., 1758)	Large Blue	V
M. arionides Stgr, 1887	Greater Large Blue	V
M. teleius Bergstrasser, 1779	Scarce Large Blue	V
M. nausithous Bergstrasser, 1779	Dusky Large Blue	E

Phylum	ARTHROPODA	Order	LEPIDOPTERA
Class	INSECTA	Family	LYCAENIDAE

SUMMARY Large blue butterflies (genus Maculinea) are bigger than most other lycaenids and are among the most attractive of the blues. They have highly specialized life histories, involving ants. The larvae are initially herbivorous, but prey on ant brood after Myrmica ants carry them into their underground nests. Large blues occur only in small colonies in special habitats, and are always uncommon. However, destruction and modification of these habitats make the large blues some of the most rapidly declining butterflies in Europe, and probably in Asia too.

DESCRIPTION Maculinea blues, as their common names suggest, are relatively large for lycaenid butterflies, with wingspans ranging from 30 mm to 45 mm (M. arionides) (22). Most species and subspecies are heavily marked with black spots and borders. However, M. alcon and M. teleius males may be entirely blue and free from spotting, while M. nausithous may be almost wholly covered with dusky brown scales, showing little blue. The underside of all species is a shade of brownish grey, crossed by arcs of white-circled black dots. Descriptions and illustrations are available for M. arionides (23,37) and for the other species (13,22,36). Markings vary greatly within and between colonies, and many subspecies have been named. These are not considered here, except to note that M. alcon rebeli is probably a distinct species (13), and is certainly vulnerable.

The Large Blue (Maculinea arion) caterpillar is collected
by the ant Myrmica sabuleti

DISTRIBUTION The genus is restricted to the Palearctic region (13). M. arion occurs from western Europe (from south of 62°N to central Spain, Pyrenees, northern Italy, Alps, Balkans, Corsica and Greece) to southern Siberia, Armenia, Mongolia and China. Recently extinct from much of northern Europe including U.K. (1979), Netherlands (1950), Belgium and northern France (32). M. alcon is found from northern Spain and France through central Europe to central Asia (Tibet). It is extinct in many former European localities (32). M. teleius extends from Spain through central and eastern Europe to China, Korea and Hokkaido, Honshu and Kyushu, Japan (23). It is very rare and local, with only about 30 colonies known in Europe, and is apparently in decline throughout its range (13,22,23,32). M. teleius is extinct in Belgium and the Netherlands (32), as is M. nausithous. The latter occurs sparingly in northern Spain and central Europe but local extinctions are occurring (17,32). M. arionides is found in Amurland (China and U.S.S.R.) and the alpine forests of Hokkaido and Honshu, Japan (22,23,37), and is the only large blue absent from Europe.

POPULATION The populations and abundance of colonies of large blues are only poorly known. Nearly all species and races form small isolated colonies whose populations fluctuate annually. Estimates for M. arion, M. nausithous and M. teleius colonies indicate adult populations ranging from less than 50 up to 500, or exceptionally several thousand. Casual observations of other species suggest colony populations of similar magnitudes. Colonies are usually isolated from neighbours and very slow to disperse. Local extinctions are known to be taking place, and recolonization is a most unlikely event (25-34).

HABITAT AND ECOLOGY Thirteen Maculinea arion sites analysed in Europe (U.K., Sweden, and France) were dry grasslands where the ant Myrmica sabuleti was abundant (32). In northern Europe (i.e. the U.K., Scandinavia) and at high altitudes elsewhere M. sabuleti only occurs in sufficient densities on the warmest south-facing slopes, and where the sward is cropped very short. In these regions, any relaxation of the grazing regime proves deleterious because Myrmica sabuleti (unlike the caterpillar's hostplant Thymus) is rapidly excluded by taller or denser grass. Myrmica scabrinodis may replace it at similar densities, but this ant cannot support a population of Maculinea arion due to low survival in the nest. Farther south in warmer conditions where the hostplant is Origanum, both butterfly and ant can survive in less intensively managed, rough grassland, so long as scrub succession is prevented (32). M. teleius and M. nausithous occur in wetlands on the edges of bogs, marshes and in swampy fields where extremely high density of Myrmica scabrinodis nests (0.5 per m^2) and slightly lower densities of the larger M. rubra nests occur. The best site known for these two species was 1 ha in size, and contained about 2500 M. rubra and 4800 M. scabrinodis nests, among 759 Sanguisorba plants (the larval hostplant). The few sites known for this species lie on farmland near Phragmites beds, but are not intensively managed. However, periodic clearance of Phragmites and other coarser vegetation apparently has occurred (probably mainly by cutting) and this seems to be essential to maintain semi-open conditions (32). Maculinea alcon also favours damp meadows and marshy places where Gentiana pneumonanthe, the larval hostplant, grows in conjunction with high densities of M. scabrinodis and/or M. rubra ants. However, French populations of M. alcon rebeli breed on dry meadows and grassy slopes, using a different Gentiana hostplant, and its ant host would probably be M. labicornis, M. sabuleti, or one of the rarer Myrmica species (32). This makes M. alcon rebeli likely to be a separate species from the nominate subspecies, which almost certainly uses different host ants. The habitat of Maculinea arionides is alpine forest, especially along stream edges (23). It is known from south-west Hokkaido and north-east and central Honshu (23).

The life cycles have been described with varying completeness for M. arion (1-6,8,11,20,26,27), M. alcon (5-7,9,10,16,19), and M. teleius (6,18,21,23). M. nausithous was the subject of an early paper (35), but this failed to establish the link with ants (14), and the assumed host ant, Formica fusca, is incorrect (32). The host is now known to be Myrmica rubra (33). The early stages of M. arionides have been described and illustrated (23). The following observations are from J.A. Thomas' work on the continental and former British Large Blue (M. arion), and the continental Scarce (M. teleius) and Dusky (M. nausithous) Large Blues (25-30,33,34). Maculinea are unique among western palearctic butterflies in being phyto-predaceous (feeding on plants for part of their lives, animals for the rest). All species have a single brood per year and generally fly for 3-4 weeks between late May and August, depending more on the locality than the species. Eggs are laid on flower buds of thyme or oregano (Thymus drucei, Origanum vulgare) by M. arion; gentians (Gentiana pneumonanthe, G. cruciata, G. germanica) by M. alcon; Great Burnet (Sanguisorba officionalis) by M. nausithous and M. teleius. Eggs are usually laid singly by M. arion and M. teleius, in groups by the others. First instar larvae of at least M. arion may be cannibalistic. Young caterpillars burrow into the flowerhead of the hostplant and feed on the developing structure. After the third moult (the second of M. alcon) they fall to the ground and wait to be found by a passing ant of the genus Myrmica. The ant will attend the larva closely for the honeydew it secretes from a posterior gland. After being milked for one quarter to four hours, Maculinea arion rears up and mimics an ant grub. This elicits a response in the ant whereby the larva is carried off to the ant nest and placed among the brood. M. teleius behaves in a similar way, but M. nausithous contorts differently and is adopted much more rapidly, and M. alcon has not been observed to rear. One or both species may possess an ant attractant pheromone. The larva may be adopted by any Myrmica species that finds it. Early reports of the caterpillar wandering to find a particular ant nest were mistaken (11). A limited number of Myrmica species is likely to coexist with a given large blue species and its hostplant, and the behaviour of the blue further limits the possibilities. Thus Maculinea arion and M. teleius fall at dusk and crawl under plant debris, corresponding with both the peak foraging time and strategy employed by Myrmica sabuleti and M. scabrinodis. M. nausithous rests on top of ground vegetation and is much more likely to be found by M. rubra, even where it coexists with M. teleius. In the more northern parts of its range and at higher altitudes, M. arion oviposits on thyme, which grows in the warmer microclimates occupied by M. sabuleti. However, in the generally warmer south this ant prefers locally cooler situations where oregano grows. M. arion therefore emerges later in such places, to coincide with the budding of oregano, but too late for the thyme on the hotter, antless hills.

Large blues spend the next nine months among the ants, overwintering in their nests. Each species behaves differently but all feed on ant eggs, grubs and prepupae. The Large Blue, Scarce Large Blue and Dusky Large Blue feed on nothing else, so that one ant nest can usually support only two to three larvae to adulthood (fewer farther north). The Alcon Large Blue may be more efficiently fed by worker ants so that many more be supported by one nest (up to 20 in captivity). Pupation occurs about four weeks prior to emergence, inside the ant nest. Pupae, also attractive to the ants, are enclosed in cells. The adults emerge at dawn, crawl out, and expand their wings once they have reached the surface.

Maculinea butterflies are best regarded as parasites of ants. However, unlike most parasites they lack mobility, the ability to seek hosts actively, and a high reproductive capacity. There is no room for the mass wastage that is the strategy of such parasites as bacteria, viruses, and flukes. Large blues require a high density of their host ants as well as their host plants, in propinquity if not together. English M. arion needed up to one nest per m^2 over a land area of 1-5 ha. Southern European sites tend to have fewer but larger nests, although the best blue colonies occur where both density and size of nests are great. The species of ant capturing the larva seems to be critical as well. Any Myrmica will adopt M. arion, but it survives well only in M. sabuleti nests. Scarce Large Blues and Alcon Large Blues seem to favour M. scabrinodis while the Dusky Large Blue prefers M.

rubra (33,34). A great deal more work is urgently required on the ecology of these insects if the progressive decline of large blues is to be reversed (25-32).

SCIENTIFIC INTEREST AND POTENTIAL VALUE Long before its unusual life cycle was known, M. arion was a great prize to entomologists because of its beauty, rarity, unpredictable appearance and remote habitats. The inability to breed in captivity added to its capture value, as did the discovery of its remarkable relationship with ants. Indeed, the life history of M. arion brought it to the attention of other naturalists and the general public. It has been adopted as the symbol of two conservation bodies (Cornish Naturalists' Trust, British Butterfly Conservation Society), appeared on the stamps of at least three countries, and its peril and later extinction in the U.K. brought wide coverage in the media. The other species of the genus, much rarer even than M. arion, arouse similar enthusiasm on the part of entomologists but have yet to receive widespread publicity. They are likely to attract much more conservation attention in future (32). No colonies of large blues have been exploited for tourism yet, but the potential exists. It will probably be on a limited scale, especially in more fragile wetlands, but dry sites could accommodate more public pressure (32). The biology, ecology and genetics of Maculinea are of the highest scientific interest in their own right. Their extreme variability makes these butterflies an elegant research tool for studies in microevolution, and this is enhanced by their extreme localization. Being so specialized, large blues are highly sensitive to change and may be useful as early indicators of environmental deterioration. In terms of cash value, variants and the rarest races have sold on the open market to ready buyers for many years. Fairly high prices (U.S. $10 to $200 per adult) (32) are still recorded. While collecting is a palpable threat to small, isolated colonies, well managed habitats can support robust populations capable of sustaining collecting. It is conceivable that this could help to pay for habitat upkeep in the future. In another way the large blues have materially aided conservation; prior to its extinction in Britain several British entomologists subscribed substantial sums (£10 to £1000) to help maintain M. arion, without requiring to see or collect it (32).

THREATS TO SURVIVAL Most Maculinea populations breed on commercially managed private land and are being eliminated by intensive land use practices (32). All large blues require unimproved land which has not been treated with fertilizer or herbicide, but management may be necessary in the form of grazing or scrub removal. At least half of the M. arion sites in central and northern Europe, and many in southern Europe, have been destroyed by conversion to arable fields, or improvement of pastures by ploughing, seeding, drilling or applying fertilizers or herbicides. Most other losses from northern Europe have been from grassland that still appears to be semi-rough with abundant thyme, but small management changes have occurred. Often it is no longer economical to graze such land as closely and regularly as before. Death of rabbits through myxomatosis in the 1950s exacerbated the adverse changes in grazing regime.

Fewer details exist for other Maculinea species. Wetland species (M. teleius, M. nausithous, M. alcon in places) are declining very rapidly through the development and drainage of wetlands (17,32). Sometimes the main reed marshes survive but lose their drier fringes, where breeding occurs. As with M. arion, land use changes are causing losses among the other species. For example, all known M. teleius and M. nausithous sites in the Rhône Valley were destroyed in 1981 when a very large reservoir was constructed (32). Like M. arion, these species suffer losses from sites that appear superficially unchanged. No analysis has been made, but the reason may be loss of the inconspicuous ant nests. The abandonment of traditional practices such as periodical burning or cutting of reed beds could account for such declines (32). In Honshu, Japan, M. teleius is widely but patchily distributed. It occurs on the northern plains of Aomori Prefecture, in the central alpine area, and in the south-west on Mt Hôki-Daisen (23). On Kyushu it is found locally on the plain around the top of the volcano Mt Aso (23). M. arionides is generally rare but locally abundant in Japan, particularly in Nagano Prefecture in the alpine zone of central Honshu, but its range and numbers are declining because of development pressure (23).

The well-publicized extinction of M. arion in Great Britain (29) followed changes in the land use regime unfavourable to Myrmica sabuleti, but the proximal cause was ignorance of necessary management measures on existing reserves (30). Such absence of facts will certainly contribute to further national, if not species, extinctions unless support can be found for the painstaking, rigorous research needed for informed management. Since large blues can be caught easily and have a low reproductive rate, collecting is a conceivable threat to populations already reduced, stressed or limited (32).

CONSERVATION MEASURES TAKEN Numerous attempts were made to conserve the M. arion in Britain, beginning in the 1920s. In the early 1960s a committee for the conservation of M. arion was established to co-ordinate the activities of the many conservation bodies working to save the species in the U.K. Surveys were organized, sites were wardened to deter collectors, five sites were obtained as nature reserves, and numerous costly projects were undertaken (15,25,29-31). All of these actions eventually failed because nothing was known about the ecology of the butterfly except for its basic life cycle. Full-time ecological research was begun in 1972, when the species had been reduced to about 250 individuals on two sites in England. The studies soon revealed that M. arion required high densities of ants of the species Myrmica sabuleti and that these had largely disappeared even from nature reserves because heavy grazing had been abandoned. It proved a straightforward matter to recreate suitable conditions, but only after it was too late to save the British race of M. arion. The numbers had declined to the point where a fervent effort had to be made to rear the offspring of the last remaining female. Although this succeeded, the population became extinct when the females all died before any males emerged to mate with them (29,30). In retrospect, the British extinction almost certainly could have been prevented with the available resources had the true nature of the threat been appreciated five to ten years earlier (29,30).

No other large blue colonies are known to occur on reserves, except for one Dusky Large Blue population, in a 'protected area' of Switzerland (12), which may now be extinct (17). The following countries have banned the collecting of Maculinea by law, in some cases only specified subspecies: M. arion (U.K., France), M. alcon (France) (12), M. teleius (France) (12).

CONSERVATION MEASURES PROPOSED A recent survey of threatened butterflies in Europe (12) included all four European large blues among the eleven endangered species of the continent. M. arion might be more properly considered vulnerable, as in this account, although many individual races are certainly endangered. Little is known of the status of Maculinea in Asia, but they are all extremely local and it is likely they are experiencing similar declines to those apparent in Europe (23,32). Surveys of M. teleius and M. arionides are needed in the Far East. M. nausithous, the only species of large blue confined to Europe, is globally endangered. Additional information is expected to confirm endangered or vulnerable status for all of the large blues globally. Lepidopterist representatives to the Council of Europe, in considering the content and implications of their survey (12), unanimously considered the highest butterfly conservation priority in Europe to be its eleven endangered species, including the large blues. Recovery recommendations include surveys to discover surviving colonies, the establishment of nature reserves where suitable, and ecological research to discover how these may be appropriately managed (12). All three measures are considered essential by prominent students of the problem (32). Inadequate knowledge in the past has led to ill-informed management and unsuitable siting of reserves. There is an urgent need to establish reserves for all the species, but especially for M.

nausithous, M. teleius and M. alcon in that order, as they are unlikely to survive elsewhere (32). Some sites in eastern France support both Scarce and Dusky Large Blues, and these would make good reserves (32). In the U.K., apparently suitable conditions for M. arion have been re-established on one site and are being attempted on two others. A reintroduction using continental stock should take place in the mid-1980s (32). It has proved almost impossible to breed Maculinea in captivity on a substantial scale, certainly beyond one generation (8,10,11). Rearing large blues on artificial media has failed, although new efforts should be made (32). Captive breeding does not offer a viable means of perpetuating any species of Maculinea in the foreseeable future. There is no suitable alternative to maintaining wild habitats in the proper condition, with both the food plants and the host ants on which the butterflies depend. Additional collecting bans will probably be implemented in Europe following the Council of Europe report (12). These are likely to be largely ineffectual, due to the nature of the threat, and they should not be allowed to eclipse necessary habitat protection and management efforts.

The broad ranges of the large blues means that it is still possible to prevent their overall extinction. This aim should be pursued fervently before they shrink to the status of isolated relics, beyond the reach of all but the most expensive and remedial conservation action.

REFERENCES
1. Blab, J. and Kudrna, O. (1982). Naturschutz Aktuell: Hilfsprogramm für schmetterlinge. Kilda-Verlag, Greven. 135 pp.
2. Chapman, T.A. (1914). The mystery of Lycaena arion. Ent. Rec. 26: 245-246.
3. Chapman, T.A. (1916). Observations completing an outline of the life of Lycaena arion L. Trans. Ent. Soc. Lond. 1916: 298-312.
4. Chapman, T.A. (1917). The evolution of the habits of the larva of Lycaena arion L. Trans. Ent. Soc. Lond. 1916: 315-321.
5. Chapman, T.A. (1918). On the life history of Lycaena alcon F. in Oberthur. Etud. Lep. Comp. 16: 277-306.
6. Chapman, T.A. (1920). Contributions to the life history of Lycaena euphemus Hb. Trans. Ent. Soc. Lond. 1919: 443-449.
7. Chapman, T.A. (1920). Notes on Lycaena alcon F. as reared in 1918-19. Trans. Ent. Soc. Lond. 1919: 443-449.
8. Clarke, C.A. (1954). Breeding the Large Blue butterfly in captivity. Ent. Rec. 60: 210.
9. Diehl, F. (1930). Die erste erfolgreiche Zucht von Lyc. alcon F. und Beobachtungen über die Biologie der ersten Stände dieses Schmetterlings. Int. Ent. Z. (Guben) 24: 35-42.
10. Elfferich, N.W. (1963). Kweekervaringen met Maculinea alcon Schiff. Ent. Ber. 23: 46-52.
11. Frohawk, F.W. (1924). The Natural History of British Butterflies, Vols. 1 and 2. London.
12. Heath, J. (1981). Threatened rhopalocera (butterflies) in Europe. Nature and Environment Series No. 23. Council of Europe, Strasbourg. 157 pp.
13. Higgins, L.G. and Riley, N.D. (1980). A Field Guide to the Butterflies of Britain and Europe. 4th ed. Collins, London. 384 pp.
14. Hinton, H.E. (1951). Myrmecophilous Lycaenidae and other Lepidoptera - a summary. Proc. S. Lond. Ent. Nat. Hist. Soc. 1949-1950: 111-175.
15. Howarth, T.G. (1973). The conservation of the Large Blue butterfly (Maculinea arion (L)) in west Devon and Cornwall.

Proc. Trans. Brit. Ent. Nat. Hist. Soc. 5: 121-126.

16. Janmoulle, E. (1960). Quelques observations sur la chenille de Maculinea xerophila Bger (Lycaenidae) Lambill 60: 5-7.
17. Mattoni, R. (1981). In litt.
18. Powell, H. (1920). Lycaena euphemus in Oberthür. Etud Lep. Comp. 17: 85-173.
19. Powell, H. (1920). Suite aux observations sur les premiers états de Lycaena alcon in Oberthür. Etud Lep. Comp. 17: 25-44, 393-419.
20. Purefoy, E.B. (1953). An unpublished account of experiments carried out at East Farleigh, Kent, in 1915 and subsequent years in the life history of Maculinea arion, the large blue butterfly. Proc. R. Ent. Soc. Lond. (A) 28: 160-162.
21. Schepdael, J. (1958). Le cycle biologique et la mymécophilie de Maculinea telejus Bergstr. (= Lycaena euphemus Hbn). Linn. Belg. 1, 17-22.
22. Seitz, A. (1907). The Macrolepidoptera of the World. Stuttgart. Pp. 320-321.
23. Shivozu, T. and Hara, A. (1962). Early Stages of Japanese Butterflies in Colour, 2. Hoikusha, Osaka.
24. Spooner, G.M. (1963). On causes of the decline of Maculinea arion (L.) (Lep., Lycaenidae). Entomologist 94: 199-210.
25. Thomas, J.A. (1976). The ecology and conservation of the large blue butterfly. Unpublished report, Institute of Terrestrial Ecology.
26. Thomas, J.A. (1977). The ecology of the large blue butterfly. A. Rep. Inst. Terr. Ecol. 1976: 25-28.
27. Thomas, J.A. (1977). Ecology and conservation of the large blue butterfly - second report. Unpublished report, Institute of Terrestrial Ecology.
28. Thomas, J.A. (1978). Report on the large blue butterfly in 1978. Unpublished report, Institute of Terrestrial Ecology.
29. Thomas, J.A. (1980). Why did the large blue butterfly become extinct in Britain? Oryx 15: 243-247.
30. Thomas, J.A. (1980). The extinction of the large blue and the conservation of the black hairstreak (a contrast of failure and success). A. Rep. Inst. Terr. Ecol. 1979: 19-23.
31. Thomas, J.A. (1981). Insect conservation in Britain: some case histories. Atala 6 (1978): 31-36.
32. Thomas, J.A. (1982). In litt., 5 January.
33. Thomas, J.A. (in press). The behaviour and habitat requirements of Maculinea nausithous (the dusky large blue) and M. teleius (the scarce large blue) in France. Biol. Conserv.
34. Thomas, J.A. (in press). The ecological segregation of two species of large blue (Maculinea) butterfly. A. Rep. Inst. Terr. Ecol. 1982.
35. Viehmeyer, H. (1907). Vorlaufige Bemerkungen zur Myrmkophilie der Lycaenidenraupen. Ent. Wochenblatt 24: 43, 50.
36. Whalley, P. (1981). The Mitchell Beazley Pocket Guide to Butterflies. Mitchell Beazley, London. 168 pp.
37. Yokoyama, M. (1964). Coloured Illustrations of the Butterflies of Japan. Hoikusha, Osaka

We are very grateful to J.A. Thomas for compiling the draft of this review, and to T. Abe, D.M. Foulkes, J. Heath, T. Matsumoto and R. Mattoni for further comment.

Strymon avalona Wright, 1905

Phylum	ARTHROPODA	Order	LEPIDOPTERA
Class	INSECTA	Family	LYCAENIDAE

SUMMARY One of the world's most narrowly endemic butterflies, the Avalon Hairstreak Strymon avalona is confined to Santa Catalina Island, California, U.S.A. The related Gray Hairstreak (S. melinus) occupies the other seven California Channel Islands, but until 1978 was not recorded on Santa Catalina. However, the Gray Hairstreak was recently found on the island and it has been claimed that the two species hybridize, although this is disputed. If hybridization does occur, then it could threaten to swamp Strymon avalona and cause its extinction. The hybridization claim requires further investigation.

DESCRIPTION Strymon avalona is a small (19-25 mm wingspan) hairstreak. The wings are largely blue-grey above with sooty scaling, becoming browner with age, the female deeper in tone than the male. The wings are light blue-grey to sandy buff beneath, crossed by an irregular light and dark line beyond which the wings are frosted with whitish scales. A vague orange and black spot lies above the short tail on the hindwing. The butterfly differs from the Gray Hairstreak (S. melinus) by lacking a strong black and white band across the ventrum, and being generally duller (9). The eggs are pale green, the caterpillar dusky grey-green with lateral stripes, and the chrysalis light brown with dark flecks (1,9).

DISTRIBUTION The Avalon Hairstreak is endemic to Santa Catalina Island, Los Angeles County, U.S.A. (11), the largest (20 000 ha) of the eight California Channel Islands. While occurring all over the island, the Avalon Hairstreak seems to prefer southern (leeward) canyons (5) and bluffs south and east of the major town, Avalon (3). Strymon melinus, the Gray Hairstreak, occurs in southern U.S.A. and Central America and is an excellent colonizer (13). Each season it establishes itself far north of where it can survive hard winters and small, unrecorded numbers probably reach Catalina Island almost every year (13).

POPULATION The observable population of the Avalon Hairstreak fluctuates markedly. Lepidopterists have reported from less than 30 to more than 1000 butterflies in a single day in comparable sites (5). Reports conflict over the relationship between abundance and rainfall, with observers linking both dry (5) and wet years (3) to high numbers of the butterfly. Males appear to outnumber females by at least two to one (5) with males much more numerous on hilltops, where they establish territories, and females common in canyon bottoms, where they seek hostplants.

HABITAT AND ECOLOGY The Avalon Hairstreak occurs in most of the available habitats on Santa Catalina, particularly in the coastal sage scrub community (12). Spring hostplants are Lotus argophyllus ornithopus and L. scoparius dendroideus. Later in the summer, when these legumes wither in the dry heat, Eriogonom giganteum giganteum and E. grande are utilized by larvae. All four hostplants are island endemics preferring dioritic, granitic reddish soil along south-facing slopes in the hottest and driest parts of the island (5). The hosts tend to grow together and in association with the dominant shrubs of the coast, Heteromeles arbutifolia, Rhus integrifolia and Artemisia californica (5). Adults fly in multiple, indistinct, overlapping and possibly continuous generations from February to October, peaking in March (9). They take nectar from the flowers of Rhus laurina and Eriogonum giganteum, among others (2). Both birds and spiders prey on adults (5).

459

SCIENTIFIC INTEREST AND POTENTIAL VALUE The extremely restricted distribution of Strymon avalona is of great interest to ecologists and island biologists (10). Determination of the butterfly's origin, whether as a mainland relict or autochthonously evolved on Santa Catalina, would shed light on the biogeography and history of the Channel Islands (7). The relationships between degrees of endemism, diversity, and patterns of colonization and extinction on all California offshore islands are the subject of scientific investigation. The Avalon Hairstreak is an integral and unique component of this programme of evolutionary biology (3,10). Investigation of the ecological and biological relationships between S. avalona and S. melinus could give insight into niche selection, interspecific competition and hybridization genetics (4). The hairstreak's nectar-seeking activity serves to pollinate many flowering plants on the island (5).

THREATS TO SURVIVAL Perhaps as remarkable as its limited distribution is the fact that this butterfly thrives so near one of the densest conurbations in the world. The water gap (30 km) and history of the island have protected the species by preventing development. Two possible threats are apparent. First, some commercial development is planned for parts of the island. The boundaries of Avalon may be extended along the shore, and a major housing development is planned for the Isthmus area (7). These activities and ancillary pressures from new residents may prove harmful to the Avalon Hairstreak, but its habitat does not seem to be seriously threatened at present (13). Its hostplants, being adventive, actually seem to spread due to certain human disturbances including firebreak construction and grazing (5).

The second possible threat comes from the butterfly's nearest, but quite distinct, relative, the Gray Hairstreak (Strymon melinus Huebner) (9,10,11). S. melinus ranges over much of North and Central America, and occupies the other seven California Channel Islands, but it was never found on Santa Catalina until 1978, despite intensive sampling (4,6). During 1978/79, the Gray Hairstreak invaded Santa Catalina Island for the first recorded time (4). The invaders have not been seen on Catalina since, but it has been claimed that hybridization took place between the two hairstreaks during that period. The data suggest that the invaders came from the mainland rather than other islands (4). Adult Gray Hairstreaks could reach Avalon artificially, on ferry or air connections, or naturally, by flying with the westerly Santa Ana wind. Pupae could arrive among potted or cut plants (4). The butterfly is highly adaptive, prolific in Los Angeles, and polyphagous. Once on the island, the Gray Hairstreak could possibly jeopardize the Avalon Hairstreak through competition for larval hostplants or genetic swamping (4,10). The threat to the endemic from the new arrival has been explored and claimed to be substantial (4). It has been suggested that S. melinus may have excluded S. avalona sequentially from the other Channel Islands, but this cannot be tested.

Recently, the claims that the two species have hybridized on Santa Catalina have been disputed (13). It is maintained that the evidence for hybridization is very weak, and that S. melinus occurs in the tropics in sympatry with closer relatives than S. avalona, with no sign of hybridization. Although S. melinus is S. avalona's closest relative, the reverse is not true (13). Until further evidence is forthcoming there is insufficient information properly to assess the threat to the Avalon Hairstreak. No individuals of the Gray Hairstreak were found on Catalina between 1980 and 1982 (3,4).

CONSERVATION MEASURES TAKEN Ownership and management of much of the island was recently acquired by the Santa Catalina Island Conservancy, a private trust. Extensive commercial development of Avalon Hairstreak habitats is therefore unlikely, although limited development will proceed (3). No direct action has been taken on behalf of S. avalona, with the exception of recent research (3,4,5,7).

CONSERVATION MEASURES PROPOSED Planners and lepidopterists should promote the survival of this species over as much of the island as possible. S. melinus readily colonizes urban and suburban secondary habitats (unlike S. avalona) and development could augment any threat from hybridization. The first priorities are to study the larval biology of both species, to attempt artificial hybridization experiments, and to assess the ability of adults to recognise mates of their own species. Such information would make it possible to predict the effect on S. avalona, should a re-invasion by S. melinus occur.

REFERENCES 1. Comstock, J.A. and Dammers, C.M. (1933). Metamorphoses of five California diurnals (Lepidoptera). Bull. So. Ca. Acad. Sci. 31: 33-45.
2. Emmel, T.C. and Emmel, J.F. (1973). The butterflies of southern California. Nat. Hist. Mus. L.A. Co., Sci. Ser. 26: 1-148.
3. Gall, L.F. (1981). In litt., 20 December.
4. Gall, L.F. (in press). The avalon hairstreak and its interaction with the recently introduced gray hairstreak, Strymon melinus on Santa Catalina Island.
5. Gorelick, G.A. (1981). In litt., 1 June.
6. Meadows, D. (1936). An annotated list of the Lepidoptera of Santa Catalina Island, California. Part 1 - Rhophalocera. Bull. So. Calif. Acad. Sci. 35: 175-180.
7. Nagano, C.D. (1982). In litt., 20 January.
8. Power, D.M. (Ed.) (1980). The California Islands: Proceedings of a Multidisciplinary Symposium. Santa Barbara Mus. Nat. Hist., California. 787 pp.
9. Pyle, R.M. (1981). The Audubon Society Field Guide to North American Butterflies. Knopf, New York. 916 pp.
10. Remington, C.L. (1971). Natural history and evolutionary genetics of the California Channel Islands. Discovery 7: 3-21.
11. Scott, J.A. (1975). Genus Strymon Hubner. In Howe, W.M. (Ed.), The Butterflies of North America. Doubleday, New York. Pp. 303-306.
12. Thorne, R.F. (1967). A flora of Santa Catalina Island. Aliso 6: 1-77.
13. Opler, P.A. and Robbins, R.K.(1982). In litt., 13 December.

We are grateful to L.F. Gall, G.A. Gorelick and C.D. Nagano for information in this review, and to J. Donahue, P.A. Opler and R.K. Robbins for further comments.

MONARCH BUTTERFLY:
MEXICAN WINTER ROOSTS

THREATENED PHENOMENON

Danaus plexippus (L., 1758)

Phylum	ARTHROPODA	Order	LEPIDOPTERA
Class	INSECTA	Family	DANAIDAE

SUMMARY The North American Monarch Butterfly overwinters in California and Mexico and is the most notable north-south annual migrant among insects. The much larger Mexican sites contain many millions of individuals. All known roosts are under immediate or potential threat from logging, tourism or other factors. Comprehensive, co-operative efforts are required to conserve this spectacular phenomenon.

DESCRIPTION The Monarch, a large (wingspan 89-102 mm), orange, black-veined butterfly, undergoes true migration in North America (17). The largest part of the population, which breeds in the eastern and midwestern states, migrates to montane Mexico and overwinters there (5,19). Within the overwintering sites, which are in mixed coniferous forests, the adult butterflies cluster in vast numbers during the winter months. The clusters unquestionably represent the most striking and spectacular butterfly phenomenon on earth (3).

DISTRIBUTION All known sites lie in the volcanic ranges of Michoacan and Mexico States, Mexico (10). The exact locations of overwintering sites will not be given here since arrangements for absorbing the impact of visitors have not yet been implemented.

POPULATION The number of butterflies clustering at a single site has been conservatively estimated at 14.25 million (5). In contrast, the Californian overwintering colonies contain a maximum of about 100 000 individuals (16).

HABITAT AND ECOLOGY The ecology of the Monarch Butterfly has been the subject of several studies, and is well described in the literature (3,5,8-11). Those insects which survive the winter and the spring migration begin the summer in North America, multiplying through two to four generations. The larvae feed on plants in the Asclepiadaceae, from which they sequester heart poisons for use in their own defences against bird predators (2). In September the autumn generation begins the southward migration. Western populations migrate to a number of winter roosts along the California coast of the U.S.A. and Mexico (16,17). By November millions of Monarchs are aggregating in mature fir, pine and cypress forests. In one well-documented Mexican locality the butterflies occupied a cool, mesic site at 3050-3150 m, and had an estimated population of 14.25 million individuals (5). Colonies tend to shift the altitude and location of roosts slightly from year to year, apparently in response to weather conditions (9,10). In 1980-1981 snowfall and cold weather caused heavy mortality (9,10). Most of the winter is spent in clusters in the trees, but as spring approaches individuals will take nectar, drink at wet earth, soar, bask, and finally court and mate prior to their departure in late February and March (3,5). The toxicity of the Monarchs to predators is variable; orioles and grosbeaks are resistant to the poison and able to prey extensively upon the migrating butterflies (4,7). In addition to the overwintering sites, Monarchs need nectar sources along the migratory route and larval hostplants (milkweeds) in their summer haunts.

SCIENTIFIC INTEREST AND POTENTIAL VALUE Although many species of insects emigrate periodically, the North American Monarch is the only species known to accomplish an annual, two-way migration in large numbers. The evolutionary and biological signficance of this is enormous. The mechanism by which the Monarchs locate their roosts and by which the offspring return to their ancestral breeding grounds are still unknown (17,21). Magnetic orientation has been suggested (13), and is beginning to be investigated. Few insect phenomena elicit greater scientific interest than the Monarch, in all its aspects. Extensive past and current researches relating to Monarch toxicity and defence systems (2,4) enhance the species' scientific importance. It is one of the most important experimental animals in the fields of animal mimicry, aposematism and chemical ecology.

The economic value of tourism associated with the winter roosts of Monarchs has already been demonstrated in California, where clusters are much smaller and less impressive (12). Mexican conservationists and government officials agree that properly managed tourism could bring in sufficient revenues to improve the local economic situation, which is not affluent. This could easily offset losses of timber revenues resulting from protection of Monarch roosting trees. The Monarch is the de facto national butterfly of all three North American countries, furnishing a strong basis for cross-border co-operation.

THREATS TO SURVIVAL The species Danaus plexippus, indigenous to the New World and extensively established in Australasia, Hawaii and elsewhere, is not endangered. However, the phenomenon of the migration of the North American population may be seriously threatened. The main threat is from logging of the conifers in which the winter roosts occur. Timber rights owned by private interest and local communities are being exercised, and a sawmill and new chipboard factory are situated near the site where most research has taken place. Logging is encroaching upon the Monarch sites, some of which have already been heavily damaged (14). Although much of the logging is selective rather than involving clear-felling, recent WWF-sponsored research has shown that thinning the forest has a highly detrimental effect on the butterfly aggregations. It permits temperatures to fall below tolerable limits at night, causing high mortality by freezing. Logged forests will not be suitable for future roosting, and there appear to be no alternative roosting sites (6,8,9,10).

The second major threat is tourism. Wide publicity and heavy visitor pressure inevitably followed the momentous discovery of the Mexican roosts (18). Magazine articles, films and television specials have reinforced this effect. Tourists, mostly from Mexico City, are already arriving in hundreds every year and the number will probably increase to thousands in the near future (14). Visitors threaten the butterflies in two major ways, by disrupting the swarms, causing butterflies to fly unnecessarily and use up their fat reserves, and by bringing the constant threat of forest fire. Unsympathetic development of tourist facilities, as in California, could be another threat (12). Grazing and trampling by cattle may be a threat to the Monarchs but this is now considered less significant than when originally suggested (5).

CONSERVATION MEASURES TAKEN Following his location of the Mexican Monarch overwintering sites, F.A. Urquhart and his associates (18,19) spoke out for conservation of the phenomenon. Films produced by Mexican and American television networks were used to raise conservation awareness. In 1976 the IUCN/SSC Lepidoptera Specialist Group declared this issue to be their first priority in world butterfly conservation; this was reaffirmed in 1981, with the California Monarch overwintering sites added. In 1980 a Decree from President Lopez Portillo of Mexico said in part: "Because of public interest, those areas in which the butterfly known as the Monarch hibernates and reproduces are made a sanctuary and reserve" (1). Capture of the Monarch without permission was expressly forbidden. The Decree was a milestone and provided the basis for Monarch conservation, but did not end the threats of logging and tourist pressure to the sites. Several delegations of scientists and conservationists have discussed these problems with federal, state and local officials in Mexico. These have involved representatives from IUCN, WWF-U.S., Mexican and American universities, lepidopterists' societies, private conservation groups, and the Mexican Institute of Biology, among others. Officials from Fauna Silvestre of the Ministry of Agriculture and Water Resources, the Ministry of Public Works and Human Settlement (involving national parks), the State of Michoacan and local towns have been consulted. A symposium on Monarch biology and conservation was held at the first Mexican-American Congress of Lepidopterology in Cocoyoc, Morelos, in August, 1981 (22). The proceedings will be widely disseminated in Mexico and North America. Research scientists from the University of Florida have been pursuing conservation goals as well as their research programme on Monarchs since 1976 (3,5-11). This work has included aerial surveys to locate additional sites, as well as basic research on all aspects of Monarch biology.

Pro Monarca, A.C., a private Mexican organization devoted to Monarch conservation, commissioned a detailed plan for dealing with tourist impact at the Monarch sites most likely to be visited (13). The group is working in conjunction with Fauna Silvestre which has already hired guards to patrol the most heavily visited areas, built a guard station, and assigned personnel to oversee Monarch activities (15). In 1982 the first organized tour to be led by a Monarch biologist was undertaken, to test the feasibility of generating income without causing undue disturbance (22).

CONSERVATION MEASURES PROPOSED In November 1981 an international group of biologists and conservationists wrote to the Minister of Agriculture and Water Resources, expressing appreciation for what had been done, and suggesting what remains to be accomplished in order to realize the intent of the Presidential Decree (14). Attached to the letter were three alternatives for reserve designation in the area. The plan calls for either a large reserve encompassing the Sierra Chinqua mountain range, an archipelago of smaller reserves linked by a protective corridor at higher altitudes, or at least three reserves of approximately 20 km^2 each, surrounded by larger buffer zones. The interior reserves would be strictly managed for conservation and research, with no logging and with carefully controlled access. One reserve should be equipped with a biological field station (20). In the buffer zone, careful forestry and tourism could be accommodated, but with emphasis on protecting the butterflies. Any reserves must incorporate sufficient geographical relief to protect both the higher ridges in which the Monarchs gather and the lower valleys in which they disperse. Damage to any one part of a site could render it unsuitable for overwintering Monarchs. Further plans will be necessary to arrange for the absorption of tourists, for the generation of local income to start co-operative approaches to local timber and chipboard firms, and to develop public relations with local communities. Means of augmenting the local economy will be essential. Additional surveys should attempt to locate any hidden roosting sites in Michoacan, Mexico, or other states. All Mexican efforts to conserve the Monarchs and their critical winter habitats should be supported vigorously by international bodies.

REFERENCES   1.   Lopez Portillo, President (1980). Presidential decree protecting the monarch in all parts of Mexico. Diario Oficial Miercoles 9 de abril de 1980, Mexico D.F.
2.   Brower, L.P. (1969). Ecological chemistry. Scientific Am. 220(2):22-20.
3.   Brower, L.P. (1977). Monarch migration. Natural History 86(6): 40-53.
4.   Brower, L.P., McEvoy, P.B., Williamson, K.L., and Flannery, M.A. (1972). Variation in cardiac glycoside content of monarch butterflies from natural populations in eastern North America. Science 177:426-29.
5.   Brower, L.P., Calvert, L.E., Hedrick and Christian, J. (1977). Biological observations on an overwintering colony of monarch butterflies (Danaus plexippus, Danaidae) in Mexico. J. Lepid Soc. 31: 232-42.

6. Calvert, W.H. (1981). Project 1958, Conservation of Mexican monarch butterfly. In Farrell, A. (Ed.), World Wildlife Fund Yearbook 1980-81. Gland, Switzerland. Pp. 142-43.

7. Calvert, W.H., Hedrick L.E. and Brower, L.P. (1979). Mortality of the monarch butterfly (Danaus plexippus L.): Avian predation at five overwintering sites in Mexico. Science 204: 847-851.

8. Calvert, W.H. and Brower, L.P. (in press). The importance of forest cover for the survival of overwintering monarch butterflies (Danaus plexippus, Danaidae). J. Lepid. Soc.

9. Calvert, W.H., Zuchowski W. and Brower, L.P. (in press) The effect of rain, snow and freezing temperatures on overwintering monarch butterflies (Danaus plexippus L.) in Mexico. Biotropica.

10. Calvert, W.H., Zuchowski, W. and Brower, L.P. (in press). Monarch butterfly conservation: interactions of cold weather, forest thinning and storms on the survival of overwintering monarch butterflies (Danaus plexippus L.) in Mexico. Atala.

11. Fink, L.S. and Brower, L.P. (1981). Birds can overcome the cardenolide defence of monarch butterflies in Mexico. Nature 291: 67-70.

12 Lane, J. (in press). The status of overwintering sites of the monarch butterfly in Alta California. Atala.

13. Monasterio, F.O., Venegas, M., Sanchez, V. and Liquidano, H.G. (in press). Magnetism as a complementary explanation for the orientation of migrant monarchs to their overwintering sites. Atala.

14. Monasterio, F.O., Liquidano, H.G., Sanchez, V., Zoreda, J.J., Araiza, G., Morones, E. and Hidalgo, J.A.L. (1980). Primera entrega: invetigacion y analisis del plan de urgencia para la proteccion de la mariposa monarca (Danaus plexippus) durante el periodo de hibernacion 1980-1981 en la zona de Angangueo, Michoacan. Aleph, Consultores Asociados, S.A. Mexico D.F. 194 pp + appen.

15. Ogarrio, R. (in press). Development of the civic group, Pro Monarca, A.C. for the protection of the monarch butterfly wintering grounds in the Republic of Mexico. Atala.

16. Tuskes, P.M. and Brower, L.P. (1978). Overwintering ecology of the monarch butterfly, Danaus plexippus L., in California. Ecol. Ent. 3: 141-153.

17. Urquhart, F.A. (1960). The Monarch Butterfly University of Toronto Press, Toronto. 361 pp.

18. Urquhart, F.A. (1976). Found at last: the monarch's winter home. Nat. Geographic Mag. 150: 160-173.

19. Urquhart, F.A. and Urquhart, N.R. (1976). The overwintering site of the eastern population of the monarch butterfly (Danaus p. plexippus: Danaidae) in southern Mexico. J. Lepid. Soc. 30: 153-158.

20. Vazquez, L., Perez, R.L. and Perez, H. (in press). The monarch as a resource for ecological investigation in Mexico. Atala.

21. Williams, C.B. (1930). The Migration of Butterflies. Oliver and Boyd, Edinburgh. 473 pp.

We are grateful to L.P. Brower and W.G. Calvert for information in this review, and to P. Ackery, R. Ogarrio, R.I. Vane-Wright and L. Vazquez for further comment. We apologize that, for technical reasons, it has not been possible to include the accents on Spanish words.

MONARCH BUTTERFLY:
CALIFORNIAN WINTER ROOSTS

Danaus plexippus (L., 1758)

Phylum	ARTHROPODA		Order	LEPIDOPTERA
Class	INSECTA		Family	DANAIDAE

SUMMARY The Monarch butterflies of western U.S.A. and Canada migrate to the California coast where they overwinter in dense roosts. All California overwintering sites are in the coastal strip, and virtually all are at risk from development or other factors. Comprehensive, co-operative efforts are required to conserve them.

DESCRIPTION The Monarch is a large (89-102 mm wingspan), black-veined butterfly, tawny below and rich orange above. California's overwintering Monarchs are thought to breed during the summer months over much of the region between the Pacific and the Rocky Mountains (9) The California winter roosts are not as large as those in central Mexico (1), but a colony can comprise tens of thousands of individuals in spectacular roosting clusters among their chosen trees (8). All California overwintering sites are in groves of trees near the Pacific shore (4). Colonies form in September or October and disperse in February or March, depending upon the weather (4,8). The preceding review of the Mexican Monarch roosts includes further information.

DISTRIBUTION Roosts have been reported from Bodega Bay (north of San Francisco) south to San Diego. Perhaps 40 significant sites occur today. Most of the colonies, including the largest ones, are found around Santa Cruz Bay south to Santa Barbara. Some large colonies north of the San Francisco Bay do not persist throughout the winter in every year. Significant colonies, which form every year and persist throughout the winter are typically situated within 1 km of the shoreline (4).

POPULATION The total number of Monarchs overwintering in California varies between years, but the same colonies seem to remain consistently large. Population estimates have been obtained for three colonies (8), the largest exceeding 90 000 individuals. This colony was subsequently re-surveyed and the numbers fell to about half, then increased to about 150 000 (12). This is still an order of magnitude smaller than one Mexican cluster with which it was compared (1). Californian clusters may not always have been smaller than those in Mexico.

HABITAT AND ECOLOGY All overwintering habitats of Californian Monarchs are in groves of trees near the coast, typically in topographically sheltered locations. Most of them employ the introduced Blue Gum (Eucalyptus globulus) as roost trees. Prior to the introduction of Eucalyptus as a source of timber in the last century, it seems that few native tree species occurred in areas where overwintering could have occurred. Monterey Pine (Pinus radiata) and Monterey Cypress (Cupressus macrocarpa) are both utilized as roost trees today and occurred in suitable areas prior to European colonization. The pine originally occurred near Ano Nuevo Point (north of Santa Cruz), on the Monterey Peninsula, on Point Lobos (11), and near Cambria, north of Morro Bay. The cypress was known only from the Monterey Peninsula and Point Lobos (11). Bishop Pine (Pinus muricata) on Santa Cruz Island is reported to support Monarch clusters (13) and this tree might support overwintering clusters on the nearby southern California coast, but no reports of such colonies are known. If the present numbers of overwintering Monarchs within the state were aggregated within only a few cluster sites, the overwintering colonies could have been much larger then than any known site today. With the introduction and spread of Eucalyptus, there has been an historic trend towards increased numbers of overwintering colony sites, and perhaps also a decrease in the size of individual overwintering colonies. At the same time there has been degradation and loss of overwintering sites, both presumed original ones with native roost trees and newer Eucalyptus roost sites.

The exact locations of roosts change from year to year, as well as during each winter. Storms and winds apparently cause some movement, and the colonies move higher into the trees as the winter progresses, probably responding to lowered nightly temperatures (1,3,4,8). Many small, temporary bivouac roosts appear each year in the autumn, but most of these break up in bad weather and join the permanent overwintering roosts. Movement occurs between large colonies provided good flying weather persists (at least in the Santa Cruz area) (4). The literature on the phenomenon often confuses temporary bivouacs with permanent winter colonies. Several lists of California sites have been published (9,10) but a complete survey has never been undertaken.

Wind can dislodge individual roosting Monarchs, subjecting them to increased mortality from predation and freezing (8), but rain does not seem to disturb them. Daily behaviour patterns such as drinking, taking nectar, basking, soaring, and courting and mating just before colony dispersal in the spring, parallel these activities in Mexican sites (1,3,8). Chestnut-backed Chickadees (Parus rufescens) have been observed preying upon Californian Monarchs (8).

SCIENTIFIC INTEREST AND POTENTIAL VALUE The annual, two-way migration of North American Monarch butterlies is a unique phenomenon of great biological interest. Many of the mechanisms governing this behaviour are not yet understood. The Californian overwintering population is probably genetically different and has made separate adaptations from the Mexican/Midwest/East component. Monarchs are perhaps the best known and most beloved butterflies in North America. The early stages are used in American schools to demonstrate complete metamorphosis, and participation in Monarch tagging programmes gives pleasure and satisfaction to many people (9). The value of Monarchs as a tourist resource in California has long been known and exploited. Pacific Grove has identified itself with the Monarchs through a Monarch Festival and parade, a large granite sculpture of a Monarch, suitably named motels, and the self-designation as 'Butterfly Town, U.S.A'. For many years, visitors to the Monarch groves have provided substantial income to Pacific Grove and other towns (4). The combined scientific, cultural and economic values make this a most remarkable insect phenomenon.

THREATS TO SURVIVAL Due to their narrow requirements and the high land values along the Californian coast, Monarchs are vulnerable to development pressures. Even in Pacific Grove, where the Monarch is protected, its habitat is not, and colonies have been lost as additional apartment units are built within the groves (4). Several colonies were lost early in this century (7) and blasting has been known to disperse a colony in midwinter (7). Forest clearing for farming around Point Ano Nuevo may have extirpated a colony (4). The greatest impediment to conserving the California Monarchs is ignorance of their importance, their exact locations and their needs. Even where they occur on public lands in parks and reserves no cohesive management plan for their conservation exists. Where parks border forests of original roost trees, the significance to Monarchs is not always appreciated by park planners. Given the pressure to develop along the coast, sites may easily be lost before they have been properly documented (4). Unsupervised tourism may disrupt colonies (7). Eucalyptus management within state parks can affect Monarchs positively or negatively, depending on the information available to park planners. Lack of awareness and planning must be considered the greatest overall threat to California Monarch colonies.

CONSERVATION MEASURES TAKEN Several cluster sites have been incidentally included within public lands. George Washinton Park is a public park in Pacific Grove. Point Lobos Reserve (near the Monterey Peninsula) and several state parks have roosts, but only one colony, at Natural Bridges Beach State Park, near Santa Cruz, has the roosting site set aside with a marked trail and an observation area. With the exception of Natural Bridges the state parks and reserves lack a comprehensive management plan for Monarch overwintering areas (4). In one park low roost branches were trimmed from trees within a campsite; park personnel apparently did not know of wintering by Monarchs in that location (4). For many years it has been a civil offense to molest Monarchs in Pacific Grove, a fine of $500 supposedly being levied upon offenders. However, this protection does not apply to roost trees or overwintering habitat, and alteration of Monarch sites in Pacific Grove has in some instances gone beyond acceptability to the insects (4). The IUCN/SSC Lepidoptera Specialist Group has given the protection of overwintering Californian and Mexican Monarch colonies the highest priority in world Lepidoptera conservation (6). The Xerces Society has begun to investigate the feasibility of opening a co-ordination and fund-raising office for Monarch conservation and research (6). The Xerces Society appealed to the California State Coastal Commission to fund a statewide survey of Monarch colonies to allow local communities to include data on colonies in their local coastal plans, in the hope that this would allow local planners to protect permanent overwintering sites. This appeal was not successful. However, at the local level, the City of Santa Cruz and the Regional Coastal Commission did mandate a seven year programme of monitoring to be carried out at Natural Bridges Beach State Park to evaluate the effects of building and operating a semiconductor factory adjacent to the park and close to the overwintering butterflies. From this study, future planners may have better data on which to decide about operations around Monarch wintering sites (4). Interpretation of overwintering Monarchs is offered at one site (Natural Bridges) and many other state parks could do the same (4).

CONSERVATION MEASURES PROPOSED Conservation measures proposed at a symposium on Monarch biology and conservation (4) called for an intensive survey of the location, condition and conservation status of all major overwintering sites. All public land agencies in control of Monarch sites, particularly state parks, were urged to protect and patrol Monarch roosts. The California Department of Parks and Recreation already have plans to make several Monarch winter roosts into Natural Preserves (11). The importance and fragility of the Monarch phenomenon needs to be emphasized to the visiting public and local planning agencies. The Monarch protection enunciated in Pacific Grove's ordinance should include roosting groves as well as individual butterflies. Important Monarch sites might be candidates for National Natural Landmarks, administered by the National Park Service and thereby afforded an added degree of protection. A full programme of basic and management research should be encouraged.

REFERENCES   1. Brower, L.P., Calvert, W.H., Hedrick, L.E. and Christian, J. (1977). Biological observations on an overwintering colony of monarch butterflies (Danaus plexippus, Danaidae) in Mexico. J. Lepid. Soc. 31: 232-242.
          2. Griffin, J.R. and Critchfield, W.B. (1972). The distribution of forest trees in California. USDA Forest Service Research Paper PSW-82/1972. 114 pp.
          3. Hill, F.H., Jr., Wenner, A.M. and Wells, P.H. (1976). Reproductive behavior in an overwintering aggregation of monarch butterflies. Am. Midl. Nat. 95: 10-19.
          4. Lane, J. (in press). The status of Monarch butterfly overwintering sites in Alta California. Atala.
          5. Opler, P.A. (1978). In litt., 12 September.
          6. Pyle, R.M. (in press). International monarch protection efforts. Atala.
          7. Shepardson, L.D. (1914). The Butterfly Trees. Herald Printers, Monterey. 32 pp.
          8. Tuskes, P.M. and Brower, L.P. (1978). Overwintering ecology of the monarch butterfly, Danaus plexippus L., in California. Ecol. Ent. 3: 141-153.
          9. Urquhart, F.A. (1960). The Monarch Butterfly. University of Toronto Press, Toronto. 361 pp.
         10. Williams, C.B., Cockbill, G.F., Gibbs, M.E. and Downes, J.A. (1942). Studies in the migration of Lepidoptera. Trans. R. Ent. Soc. Lond. 92: 101-283.
         11. Hiehle, J.L. (1982). In litt., 29 June.
         12. Dayton, J. (1982). Unpubl. data for 1982-1983 season, pers. comm. to J. Lane.
         13. Powell, J. (1982). Pers. comm. to J. Lane.

We are grateful to J. Lane for providing material used in compiling this review, and to P. Ackery, L.P. Brower, J.L. Hiehle, P.A. Opler, P. Tuskes and R.I. Vane-Wright for further comments. We wish to apologize that, for technical reasons, it has not been possible to include accents on Spanish words.

SULAWESI TREE NYMPH BUTTERFLY

Idea tambusisiana Bedford Russell, 1981

Phylum	ARTHROPODA	Order	LEPIDOPTERA
Class	INSECTA	Family	DANAIDAE

SUMMARY Although a very large and conspicuous butterfly, the Sulawesi Tree Nymph escaped detection until 1980. It is probably confined to high ground in eastern central Sulawesi, Indonesia. The type locality is in a nature reserve, but the actual degree of protection is uncertain.

DESCRIPTION The single specimen collected is a very large female with a fore-wing length of 86 mm. Fairly typical of the genus Idea (tree nymphs), it has a thinly scaled whitish ground colour with heavy black spotting. The male is likely to be similar but smaller (2). I. tambusisiana differs from the only other known species from Sulawesi, I. blanchardii, in being much more heavily marked, the outer half of the wing almost wholly blackened. The completely black body and appendages differentiate it from any other Idea species (3).

DISTRIBUTION The genus Idea is restricted to the Indo-Oriental region. I. tambusisiana is known only from the type locality at 1300 m on the south-western slopes of Gunung (Mt) Tambusisi in eastern Sulawesi Tengah (central Sulawesi), Indonesia (2).

POPULATION Six sightings including both sexes were made within a small area in March 1980. The total population is unknown.

HABITAT AND ECOLOGY The type locality is 300 m above the lower limit of montane moss forest. Two females were seen fluttering among saplings at 2-4 m, perhaps seeking an oviposition site (2). Two other sightings were of butterflies at 20-30 m in the forest canopy (2). The climate of Mt Tambusisi is monsoonal, the steep slopes are thickly wooded, and the bedrock is ultrabasic (2). The area has a lower diversity of butterflies than have limestone areas of Sulawesi (2).The genus Idea is one of the danaid groups nauseous to predators (3).

SCIENTIFIC INTEREST AND POTENTIAL VALUE Although few collections have been made in the high altitude areas of Sulawesi, it is unusual that such a large and conspicuous butterfly should have remained undescribed until 1980. The smaller, paler species I. blanchardii is widely distributed and well known in Sulawesi and surrounding islands, but may be restricted to coastal areas below 300 m (2,3). I. tambusisiana cannot be confidently placed into either the I. idea or I. lynceus groups into which the genus is divided, and is therefore of some phylogenetic interest (1). Sulawesi is known for the unique fauna which is a result of isolation from both the Sunda and Sahul shelves (3). I. tambusisiana may be a primitive species now in competition with I. blanchardii and restricted to the montane habitats of a formerly large range (3).

THREATS TO SURVIVAL The type locality is in a reserve, eight hours' walk from the nearest village. Nevertheless, demand for a new species could stimulate determined efforts by dealers to procure specimens. If the population is small and localized then heavy collecting could jeopardize the species (3). Mining or forest clearance would also threaten I. tambusisiana. Local people collect damar gum from the forest trees and clear lowland forest for shifting cultivation, but neither of these are likely to affect the habitat of this butterfly (3).

CONSERVATION MEASURES TAKEN The type locality on Gunung Tambusisi lies in the Morowali Nature Reserve. This was gazetted by the Indonesian Government as a measure of protection against timber operations (3).

CONSERVATION MEASURES PROPOSED The best protection for the Sulawesi Tree Nymph is that it is known only from a remote reserve. As long as no road is built into the area and the advances of timber and mineral extraction interests are resisted, it should be safe. These potential threats should be monitored, and the appropriate authorities clearly advised of the endemic biological resource at stake. A survey should be undertaken to locate additional populations of the butterfly and define their status and habitat requirements (3). Research on its population ecology would anticipate possible future conservation and management considerations. The insect could be considered for protection by the Indonesian Government, as have several birdwings and other papilionids.

REFERENCES   1.   Ackery, P.R. and Vane-Wright, R.I. (In press). The phylogeny and biology of danaid butterflies.
2.   Bedford Russell, A. (1981). A spectacular new Idea from Celebes (Lepidoptera, Danaidae). Syst. Ent. 6: 225-228.
3.   Bedford Russell, A. (1982). In litt., 8 February.

A. Bedford Russell kindly helped to compile this account and P.R. Ackery and R.I. Vane-Wright added helpful comments.

Tribe Pronophilini Reuter, 1897

Phylum	ARTHROPODA	Order	LEPIDOPTERA
Class	INSECTA	Family	SATYRIDAE

SUMMARY The Pronophilini is a tribe of the brown butterflies (Satyridae) and is especially richly represented in the high altitude cloud forests of the tropical Andes. It owes its high diversity to the restricted distributions of many of its species, to the extent of allopatric speciation in isolated montane locations, and to the unique way in which sister species have re-established contact after speciation and partitioned themselves into clearly defined altitudinal bands. Although no species are known to be in immediate danger, they are rare on a world scale, and their very restricted ranges renders them vulnerable to destruction of their montane habitats.

DESCRIPTION The Pronophilini comprises more than 400 species in about 66 genera. The butterflies vary in size from small to moderately large (25-90 mm wing span). The ground colour of the majority is dark brown, but some are white, silvery or blue. Upperside markings on the brown species are white, yellow, orange, chestnut or various shades of blue. Underside hindwings mostly have dark, mottled camouflage patterns or disruptive wedges or stripes of a lighter colour. All species have the M1-M2 cross vein on the hindwing strongly curving towards the wing base, and have a completely fused vein Sc+R1. The most useful generic characters for taxonomic purposes are in the male genitalia (2,3,4).

DISTRIBUTION The tribe's distribution extends from Arizona, U.S.A., to Patagonia, and includes many West Indian islands (1). In the north and south of its range (U.S.A., Mexico, Central America, Chile and Argentina), it occurs at a low species density. The greatest concentration of species is in the Andes of Colombia, Venezuela, Ecuador, Peru and Bolivia, and these are the countries where conservation action may be most needed. Above 2500 m in the tropical Andes they are the most diverse group of butterflies (5). In northern South America the Pronophilini is exclusively montane, occurring between 1100 m and the paramo at 4000 m and above (3). In the whole of the tropical Andes, only a few species live at altitudes below 1000 m. Many genera and species occur only above the tree-line, from 3000 to 4500 m, and probably higher, but maximum diversity is found in montane cloud forest (2500-2700 m) in Colombia and Venezuela (4,5). Many species are restricted to single mountain ranges or even single massifs within those ranges, and to very narrow belts of altitude (e.g. 100 or 200 m) (1).

In general, the smaller and more isolated the mountain range, and the higher the locality, the fewer pronophiline species are present, but the greater is the proportion of endemic species. Thus, while it is possible to find 32 species in one site at 2500 m in the Colombian Central Cordillera, numbers diminish rapidly above 3000 m, and the total pronophiline fauna of the very isolated Sierra Nevada de Santa Marta, Colombia, is only 22 species. Yet, of the Santa Marta species, 12 are endemic (54.5 percent), three at the generic level. The Sierra has a basal area of only about 15 000 km², but it is the highest summit between the Canadian Rockies and the Ecuadorian Andes (2). The endemics are Arhuaco ica, (rare monospecific genus, 2600-3100 m), Lymanopoda nevada (2600-3500 m); L. caeruleata (uncommon, 1100-2600 m), Manerebia nevadensis (2500-3000 m), Paramo oculata (monospecific genus abundant in paramo, 2800-4200 m), Pedaliodes leucocheilus (locally common, 1500-2500 m), P. phazania (1800-2800 m), P. tyrrheus (2300-3000 m), P. cebolleta (extremely rare, 2900-3000 m),

Physcopedaliodes symmachus (abundant, 1300-2200 m), Pronophila juliani (1800-2700 m), and Sierrasteroma polyxo (monospecific genus, above 2000 m) (2). Three of these exist in two distinct geographical races within the range, and this phenomenon is common in the tribe, e.g. in Eretris porphyria and E. apuleja in the Colombian Eastern Cordillera.

The Serrania de Valledupar mountain range is about 70 km long and 60 km wide, spanning the border between Colombia and Venezuela (where it is known as the Sierra de Perija (3,4)). Of the 35 species believed to occur in the range, only ten are also found in the Sierra Nevada de Santa Marta, whose 1500 m contour is only 40 km away (3). Eleven species (37.1 percent of the total) are endemic to the Serrania. Three of these have sister species in the Sierra Nevada de Santa Marta: Pedaliodes suspiro (P. leucocheilus in Santa Marta), P. tyrrheoides (P. tyrrheus) and Lymanopoda paramera (L. nevada) (3). Six other endemics have close allies in the Eastern Cordillera, with the exception of Dangond dangondi, which belongs to an endemic genus. Twenty of the species in Serrania de Valledupar belong to replacing series of closely allied congeners up the altitude gradient (3).

The main stem of the Andes divides into three parallel ranges in southern Colombia, one of which divides again in the north to become the Sierra de Perija in Colombia and the Cordillera de Merida in Venezuela. The latter is about 360 km long and 100 km broad. Only three species of Pronophilini have been recorded below 1800 m on the Cordillera de Merida, and maximum diversity is at 2500 m, where 21 species have been found. Of the total of 36 species, 12 are endemic (4). These include Altopedaliodes albarregas, A. albonotata, Corades pax, Diaphanos huberi, Lymanopoda dietzi, L. marianna, Pedaliodes japhleta, P. ornata, Penrosada franciscae, Pronophila epidipnis and Redonda empetrus (4). In addition, Muscopedaliodes jephtha is a great rarity, but is not endemic.

Of the three main Colombian cordilleras, the eastern (in which Bogota lies) is the most diverse, with 92 recorded species. About 27 of these are endemic (29.3 percent) and there are nine pairs of sister species separated from each other on the eastern and western slopes of the range, e.g. Pedaliodes phaea and P. ochrotaenia, Lymanopoda schmidti (7) and L. lactea, and Steroma andensis and S. bega (western species first in each case). The great majority of the endemics live in the uppermost cloud forest, at 2800 m and above, and the paramo.

The Central and Western Cordilleras are joined at higher elevations to the main Andean chain and have correspondingly fewer endemics: about 11 out of 80 (12.8 percent) in the Central and four out of 57 (7.0 percent) in the Western Cordillera. Yet many of their pronophilines are very rare, e.g. Altopedaliodes flavopunctata, Catargynnis loxo, C. ilsa, C. pholoe, Eretris centralis, Panyapedaliodes tomentosa, Parapedaliodes margaretha (7), P. nora (7), Pedaliodes spina, Penrosada inderena (7), Proboscis orsedice and Lymanopoda pieridina.

POPULATION There are no population estimates for the Andean browns; many of the new species are only known from a few collected specimens (2,3). In some cases populations may be quite large, but these could be quickly reduced by the various threats to these high altitude habitats.

HABITAT AND ECOLOGY All the Andean species require a moist, montane habitat. Of those within the forest belt, the great majority only survive in primeval forest and rarely even cross open pasture; secondary forest is also unsuitable. They prefer well-drained soils, e.g. on slopes, where species of the bamboo Chusquea grow (2). Narrow roads through cloud forest often provide a good habitat along the roadsides where Chusquea flourishes in relatively unshaded conditions. Adults of most species feed on organic matter in leaf litter, bird lime or other animal excrement (6). When not feeding, each species has a preferred stratum above ground level, and oviposition sites of different species also occur at different heights (6). The species dwelling above the treeline inhabit the cool grassland community known as paramo, which is analogous to alpine tundra. Larval hostplants are known in few cases, but are generally presumed to be species of <u>Chusquea</u> bamboos, or other grasses.

SCIENTIFIC INTEREST AND POTENTIAL VALUE The tribe is of scientific interest for a number of reasons (1,2,3,4,5):
1. It has been hypothesized that the glacial and interglacial periods of the Pleistocene resulted in a cycle of spread and isolation of pronophiline populations horizontally along the Andes chain, and vertically along altitude gradients (1,4). In the current interglacial period, this has resulted in the distributions of most species, especially those of higher altitudes, being very restricted both in altitude and geographical area. Therefore, for example, the volcanic regions of Guatemala, Costa Rica and Panama, the Colombian Sierra Nevada de Santa Marta, the Venezuelan Cordillera de Mérida, the Sierra de Perija on the northern borders of Colombia and Venezuela, and the three main Colombian Cordilleras all have high proportions of endemic species. In addition, the western and eastern slopes of the Colombian Central and Eastern Cordilleras each harbour different subspecies and species, as do the Tolima and Huila regions of the Central Cordillera. The tribe is therefore of considerable biogeographical interest (2,4).
2. Most of the species belong to assemblages of closely related, allopatric congeners (members of the same genus not living together), represented at similar elevations in different mountain ranges, and displaying the effects of geographical speciation in isolated montane locations. They are thus of interest to the evolutionary biologist (4).
3. Most of the species are parapatric, i.e. they have a close congeneric relative living above or below them (or both) on the altitude gradient, with only slight overlap. These 'replacing series' of two or three parapatric species, existing on such a large scale within one group of butterflies, are unique. It would be of interest to the ecologist to study how the members' altitude zones are so neatly maintained.
4. The colour patterns of both allopatric and parapatric sister species often differ in very clear ways, particularly among the latter. The adaptive significance of the changes that have taken place during evolutionary divergence is not always clear, but there are several cases of mimicry within the tribe (4,5).

The main potential value of the butterflies is as indicators, in cloud forest and paramo habitats, of both general ecological diversity and probable degree of faunal endemism: the former is related to the total numbers of pronophiline species recorded in the region, and the latter to the proportion of its species which are endemic to isolated montane locations.

THREATS TO SURVIVAL Felling of the cloud forests, for agriculture, logging or merely for access, is the main threat throughout the range of the tribe in the tropics. For example, on the Colombian side of the Serrania de Valledupar, extensive burning of the mountainsides has severely fragmented the forest (3). Below 1850 m there is no forest at all, except along the banks of the Rio Manaure (3). Burning at altitudes above 2500 m is likely to endanger the greatest proportion of rare and endemic species (1,6). Furthermore, agriculture and ranching can severely damage the paramo grasslands, such as has occurred in the Serrania de Valledupar between 2800 and 3100 m (1,3). No species of Pronophilini are known to be in immediate danger of extinction, but they are classified as rare on the grounds of their very restricted world distributions, and their vulnerability to habitat destruction.

CONSERVATION MEASURES TAKEN Legislation in 1964 and 1977 formulated the Sierra Nevada de Santa Marta National Park in Colombia, with an area of 383 000 ha and altitudes from sea level to 5800 m (8). Because montane forests are essential in the maintenance of water supplies, some governments are trying to preserve forest at least along water courses. No specific measures have been taken with the Pronophilini in mind.

<u>CONSERVATION MEASURES PROPOSED</u> Attempts should be made to protect representative areas of cloud forest and paramo from 1500 m to 4000 m in each main mountain range. Special emphasis should be laid on the tropical Andean countries and on elevations above 2500 m, although it is important to maintain continuity along the altitudinal gradient (6).

<u>REFERENCES</u>
1. Adams, M.J. (1977). Trapped in a Colombian sierra. <u>Geogrl Mag</u>. 49: 250-254.
2. Adams, M.J. and Bernard, G.I. (1977). Pronophiline butterflies (Satyridae) of the Sierra Nevada de Santa Marta, Colombia. <u>Syst. Ent</u>. 2: 263-281.
3. Adams, M.J. and Bernard, G.I. (1979). Pronophiline butterflies (Satyridae) of the Serrania de Valledupar, Colombia Venezuela border. <u>Syst. Ent</u>. 4: 95-118.
4. Adams, M.J. and Bernard, G.I. (1981). Pronophiline butterflies (Satyridae) of the Cordillera de Mérida, Venezuela. <u>Zool. J. Linn. Soc</u>. 71: 343-372.
5. Adams, M.J. and Bernard, G.I. (1982). In manuscript. Pronophiline butterflies (Satyridae) of the three main Colombian cordilleras.
6. Adams, M.J. (1982). In litt., 27 March.
7. Adams, M.J. (1982). In litt., 7 January.
8. IUCN (1982). <u>IUCN Directory of Neotropical Protected Areas</u>. Tycooly Int. Pub. Ltd., Dublin. 436 pp.

We are grateful to M.J. Adams for providing the orginal draft for this sheet, and to F. Medem for further comment. The compilers wish to apologize for the lack of accents on Spanish names, which is due to the limitations of the printing equipment.

UNCOMPAHGRE FRITILLARY BUTTERFLY   VULNERABLE

Boloria acrocnema Gall & Sperling, 1980

Phylum	ARTHROPODA	Order	LEPIDOPTERA
Class	INSECTA	Family	NYMPHALIDAE

SUMMARY First discovered in 1978, this presumed post-glacial relict is endemic to a small part of the San Juan Mountains, Colorado, U.S.A. Its known range is smaller than that of any other North American butterfly. Long-term survival of B. acrocnema is probably controlled by local extinctions counterbalanced by colonization of new habitat. Due to its rarity and high value, commerical and private collecting may threaten the fritillary. Sheep grazing, if uncontrolled, may also be a threat.

DESCRIPTION The Uncompahgre Fritillary is the smallest United States member of the Holarctic genus Boloria, with a wing span of 2-3 cm (1). The wings above are warm brown criss-crossed with black bars; those of females are somewhat lighter. Underneath, the forewing is light ochre and the hindwing has a bold white, jagged bar dividing the crimson brown inner half from the purple-grey scaling on the outer wing surface. In general appearance B. acrocnema most closely resembles the circumpolar butterfly B. improba, an Arctic species not occurring in the U.S.A. and with its nearest colonies in central Alberta, Canada (6). B. improba is darker and duller than B. acrocnema, lacking the contrasting markings (6). It has been claimed that B. acrocnema is a sub-species of B. improba (9), but this is not widely accepted. The Uncompahgre Fritillary was found in 1978 and described in 1980 (6).

DISTRIBUTION The type locality of the Uncompahgre Fritillary is an alpine meadow below Mt Uncompahgre, Hinsdale County, Colorado, U.S.A., at an elevation of approximately 4100 m. In July 1982 another colony was found, on Redcloud Mountain in the San Juan Mountains several kilometres south of Mt Uncompahgre (5). It is not now expected to occur widely outside the San Juans, although certain other ranges should be explored more intensively in order to find the butterfly should it be present (1,7). A report circulated in 1980 referred to five less dense colonies found elsewhere in the San Juans, but the sites remain unsubstantiated (9) .

POPULATION Detailed population structure studies have been carried out at the type locality (6,7). Maximum daily population sizes during the period 1978-1981 were in the order of 150-200 (6,7), with most being considerably smaller. The brood number (total adults maturing in a year) was judged to be certainly less than 1000 individuals in 1980, and probably between 650 and 750. This was considered the largest brood size in the four-year period of the observations (7). An independent survey using a different census method gave a larger estimated brood number of about 2000 in 1980 (9).

HABITAT AND ECOLOGY There is one generation per year, adults flying principally from mid- to late July. The timing of yearly broods varies somewhat with the extent of winter snow cover, and conditions in spring. The flight season is about two weeks. The preferred female ovipositional substrate and larval foodplant is Snow Willow, Salix nivalis, a dwarf mat-forming alpine plant (6,7). The immature stages have been reared and the larvae are typical for the genus (9). Hibernal diapause seems to be as a partially grown larva. Individual dispersal radii are less than 100 m; residence rates are less than two days. Age-specific movement occurs in females, with old individuals emigrating from the colony site. This is viewed as an adaptive compromise between diminishing reproductive success in situ and the promise of founding a new colony in an unutilized patch of the larval foodplant. The alpine meadows in the area are lush and moist since the snow lingers late, with a rich southern Rocky Mountain flora (8). A colony of the endemic, local satyrid Erebia theano demmia also occurs several hundred metres below and to the east of the B. acrocnema type colony.

SCIENTIFIC INTEREST AND POTENTIAL VALUE Boloria acrocnema has the narrowest known range of any North American butterfly, and it was the first new species to be found in the contiguous 48 states for over 20 years. Its presence in healthy numbers serves to gauge the endemism and wilderness qualities of the San Juan Mountains. The Uncompahgre Fritillary is being used as a component both in theoretical and applied research on butterfly numerical systematics (3,4,6), and in population structure theory and management (7,11). The commercial value of the butterfly has been demonstrated, with specimens offered by dealers at prices exceeding $100 for males and even higher for females. The genus Boloria is extremely popular with collectors, and rare species are sought avidly; reared specimens could furnish a source of income for management projects if trade could be regulated.

THREATS TO SURVIVAL The Uncompahgre type colony lies within the boundaries of the Uncompahgre National Forest, and, barring development of mining interests, which are not prevalent at present, there is no direct threat from habitat destruction (7). The second colony is on land belonging to the Bureau of Land Management (5) and no information on threats is available. Sheep eat the larval foodplant, but are now restrained from disturbing the B. acrocnema type colony at Mt Uncompahgre. Nevertheless, sheep browsing clearly presents the most serious general threat to this butterfly (6), especially while its precise local distribution remains unclear. Indiscriminate collecting could perhaps adversely affect survival of the Uncompahgre colonies (3,7) and commercial collecting at the type locality has already been alleged (2). Recreational pressure could represent a threat if the number of summer hikers increased substantially.

CONSERVATION MEASURES TAKEN In 1979, the U.S. Fish and Wildlife Service was petitioned for a status assessment of B. acrocnema within the context of the U.S. Endangered Species Act, and a notice of review was published (10). Regulations proposing Threatened status are still pending at the U.S. Office of Endangered Species (7). In 1980, the Mt Uncompahgre type locality was registered as a Natural Area by the Colorado State Natural Area Council (7). The U.S. Forest Service has local jurisdiction at the site. That agency conducted a survey of the San Juan Mountains in 1981 for other colonies of B. acrocnema, failing to locate any. A continuation of this survey in 1982 showed an expansion of the type colony to a small site c. 1 km from the original locality (12). The Forest Service has increased its policing of collecting activity at Uncompahgre. The vulnerability of B. acrocnema to commercial collecting has been publicized to lepidopterists (2), with an appeal to refrain from purchasing such specimens when offered for sale.

CONSERVATION MEASURES PROPOSED Additional surveys of the San Juan Mountains and other ranges where B. acrocnema could occur are being organized by the U.S. Forest Service and others. Results of investigations of population ecology already obtained (7) are being incorporated into management plans for the area. As an alternative conservation measure, introduction of the fritillary into alpine areas distant from the San Juans is being contemplated (7). The monitoring of population parameters at the Uncompahgre colonies should continue, along with work on the larval ecology. Present management should be adequate, but escalation of the threats from collecting or recreation could require more severe regulations or management at the state or federal level. Forest Service personnel should be encouraged to continue a close watch on all human activities and livestock management in the area, in order to prevent damage to the butterfly or its habitat. Research which will lead to greater understanding of the needs of B.

<u>acrocnema</u> should be facilitated. At present there does not appear to be a need for a total collecting ban, but commercial collecting of this taxon should not be permitted (7). Collectors should be encouraged to inform conservation agencies of the location of any new colonies found.

REFERENCES

1. Ferris, C.D. and Brown, F.M. (1981). <u>Butterflies of the Rocky Mountain States</u>. University of Oklahoma Press, Norman. 442 pp.

2. Gall, L.F. (1981). Re. commercial sales of <u>Boloria acrocnema</u>. <u>News</u> (Lepidopterists' Society) No. 3, May-June: 40-41.

3. Gall, L.F. (Unpubl.). The effects of marking and releasing on subsequent activity in <u>Boloria acrocnema</u> (Lepidoptera: Nymphalidae).

4. Gall, L.F. (Unpubl.). Indexing lepidopteran age-structure by wing-wear: theoretical and applied considerations.

5. Gall, L.F. (1982). In litt., 1 August.

6. Gall, L.F. and Sperling, F.A.H. (1980). A new high altitude species of <u>Boloria</u> from southwestern Colorado (Nymphalidae), with a discussion of phenetics and hierarchical decisions. <u>J. Lepid. Soc.</u> 34: 230-52.

7. Gall, L.F. and Sperling, F.A.H. (Unpubl.). Population structure and conservation of the alpine butterfly, <u>Boloria acrocnema</u>.

8. Harrington, H.D. (1954). <u>Manual of the Plants of Colorado</u>. Sage Book, Denver. 666 pp.

9. Scott, J.A. (1982). The life history and ecology of an alpine relict, <u>Boloria improba acrocnema</u> (Lepidoptera: Nymphalidae), illustrating a new mathematical population census method. <u>Papilio New Ser.</u> 2: 1-12.

10. U.S.D.I. Fish and Wildlife Service (1980). Endangered and threatened wildlife and plants review of the status of the Uncompahgre fritillary butterfly. <u>Federal Register</u> 45 (26): 8029-8030.

11. Zieroth, E. (1981). Mt Uncompahgre Fritillary Butterfly: 1981 status report. Technical report to U.S. Forest Service, Cebolla District, Uncompahgre National Forest. 7 pp.

12. Zieroth, E. (1982). Mt Uncompahgre Fritillary Butterfly 1982 status report. Technical report to U.S. Forest Service, Cebolla District, Uncompahgre National Forest.

We are grateful to L.F. Gall, who prepared a draft for this review, and to C. Ferris, B. Lapin and J. Shepard for further comment.

BAY CHECKERSPOT BUTTERFLY   ENDANGERED

Euphydryas editha bayensis Sternitzky, 1937

Phylum	ARTHROPODA	Order	LEPIDOPTERA
Class	INSECTA	Family	NYMPHALIDAE

SUMMARY The subject of one of the world's most comprehensive and long-term studies of insect population biology, the Bay Checkerspot is restricted to only three colonies on the San Francisco Peninsula in California, U.S.A. The colony on Stanford University's Jasper Ridge Preserve is probably of insufficient size to protect this butterfly, and the other two colonies are on sites presently destined for development. Conservation measures are urgently required to safeguard both the butterfly and its habitat.

DESCRIPTION A medium-sized butterfly with a maximum wing expanse of 56 mm (males) or 59 mm (females) (16). In this race the predominant colour on the upper surface is black, checkered with red and ochreous yellow markings which stand out in striking contrast (16). The underside is predominantly ochreous yellow, but still with sharp patterns in black and red (16). Male and female are very similar in appearance (16).

DISTRIBUTION The Bay Checkerspot originally occurred in a large number of disjunct colonies in the San Francisco Bay region between the Santa Cruz Mts and the Coast Range bordering the San Joaquin Valley, California, U.S.A. The northernmost documented occurrence was Twin Peaks in San Francisco and the southernmost colony was at Coyote Reservoir near San José (13,15). This small range has now been further reduced, and the butterfly is restricted to three localities on the San Francisco Peninsula.

POPULATION The total number of Bay Checkerspots has never been estimated. However, population parameters of Euphydryas editha bayensis on Jasper Ridge Preserve have been the subject of a major research programme by a team at Stanford University (7,15). Three separate populations were found to exist there, and each fluctuated widely in numbers, and independently of its neighbours (1,3). The biggest population reached a maximum of about 5000 individuals, while the smallest went extinct and was re-established from another population (2,3). The California drought of 1976-77 caused dramatic reductions in numbers, but the effects were somewhat mitigated by faster larval growth due to the additional sunlight, and the patchy pattern of host plant senescence in time (5,13,15). Nevertheless, numbers in two areas of Jasper Ridge were reduced from 5000 and 2000 in 1976, to 800 and 400 in 1977, and 125 and 400 in 1978. Two small populations in another area became extinct because of the lack of food for the larvae (5). Dramatic annual fluctuations in populations with frequent local extinctions are particularly characteristic of the Bay Checkerspot. Investigation of different races of E. editha at a number of sites has shown that the factors controlling populations are numerous, complex, and variable in space and time (15).

HABITAT AND ECOLOGY The Bay Checkerspot is patchily distributed, occurring only on isolated areas of grassland on serpentine outcrops (8,18). The reason for this limitation was obscure, since the known food plant Plantago erecta was not restricted to the serpentine soils (3). Research showed that larval mortality largely resulted from desiccation of the Plantago before the larvae were large enough to diapause. However, those that were able to find Orthocarpus densiflorus (Owl's Clover) could continue feeding until they reached diapause size (3,15). It is this species which is restricted to serpentine soils, and which determines the size and location of E. e. bayensis populations (3,7,15,18). The

481

adults emerge in early spring and females lay eggs in batches of 21 to 75 on either host plant (10). Some females lay as many as 1200 eggs, which is the largest number reported for any butterfly (10). The larvae enter diapause by early summer and remain inactive until the autumn rains when they begin to feed on new plant growth. Post-diapause larvae appear simultaneously with the new annual plants in early spring, feed for several weeks, and pupate. Adult emergence occurs about two weeks later. The adults feed on the nectar of several key plants. An increase in fecundity associated with specific nectar sources appears necessary for population survival in marginal years (14).

SCIENTIFIC INTEREST AND POTENTIAL VALUE The Bay Checkerspot has been studied extensively by P.R. Ehrlich and his associates since 1959 (7). The resultant accumulation of data on population dynamics (15), genetics (4,6,11,12) and reproductive biology (9,10) makes this one of the insects best known to science. Perhaps no other colony of butterflies is as well known in all its particulars as the Jasper Ridge populations of E. e. bayensis. Given the enormous investment represented by the body of knowledge already in print, this species is an extremely valuable resource for future research.

THREATS TO SURVIVAL Of the three extant colonies of the Bay Checkerspot Butterfly only the Jasper Ridge colony is presently safe from immediate habitat destruction. Yet this colony is too small to ensure perpetuation, given its characteristic population fluctuations (15). The severe impact of the California drought of 1976-78 emphasized the vulnerability of the Bay Checkerspot to extinction (13). Aerial spraying of pesticides threatens the Jasper ridge colony as well as others. The largest colony, located at Edgewood, is threatened by plans for a golf course on land belonging to the San Mateo County Regional Park District. The third colony, which is on San Bruno Mountain, the site of several other endangered butterflies, is also under threat (see separate review). Surveys since 1977 failed to detect a single individual there, but adults were again seen in moderate numbers in 1981.

CONSERVATION MEASURES TAKEN The habitat and colony at Jasper Ridge are well-protected by a research preserve owned by Stanford University. The Edgewood colony is publicly owned by the San Mateo County Regional Park District, and should therefore be protected. However, while the County of San Mateo is making attempts to protect the serpentine grassland habitat at Edgewood, they are also under pressure to build a golf course on the site, which is probably of inadequate size to serve both purposes (see San Bruno review). The U.S. Fish and Wildlife Service has been petitioned to list the Bay Checkerspot as an Endangered species, and a review of status was duly published (17). Regrettably, the Fish and Wildlife Service has since been unable to take any action in response to the overwhelming public support which the review evoked. A new review and emerging listing are presently being considered, as is a lawsuit to compel listing.

CONSERVATION EFFORTS PROPOSED Detailed population monitoring should continue at the Jasper Ridge and Edgewood sites with annual surveys at San Bruno Mountain. The County of San Mateo should be encouraged to abandon plans for a golf course at the Edgewood site and set up a serpentine grassland preserve instead. Efforts to preserve habitat on San Bruno Mountain should be planned in conjunction with conservation measures aimed at other endangered butterflies there. Of most immediate importance, the U.S. Fish and Wildlife Service should be strongly encouraged to act on the petition for federal Endangered status.

REFERENCES   1.   Brussard, P.F., Ehrlich, P.R. and Singer, M.C. (1974). Adult movements and population structure in Euphydryas editha. Evolution 28: 408-415.
2.   Ehrlich, P.R. (1965). The population biology of the butterfly

Euphydryas editha. II. The structure of the Jasper Ridge colony. Evolution 19: 327-336.

3. Ehrlich, P.R. (1979). The butterflies of Jasper Ridge. Co-evolution Quart. Summer 1979: 50-55.

4. Ehrlich, P.R. and Mason, L.G. (1966). The population biology of the butterfly Euphydryas editha. III. Selection and the phenetics of the Jasper Ridge colony. Evolution 20: 165-173.

5. Ehrlich, P.R., Murphy, D.D., Singer, M.C., Sherwood, C.B., White, R.R. and Brown, I.L. (1980). Extinction, reduction, stability and increase: the responses of checkerspot butterfly (Euphydryas) populations to the California drought. Oecologia 46: 101-105.

6. Ehrlich, P.R. and White, R.R. (1980). Colorado Checkerspot butterflies: isolation, neutrality and the biospecies. Am. Nat. 115: 328-341.

7. Ehrlich, P.R., White, R.R., Singer, M.C., McKechnie, S.W. and Gilbert, L.E. (1975). Checkerspot Butterflies: a Historical Perspective. Science 188: 221-228.

8. Johnson, M.P., Keith, A.D. and Ehrlich, P.R. (1968). The population biology of the butterfly, Euphydryas editha VII. Has E. editha evolved a serpentine race? Evolution 22: 422-423.

9. Labine, P.A. (1966). The population biology of the butterfly Euphydryas editha. IV. Sperm precedence - a preliminary report. Evolution 20: 580-586.

10. Labine, P.A. (1968). The population biology of the butterfly, Euphydryas editha. VIII. Oviposition and its relation to patterns of oviposition in other butterflies. Evolution 22: 799-805.

11. Mason, L.G., Ehrlich, P.R. and Emmel, T.C. (1967). The population biology of the butterfly, Euphydryas editha. V. Character clusters and asymmetry. Evolution 21: 85-91.

12. Mason, L.G., Ehrlich, P.R. and Emmel, T.C. (1968). The population biology of the butterfly Euphydryas editha VI. Phenetics of the Jasper Ridge colony, 1965-66. Evolution 22: 46-54.

13. Murphy, D.D. and Ehrlich P.R. (1980). Two California checkerspot subspecies: one new, one on the verge of extinction. J. Lepid. Soc. 34: 316-320.

14. Murphy, D.D. (1981). The role of adult resources in the population biology of Euphydryas butterflies. Unpublished Ph.D. thesis, Stanford University. Stanford, California.

15. Singer, M.C. and Ehrlich P.R. (1979). Population dynamics of the checkerspot butterfly Euphydyras editha. Fortschr. Zool. 25: 53-60.

16. Sternitsky, R.F. (1937). A race of Euphydryas editha (Lepidoptera). Can. Ent. 69: 203-205.

17. U.S.D.I. Fish and Wildlife Service (1981). Review of the status of the bay checkerspot butterfly. Federal Register 46 (30): 12214-12215.

18. White, R.R. and Singer, M.C. (1974). Geographical distribution of hostplant choice in Euphydryas editha (Nymphalidae). J. Lepid. Soc. 28: 103-107.

We are very grateful to B.A. Wilcox for information on this species, and to P.R. Ehrlich, D.D. Murphy and P.A. Opler for further comment.

NATTERER'S LONGWING BUTTERFLY

ENDANGERED

Heliconius nattereri Felder & Felder, 1865

Phylum	ARTHROPODA	Order	LEPIDOPTERA
Class	INSECTA	Family	NYMPHALIDAE

SUMMARY Heliconius nattereri from Brazil is a key primitive member of its genus, presently nearing extinction accelerated by human disturbance. Most of its potential habitat has been extensively altered or taken over by competitors. Only a single large, permanent colony has been discovered in spite of extensive searches over 15 years. Although formally protected, this colony is subject to invasion, burning and collecting.

DESCRIPTION Heliconius nattereri is a medium-large, long-winged butterfly (70-90 mm wingspan) with long antennae and abdomen. The male has three broad, longitudinal bars on each pair of wings, bright yellow over the black-brown ground colour; the female has the middle bar orange and an additional orange bar on the hindwing (1). The female was once described as a different species, H. fruhstorferi, since no other longwings (genus Heliconius) are sexually dimorphic (1).

DISTRIBUTION Endemic to Brazil, Natterer's Longwing was once known at least from central Bahia (Salvador) to southern Espirito Santo (Santa Leopoldina), and may have formerly reached Pernambuco judging from the extent of its co-mimics (6). In the past 50 years it has been seen only in the immediate region of Santa Teresa, Espirito Santo (19° 56'S, 40° 36'W) where there is a single large permanent colony and a number of subsidiary populations which are apparently established, on an irregular basis, from the central colony. Extensive searching in all former collecting sites and in many additional localities within its 19th century range has failed to disclose any additional colonies, although abundant suitable habitat was found in some areas (1).

POPULATION Unknown in detail, but believed to be low and fluctuating (3,6).

HABITAT AND ECOLOGY The species lives in steep, humid, primary rain forest where infrequent natural disturbance permits its vine food plant (Tetrastylis ovalis, a primitive passifloraceous woody perennial) to produce new growth. Since males promenade daily over several hundred metres, a dispersed distribution is probably normal for the species, and a single colony would need at least 1 km^2 of ideal habitat to persist, in the absence of competitors. Nearby human disturbance or other major habitat alteration encourages the butterfly's foodplant competitors (Heliconius numata, H. ethilla, H. melpomene, H. erato and H. sara, more aggressive as larvae than H. nattereri), without increasing the foodplant itself. The vine is very slow to reproduce and grow. No new saplings have ever been located in any of the dozens of areas where the plant is common. Thus, in areas where more adaptable competitors invade and increase by feeding on prolific passionflower vines, H. nattereri cannot persist (3,6). This helps to explain the apparent slow, natural decline which the Natterer's Longwing is suffering. The species is primitive and inflexible and its decline is accelerated by the human influences which favour its competitors.

The last known colony is in an area unusually rich in plants, insects and vertebrates, made famous by the studies of Augusto Ruschi on orchids and hummingbirds. The area is peripheral to two centres of evolution for butterflies (6) and shows a very complex topography and microhabitat structure. This greatly enhances its species diversity and permits the presence of many other species which are absent from more homogenous habitats to the north and south (6).

SCIENTIFIC INTEREST AND POTENTIAL VALUE As the most primitive member of a very widely studied genus (5), H. nattereri has considerable potential value in studies of genetics (7), physiology, behaviour and ecology. Its preservation will permit many existing and future questions about the evolution of this genus to be studied and perhaps answered. The understanding of the process of marginalization and perhaps natural extinction of primitive species, exemplified by this insect, is of central value to conservation theory and practice. The species is very attractive and has a commercial value due to its rarity and the popularity of the genus with collectors (6). One of its most remarkable traits is its sexual dimorphism, a feature long considered to be impossible in Heliconius. Longwings are routinely involved in complex mimicry rings. The female of H. nattereri is a member of a mimetic complex involving a number of other longwings and ithomiines, with pierids and nymphalines as mimics (3,8). The species combines characters from a variety of related taxa. Although this butterfly possesses both commercial and aesthetic worth, its scientific interest is certainly its greatest value.

THREATS TO SURVIVAL The primary forest habitat of H. nattereri has been almost totally destroyed by human activities, placing at risk a large number of little-known 'Bahia-species' of plants and animals, many of which were described in the middle of the last century and have never been seen since (3,4). Even where the habitat is reduced in extent, with major disturbance around the edges, this longwing will not survive due to invasion by its more aggressive foodplant competitors. It is very unlikely that it could be established in captivity, although reintroduction into some very large parks or reserves might be possible (3,6). Only a single permanent colony is known at present, with large variations in size during and between years. The entire species would be in danger of extinction should this colony be destroyed in a forest fire (6). Collecting is a potential threat but has not been demonstrated to have a substantial impact at current levels (3,6).

CONSERVATION MEASURES TAKEN In 1970, it was proposed that the area of the permanent colony be designated as a reserve as soon as possible, not only for H. nattereri but also for its other rare plants and animals (1,2,3). This was accomplished only in 1979, with the area being added to the rather distant Lombardia Biological Reserve (in which the butterfly did not occur), thus forming a discontinuous bridge or stepping stone between that Federal Reserve and a small, informal reserve of the Museu Mello-Leitao to the south (in which H. nattereri is present, perhaps resident, but very scarce) (6).

CONSERVATION MEASURES PROPOSED A visit to the colony district in May 1981 (6) showed that it was being used for shifting agriculture, destroying the forest and the adult flowers most used as sources of nectar in 1968-1973. The area must be effectively protected by fencing and probably patrolling if this colony, and perhaps the species itself, is to persist. The longwing was not seen in over 75 per cent of its usual area in the 1981 visit, when forest felling was seen and signs of further disturbance evident (6). Additional options for conserving H. nattereri need to be explored for scientific and management feasibility.

REFERENCES
1. Brown, K.S., Jr. (1970). Rediscovery of Heliconius nattereri in eastern Brazil. Ent. News (Philadelphia) 81: 129-140.
2. Brown, K.S., Jr. (1970). In litt. to A. Ruschi, Santa Teresa, Espirito Santo.
3. Brown, K.S., Jr. (1972). The heliconians of Brazil (Lepidoptera: Nymphalidae). Part III. Ecology and biology of Heliconius nattereri, a key primitive species near extinction, and comments on the evolutionary development of Heliconius and Eueides. Zoologica, N.Y. 57: 41-69.
4. Brown, K.S., Jr. (1975). The ithomiines of Brazil (Lepidoptera: Nymphalidae), part III. Rediscovery and

systematic position of <u>Napeogenes</u> <u>xanthone</u>. <u>Ent. News</u> <u>(Philadelphia)</u> 85: 265-274.

5. Brown, K.S., Jr. (1981). The biology of <u>Heliconius</u> and related genera. <u>A. Rev. Ent.</u> 26: 427-456.

6. Brown, K.S., Jr. (1982). In litt., 19 January.

7. Sheppard, P.M., Turner, J.R.G., Brown, K.S., Jr., Benson, W.W. and Singer M.C. (in press). Genetics and the evolution of Muellerian mimicry in <u>Heliconius</u> butterflies. <u>Phil. Trans.</u> <u>R. Soc. Ser. B.</u>

8. Watson, A., Walley, P.E.S., and Duckworth, D.W. (1975). <u>A</u> <u>Dictionary of Butterflies and Moths.</u> McGraw Hill, N.Y. 296 pp.

We thank K.S. Brown Jr. for information on this subject, and L. Gilbert and O. Mielke for further comment.

WIEST'S SPHINX MOTH   ENDANGERED

Euproserpinus wiesti Sperry, 1939

Phylum   ARTHROPODA   Order   LEPIDOPTERA

Class   INSECTA   Family   SPHINGIDAE

SUMMARY Among the rarest members of the Sphingidae (sphinx or hawk moths), this species remained virtually unknown for over 40 years following its original capture. In 1979 a population was found in eastern Colorado, U.S.A., but in the same year it was threatened by aerial application of Malathion for grasshopper control. A research programme is under way to develop conservation and management priorities for the moth's small known habitat.

DESCRIPTION Euproserpinus wiesti is a small, diurnal sphingid moth; the female wingspan is 38-55 mm, males are 5-6 mm smaller (3). Both sexes have grey-brown, mottled forewings, broken by a transverse grey band on the dorsum. The male generally has more brown on the forewing and thorax while the female has more grey. In both sexes, the hindwings are pale yellow dorsally with black bases and outer margins, and the ventrum is white with black scaling along the outer margin of the forewings and a black marginal band on the hindwings. Standard key characteristics for the species include four or five very stout spines on the outer surface of the first tarsal segment of the foreleg and a continuous pale yellow or white dorsal abdominal band (5,6,8). The emerald green eggs hatch into brown and white caterpillars (3). Green morphs are found in a proportion of the first four instars, but even these revert to brown in the fifth instar (3).

DISTRIBUTION The small colony in a sand wash c. 48 km south-east of Greeley, in Weld County, Colorado, is the only known population. The type locality near Kersey, Colorado (7), and a second traditional locality near Roggen, Colorado, have long since been altered by cultivation and rendered unsuitable for the moth. One worn specimen was taken near Albuquerque, New Mexico (approximately 725 km south-west of the known colony) on 6 May, 1945 (5). Four additional, but unconfirmed Colorado records are listed for Lamar, Walsenburg, and Great Sand Dunes National Monument in the 1950s and 1967 (4).

POPULATION Following the rediscovery of E. wiesti in 1979, approximately 200-300 adults were counted. In 1980, only 40-50 adults were seen, and in 1981, the number was reduced to about 25. However, most larvae were killed in 1980 by a Weld County grasshopper spraying programme, thus explaining the reduced number of adults in 1981. It is estimated that about 200 larvae were able to pupate in 1981, thus improving prospects for the 1982 emergence (3,4).

HABITAT AND ECOLOGY The colony is on a 600 m long, 25-75 m wide sand wash which supports a heavy concentration of the larval hostplant, Prairie Primrose (Oenothera latifolia). The adults and larvae of E. wiesti appear to be restricted to the sand wash although the food plant is more widely distributed in Colorado. The wash is a harsh environment with air and ground temperatures exceeding 60°C in mid-summer (3). Plant and animal species associated with E. wiesti are typical of desert and dry, high plains habitats (3). The surrounding area represents one of the last undisturbed tracts of high plateau prairie in eastern Colorado, and the diversity of indigenous plants and animals is very high (3). Detailed studies have been made of the population biology, autecology and community ecology of E. wiesti (3,4). The adults fly diurnally in May and June, usually only in clear weather. Eggs are laid singly near the base of the foodplant leaves, and hatch in 7-10 days. By mid-July all larvae have pupated below ground, and all are believed to emerge in the following spring (3). Adults do not take nectar or other nourishment.

SCIENTIFIC INTEREST AND POTENTIAL VALUE Euproserpinus is a small genus with widely separated species. One of these, E. euterpe, has already been listed by the U.S. Fish and Wildlife Service as a Threatened species (9,10). Euproserpinus species represent an intriguing set of biogeographical circumstances. E. euterpe and E. wiesti haunts may be relics of recently broad ranges and the species are probably the two most narrowly endemic hawk moths in North America (5). Their conservation and management would ensure continued existence of small examples of these habitats and their component plant and animals communities. The monetary value of rare sphinx moths has been high because collectors have concentrated on this popular group. Prices of up to $1000 have been offered for specimens of E. wiesti, and the level could go higher on the international market (3).

THREATS TO SURVIVAL The high monetary value of the moth represents a real threat. Collecting often does not jeopardize insect populations because of their great capacity for replacement, but the combination of small population size and high value could result in extinction of the E. wiesti colony if its exact location was known to collectors. The precise position of the sand wash has so far remained secret from the public, but certain collectors are known to be searching for it (7). Apart from unscrupulous collectors, the primary threat to the species is continuing grasshopper control programs. In 1980, the entire county was aerially sprayed with Malathion to reduce numbers of several pest grasshopper species. An attempt was made to avoid the sand wash but drift caused it to be sprayed anyway and heavy mortality resulted (1,2). Spraying activities will probably also occur in future outbreak years. In 1981, oil and gas exploration resulted in the placement of a pumping station and three storage tanks near the edge of the critical habitat. Their influence has apparently not been deleterious so far, but expansion of the facilities could threaten the site. Grazing has occasionally occurred on the site, but has not been a serious problem due to the co-operation of the landowner (4). In 1981 competition with increasing populations of the larvae of the White-lined Sphinx (Hyles lineata) posed a further threat. Caterpillars stripped the Prairie Primrose plants and maimed and killed E. wiesti caterpillars by slashing them behind the head with their mandibles (3).

CONSERVATION MEASURES TAKEN The U.S. Office of Endangered Species was formally petitioned to conduct a review of status of E. wiesti, with a view to listing it as Threatened or Endangered under the Endangered Species Act. An advance notice of proposed rule was subsequently published (11), but no further action has been taken. Grasshopper spraying of the site was avoided in 1981 through the co-operation of the landowners, Weld County officials and scientific investigators. Research support has been provided to a University of Wyoming team led by K. Bagdonas since shortly after the team's rediscovery of the moth in 1979. As a stated priority of the IUCN/SSC Lepidoptera Specialist Group, financial grants have been awarded by WWF-U.S. for studies on the ecology and life history of the moth. Additional co-operation has come from the ·Colorado Natural Heritage Inventory and the Colorado Division of Wildlife. The landowners are committed to the conservation of the moth and its habitat and have helped in many ways, including periodic patrolling of the wash. A group of WWF-U.S. and IUCN representatives visited the E. wiesti site in August 1981 to discuss protection with the landowners and study the moths' requirements. This resulted in extensive media coverage (7) which enhanced public concern for conservation of the site.

CONSERVATION MEASURES PROPOSED Continued close co-operation with the landowners and Weld County officials will be necessary to avoid spraying at the site. Discussions are being held to determine the feasibility of a biological control programme using parasitic protozoans to reduce the numbers of destructive grasshoppers (3). This alternative to spraying offers a realistic protective measure which should be promoted if funding can be found (3,4). Local, state, federal and international support will be sought to ensure the survival of the colony and the species. The exact locality of the site will remain confidential, at least until protected status is accorded by the federal government. The Office of Endangered Species should be permitted to list the moth as soon as possible. Meanwhile, the landowners will continue to patrol the land to prevent criminal trespass by collectors. The University of Wyoming research team has already conducted extensive surveys of the area in order to locate any other possible colonies, with negative results. However, such field activity should be vigorously pursued, not only in north-east Colorado but also in the San Luis Valley and the New Mexico area where single records have originated. The Wyoming team, in conjunction with the Colorado Natural Areas Program and with funds from WWF-U.S., is using air and ground searches to find potential transplant sites.

REFERENCES   1. Bagdonas, K. (1980). The life history and ecology of Euproserpinus wiesti Sperry. 1980 Progress report for IUCN/WWF (U.S.) Project No. 1793. 16 pp.
              2. Bagdonas, K. (1981). Project 1793. Life history and ecology of rediscovered prairie sphinx moth, U.S. In Farrell, A. (Ed.), World Wildlife Fund Yearbook 1980-81, Gland, Switzerland. p.145.
              3. Bagdonas, K. (1982). The life history and ecology of Euproserpinus wiesti Sperry. 1981 Final report for IUCN/WWF Project No. 1793. 70 pp.

4. Bagdonas, K. (in press). The rediscovery and current status of Euroserpinus wiesti Sperry, a "lost" prairie sphinx moth, in eastern Colorado. Atala.

5. Hodges, R.W. (1971). Sphingoidea of North America. Fasc. 21, Moths of North America North of Mexico. E.W. Classey and R.D.B. Publ. Inc., London. 170 pp.

6. Holland, W.J. (1904). The Moth Book. Doubleday, N.Y.

7. Schmidt, W.E. (1981). Scientists dodging collectors on the trail of elusive moth. New York Times 6 September, pp. 1, 28.

8. Sperry, J.L. (1939). Two apparently new western moths. Bull. S. Cal. Acad. Sci. 38: 126.

9. Tuskes, P.M. and Emmel, J.F. (1981). The life history and behavior of Euroserpinus euterpe (Sphingidae). J. Lepid Soc. 35: 27-33.

10. U.S.D.I. Fish and Wildlife Service (1980). Determination that the kern primrose sphinx moth (Euroserpinus euterpe) is a Threatened species. Federal Register 45(69): 24088-24090.

11. U.S.D.I. Fish and Wildlife Service (1981). Endangered and Threatened wildlife and plants; petition acceptance and status review for Wiest's Sphinx Moth. Federal Register 46(123): 33063.

We are grateful to K. Bagdonas for his help in the compilation of this review, and to B. Lapin, P.A. Opler and P.M. Tuskes for further comment.

Aneuretus simoni Emery, 1893

| Phylum | ARTHROPODA | | Order | HYMENOPTERA |
| Class | INSECTA | | Family | FORMICIDAE |

SUMMARY Aneuretus simoni is the only living representative of the subfamily Aneuretinae, a link between the very primitive Nothomyrmeciinae and the more advanced Dolichoderinae. Aneuretus is found only in southern Sri Lanka, its range appears to be shrinking and its ecological requirements are poorly known.

DESCRIPTION Light yellow to yellow-orange in colour, Aneuretus simoni is unusual in having a dimorphic worker caste (1,2). The eyes are small but the sting well-developed (2). All castes and immature stages have been described in detail (1,2).

DISTRIBUTION The ant is endemic to Sri Lanka and probably confined to the more humid southerly areas. A. simoni was originally described from the Kandy and Peradeniya area in 1892, but was not found there in 1955 (1,2). However, in 1955 A. simoni was one of the dominant ants in the area between Ratnapura, Gilimale and Adam's Peak, south of the original collecting area (1). Recent research indicates that the range has contracted still more during the past 25 years (3).

POPULATION Total populations and full distribution are unknown. However, a sample of 20 nests contained two to over 100 individuals. Nearly all nests contained brood, but only eight contained major workers, with never more than two per nest (2). Where it presently occurs, Aneuretus is among the most abundant genera, possibly only less common than Ponera, Pheidole and Myrmicaria (2).

HABITAT AND ECOLOGY At Ratnapura and Gilimale A. simoni was abundant on the edge of clearings in primary rain forest and in secondary growth, perhaps being adapted to open woodland rather than mature rain forest (2). The ants forage for collembolans and other small arthropods on the ground and in low vegetation (2). Nests are built in fallen rotten wood that ranges in size from twigs to logs (2). The close relatives of Aneuretus have long been extinct elsewhere, and the decline of the Aneuretinae is correlated with the recent evolutionary rise of Pheidole and Crematogaster (4). These latter genera occur in the same forests and tend to compete with Aneuretus for both food and nest sites. In many tropical areas Pheidole and Crematogaster overwhelmingly dominate the ant fauna, but in these Aneuretus habitats they inexplicably assume lower populations (2). Pheidole is nevertheless quite abundant, but Crematogaster is sparsely represented both in terms of species and abundance (2).

SCIENTIFIC INTEREST AND POTENTIAL VALUE Fossil aneuretines show that the group was once widespread. Three extinct genera of Oligocene age are known from Europe and North America. A. simoni is thus the only living relict of a subfamily of ants considered to be of great phylogenetic interest. The aneuretines may furnish the link between the most primitive ants of the subfamily Nothomyrmeciinae and the advanced subfamily Dolichoderinae. As such, Aneuretus simoni is of great significance for further studies (2).

THREATS TO SURVIVAL It is not clear to what extent Aneuretus is threatened. In 1955 the ant was not found around Kandy and Peradeniya, its two former localities (1,2). This area includes the Udawaddatekele Sanctuary, where a scarcity of ants may be partly due to the lack of rotting wood on the ground. Singhalese wood-gatherers there had depleted the supply of wood litter and dense populations of termites further reduced the number of available nest sites (2). Since rain forest clearance has continued on Sri Lanka, and wood collection has probably become intense over a larger area than before, A. simoni may be under survival pressure in parts of its limited range. The fact that A. simoni was prolific in scrubby secondary growth at Ratnapura in 1955 (2) suggested an apparent adaptiveness. Unfortunately this has not been substantiated since the ant had disappeared from Ratnapura in 1979 (3). It is still moderately common in less disturbed rain forest at Gilimale (3), but a further reduction of range seems inevitable.

CONSERVATION MEASURES TAKEN One of the old known localities lies within the Udawaddatakele Sanctuary, the remnants of the royal gardens of the Kandy kings (1). Unfortunately, the ant was apparently absent from this reserve in 1955 (2).

CONSERVATION MEASURES PROPOSED Periodic surveys are needed to monitor Aneuretus distribution. Much more information on the ecological needs of the species are required in order to plan conservation measures. It is not clear whether Aneuretus prefers primary or secondary forest, open or closed canopy. Competition with the wood-feeding termites (particularly the voracious Macrotermitinae) should be investigated. Management of this ant might include conservation or enhancement of the dead wood which constitutes its nest sites.

REFERENCES 1. Emery, C. (1893). Voyage de M.E. Simon à l'île de Ceylan, Jan-Feb 1892. 3rd Memoir No (1): Formicides. Ann. Soc. Ent. Fr. 62: 239-258.
2. Wilson, E.O., Eisner, T., Wheeler, G.C. and Wheeler, J. (1956). Aneuretus simoni Emery, a major link in ant evolution. Bull. Mus. Comp. Zool. Harvard 115: 81-99 + 3pl.
3. Wilson, E.O. (1981). In litt., data of A. Jayasuriya, 21 April.
4. Taylor, R.W. (1982). In litt., 21 January.

We thank E.O. Wilson and R.W. Taylor for information on this species, and B. Bolton and D. Cherix for further comment.

CAUCASIAN RELICT ANT INSUFFICIENTLY KNOWN

Aulacopone relicta Arnoldi, 1930

Phylum   ARTHROPODA   Order   HYMENOPTERA

Class   INSECTA   Family   FORMICIDAE

SUMMARY The only ponerine ant genus with known distribution limited to some part of the holarctic region, and one of only two ant genera endemic to the Soviet Union, Aulacopone is a significant biogeographic relict, apparently related to the genus Heteroponera, of Australia, New Zealand and South America. The monotypic genus is known only from two dealate queen specimens collected in Azerbaydzhan, U.S.S.R.

DESCRIPTION The Caucasian Relict Ant was originally described and illustrated by Arnoldi (I), and redescribed in detail, with scanning electron micrograph illustrations, by Taylor (3). Aulacopone has a peculiar head structure with a pointed process extending forwards over the jaws and very deep lateral longitudinal fossae to receive the folded antennae. A series of minute pits are present on each pronotal shoulder. These are unknown elsewhere among the ants and probably receive ducts from a gland which produces pheromones (3).

DISTRIBUTION The type specimen, a dealate female, was collected in 1928 at Alazapin near Lenkoran in Azerbaydzhan S.S.R., one of the southernmost states of the U.S.S.R. on the Iranian border in the Caucacasus. The second collection, also a dealate female, was taken in 1936 on Mt Gugljaband near Alekseyevka, also in Azerbaydzhan S.S.R. (3).

POPULATION Unknown.

HABITAT AND ECOLOGY The first collection was possibly associated with a colony of the formicine ant Lasius emarginatus in an oak stump (? Quercus castaneifolia) in mid-montane forest (I). Despite being found associated with another ant's nest, there is little reason to suppose that A. relicta is a social parasite, although it could be cleptobiotic or lestobiotic (3). There is no ecological information associated with the second collection.

SCIENTIFIC INTEREST AND POTENTIAL VALUE The reclusive habits of Aulacopone might explain its survival in areas which now lack, and perhaps have lost, surface-active ectatommine ants. These were certainly once present in Eurasia and North America, as the fossil record shows. Aulacopone is an ancient biogeographic relict and can be directly likened to various extinct European Baltic Amber and American Florissant Shale fossil ectatommine ants (3). Amongst the extant ants, Aulacopone most resembles the ectatommine non-palearctic genus Heteroponera, which is found in Australia (with at least nine known species), New Zealand (one endemic species) and South America (13 known species from Panama south to Uruguay and Chile) (2,3).

THREATS TO SURVIVAL Since the discovery of this species in 1928, Lenkoran city has expanded considerably and the mid-montane forest in the region has been disrupted (4). It is unclear whether the original type locality has been affected. So little is known of Aulacopone relicta that no assessment of the threats to its survival can be made.

CONSERVATION MEASURES TAKEN No specific conservation measures have been taken for the ant, but situated near Lenkoran is Gircanskiy State Nature Reserve, which includes stands of mid-montane forest with the oak Quercus castaneifolia (5). If the ant is still present in the area, it may be protected within the boundaries of the Reserve.

CONSERVATION MEASURES PROPOSED An initial contribution would be to survey the mid-montane forest in the Gircanskiy State Nature Reserve and around the second locality on Mt Gugljaband. When the species has been located, research will be needed to ensure proper management and protection. There is far too little information on this important species and the collection and study of the workers, males and immature forms of A. relicta is a significant challenge to Soviet entomologists.

REFERENCES 1. Arnoldi, K.V. (1930). Studien über die Systematik der Ameisen. IV. Aulacopone, eine neue Ponerinengattung in Russland. Zool. Anz. 89: 139-144.
2. Brown, W.L., Jr. (1958). Contributions toward a reclassification of the Formicidae. II. Tribe Ectatommini (Hymenoptera). Bull. Mus. Comp. Zool. Harvard. 118: 175-362.
3. Taylor, R.W. (1979). Notes on the Russian endemic ant genus Aulacopone Arnoldi. Psyche 86: 353-361.
4. Taylor, R.W. (1980). In litt., 2 June.
5. Radchenko, A.G. (1983). In litt., 3 January.

We are grateful to R.W. Taylor for a preliminary draft on this species, and to B. Bolton, D. Cherix and A.G. Radchenko for further comment.

Epimyrma ravouxi (André, 1896)

Phylum	ARTHROPODA	Order	HYMENOPTERA
Class	INSECTA	Family	FORMICIDAE

SUMMARY The biology of this slavemaker ant has only recently been elucidated. Its European range is extensive, but populations are sparsely distributed, and endangered through alteration of agricultural methods. The species is of great evolutionary interest.

DESCRIPTION The alate female Epimyrma ravouxi was first described at the end of the last century (1). All three castes were described under the name Epimyrma goesswaldi (9), but were later synonymized with E. ravouxi (3).

DISTRIBUTION Epimyrma ravouxi is known from numerous localities in Spain (6), southern France (type locality: Nyons, Drôme) (4), northern Italy, Switzerland (2), Austria (6), southern West Germany, East Germany (6) and Yugoslavia.

POPULATION Unknown. Despite the wide distribution, populations appear to be only in restricted areas, and may therefore be quite low.

HABITAT AND ECOLOGY Like the host species Leptothorax unifasciatus and L. nigriceps, E. ravouxi prefers warm and dry submediterranean localities with poor vegetation. Nests are found in crevices between flat stones, as well as in dry branches or bark. Comparatively dense populations exist in walls of terraced vineyards and fruit orchards (6). A dense population of the host species is a prerequisite for the survival of this slavemaker ant. The biology (7,8) and slave raids (10) of E. ravouxi have been described in some detail.

SCIENTIFIC INTEREST AND POTENTIAL VALUE E. ravouxi represents an early evolutionary stage in the development of dulosis, or slave-making, within the genus. Other Epimyrma species have reduced numbers of workers and depend increasingly heavily on captured workers, while some have no workers of their own at all (5). It is not yet known whether all the described populations belong to the same subspecies. In Austria E. ravouxi was found with Leptothorax affinis as its host, rather than the usual L. unifasciatus or L. nigriceps (6). Questions of host specificity of slavemaker ants could be usefully investigated with E. ravouxi (6).

THREATS TO SURVIVAL In near-natural habitats such as rocky slopes in the southern Alps, the survival of E. ravouxi is not seriously threatened. However, in large parts of the area, this ant prefers to dwell within walls of terraced vineyards. Recent developments of viticulture, particularly in southern Germany, have led to the systematic destruction of walls in order to enlarge the vineyards and to facilitate mechanical cultivation (6).

CONSERVATION MEASURES TAKEN No specific conservation measures have been taken for E. ravouxi.

CONSERVATION MEASURES PROPOSED Several localities with dense populations of E. ravouxi, e.g., in the Tauber Valley of Bavaria, the Swiss Valley, and in the vicinity of Nyons in Drôme, France, should be surveyed for reserves that would meet the requirements of this rare ant, with special emphasis on protection of terrace walls (6).

REFERENCES 1. André, E. (1896). Description d'une nouvelle fourmi de

France. Bull. Soc. Ent. France 2: 367-368.

2. Buschinger, A. (1971). Zur Verbreitung und Lebensweise sozialparasitischer Ameisen des Schweizer Wallis (Hym., Formicidae). Zool. Anz. 186: 47-59.

3. Buschinger, A. (in press). Epimyrma goesswaldi Menozzi 1931 = Epimyrma ravouxi (André 1896) - Morphologischer und biologischer Nachweis der Synonymie (Hym., Formicidae). Zool. Anz.

4. Buschinger, A., Ehrhardt, W. and Fischer, K. (1981). Doronomyrmex pacis, Epimyma stumperi und E. goesswaldi (Hym., Formicidae) neu für Frankreich. Insectes Soc. 28: 67-70.

5. Buschinger, A., and Winter, U. (in press). Population studies of the dulotic ant, Epimyrma ravouxi, and the degenerate slavemaker, E. kraussei (Hymenoptera: Formicidae). Ent. Gen.

6. Buschinger, A. (1982). In litt., 4 February.

7. Gösswald, K. (1930). Die Biologie einer neuen Epimyrmaart aus dem mittleren Maingebiet. Z. Wiss. Zool. 136: 464-484.

8. Gösswald, K. (1933). Weitere Untersuchungen über die Biologie von Epimyrma gösswaldi Men. und Bemerkungen über andere parasitische Ameisen. Z. Wiss. Zool. 144: 262-288.

9. Menozzi, C. (1931). Revisione del genere Epimyrma Em. (Hymen. Formicidae) e descrizione di una specie inedita di questo genere. Mem. Soc. Ent. Ital. 10: 36-53.

10. Winter, U. (1979). Epimyrma goesswaldi Menozzi eine sklavenhaltende Ameise. Naturwissenschaften 66: 581.

We are grateful to A. Buschinger for the original draft on this species, and to B. Bolton and D. Cherix for further comment.

Formica aquilonia Yarrow, 1955
F. lugubris Zetterstedt, 1838
F. polyctena Förster, 1850
F. pratensis Retzius, 1753
F. rufa Linnaeus, 1758

Phylum	ARTHROPODA	Order	HYMENOPTERA
Class	INSECTA	Family	FORMICIDAE

SUMMARY The wood ants of Europe include five closely related species. Highly valued for a number of services they perform in ecosystems, including control of pest insects and soil formation, all of these ants are declining over much of their range. Already subject to a great deal of conservation attention, they still require care to ensure that their numbers remain high enough to be of value.

DESCRIPTION Wood ants are rather large (6-11 mm), varying in colour from black-auburn to bright reddish brown. There are three castes, workers, males and females (queens). Keys, descriptions and related data for the various component species may be found in several sources (3,10,11,13,14,17,18).

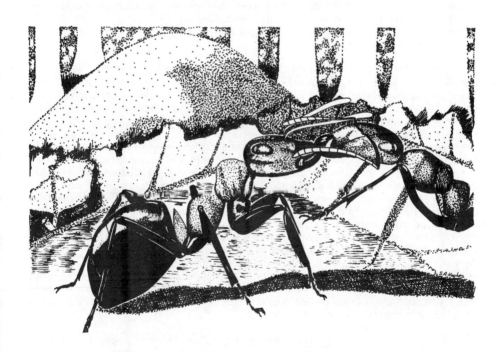

Red Wood Ant (Formica rufa) workers
outside their nest on the forest edge.

DISTRIBUTION While all five species of wood ants overlap (except in Britain, where F. polyctena is absent), each has its own specific range. Details and maps of these have been published (2,16). The distribution of wood ants includes much of northern and central Europe, the Caucasus, Siberia and North America.

POPULATION Thought to be declining in most areas, each of these species can be very numerous under the proper conditions. F. lugubris in particular is known to form enormous supercolonies involving hundreds of nests distributed over many hectares (5). Normal colonies are much smaller, but even an average nest contains 200 000 to 1 000 000 ants depending on the species (1).

HABITAT AND ECOLOGY There is a large body of literature on these ants, and much debate over the characterization of the individual species (7). F. rufa occurs generally in lowlands within or along the edges of forest. F. polyctena is also essentially a lowland forest dweller. F. lugubris lives at higher altitudes, e.g. at 1200-1400 m in the Swiss Jura, where it can occur in high density (5). F. aquilonia is another highland species, restricted in range in central Europe, but abundant in Fennoscandia and northern U.S.S.R. F. pratensis occurs in low altitude grasslands and meadows. F. polyctena, F. lugubris and F. aquilonia normally have many queens in each nest and are able to build polydomous colonies (many colonies bound together by ant tracks). The nests usually consist of large mounds of pine needles, twigs, moss or other debris situated above the underground living area. Each species has peculiarities of colony structure and behaviour, but these differences are not always obvious (6). All species feed on terrestrial and arboreal insects and on aphid honeydew, activities that relate to their economic and ecological value (1). Less attention has been paid to the ecology of these ants than to their systematics, although ecology and ethology give important clues to species relationships (17).

SCIENTIFIC INTEREST AND POTENTIAL VALUE Where they occur, wood ants are dominant predators. This role is beneficial to human needs, as has been asserted and elaborated by many workers (7,8,15). The action of wood ants in the forest ecosystem has been summarized (1) stressing these factors: a) maintenance of biological equilibrium of the forest - a large proportion of wood ants' prey consists of pest insects and if any species undergoes a population eruption, wood ants are more likely to prey upon it heavily and thereby help to restore balance; b) improving soil quality - as a result of their numbers and incessant activity, wood ants contribute to aeration, drainage, humidification and enrichment of forest and grassland soils; c) dispersal of plants - the ants collect seeds, many of which are lost or discarded from the nest; d) increasing the yield of honey - wood ants tend aphids and subsequently collect the honeydew they produce; a great surplus is left and is used to advantage by bees. Forest honey production can be 2.5 times greater in a wood with these ants than in one without them (1). The magnitude of these useful activities was suggested by one study of a large nest of F. rufa in Germany (1). On an area of 0.5 ha, the ants collected 6 000 000 insects, of which 400 000 could be considered pests in that situation, 200 litres of aphid honeydew, and 50 000 seeds. Clearly the biocontrol activity of wood ants furnishes a cheap and preferable alternative to heavy application of chemical insecticides in areas where their numbers remain high.

The supercolonies of F. lugubris known from the Jura (5) are of special scientific interest due to their previously unsuspected structure and enormous proportions. The largest of these involves some 1200 nests intercommunicating over an area of 70 ha by over 100 km of ant tracks.

THREATS TO SURVIVAL Many studies and reports show that wood ants are declining in many parts of their range (1,4,7,13). The large-scale disturbances usually involve urban expansion and pressure, land use change and forest exploitation (6). Specific destructive actions that have been observed include the removal of ant brood from the nests by pheasant and fish breeders, inadvertent destruction of nests by walkers with sticks, who do not realize the importance of thermal equilibrium in the nests; careless forestry operations, and the use of insecticides near ant colonies (1). Afforestation with conifers in British woods containing ants could threaten the already sporadic occurrence of wood ants by failing to support the necessary aphids (6). Additional factors, not fully understood, are probably involved in the overall retreat of these species.

CONSERVATION MEASURES TAKEN Wood ants have been protected by law in a number of countries, such as Switzerland (4), and included in the national red data books in others, including Belgium (19) and Luxembourg (12). A major project 'Ant Conservation in Central Europe', organized in 1977 by WWF-Switzerland, was aimed chiefly at education of the public and training of resource managers and teachers in the methods of practical wood ant conservation. A booklet on wood ant conservation, containing instructions for artificial nest emplacement and other methods, was one tangible result of this project (4). The originator of the project, Dr. K. Gosswald, has also established the Ameisenschutzwarte (Ant Protection Trust) in Wurzburg, B.R.D. This centre is engaged in studies of the reasons for disappearance of wood ants, and protection and translocation of nests (1,6,9). Nests are actively protected and information distributed to foresters and beekeepers in Italy (6). In 1958, the International Organization for Biological Control, Western Palearctic Regional Section, established a working group to study all aspects of wood ant ecology and conservation. Meetings of the working group are held periodically to co-ordinate the research and disseminate new findings (15). In 1963 the group noticed that in many areas the ants were declining and threatened, and in 1964 the Council of Europe encouraged its member countries to protect their wood ants from damage or destruction (1). The same goal has recently been named a high priority of the IUCN/SSC Ant Specialist Group (6).

CONSERVATION MEASURES PROPOSED All of the measures already begun need to be intensified. Additional research is required on distribution, relationships among the species, ecological requirements and behaviour. Species-specific survival factors require better understanding. Colonies subject to heavy human pressure should in all cases receive protection, and the notable Swiss supercolonies should be designated as reserves. Ant awareness on the part of foresters, wildlife services, beekeepers and the public should be raised through a constant flow of accurate information that appeals to their interests. The involvement of the public in practical ant conservation, as already demonstrated in Switzerland and West Germany, should be emulated elsewhere and accelerated wherever the wood ants are under threat (1,4,6).

REFERENCES
1. Anon. (1977). Wood ants. Protection de la Nature (4/77). Ligue Suisse pour la Protection de la Nature.
2. Barrett, K.E.J. (1968). A survey of the distribution and present status of the wood-ant, Formica rufa L. in England and Wales. Trans. Brit. Ent. Nat. Hist. Soc. 17: 217-233.
3. Betrem, J.G. (1960). Uber die Systematik der Formica rufa Gruppe. Tijdschr. Ent. 103: 51-81.
4. Cherix, D. (1977). Les fourmis des bois et leur protection. WWF Suisse/CSEE, Zurich. 32 pp.
5. Cherix, D. (1980). Note preliminaire sur la structure, la phenologie et le regime alimentaire d'une super-colonie de Formica lugubris Zett. Insectes Soc. 27: 226-236.
6. Cherix, D. (1982). In litt., 4 February.
7. Cotti, G. (1963). Bibliografia ragionata 1930-1961 del gruppo Formica rufa. Collona Verde 8: 413 pp.
8. Gosswald, K. (1951). Die Waldameise im Dienste der Waldhygeine. Metta Kinau, Luneburg. 160 pp.

9. Gosswald, K. (1971). Ueber den Schutz von Nestern der Waldameisen. Merkbl. z. Waldhygeine Nr. 4 (2.Auflage). Verl. Arbeitsgemeinhchaft z. Forder. Waldhygeine. Würzburg. 34 pp.

10. Kutter, H. (1975). Die Waldameisen der Turkei. Bull. Soc. Ent. Suisse 48: 159-163.

11. Kutter, H. (1977). Formicidae-Hymenoptera, Insecta Helvetica 6. Soc. Ent. Suisse, Zurich. 298 pp.

12. Mousset, A. and Pelles, A. (1975). Wirbellose Tiere: Insekten, Ameisen. Rote Waldameise. Luxembourg Red Data Book, pp. 51-52.

13. Pamilo, P., Rosengren, R., Vepsalainen, K., Varvio-Aho, S., and Pisarski, B. (1978). Population genetics of Formica ants. I. Patterns of enzyme gene variation. Hereditas 89: 233-248.

14. Pamilo, P., Rosengren, R., Vepsalainen, K., Varvio-Aho, S., and Pasarski, B. (1979). Population genetics of Formica ants. II. Genic differentiation between species. Ann. Ent. Fenn. 45: 65-76.

15. Pavan, M. (1979). Comptes rendus de la réunion des groupes de travail "Formica rufa" et vertébrés prédateurs des insectes de l'OILB (Varenna, Italie, 1978). Bull. SROP II-3, 514 pp.

16. Ronchetti, G. (1980). Distribution des fourmis du groupe Formica rufa en Europe (5 cartes), deuxième édition, OILB SROP group de travail "Formica rufa".

17. Rosengren, R. and Cherix, D. (1981). The pupa-carrying test as a taxonomic tool in the Formica rufa group. In Howse, P.E. and Clement, J.L. (Eds), Biosystematics of Social Insects. Academic Press, New York and London. Pp. 263-281.

18. Yarrow, R.H. (1955). The British ants allied to Formica rufa L. Trans. Brit. Ent. Nat. Hist. Soc. 12: 1-48.

19. Leclercq, J., Gaspar, C., Marchal, J-L., Verstraeten, C. and Wonville, C. (1980). Analyse des 1600 premières cartes de l'atlas provisoire des insectes de Belgique, et première liste rouge d'insectes menacés dans la faune Belge. Notes Fauniques de Gembloux No. 4: 1-104.

We are grateful to D. Cherix for material used in this review, and to B. Bolton and C.A. Collingwood for further comment.

JAPANESE WOOD ANT SUPERCOLONY   THREATENED PHENOMENON

Formica yessensis Forel, 1901

Phylum   ARTHROPODA   Order   HYMENOPTERA

Class   INSECTA   Family   FORMICIDAE

SUMMARY A huge supercolony of 45 000 closely related Formica yessensis nests exists on the Ishikari Coast, Hokkaido, Japan. Such a colony type is extremely aberrant from the basic life form of ants, introducing some interesting sociobiological questions. The colony may not survive the building of an industrial area and major harbour nearby.

DESCRIPTION Formica yessensis is a wood ant, fairly large in total length (6-10 mm) and varying in colour from orange to brown. As in the wood ants of Europe, the colonies consist of several thousand individuals divided into three castes, workers, males and females (queens). This species is similar in general appearance to Formica truncorum, but has fewer hairs on the body (2).

DISTRIBUTION F. yessensis occurs from mid- to northern Japan, and in Korea. This particular supercolony is situated on the Ishikari Coast, Hokkaido, Japan.

POPULATION There are an estimated 45 000 nests in the supercolony, containing 306 million workers and 1.1 million queens (1). The nests are spread over an area of about 270 ha (1).

HABITAT AND ECOLOGY The term supercolony refers to the fact that all the individuals in the estimated 45 000 nests are very closely related, and interchangeable between nests. This situation comes about by a gradual expansion from a founding colony. In the first instance, queenless branch colonies, or way-stations, may be built on foraging routes. These queenless offshoots house a number of workers, which may adopt a young queen after her mating flight from the founding colony, or an older queen which has migrated from the parent colony. Workers may continue to be interchanged between the colonies, and the supercolony gradually expands. The usual habitat of Formica yessensis is mountain woodland, but this supercolony has over 90 per cent of its nests in a sandy coastal strip between 100 and 250 m from the shore (1). In this zone Rosa rugosa, Calystegia kobomugi, Equisetum hiemale and the grass Miscanthus sinensis are prevalent (1). At this site F. yessensis prefers microhabitats with Poa pratensis, E. hiemale, Lathyrus maritimus and M. sinensis. The habitat is sunny, and the main food source for the ants is honeydew from aphids proliferating on M. sinensis and the occasional Quercus dentata sapling encroaching from the oak woods further inland. There are 16 species of ants in this coastal area, and there is evidence that the aggressive F. yessensis prevents intrusion of certain species into the coastal grassland (1).

SCIENTIFIC INTEREST AND POTENTIAL VALUE Most ant societies are monogynous, i.e. they have only one queen (3). However, this supercolony is one of the biggest ant colonies in the world, believed to contain over a million queens laying eggs. This raises some sociobiological questions, especially in relation to the evolution of polygyny and to kinship among the individuals forming the same society (2,3).

THREATS TO SURVIVAL The species Formica yessensis is not under threat, but the unusual phenomenon of this enormous supercolony may be threatened by an industrial area involving a major harbour which is under construction on this coast, scheduled for completion around 1990 (2). If the plans go ahead, the development will destroy the majority of nests in the supercolony (1).

CONSERVATION MEASURES TAKEN According to the development project, the forest of Quercus dentata, in which some nests of F. yessensis are to be found, will be left intact after the completion of the industrial area (2).

CONSERVATION MEASURES PROPOSED Not only the oak forest but also the grassland should be protected as a reserve. A detailed survey of management needs should be carried out, and the authorities responsible for the development made aware of them. If a co-operative agreement to protect this phenomenon cannot be reached, legislative means should be sought. Designation of the supercolony as a national monument seems an appropriate step (2).

REFERENCES 1. Higashi, S. and Yamauchi, K. (1979). Influence of a supercolonial ant Formica (Formica) yessensis Forel on the distribution of other ants in Ishikari Coast. Jap. J. Ecol. 29: 257-264.
2. Higashi, S. (1982). In litt., 27 January.
3. Hölldobler, B. and Wilson, E.O. (1977). The number of queens: an important trait in ant evolution. Naturwissenschaften 64: 8-15.

We are grateful to S. Higashi for providing this information, and to B. Bolton and D. Cherix for further comment.

Leptothorax goesswaldi Kutter, 1967

Phylum	ARTHROPODA	Order	HYMENOPTERA
Class	INSECTA	Family	FORMICIDAE

SUMMARY Leptothorax goesswaldi is one of three closely related inquiline ants parasitic on colonies of Leptothorax acervorum. The species is known from an extremely limited area within the Swiss Valley (Valais), and is of interest with respect to the evolution of social parasitism and species segregation.

DESCRIPTION Leptothorax goesswaldi is a minute monogynous ant which lacks a worker caste and lives in a polygynous host society. Young females of L. goesswaldi produce a sexual pheromone in their poison glands which attracts and stimulates the males (4). The sexuals are morphologically very similar to the host species (8).

DISTRIBUTION Known from only two localities in the Swiss Alps, Saas-Fee and Bellwald, both in the canton of Valais. The species may be more widespread, but searches in other areas where the related inquilines Leptothorax kutteri and Doronomyrmex pacis are quite commonly found, did not reveal L. goesswaldi (2,7).

POPULATION Unknown.

HABITAT AND ECOLOGY L. goesswaldi was found in 1970, 1973 and 1979 in the close vicinity of Bellwald at an altitude of 1600 m (2,4,6). The type material was found in 1950 at Saas Fee at 1800 m (8). The ant lives in large colonies of Leptothorax acervorum in the bark of pine and larch, preferably in more or less rotten stumps of these trees (4). Colonies of L. goesswaldi have been kept alive in culture for up to several months (4).

SCIENTIFIC INTEREST AND POTENTIAL VALUE L. goesswaldi, together with L. kutteri and Doronomyrmex pacis, form a group of closely related and often sympatrically occurring monogynous inquilines which live in polygynous colonies of the common host species L. acervorum. Their biology suggests that they are phylogenetically derived directly from a common, polygynous, independent ancestor (1). They are so closely related to each other that under experimental conditions cross-breeding was possible between all three species, resulting in intermediate females (3,5). In the field, cross-breeding seems to be prevented exclusively by the fact that the three species exhibit sexual behaviour during different hours of the day. A further interesting feature is that cross-breeding is possible despite the karyotypes being different (5). Finally it is an example of three species competing within the same ecological niche in the two localities where L. goesswaldi is known to occur (6).

THREATS TO SURVIVAL It is impossible to say to what extent Goesswald's Inquiline Ant is threatened. A much larger range is known for L. kutteri and D. pacis, but the type locality of L. goesswaldi is covered by a Saas Fee car park, and the second locality at Bellwald is seriously at risk from the construction of holiday homes in all the sites where it has been found. Since L. goesswaldi seems to prefer the open shoulders of southern mountain slopes, which are also the preferred sites for human settlement, it seems likely that the species is seriously endangered (6).

CONSERVATION MEASURES PROPOSED A systematic search for this ant in appropriate areas of the Alps should be undertaken. Further sites, if they exist, should be incorporated within protected areas. The Bellwald colony should be monitored and, if possible, development there steered away from the critical habitat. The occurrence of L. goesswaldi as well as of L. kutteri or Doronomyrmex pacis is an indicator of long-term undisturbed and near-natural habitat. Their protection, therefore, would also favour numerous other uncommon invertebrates (6).

REFERENCES 1. Buschinger, A. (1970). Neue Vorstellungen zur Evolution des Sozialparasitismus und der Dulosis bei Ameisen (Hym., Formicidae). Biol. Zbl. 88: 273-299.

2. Buschinger, A. (1971). Zur Verbreitung und Lebensweise sozialparasitischer Ameisen des Schweizer Wallis (Hym., Formicidae). Zool. Anz. 186: 47-59.

3. Buschinger, A. (1972). Kreuzung zweier sozialparasitischer Ameisenarten, Doronomyrmex pacis Kutter und Leptothroax kutteri Buschinger (Hym., Formicidae). Zool. Anz. 189: 169-179.

4. Buschinger, A. (1974). Zur Biologie der sozialparasitischen Ameise Leptothorax goesswaldi Kutter (Hym., Formicidae). Insectes Soc. 21: 133-144.

5. Buschinger, A. (1981). Biological and systematic relationships of social parasitic Leptothoracini from Europe and North America. In Howse, P.E. and Clement, J.-L. (Eds), Biosystematics of Social Insects, The Systematics Association Special Volume 19. 346 pp.

6. Buschinger, A. (1982). In litt., 4 February.

7. Buschinger, A., Ehrhardt, W. and Fischer, K. (1981). Doronomyrmex pacis, Epimyrma stumperi and E. goesswaldi (Hym., Formicidae) neu für Frankreich Insectes Soc. 28: 67-70.

8. Kutter, H. (1967). Beschreibung neuer Sozialparasiten von Leptothorax acervorum F. (Formicidae). Mitt. Schweiz. Ent. Ges. 40: 78-91.

We are grateful to A. Buschinger for the original draft on this species, and to B. Bolton and D. Cherix for further comment.

Nothomyrmecia macrops Clark, 1934

Phylum	ARTHROPODA	Order	HYMENOPTERA
Class	INSECTA	Family	FORMICIDAE

SUMMARY The only living representative of the subfamily Nothomyrmeciinae, Nothomyrmecia macrops is arguably the most primitive living ant. It is now known only from several sites within an area of less than 1 km^2 in South Australia.

DESCRIPTION The original description of the worker in 1934 (3) was amended (2) and considerably expanded in the light of new material (6). Significant features of the queen, male, larval forms (6,9) and eggs (9) have been reviewed and described. Workers are about 1 cm long, golden yellow, with large dark eyes but vestigial ocelli. They have long jaws, a single waist node, and a strong and effective sting. The stridulatory organ, which is capable of producing a barely audible chirping sound, is ventrally placed on the abdomen. This is a very peculiar feature, since in all other stridulating Hymenoptera such ' organs are dorsal. The ants have 92 chromosomes, the highest number known in the Hymenoptera.

The queens resemble the workers, but are slightly larger. They have ocelli, and the complex thoracic structure typically associated in ants with the deciduously winged condition of this caste. However, the wings are very short and peculiarly trimmed; they are not capable of maintaining the ants in flight. In most other fundamental details Nothomyrmecia resembles species of Myrmecia. These two Australian genera are believed to represent relatively unchanged survivors of the basal stocks of two lines of ant evolution which diverged in the late Mesozoic or early Tertiary times. The most important features distinguishing these lineages involve details of the form and fusion of the skeletal plates of the fourth abdominal segment (6,7).

DISTRIBUTION Restricted to South and Western Australia (6). The types were collected in Western Australia, along the track between Balladonia and Thomas River near Mt Ragged, at the west end of the Great Australian Bight (2). The significance of Nothomyrmecia as possibly the most primitive living ant (1) triggered many searches (2) and eventual re-discovery in 1977 near Ceduna, in the Eyre Peninsula, South Australia (6). The first two collections were therefore 1000 km and 46 years apart. No rediscoveries have been made in other areas, but Nothomyrmecia could be widespread in areas of mallee vegetation (dominated by Eucalyptus spp.) across southern Australia (7).

POPULATION Unknown. The reduced flight capacity of queens might be an adaptation related to distribution in small localized populations (6).

HABITAT AND ECOLOGY The sites presently known to be occupied by N. macrops are in tall mallee woodland dominated by Eucalyptus oleosa. The rather sparse canopies of these trees are virtually contiguous, and there are few herbs or grasses among the thin, but continuous, leaf litter layer (7). The sites are separated by roads, a railway line, a small settlement and wheatfields, indicating disruption of a continuous local population by European settlers since the turn of the century (7). Nothomyrmecia nests are in the soil and have obscure non-specific entrances. Above-ground activity is nocturnal, and there is no certain daytime evidence of the presence of the ants. In this area the nights are very cold relative to the days, and Nothomyrmecia forages while most other local ants are inactive. Foraging workers ascend Eucalyptus trees within a few metres of their nests and hunt cold-torpid insects among the vegetation. They leave the nests shortly after nightfall and return just before dawn (6,7). The ants apparently navigate while on the ground by using the tree canopy silhouette as a map (7). They do not appear to use scent cues or trails, even when near the nest entrance (7). Navigation while on the trees seems to be entirely geotactic (7). Impounded workers retain for up to three days the ability to return to their nest from the point of capture (7). Mating activity has not yet been observed. Winged virgin queens and males are produced in colonies in late spring and early summer and are probably released in late summer. Founding queens forage like workers while rearing their first brood, and several may co-operate in colony founding. Mature colonies excavated to date, however, contained only single mother queens (6,7).

SCIENTIFIC INTEREST AND POTENTIAL VALUE Given its generally accepted status as the the most primitive living ant, the scientific interest of Nothomyrmecia is self-evident (6). In addition the ant has proved to be easily cultured, and could become a useful experimental animal, especially for studies of learning in insects, and of the physiology of nocturnal vision. An intensive and continuous programme of studies on Nothomyrmecia is in progress, involving Australian, American and European ant specialists (4-9).

THREATS TO SURVIVAL Habitat destruction has almost certainly fragmented the local distribution of Nothomyrmecia in the Ceduna area. The site of rediscovery was bulldozed and burnt during the installation of an underground telephone line two years after studies began. The population of Nothomyrmecia at this locality was almost wiped out. Three other nearby sites are now known, two of them larger and more populous than the original. All are under the surveillance of the land owner, and one has been fenced as a reserved study area. Nothomyrmecia nests are always found beneath Eucalyptus trees in fairly dense stands. The ants may depend on overhead canopy for navigation and thus might not survive tree-clearing. They would be safe from fire while in the nests, but bush fires at night could destroy colonies indirectly by killing large numbers of foraging workers, and especially colony founding queens (7). Fires large enough to kill the trees would probably wipe out the whole ant population (7).

CONSERVATION MEASURES TAKEN N. macrops, in absentia since 1931, has been listed as an endangered species in Western Australia. No public conservation measures have been taken elsewhere, though a strong measure of protection is afforded by the interest and co-operation of local residents in South Australia (7).

CONSERVATION MEASURES PROPOSED Formal protection of the presently known habitat would not be practicable. It is hoped that further populations of Nothomyrmecia will be located in less heavily settled areas, where protection from habitat destruction and aerial insecticide spraying can be assured. Much nocturnal search for new sites has been done in a continuing programme of study (7).

REFERENCES   1. Brown, W.L. Jr. (1954). Remarks on the internal phylogeny and subfamily classification of the family Formicidae. Insectes Soc. 1: 21-31.
2. Brown, W.L. Jr. and Wilson, E.O. (1959). The search for Nothomyrmecia. West. Aust. Nat. 2: 25-30.
3. Clark, J. (1934). Notes on Australian ants, with descriptions of new species and a new genus. Mem. Natn. Mus. Victoria. 8: 5-20.
4. Hölldobler, B. and Engel, H. (1978). Tergal and sternal glands in ants. Psyche 85: 285-330.
5. Kugler, C. (1980). The sting apparatus in the primitive ants Nothomyrmecia and Myrmecia. J. Aust. Ent. Soc. 19: 263-267.
6. Taylor, R.W. (1978). Nothomyrmecia macrops: a

living-fossil ant rediscovered. Science 201: 979-985.

7. Taylor, R.W. (1982). In litt. 2nd June.
8. Ward, P.S. and Taylor, R.W. (1981). Allozyme variation, colony structure and genetic relatedness in the primitive ant Nothomyrmecia macrops Clark. J. Aust. Ent. Soc. 20: 177-183.
9. Wheeler, G.C. Wheeler, J. and Taylor, R.W. (1980). The larval and egg stages of the primitive ant Nothomyrmecia macrops Clark. J. Aust. Ent. Soc. 19: 131-137.

We are very grateful to R.W. Taylor for the original draft on this species, and to B. Bolton and D. Cherix for further comment.

WALLACE'S GIANT BEE   INSUFFICIENTLY KNOWN

Chalicodoma pluto (Smith, 1861)

Phylum   ARTHROPODA   Order   HYMENOPTERA

Class   INSECTA   Family   MEGACHILIDAE

SUMMARY Chalicodoma pluto is the world's largest bee. It is especially
interesting because it nests communally, apparently always in inhabited arboreal
termite nests. A.R. Wallace collected the type of C. pluto in 1858, from Bacan,
Indonesia. Thereafter it remained unknown and was presumed extinct until
rediscovered on other Moluccan Islands in 1981.

DESCRIPTION Chalicodoma pluto has the general appearance of a bumble bee.
The original description was for a female (5), distinguished from the male by large
size (c. 39 mm long), enlarged mandibles, white pubescence on the genae and a
tooth on the hypostomal carina. The smaller male (c. 24 mm long) was unknown
until 1981 and has normal megachilid mandibles and rufous pubescence on the head
(2).

DISTRIBUTION Bacan, an island in the north Moluccas of Indonesia, is the type
locality for C. pluto (5). A single female was collected there by Wallace in 1858
(7). The range of C. pluto also includes Soasiu (Tidore) and Halmahera, also in the
Moluccas, where it was recently rediscovered (2).

POPULATION Unknown, but uncommon throughout its range (2).

HABITAT AND ECOLOGY The female C. pluto uses her large mandibles to gather
resin from fissures in the trunk of dipterocerp trees, including 'o boulamo', a
Tobelorese name probably for Anisoptera thurifera (6). The resin is mixed with
wood chips and used to line tunnels in the inhabited arboreal nests of a termite,
Microcerotermes sp. (Termitidae). Nest dissections have shown no physical

communication between the termite and the bees (2). The nectar and pollen sources for C. pluto are not known, but other congeneric species from the area are said to visit Crotalaria (Leguminosae) (3). The meloid beetle Zonitoschema sp. nov. parasitizes C. pluto larvae and pupae (2).

C. pluto is found in lowland (altitude less than 200 m) primary and secondary forest where Microcerotermes nests are abundant. The bee was not found nesting elsewhere, and the association with termites is probably obligatory. Despite the abundance of the termite nests, the bee appears to be rare. Only seven nests were found during ten months of field work in the north Moluccas (2). Local people had never seen the bee prior to its rediscovery, yet they claimed a specific folk name, 'o ofungu-ma-koana' (2).

SCIENTIFIC INTEREST AND POTENTIAL VALUE In addition to its unique status as the world's largest bee (3), C. pluto is the only megachilid yet reported to live in association with termites. However, termite nests are also used by the solitary bee Centris derasa (1), and by various species of the meliponine bees Trigona (4). The communal social organization of Wallace's Giant Bee is also unusual among megachilids, where most species are strictly solitary nesters. Further study is needed to establish the precise level of sociality in C. pluto.

THREATS TO SURVIVAL It is not clear to what extent C. pluto is threatened. The lowland forest is under pressure from timber operations as well as from local shifting agriculture (2). Depletion of the dipterocarp resin sources seems to be the most immediate danger.

CONSERVATION MEASURES TAKEN No specific efforts have been made on behalf of this bee. Nature conservation initiatives on the islands are still in a formative stage, but some of them may help incidentally to conserve the species (2).

CONSERVATION MEASURES PROPOSED A thorough survey is needed to determine the range of C. pluto, and to assess its habitat requirements. If land use appears to threaten the bee it may need to be protected in some way.

REFERENCES 1. Bennett, F. D. (1964). Notes on the nesting site of Centris derasa. Pan-Pacif. Ent. 40: 125-128.
2. Messer, A.C. In litt., 8 April
3. Michener, C.D. (1965). A classification of the bees of the Australian and South Pacific regions. Bull. Am. Mus. Nat. Hist. 130: 1-362, plates 1-15.
4. Michener, C.D. (1974). The social behavior of the bees: a comparative study. Belknap Press, Harvard Univ.
5. Smith, F. (1861). Catalogue of hymenopterous insects collected by Mr A. R. Wallace in the islands of Bachian, Kaisaa, Amboyna, Gilolo and at Dory in New Guinea. J. Proc. Linn. Soc. Zool. 5: 93-143.
6. Taylor, P.M. (1980). Tobelorese ethnobiology: the folk classification of biotic form. Doctoral dissertation, Yale University Department of Anthropology.
7. Wallace, A.R. (1869). The Malay Archipelago. 10th edition. Dover Reprints, Dover.

We are grateful to A.C. Messer for his contribution of these data, and to G.R. Else and C. Vardy for further comment.

'MYRIAPODA'

Centipedes, millipedes, pauropods and symphylans.

INTRODUCTION Four of the five classes of the subphylum Uniramia, the Chilopoda (centipedes, 3000 spp.), Diplopoda (millipedes, 7500 spp.), Symphyla (120 spp.) and Pauropoda (380 spp.) have a body composed of a head and an elongated trunk with many leg-bearing segments (1,10). Because of this similarity they were formerly known as the Myriapoda, a term now used only as an informal collective name since it is recognized that the four groups exhibit marked differences (1,10). Myriapods lack a waxy cuticle to the exoskeleton and live in moist conditions in soil and leaf litter and beneath stones, logs and bark. Centipedes have one pair of legs per segment and prey on other arthropods using a pair of anterior poison claws. Most are 3-6 cm long, but the largest species, the tropical American Scolopendra gigantea, may reach a length of 26 cm (1,7). Millipedes have two pairs of legs per segment and feed on decomposing vegetation. They vary greatly in the number of segments and in size, from minute species only 2 mm long to the giant Spirostreptidae which may reach 28 cm. The Symphyla and Pauropoda feed on leaf litter and soil. Symphyla are 2-10 mm long and have 12 leg-bearing segments covered by 15-22 tergites. Pauropoda are less than 2 mm long and have 11 segments, nine of them with legs (1,10).

SCIENTIFIC INTEREST AND POTENTIAL VALUE Millipede cuticle is hardened not only by tanning, as in insects, but also by the deposition of calcium compounds, as in Crustacea. Local climatic or microclimatic changes may induce mass movements of millipedes and other myriapods and in some cases may result in pest outbreaks. Blaniulus guttulatus (Spotted Snake Millipede) causes damage to potatoes, sugar beet and other field and greenhouse crops in Europe. Certain millipedes are luminous, including Luminodesmus sequoiae of the Sequoia National Forest, Canada. Many millipedes have repugnatorial glands which exude irritant chemicals, sometimes in a jet or spray. Constituents include hydrocyanic acid, iodine and quinine, which may be caustic or cause blindness in small mammals and birds (2).

Centipedes in the Scutigeromorpha (Scutigerida) readily shed their legs if attacked, and in some species the detached limbs continue to stridulate. When disturbed, many Geophilomorpha (Geophilida) exude a phosphorescent fluid from glands near the leg coxae. Scolopendromorpha (Scolopendrida) produce poison from similarly situated glands. This group also shows some parental care of the young by the female, although she may devour them when stressed. Many centipedes are cave-dwellers and a few of the Geophilomorpha are marine, living under stones and seaweed at low tide (2,5,6). The Symphyla are of interest in displaying a number of features characteristic of insects. Scutigerella immaculata is a pest of field and greenhouse crops in Europe and South America (2).

THREATS TO SURVIVAL The conservation requirements of myriapods are very poorly known. Because of their cryptic habits and the paucity of studies on their taxonomy and distribution, it is difficult to assess the range of many Diplopoda and Chilopoda (7). Collections in virtually any tropical area will yield new distribution records, if not new species, and it is therefore difficult to know whether an apparently limited distribution is genuine, or due to a lack of study in adjacent areas. For example, the centipedes Arrhabdotus octosulcatus and the brilliantly coloured arboricolous Scolopendra arborea (Scolopendromorpha) are believed to be rare in Sarawak (8), but very few collectors have visited the area.

Millipedes have limited dispersal abilities and there is evidence that they are able to speciate within very restricted areas. Numerous endemics have been discovered in the Usambara Mountains, Tanzania, where 26 out of 41 species found in the Amani region are endemic (see review of Usambara Mts). Widespread speciation is also evident on Madeira (4), and perhaps occurs on many other islands. As a result of their often small ranges, millipede species may be subject to the threat of very localized environmental change. In North Carolina, U.S.A., 20 species of millipedes have been listed as threatened, 14 of which are endemic to the state (3). In addition, a recently described and possibly threatened North Carolina species, Sigmoria areolata, is known only from a very restricted area in Buncombe County despite extensive searches elsewhere (12,13). Millipedes occur frequently in caves and a number of cave endemics have been described (11). In Virginia two threatened millipedes have been listed, both from caves (9). Pseudotremia cavernarum (Cleidogonidae) may be extinct after its single cave habitat at Ellett in Montgomery County was destroyed by a quarrying operation (9). The species may survive in adjacent underground habitats.

CONSERVATION The taxonomy of all the myriapod classes is somewhat neglected (7). Without data on the distribution and abundance of species it is impossible to assess threats or conservation requirements adequately. In view of the fact that millipedes appear to speciate rather readily, they should not be overlooked in local invertebrate surveys, particularly in caves. As is the case for insects and other invertebrates, information from tropical areas of the world is particularly lacking

REFERENCES 1. Barnes, R.D. (1980). Invertebrate Zoology. Saunders College, Philadelphia. 1089 pp.
2. Cloudsley-Thompson, J.L. (1968). Spiders, Scorpions, Centipedes and Mites. Pergamon Press, Oxford. 278 pp.
3. Cooper, J.E., Robinson, S.S. and Funderburg, J.B. (1977). Endangered and Threatened Plants and Animals of North Carolina. North Carolina State Museum of Natural History, Raleigh, North Carolina.
4. Enghoff, H. (1982). The millipede genus Cylindroiulus on Madeira: an insular species swarm (Diplopoda, Iulida, Julidae). Entomologica Scandinavica Suppl. 18: 1-142.
5. Lewis, J.G.E. (1961). The life history and ecology of the littoral centipede Strigamia (=Scolioplanes) maritima (Leach). Proc. Zool. Soc. Lond. 137: 221-248.
6. Lewis, J.G.E. (1962). The ecology, distribution and taxonomy of the centipedes found on the shore in the Plymouth area. J. Mar. Biol. Ass. U.K. 42: 655-664.
7. Lewis, J.G.E. (1980). The Biology of Centipedes. Cambridge University Press, Cambridge. 476 pp.
8. Lewis, J.G.E. (1982). In litt., 1 September.
9. Linzey, D.W. (Ed.) (1979). Endangered and Threatened Plants and Animals of Virginia. Center for Environmental Studies, Blacksburg, Virginia.
10. Parker, S.P. (1982). Synopsis and Classification of Living Organisms. McGraw Hill, New York. 2 vols., 1232 pp.
11. Peck, S.B. (1981). Zoogeography of invertebrate cave faunas in southwestern Puerto Rico. Nat. Spel. Soc. Bull. 43: 70-79.
12. Shelley, R.M. (1982). In litt., 19 January.
13. Shelley, R.M. (1981). A revision of the milliped genus Sigmoria (Polydesmida: Xystodesmidae). Am. Ent. Soc. Mem. 33: 1-139.

PERIPATUS   VULNERABLE

Phylum   ONYCHOPHORA   Order   ONYCHOPHORA

                                         Family   PERIPATOPSIDAE
                                                    PERIPATIDAE

SUMMARY This group of terrestrial 'living fossils' has been described as the 'missing link' between annelids and arthropods, but recent research questions the validity of such a statement. It is now thought more likely that the Onychophora constitute a separate phylum, as the group has many unique characteristics. A few fortunate findings of fossils indicate that representatives of the taxon were probably widespread millions of years ago. Recent Onychophora are found in tropical and subtropical regions, especially in the southern hemisphere, and have a patchy discontinuous distribution typical of a relict fauna. They are usually restricted to humid cryptic habitats such as rotting wood and leaf litter in primary forests and are therefore highly vulnerable to habitat disturbance and in particular to deforestation. Their relatively small populations could be endangered by excessive collecting.

Macroperipatus geayi

DESCRIPTION Representatives of this small group of terrestrial caterpillar-like invertebrates were mistaken for molluscs when first discovered in 1826 (9). In 1853 they were given the name Onychophora, which refers to the paired terminal claws on their appendages (8). Since then about 100 species have been described, divided into two families, the Peripatidae and the Peripatopsidae. The main morphological differences between these taxa are (a) the number of appendages (b) the position of the genital pore and (c) the pigmentation (23).

Onychophorans have a markedly uniform appearance. Their bodies are vermiform, semicylindrical and flat on the ventral side. They range from 0.5 cm to 15 cm in length and have 13-43 pairs of uniform stumpy legs which are inserted laterally (6,17,27). Body segments are not distinctly defined, the surface being covered with a soft, velvety, transversely wrinkled skin with many large and small papillae. The larger papillae terminate in sensory bristles, most of which can be regarded as chemo- or mechanoreceptors (26). Two soft and mobile but non-retractile frontal antennae serve as sensory organs. A simple eye lies at the base of each. The mouth is located ventrally and is equipped wih a rasp-like 'tongue' and horny teeth-bearing jaws (6). A pair of slime-producing glands is situated on either side of the head and is used offensively as well as defensively and to capture food. Slime threads are ejected with great force to entangle an attacker and hold it to the ground (27). Many onychophorans are highly coloured and have orange, brown, blue, green, red or black pigments and are often characteristically patterned. The chitinous cuticle is only 1 μm thick and, unlike that of most arthropods, is flexible. It is moulted frequently and Peripatopsis, for example, has been recorded moulting every 2-3 weeks. Peripatopsis reaches maximum size between its third and fourth year and has a life span of nearly seven years (14).

DISTRIBUTION Onychophorans have a relatively restricted distribution and are found only in the tropics and subtropics, especially in the southern temperate regions (7). Although some species have fairly wide ranges, the majority have very restricted distributions and some are known only from their type localities. The two families never coexist. The Peripatidae are equatorial, occurring between latitudes 23°S and 27°N, while the Peripatopsidae live in the southern hemisphere between latitudes 0° and 47°S especially at the tips of the continents (6,18).

The taxonomy of the Peripatidae is still in some doubt. No species are found further north than the Himalayas, Caribbean and Central Mexico. Seven genera occur in the New World from Mexico and the Caribbean islands (except Cuba) south to Bolivia and Brazil. Peripatus, Oroperipatus, Heteroperipatus, Macroperipatus and Epiperipatus are fairly widespread (29). Three species are endemic to Jamaica, Plicatoperipatus jamaicensis, Macroperipatus insularis and Speleoperipatus spelaeus, the latter restricted to Pedro Great Cave (2,20). Plicatoperipatus and Speleoperipatus are endemic monotypic genera. Mesoperipatus is a monotypic genus restricted to the Congo (5). Typhloperipatus williamsoni is one of the few species known from north of the equator and is found near Rotang on the Dihang River, Assam, in the Himalayan foothills of eastern India (33). Eoperipatus occurs in the Indo-Malaysian region (29).

The family Peripatopsidae has twelve genera (23). South Africa has two endemic genera (Peripatopsis and Opisthopatus); Peripatopsis alba is known only from Wynberg Cave on Table Mountain, P. clavigera only from its type locality in the Knysna region and Opisthopatus roseus from its type locality in the Ngeli Forest (5,18). Chile has two; Australia five and New Guinea one (Paraperipatus) (23). Peripatoides is endemic to New Zealand and includes species with very restricted ranges such as a new species known only from the Twin Forks Cave, Paturau District, and P. suteri from Taranaki (18,23). The twelfth genus, still be be described, contains two species, fairly widespread in Tasmania and New Zealand (6).

POPULATION No data.

HABITAT AND ECOLOGY The nature of the onychophoran respiratory system exerts a strong influence on the type of habitat in which these invertebrates can live (10). Numerous microscopic pores, from which the trachea originate, are scattered all over the surface of the body. Unlike arthropod trachea, onychophoran tracheal spiracles lack closing mechanisms which means that under dry conditions the animals desiccate rapidly (5,18). Humidity is probably the most important single factor in the survival of onychophorans and it is for this reason that they are mostly found in high rainfall regions and are restricted to areas of indigenous forest. It is clear that a rather narrow dry belt could act as an ecological and zoogeographical barrier to these moisture-seeking animals and may explain why some species have such limited distributions (5,18).

Often Onychophora are associated with forests with well established layers of humus, which suits their preference for uniform conditions. They may be found in old, flat, rotting logs, under watersoaked moss covering wet rock walls, in leaf litter and under stones along stream banks (5,31,32,33). In South America some species have been found to inhabit detritus in the lower leaves of bromeliads and hollows in trees in the subtropical cloud forest at high altitude (4,22). Typhloperipatus williamsoni in the Himalayan foothills was reported to be abundant in leaf litter in scrub jungle at 1200-2000 feet (400-700 m) (33).

A high and fairly constant humidity appears to be a key factor although some interesting exceptions are known. Three troglobitic species have been described (Peripatopsis alba from South Africa, Peripatoides n.sp. from New Zealand and Speleoperipatus spelaeus from Jamaica) (5,20). Some species like Peripatoides novaezealandiae are trogloxen: during dry periods they search for caves, where conditions are more attractive at that time than in normal bush habitat (11,16).

In New Zealand Ooperipatus viridimaculatus lives in areas where winters are exceptionally severe with heavy frosts and snow (23). Under adverse conditions, such as drought or cold temperatures, animals are very sluggish or completely inactive, slowing down all body processes and surviving in fine crevices (absence of a rigid exoskeleton enables them to squeeze their bodies into confined spaces) or deep in the soil. They are able to withstand cold or extremely dry periods by becoming torpid for up to three months, returning to normal activity when rain wets the ground or the temperature rises (10,31,32). In South Africa significant numbers of Peripatopsis moseleyi and P. capensis have been found recently in grassveld habitats and it is thought that they pass the dry season, as well as the day time, by going deep into the soil (21). Many of the New World species occur in open-canopy, seasonally deciduous forests or savannas in regions with a pronounced dry season (20,31,32). It is now thought that onychophorans may be as much a part of the fauna of some soils of open habitats and of seasonal forest as they are of the litter of permanently moist forests (21).

Onychophorans are sensitive to light and air currents, showing negatively phototropic reactions which make observation of even their more ordinary activities difficult. Furthermore they are nocturnal, seeking food at night when the air is cool and moist. They feed on crickets, beetle-larvae, woodlice and grasshoppers (24). Some species display a particular preference for termites (32). The cave-dwelling species Speleoperipatus spelaeus from Jamaica was seen feeding on a blind, troglobitic Nelipophygus roach (Blattelidae) (2,20).

Sexes are always separate. Females are larger and stouter compared to males of the same age and often have more legs. The majority of onychophoran species are viviparous and the embryo is nourished by means of a special 'placental' connection to the uterine wall (3). Some species are ovoviviparous and a few species from Australasia are oviparous. The latter deposit their sculptured eggs in damp habitats, while the viviparous species keep the fertilized eggs in their genital tracts where gestation takes from 6 to 13 months (3,18).

SCIENTIFIC INTEREST AND POTENTIAL VALUE The Onychophora are excellent tools for biological research, being especially valuable for the study of evolution, phylogenetic linkages (Annelida-Arthropoda), reproductive biology, comparative physiology, discontinuous distribution and continental drift (21,24). Their distribution has been used to support the belief that American, Australian and African land masses were connected during past geological ages (15), although other factors such as dispersal across water, or the possibility of a wider original distribution during different climatic periods, may also have been involved (6). A fossil dating from the Cambrian some 550 million years ago has been suggested as being a marine onychophoran (3,15).

THREATS TO SURVIVAL   Many onychophorans are dependent on constant humidity, and are limited to forest leaf litter or caves, which are extremely vulnerable habitats. It has been suggested that the present discontinous distribution of the group indicates that it was once more widespread and is now disappearing (6). The restricted distributions of many species also make them very vulnerable to habitat alteration and if indigenous forests are destroyed a number of species will certainly become extinct. For example, in Australia a swamp tea-tree area on the outskirts of Brisbane where onychophorans were collected thirty years ago is now an industrial site (15). Typhloperipatus williamsoni has not been recorded since 1911, when it was first collected. It is now threatened by the spread of 'jhum' (slash and burn) cultivation in the areas where it occurs (3,30).

The rich South African onychophoran fauna is distributed in the narrow coastal belt below the eastern escarpment and south of the Cape fold mountains. Extensive development has occurred in these areas and is threatening the continued existence of some species (18). Forest fires in the introduced pine plantations of Signal Hill near Capetown had, by 1931, caused the virtual extinction of Peripatopsis leonina, known only from this site (13) and there are no recent records of this species (18). Opisthopatus roseus and Peripatopsis clavigera may also be extinct; P. clavigera could not be found at its previously recorded locality at Knysna in 1933 (14,18). P. moseleyi is apparently the only species which has managed to spread from forest or to persist when habitats are altered to plantations or cultivated ground presenting similar physiographic conditions (5).

Onychophora are avidly sought by invertebrate zoologists on account of their scientific interest, and over-collecting could be a threat to a number of species, particularly troglobitic species with small populations such as Peripatopsis alba (12,18,20). Furthermore cave species could be highly susceptible to factors such as excessive pollution or drying out (18).

CONSERVATION MEASURES TAKEN   In South Africa all onychophorans are protected, and may not be collected or exported without a permit (12). A population of Peripatopsis moseleyi occurs in the Cathedral Peaks Forestry Research Reserve in Natal (21). In Australia, although many habitats are being destroyed, a number are probably protected in national parks (15). In Brazil, several species of Peripatus occur in the Federal Biological Reserve of Corrego do Veado in the low altitude coastal region of northern Espirito Santo (19,28). Species have also been recorded from near Rancho Grande in the Aragua National Park in north central Venezuela (4). A subspecies of Peripatus, known only from mountains on St John, U.S. Virgin Islands, occurs within the U.S. Virgin Islands National Park (33).

CONSERVATION MEASURES PROPOSED   Further information on the ecology and distribution of the Onychophora is required. As with many invertebrates small reserves, or extensions to presently existing reserves, may well be sufficient to ensure the survival of many species. Countries where onychophorans occur should control and monitor collecting and encouragement should be given to maintaining captive populations of these animals (24). Captive colonies have been maintained and juveniles born in such cultures, but breeding in captivity has never been observed (32).

REFERENCES   1.   Anon. (1977). Nature Conservation in the Cape. The Dept. of Nature and Environ. Conserv., Cape Provincial Admin., Cape Town. 60 pp.

2. Arnett, R.G. (1961). The Onychophora of Jamaica. Ent. News 72: 213-220.
3. Biswas, S. and Ghosh, A.K. (1976). Impact of shifting cultivation on wildlife in Meghalaya. In Shifting Cultivation in Northeast India. Proceedings of a seminar on the socioeconomic problems of the shifting cultivation in Northeast India with special reference to Meghalaya. North east India Council for Social Sciences Research, Shillong. Pp. 77-79.
4. Beebe, W. and Crane, J. (1947). Ecology of Rancho Grande, a subtropical cloud forest in Northern Venezuela. Zoologica 32(5): 43-59.
5. Brinck, P. (1956). Onychophora, a review of the South African species with a discussion on the significance of the geographical distribution of the group. In Hanström, B., Brinck, P. and Rudebeck, G. (Eds), South African Animal Life: Results of the Lund University Expedition 1950-51. Almquist and Wiksells, Uppsala. Pp. 7-32.
6. Rhuberg, H. (1982). In litt., 30 September.
7. Darlington, P.J. (1965). Biogeography of the Southern End of the World. Harvard University Press, Cambridge, Mass.
8. Grube, E. (1853). Über den Bau von Peripatus edwardsii. Müller's Archiv. Pp. 322-360.
9. Guilding, L. (1826). Mollusca caribbeana, No.2. An account of a new genus of Mollusca. Zool. Jb., Jena 2(11): 437-444.
10. Hardie, R. (1974). The riddle of Peripatus. Austr. Nat. Hist. 18(5): 180-185.
11. Harrison, R.A. (1969). Forest invertebrates. Chapter 19. In Knox, G.A. (Ed.), The Natural History of Canterbury. Reed, Wellington, New Zealand.
12. Hey, D. (1978). The Fauna and Flora of Southern Africa and their Conservation. Dept. of Nature and Environ. Conserv., Cape Provincial Administration, Cape Town, South Africa.
13. Lawrence, R.F. (1931). A new peripatopsid from the Table Mountain caves. Ann. S. Afr. Mus. 11: 165-168.
14. Manton, S.M. (1938). Studies on the Onychophora. 6: The life-history of Peripatopsis. Annls. Mag. Nat. Hist. 1(11): 515-529.
15. Marks, E.N. (1969). The Invertebrates. In Webb, L.J., Whitelock, D. and Brereton, J.L.G. (Eds), The Last of Lands. Milton, Jacaranda.
16. May, B.M. (1963). New Zealand cave fauna. Trans. Roy. Soc., N.Z. 3(19): 185.
17. Meglitsch, P.A. (1972). Invertebrate Zoology. Phylum Onychophora. Oxford Univ. Press. New York, London, Toronto, 2nd Ed. Pp. 440-445.
18. Newlands, G. and Ruhberg, H. (1978). Onychophora. In Werger, M.J.A. (Ed.), Biogeography and Ecology of Southern Africa. W. Junk, The Hague. Pp. 679-684.
19. Padua, M.T.J., Magnanini, A. and Mittermeier, R.A. (1974). Brazil's National Parks. Oryx 12(4): 452-464.
20. Peck, S.B. (1975). A review of the New World Onychophora with the description of a new cavernicolous genus and species from Jamaica. Psyche 82 (3-4): 341-260.
21. Peck, S.B. and Endrody-Youngo, S. (in press). The onychophorans Peripatopsis capensis and P. moselyi in South African grassveld habitats (Onychophora, Peripatopsidae). Ann. Trans. Mus. Pretoria, S. Africa.
22. Picado, M.C. (1911). Sur un habitat nouveau de Peripatus.

Bull. Mus. Nat. Hist. Paris. Pp. 415-416.

23. Ruhberg, H. (in prep.). Die Peripatopsidae (Onychophora). Systematik, Chorologie und phylogenetische Aspekte. Verh. Naturwiss. Ver. Hamburg (NF), Dissertation.

24. Ruhberg, H. and Nutting, W.B. (1980). Onychophora: feeding, structure, function, behaviour and maintainence (Pararthropoda). Verh. Naturwiss. Ver. Hamburg (NF) 24(1): 79-87.

25. Ruhberg, H. and Storch, V. (1977). Über Wehrdrüsen von Peripatopsis moseleyi (Onychophora). Zool. Anz. Jena. 198 (1/2): 9-19.

26. Storch, V. and Ruhberg, H. (1977). Fine structure of the sensilla of Peripatopsis moseleyi (Onchophora). Cell Tiss. Res. 177: 539-553.

27. Tryon, H. (1887). On Peripatus and its occurrence in Australia. Proc. Roy. Soc. Queensland 4: 78-80.

28. IUCN (1982). IUCN Directory of Neotropical Protected Areas. Tycooly International Publishing Ltd., Dublin. Pp. 94-95.

29. Peck, S.B. (1982). Onychophora. In Parker, S.P. (Ed.), Synopsis and Classification of Living Organisms. McGraw Hill Co. Pp. 729-730.

30. Ghosh, A.K. (1981). In litt., 30 Dec.

31. Lavallard, R., Campiglia, S., Alvares, E.P. and Valle, C.M.C. (1975). Contribution à la biologie de Peripatus acacioi Marcus et Marcus (Onychophore) III. Etude descriptive de l'habitat. Vie et Milieu 25: 87-118.

32. Van der Lande, V. (1978). The occurrence, culture and reproduction of Peripatoides gilesii Spencer (Onychophora) on the Swan coastal plain. West Aust. Natur. 14: 29-36.

33. Canoy, M. (1981). In litt., 7 April.

We are very grateful to H. Ruhberg who wrote a substantial part of this data sheet and to A.K. Ghosh and S.B. Peck for providing data and helpful comments.

TARDIGRADA

Water Bears

INTRODUCTION Currently 514 tardigrade species are known, ranging in size from 50 to 1200 μm in length. These bilaterally symmetrical animals generally have four pairs of legs, each of which terminates with a claw or disc-shaped appendage. When observed in water they demonstrate a lumbering movement which has earned them the name Water Bears or tardigrades (tardus, L = slow; gradus, L = step). The body is covered with a cuticle which in many cases has taxonomically significant variations (such as plates, papillae, tubercles, spines, cirri) in its surface. Pigmentation varies from pale, colourless juveniles to dense, darkly coloured adults (9,16,22). Growth is through a series of 12 successive moults in a life-span. The eggs are either laid in the female's moulted cuticle or shed freely into the surrounding environment. Many of the eggs have particular surface characteristics which vary even within species (9,22,24,26). In non-marine tardigrades, especially Echiniscus, males are frequently absent and the females reproduce by parthenogensis (16,26).

Echiniscus sp.

While many tardigrade species are considered cosmopolitan (e.g. Milnesium tardigradumm and Macrobiotus hufelandi), others possibly may have more restricted distributions. Echiniscus tympanista, for example, is currently known only from the British Isles although it may well occur elsewhere (16) and the rare Angursa bicuspis is known only from two individuals collected at two separate distinct sites (20). There are probably many more examples but this disparity in apparent distribution may largely reflect sparse collection. Experiments have been carried out to determine if distribution is by wind dispersal but these were inconclusive (23). Extremely high population densities are found in some cases. For example, in a Danish beech forest 1000-12 500 tardigrades were found in a one square metre (12) and in another study in the U.K. a density of 2 287 000 per m^2 was recorded (15).

Tardigrades are found in terrestrial, marine and freshwater environments. The terrestrial forms usually inhabit moss and lichen (2), but have also been found in soil and leaf litter and a variety of other habitats (13,14). Numerous factors have been examined to establish and understand habitat preference. The precise living conditions of most tardigrades are still unknown, and some clearly have very broad requirements (17). Moisture is almost certainly the most important factor affecting distribution (13,17). Moss cushions in general harbour the greatest tardigrade populations, the compass orientation of the cushion, its altitude and its height on the supporting substratum all influencing abundance (17). A spatial preference within the moss cushion itself has been suggested, with certain species inhabiting a specific microhabitat (10,11), but there is no correlation between moss species and tardigrade species (17). Population densities show a seasonal variation, with daylight, temperature, level of precipitation, relative humidity and fungal attack being the major determinants (8,12,15). In freshwater (lakes, ponds, rivers), tardigrades are found in the substratum of sandy shoreline sediment where there is constant wave action. Important factors contributing to their occurrence are depth, current, presence of detritus and concentration of oxygen (14).

Marine tardigrades also largely inhabit beaches, although occasionally individuals are found at depths of 4700 m (21) or living in association with marine flora and invertebrates. Beaches with fine sand grain sizes (less than 300 μm) and with large grains whose pores are easily clogged with silt have lower tardigrade numbers (19). Marine species are zoned on the beach surface as well as within the sediment, with water content, temperature and light affecting this zonal distribution (18).

Most tardigrades are herbivores but some species, such as Milnesium tardigradum, are carnivorous and have been known to prey on protozoans, rotifers and nematodes (5,12). This establishes a soil invertebrate predator-prey food chain that could be disrupted by herbicides, pesticides and pollution, although no work has been carried out to confirm this yet. Obligate freshwater tardigrades and some soil-dwelling species form cysts, in which they survive adverse conditions (9,22). Even more remarkable is the fact that many terrestrial species can enter a state of cryptobiosis (see next paragraph).

SCIENTIFIC INTEREST AND POTENTIAL VALUE Of major interest is the ability of some invertebrates to achieve the state of cryptobiosis, a form of suspended animation. Terrestrial tardigrades, along with rotifers and nematodes, have this ability and have been recovered in a living condition from 120-year-old dried moss samples (4). With periodic entrances into this dormant state the average one year life expectancy can be greatly extended with a total suspension of the growth and ageing process (3). The mechanism for this remarkable property is the gradual removal of surrounding moisture, which reduces the size and configuration of the Water Bear to a protective state called the tun. Water normally makes up 85 per cent of the weight of tardigrades but in cryptobiosis the water content drops to only 3 per cent. In the cryptobiotic state they are able to withstand extremes in temperature, radiation, oxygen and water concentrations (3). Tardigrades thus provide a unique experimental living system for use in the study of basic biological phenomena including 1) the role of water in the maintenance of macromolecular and membrane structure, 2) the characteristics of life itself, and 3) the events which result in senescence (3).

THREATS TO SURVIVAL An organism which can undergo cryptobiosis and endure chemical and physical extremes would appear to be virtually indestructable. However, rapid removal of water prevents tardigrades from entering a cryptobiotic state (3) and there is evidence that they could be vulnerable to a variety of factors. Elimination of habitat is probably the greatest threat to many species on a local basis (25). Urban areas are generally poor in terms of tardigrade habitat, because of soil disturbance and the scarcity of lichens and mosses due to atmospheric pollution and lack of suitable substrate. Work in Hungary has suggested that tardigrades are scarce in soils treated with herbicides and pesticides (14). A study in the U.S.A. showed that the use of DDT was found to greatly reduce both the density and diversity of tardigrades (1). Following treatment, a sample area that once contained 97 Macrobiotus islandicus, Macrobiotus areolatus and Milnesium tardigradum only supported four individuals of M. tardigradum (1). However another study has shown tardigrades to be fairly resistant to insecticides (25). Substrate structure disturbances can also reduce tardigrade populations and the disruption of soil structure by cultivation has a detrimental effect on tardigrade abundance (8). Finally, natural predators and fungal parasites can decrease tardigrade numbers (6).

CONSERVATION The lack of detailed ecological and environmental studies allow only speculation in this area. Increased studies of the factors that affect species diversity and density, and development of comparative faunal lists to monitor population changes, are vital. Greater care must also be directed to the supporting substrates, as these are delicate microhabitats which have a direct influence on tardigrades as well as on many other micro-organisms (25).

REFERENCES   1. Barrett, G. W. and Kimmel, R. G. (1972). Effects of DDT on the density and diversity of tardigrades. Proc. Iowa Acad. Sci. 78: 41-42.
2. Bartos, E. (1950). Additions to knowledge of moss-dwelling fauna of Switzerland. Hydrobiol. 2: 285-295.
3. Crowe, J. H. (1975). The physiology of cryptobiosis in tardigrades. Mem. Ist. Ital. Idrobiol. 32. Suppl. Pp. 37-59.
4. Crowe, J. H., and Cooper, A.F., Jr. (1971). Cryptobiosis. Sci. Am. 225(6): 30-36.
5. Doncaster, C.C. and Hooper, D.J. (1961). Nematodes attacked by protozoa and tardigrades. Nematolog. 6: 333-335.
6. Drechsler, C. (1951). An entomophthoraceous tardigrade parasite producing small conidia on propulsive cells in spicate heads. Bull. Torrey Bot. Club. 78(3): 183-200.
7. Martin, N.A. and Yeates, G.W. (1975). Effect of four insecticides on the pasture ecosystem. III. Nematoda, rotifers and tardigrades. New Zealand Journal of Agricultural Research 18: 307-312.
8. Fleeger, J.W. and Hummon, W.D. (1975). Distribution and abundance of soil Tardigrada in cultivated and uncultivated plots of an old field pasture. Mem. Ist. Ital. Idrobiol. 32 Suppl. Pp. 93-112.
9. Ramazzotti, G. (1972). Il phylum Tardigrada. Mem. Ist. Ital. Idrobiol. 28:1-732.
10. Hallas, T.E. (1975). Interstitial water and Tardigrada in a moss cushion. Ann. Zool. Fenn. 12: 255-259.
11. Hallas, T.E. (1978). Habitat preference in terrestrial tardigrades. Ann. Zool. Fenn. 15: 66-68.
12. Hallas, T.E. and Yeates, G.W. (1972). Tardigrades of the soil and litter of a Danish beech forest. Pedobiol. 12: 287-304.
13. Iharos, A. (1947). The Tardigrada fauna of the Tihany Peninsula. Arch. Biol. Hung. 17: 38-43.
14. Iharos, G. (1975). Summary of the results of forty years of research on Tardigrada. Mem. Ist. Ital. Idrobiol. 32 Suppl. Pp. 159-169.
15. Morgan, C.I. (1977). Population dynamics of two species of Tardigrada, Macrobiotus hufelandii (Schultze) and

Echiniscus (Echiniscus) testudo (Doyere), in roof moss from Swansea. J. Anim. Ecol. 46: 263-279.

16. Morgan, C.I. and King, P.E. (1976). British Tardigrades. Synopses of the British Fauna (NS) 9. Academic Press, London.

17. Nelson, D.R. (1975). Ecological distribution of tardigrades on Roan Mountain, Tennessee. North Carolina. Mem. Ist. Ital. Idrobiol. 32 Suppl. Pp. 225-276.

18. Pollock, L.W. (1975). The role of three environmental factors in the distribution of the interstitial tardigrade Batillipes mirus Richters. Mem. Ist. Ital. Idrobiol. 32 Suppl. Pp. 305-324.

19. Pollock, L.W. (1976). Marine flora and fauna of the north eastern United States. Tardigrada. NOAA Technical Report NMFS CIRC-394.

20. Pollock, L.W. (1979). Angursa biscuspis n.g., n.sp., a marine Arthrotardigrade from the western North Atlantic. Trans. Am. Micro. Soc. 98(4): 558-562.

21. Renaud-Morant, J. (1974). Une nouvelle famille de tardigrades marine abyssaux: les Coronarctidae fam. nov. (Heterotardigrada). C.R. Acad. Sci. Paris 278: 3087-3090.

22. Marcus, E. (1929). Tardigrada. In Bronn, H.G. (Ed.), Klassen and Ordnungen des Tierreichs Bd. 5, Abtlg. 4, Buch 3. Akademische Verlagsgesellschaft, Leipzig.

23. Sudzuki, M. (1972). An analysis of colonization in freshwater micro-organisms. Two simple experiments on dispersal by wind. Jap. J. Ecol. 22(5): 222-225.

24. Toftner, E.C. Grigarick, A.A. and Schuster, R.O. (1975). Analysis of scanning electron microscope images of Macrobiotus eggs. Mem. Ist. Ital. Idrobio. 32. Suppl. Pp. 393-411.

25. Morgan, C. (1982). In litt., 1 October.

26. Bertolani R. (1982). Cytology and reproductive mechanisms in tardigardes. Proc. Third Int. Symp. on Tardigrada. Johnson City, Tennessee, U.S.A., August 1980, pp 93-114.

We are very grateful to D. Prochnow for preparing the draft of this account and to R. Bertolani, H. Greven, T. Hallas and C. Morgan for helpful comments.

BRYOZOA

Moss Animals

INTRODUCTION The Bryozoa contains over 4000 living species which were once regarded as plants (the 'moss animalcules'). They form small (1 mm-50 cm) sessile colonies displaying a wide range of form, including encrusting calcareous or gelatinous patches, long branching chains, apparently jumbled heaps, or bushy, branching growths, some of which superficially resemble corals. The colonies are formed by the asexual budding of a few to many thousand zooids which are small (average length 0.5 mm) and regularly patterned with chitinous, calcareous or gelatinous body walls. Bryozoans feed on small plankton and organic particles which the zooids filter from the water using the lophophore a structure which bears ciliated tentacles and which can be retracted for protection. Within the microscopic scale zooids display a wide diversity of form and some are modified for functions other than feeding such as attachment, support or protection. Many of their specializations are still not fully understood because, although bryozoans are often common and abundant, their study has been neglected until recently. They are often overlooked and may be mistaken for seaweeds or coral (6,13,28).

A shelf fauna anywhere in the world will have from 50 to 500 species, and an increasing number of species are being discovered in abyssal oozes down to more than 8000 m (5,9,14). Although many species appear to have a virtually worldwide distribution (17), biochemical studies on the intertidal genus Alcyonidium have shown the number of species to be much greater than was previously realised. Sympatric species have been found which are morphologically indistinguishable, and genetic speciation may occur over far shorter geographical distances than was previously supposed (29). Only about 50 freshwater byozoans are known. These are usually ubiquitous, occurring in all types of water bodies and having very wide distributions.

Marine bryozoans are most abundant from sublittoral to shallow-shelf (down to 200 m) depth, in rocky caves, shell beds and kelp beds, species diversity and abundance declining steadily from the shelf into coastal waters and from the edge of the shelf into deeper waters (9). In some areas, a forest 1 m in height or more, consisting of bryozoans, sponges and hydroids, covers the sea bottom, probably for square miles (1,24). In southern Australia for example, Bryozoa make up a large portion of the living fauna on the coastal shelf (39).

Colonies are found attached to almost any kind of firm substratum including rocks, shells, gravel or seaweeds (35). The most important limiting factor seems to be availability of substrate. A study in the Bay of Biscay showed that erect, rigid forms predominated between 50 m and 200 m, but were poorly represented on the slope where they were outnumbered by encrusting species. Flexible erect anascans became predominant below 1000 m, and below 2000 m mainly erect, lightly calcified anascans or ctenostomes were found (9), although some encrusting species are found down to 4690 m (14,42). Sedimentation may limit bryozoan distribution near estuaries and deltas or where currents cause turbidity (24). Temperature and the presence of phytoplankton may also be important factors limiting the distribution of some species (40). Colonies are hermaphroditic, although some zooids may be male or female. Food availability may be important in initiating reproduction (41). In most species the fertilized eggs pass into a brood chamber; brooded larvae have shorter free-living stages and do not feed. Most larvae attach themselves to a suitable substrate and change into a zooid from which the colony develops by asexual budding (6).

SCIENTIFIC INTEREST AND POTENTIAL VALUE Bryozoans, with sponges, ascidians and some hydroids, are probably the 'fixers' of phytoplankton as a result of their filter-feeding activity, providing the basis for food webs leading to some of the larger fish and crustacea. Since a colony rarely dies or is killed by predation (bryozoans can transfer coelomic tissues and repair damage at any place in the colony) it can support a fauna of predators such as nudibranchs and still continue to grow (24,33). Although individual colonies are usually small they are very numerous, and the erect and foliaceous forms provide a substratum for other sessile animals and shelter for mobile animals, particularly the young stages of molluscs, fish and crustacea (24). A study of two species in India (Thalamoporella gothica var. indica and Pherusella tubulosa) showed that a large number of species of almost every invertebrate group depend on bryozoans for their existence; for example 100 g of T. gothica colony supported 333 amphipods, 266 bivalves, 300 serpulid worms, 230 nematodes, 220 copepods and 240 chironomid larvae among other species (25). On parts of the north-western coast of New Zealand's South Island grounds of bryozoan 'mounds' occur in such height and density that they support a variety of associated invertebrates, the whole community serving as a nursery ground for juveniles of commercial fish species. The two main bryozoan species found there are Celleporaria agglutinans and Hipponenella vellicata, both of which are endemic to New Zealand (32).

Unfortunately bryozoans are important fouling organisms and some species have been found to be very resistant to anti-fouling paints (6). Alcyonidium gelatinosum and A. hirsutum in the North Sea cause a skin allergy known as 'Dogger Bank Itch' which affects fishermen (3,36). However bryozoans have some useful properties for man as well. Besides providing attractive curios for sport-divers and contributing to the aesthetic appeal of the underwater environment, which is now recognized as an important element in tourism, bryozoans may be of medical value. An extract of Bugula neritina was found to extend the lives of mice with lymphatic leukaemia by 68-100 per cent (26).

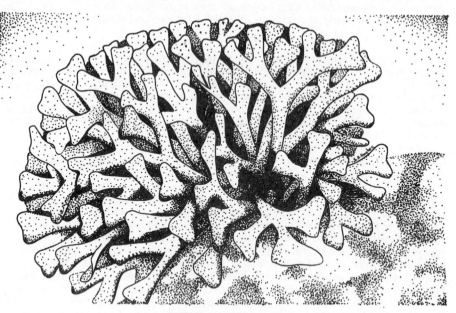

Pentapora fascialis

THREATS TO SURVIVAL As little research has been carried out on the ecology of this group (5), there is no literature available to indicates whether any particular species is under threat. However, given the habitats and ecology known so far for this phylum, it can be inferred that many species could be vulnerable to a variety of adverse factors (12). Pollution, increased turbidity (which is fatal to filter-feeders), sedimentation and siltation, changes in salinity and temperature, and loss of substrate e.g. by dredging and kelp harvesting, are potential hazards to whole bryozoan faunas and to the animals that directly or indirectly depend on them as a food supply (4,24). Since all species have mobile, sometimes planktonic larvae, recolonization of locally destroyed populations may take place fairly easily; studies off Brittany have shown that species may be present in normal abundance only a year after the incidence of heavy pollution (15). Examples are given below of species that have either been shown to be, or which could be, susceptible to particular human activities.

1. Marine Pollution In heavily polluted areas e.g. the Gulf of Fos near Marseilles, France, species diversity was found to decrease progressively along a gradient of increasing pollution. The most susceptible species were Entalophoroecia robusta, Plagioecia sarniensis, Microecia occulta, Lichenopora radiata, Cellaria sp., Cribilaria innominata, Synnotum aegyptiacum, Escharina vulgaris, Arthropoma cecilii, Smittina marmorea, Reptadeonella violacea, Schizotheca fissa and Cellepora sp., which only occurred in areas where extreme siltation was not apparent. Throughout this region the large species Pentapora fascialis, Turbicellepora avicularis, Adeonella calveti and Sertella spp., which are usually abundant, were absent (8). Off La Ciotat, east of Marseilles, populations of large species (Turbicellepora spp., Pentapora spp., Porella cervicornis), which are free-living on detritic sands, were reduced as a result of increased pollution and fine particle deposits (22). Bryozoan populations suffered heavy losses off Roscoff, on the west coast of France, when the oil tanker the Amoco Cadiz foundered in 1978 causing large scale hydrocarbon pollution. The only arborescent species in this area, Scrupocellaria reptans, was almost exterminated, and recolonization did not begin until two years later. This species could therefore be seriously affected by widespread pollution (15). In Willapa Bay, Washington, U.S.A., colonies of Schizoporella unicornis were found with abnormal ovicells in areas where petroleum derivatives were evident in the water. These presumably came from several sources including small craft, engine oil from boat bilges and creosote leaking from dock pilings (23). However, the correlation between petroleum derivatives and cell abnormalities has yet to be confirmed in other studies.

Estuarine forms are most likely to be endangered by pollution, since they are found in harbours and along shorelines in the path of offshore pollution spills. Since there are usually large estuarine areas that do not have major harbours nearby, recruitment to the damaged area should be possible and species with relatively long-lived planktonic larvae should be able to recolonize such areas fairly rapidly. However, the majority of estuarine species, particularly those of the east coast of the United States, have very short-lived larvae which tend to settle out within a few hours of leaving the parent stock. These species would be much slower recolonizers than those with long-lived larvae. Fortunately they tend in general to be widespread, temperature- and salinity-tolerant species which are opportunists in travelling with man's aid on coastal vessels and debris, making extinction unlikely (18). However, research is revealing that species may have different tolerances to temperature and dissolved oxygen levels, and many bryozoans are unable to tolerate oxygen levels below 1.0 ppm (40). Other species which could be susceptible to such pollution include (13,28):

Conopeum reticulum (Linnaeus). Encrusts hard substrata in coastal and estuarine environments around most of the British Isles, the southern shores of the North Sea and the north-east U.S.A. The larvae are planktonic, in inshore plankton; both stages of the life cycle are therefore possibly susceptible to coastal and estuarine pollution.

Conopeum seurati (Canu). Habitat limited to brackish lagoons and riverine areas;

frequently on plant stems (e.g. Ruppia). Larvae planktonic. Distribution imperfectly known: East Anglia and Barrow-in-Furness, U.K., southern coasts of North Sea, south coast of France, Tunisia, Italy. Vulnerable to wetland drainage, pollution etc.

Electra crustulenta (Pallas). A brackish water species, but more euryhaline than C. seurati. Larvae planktonic. Known from estuaries of southern and south-east England, Northern Ireland and the Baltic Sea.

Electra monostachys (Busk). An estuarine species usually associated with hard substrata such as shells. Distribution imperfectly known; present in estuaries and along coast of North Sea, Irish Sea, English Channel. Larvae planktonic.

Aspidelectra melolontha (Landsborough). Distribution very limited: estuaries of the Thames, Crouch, Roach, and in the Dover Straits, U.K., and a few similar localities on Danish, Belgian and German coasts. Life history incompletely known but larvae assumed to be planktonic. Possibly susceptible to coastal pollution.

Phaeostachys spinifera (Johnston). Probably limited to the British Isles. Found from the lower shore down to about 100 m. At present common but the unusually small geographical range may be a vulnerable feature. Larvae mobile.

Victorella pavida Kent and Bulbella abscondita Braem. These two species are still taxonomically confused. Both are known only from the London docks, and a few similar localities in Germany and the low countries, but V. pavida has been reported from southern Europe and the eastern U.S.A.

Farrella repens (Farre). Probably limited to the British Isles and the southern coast of the North Sea. Coastal or estuarine, often epizoic.

2. Freshwater Pollution Although most species have apparently cosmopolitan distributions, some appear to have fairly circumscribed ranges and could be threatened by pollution (2,37).

Lophopus crystallinus Has been recorded from only two rivers (the Illinois river and the Schuylkill River) in the U.S.A. and from Lake Erie, and there have been no new published occurrences for 30 years. It is more widespread in Europe but recent records are scanty; there are old records of its presence in high mountain lakes e.g. in the Pyrenées, which are often refuge habitats for species sensitive to pollution. Unlike many freshwater bryozoans, it has never been found abundantly, and it also has a much lower statoblast production rate than many species. (These are asexual dormant structures which germinate to form new colonies when favourable conditions prevail) (2,20).

Cristatella mucedo has a holarctic distribution, having been recorded in south-west Russia, the U.S.A., central Europe, Iran and the U.K. Many areas where it occurs are free from severe pollution but in the U.K. a decline was noted in the Norfolk Broads, which was accounted for by pollution and the increasing turbidity of the broadland waterways during the summer months (2,27,34). Studies in progress suggest that this species is also very sensitive to heavy metal pollution (especially copper). However statoblasts of members of the class Phylactolaemata appear to be much more resistant to heavy metals than adult colonies (21).

Other freshwater species which could be vulnerable to pollution include Fredericella sultana, Plumatella repens, P. fungosa, P. emarginata, P. fruticosa and Paludicella articulata (12,19).

3. Loss of Substrate a) Epiphytic species A number of bryozoans are restricted either wholly or partly to living on coastal seaweeds such as Laminaria spp., Fucus

serratus, Gigartina and Chondrus (10,12). Recent research has shown that there is enormous genetic diversity in these species, and has furnished evidence of cryptic speciation, but most are still poorly known or even undescribed (29,11). In California 100 m long fronds of Macrocystis may become entirely covered with bryozoans (30). Such species may be vulnerable through loss of seaweeds as a result of pollution or commercial exploitation for food and substances such as alginates, agar and carageenan. Seaweeds have been found to contain a wide variety of compounds useful to the drug industry and for medical research; it is therefore to be expected that commercial exploitation will increase (26). The following epiphytic bryozoans are still generally widespread in the North Atlantic region: Membranipora membranacea, Flustrellidra hispida, Alcyonidium spp. (all wholly epiphytic), Electra pilosa, Puellina gattyae, Haplopoma impressum, H. bimucronatum and Plesiothoa gigerium (partly epiphytic) (10,13,28).

b) Dredging Dredging for commercial organisms, such as scallops (Pecten) at Roscoff, may cause loss of habitat for bryozoans (15). Freshwater species such as the Plumatellidae, the Fredericellidae and Paludicella articulata, which are fairly tolerant of pollution in general, are nevertheless susceptible to loss of substrate (38). One notable case of the destruction of a bryozoan community took place off New Zealand. With the development of special techniques, it became possible to fish the bryozoan mounds in the Tasman Bay-Golden Bay area. Extensive destruction of the mounds resulted, and by the late 1970s, the Torrent Bay ground was virtually destroyed. Where the mounds still persisted they were markedly reduced in size and density. With loss of shelter and food organisms, there has been a concomitant decrease in stocks of commercial fish (32).

4. Exploitation A number of bryozoans form delicate, attractive, easily preserved structures which are conspicuous underwater and are collected by sport-divers as curios. Although this is not a major threat, low-level exploitation could cause damage to local populations as they are probably slow-growing, occur in low densities and may have poor recruitment (16). The species which are collected tend also to be the large branching forms which provide protection and food for a large number of other invertebrates (4). The most commonly collected species in Atlantic and Mediterranean waters are:

Myriapora truncata (False Coral) limited to the Mediterranean (6,7).
Pentapora foliacea (Rose Coral) north-east Atlantic, including English Channel. Forms large perennial colonies in sublittoral habitats characterized by strong currents and a hard substrate. Colonies probably live for 15-20 years and may reach over 1 m in diameter; one small colony (e.g. 30 cm in diameter) can support thousands of small animals and their eggs, and provides refuge for several hundred mobile ones. Deeper water colonies are reasonably safe (except from dredging). The removal of this species down to about 30 m from the south and south west-coast of the U.K. could affect a very large number of animals (4,6,7,12).
Pentapora fascialis (Rose Coral) similar to P. foliacea but limited to Mediterranean.
Reteporids (e.g. Sertella beaniana, Reteporella sp.) (7).
Hornera spp. Mediterranean (6,7).

Ornamental species collected in New Zealand waters include Hippellozoon novaezelandiae and Hippomenella vellicata; the latter has a wide distribution and is unlikely to become threatened (31).

CONSERVATION In view of the lack of information, no precise recommendations can be made, but the data available suggest that the following points should be considered:

a) Further research into the effects of pollution and sedimentation and the distribution and ecology of bryozoans is required.

b) Pollution control measures should be improved and strictly enforced.
c) Measures should be taken to ensure that over-collection of popular ornamental species does not occur. For example, one of the best-known collecting grounds of Hippellozoon novaezelandiae in New Zealand has now been protected through the creation of Poor Knights Islands Marine Reserve (31).

Possibly the only example of bryozoans being specifically protected, in this case to conserve a commercial fishery, comes from New Zealand. In December 1980, an area of sea bed off Separation Point, South Island, delimiting bryozoan coral-like mounds, was closed to power-fishing methods such as trawling and dredging which were likely to continue destroying the mounds. Restoration of the habitat and fish numbers is being monitored (32).

REFERENCES
1. Bullivant, J.S. (1959). Photographs of the bottom fauna in the Ross Sea. N.Z. Journ. Sci. 2(4): 485-497.
2. Bushnell J.H. (1974). Bryozoans (Ectoprocta). Chap. 6. In Hart, C.W. and Fuller, S.L.H. (Eds), Pollution Ecology of Freshwater Invertebrates, Academic Press, London.
3. Carlé, J.S. and Christophersen, C. (1980). Dogger Bank Itch. The allergen is (2-Hydroxyethyl) dimethylsulfonium ion. J. Am. Chem. Soc. 102: 5107-5108.
4. Cook, P.L. (1980). In litt., 2 April.
5. Cook, P.L. (in press). Bryozoa from Ghana - a preliminary survey. Ann. Mus. and Afr. Cent.
6. George D. and George, J. (1979). An Encyclopaedia of Marine Life. Harrap & Co Ltd., London.
7. Harmelin J.G. (1980). In litt., 16 July.
8. Harmelin, J.G. and Hong, J.S. (1979). Données preliminaires sur le peuplement d'un fond de concretionnement soumis a un gradient de pollution. 2 - Faune Bryozoologique. Rapp. Comm. Int. Mer. Médit. 25/26 (4): 175-177.
9. Hayward, P.J. (1978). Bryozoa from the west European continental slope. J. Zool. Lond. 184: 207-224.
10. Hayward, P.J. (1980). Invertebrate epiphytes of coastal marine algae. In Price, J.H., Irvine, D.E.G. and Farnham, W.F. (Eds), The Shore Environment Vol.2. Ecosystems. Systematics Assn. Special Vol. 17(b): 761-787. Academic Press, London and New York.
11. Hayward, P.J. (1980). In litt., 14 July.
12. Hayward, P.J. (1981). In litt., 10 February.
13. Hayward, P.J. and Ryland, J.S. (1979). British Ascophoran Bryozoans. Synopses of the British Fauna (NS) 14. Academic Press.
14. d'Hondt, J.-L. (1975). Bryozoaires Cténostomes et Cheilostomes (Cribrimorphes et Escharellidae exceptes) prevenant des dragages de la campagne océanographique Biacôres du 'Jean Charcot'. Bull. Mus. Natn. Hist. Nat. Paris 3(229) (Zool. 209): 553-600.
15. d'Hondt, J.-L. (in press). Incidence de la pollution par les hydrocarbures de l' "Amoco Cadiz" sur les populations de Bryozoaires épibiontes du Chenal de l'Ile Verte (Roscoff, France). Trav. Stat. Biol. Roscoff.
16. Hunnam P.J. (1980). Mediterranean marine species in possible need of protection. UNEP/UNESCO, UNEP/IG/INF.6.
17. Lagaaij, R. and Cook, P.L. (1973). Some tertiary to recent Bryozoa. In Hallan, A. (Ed.), Atlas of Palaeobiogeography. Elsevier.
18 Maturo, F. (1981). In litt., 4 June.

19. Mundy, S.P. (1980). A key to the British and European freshwater bryozoans. Scient. Publs. Freshwat. Biol. Ass. 41: 1-30.
20. Mundy, P.S. (1980). Stereoscan studies of phylactolaemate bryozoan statoblasts including a key to the statoblasts of the British and European Phylactolaemata. J. Zool. Lond. 192: 511-530.
21. Mundy, S.P. (1981). Some effects of low concentrations of copper ions on Cristatella mucedo (Bryozoa: Phylactolaemata). In Larwood, G.P. and Nielsen, C. (Eds), Recent and Fossil Bryozoa. Olsen and Olsen, Fredensborg.
22. Picard, J. and Bourcier, M. (1976). Evolution sous influences humaines des peuplements benthiques des parages de La Ciotat entre 1954 et 1972. Téthys 7(2-3): 213-222.
23. Powell, N.A. et al (1970). Hyperplasia in an estuarine bryozoan attributable to coal tar derivatives. J. Fish Res. Bd. Canada 27: 2095-2096.
24. Probert, P.K., Batham, E.J. and Wilson, J.B. (1979). Epibenthic macrofauna off southeastern New Zealand and mid-shelf bryozoan dominance. N.Z. J. Mar. Freshwater Res. 13(3): 379-392.
25. Rao, K.S. and Ganapati, P.N. (1980). Epizoic fauna of Thalamoporella gothica var. indica and Pherusella tubulosa (Bryozoa). Bull. Marine Sci. 30(1): 34-44.
26. Ruggiere, G.D. (1976). Drugs from the Sea. Science 194: 491-497.
27. Ryland, J.S. (1970). Bryozoans. Hutchinson, London.
28. Ryland, J.S. and Hayward, P.J. (1977). British Anascan Bryozoans. Synopses of the British Fauna (NS) 10. Academic Press.
29. Thorpe, J.P. and Ryland, J.S. (1979). Cryptic speciation detected by biochemical genetics in three ecologically important intertidal bryozoans. Estuar. Coast. Mar. Sci. 8: 395-398.
30. Wing, B.L. and Clendenning, (1971). Kelp surfaces and associated invertebrates. In North, W.J. (Ed.), The biology of the giant kelp beds (Macrocystis) in California. Beih. Nova Hed. Wigia. 32: 319-341.
31. Gordon, D.P. (1982). In litt., 16 April.
32. Saxton, F.L., Bradstock, M. and Gordon, D.P. (in press). The protection of bryozoans as a conservation measure for commercial fish stocks. N.Z. J. Mar. Freshwater Res.
33. Clark, K.B. (1975). Nudibranch life cycles in the northwest Atlantic and their relationships to the ecology of fouling communities. Helgoländer Wiss. Meeresunters 27(1): 28-69.
34. Ellis, E.A. (1965). The Broads. Collins, London.
35. Eggleston, D. (1972). Factors influencing the distribution of sub-littoral ectoprocts off the south of the Isle of Man (Irish Sea). J. Nat. Hist. 6: 247-260.
36. Dubos, M. (1980). Etude expérimentale du pouvoir sensibilisant d'Aleyonidium gelatinosum (L.) (Bryozoaire marin.). Arch. mal. prof. 41(1): 9-13 and (2): 75-77.
37. Lacourt, A.W. (1968). A Monograph of the Freshwater Bryozoa - Phylolactolaemata. Zool. Verh. Uitgegeven Rijksmuseum Natuur. Histor. Leiden no. 93: 1-59.
38. Mundy, S.P. (1981). In litt., 24 February.
39. Wass, R.E. (1970). Bryozoan carbonate sand continuous along southern Australia. Marine Geology : 63-73.
40. Soule, D.F., Soule, J.D. and Henry, C.A. (1979). The

influence of environmental factors on the distribution of estuarine bryozoans, as determined by multivariate analysis. In Larwood, G.P. and Abbott, M.B. (Eds), <u>Advances in Bryozoology</u>. Academic Press, London. Pp. 347-372.

41. Gordon, D.P. (1970). Reproductive ecology of some northern New Zealand Bryozoa. <u>Cah. Biol. Mar.</u> 11: 307-323.
42. d'Hondt, J-L. (1982). Les bryozoaires eurystomes abyssaux. <u>C.R. Soc. Biogéogr.</u> 58(1): 30-48.

We are extremely grateful to P. Cook, for assistance in the compilation of this data sheet, and to J. Chimonides, D.P. Gordon, J.G. Harmelin, P.J. Hayward, J-L. d'Hondt, F. Maturo, P.S. Mundy and J.S. Ryland for additional information and helpful comments.

BRACHIOPODA

Lamp Shells

INTRODUCTION Although there is a large and diverse fossil brachiopod record (about 30 000 fossil species have been described) only about 300 species are known today (28). Most belong to the class Articulata, with about 40 species in the class Inarticulata. Both classes bear a superficial resemblance to bivalve molluscs. The body is enclosed within a shell, made of two valves which cover the dorsal and ventral surfaces of the animal, and ranges in size from a few mm to 10 cm in length. The two classes differ in many important respects, to the extent that many authors argue for a polyphyletic origin of the phylum (9). The class Articulata has a hinge of interlocking teeth and sockets between the calcareous valves, and the stalk or pedicle, when present, emerges through the ventral valve. In the class Inarticulata the valves are held together by complex muscle systems and the pedicle either emerges from between the valves or through an opening in the ventral valve. Two families (Lingulidae and Discinidae) have valves composed of a chitino-phosphatic material, and the third family (Craniidae) has calcareous valves. In both classes the body of the animal occupies only a small portion of the shell cavity at the posterior end, but extensions of the body wall (mantle lobes) line the internal surfaces of the valves forming a large mantle cavity which contains a filament-bearing structure (lophophore) used for feeding and respiration (5).

Articulate brachiopods have a worldwide marine distribution, occurring in greatest numbers in the shallow seas surrounding the continents (5). Although fossil species were once widely distributed and abundant on coral reefs, modern articulates are largely inhabitants of temperate waters and have a rather patchy distribution (1). They are abundant in only a few places, such as round the Japanese islands and in the Antarctic, where they have been reported to be the most abundant part of the megafauna in a few local areas (3,29). Densities of up to 400 per m^2 have been recorded (2). SCUBA diving has shown that brachiopods may be important members of the cryptic brachiopod-coralline sponge community in the Caribbean (Jamaica, Curaçao), Red Sea and Pacific. High densities are found on the undersides of foliaceous corals or overhangs and in the interiors of crevices or caves (6). A variety of rarer forms are known from depths of between 500 m and 6000 m (3). Examples are the relict species Sphenarina ezogremena and Septicollarina hemichinata, known only from single specimens collected at a depth of 240 m in the Bali Sea (27). Probably the richest brachiopod area in the northern hemisphere are in Puget Sound, around the San Juan Islands (Washington State, U.S.A.) and off the southern coast of British Columbia, Canada (30). Brachiopods are rarely found washed up on the sea shore since the shell breaks easily once the animal dies (2).

In the inarticulate brachiopods the Lingulidae and Discinidae occur in tropical and subtropical regions. The two genera in the family Lingulidae are geographically disjunct: Lingula is found in shallow inshore waters in the east Atlantic and Indo-West Pacific regions, while Glottidia is confined to both coasts of the Americas. Representatives of the Discinidae are restricted to shelf waters in all oceans, except for a single cosmopolitan deep-sea species. The Craniidae occur worldwide, but mainly in temperate shelf waters. All groups may be exceedingly abundant, and can be found at densities of up to 1000 per m^2 (17,18).

Most articulate brachiopods live permanently attached to hard material on the sea floor either by a fleshy stalk or by cementing the ventral valve directly to the surface, but a few species are able to attach to aggregated sediment grains and other soft substrates (9,31). They feed mainly on plankton and other organic material collected by the filaments on the lophophore, which create a current of water through the mantle cavity; they may also feed directly on dissolved nutrients (2,5). They are not found in areas of soft, fine sediment with a high rate of sedimentation, since such conditions would clog the lophohore, but they can survive strong water currents. They do not generally tolerate salinities deviating more than a few per cent from that of normal sea water (2). Sexes are normally separate although there are a few hermaphroditic species. Eggs and sperm are usually shed into the water, where fertilization occurs, but a few warm-water species brood their eggs. The fertilized egg develops into a free-swimming larva which, after a short planktonic existence, sinks to the bottom and develops into the adult form (5).

In the Inarticulata, lingulids live in semi-permanent vertical burrows in sandy sediments (19). They occur in areas that may be subjected to lowered salinities for relatively prolonged periods. Craniids require a hard substratum for attachment, as do the discinids, which are often found in clumps attached to each other. Little is known of the salinity tolerance of these two families. Larval development patterns appear to enhance wide dispersal of lingulids and discinids (14,15) but craniids have a short larval life and only limited potential for dispersal (20).

SCIENTIFIC INTEREST AND POTENTIAL VALUE   The phylum is particularly important in the study of evolutionary history, as brachiopod remains are often abundant in fossil-bearing rocks of almost every geological age. Brachiopods have one of the longest observable evolutionary histories, having existed for at least 550 million years (2); the continuity of their fossil record is extremely good (9). The genus Lingula is the classic 'living fossil', remaining unaltered in external appearance and probably in life habit for about 500 million years. It is the most studied of all brachiopod genera.

Lingula has been utilized as a food item in Japan (26), the Philippines, Burma, Thailand and by Australian aborigines. Some discinids have been recorded as fouling ships' hulls (12), but they are likely to be of only minor significance in this regard. No other species of commercial importance is known.

THREATS TO SURVIVAL   Although there is no literature available indicating particular species as being endangered, brachiopods, as sessile suspension feeders, are probably subject to the same adverse factors as other invertebrates living under similar circumstances. All reef brachiopods face the threat of losing their attachment base if the host corals are destroyed by dredging, anchor damage, starfish predation, disease, hurricanes and pollution (7). Populations on the north-west coast of the U.S.A. are said to be being greatly reduced through pollution (particularly in Puget Sound), commercial collecting for sale to schools throughout the U.S.A., collecting for scientific purposes and teaching, and accidental killing in other bottom fishing operations. The species involved include Terebratalia transversa, Hemithiris psittacea, Terebratulina unguicula and Laqueus californianus (30).

1. Loss of substrate Scallop dredging in the Bay of Fundy, Canada, could effect populations of Terebratulina septentrionalis (Articulata) (7). This species occurs mainly under boulders but it may also be attached to the vagile sea scallop, Placopecten magellanicus. This method of attachment may be important for the dispersal of the brachiopod since research has shown that it has a shorter planktonic stage than other marine species by virtue of brooding its larvae. Although this provides protection for the larvae, it could restrict its geographic dispersal which may be counteracted by mobile vectors such as sea scallops carrying brachiopods out of the main population area (11).

In Bermuda, dredging may be affecting Argyrotheca bermudana (Articulata). In Castle Harbour there are no living populations although there is evidence of their former existence; dredging could have been the cause of their disappearance (7). It was a favourite collecting ground for naturalists who visited Bermuda as the water was relatively clear and there was a diverse marine community. However it was significantly altered by dredging from 1941-43 for the construction of U.S. Naval Air Station Bermuda, and an estimated 12-15 million cubic metres of substrate was removed for infilling, reducing the area of the harbour by about one fifth. The water is now turbid, due to the large amount of sediments churned up during dredging, reduced circulation caused by infilling, and possibly from washings from the Government Quarry where crushed limestone is produced for building purposes. Evidence from other studies suggests that high turbidity and sedimentation rates may be responsible for decreased coral abundance and species diversity as a result of decreased light penetration (4).

Removal of sediments for landfill in Tampa Bay, Florida, resulted in the elimination of Glottidia pyramidata (Inarticulata: Lingulidae) from the affected area (21). This was attributed to the development of soft, fine sediments, combined with low oxygen levels. The species is adapted to sand-sized sediments (19) but little is known of its oxygen requirements.

2. Pollution This is a potential widespread danger. Articulate brachiopod populations in the Mediterranean may be threatened, especially in the area off Marseilles (7). The easily accessible intertidal articulate population of the Straits of Georgia and Puget Sound on the west coast of North America could be threatened by local oil refineries and associated tanker traffic bearing Alaskan crude oil. The same taxa are present subtidally but are less abundant (10). At least three examples indicate that highly eutrophic conditions are inimical to lingulids probably through the reduction of available oxygen. This occurs despite tolerance by some species of anoxic conditions for long periods (13).

(a) Twenty years ago, Glottidia pyramidata was known from an area of Kingston Harbour, Jamaica (16), that has since become virtually anoxic, following heavy organic pollution. Subsequent extensive surveys of the benthos (23) have not yielded further specimens.

(b) On the southern California coast, Glottidia albida was found to be absent from sites influenced by municipal wastewater outfalls (22,24), despite being recorded at extremely high abundances (up to 1000 per m^2) in an earlier study of the same area (17). It is considered an indicator species in comparative studies of infaunal communities (22).

(c) In Kaneohe Bay, Hawaii, Lingula reevii, a species possibly endemic to the Bay, is absent from zones of high eutrophication, and populations may be declining (25).

3. Exploitation Exploitation of Lingula is unlikely to be a threat to the survival of the genus, although heavy collecting may have depleted some local populations in Japan (26). A rock pool near Lyttleton Harbour, Christchurch, New Zealand, is the only known locality in the world with three different brachiopod species (Terebratella (Waltonia) inconspicua, Notosaria nigricans and Pumilus antiquatus). On account of its biological interest it is being disturbed greatly by visiting scientists and students, and populations are being reduced through over-collecting (30).

4. Exploitation of other marine species In New Zealand, Paterson Inlet (Stewart Island) has been proposed as a site for farming salmon and oysters. This is one of the few known localities where brachiopods are found on a soft substrate. Furthermore, within the area brachiopods, like their Palaeozoic counterparts, are dominant members of the benthos and subtidal and intertidal rock habitats. The inlet provides apparently optimum conditions for growth and diversification of brachiopods, whose densities far exceed that of bivalve molluscs (31). Neither salmon nor oysters are found within the inlet and so if permission is granted to the promoters, they will attempt to modify the environment to suit the product they wish to farm. The changes that may occur through farming are unknown, but are likely to be damaging to the rare articulate brachiopod taxa (Notosaria nigricans, Neothyris lenticularis, Terebratella sanguinea, Terebratella inconspicua) found there. The whole area, including the inlets of Stewart Island, Foveaux Strait and the fjords in which Crania huttoni, Terebratulina sp., Amphithyris richardsonae and Liothyrella neozelanica are found, holds considerable promise as a genetic and ecological resource on account of its fauna. It also provides a comparative framework of different and isolated marine environments of known age (8). Populations of Laqueus californianus on rocky bottoms off Morro Bay, California, are being reduced by fishing operations which accidentally bring them up when dredging for Rock Cod. It is suggested that populations of deep water brachiopods such as Terebratulia transversa, Dallinella obsoleta and Terebratulina unguicula may be threatened off southern California by pollution and by accidental killing in other dredging operations (30).

CONSERVATION Since there is very little information available no precise recommendations can be made. Further research into the distribution and ecology of brachiopods is required, and consideration should be given to such benthic groups in areas of dredging activity, high pollution, or commercial exploitation of marine organisms.

REFERENCES
1. Barnes, R.D. (1980). Invertebrate Zoology. 4th Ed. Saunders College, Philadelphia.
2. Brunton, C.H.C. and Curry, G.B. (1979). British Brachiopods. Synopses of the British Fauna N.S. 17. Linn. Soc. Lond., Academic Press.
3. Cooper, G.A. (1977). Brachiopods from the Caribbean Sea and Adjacent Waters. Studies in Tropical Oceanography, 14. Univ. of Miami Press, Florida.
4. Dryer, S. and Logan, A. (1978). Holocene reefs and sediments of Castle Harbour, Bermuda. Journal of Marine

Research 36(3): 399-425.

5. Moore, R.C. (Ed.) (1965). Treatise on invertebrate palaeontology. Pt. H. Brachiopoda. Geological Society of America and University of Kansas Press.

6. Logan, A. (1975). Ecological observations on the recent articulate brachiopod Argyrotheca bermudana Dall from the Bermuda Platform. Bull. Mar. Sci. 25(2): 186-204.

7. Logan, A. (1980). In litt., 30 April.

8. Richardson, J. (1981). In litt., 11 February.

9. Rudwick, M.J.S. (1970). Living and Fossil Brachiopods. Hutchinson University Library, London.

10. Thayer, C.W. (1981). In litt., 1 April.

11. Webb, G.R., Logan, A. and Noble J.P.A. (1976). Occurrence and significance of brooded larvae in a recent brachiopod, Bay of Fundy, Canada. Journal of Palaeontology 50(5): 869-871.

12. Dall, W.H. (1920). Annotated list of the recent Brachiopoda in the collection of the United States National Museum, with descriptions of thirty-three forms. Proc. U.S. Nat. Mus. 57(2314): 261-377.

13. Hammen, C.S. (1977). Brachiopod metabolism and enzymes. Am. Zool. 17(1): 141-148.

14. Hammond, L.S. (1980). The larvae of a discinid (Brachiopoda: Inarticulata) from inshore waters near Townsville, Australia, with revised identifications of previous records. J. Nat. Hist. 14(6): 647-661.

15. Hammond, L.S. (1982). Breeding season, larval development and dispersal of Lingula anatina (Brachiopoda, Inarticulata) from Townsville, Australia. J. Zool. 198(2): 183-196.

16. Two specimens, Acc. Nos. M2448 and M2449, Institute of Jamaica, Kingston, Jamaica. Collected 17 September 1960, Kingston Harbour.

17. Jones, G.F. and Barnard, J.L. (1963). The distribution and abundance of the inarticulate brachiopod Glottidia albida (Hinds) on the mainland shelf of southern California. Pac. Nat. 4(2): 27-52.

18. Kenchington, R.A. and Hammond L.S. (1978). Population structure, growth and distribution of Lingula anatina (Brachiopoda) in Queensland, Australia. J. Zool. 184(1): 63-81.

19. Paine, R.T. (1970). The sediment occupied by recent lingulid brachiopods and some paleoecological implications. Palaeogeog. Palaeoclimatol. Palaeoecol. 7(1): 21-31.

20. Rowell, A.J. (1960). Some early stages in the development of the brachiopod Crania anomala (Muller). Annu. Mag. Nat. Hist. Ser. 13(3): 35-52.

21. Taylor, J.L. (1970). Coastal development in Tampa Bay, Florida. Mar. Pollut. Bull. 1: 153-156.

22. Thompson, B.E. (1981). Pers. comm. to L. Hammond.

23. Wade, B.A., Antonio, L. and Mahon, R. (1972). Increasing organic pollution in Kingston Harbour, Jamaica. Mar. Pollut. Bull. 3: 106-110.

24. Ward, J.Q. and Mearns, A. J. (1979). Sixty-meter control survey off Southern California. Southern California Coastal Water Research Project, TM 229, 58 pp.

25. Worcester, W. (1969). On Lingula reevii. M.Sc. thesis. University of Hawaii, 49 pp.

26. Yatsu, N. (1902). On the habits of Japanese Lingula. Annot. Zool. Japonenses 4(2): 61-67.

27. Zezina, O.M. (1981). Recent deep-sea Brachiopoda from the Western Pacific. Galathea Report 15: 7-20.
28. Foster, M.W. (1982). Brachiopoda. In Parker, S.P. (Ed.), Synopsis and Classification of Living Organisms, McGraw-Hill, New York Pp. 773-780.
29. Foster, M.W. (1974). Recent Antarctic and subantarctic Research. American Geophysical Union Antarctic Research Series 21. 189 pp.
30. Foster, M.W. (1982) In litt., 26 March.
31. Richardon, J.R. (1981). Recent brachiopods from New Zealand - background to the study cruises of 1977-79. New Zealand Journal of Zoology 8(2): 133-143.

We are very grateful to L. Hammond for writing a substantial part of the draft and to C.H.C. Brunton, M.W. Foster, A. Logan, J. Richardson, C.W. Thayer and O. Zezina for information provided for this data sheet.

<u>INTRODUCTION</u> The echinoderms include starfish, brittle stars, feather stars, sea-lilies, sand dollars, sea urchins, and sea cucumbers, and comprise about 6000 living species. All are marine and the adults are usually benthic. Their unique characteristic is their basically radial 5-rayed symmetry, although sea cucumbers have some superimposed bilateral symmetry. Echinoderms have an internal skeleton lying just below the external surface, made of calcareous ossicles which may articulate with each other as in starfish or be sutured together to form a 'test' as in sea urchins. The word echinoderm means spiny skin, which aptly describes the external surface of most species. Each plate or spine is a single calcite crystal, and although essentially simple in starfish, these reach a high degree of complexity in sea urchins.

Echinoderms are also characterized by their tube feet which project through the body wall and are connected to an internal water vascular system which acts hydraulically. The tube feet are used for locomotion, to maintain adhesion, and to manipulate edible matter, but also play an important role in gas exchange and sensory reception. Sexes are usually separate and, as in many marine invertebrates, eggs and sperm are shed into the sea, where fertilization takes place. The egg develops into a larva called a 'pluteus' which swims in the plankton for about two months. After a complex metamorphosis it settles on a suitable substrate and takes on the adult form.

Four classes are traditionally recognized: Crinoidea, Stelleroidea, Echinoidea, Holothuroidea. The Stelleroidea include starfish, brittle stars and basket-stars. Starfish have five or more stout arms and generally occur on rocky bottoms, feeding on a variety of foods ranging from algae, sponges and corals to molluscs and even other echinoderms. Brittle and basket stars have long slender arms and may occur in dense mats. European brittle stars can attain densities as great as 1000-2000 per m^2 in suitable localities. The Crinoidea include sea lilies and feather stars and are the most primitive echinoderms. Most sea lilies live at great depths and are seldom encountered. Feather stars extend into shallow waters and are found in great variety on coral reefs. The Holothuroidea, or sea cucumbers, unlike other echinoderms are elongated and lie on their sides rather than on the oral or aboral surface. The skeleton is reduced to microscopic ossicles in a flexible body wall, and the tube feet around the mouth are modified to form a circle of retractile tentacles for deposit or suspension feeding.

Since the two data sheets which follow are for sea urchins, the Echinoidea are discussed in greater detail. They include sea urchins, heart urchins and sand dollars and are characterized by the fusion of the ossicles to form a rigid 'test'. The organisms usually have a diameter of 2-12 cm and may be very colourful. Greens and browns probably predominate but some are red, purple or black or else so pale as to be nearly white. The shape of the spines may be adapted to the habitat of the species; for example on surf-beaten shores urchins tend to have very short stout spines. Within the individual, different spines serve different functions such as defence, protection against desiccation, clinging to rocks and, in some species, burrowing. Situated among the spines are minute pincer-like organs called pedicellariae which are especially adapted for grasping and may be used for feeding and defence. Five double rows of tube-feet protrude through the test and are almost exclusively used for locomotion. Sea urchins have a unique complex scraping apparatus called Aristotle's lantern which consists of a framework of skeletal bars supporting five large vertically aligned teeth together with powerful muscles to operate them (7,8). Most species live on rocks and other hard substrates, feeding on algae, barnacles, hydroids, tube worms and sponges.

<u>SCIENTIFIC INTEREST AND POTENTIAL VALUE</u> Certain aspects of echinoderm

biology such as the structure of the larvae and development of the egg provide strong evidence that echinoderms may be closely related to the group which gave rise to the vertebrates. The eggs of sea urchins have been used as subjects for fundamental research on cell structure, fertilization and experimental embryology (16,17). In recent years sea-urchin eggs have even been shot into space in rockets to find out whether cosmic rays or other such phenomena have an effect on living organisms. Their relatively short development period allows researchers to observe the teratological effects of drugs, since adverse reactions can be noted in a matter of days instead of weeks or months (7), and there is potential for the use of sea urchin larvae in bioassay techniques (1). Interest has been shown recently in the large spines of the Slate-pencil Urchin, Heterocentrotus mammillatus, since the structure of the material of which they are composed may provide a template for the synthesis of artificial bone (13).

Several echinoderms have been found to contain active compounds that may be of use medicinally. Holothurin, a steroid saponin isolated from the Bahamian sea cucumber Actinopyga agassizi suppresses the growth of tumours in mice, and holothurin-like substances from other sea cucumbers and sea stars elicit a variety of effects on a number of biological systems. Crude extracts of the sea stars Asterias forbesi, Acanthaster planci and Asterina pectinifera are effective against influenza B virus in embryonated chicks, and the sea cucumber Stichopus japonicus is the source of two extracts, one of which shows antitumour activity, the other exhibiting high antifungal activity (20).

Echinoderms have been used for food since before the time of Aristotle (16). Currently 50 000 tonnes of echinoderms are harvested annually, most of which are sea urchins, although in the South Pacific holothurians have a greater relative importance (19). Sea urchins of various species are marketed in a number of countries, largely as a luxury food. Paracentrotus lividus is collected in Ireland, the Mediterranean and Brittany (see data sheet), Tripneustes ventricosus in Barbados and Loxochinus albus in Chile (9,12). Strongylocentrotus franciscanus is collected on the west coasts of the U.S.A. and Canada and S. droebachiensis on the east coasts. The gonads are sold fresh or frozen, and are particularly popular with Italians and other immigrants (16). In Japan S. intermedius, Heliocidaris and Hemicentrotus are fished commercially, the former being the most important. In Australia surveys are in progress and a fishery for Heliocidaris erythrogramma, Tripneustes sp., Centrostephanus and Phyllacanthus is being considered (31).

Dried sea cucumbers, known as trepang or bêche-de-mer, were among the earliest products of the South Pacific to enter international commerce. In the Far East they are highly regarded as an aphrodisiac, a tonic and as a cure for various illnesses, particularly high blood pressure. However, their most important use is as a food, their protein content being as high as 43 per cent. The main consumers are the Chinese, both in China and in countries throughout South East Asia. Hong Kong and Singapore (the main trading countries) both import more than 450 000 kg annually; imports into Japan, Taiwan, the U.S.A. and France have increased recently and Australia also has a small market. The trade has always been run on a cottage-industry scale, and local processing on coral cays and in coastal towns has been found to be more economically feasible than large central processing factories. Sources of supply are scattered from the east coast of Africa to the South Pacific, the main exporters being Fiji, the Solomon Islands, Papua New Guinea, Palau, Sri Lanka, Indonesia and the Philippines. Only a few species are commercially valuable, the most important being Microthele nobilis and Actinopyga miliaris, and the best quality coming from the South Pacific. Sea cucumbers are easy to catch; in the Solomon Islands, over a period of nine days, 7122 specimens of H. nobilis were caught by local fishermen, an average of 11 specimens per man per hour (19,25,27,28,29,30). In the U.K. sea urchins are collected for the curio trade (see Echinus esculentus data sheet), and dried starfish and sand dollars are collected in other parts of the world for ornamental purposes.

The spines of the Slate-pencil Urchin are used in the shellcraft trade in large quantities particularly in the Philippines (18).

Sea urchins are the main grazing element in shallow waters and play an important role in controlling algal growth. Off Plymouth, U.K., at least one third of the rock is cleared of its seaweeds and encrusting algae each year by Echinus esculentus (see data sheet). Under certain conditions, urchin populations may reach pest proportions and cause overgrazing of kelp beds. This has occurred in some areas of Canada, possibly as a result of the decline in abundance of the lobster, Homarus americanus which is a known predator of the urchin Strongylocentrotus droebachiensis (32). The starfish Asterias of northern temperate seas may be a pest of commercial shellfish beds. The Crown-of-thorns Starfish Acanthaster planci of the Indo-west Pacific has gained a certain notoriety on account of the damage it can cause to coral reefs. It feeds on living corals, and in some areas where major population outbreaks have occurred up to 95 per cent destruction has been recorded. There has been considerable controversy over the causes of such plagues. It is not clear to what extent this species may have a natural fluctuating population, or whether the current outbreaks are due to human activities (see section on Cnidaria). No satisfactory methods have been found to control the outbreaks (22).

THREATS TO SURVIVAL 1. Pollution Echinoderms are nearly all sensitive indicators of clean water and may be killed in large numbers by toxic oil. It has been found that a 0.01 per cent oil emulsion inactivated the tube-feet of urchins and was lethal within one hour of exposure (2). The developing eggs of Strongylocentrotus purpuratus have been found to be very sensitive to oil pollutants, particularly crude and heavy bunker oils (3). In a study of urchins on the coast of Italy it was found that there was a much lower density of Paracentrotus lividus and Arbacia lixula near Naples than further south in unpolluted areas, and that these two species were absent at Sorrento and Naples wherever the sewage loads were very high. Furthermore, the dry weight of urchins was found to be much smaller in the Bay of Naples than further south (4). However, a study on the French coast near Marseilles revealed that domestic pollution could lead to huge population oscillations in sea urchins with densities as high as 400 per 10 m^2 being reached (5). The same study illustrated the sensitivity of three species (Paracentrotus lividus, Arbacia lixula, Sphaerechinus granularis) to industrial pollution, which caused a reduction in the size of individuals as well as a decrease in numbers. The sea cucumbers Holothuria tubulosa and H. forskali were found to be even more sensitive and disappeared in areas of high pollution. This is probably due to their feeding method since they ingest sediment and pollutants tend to become concentrated in this part of the benthos (5). Two populations of highly deformed sea urchins of the species Tripneustes cf. gratilla were found in the vicinity of a combined power and desalination plant in the Gulf of Eilat (Aqaba), Red Sea, an area which is highly polluted by thermohaline and heavy metal ion effluents. More than 60 per cent of the urchins showed irregular bulging of the aboral half of the test and under unpolluted conditions in aquaria the growth rate of such urchins was very slow (14). A large concentration of Lytechinus variegatus urchins were found in Cartagena Bay, Colombia with skeletal deformations which were probably caused by oil and waste water pollution (15). Studies following the wrecking of the 'Torrey Canyon' tanker showed that oil had affected a number of echinoderms including the starfish Marthasterias glacialis, the burrowing heart urchin Echinocardium cordatum, and Echinus esculentus, although some species (e.g. the starfish Asterias rubens and the brittle star Acrocnida brachiata) appeared to be little affected (23). The effects of pollution could be particularly serious in places such as the Red Sea where echinoderm endemicity has been estimated at as high as 12 per cent (24).

2. Exploitation Although small scale artisan fisheries probably have little effect

on sea urchin populations, there are numerous reports of local population declines through large scale or commercial collection. Over-exploitation of Echinus esculentus (for curios) and Paracentrotus lividus (for food) are discussed in the data sheets which follow. The starfish Solaster papposus, Marthasterias glacialis, Luidia sp. and Henricia sanguinolenta (21) have been recorded as declining around Milford Haven, U.K., perhaps due to overfishing by divers. Loxechinus albus has been grossly over-exploited in Chile particularly in easily accessible areas (9,12). Catches, which are mainly consumed locally, declined from 4200 tonnes in 1971 to 2500 tonnes in 1973, (although FAO records a dramatic increase in catch from 3200 tonnes in 1970 to 13 649 tonnes in 1980 (25)), and the fishery in the north and centre of the country has come to an end (12). In Japan, beds of Strongylocentrotus intermedius are said to be depleted (6), and Heliocidaris and Pseudocentrotus are reported to have been overfished (26). In the 1940s stocks of Strongylocentrotus off the west coast of the U.S.A. were reported to be depleted, and the heavy demand for Tripneustes in Barbados was said to be causing over-collection (16). The heavy exploitation of sea cucumbers for the trepang trade has resulted in signs of over-collection. Populations of Microthele nobilis and Actinopyga miliaris in the Ongtong Java atoll in the Solomon Islands have decreased since the 1960s, especially in the shallower parts of the lagoon. Fishermen are reported to have to travel greater distances and to fish in deeper water in order to find substantial catches (29). Some of the rich beds in Fiji are also said to be showing signs of depletion (28).

CONSERVATION As far as is known there is no immediate likelihood of any species of echinoderm becoming extinct in the near future. However, recent studies of the effects of industrial land-fill and pollution and reports of over-exploitation of urchins give cause for concern. The two accounts which follow are for European sea urchins, a bias which is due simply to paucity of data on other species and from other parts of the world. The increasing number of marine reserves being established in many countries will help to protect breeding stocks of echinoderms, which can then replenish over-exploited areas and damaged populations. However, the evidence provided by studies on the effects of pollution on sea urchins strongly emphasizes the need for much stricter controls on the siting of coastal developments and disposal of the industrial wastes.

Commercial fisheries clearly need to be managed on a sustainable yield basis. In 1956 it was reported that Barbados had laws regulating the gathering and sale of Tripneustes (16). There is a minimum size limit of 80 mm for collection of Loxechinus albus in northern Chile and 100 mm in central and southern Chile. However, in a study in northern Chile only 37.6 per cent of individuals taken were of the correct size. It has been recommended that the size limit should be raised to 100 mm for the whole country, and clearly better enforcement is required (9). In efforts to devise management schemes for this species, studies have been carried out on growth rates of individuals raised in captivity, and their maximum food requirements have been determined (10,11).

Fukui Prefectural Fisheries Research Station in Japan has recently successfully developed and tested an artificial substrate for growing juvenile sea urchins, which is expected to be a turning point in a huge restocking programme for Pseudocentrotus and Heliocidaris. Nylon fibres are set into concrete blocks to form nests on which juvenile urchins settle. The highest mortality in natural populations occurs at this stage, after metamorphosis, when juveniles normally settle into algal beds (26).

Efforts are also being made to improve the sea cucumber fishery. Several organizations including FAO, the South Pacific Commission (SPC), the South Pacific Islands Fisheries Development Agency and individual governments have participated in a programme to revive the bêche-de-mer fishery in the South Pacific (30). In Fiji a basic biological research programme has been initiated on commercial species to study growth, natural mortality, habitat requirements, abundance, juvenile ecology and reproduction with a view to establishing levels of harvest. The SPC has sponsored courses in trepang processing and trainees from these courses are helping to establish fisheries in New Caledonia, French Polynesia and Tonga (28). In China a sea cucumber farm has reportedly been established in Shantung. Larval sea cucumbers are reared in a hatchery until several centimetres long and are then released into a sheltered bay, and it is claimed that this process has increased the harvest (28).

REFERENCES

1. Bougis, P., Corre, M.C. and Etienne, M. (1979). Sea-urchin larvae as a tool for assessment of the quality of sea water. Ann. Inst. Océanogr. Paris (N.S.) 555(1): 21-26.

2. North, W.J., Neushul, M. and Clendenning, K.A. (1965). Successive biological changes observed in a marine cove exposed to a large spillage of oil. Symp. Comm. Int. Explor. Scient. Mar. Medit. Monaco 1964: 335-354.

3. Allen, H. (1971). Effects of petroleum fractions on the early development of a sea urchin. Mar. Poll. Bull. 2(9): 138-40.

4. Sheppard, C.R.C. and Bellamy, D.J. (1974). Pollution of the Mediterranean around Naples. Mar. Poll. Bull. 5(3): 42-44.

5. Harmelin, J.G., Bouchon, C. and Hong, J.S. (1981). Impact de la pollution sur la distribution des échinodermes des substrats durs en Provence (Mediterranée Nord-Occidentale). Tethys 10(1): 13-36.

6. Southward, A. and Southward, E. (1975). Endangered urchins. New Scientist 66: 70-72.

7. Clark, A.M. (1977). Starfishes and Related Echinoderms. 3rd Ed. British Museum (Natural History).

8. Nichols, D. (1962). Echinoderms. Hutchinson Univ. Lib., London.

9. Deppe, R. and Viviani, C.A. (1977). La pesqueria artesanal del erizo comestible Loxechinus albus (Molina) (Echinodermata, Echinoidea, Echinidae) en la region de Iquique. Biol. Pesq. Chile 9: 23-41.

10. Buckle, F.R., Guisado, C.A., Serrano, C.T., Cordova, L., Pena, L. and Vasquez, E. (1977). Estudio de crecimiento en cautiverio del erizo Lóxechinus albus (Molina) en las costas de Valparaiso y Chiloé, Chile. An. Centro Cienc. del Mar y Limnol, Univ. Nac. Auton. México 4(1): 141-152.

11. Buckle, F.R., Guisado, C.T., Serrano, C.T., and Vasquez, E. (1977). Estudio de la alimentacion en cautiverio del erizo Loxechinus albus (Molina) en las costas de Valparaiso y Chiloé, Chile. An. Centro. Cienc. del Mar. y Limnol. Univ. Nac. Auton. México 4(1): 153-160.

12. Castilla, J.C. and Becerra, R.M. (1975). The shellfisheries of Chile: an analysis of the statistics 1960-1973. Proc. Int. Symp. Coastal Upwelling. Coquimbo, Chile. Nov. 18-19, 1975. Universidad del Norte, Centro de Investigaciones Submarinas, Coquimbo.

13. Weber, J.N., White, E.W. and Lebiedzik, J. (1971). New porous biomaterials by replication of echinoderm skeletal microstructures. Nature 233: 337-339.

14. Dafni, J. (1980). Abnormal growth patterns in the sea urchin Tripneustes cf. gratilla (L.) under pollution (Echinodermata, Echinoidea). J. Exp. Mar. Biol. Ecol. 47: 259-279.

15. Allain, J.Y. (1978). Déformations du test chez l'oursin Lytechinus variegatus (Lamarck) de la baie de Carthagene. Caldasia 12: 363-375.

16. Harvey, E.B. (1956). The American Arbacia and Other

Sea-urchins. Princeton University Press.

17. Hörstadius, S. (1973). Experimental embryology of Echinoderms. Clarendon Press, Oxford.

18. Wells, S.M. (1982). Marine conservation in the Philippines and Papua New Guinea with special emphasis on the ornamental coral and shell trade. Report to Winston Churchill Memorial Trust, London.

19. Salvat, B. (1980). The living marine resources of the South Pacific past, present and future. In Population environment relations in tropical islands: the case of eastern Fiji. Technical Notes 13, UNESCO.

20. Ruggieri, G.D. (1976). Drugs from the sea. Science 194: 491-497.

21. Natural Environment Research Council (1973). Marine Wildlife Conservation Publications Series 'B', No. 5.

22. Kenchington, R.A. (1978). The crown-of-thorns crisis in Australia: a retrospective analysis. Environmental Conservation 5(1): 11-20.

23. Smith, J.E. (Ed.) (1970). 'Torrey Canyon' Pollution and Marine Life. Cambridge University Press.

24. Price, A. (in press). Comparison between echinoderm faunas of Arabian Gulf, south-east Arabia, Red Sea and Gulfs of Aqaba and Suez. Fauna of Saudi Arabia 4.

25. FAO statistics, (1982). Rome (provided by E. Aykyuz).

26. Anon. (1981). Urchins grown on sea nests. ICLARM Newsletter 3(4): 16.

27. Parrish, P. (1978). Processing guidelines for bêche-de-mer. Australian Fisheries 37(10): 26-27.

28. Gentle, M. (1979). The fisheries biology of bêche-de-mer. South Pacific Bulletin 4th Quarter: 25-27,50.

29. Crean, K. (1977). Some aspects of the bêche-de-mer industry in Ongtong Java, Solomon Islands. South Pacific Commission Fisheries Newsletter 15: 36-48.

30. Sachithanathan, K. (1972). South Pacific Islands Bêche-de-Mer Industry. FAO Report FI: DP/RAS/69/102/11.

31. Nichols, D. (1981). The Cornish sea-urchin fishery. Cornish Studies 9: 5-18.

32. Mann, K.H. (1977). Destruction of kelp-beds by sea-urchins: a cyclical phenomenon or irreversible degradation? Helgoländer wiss. Meeresunters 30: 455-465.

We are very grateful to A.M. Clarke and A.R.G. Price for assistance with this section.

EUROPEAN EDIBLE SEA URCHIN   INSUFFICIENTLY KNOWN

Echinus esculentus Linnaeus, 1758

Phylum   ECHINODERMATA   Order   ECHINOIDA

Class   ECHINOIDEA   Family   ECHINIDAE

SUMMARY Echinus esculentus is found on the north-east Atlantic coast of Europe
where it is an important element in the sublittoral community. The collection of
this species for the curio trade has led to local depletions in some areas around the
British coast. Recolonization occurs fairly rapidly since the larvae are planktonic,
and there is little danger of extinction, but large-scale commercial exploitation
for food could have a serious effect on populations. The species should be
protected in order to avoid this happening.

DESCRIPTION Colours vary from violet to rusty red or pale brown (11). There is
considerable variation in size and rate of growth. Specimens from Plymouth
measured up to 15 cm in diameter (11) whereas those from the Clyde sea area,
Scotland, reached a maximum diameter of 12 cm (2).

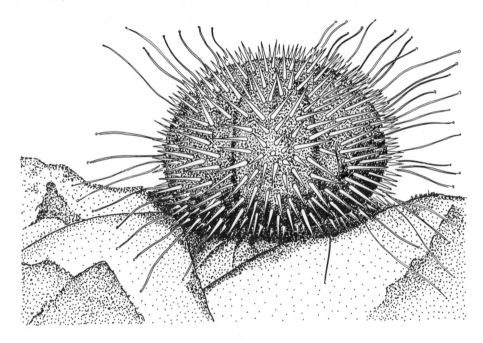

DISTRIBUTION North-east Atlantic coast of Europe from northern Norway to
northern Portugal and the Canaries; does not occur in the Baltic, on the German
coast, or in the Mediterranean (18). Found on Swedish and Danish coasts and on
offshore banks in Dutch and Belgian waters (18). In the U.K. found around all
shores except the southern North Sea and the eastern English Channel (15). It is
absent from much of the north coast of Spain and the west coast of France (3)
where it is replaced by Paracentrotus lividus but it is found on the north and
south-west coast of France and on parts of the Spanish Atlantic coast (18). Its
distribution may be influenced by large scale water movements in connection with
the North Atlantic Drift (18).

POPULATION Unknown. Its density varies significantly over its geographical range, but since the range is large the total population is probably large. Off Plymouth, density has been estimated at one urchin per 4.6 m^2, and in 1959 a population of 1.4 million urchins was estimated in a 2 km square between the 13 m and 31 m depth contours (3). At Port Erin the density varied from two to five per m^2 (5). Around Lundy Island densities have been estimated at 0.05-0.27 per m^2 (22). It is abundant in several sites in Scotland on the west coast (25,26).

HABITAT AND ECOLOGY It is most dense below the low tide mark down to about 20 m depth (16). Has been reported as deep as 1264 m (18) but these records may be based on incorrect identification. Larger urchins tend to inhabit shallower water possibly because there is a better food supply there (16,11), but it was found that off the Swedish coast the species tended to migrate to deeper water with age (7). Algae form the principal part of the diet, particularly for shallow-living individuals, but urchins take other bottom living organisms such as barnacles, bryozoans and tunicates, especially if they live at depths below which weed cannot survive (5,8,11,12,16,18). In a study off Plymouth it was calculated that at least one third of the rock is cleared of its algae and encrusting organisms each year as a result of grazing (3). Around Heligoland the main prey of Echinus is the Bridle Worm Polydora; urchins remove layers of rock with their rasping apparatus and it has been calculated that about 0.2-0.5 cm of rock are eroded per m^2 annually (6). However, this method of feeding has not been reported elsewere. Dispersal is not fully understood but adults may be moved by currents across the sea bed, even across silt, either as free individuals or wrapped in bundles of weed (13).

Spawning begins in February or March when the sea temperature is at its lowest and continues to May or later, but it is not known how frequently it occurs. A large mature urchin is estimated to contain 20 million eggs but probably only a proportion of these are shed at any one time (2,9). The maximum life span appears to be about 8-10 years (10).

SCIENTIFIC INTEREST AND POTENTIAL VALUE Like other urchins E. esculentus plays a very important role in preventing the overgrowth of kelp and other bottom-living organisms (5). The activities of divers in removing large numbers from localized areas e.g. Martin's Haven, Pembrokeshire, U.K., may have already resulted in increased algal cover (20). The roes are a culinary delicacy but the species is not yet exploited in large numbers for food as it reaches its prime condition in the winter months when collecting by divers is difficult (23). Its eggs provide a convenient source of large cells for cytologists and developmental biologists (17,24). In recent years the market for the curio and ornament trade has built up to such an extent that a small fishery has developed in Cornwall at Penzance to supply specimens for an industry which processes the shells for sale in the U.K. and abroad (23). In the mid-1970s a single diver could expect to sell up to 50 000 cleaned urchin tests every year. In 1978 one firm was fishing sea urchins at the rate of 10 000 a week for turning into table lamps (4). Currently singly divers manage to collect only 6000 annually, and the overall annual catch may be just below half a million specimens. All urchins of marketable size are collected from between 10 and 20 m depth, each site being visited about once a year. The tissue removed from the test during processing is used for fertilizer. A local trade has existed for many years but now the majority of tests collected are exported to France, Spain, Italy, Japan and the U.S.A. In the U.K., considerable quantities are incorporated into shellcraft articles. A dealer in California imports 6000 a year for sale in California, Hawaii, elsewhere in the U.S.A. and for re-export to Mexico (23).

THREATS TO SURVIVAL Since the population of this species is still large, present diver pressure is not likely to endanger it. However, in the U.K., urchins are usually collected from a few easily accessible and well-known sites which have in the past become seriously depleted. Divers who knew such sites over a number of years reported that they were seeing fewer large urchins (16). Samples from popular collecting sites, such as Lamorna Cove, Cornwall, showed a paucity of large specimens from depths of between five and fifteen metres, which may have been due to over-collecting for the curio trade (15). The species is reported to have declined or disappeared from Martin's Haven, South Wales, and from sites in North Wales. A reduction in the average size of the test diameter has also been reported. These effects are said to be due to over-collection by skin divers and possibly by collecting for educational and commercial purposes (20). If collecting pressure is reduced, recolonization may take place fairly quickly. However if large scale commercial harvesting started this could rapidly lead to a decline, as has happened with other urchin species (19). The Japanese have shown an interest in starting an industry based on sea urchin roes which are a delicacy in many parts of the world. If they turn to E. esculentus, serious overfishing could result if good husbandry methods are not adopted (23).

In some areas in the U.K. urchins have been found with their apical spines missing and with accompanying blackening of the epithelium. It is thought that this could be caused by weak detergent from the treatment of oil spillages although too few data have been collected to be certain (14).

No information is available on the conservation status of this species in other parts of its range.

CONSERVATION MEASURES TAKEN In the U.K., an integrated project was launched in connection with Underwater Conservation Year 1977 to study this species, making use of the diving expertise of interested amateurs and in particular, members of the Underwater Conservation Society (15, 21). A study of the reproductive cycle of the species was also undertaken under a contract from the U.K. Department of the Environment which is interested in the fisheries implications (23).

In 1979 the Cornwall Sea Fisheries Committee was advised that further exploitation of the species in U.K. waters should not be permitted. A commercial processing firm would probably require a catch of some 40 000 mature individuals a day which is many more than the populations could withstand (23). No information is available on conservation measures taken in other countries.

CONSERVATION MEASURES PROPOSED There is no biological evidence so far available that the present catch level is having a serious effect on the total population size. There appear to be large stocks in waters deeper than those normally exploited by professional collectors; individual urchins from these deeper stocks tend to migrate into shallower water under certain influences; and thirdly urchins will establish and prosper on shallow reefs if there are food resources available following depletion (1,21,23). However, it is strongly recommended that urchins are never completely removed from any reef (23). Investigations should be made into methods of combining the requirements of the curio trade with those of industries involving use of the roe, by removing the gonads from the rest without damage to either. Finally, biological surveys at key sites should be maintained to ensure that there is adequate monitoring of populations (23).

REFERENCES   1.   Lewis, G.A. and Nichols, D. (1981). Preliminary experiments on the colonisation of artificial reefs and a cleared site by the European sea-urchin Echinus esculentus (Echinodermata: Echinoidea). Progress in Underwater Science, 6 (N.S.): 29-35.
2.   Elmhirst, R.E. (1922). Notes on the breeding and growth of marine animals in the Clyde Sea area. Ann. Rep. Scottish Mar. Biol. Ass. 1922: 19-43.
3.   Forster, G.R. (1959). The ecology of Echinus esculentus L.: quantitative distribution and feeding. J. Mar. Biol. Ass. U.K.

38: 361-367.

4. Grimes, J. (1978). In litt., 28 September.
5. Jones, N.S. and Kain, J.M. (1967). Subtidal algae colonisation following the removal of Echinus. Helgol. wiss. Meeresunters 15: 460-466.
6. Krumbein, W.E. and Pers, J.N.C. (1974). Diving investigations on biodeterioration by sea-urchins in the rock sublittoral of Heligoland. Helgol. wiss. Meeresunters 26: 1-17.
7. Larsson, B.A.S. (1968). Scuba-studies on vertical distribution of Swedish rocky bottom echinoderms, a methodological study. Ophelia 5: 137-156 (quoted in 16).
8. Lawrence, J.M. (1975). On the relationship between marine plants and sea urchins. Oceanogr. Mar. Biol. Ann. Rev. 13: 213-286.
9. Macbride, E.W. (1906). Echinodermata. In Cambridge Natural History, Macmillan, London 692 pp.
10. Macbride, E.W. (1914). Textbook of Embryology Vol. I. Invertebrata Macmillan, London 692 pp.
11. Moore, H.B. (1934,1935,1937). A comparison of the biology of Echinus esculentus in different habitats. Parts I, II and III. J. Mar. Biol. Ass. U.K. 19: 869-885; 20: 109-128; 21: 711-720.
12. Nichols, D. (1978). Can we save the sea urchin? Diver October 1978: 490-491.
13. Nichols, D. (1979). The Echinus project - three years of results. Underwater Conservation Society Newsletter. September 1979.
14. Nichols, D. (1979). Aboral spine loss in the European sea urchin Echinus esculentus. In Earll, R.C. and Jones, H.D. (Eds), Observation Scheme Report 1977/78, Underwater Conservation Society. Pp. 43-46.
15. Nichols, D. (1979). A nationwide survey of the British sea-urchin, Echinus esculentus. In Gamble, J.C. and George, J.D. (Eds), Progress in Underwater Science 4(NS): 161-187. Pentech press, Plymouth.
16. Nichols, D. (1979). The biology of the edible sea-urchin. Echinus esculentus, a brief review. Appendix II of (15).
17. Hörstadius, S. (1975). Experimental Embryology of Echinoderms. Clarendon Press, Oxford.
18. Reid, D.M. (1935). The range of the sea-urchin Echinus esculentus. J. Anim. Ecol. 4: 7-16.
19. Southward, A.J. and Southward, E.C. (1975). Endangered urchins. New Sci. 66: 70-72.
20. NERC (1973). Marine Wildlife Conservation. Publications of Environmental Research Council, U.K. Series B. 5 January.
21. Lewis, G.A. and Nichols, D. (1979). Colonisation of an artificial reef by the sea-urchin, Echinus esculentus. Progress in Underwater Science 4(N.S.): 189-195.
22. Rodhouse, P.G. and Tyler, I.D. (1978). Distribution and production indices of the sea-urchin Echinus esculentus L. in the shallow sublittoral around Lundy. Progress in Underwater Science 3 (N.S.): 147-163.
23. Nichols, D. (1981). The Cornish sea-urchin fishery. Cornish Studies 9: 5-18.
24. Harvey, E.B. (1956). The American Arbacia and Other Sea-urchins. Princeton University Press.
25. Dipper, F. (1981). Report of a sublittoral survey of South Skye, Inner Hebrides. CST Report 342, Nature Conservancy Council, U.K.

26. Dipper, F. (1982). Sublittoral survey of habitats and species around the Summer Isles, Ross and Cromarty, CST Report 365, Nature Conservancy Council, U.K.

We are very grateful to D. Nichols and E. Southward for information provided for this data sheet, and to A. Clark for comments.

PURPLE URCHIN   COMMERCIALLY THREATENED

Paracentrotus lividus (Lamarck, 1816)

Phylum   ECHINODERMATA   Order   ECHINOIDA

Class   ECHINOIDEA   Family   ECHINIDAE

SUMMARY   The Purple Urchin still has a wide range, occurring in the Mediterranean and along the Atlantic shores of France and Britain. Its roe has long been regarded as a delicacy in France and this has led to local over-exploitation in some areas. Efforts should be made to implement some form of control over commercial fisheries. Research is revealing that the species can also be affected by pollution, which could further aggravate the effects of over-collection.

DESCRIPTION   Colour varies from brown or green to dark violet; for example, the Maltese population is predominantly blue, brown or sepia coloured (7). The suckered tube feet often hold pieces of algae and shells on the upper surface (2). Significant size differences appear to occur between populations. Specimens are larger on the north Brittany coast (8.5-9.0 cm in diameter) (14) than on Atlantic coasts (6.2 cm diameter) (15) and in the Mediterranean (6.5 cm) (16). In Maltese waters most individuals in the population were less than 2 cm in diameter and were presumably juveniles since some adults reached 6 cm in diameter (7). Within populations there are again considerable size differences; in Brittany the largest specimens are found in the bays of Morlaix and Lannion (17).

DISTRIBUTION   Found throughout much of the Mediterranean and on the Atlantic coasts of Ireland, France and Spain. Very rare in Britain although a few are found in south-west Scotland mainly around Skye. They have been found occasionally near Plymouth although there are no records from this area since 1944. They are still found in Guernsey, and probably Jersey but are generally rare in the Channel Islands. Specimens have been found in the Isles of Scilly. Specimens on the English side of the Channel are probably the result of planktonic larvae drifting over from the French population (1,21).

POPULATION   Unknown. Urchin density is high around the Maltese islands (6) and down to a depth of 30 m on Algerian coasts (18). Near Marseilles, France, densities of 40-110 individuals per m^2 are found (27). Around Corsica densities are lower and have been estimated at 0.5-2.0 per m^2 (27). The species is said to be abundant in Ireland (10).

HABITAT AND ECOLOGY   Occurs in shallow water and in tide pools on the shore and to a depth of 30 m, sometimes in cavities excavated into soft rock which protect it from wave action. The animal burrows by clinging to the rock surface with its tube feet and abrading the rock surface with its spines and teeth. Populations on the coast of Ireland and north-west France burrow. Urchins in the Mediterranean rarely do so, even if similar rock is available (2,5), but in Algerian waters, individuals living in the first 50 cm have been found in cavities (18). In the Mediterranean Paracentrotus lividus is usually found on rocky bottoms and slopes, especially in areas sheltered from light and rich in vegetation, and it is often an important species in the Posidonia community (24). In Algeria, Urchins living down to one metre are found on rocky bottoms, whereas those living deeper are found mainly among Posidonia vegetation (18). In Malta smaller individuals were found mainly at 10 m, where seaweed cover is sparse, but larger individuals were most abundant in shallow water (7).

Although one study has shown individuals to exhibit few day to day movements (8),

551

other studies suggest that daily and seasonal migrations occur. Individuals in Lough Ine, Ireland were found to show a diurnal migration, moving on to the tops of boulders in daylight hours (15). In the Mediterranean at Port-Cros, France, individuals in shallow water showed activity peaks at sunrise and sunset (25), and small individuals, of unknown origin, have been found to migrate into the <u>Posidonia</u> beds in spring (27). Purple Urchins feed primarily on algae, particularly large brown species, and other vegetation, mainly at night (6,8,20). Individuals in <u>Posidonia</u> beds have been calculated to consume 240-490 mg dry weight per day, mainly of leaves and epiphytes (27). Burrowing individuals may feed on particles washed to the bottom of the cavity. Individuals probably live for about 7-9 years (14). Crabs and wading birds seemed to be the main predators of Urchins at Lough Ine in Ireland (15).

SCIENTIFIC INTEREST AND POTENTIAL VALUE Purple Urchins have been collected for food in the Mediterranean for centuries and the roes are regarded as a delicacy in many countries (4). The fishery consists of a shallow water artisan fishery, incidental catches in prawn and scallop fishing nets, and a deeper water fishery (3). Annual catches in France between 1974 and 1980 ranged from 233 tonnes to 470 tonnes, with an average of 351 tonnes (11). The Brittany fishery provided a winter livelihood for the poorer fishing families and older fishermen, until the development of large scale commercial operations led to depletion of the Urchin population. In the 1960s the species was being collected in Ireland, mainly by divers by hand, for export to France (1). Between 1930 and 1960 it was collected on a large scale in Algerian waters by European fishermen, but it is now only collected there in small numbers for local consumption. However, population levels are high enough to support a commercial fishery (18).

The eggs of the Purple Urchin have been used extensively in laboratory studies by biochemists and physiologists, as they have particularly useful characteristics (12). It is thought that the larvae could be used in bioassay tests to estimate biologically active copper and other metals in sea water (23). Like other sea urchins, <u>P. lividus</u> plays an important role in controlling algal growth (22,24).

THREATS TO SURVIVAL Mediterranean sources of this species for the luxury food market became inadequate at the turn of the century and the trade turned to Brittany where the use of the drag net caused a drastic depletion of the population. Large urchins were collected and the small ones were lost by being smashed during the net cleaning process (1,19). The drag net also destroyed juveniles on the sea bed and the habitat in general. In the Morlaix region of Brittany, where only a few tonnes were sold each year at the beginning of the century, 150 tonnes were being landed in 1947. By the early 1970s, when only a few adults were left in inaccessible areas, the catch had declined considerably and the market was again dependent on supplies from the Mediterranean (1,14). There have been no recent reports of over-collection in the Mediterranean and a commercial fishery was still in existence around Marseilles in the 1960s (20).

Some studies have shown <u>P. lividus</u> to be very sensitive to oil pollution. Exposure of individuals to crude oil results in 60 per cent mortality within five days and 90 per cent within nine days, one of the reasons being that the animal finds it difficult to right itself. The righting response is an essential part of the animal's behaviour in its rocky and uneven environment (6). A study in the region of Marseilles has shown that <u>P. lividus</u> is one of the urchins more tolerant of industrial pollution but that even this species disappears in high levels of pollution in the Gulf of Fos. In heavy domestic pollution around Marseilles, <u>P. lividus</u> shows large oscillations in population size and individuals tend to be much smaller in size than in unpolluted waters. However, densities may be as high as 100-400 individuals per 10 m^2 in such areas compared with 10-50 per 10 m^2 in Port-Cros National Park. High densities in areas of domestic pollution are partly explained by the Urchins' ability to feed by absorbing microparticles of organic matter (9).

P. lividus has also been found to be an indicator of mercury pollution (13).

A necrotic disease, which causes loss of spines, affects P. lividus as well as other sea urchins. Its cause is unknown and it is equally likely to occur in clean or polluted waters. At Port-Cros in France, the disease caused a decline of 60 per cent in the Purple Urchin population in only six months (26,27).

CONSERVATION MEASURES TAKEN A distribution survey is being carried out for the British Isles, English Channel and North Sea (10). The species occurs in a number of marine parks in the Mediterranean such as Port-Cros National Park, France (9), and El Kala Park, Algeria (18).

CONSERVATION MEASURES PROPOSED Close seasons should be instituted and catch limits controlled for commercial fisheries (1). Artificial reseeding would be necessary to restore populations in Brittany to their former economically valuable size. This is thought to be feasible and it has been suggested that reseeding should start in those areas where urchins are known to grow to particularly large sizes. No attempts have yet been made. Culturing would probably not be economically viable since the species does not reach marketable size until 3-4 years old (17). Populations should be monitored for effects of pollution in industrialized areas, and these effects should be borne in mind when plans are being made for coastal developments which could lead to high levels of industrial pollutants, particularly in the Mediterranean.

REFERENCES
1. Southward, A. and Southward, E. (1975). Endangered urchins. New Scientist 66: 70-72.
2. George, J.D. and George J.J. (1979). Marine Life. Harrap, London.
3. Hunnam, P.J. (1980). Mediterranean species in possible need of protection. Report to IUCN, UNEP/UNESCO UNEP/16.20/INF.6.
4. Clark, A.M. (1977). Starfishes and related echinoderms. 3rd ed., British Museum (Natural History), London.
5. Nichols, D. (1962). Echinoderms. Hutchinson Univ. Lib., London.
6. Axiak, V. and Saliba, L.J. (1981). Effects of surface and sunken crude oil on the behaviour of a sea urchin. Mar. Poll. Bull. 12(1): 14-18.
7. Gamble, J.C. (1967). Ecological studies on Paracentrotus lividus (Lmk). Underwater Ass. Rep. 1966-67: 85-88.
8. Neill, S.R. and Larkhum, H. (1965). Ecology of some echinoderms in Maltese waters. Symp. Underwater Ass. Malta Pp. 51-55.
9. Harmelin, J-G., Bouchon, C. and Hong, J-S. (1981). Impact de la pollution sur la distribution des echinodermes des substrats durs en Provence (Mediterranée Nord-Occidentale). Tethys 10(1): 13-36.
10. Southward, E.C. and Holme, N.A. (1973). Survey of British Echinoderms. Unpub. Report, Marine Biological Association and British Records Centre.
11. FAO statistics, Rome (provided by E. Aykyz).
12. Hörstadius, S. (1973). Experimental embryology of echinoderms. Clarendon Press, Oxford.
13. Augier, H., Gilles, G. and Ramonda, G. (1979). Recherche sur la pollution mercurielle du milieu maritime dans la region de Marseille (Méditerranée France). Partie 3- Degré de contamination par le mercure des Echinodermes prelevés dans l'herbier de Posidonies à proximitié des ports et du rejet du grand collecteur d'égout de la ville de Marseille.

Env. Pollut. 18(3): 179-185.

14. Allain, J.Y. (1978). Age et croissance de Paracentrotus lividus (Lamarek) et de Psammechinus miliaris (Gmelin) des côtes nord de Bretagne (Echinoidea). Cah. Biol. Mar. 19(1): 11-21.

15. Eblings, F.J., Hawkins, A.D., Kitching, J.A., Muntz, L. and Pratt, V.M. (1966). The ecology of Lough Ine. 16. Predation and diurnal migration in a Paracentrotus community. J. Anim. Ecol. 35: 559-566.

16. Regis, M.B. (1969). Premières données sur la croissance de Paracentrotus lividus (Lamarck). Téthys 1(4): 1049-1056.

17. Allain, J.Y. (1972). Sur les populations de Paracentrotus lividus (Lamarck) et de Psammechinus miliaris (Gmelin) de Bretagne nord (Echinodermes). Bull. Mus. Hist. Nat. Series 3, 32(26): 305-315.

18. Allain, J.Y. (1982). In litt., 15 October.

19. Allain, J.Y. (1971). Note sur la pêche et la commercialisation des oursins en Bretagne nord. Trav. Lab. biol. Halient., Univ. Rennes 5: 59-69.

20. Kempf, M. (1962). Recherches d'écologie comparée sur Paracentrotus lividus (Lmk) et Arbacia lixula (L.). Rec. Trav. Sta. mar. Endoume 39(25): 47-116.

21. Southward, E. (1982). In litt., 15 June.

22. Kitching, J.A. and Ebling, F.J. (1961). The ecology of Lough Ine. 11. The control of algae by Paracentrotus lividus (Echinoidea). J. Anim. Ecol. 30: 373-383.

23. Bougis, P., Corre, M.C. and Etienne, M. (1979). Sea-urchin larvae as a tool for assessment of the quality of sea water. Ann. Inst. Oceanogr. Paris (N.S.) 55(1): 21-26.

24. Nedelec, H. and Verlaque, M. (1982). Quantification in situ de la consommation par Paracentrotus lividus dans un herbier à Posidonia oceanica en Corse (Méditerranée, France). 2éme Séminaire International sur les Echinodermes. 15-16 September, Abstract.

25. Shepherd, S.A. and Boudouresque, C.F. (1979). A preliminary note on the movement of the sea urchin Paracentrotus lividus. Trav. Sci. Parc nation. Port-Cros 5: 155-158.

26. Boudouresque, C.F., Nedelec, H. and Shepherd, S.A. (1980). The decline of a population of the sea urchin Paracentrotus lividus in the bay of Port-Cros (Var, France). Trav. Sci. Parc nation. Port-Cros 6:243-251.

27. Harmelin, J.G. (1983). In litt., 13 January. (including results of work by H. Nedelec and M. Verlaque).

We are very grateful to J.Y. Allain, J.G. Harmelin and E. Southward for assistance with this data sheet.

INTRODUCTION Many of the invertebrate phyla in this section are poorly known and will be unfamiliar to some readers. The majority are aquatic and include species with cosmopolitan distributions or parasitic life styles which are unlikely to be seriously affected by human activities. However, it can be deduced, from the material gathered on other groups, that species with small ranges or restricted to vulnerable habitats may be under pressure from man.

Planktonic animals.

The CTENOPHORA and CHAETOGNATHA are found in the surface plankton of all seas, sometimes in large swarms. The ctenophores or comb jellies are thought to be an offshoot of the Cnidaria, as many bear a superficial resemblance to jellyfish. Fewer than 100 species have been described, perhaps partly because it is very difficult to collect undamaged specimens. Ranging in size from a few mm to over a metre in width, most of them are delicate, transparent, gelatinous animals with eight characteristic bands of cilia which propel them through the water. The chaetognaths or arrow worms, of which about 50 species are known, are probably related to the higher invertebrates such as echinoderms. They have torpedo-shaped bodies bearing lateral and caudal fins and range in length from 3 to 10 cm. Although several species are cosmopolitan, many are restricted to particular types of water. Cold or warm water and coastal or open water species can be distinguished, which enable arrow worms to be used as indicators of the movements of different water masses in the world's oceans (5,6).

Burrowing and tube-dwelling animals

Three phyla include species resembling tiny flatworms which live in the interstitial spaces of marine and freshwater sediments. The GNATHOSTOMULIDA, with about 80 species, and the KINORHYNCHA, with about 100 species, are entirely marine. About 400 species have been described in the phylum GASTROTRICHA from sediments in the sea and fresh waters, in water films on soil particles, and from the surfaces of submerged plants and animals (5,6).

Several wormlike groups of animals are burrowers. The phylum PRIAPULIDA comprises about ten known species, mainly from cold seas off North America and Siberia, the Baltic Sea, and the Antarctic. Individuals range in size from a few mm to over 8 cm. The 325 known species in the phylum SIPUNCULA are called peanut worms and range in size from a few mm to over 50 cm. The 100 known species of spoon-worm in the phylum ECHIURA, named after the shape of their proboscis, used to be grouped with the sipunculans, which they closely resemble. Both groups are marine (although some brackish water echiurans are known) and include species with cosmopolitan distributions. Sipunculans often live in protected situations such as crevices, empty shells, mussel beds, mangrove roots or seaweed holdfasts. Echiurans and some sipunculans burrow into sand or mud. Many sipunculans are found in tropical reef limestone and densities of up to 700 individuals per m^2 of coralline rock have been reported from Hawaii (1). The POGONOPHORA are a phylum of about 100 species known mainly from deep waters in the north-west Pacific, although this may reflect collecting effort more than actual distribution. They live in tightly fitting cylindrical chitinous tubes which are usually buried vertically in mud sometimes in dense aggregations of about 200 individuals per m^2. The body is often at least 80 cm long and individuals over 2 m in length have recently been found at 9000 m on the floor of the Galapagos rift. They are remarkable in having neither mouth nor digestive tract, and probably feed by absorbing dissolved organic matter (5,6).

Only two genera and about 15 species have been described in the phylum

PHORONIDA. Commonly known as horseshoe worms, they bear a superficial resemblance to tube-dwelling polychaetes as they live in chitinous tubes, either buried, or attached to rocks and other hard substrates in intertidal and shallow seas. The body is usually less than 20 cm long and is characterized (like the bryozoans) by a horseshoe-shaped row of cilated tentacles (the lophophore) which surrounds the mouth and is used for filter feeding. The HEMICHORDATA, comprising 90 known species, were once considered a subphylum of the Chordata. They include the acorn worms which live in U-shaped burrows in shallow waters particularly in tropical seas, and a second group of benthic species which may live colonially and are found mainly in deep waters in the Southern Hemisphere. Hemichordates are wormlike, sluggish animals usually between 9 and 45 cm in length, although some reach 2 m. They are very fragile and are difficult to collect intact (5,6).

Parasites, parasitoids and other groups.

The unsegmented, colourless roundworms in the phylum NEMATODA (NEMATA), include some of the most widespread and numerous of all multicellular animals. Some 15 000 species have been described from soil, fresh water and the sea, in habitats as diverse as polar regions, deserts, hot springs, high mountains and ocean depths. Many species are parasitic in plants and animals, including some which cause diseases in man. The majority are free-living and many have cosmopolitan distributions as they are easily dispersed. They often occur in enormous numbers and play a key role in the breakdown and recycling of organic matter; in organically rich sediments as many as 100 million per m^2 have been found. Most free-living forms are only 1-3 mm in length, but marine species reach lengths of several cm. A characteristic writhing, sinusoidal motion enables nematodes to glide between sediment grains and through organic debris. The 230 known species of horsehair worms or hairworms in the phylum NEMATOMORPHA have long hairlike bodies. The adults are free-living, generally in fresh water or damp soil although a few pelagic marine species are known, and the juveniles are parasitic in arthropods (5,6).

The 1800 species in the ROTIFERA are about the size of ciliated protozoans and are mainly found in fresh water although some marine species are known. They are characterized by a structure on the head known as the wheel organ, on which cilia beat rhythmically to propel the animal. Dormant eggs which can withstand adverse conditions enable some species to live in temporary pools. The ACANTHOCEPHALA and MESOZOA consist of parasitic, wormlike animals. The PLACOZOA is a phylum with only one known species of minute, flattened and flagellated organisms. The ENTOPROCTA, comprising some 60 species, was formerly included in the Bryozoa. They are small, sessile, solitary or colonial animals, rarely exceeding 5 mm in length, and bear a superficial resemblance to hydroid polyps. Most are marine and are widely distributed in coastal waters, attached to rocks, shells, sponges and even crabs (5,6).

Protochordates

The phylum CHORDATA includes the Vertebrata, that is, animals with backbones. Two other subphyla in this group are generally considered invertebrates as they have no backbone as such, but at some stage of their life cycle have the chordate characteristics of a skeletal rod of notochord-like cells lying beneath a dorsal tubular nerve cord, and a pharynx perforated by gill slits.

The largest group is the Urochordata or tunicates, containing about 1300 marine species, the bodies of which are enclosed within a gelatinous or leathery tunic and have a perforated pharynx through which a current flows, permitting respiration and feeding. The notochord and nerve chord are found only in the larval stages which look like small tadpoles. Over 1200 tunicate species, called sea squirts, are in the class Ascidacea. These are barrel-shaped, solitary or colonial animals which live attached by their tunics to the sea bottom, submerged objects or other organisms. They are usually found in shallow waters, and a particularly high species diversity is found on tropical reefs. Many are very colourful. Solitary forms reach a length of over 20 cm but individuals within colonies are smaller, although colonies may extend for more than 50 cm. Two openings at the unattached end permit a flow of water through the pharynx, huge quantities often being strained in a day. The two other classes of tunicates are free-swimming planktonic forms found mainly in the surface waters of the open oceans. The Thaliacea include the transparent salps of tropical and subtropical waters. The Larvacea are minute animals, often found in dense aggregations, that feed on the fine phytoplankton unavailable to many other planktonic filter feeders. Their bodies are enclosed in unique transparent mucous 'houses' which are constantly shed and replaced (5,6).

The second protochordate group is the Cephalochordata which contains only about 25 species, commonly known as amphioxus or lancelets. These are translucent fishlike animals found in coarse sands and shell gravels in shallow seas of both tropical and temperate regions (6).

CONSERVATION Surprisingly, even such comparatively obscure animals may be of value to man. For example, adult amphioxus are fished commercially in areas where they congregate in large numbers (6). Sea squirts are also collected for food and FAO recorded a world catch of 5527 tonnes in 1980. Pyura stolonifera is collected in South Africa, P. chilensis in Chile, Halocynthia roretzi in Japan, Microcosmos sulcatus (the Grooved Sea-squirt) in France and Ireland, and other species in Korea (8). Some species may prove to be of use in medical research. Extracts from adult females of the echiuran Bonellia viridis and of a number of ascidians including Ecteinascidia turbinata, Molgula occidentalis, Clavelina picta, and Aplidium sp. have been found to have an anticancer activity (2).

Many species in these phyla are under no immediate threat because of their cosmopolitan distribution or their inaccessibility in deep waters and they therefore do not have a high priority for conservation activities. Over-exploitation is unlikely to affect many species although there has been a significant but unexplained decline in the French catch of the edible sea squirt Microcosmos sulcatus in recent years (11).

However, pollution could affect aquatic species in coastal habitats. Thalassema hartmani, an estuarine echiuran, is designated as of Special Concern in California (3) and T. neptuni is reported to have declined in the Plymouth area, U.K. through unknown causes, although pollution has been implicated (4). The ascidian Ascidia conchilega has become very rare in polluted areas near Marseilles, France, although it still occurs abundantly in the clean waters of Port-Cros National Park (9). Pollution could also affect species with restricted distributions. For example, the sea squirt Diplosoma multipapillata currently is known only from the cascades of the south-western fringing reef of Viti Levu, Fiji (10) and could be vulnerable to damage to the reef. The deep sea trenches from which unique faunal communities including pogonophorans have been described are largely protected through their inaccessibility, but pollution is a potential threat. The importance of these communities is discussed in (7).

Large declines in populations of the sea squirt Ciona intestinalis have been reported in several countries but it is not clear whether these are due to human activities or to natural fluctuations in population levels. Its disappearance off Marseilles. France, in the 1950s may have been associated with pollution from detergents in sewage effluent (13). In Australia, previously abundant populations disappeared from many harbours in the 1970s. The species may have been introduced to Australian waters from its natural range on the Atlantic coast of the

U.S.A. It has been postulated that more effective anti-fouling paint and faster shipping could cause a breakdown in gene flow between populations and a consequent decline in genetic vigour in Australian populations (13). The species has also disappeared from some localities in California, U.S.A. (12) and on the south coast of England, U.K. (4,14). However, in other U.S. localities and in some parts of Norway, Sweden and Italy, large fluctuations in population numbers have been observed (12,14) which, it is thought, may be normal for this species. This example illustrates the difficulties of interpreting changes in population levels of marine invertebrates and the need for long-term studies to determine the effects of man's activities. In conclusion, it seems that there is no immediate threat to these organisms, but it should be borne in mind that their responses to man's activities are likely to be similar to those of some of the better known invertebrates.

REFERENCES

1. Stephens, A.C. and Edmonds, S.J. (1972). The Phyla Sipuncula and Echiura. Trustees of the British Museum (Natural History), London.

2. Ruggieri, G.D. (1976). Drugs from the sea. Science 194: 491-497.

3. Porter, H.J., Johnson, C. and McCravy, A.B. (1977). Marine invertebrates. Proc. Symp. Endangered and Threatened Plants and Animals in North Carolina. State Museum of Natural History, Raleigh.

4. NERC (1973). Marine Wildlife Conservation. The Natural Environment Research Council, Publications Series 'B', No. 5.

5. Barnes, R.D. (1980). Invertebrate Zoology. 4th Ed. Saunders College, Philadelphia.

6. George, J.D. and George, J.J. (1979). Marine Life. Harrap, London.

7. Angel, M.V. (1982). Ocean trench conservation. Commission on Ecology Papers 1, IUCN, Gland, Switzerland.

8. FAO (1981). 1980 Yearbook of Fishery Statistics, Catches and Landings Vol. 50. FAO, Rome.

9. Baccar, H. (1977). A survey of existing and potential marine parks and reserves in the Mediterranean region. IUCN background paper to the Expert Consultation on Mediterranean Marine Parks and Wetlands, Hammamet, Tunisia, January 1977.

10. Kott, P. (1980). Algal-bearing didemnid ascidians of the Indo-West Pacific. Mem. Qd. Mus. 20: 29

11. Hunnam, P.J. (1980). Mediterranean species in possible need of protection. Report to IUCN by Aquatic Biological Consultancy Services Ltd., Chelmsford, U.K.

12. Lambert, C. and Lambert, G. (1977). Decline in Ciona populations in southern California. Pers. comm. in Ascidian News 7: 1.

13. Peres, J.M., Mather, P. and Gulliksen, B. (1977). Decline in Ciona populations, further observations. Pers. comm. in Ascidian News 8: 2-3.

14. Thorndyke, M. and Tursi, A. (1978). Decline in Ciona populations, further observations. Pers. comm. in Ascidian News 9: 2.

We are very grateful to P. Mather for help with the Protochordate section of this account.

Threatened Community accounts have been compiled to illustrate situations where entire invertebrate communities are in need of conservation. They emphasize that large numbers of invertebrates may become endangered through single events or through human activities over a small area, and it is hoped that the data could be used in the formulation of management plans or reserve proposals. Eleven examples have been selected to represent a range of countries and biomes. These have been chosen on the basis of their need for conservation, the number of endemic or scientifically interesting species found there and the amount of information available. As with species sheets, the examples chosen do not indicate any priority in their need for conservation, and it is fully understood that there are many other communities in equal or greater need. The word 'Threatened' is used here only to indicate that some sort of conservation action is required; we have not distinguished between 'Endangered', 'Vulnerable' or 'Rare' communities although it will be obvious from the accounts that some of the examples have a higher conservation priority than others.

Tropical forest communities are represented by the lowland and montane rain forests of Gunung Mulu in Sarawak, Malaysia and the montane rain forests of the Usambara Mountains, Tanzania. Both sites illustrate the importance and diversity of invertebrates within forest ecosystems and the high levels of endemism which are reached within a range of taxonomic groups. Similar forest communities could have been chosen from many other areas in Africa, South East Asia and South America where there are severe pressures on forests, but detailed studies of the invertebrate fauna have been made in remarkably few places. As discussed in the Insecta introduction, it is evident that the present rate of logging of tropical forests will cause the extinction of many invertebrates before they are known to science. Tropical forest conservation is now a major issue and action is being taken on many fronts to fulfil the needs of mammals, birds, man and the forests themselves. The two examples given here emphasize the need for invertebrates to be considered as well.

Xeric (dry) biomes are represented by the Dead Sea Depression in Israel and Jordan, and the El Segundo Sand Dunes in the U.S.A. Because of their apparently impoverished fauna and flora, such areas are often overlooked in conservation programmes. These two examples illustrate the specialized communities that they can support.

Two examples of cave communities are given, Deadhorse Cave in the U.S.A. and Cueva los Chorros in Puerto Rico. Very little is known about the specific requirements of most cave organisms but it is clear that highly specialized communities may evolve in such isolated localities. Most cavernicoles are adapted to the ambient conditions in their own particular cave system, and are highly sensitive to alterations. Caves are coming under increasing threat as remote areas become more accessible, interest in speleology increases and the pollution and tapping of subterranean waters becomes more widespread. In the U.S.A. many are already in need of protection, such as Malheur Cave in Oregon and Shoshone Cave in the Mojave Desert, California, and closer attention should be paid to cave habitats in other countries (see Chelicerata introduction).

San Bruno Mountain and Banks Peninsula, which for a variety of reasons contain a high diversity of invertebrates and an unusually large number of endemic species, are areas close to large conurbations. Banks Peninsula is important as the site of many invertebrate type localities, and both areas are ideal locations for people in the nearby big cities to carry out research and fieldwork. Like the El Segundo Sand Dunes, they illustrate the conflict which may arise when land needed for housing and urban development contains important endemic species and is the only readily accessible natural area available for recreational and educational purposes.

Wetland and freshwater communities are under-represented in this section. The IUCN Directory of Wetlands of International Importance in the Western Palearctic is an important contribution to the conservation of these biomes but further information on localities important for invertebrates is required. The Mires of the Sumava Mountains in Czechoslovakia provide an example of the unique invertebrate communities which can be found in peat bogs. These mires are glacial relicts with a fauna similar to that of the subarctic regions. The springs of the Dead Sea Depression and the tropical ghors in this region are also important for the relict fauna they contain. Of major importance to science are the faunas of the ancient lakes such as Lake Baikal in U.S.S.R., Lake Ohrid in Yugoslavia and Lake Malawi on the borders of Malawi, Tanzania and Mozambique. These contain large numbers of endemic invertebrates and are highly vulnerable to pollution and alterations to water levels and flow. Possibly the best example of the total alteration of a lake community, involving the extinction of several invertebrates, is the 1972 flooding of Lake Pedder in Tasmania. There are fears that the current proposals for the Lower Gordon River, also in Tasmania, could cause similar or even greater damage.

The problem of assessing the conservation status of marine species was discussed in the introduction, where it was emphasized that threats to marine communities are usually far greater than those to individual species. Taka Bone Rate Atoll in Indonesia is an example of a coral reef under threat, and more information on the importance of reef communities and their need for conservation is given in the introduction to the Cnidaria section. The second marine account is of the Roseland Marine Conservation Area, which is an example of an threatened estuary. As illustrated by this account, estuaries, like salt marshes and many other coastal biomes, are of prime importance to man as well as to the animals which live in them. Since they also tend to be areas of high population density, their faunal communities are often under considerable pressure. Estuaries, lagoons, coral reefs and marshes, unlike most marine ecosystems, may contain endemic species which cannot readily disperse between suitable localities.

SUMMARY The Usambara Mountain forests constitute what is probably one of the richest biological communities in East Africa in terms of the diversity of plant and animal species and endemic taxa. This is considered to be due to long periods of isolation and geological stability coupled with periods of species immigration during times of re-establishment of forest cover. The East Usambaras show the greatest diversity and endemism, with a unique flora and fauna deserving particular study and conservation. However, the Usambara Mountains have a very high human population density and the forests are subject to increasing pressure from legal and illegal encroachment for tea, cardamom, and subsistence agriculture as well as timber operations. The Usambaras, with the other mountains of Tanzania, have been included in the IUCN/WWF Tropical Forests and Primates Programme, and considerable efforts are being made to implement conservation measures.

DESCRIPTION The Usambara Mountains consist of two highland blocks, together comprising less than 2000 km^2, some 100 km from Tanga on the coast. The East Usambaras rise to 1500 m and are separated by the Lwengera valley from the larger block of the West Usambaras which reaches an altitude of 2250 m. They form part of a chain of islands of forested basement block mountains stretching from near the Kenya-Tanzania border to the south of Malawi and beyond. They were still largely covered with natural vegetation, mainly different types of forest, when they were first visited by botanists but destruction has been widespread since then (1).

There are at least 276 forest tree species over 10 m tall, about fifty of which are endemic and three of which belong to monotypic endemic genera (Cephalosphaera usambarensis, Englerodendron usambaranse and Platypterocarpus tanganyikensis). Endemic herbs and smaller plants are probably no less numerous. The most widespread vegetation was formerly forest: lowland and transitional rain forest in the East Usambaras and on the wetter slopes of the West Usambaras, and montane forest on the upper slopes of the West Usambaras. On the most exposed ridges forest gives way to ericaceous brushland and thicket dominated by Erica arborea or Philippia. To the east and south, rain forest is replaced by various types of dry evergreen woodland; to the north and west there is a transition to deciduous bushland via semi-evergreen scrub forest and bushland in which cactiform euphorbias are conspicuous (1). Much of the lowland forest has been destroyed and a sub-climax savannah has developed following fire in several places (6).

Climatically the rainfall of the Usambaras is transitional between that of West Africa and Zaire and that of southern Africa. Daylength regime is typical of the sub-equatorial region but, unlike anywhere else in eastern Africa, it occurs with well watered, warm lowland forests. Distribution of rainfall is highly irregular spatially, being very high on the East Usambara plateau (nearly 80 inches (2000 mm) a year), low in the central rain shadow of the Lwengera valley (less than 40 inches (1000 mm)), high at the eastern scarp of the West Usambaras and low at their extreme western end. Seasonal discontinuity with the rest of Africa is thus coupled with rainfall discontinuity within the region, encouraging what appears to be a zone with great disjunction between communities which has led to rapid speciation. Furthermore the many warm humid tropical niches of the East Usambaras, like the oceanic coastal forests, have offered relict habitats for several species which have an Indo-Oriental rather than strictly African affinity (6).

INVERTEBRATE FAUNA The Usambara Mountain forests have been

comparatively well studied as Amani, in the East Usambaras, was the site of the German and later the British Agricultural Research Programme. Nevertheless, only very limited parts of the invertebrate fauna have been studied in depth, and there are probably many species yet to be described. A survey during the late 1960s revealed a wide diversity of arthropods in the Amani West Forest Reserve (5). Collembola (springtails) and Staphylinidae (rove beetles) were particularly numerous and large numbers of other arthropods including flies (Diptera: Cyclorrhapha), ants (Hymenoptera: Formicidae), parasitic wasps (Hymenoptera) and terrestrial Isopoda and Amphipoda (Crustacea) were found. More detailed studies have been carried out on a few groups, and species lists are given in (1).

ARACHNIDA
Araneae
   Recent collections in the Usambara Mts have not yet been studied in detail, but are expected to contain a substantial proportion of endemic species (16).
INSECTA
Orthoptera
   The three genera of the Euschmidtiinae, Euschmidtia, Stenomastax and Chromomastax, contain 24 species, 17 of which are found only in the coastal forests and the Usambaras (17). The Thericleidae has a number of tribes centred on the Usambaras (18), and many species are represented by unique holotypes from there. The acridine grasshoppers Odontomelus and Parodontomelus are richly represented, with a large proportion of endemic species (19). Rhainopomma, a new genus of Lentulidae centred on the Usambaras and coastal forests, has recently been described (20). The 41 genera of Phaneropterinae with open tympana have been reviewed (21); four genera are endemic to the Usambaras and a fifth genus has four of its seven species endemic there.
Dermaptera
   The earwig fauna of the Uluguru mountains contains many endemics and a similar situation may exist in the Usambaras (14).
Coleoptera
   The Coleoptera of the Usambara Mts have not been thoroughly investigated, but in the Uluguru Mts 108 species of Carabidae, 47 Tenebrionidae and 43 Pselaphidae have been found (14). Of these, 40, 17 and 41 species respectively are endemic (14). Similar results may be expected from the Usambaras (14). Ozaniella bimaculata (Carabidae) is known only from the Usambara and Uluguru mountains (14).
Trichoptera
   As in the Araneae, recent collections are believed to contain a number of new species (16).
Lepidoptera
   The butterflies Cymothoe aurivilli and C. amaniensis (Nymphalidae) are confined to the Usambaras (11). Three endemic Lycaenidae also occur there, Sytarucus sp. nov. is found in swamps and marshes, and Spindasis collinsi and Uranothauma sp. nov. near williamsi occur in highland forest (15).
Hymenoptera
   Solitary predatory wasps of the family Sphecidae are represented by 131 species in the East Usambaras. Of these, 57 are savanna species and 74 are forest species including 42 arboricolous species, 26 terricolous species and six mud-daubers. There are possibly 27 endemic species of Sphecidae, which is remarkable since the family has conventionally been regarded as having good dispersal ability (1).
DIPLOPODA
   Millipedes are a conspicuous element in the invertebrate fauna. A total of 41 millipedes have been collected from the Amani region, of which 26 are new and still undescribed. Most of these species are either endemic to the Usambaras only or to a few of the East African rain forest localities (16). Little collecting has been carried out in the West Usambaras but it is

apparent that there are distinct differences. The Diplopoda are a useful group for zoogeographical study because of their abundance, ease of collection and the fact that they are an ancient group incapable of rapid dispersal (1).

MOLLUSCA

Gastropoda

115 molluscs have been recorded from the Usambaras, roughly ten per cent of the entire known non-marine mollusca of East Africa (Kenya, Tanzania and Uganda) (3). Molluscan assemblages in tropical African forests are markedly different from those in most of Europe owing to the large number of carnivorous species of the family Streptaxidae present. In the Usambaras there are at least 40 streptaxid species of which 62.5 per cent are endemic to the area (including the outlying Mt Mlinga and Mt Tongwe) (1). The largest species of the genus Edentulina are predators on large slugs, the smaller ones on young Achatina. The genus Gulella is particularly well represented and contains a number of interesting forms, sometimes brightly coloured and often with complex patterns of denticles around the apertures. Eighteen of the 24 Gulella species are endemic and doubtless many other minute species remain to be discovered (12). Other genera include Tayloria (three out of four species endemic); Gonaxis (two out of eight); Ptychotrema (the only species is endemic) and Streptostele (not endemic). Some gastropods are apparently endemic to specific areas of the Usambaras and others are represented by different subspecies in the East and West blocks of the mountains. The non-endemic species often show interesting zoogeographical relationships. For example, one is a subspecies of a Natal species and another is only otherwise known from Kakamega Forest in Kenya (an area which has been heavily damaged through deforestation), but it is minute and may be overlooked. There is some evidence that before destruction of the forests, many species thought to be endemic to the Usambaras extended over wider ranges (1).

SCIENTIFIC INTEREST AND POTENTIAL VALUE The Usambara Mountain forests are particularly outstanding examples of biological diversity and the intermediate rain forest of the East Usambaras is particularly rich on account of the species of lowland affinity which are found there (2). Predatory wasps and bryophytes show species affinities with those of Madagascar indicating that some elements of the Usambara forests are very ancient, and the community as a whole probably dates from the Miocene (3). The Usambaras could equal the Galapagos and the Indo-Pacific islands in providing insight into aspects of evolution, biogeography and community ecology. The existence of other nearby montane forest islands of similar (Ulugurus, Uzungwas, Ngurus, Ukagurus) or different (Kilimanjaro, Meru, Rungwe) geological history and of equally high invertebrate diversity, greatly adds to their importance as a research site (2).

In addition to the large number of endemic plants and invertebrates found in the Usambaras, there are also many interesting vertebrates. The amphibian fauna is unique in Africa in its diversity and eight of the 15 known forest species are endemic. Of the reptiles, 14 lizards, seven species and subspecies of chameleon and a number of snakes are endemic (1). The avifauna has been extensively studied and includes one endemic species and several species with very restricted ranges, many of which are threatened, and at least five of which are to be included in the ICBP/IUCN Bird Red Data Book for Africa (4,9).

Within Tanzania, the forests are valued not so much for their unique biological characteristics as for their importance as major water catchments for agricultural, hydro-electric and urban development. The Usambaras supply water for the town of Tanga and for hydro-electric stations on the Pangani River (2).

THREATS TO SURVIVAL Rapidly increasing human pressure on the land is the main threat to the forests of the Usambara Mountains. This is a result of several factors, including the increase in human population (naturally and by immigration), the development of a new cash crop (cardamom), the expansion of the timber industry and the exhaustion of land poorly farmed by subsistence cultivators who must then find new areas for food crops (1). The once continuous forest is now fragmented into a patchwork of primary and secondary forest and commercial and subsistence agriculture, seriously reducing the amount of habitat available to the unique native fauna (2).

The proportion of forest cover has declined drastically since the end of the last century, and in the Amani area a decrease of 50 per cent was recorded between 1954 and 1978 (4,10). Despite an aerial survey giving an estimate of 70 985 ha of forest, it has now been calculated that only 45 568 ha of the Usambaras are truly forested, if cardamom plantations are taken into account and when aerial surveys are backed up by ground checks (4). Generally the rate of decline has been greater in the West Usambaras than in the East, largely because the former have long been an important area of human settlement and currently have a higher population density than the East Usambaras (4). Large areas of the West Usambaras have been planted with conifers or made into tea plantations. Erosion from road construction and mechanical logging is often serious, and native trees are used for fuel in tea factories. Plantations of cardamom, a recently introduced highly priced cash crop, are increasing rapidly, particularly in the East Usambaras. Although the canopy trees of the forest are left to provide shade for the crop, the entire understorey is removed. Subsistence agriculture occurs haphazardly throughout the region, often on steep slopes which accentuate the problems of erosion. Although a number of forest reserves exist, these were set up mainly as water catchment areas and in fact are often exploited. There is no strict protection of areas known to be unique and cardamom is often illegally planted in reserves (1).

CONSERVATION MEASURES TAKEN In the East Usambaras there are more than 20 gazetted forest reserves with a total of about 16 500 ha, part of which is declared inviolate watershed reserve, part of which is open for commercial and individual logging and pit sawing, and part of which is being replanted with Tanzanian and exotic hard woods (1,2). The West Usambaras contain the Shume Nature Reserve which protects a small area of forest and Mazumbai, probably the most important reserve in the whole area. This covers an isolated area of 300 ha, from 1300 to 1900 m in altitude, and is controlled as a research area by the University of Dar es Salaam. It has been well protected in the past and has a full time Forest Guard but is under considerable pressure for logging (13).

The Usambara Mountains, along with other mountains in Tanzania have been included within the IUCN/WWF Tropical Forests and Primates Programme. Four project proposals have been drawn up. These include projects to evaluate and describe the fragmented forest resources of the Usambara and Uzungwa Mountains and to prepare for a land use plan for the Usambaras. This will involve co-ordination between land use planning authorities so that forest for conservation and watershed purposes is given a major role. A regional land planning programme (Tanga Integrated Rural Development Programme) already exists and is in a position to implement rational conservation suggestions.

CONSERVATION MEASURES PROPOSED Ideally there should be a complete halt to all further forest clearing until the situation has been assessed and action taken to implement a land use plan integrating conservation, forestry and agricultural interests. Gazetted forests should be surveyed and strict natural reserves demarcated to preserve unique areas. All forest on slopes with a gradient of over 60 per cent or within 50 m of streams and around stream heads should be preserved as catchment forest in both gazetted and ungazetted areas (1). It is still not known which forest areas are of greatest value and in most urgent need of conservation, and further research is clearly required. However, conservation action should start immediately. Since endemic species are present in both highland and lowland forest types it is evident that sufficient examples of each habitat type must be conserved. It has been stressed that the following forest types in the East Usambaras require conservation (10):

1. Higher altitude forest (900 m and above) in the north of the East Usambaras (e.g. the Mtai and Lutindi Forest Reserves).
2. Medium altitude forest (e.g. the large block composed of the contiguous Kihuhwi, Kwamsambia, Kwamkoro and Amani-Sigi Forest Reserves).
3. Low altitude forests at the eastern foot of the Usambaras. The Mwarimba Forest Reserve is a particularly important site as it not only contains a high species diversity, but also occupies a very small area and is currently extremely vulnerable (19).
4. Amani West Forest Reserve, which despite its small size (99 ha) is of extreme importance as the reserve most accessible from the Amani Research Station and the site of most ecological research in the past; this Reserve merits immediate total protection. In addition, the Amani Arboretum and Institute grounds should receive official protection (19).

A report in 1981 stressed the following areas in the West Usambaras, as being of top priority: Shume-Magamba, Shagayu, Ambangulu, Mazumbai and Kitivo South (4). Mazumbai, with its adjacent forests, is in particular need of protection and efforts should be made to ensure that this area is effectively controlled (19).

Improved management of the forestry, tea and subsistence agriculture resources would lead to greater production with higher yields and economic returns which would reduce pressure on natural forests. Estates could be improved to reduce erosion and increase tea productivity. Eucalyptus should be planted to provide fuel wood, and a system of local subsistence or co-operative tea outgrowers would increase overall tea output. Improved soil conservation measures, such as contour planting with intervening grass leys, increased fertilizer inputs and avoidance of cultivation on slopes in excess of 30 per cent where possible, are essential as far as subsistence agriculture is concerned. Livestock production and cardamom cultivation are also open to improvement (1). Land use planners should be made aware of the importance of forest conservation in the Usambara Mountains, and the need for agricultural and forestry developments to be undertaken with the aim of decreasing present land pressures on natural forest lands (2).

REFERENCES 1. Rodgers, W.A. and Homewood, K.M. (1982). Species richness and endemism in the Usambara Mountain forests, Tanzania. Biol. J. Linn. Soc. 18: 197-242.
2. Rodgers, W.A. (1982). WWF/IUCN Tropical Forests and Primates Programme, Tanzania. IUCN/WWF, Gland, Switzerland. Unpub. report.
3. Verdcourt, B. (1972). The zoogeography of the non-marine Mollusca of East Africa. J. Conch. 7: 291-348.
4. van der Willigen, T. A. and Lovett, J. (Eds) (1981). Report of the Oxford expedition to Tanzania 1979. Unpub. report. 95 pp.
5. Jago, N.D. and Masinde, S.K. (1968). Aspects of the ecology of the montane evergreen forest near Amani, East Usambaras. Tanzania Notes and Records 68: 1-30.
6. Moreau, R.E. (1935). A synecological study of Usambara, Tanganyika Territory with particular reference to birds. J. Ecol. 23: 1-43.
7. Baumann, O. (1891). Usambara und seine Nachbargebiete. Berlin, D. Reimer.

8. Enghoff, I. and Enghoff, H. (1975). Notes on myriapods observed and collected in Tanzania and Kenya during the summer 1974. Unpub. report. Zool. Mus. Copenhagen, Denmark.

9. Collar, N.J. (1982). Extracts from the Red Data Book for the Birds of Africa and Associated Islands. (International Bird Red Data Book, 3rd ed., part 1). International Council for Bird Preservation, Cambridge.

10. Stuart, S.N. and van der Willigen, T.A. (1978). Report of the Cambridge ecological expedition to Tanzania 1978. Unpub. report, Oxford.

11. Clifton, M.P. (1982). In litt., 30 August.

12. Verdcourt, B. (1957). The Gulellae (Moll. Streptaxidae) of the Usambara Mountains, N.E. Tanganyika. Tanganyika Notes and Records 47 and 48: 92-102.

13. Redhead, J.F. (1981). The Mazumbai Forest: an island of lower montane rain forest in the West Usambaras. Afr. J. Ecol. 19 (1 and 2): 195-199.

14. Scharff, N., Stoltze M., and Jensen, F.P. (1981). The Uluguru Mts., Tanzania. Report of a study tour, 1981. 51 pp.

15. Kielland, J. (unpubl.). List of butterflies taken and observed at Masumbai Tea Estate in Lushoto District, Usambara Mts. In litt., 4 pp.

16. Stoltze, M. (1982). In litt., 2 November.

17. Descamps, M. (1964). Revision préliminaire des Euschmidtiinae (Orthoptera, Eumastacidae). Mem. Mus. Natn. Hist. Nat., Paris A 30: 1-321.

18. Descamps, M. (1977). Monographie des Thericleidae (Orthoptera, Acridomorpha, Eumastacoidea). Ann. Mus. R. Afr. Centr., Tervuren 216: 1-475.

19. Jago, N.D. (1982). In litt., 26 November.

20. Jago, N.D. (1981). A revision of the genus Usambilla Sjöstedt (Orthoptera: Acridoidea) and its allies. Bull. Br. Mus. Nat. Hist. Ent. 43(1): 1-38.

21. Ragge, D.R. (1980). A review of the African Phaneropterinae with open tympana (Orthoptera: Tettigoniidae). Bull. Br. Mus. Nat. Hist. Ent. 40(2): 67-192.

We are very grateful to M.P. Clifton, N.J. Collar, K.M. Howell, N.D. Jago, J. Kielland, W.A. Rodgers, M. Stoltze, S.N. Stuart, and B. Verdcourt for assistance with this data sheet.

SUMMARY The Gunung Mulu National Park includes examples of almost all Sarawak's inland forest types. It has been estimated that the park contains more than 20 000 species of invertebrates. As in rain forests worldwide, a very large proportion of these are undescribed. Although there are a few local pressures, this particular area is in no immediate danger, being under the protection of the Sarawak Forest Department. This review presents Mulu as an unusually well-studied example of the world's tropical forests, vast areas of which are imminently threatened by short term exploitation which cannot be sustained, and whose destruction will result in the undocumented extinction of many invertebrates.

DESCRIPTION The Gunung (Mount) Mulu National Park, constituted in 1974, spans the border of Fourth and Fifth Divisions in northern Sarawak, a Malaysian State in Borneo. In 1977/78 the park was the venue for a large scientific project organized by the Royal Geographical Society of London and the Sarawak Forest Department (21). Although publication of the results is still not complete, a large body of literature has been prepared by the scientific participants (26). In particular, a management plan, believed to be the most detailed for any rain forest reserve, has recently been published by the Royal Geographical Society (3) and approved by the Sarawak Forest Department.

The comparatively small (529 000 ha) park lies between the headwaters of the Tutuh and Medalam Rivers. The slate and sandstone massif of Gunung Mulu, which extends from 60 m to 2377 m above sea level, dominates the southern half of the park. G. Mulu abuts onto the Melinau limestone massif which rises, sometimes precipitously, to c. 1700 m at the summit of Gunung Api, the highest limestone peak in the region (17). More northerly peaks in the limestone massif are G. Benarat and Bukit Buda. North-west of the limestone outcrop is an alluvial plain, particularly rich in vertebrate fauna. The northern part of the park runs along the Brunei border and consists of low, sharply dissected hills and extensive Quaternary river terraces. Accounts of the flora (2,25), climate (34), hydrology (33), geology (35) geomorphology (8) and soils (4) are given in the management plan (3) and the two volumes of scientific results in preparation (26).

The significance of the park as a conservation area and the great richness of the invertebrate fauna are due to the remarkable diversity of vegetation types (1,14). As a result of geological and altitudinal heterogeneity, all the major inland vegetation formations of Sarawak are present within the park (1), although certain facies of each formation may be missing or poorly represented. The alluvial plains are inundated periodically by flooding rivers and they support the most floristically rich vegetation type, alluvial lowland rain forest. The kerangas (tropical heath) forest grows on the humus podzols of Quaternary river terraces, and is very variable in character, dependent upon physical and chemical properties of the soil and drainage. The lower slopes of Mulu support lowland dipterocarp rain forest, which is the predominant formation in Malaysia, and is the most heavily exploited (38). On the upper slopes of Gunung Mulu the interdigitating altitudinal sequence of forests has been classified into lowland dipterocarp, lower montane and upper montane rain forest formations (1,13). Parallel but floristically different formations are found on the limestone outcrops (1,17). The park also contains a small area of peat swamp forest, an extensive formation in coastal Sarawak, much of which has now been logged (38).

The caves of the Mulu area have been described fully (9,19) but a detailed discussion of their conservation lies outside the scope of this review. However, it should be noted that the 100 km of passage so far surveyed contain a great variety of invertebrates dependent primarily or secondarily upon the huge input of guano by bats and cave swiftlets (10). The foraging area of the bats is unknown, but may extend beyond the present park boundaries. Many invertebrates are endemic to the caves, and a few of the more notable species will be described below.

A tribe of semi-settled and truly nomadic people, the Penan, hunt for game and collect wild Sago Palms (Eugeissonna utilis) in the area. There are believed to be about 140 Penan exercising their traditional privileges within the park boundaries (3).

INVERTEBRATE FAUNA The number of invertebrate species in the park has been conservatively estimated at about 20 000, including about 10 000 Coleoptera (50) and 3000 Lepidoptera (14,22,23). This is a remarkable diversity for such a relatively small park, but it may not be exceptional. It is commonly stated that rain forests harbour rich and diverse faunas, but there is a paucity of data on invertebrates to support this contention. The brief ensuing description of the invertebrates of Mulu is an example of what is almost certainly a widespread phenomenon, the extraordinary richness of rain forest invertebrate communities.
PLATYHELMINTHES
Turbellaria
    In the constant high humidity of rain forest, flatworms are found in soils and litter, as well as in their more usual aquatic habitats. A spectacular black and yellow striped planarian, Bipalium sp., is regularly found in forest litter, but could not survive the desiccation that follows forest destruction (14).
Temnocephalida
    Temnocephala semperi, an unusual commensal not previously known from Borneo, occurs on crabs in the alluvial forests.
ANNELIDA
Megascolecidae
    Perichaetine and octochaetine megascolecid earthworms are an important element of the soil fauna. An unusual arboricolous earthworm Planapheretima sera has been described from Mulu (41). This and other arboricolous species feed on decomposing epiphytic materials in the trees, and would not survive logging activities.
Enchytraeidae, Moniligastridae
    These families of worms are present in soils and await further study (13).
Hirudinea
    Two species of blood-sucking land leech, Haemadipsa zeylanica and H. picta, are common on lowland paths. Little-known predatory leeches, the Pharyngobdellida, occur in montane forest soils, feeding on worms and insect larvae (13,14).
MOLLUSCA
Gastropoda
    Snails and slugs have not been well-studied in Mulu but are present in low densities except in limestone areas, where the snail Cyclotus sp. is fairly common (13).
ARACHNIDA
Scorpionida (scorpions), Uropygi (whip scorpions), Amblypygi (tailless whip scorpions) and Schizomida (schizomids)
    All these arachnid groups are well represented at Mulu, but collections have received little study.
Araneae
    The forest spiders are diverse (about 360 species) and many new species are being described (36). These include ant mimics of the genus Myrmarachne (Salticidae), a Theridion species (Theridiidae) which builds its webs in pitcher plants and shares the insects which the plants attract, and blind Gnaphosidae, Zodaridae and Clubionidae from soils and caves. Of 32 spider families listed so

far, 18 are new records for Borneo (36). Many further discoveries are envisaged.

Opiliones

A collection of harvestmen from Mulu contains about 40 species, of which approximately one third are new to science. Descriptions are in preparation (42).

CRUSTACEA

Decapoda: Brachyura

With a density of 0.35-0.40 per m^2, freshwater crabs are a significant component of the soil fauna in the alluvial forests (11). They live in wet burrows and at night scavenge on insects and rotting organic matter (11). Four species (all Gecarcinucoidea) are found in the alluvial forest: Perbrinckia loxophthalma, Thelphusula baramensis and T. granosa (new species) are all endemic to Borneo, while Geosesarma gracillima (Grapsoidea) is also found in the Natuna Islands and on the South East Asian mainland (24). One species, Palawanthelphusa pulcherrima, was found in a stream in dipterocarp forest (24). Five new crabs endemic to the cave systems within the park have also been described: Adeleana chapmani, Cerberusa caeca, C. tipula, Sundathelphusula tenebrosa and Isolopotamon collinsi (24).

Isopoda

Six species of woodlice in six genera of Oniscoidea have been recorded from Mulu caves; five are new species, one a new monotypic genus. All are probably troglophiles and may also occur on the surface (43). The species are Armadillo solumcolus, Nagurus lavis, Paraperiscyphis platyperaeon, Selaphora parvicaputa, Tuberillo sarawakensis (gen. nov.) and Triadillo amandalei, a species collected once before in Sarawak (43)

CHILOPODA

Scolopendromorpha

The bizarre and rare Borneo centipede Arrhabdotus octosulcatus is a monotypic genus assigned to its own tribe, the Arrhabdotini, of the subfamily Otostigminae. It was described from south-east Borneo and was also found at Mulu on the trunk of a sapling (28). It is unusual in being slow-moving and exceptionally short-legged.

INSECTA

Orthoptera

Locusta migratoria manilensis was taken in a cleared area. In Sabah this locust occasionally swarms and causes considerable damage to agriculture. Its spread to Mulu gives a clear warning of the dangers of clearing large tracts of forest (14).

Blattaria

Trogloblatella chapmani is a new species of troglobitic cockroach known only from the Mulu caves. The genus was known previously only from a single species found in limestone caves in Australia (32).

Dermaptera

Arixenia sp. is a blind, supposedly ectoparasitic earwig which lives on the Hairless Bat (Cheiromeles torquatus). Despite rather low populations of this bat in the Mulu caves, spectacular populations of these earwigs infest guano heaps in certain areas. A new species of earwig, Brachylabis collinsi, has been found in the lowland forests (7).

Isoptera

Seventy two species of termites have been recorded from the forests of Mulu, an unusually high diversity for this conservative group of social insects (15). Labritermes emersoni has recently been described from Mulu material (27), and one new nasute genus and several termitine species await description (15).

Coleoptera

A variety of collecting methods produced about 4500 species of beetles, out of a probable total for the park of over 10 000. Perhaps 80 per cent of these are confined to the lowland forests. Many species are undescribed, for example 78 out of 110 species of Staphylinidae attracted to baited traps are new, as are

7-11 out of 24 Lucanidae and 2 out of 19 Cicindelidae. In the chrysomelid flea-beetle genus Clavicornaltica, previously unknown from Borneo, 72 new species have been found. Only three species were previously assigned to this genus worldwide. It has been estimated that up to 80 per cent of the beetle fauna of Mulu may be undescribed, and at least two new families are represented. Over half of the collections have been carefully studied and the degree of species overlap between forest types is remarkably small. The alluvial and dipterocarp lowland forests (which are contiguous) share only 20 per cent of their beetle species, and this is the highest value found for any comparison between forests. Only 13 per cent is shared between dipterocarp forest and limestone lowland forest, 5 per cent between dipterocarp forest and Mulu upper montane forest, and 0.1 per cent between alluvial forest and limestone lowland forest. There is virtually no overlap between the faunas of high and low altitude forests, although exposed ridges in the lowlands tend to support a fauna typical of higher altitudes. The high altitude zone revealed interesting representatives of north temperate groups such as the burying beetles, Nicrophorus spp. (Silphidae, see Nicrophorus americanus review). A new species of Dasycerus (Dasyceridae) and three species of Aspidiphorus (Sphindidae) from high on Mulu were the first records in Borneo for these essentially temperate region families. These discoveries emphasize the value of the altitudinal range within the park (50). Recent studies in South American rain forest canopies suggest that a vast undescribed beetle fauna may exist there, possibly enough to increase the estimates of the world's beetle fauna by at least an order of magnitude (51). If this claim can be further substantiated then the estimates for the world's total arthropod species will need to be re-examined.

Homoptera

Over 40 species of cicadas were taken in the park, several of which are undescribed.

Trichoptera.

A new species of caddis-fly, Polymorphanismus muluensis, has been described from a light-trap catch in Mulu (5). Other species found there are P. quadripunctatus, P. scutellatus, and P. ocularis (5).

Lepidoptera.

The butterfly (Papilionoidea) fauna of Borneo may include about 1000 species, 276 of which have been recorded from Mulu. The high altitude areas of Mulu and Api contain a number of characteristically montane species, most of which are new records for Sarawak (22). Arhopala ariel (Lycaenidae) is a new record for Borneo and three species are otherwise known only from G. Kinabalu in Sabah. A new species of Celastrina (Lycaenidae) has been found on G. Api and one other new species, Ypthima hanburyi (Satyridae) was taken in lowland forest (22). The territoriality and courtship of Rajah Brooke's Birdwing (Trogonoptera brookiana) have been studied in the park, where the butterfly is relatively common (30). This species is only found in virgin forest in Peninsular Malaysia and Borneo, and it has been listed in the Convention on International Trade in Endangered Species of Wild Fauna and Flora (CITES) in order to monitor international trade. Butterflies constitute only one of about 20 superfamilies within the Lepidoptera, the rest being moths. Over 100 new species, some of them already described (44), are present in the 2400 species of larger moths ('heterocera') collected in Mulu (23). This represents about two-thirds of the Bornean total of 3000-4000 species. Collections of tineid moths from Mulu's caves stimulated a complete revision of this important component of the fauna of caves worldwide (31).

Hymenoptera.

The Symphyta (saw-flies and wood-wasps) of Mulu are virtually unknown, and of the Apocrita, only the ants (Formicidae) have been studied well. They are extraordinarily diverse; over 450 species in 78 genera have been found, more than occur in the whole of North America. Two new genera have been described, and many others are rare or previously unrecorded from the region, including Stenamma, Eurhopalothrix, Dysedrognathus, Amblyopone, Euprenolepis, Belonopelta, Mystrium, Bregmatomyrma and Anillomyrma. The symbiotic association between Iridomyrmex species and ant-plants such as Myrmecodia and Hydnophytum is common in kerangas forest and at high altitude on G. Mulu and G. Api. The digger wasp Sphex subtruncatus has been recorded (30), but in general the aculeate (sting-bearing) hymenopterans other than ants are poorly known and in need of further study. The Parasitica are also poorly known; some descriptions of new species of Chalcidoidea (6) and Ichneumonidae (20) have already been published but many other new species await description. An interesting new species from the unusual subfamily Loboscelidiinae (Chrysididae) has been described (18).

SCIENTIFIC INTEREST AND POTENTIAL VALUE The interest and value of Mulu may be viewed on a local scale, and as a representative of the world's tropical rain forests. On a local level the park is a potential source of revenue from tourism, and of employment for the residents of local longhouses. The nomadic hunter-gatherers of the Penan tribe, whose heritage of rain forest lore should not be undervalued, depend upon the lowland forests for their livelihood. Rainfall in the headwaters of the Tutuh and Medalam is 5000-6000 mm per year. The heaviest recorded fall was on 2 November 1977, when 187 mm fell in a single day (34). The destructive force of such waters once they are channelled into rivers can be quite devastating, but the forested watersheds moderate the release of rain water. Without the forests on inland hills, siltation and flooding would occur downstream, killing people and livestock, and destroying towns and villages. For example, in 1981 logging and burning of a 10 ha area of forest on a 15° slope in Sumatra caused five landslides which ran into a river and formed a series of dams that eventually burst, destroying 17 houses and killing 13 people (45).

Such arguments give a local rationale to a worldwide need for preservation of rain forest. The value of the forests lies in watershed management, in production of medical drugs from natural sources, in storing the genetic resources of current and potential food plants (29,37) and animals (46), in meeting the needs of more than 200 million people living in forests or on their fringes, in productivity of timber for export, and probably in stabilizing the local and world climate (29,39,40,47,48,49).

This description of the invertebrate fauna in Mulu gives some measure of the levels of diversity to be found in rain forests. Probably more than 20 000 species of invertebrates occupy this relatively small National Park. The value of each individual worm or insect may appear to be slight, but together they form the communities of decomposers, herbivores, predators and pollinators that are essential to the dynamic nature and survival of the forest. Cryptic insects such as termites play a significant role in the cyclic transfer of nutrients between plants and the soil, but their diversity may be reduced by 75 per cent on forest clearance (12,16). The tiny thrips (Thysanoptera) are only 0.5-8 mm long, but are essential pollinators of some huge rain forest trees. It is still uncertain which insects pollinate Shorea curtisii, one of the more important of the dipterocarp timber trees, but stingless bees may be responsible. The Mulu forest harbours many new species of parasitic wasps and phytophagous insects of the type used in biological pest control programmes (6,18,20). Many moth species appear to be more or less specific to certain forest types and may prove to be of value as indicators in rain forest monitoring (23). These are some simple examples of the documented value of forest invertebrates; many more remain to be discovered.

THREATS TO SURVIVAL The Gunung Mulu National Park itself is not currently threatened on a large scale, and is capably protected by the Sarawak Forest Department. However the increasing pressure on the surrounding lands, and the issue of licences to log the forests surrounding the park, will eventually cause damage to the park margins and open up the possibility of illicit activities. The western edge of the park between Sungei Melinau and the Brunei border has already been logged, although it is believed that the Forest Department demanded a 20 chain (c. 400 m) buffer zone along the river bank. With the increasing population pressures and demand for land, logging roads are quickly used for access by shifting agriculturalists. At the present level of staffing (one forest guard at Long Melinau), there would be little chance of preventing park abuse by immigrants to the area. There are reports that some local residents exceed their privileges in the removal of trees, game and other forest products from the park.

The threats to rain forests worldwide are well documented (39,40,47,48,49). Deforestation of equatorial regions is widely considered to represent the greatest single threat to insect species. Timber exploitation is often short-term and not sustainable, and as a result of population pressure may be followed by shifting agriculture with insufficient periods of fallow. Levels of reforestation are inadequate, while logging accelerates at an unprecedented rate (see section 1a in Insecta). Despite warnings that at least 50 per cent of the world's insects are confined to rain forests and that extinctions could be occurring in unprecedented proportions by the end of this century (39), it is surprisingly difficult to find documented examples. This is not to imply that extinctions are not occurring, but simply that the taxonomy and distribution of forest invertebrates are so poorly known that nobody is recording the losses.

CONSERVATION MEASURES TAKEN   Malaysia contains some of the richest tropical rain forests in the world but in Peninsular Malaysia the virgin lowland forests have nearly all been logged over and are controlled under careful silvicultural practice. Sarawak still has vast tracts of virgin forest, and since the early 1960s the Sarawak Forest Department has actively pursued policies of both conservation and sustainable utilization of the forests under their jurisdiction. The Forest Department is aware of the need to give total protection to substantial areas in the face of a growing population and the rapid spread of shifting cultivation. At the same time, the Department is exceptional in its adherence to sound forest management for sustained yield. In production forests the annual harvest is strictly controlled, licensees are fined for causing unnecessary damage to the residual stand, and silvicultural practices are implemented to encourage natural regeneration. Recent changes to the Forest Ordinances provide for heavy fines, confiscation of equipment and rapid eviction and prosecution of illegal settlers. Enforcement measures are being implemented.

The full history of the Mulu National Park is given in the management plan (3), and will only be summarized here. In 1961 a botanical expedition from the Sarawak Forest Department visited Mulu and subsequently recommended the constitution of a National Park. In 1962 the National Parks Trustees and the Chief Secretary approved the proposals. In 1965 a Notice of Intention to create the park was published and in 1970 claims by the Residents, Fourth and Fifth Divisions, were settled. To meet the claims of people in the lower reaches of the Tutuh River, a substantial proportion of the proposed park was excised. The G. Mulu National Park was finally constituted by the Sarawak Government on 3 October 1974.

The Royal Geographical Society/Sarawak Forest Department expedition in 1977/78 produced a substantial amount of scientific data which have been summarized into a management plan approved by the Sarawak Forest Department (3). Since 1978 the Department has not officially opened the park to the public, and has stationed only one forest guard there. However, the Department is trying to protect a buffer zone in order to conserve larger areas of lowland forest, and to prevent damage to the park margins by logging companies.

In 1979 the International Committee on Research Priorities in Tropical Biology of the U.S. National Academy of Sciences met to choose a prime site for future biological research in tropical Asia (29). They chose Mulu for its outstanding

572

scientific interest, and urged that the area should be developed as a major ecosystem site for research purposes. No steps have yet been taken to implement this resolution, and it is hoped that research projects will soon be developed jointly with the Sarawak Forest Department.

From 1983 to 1985 WWF is funding the development of management plans for protected areas in Sarawak as part of its tropical forests and primates campaign (52). Implementation of the Mulu management plan is one of the immediate priorities of the programme.

CONSERVATION MEASURES PROPOSED It is strongly recommended in the management plan that the park be extended to include part of the mainly low-lying Medalam Protected Forest to the north (3). This is not the same area that was excised from the park on the request of the Tutuh residents in 1970 (3). The reason for this proposal is that the present boundary along the Medalam River makes it impossible to control river access to the park. Moreover, unless both banks of the river lie within the park, the river fauna cannot be protected. In addition, the area of lowland forest within the current park boundaries is inadequate to protect the sparse populations of the very rich flora and fauna, and is also inadequate to support the traditional privileges of the nomadic Penan people. If the Medalam Protected Forest was to be licensed for logging, the Penan would be forced into the area of lowland within the current park boundaries, which is too small to support their need for game and sago. This could lead to the withdrawal of their privileges, which would be disastrous for their way of life.

The proposed Beluru-Limbang highway will be an asset to the park provided that the route is suitable, that during its construction the park regulations are fully respected, and that the Forest Department prepares facilities to absorb the consequent impact of increased tourism in the area. The threat of shifting cultivators gaining access to the park via the road should not be under-estimated; regular patrols will be required.

In order to protect the park properly, the Sarawak Forest Department should develop park facilities and provide resident staff. Technical advice will be needed on management of the park and its wildlife, on protection of the flora and fauna (particularly primates), and on the scientific and touristic potential of the cave systems within the Melinau massif. The management plan considers that visitors should be encouraged, particularly in order to demonstrate the park's value to local residents. The development of an information and interpretative centre should be given priority. The recommendation of the U.S. Committee on Research Priorities in Tropical Biology, that Mulu should be developed as a centre for scientific research, should receive international funding as soon as possible. Mulu is of the status of Unesco/MAB World Heritage Sites, and should be considered for inclusion if Malaysia becomes a signatory to the World Heritage Convention. The staff of the new WWF project to develop management plans in Sarawak should be given every encouragement to implement the Mulu management plan with the co-operation of the Forest Department.

Rain forests throughout the tropics are being logged for short-term gains long before the invertebrates and other fauna and flora have been properly described. At the present rate of logging, species extinctions are inevitable, particularly amongst the flora and invertebrates. It is essential that substantial reserves of high species diversity and endemism be set aside for future research. Sustainable logging practices should be followed, with provision for corridors for species re-introductions from forest reserves. The Sarawak Forest Department and other government departments in tropical countries with rain forest should be given international support in the delicate task of conserving their natural heritage at the same time as developing methods of sustainable utilization of rain forest products.

REFERENCES
1. Anderson, J.A.R. and Chai, P.P.K. (1982). Vegetation formations. In (26), part 1, chapter 9.
2. Anderson, J.A.R. and Chai, P.P.K. (1982/83). Flowering plants. In (3), pp. 74-81, and in (26) part 2.
3. Anderson, J.A.R., Jermy, A.C. and Cranbrook, The Earl of, (1982). Gunung Mulu National Park, A Management and Development Plan. Royal Geographical Society, London. 345 pp.
4. Baillie, I.C., Tie, Y.L., Lim, C.P and Phang, C.M.S. (1982). Soils. In (26), part 1, chapter 8.
5. Barnard, P.C. (1980). A revision of the Old World Polymorphanisini (Trichoptera: Hydropsychidae). Bull. Br. Mus. Nat. Hist. (Ent.) 41: 59-106.
6. Boucek, Z. (1978). A generic key to Perilampinae (Hymenoptera, Chalcidoidea) with a revision of Krombeinius n. gen. and Euperilampus Walker. Ent. Scand. 9: 299-307.
7. Brindle, A. (1980). Dermaptera from the Gunong Mulu National Park, Borneo. Entomologist's Rec. J. Var. 92: 172-175.
8. Brook, D.B., Laverty, M. and Waltham, A.C. (1982). Geomorphology. In (3), pp. 34-42.
9. Brook, D.B. and Waltham, A.C. (Eds) (1978). Caves of Mulu. The Limestone Caves of the Gunung Mulu National Park, Sarawak. Royal Geographical Society, London. 44 pp.
10. Chapman, P. (1982). The ecology of caves in the Gunung Mulu National Park, Sarawak. Trans. Brit. Cave Res. Assoc. 9: 142-162.
11. Collins, N.M. (1980). The habits and populations of terrestrial crabs (Brachyura: Gecarcinucoidea and Grapsoidea) in the Gunung Mulu National Park, Sarawak. Zool. Meded. Leiden 55: 81-85.
12. Collins, N.M. (1980). The effect of logging on termite (Isoptera) diversity and decomposition processes in lowland dipterocarp forests. In Furtado, J.I. (Ed.), Tropical Ecology and Development. pp. 113-121. Int. Soc. Tropical Ecology, Kuala Lumpur. 1383 pp.
13. Collins, N.M. (1980). The distribution of soil macrofauna on the west ridge of Gunung (Mount) Mulu, Sarawak. Oecologia (Berl.) 44: 263-275.
14. Collins, N.M. (1982). The significance of the invertebrate fauna. In (3), pp. 104-115.
15. Collins, N.M. (in press). The termites (Isoptera) of the Gunung Mulu National Park, with a key to the genera known from Sarawak. In (26) part 2.
16. Collins, N.M. (in press). Termite (Isoptera) populations and their role in litter removal in Malaysian rain forests. In Sutton, S.C., Whitmore, T.C. and Chadwick, A.C. (Eds), The Tropical Rain Forest: Ecology and Management. Blackwells, Oxford.
17. Collins, N.M., Holloway, J.D. and Proctor, J. (in prep.). Notes on the ascent and natural history of Gunung Api, Sarawak.
18. Day, M.C. (1979). The affinities of Loboscelidia Westwood (Hymenoptera: Chrysididae, Loboscelidiinae). Syst. Ent. 4: 21-30.
19. Eavis, A.J. (Comp.) (1981). Caves of Mulu '80. Royal Geographical Society, London. 52 pp.
20. Gauld, I.D. (1981). The taxonomy, distribution and host preference of Indo-Papuan parasitic wasps of the subfamily

Ophioninae (Hymenoptera: Ichneumonidae). Commonwealth Agricultural Bureaux, Farnham Royal, U.K..

21. Hanbury-Tenison, A.R. (1980). Mulu: the Rain Forest. Weidenfeld and Nicolson, London. 176 pp.

22 Holloway, J.D. (in press). Notes on the butterflies of the Gunung Mulu National Park. In (26), part 2.

23 Holloway, J.D. (in press). The larger moths of the Gunung Mulu National Park; a preliminary assessment of their distribution, ecology, and potential as environmental indicators. In (26), part 2.

24. Holthuis, L.B. (1979). Cavernicolous and terrestrial decapod Crustacea from northern Sarawak, Borneo. Zool. Verhand. Leiden 171: 3-47.

25. Jermy, A.C., Jülich, W. and Touw, A. (1982). Cryptogams. In (3) pp. 82-84.

26. Jermy, A.C. and Kavanagh, K. (Eds) (1982/83). The Gunung Mulu National Park, Sarawak. Sarawak Mus. J. Special Issue Number 2, parts 1 (1982) and 2(in press for 1983).

27. Krishna, K. and Adams, C. (1982). The Oriental termite genus Labritermes Holmgren (Isoptera, Termitidae, Termitinae). Am. Mus. Nov, 2753: 1-14.

28. Lewis, J.G.E. (1981). Observations on the morphology and habits of the bizarre Borneo centipede Arrhabdotus octosulcatus (Tömösv'ary), (Chilopoda, Scolopendromorpha). Entomologist's Mon. Mag. 117: 245-248.

29. National Research Council (1980). Research Priorities in Tropical Biology. National Academy of Sciences, Washington. 116 pp.

30. Panchen, A.L. (1980). Notes on the behaviour of Rajah Brooke's birdwing butterfly Trogonoptera brookiana brookiana (Wallace) in Sarawak. Entomologists's Rec. J. Var. 92: 98-102.

31. Robinson, G.S. (1980). Cave-dwelling Tineid moths: a taxonomic review of the world species (Lepidoptera: Tineidae). Trans. Brit. Cave Res. Assoc. 7: 83-120.

32. Roth, L.M. (1980). Cave dwelling cockroaches from Sarawak, with one new species. Syst. Ent. 5: 97-104.

33. Walsh, R.P.D. (1982). Hydrology and water chemistry. In (26), part 1, chapter 7.

34. Walsh, R.P.D. (1982). Climate. In (26), part 1, chapter 3.

35. Waltham, A.C. and Webb, B. (1982). Geology. In (26), part 1, chapter 4.

36. Wanless, F.R. and Hillyard, P.D. (in press). Report on the spiders with a preliminary list of the harvestmen of Gunung Mulu National Park. In (26), part 2.

37. Whitmore, T.C. (1975). Tropical Rain Forests of the Far East. Clarendon Press, Oxford.

38. Whitmore, T.C. (1976). Conservation Review of Tropical Rain Forests, General Recommendations and Asia. IUCN, Gland. 116 pp.

39. Myers, N. (1979). The Sinking Ark. Pergamon Press, Oxford. 307 pp.

40. Myers, N. (1980). Conversion of Tropical Moist Forests. Report to the Academy of National Sciences. National Research Council, Washington D.C.

41. Easton, E.G. (1979). A revision of the acaecate earthworms of the Pheretima group (Megascolecidae: Oligochaeta): Archipheretima, Metapheretima, Planapheretima, Pleionogaster and Polypheretima. Bull. Br. Mus. Nat. Hist.

Zool. 35 (1): 1-126.

42. Hillyard, P.M. (1983). In litt., 6 January.
43. Schultz, G.A. (1982). Terrestrial isopod crustaceans (Oniscoidea) from Mulu caves, Sarawak, Borneo. J. Nat. Hist. 16: 101-117.
44. Holloway, J.D. (1982). Taxonomic appendix. In Barlow, H.S. An Introduction to the Moths of South East Asia. Pp. 174-271. H.S. Barlow, Kuala Lumpur.
45. Robertson, J.M.Y. and Soetrisno, B.R. (1982). Logging on slopes kills. Oryx 16: 229-230.
46. Medway, Lord (1980). Tropical forests as a source of animal genetic resources. BioIndonesia 7: 55-63.
47. Brown, S. and Lugo, A.E. (1982). The storage and production of organic matter in tropical forests and their role in the global carbon cycle. Oecologia 14: 161-187.
48. Unesco/UNEP/FAO (1978). Tropical Forest Ecosystems. Natural Resources Research 14. Unesco, Paris. 683 pp.
49. FAO/UNEP (1981). Tropical Forest Resources Assessment Project. (3 vols.). FAO, Rome.
50. Hammond, P.M. (1983). Pers. comm. 20 January.
51. Irwin, T.L. (1982). Tropical forests: their richness in Coleoptera and other arthropod species. Coleop. Bull. 36(1): 74-75.
52. World Wildlife Fund (1982). No Trees No Life. Forest Pack 5, launch edition. WWF Tropical Forests and Primates Campaign, Gland.

We are grateful to P.C. Barnard, B. Bolton, Z. Boucek, P.P.K. Chai, the Earl of Cranbrook, M.J. Day, E.G. Easton, Abang Abdul Hamid, P.M. Hammond, P.M. Hillyard, J.D. Holloway, R.W. Ingle, A.C. Jermy, J.G.E. Lewis, Abang Kassim Morshidi, N. Myers, F. Naggs, J. Noyes, J. Proctor, K. Proud, F.R. Wanless, and T.C. Whitmore for their comments on this review.

California, United States of America

SUMMARY   San Bruno Mountain is a largely undeveloped mountain habitat surrounded by the San Francisco metropolis. It harbours populations of a number of rare or federally protected vertebrates, invertebrates and plants, some of which are San Francisco Peninsula endemics now restricted to the mountain. Much of the mountain's natural habitat is included within a park boundary, but invasive non-native vegetation, recreational activities, and the unnatural suppression of fire remain potential or actual threats. Large proposed housing developments will significantly reduce the grassland habitat of some of the mountain invertebrates.

DESCRIPTION   San Bruno Mountain is a small range with a length of about 6.4 km and a maximum elevation of 401 m. It extends diagonally across much of the San Francisco peninsula in San Mateo County, just south of the San Francisco city limits (8). It is surrounded by the municipalities of Bayshore, Daly City, Colma, South San Francisco and Brisbane (7). San Bruno Mountain is comprised of two nearly parallel ranges, separated partially by Guadalupe Valley (now an industrial park) and united by a plateau known as the Saddle. The bulk of the mountain is composed of dark green or grey graywacke rocks of the Franciscan formation; one remaining enclave of serpentine rock harbours a population of a threatened butterfly, the Bay Checkerspot (4, and separate review). The climate varies greatly over this small area, playing a major role in the distribution and evolution of the flora and fauna. The mountain often experiences high winds, particularly in the summer. Heavy fog often envelops the western half of the mountain, most extensively in July and August. Four distinct vegetational communities are present, listed in order of area coverage: grassland, coastal scrub, foothill woodland and freshwater marsh (8). Further details are available (3,5,7,8).

INVERTEBRATE FAUNA   No systematic survey of the mountain's invertebrates has been carried out, but a number of rare or endemic insect species are known to be present. Many are peninsular endemics, some now restricted to the mountain due to the removal of surrounding natural habitat.
LEPIDOPTERA
Lycaenidae
   Incisalia mossii bayensis (the San Bruno Elfin butterfly). The largest known populations of, this peninsular endemic inhabit the higher reaches of north-facing slopes where the caterpillar foodplant, Sedum spathulifolium, grows. Most known colonies lie within county regional park boundaries, but some are threatened by invasive gorse (Ulex europaeus), including the type population (2,4,5). This butterfly has been designated as Endangered under the U.S. Endangered Species Act (12).
   Icaricia icarioides missionensis (the Mission Blue butterfly). The largest populations of this federally Endangered species inhabit eastern grasslands of the mountain (12); it also occurs less frequently on the mountain's coastal sage scrub sites. It is also known from San Francisco (Twin Peaks) and Marin County (just north of Golden Gate Bridge) where populations are low and unstable (1,10). Eastern grasslands may be significantly reduced by proposed housing developments (5,7,10).
Nymphalidae
   Speyeria callippe callippe (the Callippe Silverspot butterfly). Proposed for federal protection in 1978, this species is now believed to survive only on San Bruno Mountain, primarily in the eastern grasslands. It was formerly known from San Francisco. This species may also be threatened by the proposed housing developments (3,5,7,10,13).
   Euphydryas editha bayensis (the Bay Checkerspot). This butterfly is endemic to serpentine outcrops and survives as three peninsular populations, the

smallest situated near the top of the mountain's main ridge. It has been the subject of a population study by P.R. Ehrlich and his colleagues at Stanford University for over 20 years. The species continues to have high scientific value in population dynamics and plant-insect inter-relationship studies (5,9,10 and separate review).

Tortricidae

Grapholitha edwardsiana (San Francisco Tree Lupine Moth) is a peninsular endemic restricted to sandy habitats occupied by the larval foodplant, Lupinus arboreus. It occurs within state park boundaries of the Saddle on San Bruno Mountain, and on unprotected sandy remnants in lower Colma Canyon and skirting the south-western slopes. It was proposed in 1978 for federal protection but was withdrawn in 1980 because it was found to be more widespread than previously thought (5,10,13).

HYMENOPTERA

Halictidae

Duforea stagei is a recently described species of bee, once thought to be restricted to its type locality on the Saddle of San Bruno Mountain. Specimens have recently been found in outlying areas, but the species is still considered a vulnerable peninsular endemic (5,10).

SCIENTIFIC INTEREST AND POTENTIAL VALUE The presence of so many unique invertebrates, combined with the presence of many rare and possibly endangered plants, and the federally Endangered San Francisco Garter Snake makes this an exceptionally valuable habitat (5,6,10). Its location in the midst of a large metropolis and near several institutions of higher learning (including the University of California, Berkeley, Stanford University and two California state universities) make it an ideal field site for biology classes and research. The effects of the varied climate on local distribution of the mountain's unique flora and fauna make it a prime location for continued ecological and evolutionary investigations (6,8,10). Additionally, it is a site of scientific interest as the type locality for several vulnerable species. The recreational and open space value of San Bruno is apparent from the intense public resistance to further development (2,6,7).

THREATS TO SURVIVAL OF FAUNA Habitat loss on San Bruno Mountain has resulted in the past from construction of roads, utilities, homes, industrial and commercial structures, quarrying for rock and sand, grazing of livestock, diversion of groundwater, off-road vehicles and increased frequency of fires (5). Currently, much of the eastern and southern grassland areas are threatened by a proposed housing development(7,10). Long-range plans for development of the parklands are sketchy; some recreational uses would be incompatible with the unique flora and fauna. Many of the vulnerable species are somehow dependent on recurrent fires; their continued survival depends on a sound fire management policy not yet developed. Several species of non-native plants, including Gum (Eucalyptus globulus) (Myrtaceae), Gorse (Ulex europaeus) and species of Broom (Cytisus) (both Fabaceae) are extending their range on the mountain, displacing native vegetation (8,10). Off-road vehicle activities, though diminishing, continue to inflict damage on the flora, with long-term adverse effects on erosion and soil compaction. An active, expanding quarry on the mountain's north flanks has destroyed populations of the Endangered San Bruno Elfin (2,9). Some sections of the range of the Bay Checkerspot may be threatened by the Edgewood Golf Course.

CONSERVATION MEASURES TAKEN Since 1978, approximately 700 ha of San Bruno Mountain have gained protection as county regional parkland. This land is concentrated on the main ridge and adjacent slopes. Some management of non-native vegetation has been attempted within this area (6,10). Most of the Saddle has been dedicated as a state park, although much of this site has been severely disturbed by prior off-road vehicle activity and the invasion of Ulex europaeus. Local environmentalists are working to have the quarry closed and disturbed lands revegetated with native species (10). Continued research on the mountain's vulnerable and endangered butterflies is being undertaken by several lepidopterists, with a view toward refining management activities (10).

The San Bruno Mountain Recovery Plan, an intensive biological report on the mountain's vulnerable grassland butterflies, has recently been completed (5), as has a further study by a private consultant company (11). The results of the Recovery Plan in particular should have been applied to the formulation of the San Bruno Mountain Area Habitat Conservation Plan (HCP), a document which was designed to minimize the impact on the protected or vulnerable lepidopterans and the Garter Snake, while still allowing the proposed housing developments. Some authorities, including the author of the San Bruno Mountain Recovery Plan, feel strongly that that document was inadequately considered in the formulation of the HCP (15). The compiler of this review was unable to obtain a copy of the HCP before going to press. In accordance with the HCP, and under Section 10(a)(1)(b) of the Endangered Species Act, a permit application PTR 2-9818 has been lodged with the Federal Wildlife Permit Office in Arlington to take Mission Blue Butterflies, San Bruno Elfin Butterflies and San Francisco Garter Snakes incidentally to development on the mountain. Some knowledgeable scientists feel that any construction will have disastrous long-term implications, and are opposed to the granting of any licences for development (10). The Chairman of the IUCN/SSC Lepidoptera Specialist Group (LSG), on behalf of over 40 members worldwide, has communicated this view to the Federal Wildlife Permit Office (14), as has the author of the San Bruno Mountain Recovery Plan (15).

CONSERVATION MEASURES PROPOSED The situation at San Bruno Mountain is currently very volatile, and opinion in the conservation community is divided over the level of conservation measures required. The LSG is opposed to all major developments that may adversely affect rare or endangered Lepidoptera unless adequate provision for their protection can be demonstrated, and it believes that the provisions given in the HCP to restore the habitat are inadequate (14). Some other conservationists feel that the measures proposed for restoration of disturbed habitat are sufficient to safeguard the Lepidoptera. In addition, the LSG questions whether the proposed tax on local eventual residents to pay for the proposals in the HCP is a sufficiently reliable funding mechanism (14). The Group believes that the fauna is best served by preventing any further development, by further research and effort in habitat management, and by listing of the Callippe Silverspot and the Bay Checkerspot under the Endangered Species Act (14).

REFERENCES
1. Anonymous (1978). The endangered Mission Blue butterfly of San Francisco. Xerces Soc. Educ. Leaflet 1, 2 pp.
2. Anonymous (1980). The endangered San Bruno elfin butterfly of California. Xerces Soc. Educ. Leaflet 6, 4 pp.
3. Anonymous (1980). The endangered San Francisco silverspot butterfly of California. Xerces Soc. Educ. Leaflet 5, 4 pp.
4. Arnold, R.A. (1978). Survey and status of six endangered butterflies in California. Calif. Dept. Fish and Game, Non-game Wildlife Investigations E-1-1. Study V, Job 2.20. Final report, 95pp.
5. Arnold, R.A. (1982). San Bruno Mountain Recovery Plan. Technical Review Draft, U.S. Fish and Wildlife Service, Portland, Oregon. 75 pp.
6. Committee to Save San Bruno Mountain (1981). Rare and Endangered San Bruno Mountain Wildlife. 1 p.
7. King, W. (1982). N.Y. Times, February 7, p. 9.
8. McClintock, E. and Knight, W. (1968). A flora of the San Bruno Mountains, San Mateo County, California. Proc. Calif. Acad. Sci. Fourth Series. 23: 587-677.
9. Murphy, D.D. and Ehrlich, P.R. (1980). Two California

checkerspot subspecies: one new, one on the verge of extinction. J. Lepid. Soc. 34: 316-320.

10. Orsak, L.J. (1982). In litt., 20 February.

11. Reid, T. and Associates (1981). Endangered species survey San Bruno Mountain. Biological Study, 1981. Draft report to the San Mateo steering committee for San Bruno Mountain.

12. U.S.D.I. Fish and Wildlife Service (1976). Determination that six species of butterflies are Endangered species. Federal Register 41 (106): 22041-22044.

13. Arnold, R.A. (1981). Distribution, life history, and status of three California Lepidoptera proposed as endangered or threatened species. Calif. Dept. Fish and Game, Inland Fisheries Branch Investigation E-F-3, S-1620. Final report. 39pp.

14. Pyle, R.M. (1982). In litt., letter to Director, Federal Wildlife Permit Office, 14 December.

15. Arnold, R.A. (1982). In litt., letter to Director, OES/USFWS, Washington, 10 December.

We are grateful to L.J. Orsak for a preliminary draft of this review, and to R.A. Arnold, P.R. Ehrlich, D. Murphy and P.A. Opler for further comment.

SUMMARY Banks Peninsula in Canterbury, New Zealand, was once a separate island and is still geographically isolated from the South Island mountain ranges by the Canterbury Plains. It contains a high number of endemic invertebrates and type localities. Deforestation was extensive in the last century as a result of timber milling and fires, and only small remnants of forest remain. Farming and housing could encroach on these remaining areas.

DESCRIPTION The Peninsula is the visible portion (about 100 000 ha) of a basaltic lava volcano dome probably built during the Pliocene. Eruptions may have been in the form of lava flows rather than violent explosions (3). It was almost certainly once a separate island and is still geographically isolated from the mainland of South Island, New Zealand by plains of glacial outwash gravels, clays and loess (1,9). It is 96 km long and has a number of natural harbours and several towns and villages. Agriculture is the main activity (1). Along the coast, gently sloping platforms eroded by the sea give way to progressively steeper slopes that are dissected by a series of radial valleys. The climate is mild and subhumid on the coast, subhumid to humid in the valleys and inner harbours, and cooler and humid on the summits. At the time of human settlement broadleaf podocarp forest occupied many of the wetter slopes, giving way to tall tussock and subalpine vegetation on exposed south-facing ridge crests. Tussock grasses and occasional patches of broadleaf scrub forest grew on the coastal flats (11).

INVERTEBRATE FAUNA At least 800 arthropod species have been described or recorded and many are endemic to the Peninsula. Three categories of endemism can be recognised: a) species which are isolated monospecific genera or which differ greatly from their congeners, b) species or subspecies which are clearly related to others in the South Island, c) isolated populations of widespread species (4). Category (a) is particularly important for conservation as illustrated by the following (4,5):
NEMERTEA
Antiponemertes allisonae is a nemertine known only from the type specimens which came from highly susceptible second-growth scrub habitat on a farm near Menzies Bay. It has only two eyes and a mottled brown dorsal surface with a clear stripe over the proboscis (9). Its habitat could be severely altered by shrub clearance.
INSECTA
Orthoptera
Hemideina ricta (Stenopelmatidae) is an endemic weta with a small and probably stable population. It is now known only from a small area, although it has been recorded within a reserve. It lives in tree galleries but its preferred tree is not regenerating well (2,4). First seen in 1890 and not rediscovered until 1978, this weta has very high scientific value and live material is needed for current behavioural and genetic research on members of the genus (5 and see review of giant wetas).
Coleoptera
Mecodema howitti (Carabidae) is an endemic ground beetle found within several reserves and once very common in rough pastures. It is the largest arthropod on the peninsula (5), and apparently quite isolated from other species of Mecodema (10).
Omaseus pantomelas (Carabidae) is an endemic ground beetle which is widespread (occurring in several reserves) but uncommon. Six specimens were collected in 1841, one in 1964, one in 1973 and six since then (4). Although the species is currently assigned to the genus Omaseus, this is due for revision (10).

Pheloneis gratiosus (Tenebrionidae) is an isolated, very small population of a mountain species, which has not been seen on the peninsula for several years (5).

Megacolabus sculpturatus (Curculionidae) is an endemic weevil which is possibly extinct (5).

Hadramphus tuberculatus (Curculionidae) is an endemic weevil, also possibly extinct. It is known to be associated with the spiny plant _Aciphylla_ (Umbelliferae), now quite uncommon on the peninsula (5).

Diptera

Austrolimnophila n.sp. (Tipulidae) is known only from one specimen from Otepatotu Scenic Reserve. Apparently its nearest relative is another undescribed and wingless species from the subalpine tussock of eastern areas of Fiordland National Park (5).

Lepidoptera

Kupea electilis (Pyralidae) is a monotypic genus of moth endemic to the Birdlings Flat Area. Detailed biological field studies are needed in order to delineate its habitat for possible reservation (5).

Kiwaia jeanae (Gelechidae) is another monotypic moth genus which is endemic to the Birdlings Flat Area, whose biology is very poorly known. Although some areas near the known site are reserved, the habitat is very susceptible to damage or modification as it is close to areas of intensive farming or recreational use (4,5).

DIPLOPODA

A number of new millipede species are also to be described, some of which may be limited to small ranges (5).

SCIENTIFIC INTEREST AND POTENTIAL VALUE Twenty eight per cent of the species described from the peninsula have their type localities there. Many New Zealand insects were first described from Banks Peninsula, and type localities are particularly important in this country since it has so many variable or restricted species. Endemism on Banks Peninsula is very high (4). _Hemideina ricta_, for example, represents a relict stock stranded from the mainland _Hemideina_ species by geographical isolation during the Pleistocene glaciations (2).

THREATS TO SURVIVAL The mixed podocarp forest on Banks Peninsula was subjected to a deliberate policy of milling and clearing for its valuable timber in the last century. By 1880 substantial inroads had been made and by 1900 the area had been considerably deforested, exacerbated by a series of accidental and even deliberate fires. In 1849 the Banks Peninsula forest was estimated at 54 000–81 000 ha but there is now no large remnant; in the main it was replaced by artificial grassland produced mainly by seeding with cocksfoot (6). It is calculated that 93 per cent of native forest cover has been lost. The area thus cleared has passed through various cycles of farming type and intensity and today about 5 per cent of the cleared area has reverted to unproductive, mixed native and exotic shrubland or gorse. Where the mature forest was burnt, the large stumps were left to rot in the paddocks. It is only since the 1960s that the use of heavy machinery to withdraw these stumps has allowed the development of uniform pastures at higher altitudes (5), accelerating destruction of the invertebrate fauna.

Roadside banks and gullies within paddocks have tall shrub vegetation (_Fuchsia_, _Leptospermum_, _Pittosporum_) and the latter even have some areas of forest. Where such areas are fenced, the vegetation is in good health, but where cattle and other stock roam freely the understorey is badly damaged. If this occurs where the canopy-feeding Australian possum (_Trichosurus vulpecula_) is abundant, damage may be severe. The combination of farming, fire and stock damage has now reduced the vegetation to very small areas and only a few of those that are not already protected are suitable refugia (5). The Peninsula is also under threat from the expansion of Lyttelton and from recreational activities including the building of holiday homes (1,2). The 'Dry Bush' type locality for many species has gone as a result of farming, fire and housing. The northwestern dry and sunny side of the Peninsula has easy access from Christchurch and only a little of the natural vegetation is left, confined to steep, rocky slopes (4).

CONSERVATION MEASURES TAKEN The western hills of Banks Peninsula are subject to a preservation order, but development projects from Christchurch are still encroaching (1). Most forest patches on the Peninsula have some type of reserve status. Forest vegetation in high rainfall areas has been reserved and it is hoped that the podocarp trees will regenerate. One of the largest forest remnants on the Peninsula was reserved in 1980 (4). At present ten of the established reserves are rated at six or higher on a scientific value scale of zero to ten; these areas are still close to their original state before human colonization (7). About 1.1 per cent of the total area has been set aside in 35 Scenic Reserves, just over 0.043 per cent in three Scientific Reserves, and a smaller percentage in three highly modified Recreational Reserves or Domains. The Scientific Reserves have strong restrictions on public access but for all other reserves the public has right of access unless such access is over private land, when the landowner's permission is needed, as is the case for three of the Scenic Reserves. A survey of the invertebrate fauna of the Peninsula has been carried out (5).

CONSERVATION MEASURES PROPOSED The results of the invertebrate survey will be used to regrade current reserves and suggest further areas for reservation (5). Of prime importance is the completion or strengthening of fencing surrounding the reserves so as to exclude stock. Funds are needed to buy land and critical habitats for endemic species. Since many are confined to limited localities small reserves would be sufficient (1). A proposal has been submitted to study the biology, bioacoustics, distribution and habitat requirements of the weta *Hemideina ricta*. Further efforts will establish awareness of its importance and the need for local protection of the species. The information obtained from the study will be used to inform farmers of the need to conserve certain flora of their property and to emphasize to local residents that wetas should not be killed in spite of their apparent ferocity (2).

REFERENCES 1. Field, L. (1980). In litt.
2. Field, L. (1980). Conservation of endangered species of the New Zealand weta fauna (Insecta, Orthoptera, Stenopelmatidae) with special reference to Hemideina ricta. Project proposal to IUCN/WWF.
3. Gage, M. (1969). Rocks and Landscape. In ref. (8).
4. Johns, P. (1980). In litt., 29 August.
5. Johns, P. (1981). In litt., 29 May.
6. Johnson, W.B. (1969). Modification of the natural environment by man. In: (8).
7. Kelly, G.C. (1972). Scenic Reserves of Canterbury. Botany Division, Department of Scientific and Industrial Research: Biological Survey of Reserves. Report No. 2: 30, 390 pp. Christchurch.
8. Knox, G.A. (Ed.) (1969). The Natural History of Canterbury. Reed, Wellington.
9. Moore, J. (1973). Land nemertines of New Zealand. J. Linn. Soc. (Zool.) 52: 293-313.
10. Scott, R.R. and Emberson, R. (1982). In litt., 15 June.
11. Vucetich, C.G. (1969). Soils of Canterbury. In ref. (8).

We are very grateful to P.M. Johns, R. Emberson and R.R. Scott for information on this subject, and to J.C. Watt and G.R. Williams for comments on the review.

California, United States of America

SUMMARY The El Segundo coastal dunes of Los Angeles County, California, U.S.A., constitute an isolated ecosystem reduced to two small remnants which have been disturbed moderately or severely. Several endemic invertebrates and populations of vulnerable plants, reptiles and mammals still remain. Elimination of the remaining open habitat and continued spread of non-native vegetation threaten the dune organisms.

DESCRIPTION The El Segundo dunes, of Pleistocene origin, once covered c. 9600 ha along the Pacific coast of the Los Angeles basin, U.S.A. (12). Originally the sand hills reached 30.5-61.0 m in height. Two remnants survive, the smaller of which covers 0.6 ha and is surrounded by an oil refinery and housing. The larger, sandwiched between the Pacific Ocean and Los Angeles International Airport, contains 32 ha of moderately disturbed habitat and 90 ha of severely disturbed habitat. The unique dune organisms are generally restricted to the moderately disturbed portion; the severely disturbed site was once covered with houses which were razed because of their close proximity to the airport. Dune organisms and sand are slowly re-invading this area. These small remnants represent the last coastal dune habitat of significant size along the California coast between Point Dume and San Clemente (1,2,3,6).

INVERTEBRATE FAUNA The El Segundo dunes are the type localities for a number of arthropod species. Many of these are endemic or restricted to this and other disappearing coastal sand dune habitats.
ARACHNIDA
Acarina
   The mite Eremobates sp. nov. (Cryptostigmata: Oribatidae) was recently discovered in a nearby much smaller dune remnant, but is also expected to occur at El Segundo (7).
   Erythraeus tuberculatus (Prostigmata: Erythraeidae) is another mite, known only from the El Segundo dunes (4).
INSECTA
Coleoptera
   Trigonoscuta dorothea dorothea (Curculionidae) is a weevil which inhabits coastal sand dunes from Point Dume south to Orange County (11).
   Onychobaris langei is another weevil. It is known only from seven specimens and is believed to be endemic to the dunes (4).
Diptera
   Brennania belkini (Belkin's Dune Tabanid Fly, Tabanidae) is only recorded from seven adult specimens collected in coastal dunes between Los Angeles and Ensenada (Mexico). It has been collected from El Segundo dunes and may still reside there since it occurs in a smaller remnant nearby. It was recently proposed for federal protection (2,5, and see separate review in this volume).
Lepidoptera
   Euphilotes battoides allyni (the El Segundo Blue butterfly, Lycaenidae) was described from, and is restricted to, the El Segundo dunes, where Eriogonum parvifolium blossoms are the sole larval and adult food sources. The species is federally protected (1,2,5,9,10,13).
   Lorita abornana (Tortricidae: Olethreutinae). Both the genus and species of this moth are endemic to the dunes, and the species has been nominated for federal protection (4).
   Eucosma hennei (Tortricidae: Olethreutinae) is a moth which is endemic to the dunes. It has been nominated for federal protection, (4).
   Carolella busckana (Cochylidae) is a moth known from only a few localities in Los Angeles County. It has been nominated for federal protection (4).

Psammobotys fordi (Pyralidae: Pyraustinae) is a moth endemic to the dunes (4).

SCIENTIFIC INTEREST AND POTENTIAL VALUE   The dune organisms have considerable potential in studies of island biogeography (species diversity in relation to dune size), in adaptation studies, and in various ecological studies. The dunes are the type localities of several arthropod species collected by members of the Los Angeles County Museum in the 1930s. As a unique habitat surrounded by a major metropolis and numerous institutes of higher learning, the dunes could become a valuable outdoor teaching and research laboratory (2,4,6).

THREATS TO SURVIVAL   The major short-term threat is the invasion of non-native plants such as the Ice-plant Mesembryanthemum sp. (2). Long-term threats come from the proposed conversion of the 90 ha severely disturbed site to tennis courts and a 27-hole golf course (3). This site, if managed as a reserve, offers considerable potential for reclamation by the endemic dune organisms, increasing populations and making their status considerably less vulnerable (6).

CONSERVATION MEASURES TAKEN   Standard Oil of California fenced their 0.6 ha site and declared it as a preserve (8). In 1976 the El Segundo Blue butterfly was given federal protection as an Endangered species (13). Detailed studies of the butterfly have been sponsored by the State of California (1) and these are incorporated into a recovery plan issued by the U.S. Fish and Wildlife Service (2).

CONSERVATION MEASURES PROPOSED   Los Angeles International Airport officials propose a maximum reserve of 32 ha (3). Conservationists and nearby residents favour preservation of the entire 122 ha site (6). Property access restrictions on airport land have prevented an updated survey of the flora and fauna, which would help determine which endemic species still survive and the extent of their distribution. Elimination of invasive, non-native Mesembryanthemum would be a necessary component of any habitat management plan (2). Propagation of dune plants and dissemination within the severely disturbed remnant has been contemplated (2). The Airport Dunes are listed as a Significant Ecological Area by the County of Los Angeles. Development having been approved by the Airport Dunes Authority, the last legal chance for protection short of legal appeal lies with the California Coastal Commission, who must issue a permit before development can begin. The Friends of the Dunes is an active local group working with biologists towards the protection of the unique habitat (6).

REFERENCES   1. Arnold, R.A. (1978). Status of six endangered California butterflies, (1977). Report to Calif. Dept. Fish and Game. 93 pp.
2. Arnold, R.A. (1981). El Segundo blue recovery plan. U.S. Fish and Wildlife Service Tech. Review Draft. 42 pp.
3. City of Los Angeles. (1980). Airport Dunes Study. Project No.6, local coastal program of the City of Los Angeles. 109 pp.
4. Donahue, J.P. (1979). In litt. to R. Beard, Los Angeles Office of Environmental Planning, 26 November.
5. Donahue, J.P. (1975). A report on the 24 species of California butterflies being considered for placement on the federal lists of Endangered or Threatened species. Nat. Hist. Mus. of Los Angeles Co., Calif. 58 pp.
6. Friends of the Dunes (1981). A specific plan for a dunes nature refuge, preserve, and education centre. Playa del Rey, Calif. 25 pp. + appendices.
7. Nagano, C. (1981). In litt. to P. Brown, Los Angeles City Planning Department, 20 January.
8. Oppewall, J.C. (1976). The saving of the El Segundo blue.

Atala 3: 25-28.

9. Orsak, L.J. (in press). The endangered El Segundo blue butterfly and its dune habitat. Xerces Soc. Educ. Leafl. 7. 6 pp.

10. Shields, O. (1975). Studies on North American Philotes (Lycaenidae). IV. Taxonomic and biological notes, and new subspecies. Bull. Allyn Mus. (3 September).

11. Pierce, W.D. (1975). The sand dune weevils of the genus Trigonoscuta Motchulsky (Coleoptera: Curculionidae). Privately published, Los Angeles. 164 pp.

12. Cooper, W.S. (1967). Coastal dunes of California. Geol. Soc. Am. Mem. 104: 95-99.

13. U.S.D.I. Fish and Wildlife Service (1976). Determination that six species of butterflies are Endangered species. Federal Register 41(106): 22041-22044.

We are grateful to L.J. Orsak for providing a preliminary draft of this review, and to C.D. Nagano, P.A. Opler and J. Oppenwall for further comment.

SUMMARY Parts of the Dead Sea Depression contain a significant isolated tropical flora situated further north than that ' of any similar area and a remarkable freshwater spring and terrestrial fauna. Contact with the main tropical areas of Arabia and Africa is almost lost. The ecosystems under discussion are limited to a number of clearly defined pockets in the middle of typical Saharo-Sindian associations. They are limited in size, extremely fragile and in need of conservation.

DESCRIPTION The Dead Sea Depression is part of the Jordan Rift Valley and consists essentially of flat, open land, located mainly in Jordan, the south-western portion belonging to Israel. The West Bank has been under Israeli military administration since 1967. Much of the area is heavily farmed and grazed by livestock (camels and goats) belonging to both sedentary and semi-nomadic farmers. It is particularly interesting for two types of habitat found there.

Firstly, most of the area belongs to the Saharo-Sindian vegetation zone (1), but there are isolated and well defined pockets of tropical vegetation classed as Sudanian by Zohary (4). Locally such pockets are known as ghors - although the term ghor has also been applied to the whole of the Dead Sea Depression. The ghors, small as they are, are fully tropical ecosystems quite distinct from the surrounding areas. It seems that special combinations of soil and microclimate are responsible for exact siting; capacity for regeneration in the case of destruction must be very limited as they would be taken over by the Saharo-Sindian vegetation. These tropical sites are all situated below sea level, in some cases on the banks of the Dead Sea, and comprise:

Ghor Feifa   Shunit Nimrin
Ghor es Safi   Ein Gedi
Ghor edh Dhura   Wadi Farra, Jericho
Ghor en Numeira

Secondly, many of these pockets, as well as some non-tropical sites, contain springs and small streams that give rise to comparatively luxuriant vegetation between the coast and rocky heights bordering the Dead Sea. The springs occur at an altitude of about 300 m below sea level and on the western border are fed by water from the Judaean Hills and on the eastern side by water from the escarpment abutting the Jordanian Plateaux. Most of these springs and spring complexes form striking contrasts with the surroundings of the Dead Sea and rocky desolate heights. The most important springs on the western border are as follows: -

1.   Ein Nueima (En Duyuk) on the lower course of the Jordan River, West Bank.
2.   Ain Fashkha (En Feshkha), West Bank: an important group of springs and streams giving rise to a brackish marshy area on the coast, with a rich vegetation of reeds and other aquatic plants. The streamlets are fast flowing with gravel and rubble substrates.
3.   Ein Turaba (En Turabe), West Bank: water in this small complex of freshwater springs rises from a gravelly beach on the coast. The springs are richly vegetated, mainly with reeds and, in areas of still water, with abundant filamentous algae.
4.   Ein Ghuweir (En Uver) north of Ein Turaba, West Bank.
5.   Ein Gedi (Nahal David), Israel: the most important complex of springs on the western shore, with good fresh water rising on the slopes at some distance from the shore. Most of the water from the springs flows out through the

Ein Gedi (Nahal David) stream. Diverse aquatic habitats and rich tropical vegetation are found along this watercourse.

6.   Nahal Arugot south of Ein Gedi, Israel.
7.   Ein Mishmar on the shore south of Nahal Arugot, Israel.
8.   En Boqeq sulphur springs at the south end of the Dead Sea, Israel.
9.   Hamei Zohar not far from the south end of western shore, Israel; thermal and strongly mineralized.
10.   Neot (Ein) Hakikar, Israel, south of the Dead Sea.

FLORA AND INVERTEBRATE FAUNA   Published accounts of the fauna and flora of this area are very incomplete. The best indicators for the habitat are the plants of Sudanian origin, prominent among which are:

Maerua crassifolia (Capparaceae)   Grewia villosa (Tiliaceae)
Acacia tortilis (Leguminosae)   Abutilon fruticosum (Malvaceae)
A. raddiana (Leguminosae)   A. muticum (Malvaceae)
A. laeta (Leguminosae)   Solanum incanum (Solanaceae)
Hyphaene thebaica (Palmae)   Cassia obovata (Leguminosa)
Calotropis procera (Asclepiadaceae)   Salvadora persica (Salvadoraceae)
Loranthus acaciae (Loranthaceae)   Senecio flavus (Compositae)

The tropical flora comprises a total of about 60 species. They generally do not penetrate the adjacent Saharo-Sindian vegetation or the Irano-Turanian vegetation of the surrounding hills.

Endemism at the species and subspecies level is high in the southern Rift Valley. The ghors have their associated terrestrial invertebrate fauna, found nowhere else in Jordan, Israel, Syria or Lebanon, and usually not on the Lower Nile. Many endemic insect species are chiefly associated with the African vegetation of the area (9). Thus five out of about 70 Jordanian butterflies are wholly linked to the tropical habitats in the Dead Sea Depression (2):

Colotis chrysonome (Klug, 1829) (Pieridae) (probably extinct on West Bank) - Mauritania to Arabia
C. phisadia (Godart, 1819) (Pieridae) - Mauritania to Arabia and India
Epamera glaucus (Butler, 1886) (Lycaenidae) - Arabian-Jordanian endemic
Anthene amarah (Guérin, 1849) (Lycaenidae) - all Africa and Arabia
Azanus ubaldus (Cramer, 1872) (Lycaenidae) - all Africa and Arabia

These species are all linked to plants from the tropical associations, are very local and are not all found in any one locality. A number of other tropical butterflies occur in the Dead Sea area, but are not wholly limited to the tropical plant associations. At Ein Gedi, two African species reach their northern limits, with endemic subspecies: Colotis phisadia palaesinensis Staudinger and Iolus glaucus jordanus Staudinger (3). Extrapolating from butterflies it is reasonable to conclude that at least a hundred terrestrial invertebrates owe their presence so far north to this habitat.

The highest level of endemism is found in the freshwater fauna and has probably been favoured by the endorrheic isolated conditions of the Rift Valley (9). The fauna of the springs has a relatively low number of species. A survey carried out in 1971 revealed fish, flatworms, various gastropods, Ephemeroptera (mayflies), Diptera: Chironomidae (midges) and Trichoptera (caddis-flies) (6). The flatworms, mayflies and chironomids have not been described but a brief outline of some of the other groups follows:

MOLLUSCA
Gastropoda (6)
    Theodoxus jordani (Sowerby, 1836): endemic to the Jordan Rift Valley.

Melanoides tuberculata (O.F.Müller, 1774)
M. praemorsa (Linnaeus, 1758)
Pseudamnicola solitaria Tchernov, 1971: known only from the following springs: Ein Hakikar, En Tamar, Ein Turaba, Ein Ghuweir, Ein Fashkha, and springs in the Wadi Zin, south of the Dead Sea (7).
Bythinella sp.
Hydrobia sp. discovered in springs at En Feshkha.
CRUSTACEA (9,10)
Copepoda
Nitocra incerta (Toalnearia Por, 1964) a harpacticoid copepod found at En Gedi and Hamei Zohar.
Ostracoda
Darwinula stevensoni (Br. et Rob.) an ostracod found at Ein Gedi.
Both the above species are considered to be old lacustrine relicts in the area (9).
Thermosbaenacea
Monodella relicta Por, 1962: known only from the thermal and strongly mineralized spring at Hamei Zohar, on the south-western shore of the Dead Sea (10), and from a spring on the northern shore of Lake Tiberias.
Isopoda
Typhlocirolana reichi Por, 1962: a species in the family Cirolanidae known only from the springs at Ein Hakikar, and from two localities south of the Dead Sea (10).
Amphipoda
Bogidiella hebraea Ruffo, 1963: known only from Ein Hakikar, and from a spring on the northern shore of Lake Tiberias.
The above three species are true stygobionts (i.e. they are restricted to underground water), are found only in the Jordan Rift area, and are relict species with a marine origin (10). Other crustaceans discovered recently include two amphipods (Echinogammarus sp. and Metacrangonyx sp.) from south of the Dead Sea, and a cyclopoid copepod (Halicyclops) from springs at En Feshkha.
INSECTA
Trichoptera (8):
Stactobia margalitana Botosaneanu, 1974: the larvae build minute barrel-shaped cases and exclusively inhabit hygropetric (madicolous) waters; known only from the springs of Nahal Arugot.
Ithytrichia dovporiana Botosaneanu, 1980 is also known only from Nahal Arugot; it is a minute species, whose larvae spin transparent cases, with the characteristic shape of a pumpkin seed.
The genus Hydroptila: larvae build cases shaped like a spectacle case and covered with fine sand or algae. Three species have been described from this area. H. hirra Mosely, 1948 is known from Ein Nueima, En Uver, En Turabe, Nahal Arugot, Ein Mishmar. It has a wide distribution from south-west Arabia, Sinai and northern Galilee, to the Air (Azbine) mountains of Niger. Its larvae are unusual, their cases being sheltered under delicate cupolas built from sand and fastened to the stones. H. adana Mosely, 1948 was found at En Turabe and Nahal Arugot; the species is otherwise known from south-west Arabia and Sinai. H. angustata Mosely, 1939 first described from the Zeitoun Oasis (Egypt) and widely distributed in the Near East was found in En Feshkha, Nahal David, and Nahal Arugot. Undetermined larvae of Hydroptila are also known from En Boqeq.
Chimarra lejea Mosely, 1948 is a much larger species than those already mentioned. Its larvae, which are carnivorous and very active, do not build mobile cases, but secrete nets, which are used as traps to catch live prey. The nets are fastened under or among boulders in relatively fast current. Caught at En Feshkha, En Turabe, Nahal David, Neot Hakikar; was already known from south-west Arabia and Ethiopia.
Tinades negevianus Botosaneanu & Gasith, 1971: larvae build characteristic serpentiform galleries fastened to the surface of boulders and pebbles. It is known from Nahal David but also inhabits springs in the Judaean Hills and in

Lebanon.

Setodes alalus Mosely, 1948 is known from En Turabe; the species was previously found in south-west Arabia, Hedjaz, Sinai, as well as from the southern Iran. Larvae and pupae of this species build delicate mobile cylindrical cases from sand.

SCIENTIFIC INTEREST AND POTENTIAL VALUE The Jordan Rift Valley is one of the sections of the great rift system extending from the Zambezi River in Mozambique to the Lebanon. The most interesting parts faunistically are the deepest parts, including the Dead Sea Depression which in some places is almost 800 m below sea level. The valley served as a gateway for the penetration of tropical African elements towards the north, and more recently, palaearctic freshwater fauna towards the south, and is the meeting area for three realms of the animal world: the Palaearctic, Oriental and Ethiopian regions. The oases of the Dead Sea area are of particularly outstanding zoogeographical importance (9,10).

The tropical pockets in the Dead Sea Depression are particularly interesting since, as in all marginal and more or less isolated localities, there is potential for study of speciation processes. Furthermore they are the only areas in the western Palaearctic where tropical vegetation comes so close to Irano-Turanian vegetation, and are the most northerly strongholds for many species which cannot survive even in Lower Egypt. A further interesting feature, at least in the plants but probably also in some invertebrates, is that phenology is geared to the rainfall conditions that still prevail in Sudan and Arabia, but no longer in the Dead Sea area. It would be interesting to find out why they survive nonetheless and why their time of flowering has not adjusted.

The springs, though poor in species, have a highly original faunal composition and can be considered as relict habitats and refuges for freshwater fauna and flora in a large area characterized by its aridity and extremely severe environmental conditions. The species occurring there that have a more widespread distribution (e.g. trichopterans) are always restricted to springs and streams in arid areas and are evidence of the remarkable faunal migration which has taken place along the Great Rift Valley (8). Some of the species may be true 'eremial elements' in the freshwater fauna. The relict crustacea provide evidence for the existence of a marine gulf of the Mediterranean which once extended south along the Jordan Rift Valley (10).

THREATS TO SURVIVAL The tropical associations of plants and terrestrial animals of the area are small, fragile and in actual danger. Since 1967 some of the pockets have been partly used for minefields, which afforded significant protection and the area was also fairly inaccessible for security reasons. Since 1977, minefields are being cleared and development of the area is in progress. A metalled road has been built from Aqaba to Kerak, a factory to extract minerals from the Dead Sea has been built on Ghor Numeira and there are plans for a resort near Ghor adh Dhira.

The main dangers appear to be extension of farming, over-grazing (especially by goats), fuel gathering, development of tourism, and random physical destruction (picnic sites, waste dumping, building activity, tracks). The tropical pockets in the Nimrin area have perhaps ceased to be viable ecosystems because of farming; Ghor Numeira is suffering from the effects of the factory and its associated activities. Ein Gedi has suffered much from casual tourism (3). In particular, the butterfly Iolaus glaucus jordanus is threatened by destruction of Acacia, Zizyphus and other trees that support the host plant of this species, a parasitic plant (3).

The drainage of many swamps in the Jordan Rift Valley has had a catastrophic effect on freshwater faunal diversity (9) and is likely to be a major threat to the springs of the Dead Sea Depression. In addition the development of springs for recreational activities could jeopardize many sites. At En Feshkha some of the most interesting aquatic habitats have already been disturbed and even destroyed by the construction of bathing facilities. The thermal springs at Hamei Zohar are also used regularly for bathing, a factor which could seriously jeopardize the endemic crustacean Monodella relicta.

CONSERVATION MEASURES TAKEN The springs and streams of Ein Gedi are part of a nature reserve set up to protect the rich tropical vegetation and varied wildlife found there. It has been designed to cater for a large number of visitors (11).

CONSERVATION MEASURES PROPOSED The Dead Sea Depression is becoming a major tourist attraction on account of its scenery. It should be possible to develop this potential without damaging the wildlife:

1. Four or five specially selected areas of tropical vegetation need protection so that the fully developed tropical associations can be maintained and allowed to reach climax condition. This would entail their declaration as national parks, effective fencing to stop grazing and fuel gathering and access only for bona fide scientific research. All the above mentioned springs should be designated reserves.
2. The conservation and survival of the spring complexes and oases will depend on water management as well as preservation (9). Alterations (for irrigation, provision of drinking water, or bathing or recreational use) should be made only in the lowest parts of the watercourses and conservation organizations should be consulted at the planning stage.
3. A simple but efficient system of fences and footbridges at important sites should be established to prevent erosion and trampling of the vegetation by tourists.

Jordan has an enlightened policy towards conservation as witnessed by the large reserve and wetland conservation programme near Azraq in the north of the country (5) and it is hoped that similar steps will be taken to protect the unique habitats in the Dead Sea Depression. Israel also has a strong interest in conservation as shown by the active work of the Nature Reserves Authority. The wealth of endemic and marginal populations found in the Dead Sea Depression makes protection of its species and habitats especially important (9).

REFERENCES
1. Eig, A. (1931-32). Les éléments et les groupes phytogéographiques auxiliaires dans la flore Palestinienne. Feddes Repert. Spec. Nov. Reg. Veg. Beih. 63: 1-201., 1-120. (2 vol.).
2. Larsen, T.B. and Nakamura, I. (in press). The Butterflies of East Jordan. E.W. Classey.
3. Nakamura, I. (1976). Butterfly conservation problems in the Palestinian region. Atala 4: 27-28.
4. Zohary, M. (1962). Plant life of Palestine. New York.
5. Carp, E. (1980). Directory of Wetlands of International Importance in the Western Palearctic. IUCN-UNEP, Gland, Switzerland.
6. Botosaneanu, L. (1982). In litt.
7. Tchernov, E. (1971). Pseudamnicola solitaria n.sp. a new prosobranch gastropod from the Dead Sea area, Israel. Israel J. Zool. 20: 201-207.
8. Botosaneanu, L. (1973). Au carrefour des régions orientale, ethiopienne et palearctique. Essai de reconstitution de l'histoire de quelques lignées "Cool adapted" de Trichopteres. Fragm. Ent. 9(2): 61-80.

9. Por, F.D. (1975). An outline of the zoogeography of the Levant. Zool. Scripta 4: 5-20.
10. Por, F.D. (1963). The relict aquatic fauna of the Jordan Rift Valley. Israel J. Zool. 12 (1-4): 47-58.
11. Nature Reserves Authority (1972). Nature Reserves in Israel.

We are very grateful to T.B. Larsen and L. Botosaneanu for providing drafts from which this account has been compiled, and to F.D. Por for further assistance.

SUMMARY Cueva los Chorros is one of the most interesting caves in Puerto Rico because of its rich invertebrate fauna. The cave is very small which makes it vulnerable to any kind of disturbance, particularly visits from large numbers of people. It deserves to be officially protected and managed.

DESCRIPTION Cueva los Chorros is situated on the main highway, 15 km south of Arecibo at about 35 m above sea level. There are two entrances on a cliff side; one is a permanent spring resurgence and the other is a spring only in wet weather. The cave consists of a chamber and stream passage only a few hundred metres long, with silt banks and guano patches (1).

INVERTEBRATE FAUNA The cave fauna of Puerto Rico consists mainly of wide-ranging species that frequently show an affinity for caves, but are not strictly dependent on them, being also found in moist, forest leaf litter (2). Cueva los Chorros contains an abundant and unusually varied fauna including a troglobitic millipede and cockcroach. Guano samples contained cydnid bugs, nitidulid beetles, terrestrial isopods, ants, centipedes, millipedes, 17 species of mites and abundant fly larvae, ptiliid beetles and Collembola. A complete listing of the species so far known from the cave, first described in 1981 (1), illustrates the diversity of such a cave fauna, the varying degrees of dependency of invertebrates on the cave ecosystem, and the way in which many species are adapted to living on guano.

Troglobionts (troglobites) are true cave animals and are adapted for a totally underground existence. Troglophiles are species that may prefer caves but are not specially modified for this way of life. Trogloxenes are visitors which move in and out of caves with some frequency, but which are unable to complete their life cycles there. Stygobionts are aquatic species occurring exclusively in groundwater-bearing subterranean cavities; stygophiles are species able to live in such places but may also use other suitable habitats.

MOLLUSCA
Gastropoda
    Subulina octona (Subulinidae, Stylommatophora) is a troglophile land snail, known also from Cueva El Convento. It scavenges on guano and is found fairly abundantly.
ARACHNIDA
Amblypygi
    Paraphrynus viridiceps (Phrynidae) is a predaceous troglophile tailless whip scorpion, also known from three other caves.
Schizomida
    Schizomus portoricensis (Schizomidae) is a troglophile predatory whip scorpion, known also from two other caves.
Opiliones
    Stygnomma sp. (Phalangodidae) is a possibly troglophilic, predatory harvestman.
Araneae
    Modisimus montanus (Pholcidae) and Scytodes sp. nov. (Scytodidae) are both troglophilic predatory true spiders.
Acarina
    17 species of mite have been found living in guano:
        Acotyledon sp. nr. krameri (Acaridae) is saprophagous or fungivorous.
        Rhizoglyphus sp. (Acaridae) is probably saprophagous.
        Histiostoma sp. (Aneotidae) possibly feeds by straining the micro-organisms from surface water films, perhaps on wet guano pellets. It is also found in one other cave.
        Evimirus sp. (Eviphididae); feeding habits unknown.

Galumna sp. (Galumnidae) is a saprophage and/or fungivore on decaying guano materials; known from three other caves.

Hypoaspis (Stratiolaelaps) miles and *H.* sp. nr. angusta (Laelapidae) are probably freeliving predators of small arthropods on guano.

Macrocheles sp. (Macrochelidae) is a predator on small arthropods (including eggs) and nematodes.

Malaconothrus sp. (Malaconothridae) is saprophagous.

Brachioppia sp. (Oppiidae) is fungivorous or saprophagous.

Scheloribates spp. (Oribatulidae). The two known species are saprophagous or fungivorous.

Parasitus sp. (Parasitidae) is a free-living predator on small arthropods.

Archegozetes sp. (Trhypochthoniidae) is a saprophage or fungivore in moist habitats.

Triplogynium vallei (Triplogyniidae) is probably a free-living predator.

Uroactinia sp. nr. hippocrepoides (Uropodidae) is possibly a fungivorous grazer.

Uropoda (Phaulodinychus) sp. nr. argasiformis (Uropodidae) is possibly saprophagous.

CRUSTACEA
Isopoda

An undetermined isopod is abundant on guano.

Decapoda

Xiphocaris elongata (Atyidae), is a troglophilic, detritivorous shrimp. *Epilobocera sinuatifrons* (Pseudothelphusidae) is a troglophilic, omnivorous freshwater crab found near the stream in Cueva los Chorros, and in other caves.

DIPLOPODA

Cylindromus uniporus (Eurydesmidae) is a troglobitic detritivorous millipede; this genus is known only from a single specimen collected from this cave (3). *Prostemmiulus* sp. (Stemmiulidae) was also present and is probably troglophilic.

CHILOPODA

An undetermined predatory centipede from the family Henicopidae.

INSECTA
Collembola

Paronella sp. (Entomobryidae), *Folsomides* sp. (Istomidae), *Onychiurus* sp. (Onychiuridae) and *Collophora quadrioculata* (Sminthuridae) are all troglophilic guano scavengers. The latter species was previously known only from Costa Rica; it is now known from this and two other Puerto Rican sites.

Blattaria

Nelipophygus sp. (Blatellidae) is a cockroach which probably also occurs in the Rio Camuy Cave System on Puerto Rico. It has eyes, reduced wings and elongated legs.

Orthoptera

Amphiacusta sp. (Gryllidae) is a cricket which is also found in other caves. It has not yet been found outside caves but other species from this genus occur in forest.

Hemiptera

Amnestus sp. (Cydnidae) is a bug which possibly also occurs in Cueva Tuna. It is a troglophilic guano scavenger.

Coleoptera

Stelidota sp. (Nitidulidae). This beetle and an undetermined species in the family Ptiliidae are occasionally found on carrion, and are both troglophilic guano scavengers.

Hymenoptera

Hypoponera sp. (Formicidae) is a predatory ant, possibly troglophilic, known also from other caves in Puerto Rico and Jamaica.

SCIENTIFIC INTEREST AND POTENTIAL VALUE Puerto Rico has an abundance of caves which offer excellent opportunities for detailed ecological studies and in which many future discoveries are to be expected (2). Cueva los Chorros is scientifically one of the most significant caves in Puerto Rico because of its rich invertebrate community, which includes two terrestrial troglobites (1). The guano mite fauna is surprisingly rich, and contains an unusually low proportion of predators and ectoparasites (1).

THREATS TO SURVIVAL Most of the caves studied in Puerto Rico are relatively safe from the small-scale habitat alteration and faunal destruction that may be caused by visitors. However, because Cueva los Chorros is so small, its microhabitats and their faunas cannot withstand much human impact. Visits by many or large groups, even informed student groups, would soon degrade the community by trampling it into the mud (1).

CONSERVATION MEASURES TAKEN None.

CONSERVATION MEASURES PROPOSED The cave should be given legal protection and the number of visitors should be strictly controlled (1).

REFERENCES   1. Peck, S.B. (1981). Zoogeography of invertebrate cave faunas in southwestern Puerto Rico. Nat. Speleol. Soc. Bull. 43: 70-79.
2. Peck, S.B. (1974). The invertebrate fauna of tropical American caves, Part III: Puerto Rico, an ecological and zoogeographical analysis. Biotropica 6: 14-31.
3. Loomis, H.F. (1977). Three new millipeds from West Indian Caves. Florida Entomol. 60: 21-25.

We are very grateful to S.B. Peck for commenting on this account.

SUMMARY Deadhorse Cave, Washington, U.S.A., contains the most diverse stream ecosystem of any lava tube cave in North America; the terrestrial fauna is also significant and the cave itself is of geological interest. It is particularly vulnerable to modification of the surface and to damage by careless cave explorers.

DESCRIPTION Deadhorse Cave is an unusually complex lava tube cave with about 1400 m of passage. It is situated in the Mount Adams cave area, north-west of Trout Lake in Skamania County, Washington. The lower entrance is a sink, formed when part of the ceiling collapsed. The upper entrance is an air hole, approximately 500 m uphill from the lower entrance. The cave has numerous lava features characteristic of the area, including stalactites, extensive breakdown and lava seals. It differs from other caves in the area in exhibiting a complex network of passages, many on different levels, with numerous side branches (7,8,11). 'Springs' in the cave walls allow entry of groundwater and surface water seeping from a nearby surface stream. This water reaches the cave through a permeable fractured layer at the lower contact of the lava flow forming the cave. In the cave, the water forms streams and pools which are inhabited by the aquatic fauna listed below. Inputs of organic matter to the terrestrial organisms include tree roots penetrating the cave ceiling, organic solutes in seepage water which nourish bacteria on the cave walls, and trogloxenic mammals. The largest organic input is via the 'springs' which carry much organic matter from the surface stream. Water volume varies throughout the year and increases considerably between October and March, when a small lake is formed in the cave (8,11).

INVERTEBRATE FAUNA A remarkable assemblage of invertebrates occurs in Deadhorse Cave. While none of the obligate subterranean species are thought to be restricted to this one cave, their collective diversity is very high for a lava tube system (3). The terrestrial troglobionts (the millipede, campodeid and mite) occur in caves and rock fissures throughout the 10-30 km^2 occupied by the lava flow. The described aquatic fauna (stygobionts and stygophiles) are more widely distributed. Distribution of the undescribed species is unknown, but probably similar. Definitions of biospeleological terms are given in the review of Cueva los Chorros in this volume.

PLATYHELMINTHES

Tricladida

Polycelis coronata, a stygophilic planarian flatworm, is widespread in surface waters in North America but isolated in this cave population (10). It differs from known surface populations in behaviour (9) and biochemistry (12). There is also an undescribed, stygobiotic, kenkiid flatworm in the cave (3).

ARACHNIDA

Acarina

An undescribed, stygobiotic Parasitengona water mite, and a terrestrial troglobiotic rhagidiid mite, Flabellorhagidia sp. nov., have been found (3,5). Both belong to the Prostigmata and are predaceous as adults, but the water mites are parasitic as larvae.

CRUSTACEA

Copepoda

An undescribed cyclopoid copepod, stygobiont (3).

Isopoda

The stygobiont Salmasellus steganothrix has been found. It is an eyeless, unpigmented asellid isopod, known from other groundwater habitats such as karst springs (1,14).

Amphipoda

599

Stygobromus elliotti, a gammarid amphipod, is a comparatively large stygobiont species, reaching lengths of 7-9 mm (7).
INSECTA
Diplura
Haplocampa sp. nov., a campodeid bristletail, is a terrestrial troglobiont (6).
Grylloblattaria
Grylloblatta sp. nov. is a terrestrial troglophile rock crawler. It is wingless, light yellow in colour and adapted for rapid movement (10,11,13).
DIPLOPODA
Chordeumida
Lophomus skamania, an unpigmented conotylid millipede, is a terrestrial troglobiont (2).

In addition to these invertebrates, there are numerous other species with varying dependancy upon the cave system (3).

SCIENTIFIC INTEREST AND POTENTIAL VALUE Although the diversity of this cave ecosystem would not be considered unusual in comparison to that of many limestone caves, it contains the most diverse stream ecosystem of any lava tube cave in North America, and possibly the world (3). It is also the only cave in the Cascade Mountains which contains more than one species of aquatic stygobiont (3). It contains the most accessible population of the unique isopod Salmasellus steganothrix, otherwise known only from deep limestone caves and karst springs in the Canadian Rockies (1). The cave population of the flatworm Polycelis coronata is apparently evolving into a distinct, stygobiont species, a situation rarely encountered (12). The combination of groundwater and surface water faunas in one spring-fed cave stream also affords unique research opportunities.

THREATS TO SURVIVAL The habitat of Deadhorse Cave is vulnerable to surface logging operations, which could block the cave's small upper entrance and cut off vital tree root nutrients from the cave's fauna, as well as fouling the surface stream which contributes largely to the unique aquatic ecosystem. Another serious danger is the dumping of calcium hydroxide residue from acetylene headlamps by uninformed cave explorers. This has occurred several times in the past decade and could poison the cave stream (3). In addition, cave explorers have severely damaged the exposed roots in some parts of the cave (4).

CONSERVATION MEASURES TAKEN The cave is located in the Gifford Pinchot National Forest. Cavers noticed in 1978 that an abandoned logging road passing Deadhorse Cave's upper entrance was flagged for rebuilding, and that the timber over the cave was proposed for logging. Protest letters and a report were submitted to the Mt Adams Ranger District, as a result of which the proposed roadbuilding and timber sale were abandoned. However, no irrevocable steps have been taken to protect the cave (3).

CONSERVATION MEASURES PROPOSED A formal designation of Research Natural Area should be established to protect the cave and its environs. Signs should be posted in the cave entrances (but invisible from outside) telling of the harmful effects of dumping acetylene lamp waste in the cave. The latter action is currently planned by the Cascade Grotto section of the National Speleological Society. Perhaps, as a last resort, the cave should be gated (3). Close communications must be maintained between Gifford Pinchot National Forest and biospeleologists, to ensure suitability of continuing management measures.

REFERENCES   1.   Bowman, T.E., (1975). Three new troglobitic asellids from western North America (Crustacea: Isopoda: Asellidae). Int. J. Speleol. 7: 339-356.
2.   Causey, N.B. (1972). Two new conotylid millipedes from western North America and a key to the genus Adrityla

(Chordeumida: Diplopoda). Proc. Louisiana Acad. Sci. 35: 27-32.

3. Crawford, R.L. (1982). In litt., 20 February.

4. Crawford, R.L. (1982). In litt., 3 October.

5. Elliott, W.R. (1976). New cavernicolous Rhagidiidae from Idaho, Washington, and Utah (Prostigmata: Acari: Arachnida). Occ. Papers Texas Tech. Univ. Mus. 43: 1-15.

6. Ferguson, L.M. (1981). Cave Diplura of the United States. Proc. 8th Int. Cong. Speleol. 1: 11-12.

7. Holsinger, J.R. (1974). Systematics of the subterranean amphipod genus Stygobromus (Gammaridae), Part 1: Species of the Western United States. Smithson. Contrib. Zool. 160: 1-63.

8. Nieland, J. (1971). Deadhorse Cave: just another lava tube? Nat. Speleol. Soc. News 29: 142-144.

9. Nixon, S.E. (1974). Some behavioral observations on a cave dwelling planarian. J. Biol. Psych. 16: 32-33.

10. Nixon, S.E. (1975). The effect of snowfall on cave organisms. Speleograph 11: 24-25.

11. Nixon, S.E. (1975). The ecology of Deadhorse Cave. Northwest Science, 49: 65-70.

12. Nixon, S.E. and Taylor, R.J. (1977). Large genetic distance associated with little morphological variation in Polycelis coronata and Dugesia tigrina (Planaria). Syst. Zool. 26: 152-164.

13. Kamp, J.W. (1980). Pers. comm. to R.L. Crawford.

14. Lewis, J.J. (1981). Pers. comm. to R.L. Crawford.

We are grateful to R.L. Crawford for the initial draft of this review, and to G. Magniez, J.E. Mylroie and S.B. Peck for further comment.

Czechoslovakia

SUMMARY The Sumava Mountains of south-west Bohemia, Czechoslovakia, contain scattered montane peat bogs which are refugia for relict mire, boreo-alpine and subarctic Lepidoptera and other insects. Some species are restricted or threatened in most European countries. There is an urgent need to extend the existing network of mire nature reserves to include their hydrological sources.

DESCRIPTION OF HABITAT The Sumava Mountains (Cesky les Sumava) are the southern and south-western frontier highlands between Czechoslovakia (Bohemia), Federal Republic of Germany (Bavaria) and Austria (2). On the German side, the mountains are called the Böhmerwald. Sumava is virtually covered in woodland, with Picea excelsa, Sorbus aucuparia, Acer platanus and Fagus silvatica common, and local growths of Pinus mugo (sens. lat.). The peat bogs or mires are a very characteristic component of the Sumava landscape. The total area of all mires is nearly 3500 ha and an average area of an isolated mire is 100-120 ha. Smaller localities are not taken into account here. Several of the largest, isolated localities (e.g. Mrtvy Luh near Volary (351 ha), Jezerni Slat near Kvilda (120 ha), Velka Niva near Lenora (136 ha) and the large (nearly 2000 ha) complex of Modravské Slate) are existing or proposed nature reserves of international importance and of unique value for all Europe (1,4,8).

Typical habitat for mire Lepidoptera and other specialized insects in the Sumava Mountains is montane raised bog with the characteristic plant association of the Mountain Pine Pinus mugo and Sphagnum mosses (3,6,8). The peat bog ecosystem is closely related to the subarctic forest-tundra and subalpine communities of dwarf forest near the timber line. The vegetation is sub-climax for the climate, but is maintained by waterlogging and other edaphic constraints (7,8).The large, raised peat bogs occur at altitudes between 700 and 1300 m. There are some slight faunistic and ecological differences between mires of lower altitude in the valley of the river Vltava (700-800 m altitude, e.g. Mrtvy Luh near Volary) and upland mires above 1000 m (e.g. Jezerni Slat near Kvilda, Modravské Slate) (6,8). Jezerni Slat is a typical montane raised mire and is fringed with peat bog communities of the sedges Carex nigra, C. stellulata, C. canescens, C. limosa and C. rostrata, and the moss Sphagnum recurvum (12). On drier sites Dwarf Birch Betula nana is abundant. The central part of the bog is covered by a solid stand of Mountain Pine Pinus mugo subsp. pumilio, with undergrowth of Wortleberries Vaccinium uliginosum, V. vitis-idaea, Crowberry Empetrum nigrum, Ling Calluna vulgaris etc. (12). Pinus mugo pumilio does not grow in moister sections of the bog, where the cover is composed of Cotton-grass Eriophorum vaginatum, the sedge Carex pauciflora, Cranberry Oxycoccus quadripetalus, Bog Rosemary Andromeda polifolia, Deer-grass Trichoforum (=Scirpus) caespitosum subsp. austriacum, and various mosses Sphagnum spp (12). The adjoining pastures are notable for the presence of the Brown or Hungarian Gentian Gentiana pannonica (12). All the large peat bogs in the Sumava Mountains are in good ecological condition, and have been subject to very little human impact (10).

The mountains formerly supported European Bison, Brown Bear, Wolf and Lynx, but these are now absent. The Northern Birch Mouse Sicista betulina is a glacial relict, and the Alpine Shrew Sorex alpinus pre-glacial.

INVERTEBRATE FAUNA From the point of view of nature conservation, certain Lepidoptera constitute the best known and most important relicts (2,8,9,11), although there are also notable and relict insects from the Odonata (Aeschna subarctica, A. juncea, Leucorrhina dubia), Heteroptera (Notonecta reuteri),

Coleoptera (Agonum ericeti, Carabus menetriesii, Stenus kiesenwetteri, Hydraena britteni) (9,11) and Diptera (Tipula subnodicornis, Atylotus sublunaticornis) (10,11). About 25 species of the larger Lepidoptera have isolated and relict populations in the Sumava Mountain mires, and lists of recorded species have been published (4,5,6,8,9,11). In Europe, the boreal crambine moth Pediasia truncatella is only found within the subarctic zone and in the Sumava Mountains. The mires are a southern refuge for the Fennoscandian butterfly Colias palaeno although the species is widely distributed outside Europe. Boloria aquilonaris is a skipper also dependent upon Vaccinium, but locally common in suitable habitats. Important moths include Eugraphe subrosea, Anarta cordigera, Carsia sororiata, Arichanna melanaria, Celaena haworthii and Chionides viduella (4,5,6,8,11). These few species are examples of an undoubtedly much larger relict fauna. The Microlepidoptera of the mires have been studied in detail and characterized into obligatory and opportunistic inhabitants (2). This area also contains the remaining Czechoslovakian populations of the Freshwater Pearl Mussel Margaritifera margaritifera (see separate review). More studies are needed to demonstrate further the unique fauna of these important mires.

SCIENTIFIC INTEREST AND POTENTIAL VALUE Most of the Sumava Mountains peat bogs are not seriously disturbed by man and they form an ecological archipelago of mire habitats inside zones of various montane forest types. The relict insect populations in the mires are at their southerly limit of distribution in Europe (7,10). Most relict populations of mire Lepidoptera are endangered or vulnerable in central and western Europe. The peat bogs of the Sumava Mountains are important for studies in mire conservation and for species-area studies in island biogeography (8,10).

THREATS TO SURVIVAL Outside the State Nature Reserves there are no restrictions on forestry and agriculture. The land use plan for the Sumava Mountains includes some hydrological and forest management changes, which represent the main threat to some mires. The limited exploitation of peat will probably cease. Acid rain originating from distant industries could be a threat (10).

CONSERVATION MEASURES TAKEN There are ten State Nature Reserves in the Sumava Mountains, six of which are strictly protected mires totalling an area of 950 ha (10). Some associated mires may be protected in the Böhmerwald National Park of Bavaria.

CONSERVATION MEASURES PROPOSED A joint project of the State Nature Conservancy and other institutions seeks to extend the total reserve area up to 3000 ha of mires, including safe hydrological buffer zones. It is of great importance that the hydrological origins of the mires should be protected effectively. This aim of the joint project has not yet been realized (10). The most significant parts of the proposed project concern these localities: Velka Niva (136 ha), Mala Niva (90 ha), Slat u Knizecich Plani (99 ha), Chalupska slat (137 ha) and the large complex of mires and montane wetlands known as Modravske slate (c. 2000 ha). The large mires of international importance are situated mostly in the Czechoslovakian parts of the Sumava Mountains, and their complete conservation is necessary (7,10). However, it would also be useful to protect all undisturbed and little-altered peat bogs on the Bavarian and Austrian slopes of the Sumava Mountains (Böhmerwald), as has been suggested for peat bogs in south-west Germany (13).

REFERENCES   1.   Dohnal, J. (Ed.) (1965). Czechoslovak mires and fens. Czech. Academia, Praha.

2.   Elsner, G., Krampl, F., Novak, I. and Spitzer, K. (1981). Microlepidoptera of the Sumava Mountains peat bogs. Sbor. Jihoces. Muz. v Ces. Budejovicich Prir. Vedy 21: 73-88.

3. Neuhäusl, R. (1972). Subkontinentale Hochmoore und ihre Vegetation. Studie CSAV Praha 13: 1-121.

4. Novak, I., and Spitzer K., (1972). Results of faunistic and ecological studies of Lepidoptera (Noctuidae, Geometridae) of the peat bog Mrtvy luh near Volary. Sbor. Jihoces. Muz. v Ces. Budejovicich Prir. Vedy 12 (Suppl. 1): 1-63. (In Czech with German summary).

5. Soffner, J. (1930). Zur Schmetterlingsfauna des mittleren Böhmerwaldes. Mitt. Münch. Ent. Ges. 20: 115-132.

6. Spitzer, K. (1975). Zum zoogeographisch-ökologischen Begriff der südböhmischen Hochmoore. Verh. 6th Internat. Symp. Entomofaunistik Mitteleuropa (Lunz am See, 1975). Pp. 293-298. (English summary).

7. Spitzer, K. (1980). The Sumava Mountains as an environment for montane and alpine Lepidoptera. Prace Muz. H. Kralove, Acta Musei Reginaehradecensis S. A Supplementum 1980: 114-118.

8. Spitzer, K. (1981). Ökologie und Biogeographie der bedrohten Schmetterlinge der südböhmischen Hochmoore. Beih. Veröff. Naturschutz Landschaftspflege Bad. -Württ. 21: 125-131. (English summary).

9. Spitzer, K. (1982). Evolution of the mire insect fauna associated with Pinus rotundata Link. in South Bohemia. Sbor. Jihoces. Muz. v Ces. Budejovicich Prir. Vedy 22: 93-96.

10. Spitzer, K. (1982). In litt., 18 February.

11. Novak, I. and Spitzer, K. (1982). Ohrozeny svet hmyzu. (Endangered World of Insects). Ceskoslovenska akademie ved. 139 pp.

12. Carp, E. (1980). Directory of Wetlands of International Importance in the Western Palearctic. UNEP,Nairobi; IUCN, Gland. 506 pp.

13. Meineke, J-U. (1982). Einige Aspekte des Moor-Biotopschutzes für Schmetterlinge am Beispiel moorbewohnender Grobschmetterlingsarten in Südwestdeutschland. TELMA 12: 85-98.

We are grateful to K. Spitzer for the data used in compilation of this sheet, and to J. Harrison and M.G. Morris for further comment.

SUMMARY Taka Bone Rate is the largest coral atoll in Indonesia and supports a high diversity of coral and mollusc species, including vulnerable species such as giant clams and the Giant Triton. Damaging fishing methods have destroyed large areas of reef and many species have become severely depleted through over-collection. It is feared that much of the area is now so severely damaged that little benefit would be gained from major conservation efforts but a sanctuary should be set up to protect the only known populations of two giant clam species in Indonesia. Action urgently needs to be taken in this and similar areas to ensure that the livelihood of local inhabitants dependent on reef resources is not threatened further.

DESCRIPTION Taka Bone Rate is situated in the Flores Sea east of the southern tip of Selayar Island and approximately halfway between Sulawesi and Flores. It is the largest atoll in Indonesia, having an area of 2220 km^2, comparable with the Kwajalein atoll in the Marshall Islands and Suvadiva in the Maldives which are well known as the largest atolls in the world. It comprises a complex system of patch reefs, barrier reefs and faroes with 21 sand cays, the largest of which have coconut plantations and at least nine of which have permanent inhabitants. The margin of the atoll slopes steeply to great depths but is apparently nowhere sheer. A complex system of large sea-level reefs occupies the central portion of the atoll, with extensive reef flats or central lagoons (1).

INVERTEBRATE FAUNA The atoll supports a high diversity of corals. A total of 68 genera and subgenera and 158 species of coral have been recorded from the area and are listed in (1). Two of these genera, Palauastrea and Blastomussa are new records for Indonesia. Taka Bone Rate atoll alone yielded 65 coral genera and subgenera and 149 species. In general the outer perimeter of reef is dominated by large massive boulders of Porites and some faviid corals or by tabular Acropora. In more sheltered areas staghorn or banks of leafy corals, such as Montipora foliosa, may dominate from the reef crest down the reef slope. Lagoonal reefs have essentially the same corals as outer reefs but the colonies of branching and leafy forms attain larger and more fragile proportions.

A total of 121 gastropod, 78 bivalve and one scaphopod species of mollusc have been identified and are listed in (1). Nautilus, cuttlefish, squid and octopus species have also been recorded, and probably many more remain to be found. However, the living mollusc fauna on the reefs seem to be sparse. The most commonly seen species were boring bivalves (Lithophaga spp.) and the giant clams. Large oysters (Lopha spp.) and pen shells (Pinna sp.) were seen at most dive sites but in small numbers. The most common gastropod species were of the family Strombidae, particularly Strombus luhuanus and Lambis spp. The reef-flats of uninhabited islands are particularly rich in gastropods with species of the genera Strombus, Oliva, Conus and Cymbiola being abundant. On Pulau Santigiang, typical rocky shore species of the genera Crassostrea, Drupa, Conus, Cypraea, Turbo, Trochus and Tectus occur on beach rock exposed at low tide.

The commercially valuable mother-of-pearl shells such as the commercial trochus (Trochus niloticus), the green snail (Turbo marmoratus), and the pearl oysters (Pinctada maxima, P. margaritifera and Pteria spp.) were notable for their scarcity. Only one living trochus and five living P. margaritifera were seen during the survey.

Six of the seven species of giant clams (see review) were seen living at Taka Bone Rate, the most abundant being the smaller species Tridacna crocea and T.

maxima. Living T. squamosa, T. derasa, and Hippopus spp. were seen at 100 per cent, 50 per cent and 70 per cent respectively of the snorkel and dive sites, but never in great abundance. T. gigas is reported to occur on the reefs but no living specimens were seen (1).

SCIENTIFIC INTEREST AND POTENTIAL VALUE The central location of the atoll combined with its high diversity of corals, should argue strongly for its protection as a conservation area around which to design a network of coral reef reserves, although the extensive damage that it has already undergone reduces its value. It is the first site in Indonesia with confirmed sightings of the clams Tridacna derasa and Hippopus porcellanus in substantial numbers. Other species of the vulnerable Tridacnidae are found there, and there is evidence that the rare Charonia tritonis (see review) also occurs there. The Green Turtle (Chelonia mydas) and Hawksbill Turtle (Eretmochelys imbricata) occur in the area and nest on uninhabitated cays, but are rare. Both are listed as Endangered in the IUCN Amphibia-Reptilia Red Data Book (2).

The reef resources of the area are of major economic importance to the local inhabitants. The meat of giant clams and of a variety of other molluscs, including Cassis cornuta, Chicoreus ramosus, Cymbiola vespertilio, Strombus luhuanus and Lambis lambis, is eaten by the islanders. Dried clam meat, dried octopus, sea cucumbers and mother-of-pearl shells are exported to the neighbouring islands and mainland Sulawesi. The shells of Tridacna gigas and of large T. derasa and T. squamosa are stored on Pulau Jinatu from where they are exported to Jakarta to be crushed and made into high quality floor tiles. The fishermen use shells of the Giant Triton as trumpets or export them for sale as souvenirs.

THREATS TO SURVIVAL The reefs have already been extensively devastated by explosives and the harvest of molluscs. Use of explosives for fishing continues, although islanders claim that the practice is declining because of increasing difficulty in obtaining dynamite. Local people claim that it is people from Flores who most frequently fish these reefs with explosives. Certain larger species of molluscs including the commercially valuable mother-of-pearl shells have been intensively collected and are now rare. Itinerant fishermen and teams of professional divers based far from the atoll (such as those from Ujung Pandang) place heavy pressure on populations of valuable species. The former collect any species of food or commercial value while the latter specialise in mother-of-pearl shells and species of souvenir value. The divers strip reefs of molluscs with great efficiency and reach greater depths with SCUBA equipment than others can attain. Both groups smash corals and the itinerant fishermen often fish with explosives which kill molluscs and their eggs and destroy their habitat. The combination of heavy collecting pressure and habitat destruction appears to have caused the local extinction of the larger mollusc species, notably Tridacna gigas and Charonia tritonis, and others such as Cassis cornuta and the mother-of-pearl species may well be becoming rarer (1).

CONSERVATION MEASURES TAKEN None taken.

CONSERVATION MEASURES PROPOSED The extent of damage to Taka Bone Rate emphasizes the alarming rate at which Indonesia's valuable reef resources are being destroyed by careless and unregulated fishing methods. In some respects this area has been so devastated that it is feared that a major conservation effort would not be worthwhile. However, as it still contains several species threatened on a worldwide scale and is of importance to local inhabitants, it is suggested that the atoll merits immediate attention.

Recommendations for the area have been made in (1). It is hoped that the Directorate of Nature Conservation (PPA) will discuss with the Directorate General of Fisheries ways to control destructive fishing methods. A prime requirement is the establishment of a sanctuary in the central area of the reef to protect the only known sites of T. derasa and H. porcellanus in Indonesia, until these species are located in other areas more suitable for reserve status. Once established, the sanctuary could be controlled by local people, and turtle capture and turtle egg collection should be prevented in the area. It has been suggested that there should be a national ban on the operation of boats carrying groups of divers and modern equipment from ports, such as Ujung Pandang, to distant reefs for the collection of molluscs and corals. The activities of itinerant fishermen should also be controlled in order to safeguard the reef resources for local inhabitants.

REFERENCES   1.   Salm, R.V., Usher, G., Mashudi, E. (1982). Taka Bone Rate: Assessment of marine conservation value and needs. Field Report of UNDP/FAO National Parks Development Project, INS/78/061. Bogor, Indonesia.
             2.   Groombridge, B. and Wright, L. (1982). The IUCN Amphibia-Reptilia Red Data Book, Part 1. IUCN, Gland, Switzerland. 426 Pp.

We are very grateful to G. Usher for comments on this account.

SUMMARY The Roseland Voluntary Marine Conservation Area is a small area of estuarine and coastal habitat on the south coast of Cornwall, U.K. The marine community found there is extremely diverse and the area is particularly important for the living maërl beds found on St Mawes Bank. These deposits of calcareous algae support a diverse fauna but are very vulnerable to damage on account of their very slow growth rates. Maërl is of commercial value and is collected in surrounding areas for use as soil fertilizer. In 1982 this area became a voluntary marine nature reserve and activities within it are monitored by the Roseland Marine Nature Conservation Area Advisory Group.

DESCRIPTION The Roseland Voluntary Marine Conservation Area covers 600 ha. It is bounded by the coastline from St Anthony Head, at the mouth of the Fal River, to St Just-in-Roseland (including the Percuil River within tidal influence), and on the offshore side by the eastern edge of the shipping channel. It includes two particularly important sites, St Mawes Bank and Place Cove.

The substratum of St Mawes Bank is composed of an almost pure growth of unattached red calcareous algae known as maërl over an area of about one ha. The main species is Phymatolithon calcareum (= Lithothamnium calcareum) but L. corallioides is also present where living deposits are at least 15 cm deep. Phymatolithon and Lithothamnium are genera of the Corallinaceae, a family of calcareous marine algae which encrust stone or other algae or, as in St Mawes Bank, form free 'rhodoliths'. These range from small pink to purple twigs of 2-4 cm in length to a mass of coalescing knobs up to 10 cm or more in diameter. Both species have a great range of forms and it requires a specialist to identify them. Maërl has an extremely slow growth rate, increasing in length by as little as 1 mm or less per year. It is thought that maërl banks are usually formed in situ rather than being transported by water currents (5). The formation of living maërl banks is dependent on the absence of more than very gentle water movement, which would otherwise bury the algae under fine sediment. Maërl is thus usually found in estuaries and semi-protected bays less than 20 m deep (5). Dead Lithothamnium is broken down by wave action and abrasion, and produces extensive banks of yellowish-white maërl deposit resembling coral sand and consisting of short calcareous rods (5,11). Dead deposits in the vicinity of St Mawes trap fine sediments in what is essentially a coarse open substrate, thus creating a specialized habitat for a diverse fauna of suspension and deposit feeders, as well as providing hard substrate for settlement of hydroids, bryozoans and certain molluscs (1).

Place Cove, at the mouth of the Percuil River, is an area of 5.7 ha with minimum freshwater influence. It has a salt water lagoon which is exposed at low tide. Around the lagoon are soft mud deposits with, towards low water, sandier mud mixed with a small amount of gravel and shell. Its sheltered position has resulted in an abundant and diverse soft substrate fauna. In other parts of the Conservation Area, there are a variety of marine habitats, including sublittoral beds of eel-grass (Zostera marina) which are just exposed at low water below Newton Cliff; beds of the polychaete worm Lanice conchilega, reaching a high density and associated with several species of echinoderms, molluscs, and other polychaetes; mud flats with a dense population of the sea anemone Cereus pedunculatus (estimated at 1000 per m^2) and associated with a few slime-tube worms, Myxicola infundibulum; fringing reefs of fissured Devonian rock with classic zonation of lichens, algae, barnacles and molluscs; rock pools; and beaches composed of shaled slate cobbles (which provide cover and settling space for a wide selection of fauna), shingle, shell gravel, sand and silt (1).

INVERTEBRATE FAUNA Both living and dead maërl support a very large number of plant and animal species, representing many groups and forming an unusual community (1). Crustaceans dominate the fauna of living maërl, which provides many crevices for epifauna. The decapods Porcellana longicornis, Pilumnus hirtellus and Galathea squamifera are especially common (5). Deposit and suspension feeding polychaetes are also common and over 60 species have been recorded from St Mawes Bank in either living or dead maërl (1). Of special interest is the probable sighting of the anthozoan Halcampoides purpurea. This species has a wide holarctic distribution, but in the British Isles, apart from St Mawes Bank, it is only known from maërl beds in west Ireland (12). Maërl also forms a focus for lime-boring polychaetes and the boring sponge Cliona celata (1).

The fauna of dead maërl is not as abundant as that of living maërl, and resembles that of fine shell gravel. However, the wide range of polychaetes present, from deposit feeders to suspension feeders, demonstrates the specialized ecological habitat that maërl deposits provide, namely the varying degree of fine subtrate within an otherwise coarse open matrix (1,5).

An interim faunal list has been compiled until the fully annotated list is ready (1). This includes 69 species of polychaete, 99 species of mollusc, 31 decapods, 18 echinoderms, 14 anthozoans, 12 sponges, 10 hydrozoans and 10 bryozoans, as well as species from many other marine invertebrate groups (1,5).

Place Cove has soft substrates with a great variety of species occurring under almost fully marine conditions, and every area of hard surface is covered with epifauna and flora. Many of the species are rare southern types, and a number of molluscs, polychaetes and echinoderms here extend into the littoral zone although elsewhere they are found only offshore (1,5).

SCIENTIFIC, INTEREST AND POTENTIAL VALUE The maërl that typifies St Mawes Bank is known in Britain only on the west coast of Scotland, the Irish Sea, west Ireland and south-west England (St Austell Bay, Carrick Roads and Falmouth Bay). St Mawes Bank is particularly important as it has the most extensive bed of living rhodoliths and has given rise to a unique, diverse marine community of considerable scientific value (1). Banks of predominantly living maërl are rare. In U.K. waters, because of the relatively high turbidities, they are only to be expected in depths of a maximum of 20 m in semi-protected bays and estuaries where the lack of heavy wave action allows them to develop in spite of their slow growth rate (6,7).

Maërl is used as a soil conditioner, either alone or mixed with traditional fertilizers such as phosphate or lime. The U.K. deposits are largely unexploited although there is evidence that dredging for maërl was carried out near Falmouth in the 18th century. Fresh interest in this product in the mid-1970s led to a survey of the beds by MAFF (Ministry of Agriculture, Fisheries and Food) (5). The Cornish Calcified Seaweed Company currently has a licence to dredge maërl from off Mylor, west of the shipping channel (15).

The area and its surroundings are used for a number of other purposes. Oysters are dredged in Carrick Roads and are farmed in the Percuil River; winkles and cockles are collected in some places. Anglers dig for bait in the Place Cove area (1). Part of the area is a designated water ski area (1).

The beaches of Fal Bay have been considered the richest for marine life in the entire British Isles (3) and are still very rich although human disturbance of various kinds has affected them adversely (4). Place Cove, with its particularly rich assemblage, is an important area for controlled research and has long been a focal study site for the Marine Biological Association, U.K. and for university and school groups. Otters, now an endangered and protected species, live in the area and several sites are important as feeding areas for birds (1). Nearly 120 species of marine algae have been collected and identified from the maërl beds, many of which, such as Solieria chordalis and Cruoria cruoriaeformis, were first records for the British Isles (10,17). These constitute a distinctive community, similar to that described for Brittany (16).

THREATS TO SURVIVAL The greatest threat to the area is the loss of, or damage to, the living maërl beds. Maërl is particularly vulnerable to activities such as dredging, on account of its slow growth rate and the fact that the living plants occur only as surface cover to beds that are mainly composed of dead material (9,10). The effects of maërl extraction on the marine environment are likely to be similar to those caused by marine sand and gravel extraction. These effects have been reviewed comprehensively by a Working Group of the International Council for the Exploration of the Sea (ICES) and include: alteration of bottom topography, reduction in numbers of benthic fish food species, change in species composition, damage to fish spawning grounds and increase in suspended solids, which can adversely affect fish, shellfish and other organisms or drive them away from the area (8). The extent to which these effects are applicable to Falmouth Bay and the Fal Estuary are discussed in (5) and it is concluded that limited commercial extraction of maërl would be unlikely to have a significant effect on the benthic productivity of the area as a whole. The greatest danger would be the loss of the unique habitat of St Mawes Bank, and consequent reduction in the diversity of the fauna (1).

Another possible threat is the proposed Falmouth Container Terminal which, if constructed, would increase tidal flow and siltation and lead to turbulent water, higher levels of pollution and an increase in scouring of the sea bed with the possible removal of some of the present sediment which is so important in the conservation area. This proposal has not yet gone ahead and it is presumed that detailed hydrographic work will be carried out before any further steps are taken (13).

In the past, digging by university and school parties has sometimes been excessive and has caused damage (1), and it is thought that many of the beaches of Fal Bay are over-exploited by collectors, anglers and educational groups (4). There is evidence from exploratory dives that anchors can damage living maërl and Zostera beds, but as the Bank is used as a water ski area it is largely an exclusion zone for all other activities (14).

CONSERVATION MEASURES TAKEN St Mawes Inlet is listed as a Primary Site of National Marine Biological Importance from St Anthony Head to north of St Mawes Castle, including the Percuil River within tidal limits (2). The whole area is part of an Area of Outstanding Natural Beauty. It includes several Special Sites recognized by the Cornish Naturalists' Trust and scheduled by Cornwall County Council Planning Department. The National Trust owns cliffland above St Mawes Bank (1).

The Roseland Marine Nature Conservation Group was set up in 1982 with the aim of preventing the unwitting damage which is known to result from excessive bait-digging, trench transects dug by parties of students and unnecessary over-collecting of specimens. The Group has representatives from the University of Exeter, the Cornwall Naturalists' Trust Council (CNT), the Marine Biological Association U.K., the Nature Conservancy Council (NCC) and other interested parties. The Conservation Area is a purely voluntary one with no legal powers, and its primary purpose is educational. A code of conduct is being prepared. Attempts have been made to discourage school and university parties from excessive digging. The Cornish Calcified Seaweed Company has been asked by the NCC and MAFF not to remove living maërl by dredging on St Mawes Bank, and the company is now aware of the importance of the bank (1,14). Surveys by members of the Underwater Conservation Society and MAFF have contributed to the considerable body of information available on the marine life of the area. Additional littoral and sublittoral surveys, organized by the University of Exeter, are in progress so that the present extent of the various communities can be mapped (14).

CONSERVATION MEASURES PROPOSED St Mawes Bank should remain inviolate from dredging for maërl or oysters. Place Cove and Amsterdam Point should be treated as very sensitive, with digging kept to a minimum and preferably only undertaken after consultation with organizations such as the NCC, CNT or MBA (1). The activities of the Conservation Group should be fully supported and encouraged. The example of St Mawes Bank being used as a Water Ski area, which makes it largely an exclusion zone for other activities, illustrates the manner in which the Conservation Group might reconcile potentially conflicting commercial, leisure and conservation interests.

REFERENCES
1. Turk, S.M. (1982). Consultation paper re proposal to create a marine nature reserve in Carrick Roads, Falmouth, to include St Mawes Bank and St Mawes Inlet with adjacent areas. Preliminary Draft.
2. Bishop, G.M. and Holme, N.A. (1980). The sediment shores of Great Britain: an assessment of their conservation value. Final Report, Part 1. Survey of the littoral zone of Great Britain. Scottish Marine Biological Association/Marine Biological Association of the United Kingdom, Intertidal Survey Unit.
3. Clark, J. (1907). The excursion of Gyllyngvase Beach. Rep. R. Cornwall Polyt. Soc. 25: 47-51.
4. Pitchford, G.W. Some Cornish beaches. Conchologists' Newsletter 63: 49-52.
5. Hardiman, P.A., Rolfe, M.S. and White, I.C. (1976). Lithothamnium studies off the south west coast of England. International Council for Exploration of the Sea, Shellfish and Benthos Committee.
6. Adey, W.H. and Adey, P.J. (1973). Studies on the biosystematics and ecology of epilithic crustose Corallinaceae of the British Isles. J. Phycol. 8: 343-407.
7. Adey, W.H. and McKibbin, D.L. (1970). Studies on the maërl species Phymatholithon calcareum (Pallas) nov. comb. and Lithothamnium corallioides Crouan in the Ria de Vigo. Botanica Mar. 13: 100-106.
8. ICES (1975). Report of the Working Group on effects on fisheries of marine sand and gravel extraction. International Committee for Exploration of the Sea Cooperative Research Reports 46. 57 pp.
9. Blunden, G., Farnham, W.F., Jephson, N., Barwell, C.J., Fenn, R.H. and Plunkett, B. (1981). The composition of maërl beds of economic interest in northern Brittany, Cornwall and Ireland. Proc. Intern. Seaweed Symp. 10: 651-656.
10. Farnham, W.F. and Jephson, N.A. (1977). A survey of the maërl beds of Falmouth (Cornwall). Brit. Phycol. J. 12: 119.
11. Bosence, D. (1976). Ecological and sedimentological studies on some carbonate sediment producing organisms, Co. Galway, Ireland. Ph.D. thesis, Univ. of Reading, U.K.
12. Manuel, R.L. (1981). British Anthozoa. In Kermack, D.M. and Barnes, R.S.K. (Eds), Synopsis of the British Fauna, (N.S.) 18. Linn. Soc. London, Academic Press.

13. Slater, P. (1982). The proposed Falmouth Container Terminal. Unpub. report.
14. Burrows, R. and Turk, S.M. (1982). Roseland Marine Nature Conservation Area. Information Sheet.
15. Turk, S.M. (1982). In litt., 30 December.
16. Blunden, G., Farnham, W.F., Fenn, R.H. and Blinkett, B.A. (1977). The composition of maërl from the Glenan Islands of southern Brittany. Bot. mar. 20: 121-125.
17. Farnham, W.F. (1980). Studies on aliens in the marine flora of southern England. In Price, J.H., Irvine, D.E.G. and Farnham, W.F. (Eds), The Shore Environment, Vol. 2. Ecosystems. Academic Press, London, pp. 875-916.

We are very grateful to R. Burrows, W.F. Farnham, N. Holme and S.M. Turk for assisting with this account.

Japanese Wood Ant Supercolony 503